# BIO
# LOGY

**Appleton-Century-Crofts**

Educational Division
Meredith Corporation

**New York**

# BIOLOGY

**Aaron O. Wasserman**
City College of the City University of New York

Design and photomontages by
**Ladislav Svatos**

Illustrations by
**Russell Peterson and Vantage Art, Inc.**

Picture research by
**Alycia Smith Butler**

Cover transparency: Jen & Des Bartllett — Bruce Coleman Inc.

73 74 75 76 77 / 10 9 8 7 6 5 4 3

72-90618

PRINTED IN THE UNITED STATES OF AMERICA
390-92744-9

# Contents

**Membrane systems: the cell**
    The cell theory. Studying cells.
**Cell form and function**
**The structure of the cell**
    The cell nucleus. The endoplasmic reticulum and
    ribosomes. The Golgi apparatus. Mitochondria.
    Plastids. Lysosomes. Vacuoles. Centrioles. Cilia
    and flagella. Microtubules. Cell coatings.
**Cell specialization**
    Specialized plant cells. Specialized animal cells.
    The multicellular organism.

part **2** The functioning of the organism

**Other types of muscle**
Invertebrate muscles. Vertebrate smooth muscle. Vertebrate cardiac muscle.

**Receiving information**
Information about the external environment. Information about the internal environment.
**Sending signals**
Membrane potential of the neuron. The nerve impulse. Conduction of nervous impulses. Transmission across the synapse. Information coding.
**Coordination**
Reflexes. The autonomic nervous system. The central nervous system.

**Agents of chemical control**
Hormones. Prostaglandins. The mechanism of hormone action. The role of cyclic AMP in hormone action.
**Chemical control in plants**
Plant hormones.
**Chemical control in invertebrates**
Hormones in arthropods.
**Chemical control in vertebrates**
Endocrine organs in mammals.

part 3 **Reproduction**

**Kingdom Plantae**
Phylum Rhodophyta. Phylum Phaeophyta. Phylum Chlorophyta. Phylum Bryophyta. Phylum Tracheophyta.

**Kingdom Animalia**
Phylum Mesozoa. Phylum Porifera. Phylum Cnidaria. Phylum Ctenophora. Phylum Platy-helminthes. Phylum Aschelminthes. Phylum Annelida. Phylum Sipunculoidea. Phylum Arthropoda. Phylum Pentastomida. Phylum Tardigrada. Phylum Bryozoa. Phylum Phoronidea. Phylum Brachipoda. Phylum Mollusca. Phylum Chaetognatha. Phylum Echinodermata. Phylum Pogonophora. Phylum Hemichordata. Phylum Chordata.

# Preface

## Purpose

This text is designed for introductory biology courses at the college level. It includes systematic and comprehensive coverage of basic concepts and principles, terminology, and current biological trends and issues. The book also functions as a reference, enabling the student quickly to look up basic data. For example, there is a table that summarizes geological time, and another that gives metric equivalents of inches and ounces. Taxonomy, which is introduced in Chapter 4, is covered more extensively and in easy-reference format in an appendix. Topics are summarized at the end of each chapter, and all terms are defined in a glossary at the back of the book.

Every textbook writer must start with the realization that there is no one "right" way to teach a course; a good teacher will adapt his material to suit his own special abilities and interest and his students' needs. To limit a book to only one school of biological thought—phylogenetic, behavioral, or ecological, for example—is to limit its usefulness. This book, therefore, introduces many different biological points of view. An inclusive approach seems especially valuable in a field such as biology, which is changing and developing rapidly, for it gives the student a broad framework in which to integrate the new concepts and data that will arise in the next few years. A special effort has been made to use many current resources which will help keep the reader abreast of recent developments with significance for the future. For example, we include discussions of and references to current research in genetic engineering, cloning, the energy crisis, and behavioral effects of crowding.

The full scope and variety of biological study are introduced for the beginning student. Basic concepts are presented here, in a manner that points out that not all biologists agree on how to define them. We discuss the results of studies and research projects conducted by eminent men in the field today. Charts, graphs, and tables found throughout the book present statistical data; specially-prepared photomicrographs are accompanied by explicit line drawings.

In this text we have attempted a fresh approach to current biology that emphasizes principles in a readable and exciting way. Where possible, we have included common terms as well as technical terms, and have sought to relate concepts and processes to familiar experiences. Certain underlying themes of life, such as the use and relevance of energy to organism and community, are used as recurrent threads within a tapestry that seeks to make more meaningful the universality of internal life processes and the interrelationships of organisms.

## The plan of the book

**Introduction:** This chapter introduces the student to the continuing inquiry into the methods, assumptions, and theories of biology. It includes preliminary discussions of the scientific method, the characteristics of life, and the concept of evolution, topics that are covered in greater detail in subsequent chapters.

## Part I:

**The Cell:** Part I begins with the basic physical and chemical principles that provide the background necessary for understanding the structures and processes of cells. The student becomes familiar with the challenges of existence confronted by all living organisms.

## Part II:

**The Functioning of the Organism:** A comprehensive study of the latest ideas on the anatomical and physiological processes of organisms builds on the problems introduced in Part I: nutrition, transport, gas exchange, homeostasis, chemical control, nervous control, and movement. Plants and animals are covered together to provide the student with a sense of the two related systems of biological organization.

## Part III:

**Reproduction:** This section deals with the problems of sustaining, adapting, and transmitting organization. The processes of mitosis and meiosis are introduced, as are the consequences of these processes for inheritance. In molecular genetics, the reproductive mechanisms and the control of the cell are explained biochemically. The application of this cell control is then demonstrated in the chapter on the developmental processes.

## Part IV:

**The Organism and Its Environment:** Part IV introduces the student to evolution, behavior, and ecology, the topics that help develop the student's understanding of himself and his environment. The final chapter considers the intricate relationship of living systems and the effects of man upon them.

## Appendix

**Taxonomy:** This supplement provides more extensive coverage of taxonomic principles and individual phyla. It may be used in conjunction with Chapter 4.

Each Part of the text forms an integral unit. However, the order of the Parts may be altered, permitting changes in emphasis according to the individual teacher's orientation and objectives. Five common sequences are provided here, but others may occur to the teacher as he plans his course.

## I.  Functional emphasis

**First semester**
    Introduction
    Part I
    Part II

**Second semester**
    Part III
    Part IV

## II.  Phylogenetic emphasis

**First semester**
    Introduction
    Part I
    Appendix
    Part II

**Second semester**
    Part III
    Part IV

**First semester**
    Introduction
    Part I (postponing Chapter 4)
    Part II
    Part III

**Second semester**
    Chapter 4 with Appendix
    Part IV

## III.  Biochemical-genetic emphasis

**First semester**
    Introduction
    Part I (postponing Chapter 4)
    Part III

**Second semester**
    Part II
    Part IV
    Chapter 4 (Appendix optional)

## IV.  Evolutionary emphasis

**First semester**
    Introduction
    Chapters 17, 18
    Chapter 4 (appendix optional)
    Part I

**Second semester**
    Part II
    Chapter 19
    Part III
    Chapters 20–22

## V. Ecological emphasis

## Features of the book

We are putting this book on the market because we realize that biologists require a text that satisfies their students' needs for clear, up-to-date, and relevant material. Improvement along these lines would make a book that is more interesting and useful to students and a better teaching tool for professors.

## Readability

The only good textbook is one that students can and will read, yet the complaint most frequently made by teachers and students alike about introductory textbooks is that they lack readability. A first step in producing this book, therefore, was to define "readability" in the context of the teaching situation—particularly in terms of the complex biological concepts and processes that must be explained—and to determine how it might be achieved. After much discussion and analysis, and with the help of professionals, it was decided that readability in biology depends essentially on organization and relevance.

Good organization makes a book readable. Students learn best when material is clearly presented one step at a time and the sequence of presentation indicates important interrelationships. Each Part of this book is a separate and virtually self-contained unit of study. This style of organization makes the material easier for the student to comprehend and at the same time allows the teacher considerable latitude in arranging reading assignments to suit his own curriculum.

The manner in which terms and concepts are introduced can affect readability. Students often find it difficult to learn a whole new vocabulary in order to understand their text. Each new term, concept or principle is carefully explained with concrete examples of its application. These definitions reappear in the end-of-chapter summaries and again in the glossary, which serves as a convenient reference and study guide. For immediate reference, the terms are set off in italics.

Relevance makes a book readable, because it provides a motivation to read. Fortunately, it is often possible to include interesting and possibly familiar examples of important concepts. These examples, however, are not ends in themselves; they serve primarily as the means of clarifying the concepts, processes, and principles of science, which are emphasized in this book.

## Illustrations

Modern society is increasingly oriented toward visual images. This book contains a large number of photographs and line drawings that give visual impact and support to the text. The captions assist the student in relating the meaning of the illustrations to the material in the text. Where a photograph does not sufficiently illustrate the item, a line drawing is also included to identify the different parts of the organ or organism.

## Tables and graphs

Since biology depends heavily on empirical evidence, it is important for the new student to be able to understand statistical data. All the statistical data in this text has been designed by a specialist, to make it attractive, graphic, and readily comprehensible.

## Photoessays

Several chapters are supplemented by brief, self-contained essays, each on a specialized element of the subject under discussion. These essays are heavily illustrated with photographs and clarifying line drawings. For example, Chapter 11 includes material on specific photoperiodic cycles.

## Portfolios

The book contains four color portfolios. They deal either with a classic situation that requires additional coverage or a current issue that is best understood through visual imagery. For example, an insert on the ecological crisis includes Arthur Westing's graphic photographs of cratering in Southeast Asia.

## Readings

Many teachers assign readings to supplement and illustrate the textbook they have chosen. Each chapter of this book includes a fully-annotated list of books and articles. These lists are intended to encourage the student to keep himself up-to-date, to go beyond the text and discover the exciting work that biologists are doing today.

Also, in keeping with the current trend toward interdisciplinary study, these reading lists include items by chemists and writers in related fields, such as psychology, sociology, and anthropology. Some are written in an academic style, others in a more popular vein; some concentrate on small specific studies, and others formulate broad principles and concepts. Each was chosen because of its application to the subject of the chapter in which it appears and its probable interest to introductory students.

## Supplements

Accompanying the text are a Study Guide and ACCESS Workbook, a lab manual, and an Instructor's Manual. The workbook is designed to help students interrelate concepts from various sections and chapters and to review the important points of the text. For each chapter of the text, the workbook includes a review outline, questions for study and discussion, and application questions relating biology to the student's own concerns. (The self-test utilizes ACCESS, a process in which an answer printed in invisible ink is revealed by rubbing it with the ACCESS activator. The student thus receives immediate verification of his self-test responses.)

The Instructor's Manual includes not only an overview of each text chapter, but also suggested illustrations for lecture use and topics for research projects for many of the major points of the text. Information is given on how to obtain the films listed at the end of each chapter. For each chapter there are also suggestions for essays or essay-type exam questions.

## Acknowledgments

In the final chapter on Man and His Environment, a community is depicted as a highly integrated supra-organism. Probably the best example of such a supra-organism in human society is the extremely efficient and well-organized team that has helped to produce this book. The close collaboration and cooperation between the Appleton-Century-Crofts specialists and myself has been gratifying and rewarding. It has also, I hope, resulted in a superior text. For her professional assistance in the writing, the lion's share of the credit goes to Caroline Latham, whose grasp of complex technical material and whose stamina never ceased to amaze me. To Pete Salwen, for his assistance in preparing portions of the material, I extend my appreciation. To Christine Tello, who handled the production of a complex manuscript, I extend a special thanks.

I also wish to express my everlasting gratitude and thanks to the following colleagues of the Biology Department at The City College of the City University of New York: To Professors William N. Tavolga and Robert A. Ortman for critically reading portions of the manuscript; to Professor Lawrence J. Crockett for providing source material and numerous suggestions, and to Professors Neil Grant and Jess Hanks for ideas and information in matters botanical. Teaching suggestions from Joseph Rubinstein and Vincent Chiappetta of New York University were also appreciated. Last of all, I should like to pay tribute to my wife, Solange, and to my son, Gilbert, whose patience and encouragement were a source of much needed moral support.

A.O.W.

*August 1972*

# Introduction

One of the oldest and most basic of human emotions is the feeling of reverence for life. We know from the art early man left behind that he worshipped divinities for their power to bestow life. His idols were statuettes of heavy-breasted pregnant goddesses, ripe with the promise of new life and the sustenance of life. His religious rituals were elaborated pleas for continuing fertility. The sun, the rain, the flooding river—these nourishers of life were in many cultures elevated to the status of deities. The spring equinox, which heralds the greening of new life from the earth, was the most sacred time of year.

In our modern age, the ancient worship of life continues to find expression. It is a vital part of all our religious tradition and our ethical values. It is the basis of much of our legal and moral framework. Many of our customs, such as throwing rice at weddings or decorating eggs at Easter, still reflect the urge to celebrate life. Especially in our emotions we see the power of our reverence for life; even the clearest understanding of the dangers of overpopulation cannot prevent us from responding to a newborn child with feelings of hope and joy.

Implicit in much of the worship of life has been the assumption that man will never know its mysteries. This assumption is under challenge from the activities of modern science. Physicists, chemists, and biologists have set themselves the task of investigating the mystery. Their success, while far from complete, has been notable. Many life functions can now be explained in terms of chemical and physical laws, rather than mysterious vital forces.

The heart furnishes a vivid example of science's ability to dispel certain kinds of myth and mystery. The heart, that organ favored by poets and cherished by lovers, was once thought to be the source of all emotion, a touchstone for truth, even the site of intellectual understanding. The Egyptians quite literally thought they would be judged in the hereafter by their hearts, and that only those with hearts lighter than a feather could be admitted to the pleasures of paradise. In France, it was the custom to bury the heart of a dead king separately in some special place; this burial was much more significant than the state funeral in which his body was buried at St. Denis.

Today most of us regard the heart not as a magical force but as an efficient pump. Scientists have measured the rate at which it beats, the volume of blood it can move, and the amount of hydraulic pressure it creates. To us, a broken heart is only a metaphor, and although we may still exchange valentines on February 14, they are a paper symbol of a belief in some mystery we can no longer accept.

Looking back on these vanished myths, some writers have tried to make a case for the view that science's gain has been humanity's loss. They suggest that by reducing life to an interaction of atoms and molecules, scientists have also reduced the magic and mystery that was the basis of man's reverence for life. How, they ask, can we possibly worship the statistical probability of the random movement of atomic particles?

Certainly the old view of life's mystery is gone, and even an immediate moratorium on all scientific investigation cannot bring it back. But biologists, and those who study biology, have not lost their feelings of reverence and awe toward life. A knowledge of the discoveries of biology serves to increase man's wonder.

Again, the example of the heart is instructive. Perhaps we have lost many of the romantic myths, but the scientific facts with which they have been replaced are no less evocative. For example, consider the heart's efficiency as a pump. This organ is formed by interwoven fibers of cardiac muscle. The fibers contract, in a synchronized wave, about 70 times every minute during normal activity. By the time a person reaches the age of 70, his heart has probably contracted at least 2.5 billion times, and only then does the normal heart begin to show signs of wear and fatigue. This is far greater efficiency than any man-made pump has ever displayed.

1 *This example of a fertility figure is thought to date from about 4000 B.C. At that time, a woman's ability to produce and nourish life was associated with divine powers. (Collection Musée de l'Homme)*

Another aspect of the heart's efficiency is the fact that it stimulates its own contractions, by means of the specialized tissue in the sinoatrial node, which is both a muscle and a nervous sensor. Isolated hearts of lower vertebrates, such as frogs, can go on beating in a regular rhythm for hours. Even more amazing is the way in which, through the action of the vagus nerve and a hormone called acetylcholine, the rate of heartbeat and the volume of blood pumped to the organ can be varied in response to the needs of the organism. A glimpse of an enemy or predator, a change in the external temperature, a wound — these, and thousands of other factors, can cause sudden and automatic changes in the rate of blood circulation and the extent of blood pressure.

## The study of life

In a sense, we are all students of life, for there is at least one organism about which we are intensely curious — ourselves. A person may want to know why he caught a cold when no one else in the family was ill, why he is depressed on gray winter days, what makes his hair fall out, whether or not it is true that marijuana is a sexual stimulant.

Biologists ask the same questions, but they ask them in a somewhat different way, and they are more persistent about finding out the answers. The primary difference between the kind of curiosity we all have about life and the scientific study of biology is the use of scientific method as a tool of inquiry.

## Scientific method

Scientific method begins with careful observation. An important part of any scientist's education is learning to look at things closely, accurately, and with an open mind. To a certain extent, we all see what we expect to see; our assumptions often outweigh our observations. Psychologists have demonstrated this conclusively with experiments in which a group of people observes some sequence of events and then each person recounts what he saw. These experiments always produce several very different versions of the same events, and psychologists have found that by giving observers different clues about what is going to happen, they can produce an even wider difference in the accounts.

In observing natural phenomena, the scientist must try to take the events on their own terms and resist filtering his impressions through his preconceptions. It is interesting to note, in this context, that innovative scientists, such as Charles Darwin and Albert Einstein, have been described by those who knew them as "childlike" or "innocent." This description probably refers to their ability to look at things with the same lack of preconceptions that a child might have.

On the basis of observation, the scientist can formulate a hypothesis, or statement to be tested. The scientist will probably begin testing his hypothesis by collecting all available data from other people's work in the same field. If the known facts seem consistent with his hypothesis, he will then proceed to draw up a research design for testing the hypothesis through experimentation. The basic objective of the research design is to eliminate all variables except those directly involved in the hypothesis. This is usually done through the use of a control group — organisms that are treated in exactly the same way in all respects except the one single factor being tested. For example, one group of rats is fed a certain chemical that is suspected of speeding up growth rates; a second group of rats, with a similar genetic inheritance, is kept in the same room, in the same kind of cage, under the same environmental conditions. It is then safe to assume that any difference in growth between the experimental group and the control group is due solely to the chemical the experimental group receives.

3

**2** *This obese rat began to overeat when an appetite control center in his brain was destroyed. The experimenter will record his weight gain and compare it with the weights of a control group of normal rats; in this way he can establish the exact effect of the experimental procedure. (Courtesy of Neal E. Miller)*

Before experimental evidence can be accepted by the scientific community as a whole, it must meet one important test. The results must be shown to be reproducible when other scientists repeat the same experiments. This requirement guards not only against fraud (which is actually relatively rare) but more importantly, against accident, coincidence, faulty equipment, and human failings. A thermometer may be inaccurate; a chemical sample may deteriorate, changing the structure of the compound; a tired assistant may make an error in reading a set of figures. There are hundreds of such tiny accidents and errors that may influence the outcome of an experiment. An even more subtle influence on the results may be the experimenter's desire to find a certain pattern or to establish a certain outcome. The unconscious bias of the person performing the research can create effects of which he is entirely unaware.

Once a body of experimental evidence has been established, the final step in the scientific method is the drawing of conclusions regarding the acceptability of the hypothesis. To some extent, this third step is always a subjective one; using exactly the same data, two scientists can arrive at different conclusions regarding their meaning. Most scientific controversies arise over the interpretation of data, not their accuracy.

## Certainties and probabilities

The work of scientists has led to the establishment of scientific laws, many of which are referred to in this text to explain the behavior of atoms and organisms. It is important to understand just what a scientific law is, and what it is not.

A scientific law is formulated by observing large groups of individuals or events; it deals with averages rather than with individual instances. By using scientific laws, we can predict what will probably happen in any given situation, but we cannot necessarily predict the outcome of one single event. An example of this can be seen in the law of random behavior as it applies to the throwing of dice. Using a chart like the one shown in Table 1, statisticians can calculate the number of chances there are to throw each number from 2 to 12. On the basis of this calculation, they can predict that if a 1000 people throw the dice, there will be about five times more 6s thrown than there are 2s. But they cannot predict that any one person will throw a 6 instead of a 2.

Scientific laws apply to natural phenomena in exactly the same way that the statistician's law applies to the throwing of dice. The Mendelian law in genetics states that if Mr. Brown-

Eyes marries Miss Blue-Eyes, the chances are 3 to 1 that their children will have brown eyes. In the unlikely event that this couple has 1000 children, it is fairly certain that about 750 will have brown eyes and only 250 will have blue eyes. But if the newlyweds are dedicated members of Zero Population Growth and have only one child, no geneticist can predict with certainty the color that child's eyes will be.

# 1
# Probability

|   | 1 | 2 | 3 | 4 | 5 | 6 | Odds |
|---|---|---|---|---|---|---|------|
| **1** | 2 | 3 | 4 | 5 | 6 | 7 | 2 = 1 out of 36 |
|   |   |   |   |   |   |   | 3 = 2 out of 36 |
|   |   |   |   |   |   |   | 4 = 3 out of 36 |
| **2** | 3 | 4 | 5 | 6 | 7 | 8 | 5 = 4 out of 36 |
|   |   |   |   |   |   |   | 6 = 5 out of 36 |
| **3** | 4 | 5 | 6 | 7 | 8 | 9 | 7 = 6 out of 36 |
|   |   |   |   |   |   |   | 8 = 5 out of 36 |
| **4** | 5 | 6 | 7 | 8 | 9 | 10 | 9 = 4 out of 36 |
|   |   |   |   |   |   |   | 10 = 3 out of 36 |
| **5** | 6 | 7 | 8 | 9 | 10 | 11 | 11 = 2 out of 36 |
|   |   |   |   |   |   |   | 12 = 1 out of 36 |
| **6** | 7 | 8 | 9 | 10 | 11 | 12 |  |

The same thing is true of the laws that predict the path of electrons or the direction in which molecules will move. These laws express the statistical probability that most members of the group will behave that way. They do not express the certainty that any one selected electron or molecule will behave as expected.

## The characteristics of life

It seems to be traditional to introduce students to biology by asking them the question: What is life? Oddly enough, biologists are the only people who think this is a difficult question to answer. Most of us can tell a stone from a stork, and have no trouble deciding which one is alive and which one is not. To biologists the question is more troubling, because they continually run across cases where the dividing line is not so clear-cut. Is a bouquet of roses alive? Roses in a vase have the same vivid color and sweet perfume found in roses in the garden; their buds develop and open in the same way; they show the same sensitivity to sunlight and to cold. But the roses in the vase are surely dying, and in time they will be dusty brown mummies, consigned to the wastebasket. When did they die? Was it the moment they were cut from the plant, or was it when they were finally thrown away, or at some indeterminate hour in between? And how does their death differ from the fading of the flowers that were left on the plant?

Biologists have found that it is impossible to formulate a definition of life that will provide answers to all such questions. Life cannot really be defined; it can only be described. There are certain characteristics which are typical of living things. An entity that displays all these characteristics is surely alive, and an entity that displays none of them is certainly dead or inanimate. And those that are in between—that show some of the characteristics but not all of them—will continue to be a source of question to biologists.

One important characteristic of living things is their ability to carry on metabolic functions, or functions that lead to some exchange of materials between the organism and its environment. One such function is nutrition, the taking in of food; another is respiration, in which oxygen is taken in and carbon dioxide is given off. This ability to use, and even more important, to transform, materials from the environment is found only in living things.

A second important characteristic of life is the ability to adapt to the environment. Living things show measurable responses to changes in temperature, light, chemical composition, and other aspects of the external environment. We must be careful here to distinguish between passive changes, such as those seen in a rock that is hit by a hammer, and active changes, such as those seen in an animal that runs away from fire. Living things change themselves so that their internal environment remains constant in spite of external change.

A third important characteristic is the ability to reproduce new individuals that are either exact duplicates or close copies of the original. Although we can see certain kinds of growth in the nonliving world—some mountains can be said to grow, and crystals can grow when placed in chemical solutions—this property of exact reproduction (including the reproduction of inheritable variations) is limited to living organisms.

To these functional characteristics of life, we may add a fourth distinguishing characteristic, which is structural in nature. Most living things have a structure that is not seen in the nonliving world, a structure based on tiny units called cells. To create and maintain a cell requires a constant input of energy in regular graded amounts, and only living things are capable of providing this energy. As soon as a living organism dies, its cells begin to break down, because no new supplies of energy are available.

In the last 20 years, the question, "What is life?" has ceased to be an academic exercise and has taken on practical dimensions. Since space vehicles are now looking for life on other planets, we must be able to define the object of our search; it is hard enough to find a needle in a haystack without the additional problem of not knowing what a needle is. Many of the characteristics we associate with life—cellular structure and oxidative respiration, for example—are probably found only on this planet. How will we recognize as living something without these traits? The National Aeronautics and Space Administration has enlisted the aid of biologists to help find an answer to this question, and they have formulated a working definition to guide the search for life in outer space. According to NASA, only two characteristics distinguish life from inanimate matter: the capacity for self-replication; and the inheritance of mutations, or changes in genetic material.

## Life and change

Nothing is more characteristic of life than change. The dramatic blossoming of the Christmas cactus, the metamorphosis of the caterpillar into a butterfly, the whitening of the ermine's coat in winter—it is these visible signs of constant change that reassure us life is present. As the English historian Thomas Macauley said, "In the dead there is no change."

Biochemists can demonstrate that even on the most fundamental level of organization, living things show constant change. A molecule that enters the human body will undergo many changes and transformations, and it is overwhelmingly likely that it will remain within the body for only a small fraction of the individual's total life span. The molecules that make up a boulder, on the other hand, are locked up in that form for centuries, and they undergo no change until they are finally released by the weathering action of wind or water.

In the nineteenth century, Charles Darwin observed and recorded another aspect of the change found in living things. Over long periods of time, living things change in form and function, and the direction of this change, over many generations, is toward increasing adaptation to the environment in which they live. Darwin called this process of change organic evolution.

Central to Darwin's interpretation of evolution was his conviction that the changes occur at random. Most of the changes will be harmful, for they will handicap the organism as it competes with others of its kind for food, living space, and the opportunity for sexual reproduction. But a few of these random changes will increase the organism's chances of survival, and the organism will reproduce more individuals with that advantageous trait, until eventually there is a whole group of organisms in which that trait is common. According to Darwin, the changes are random, but the perpetuation of successful changes through natural selection is more or less predictable.

It was especially this emphasis on the randomness of change that made Darwin's explanation of evolution so difficult for many of his contemporaries to accept, since it was in direct conflict with the need for order and desire for purpose that seem to be a part of human nature. Even today, when the weight of accumulated evidence makes it impossible to quarrel with Darwin's thesis, we have not yet learned to accept all the implications of change's randomness. Not only students but even teachers and practitioners of biology sometimes explain occurrences by resorting to teleology, or using an argument based on purposiveness. An example of a teleological statement is the explanation that the purpose of brightly colored flowers is to attract insects. Such a statement implies that the flower (or perhaps a deity) carefully thought the whole thing out and then selected red petals, like a woman in a Clairol advertisement. It is more accurate to say that bright colors attract insects and therefore natural selection favors this trait when it appears as a random mutation in flowers; this way of putting it eliminates the implication of purpose and reminds us of the random nature of the process.

The understanding of this process of evolution has considerably broadened man's knowledge of biology, for it has provided explanations of many puzzling events and relationships. The influence of Darwin's theory, and subsequent biological discoveries made with its help, has spread far beyond the field of biology. It has also changed the way man looks at himself and at his universe; it has changed his view of his relationship with other living things; to a certain extent, it has also colored his deepest religious and moral beliefs.

The work of scientific research may be a narrow, specialized discipline, but the discoveries of science have the broadest significance for all mankind.

# 1

## Cells and their structure

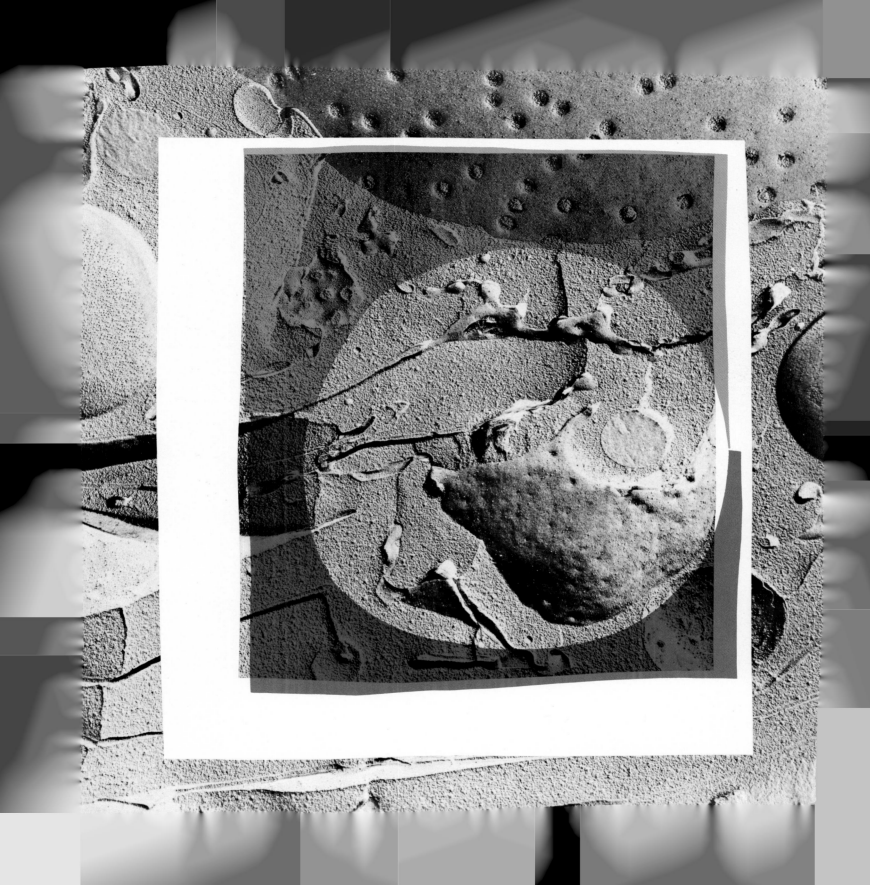

# 1
# The chemical basis of life

The images which the word "life" generally calls to mind are those of living plants and animals—a spring robin singing in the sunshine, a fern unrolling its delicate fronds. It is on the level of the organism that life is familiar, and most of us are content to look no farther. If we ask questions about the nature of life, they are questions relating to the aspects we ordinarily see and hear. We may want to know why the beaver builds dams, or how worker bees can tell when they need to start feeding a new queen. Our questions usually concern the functioning of the whole organism.

Biologists often ask these same questions, but their interest is not confined solely to the organismic level. Like the Elephant's Child, they are "full of 'satiable curiosity,'" and they want to know not only why living things act or look as they do but also how they are organized and what they are made of. Biologists question aspects of life that are less visible but perhaps more fundamental.

The question of what living things are made of was first asked thousands of years ago. Because there is such an obvious and dramatic difference between living things and the nonliving environment, early scientists assumed that there must be some corresponding difference in their composition. This possibility has been persistently investigated for hundreds of years, and in the seventeenth century, it seemed that the question might at last be settled. The invention of the microscope made it possible to examine both living and nonliving things more closely, and a major difference between the two was indeed apparent. It could be observed that living things were made of countless little structural units (given the name of cells) and that these units were not found in nonliving materials.

This apparent distinction between living and nonliving things began to lose its validity during the nineteenth century, as chemists were able to identify and name a group of basic elements found in all matter. Among the first elements to be explored were carbon, hydrogen, and oxygen, and newly devised tests for the presence of these elements showed that they were found in all cells, as well as in the earth and the atmosphere. Although scientists searched for a "mystery" element that might be present only in living things, no such element was found. Cells proved to be made of exactly the same elements as are rocks and rivers. The only difference between living and nonliving things seems to be the way these basic materials are organized.

In the twentieth century, the investigation into the composition of all matter has continued. It has been discovered that even the difference between elements is due to organization rather than content. All elements are actually composed of even smaller units, referred to as subatomic particles, and it is the pattern in which these particles are assembled that gives each element its individuality. Nearly a dozen subatomic particles have been discovered, and physicists continue to search for that hypothetical entity, the fundamental particle—the smallest unit of matter, of which all other matter is made.

On the basis of the evidence now available, we must conclude that living things are made of exactly the same materials as nonliving things. This conclusion carries with it a corollary: the chemical and physical laws that apply to these materials when they are present in nonliving things anywhere in the universe will also apply when they are present in living organisms. Thus all the characteristics of life must be explained within the confines of these laws.

## Life and energy

Perhaps the most difficult aspect of life to explain in terms of known chemical and physical laws is the seemingly inexhaustible energy of living organisms. Physicists tell us that the universe is slowly but surely running down, losing its supply of energy available to do work. Yet in the living world, or **biosphere,** we see the reverse situation. Since the dawn of life on our planet, living things have shown an enormous increase in both numbers and complexity. Although the available energy of the solar system has been declining, the energy of earth's biosphere has been increasing. In the past, some people have interpreted this contradiction to mean that the biosphere is somehow exempt from the laws governing energy in the rest of the universe. Biologists today disagree with such an interpretation; they suggest, instead, that the biosphere is a closed system into which new supplies of energy, from the larger system of the universe, are continually being injected. In order to understand the reasoning that lies behind this conclusion, we must first know something about the nature of energy.

## Free energy and entropy

Energy is impossible to define but easy to describe. Energy itself cannot be directly perceived, but it can be quite accurately measured by its effects on matter. In general, we can say that the greater the effect is, the greater is the amount of energy that caused it. For example, a football thrown by a two-year-old toddler will travel only a short distance at a relatively slow speed. When the same football is thrown by Joe Namath, it will travel much farther and at a much greater velocity. Therefore, we can safely conclude that Joe Namath transfers more energy to the ball than the child does, even though we cannot see the energy itself.

Although it is impossible to see energy, scientists can establish theoretical models that help to explain its effects. One such model envisions energy traveling through the universe in waves; this is why we speak of light waves and radio waves, both forms of energy. A second, more recent model depicts energy in the form of individual packets. According to this model, energy packets are flung from the sun and bombard the earth like buckshot out of a shotgun. In light of current knowledge, neither of these two models is more correct than the other. Each serves to explain certain observable aspects of the intangible phenomenon of energy.

Physicists say that energy is indestructible. It exists in a number of different forms, and it can be converted from one form to another. But it cannot be destroyed, no matter how it is transformed. This concept is expressed in the first law of thermodynamics. In every transfer of energy, the total quantity of energy remains unchanged; therefore in any closed system, total energy content remains constant.

Although energy cannot be destroyed, it can be changed into a form that is difficult or impossible to utilize. In order for energy to do work, it must be available in an orderly or organized form. Organized energy is called **free energy.** As the disorder of free energy increases, its ability to do work -that is, to affect or move matter—decreases. This disorderliness is called **entropy.** The term may be new, but the concept is really quite familiar. An analogy to entropy might be drawn in the organization of your college library. Suppose that the librarians worked all summer to put the books in order and then left for a Caribbean vacation the day the fall term opened. For the first few weeks of the term, the library would probably function quite well, doing the work of supplying the information you needed. But as time went by, more and more books would be left on library tables, behind filing cabinets, upside down on the wrong shelves. Eventually the disorder would become so great that you would be unable to locate the books you needed, and the library would no longer be doing its work. The books you want are still there in the building; they have not been destroyed, and each single book is still potentially as useful as it ever was. But the disorganization—the entropy—of the library has made those books useless to you.

Through observation and experimentation, physicists have formulated the second law of thermodynamics, which states that energy must always flow in the direction of increasing entropy. This is true both in the physical system of the universe and the biological system of a living organism. Everything moves in the direction of increasing disorderliness, toward the final equilibrium point of maximum disorganization or degradation, when it is no longer possible for energy to affect matter.

The basic problem that all living systems must solve is retarding the movement toward maximum disorder (which is death for the organism), or as one scholar puts it, "evading the decay to equilibrium." Only by adding more energy to the closed system of the biosphere is this possible. Through the process of photosynthesis, green plants **transduce,** or change, light energy into chemical energy, thereby increasing the total amount of free energy available in the biosphere. Thus free energy increases in this one closed system, even though the larger system of which it is a part (the universe) is moving inexorably in the direction of increasing entropy.

## The forms of energy

There is a vast quantity of energy in the universe, but it is not all equally available or useful to living systems, because energy is found in different forms and different states. We can distinguish between the **potential state** of energy, which is stored energy, and the **kinetic state** of energy, which is energy in action. For example, if you lift your textbook to shoulder height, it acquires a certain amount of potential energy. When you let go of the book and it starts to fall, that potential energy is being converted into kinetic energy, which is accomplishing the work of moving the book. By the time the book hits the ground, its total supply of potential energy has been converted, and there is no possibility of further movement. Of course, you can replenish the book's supply of energy simply by picking it up

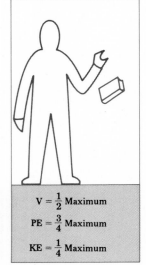

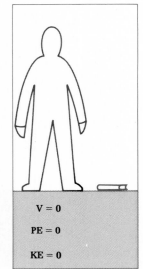

| $V = 0$ | $V = \frac{1}{2}$ Maximum | $V = $ Maximum | $V = 0$ |
| $PE = $ Maximum | $PE = \frac{3}{4}$ Maximum | $PE \cong 0$ | $PE = 0$ |
| $KE = 0$ | $KE = \frac{1}{4}$ Maximum | $KE = $ Maximum | $KE = 0$ |

$V = $ Velocity
$PE = $ Potential energy
$KE = $ Kinetic energy

**1-1** *When the textbook is raised to shoulder height, its stored or potential energy is at maximum. As the book falls, potential energy is rapidly converted into the kinetic energy of motion. By the time the book hits the ground, it contains neither potential nor kinetic energy. The man can transfer additional supplies of energy to the book by picking it up again.*

again; by doing so, you transfer some of your kinetic energy to be stored as potential energy in the book.

In living organisms, energy is constantly being changed from kinetic to potential and back again. Cells store energy and then later convert that potential energy into kinetic energy that permits the organism to undertake a variety of activities — including the storage of more potential energy.

Living organisms also continually transform energy from one form to another. According to the second law of thermodynamics, these changes must always proceed in the direction of increasing entropy, or disorder. Some of the forms of energy important to biological systems are listed below.

1. **Light energy** is the basic source of outside energy for all living systems; the addition of light energy to the closed system of the biosphere counteracts the trend toward increasing entropy. As biologist David M. Gates says:

> The radiant energy that bathes the earth builds order from disorder through the processes of life. Most events in the universe proceed toward increasing entropy, but life postpones the effect of this basic law by using the stream of sunlight to build highly complex assemblages of proteins, carbo-hydrates, lipids, and other biological molecules.*

Absorbed light energy is used by living systems to start or speed up chemical reactions that alter the energy levels of participating molecules.

2. **Chemical energy** is the means by which most life processes are carried on. Every activity of

*D. M. Gates. 1971. "The Flow of Energy in the Biosphere." *Scientific American.* 224(3): 89.

any living system can ultimately be traced to a chemical reaction taking place within that system. Some chemical reactions release energy; others store it for future use. The release of chemical energy can be controlled with great accuracy, simply by regulating the quantity of reacting substances.

3. **Heat energy** is associated with a high degree of entropy; once energy is converted to heat, it is usually lost to the biological system, because in living organisms heat energy cannot be utilized to do work. Heat affects matter by speeding up the rate at which individual particles of matter move and react with one another.

4. **Electrical energy** is the result of a flow of electrons along a conductor. Many living organisms convert chemical energy into a form of electrical energy utilizing charged particles of matter that move into and out of the cell. This electrochemical impulse can be transmitted rapidly throughout the entire organism; the nervous system of animals specializes in this conduction function.

5. **Mechanical energy** is energy directly involved in moving matter. Examples are the kinetic energy of a moving body or the energy associated with an expanding gas, as occurs in the steam engine. Organisms can convert the chemical energy contained in the molecules from which muscles are made into mechanical energy that does the work of moving the organism.

## The structure of matter

Matter can be defined as that which has mass and occupies space. At one time, physicists thought that matter was a category quite separate and distinct from energy. The work of twentieth-century scientists has proved this distinction to be artificial. Einstein's famous formula, $E = mc^2$, provides a way to calculate the amount of energy (E) obtained when mass (m) is converted into energy. The amount is huge, since the constant (c) by which the mass is multiplied is the speed

**1-2** *Rube Goldberg's not-quite-scientific studies express a profound truth basic to biology: energy may be transformed from one state to another within a system and energy from one system can trigger release of energy from another. Goldberg's drawings amuse us because he seizes on the unexpected consequence of an energy transformation, but the study of life constantly reveals subtle interactions no less unexpected and infinitely more varied and fascinating. (Drawing by Rube Goldberg.)*

PROFESSOR BUTTS FALLS ON HIS HEAD AND DOPES OUT A SIMPLIFIED CAN-OPENER WHILE HE IS STILL GROGGY. GO OUTSIDE AND CALL UP YOUR HOME. WHEN PHONE BELL RINGS, MAID (A) MISTAKES IT FOR AN ALARM CLOCK - SHE AWAKENS AND STRETCHES, PULLING CORD (B) WHICH RAISES END OF LADLE (C). BALL (D) DROPS INTO NET (E) CAUSING GOLF CLUB (F) TO SWING AGAINST BALL (G), MAKING A CLEAN DRIVE AND UP-SETTING MILK CAN (H). MILK SPILLS INTO GLASS (I) AND THE WEIGHT PULLS SWITCH ON RADIO (J). WALTZING MICE (K) HEARING MUSIC AND PROCEED TO DANCE, CAUSING REVOLVING APPARATUS (L) TO SPIN AND TURN. SPIKES (M) SCRATCH TAIL OF PET DRAGON (N) WHO IN ANGER EMITS FIRE IGNITING ACETYLENE TORCH (O) AND BURNING OFF TOP OF TOMATO CAN (P) AS IT ROTATES. WHEN NOT OPENING CANS, THE DRAGON CAN ALWAYS BE KEPT BUSY CHASING AWAY INCOME TAX INVESTIGATORS AND PROHIBITION OFFICERS.

## 1-1

## Elements of major biological importance and their natural abundance

| Element | Symbol | Atomic weight | Biological significance | Atomic % composition of: human body | earth's crust |
|---|---|---|---|---|---|
| Hydrogen | H | 1.01 | present in water and organic molecules | 60 | * |
| Oxygen | O | 16.00 | cellular respiration; present in water and organic molecules | 26 | 62.6 |
| Carbon | C | 12.01 | basic element of all organic molecules | 11 | * |
| Nitrogen | N | 14.01 | present in proteins and nucleic acids | 2.4 | * |
| Sodium | Na | 23.00 | ionic balance; nerve conduction | 0.7 | 2.6 |
| Calcium | Ca | 40.08 | important in membrane function, bone structure, blood clotting | 0.22 | 1.9 |
| Phosphorus | P | 30.98 | bone structure; nerve function; present in nucleic acids and ATP | 0.13 | * |
| Sulfur | S | 32.07 | present in some proteins | 0.13 | * |
| Potassium | K | 39.10 | ionic balance and nerve conduction | 0.04 | 1.4 |
| Chlorine | Cl | 35.46 | ionic balance | 0.03 | * |
| Magnesium | Mg | 24.32 | necessary for certain enzyme functions | 0.01 | 1.8 |
| Iron | Fe | 55.85 | respiration; present in heme molecules | 1.01 | 1.9 |
| Silicon | Si | 28.06 | found in lower forms | * | 21.2 |
| Aluminum | Al | 26.97 | unknown | * | 6.5 |

*negligible

of light, or 186,000 miles per second. According to this formula, one unit of mass yields $(1 \times 186,000)^2$ units of energy, or 34½ billion units. The successful explosion of the atomic bomb, which converts a small amount of matter into a huge amount of energy, demonstrated the validity of Einstein's equation.

Because the conversion of matter into energy rarely takes place in the biosphere, biologists continue to treat these two states separately. All matter is made up of a few basic **elements.** The characteristic quality of an element is that no ordinary physical process, such as heating, cooling, or reacting with other substances, can break the element down into simpler substances.

Ninety-two different elements have been found to occur naturally. Among these are such familiar substances as oxygen, hydrogen, and carbon. Chemists have been able to produce at least 12 more elements in the laboratory. It is possible that more man-made elements will be produced; there is a still unconfirmed report that element 105 was recently produced at the Joint Institute for Nuclear Research at Ubna, in the U.S.S.R.

Two or more elements may combine to form a substance with new chemical and physical properties, called a **compound.** Water, for example, is a compound, made of the elements of hydrogen and oxygen. This combination of elements depends on the creation of chemical bonds, which usually releases some amount of kinetic energy. To break the compound down into its elements again, this same amount of energy must be added. This additional energy is called the **energy of activation.**

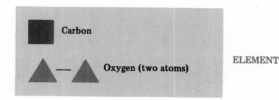

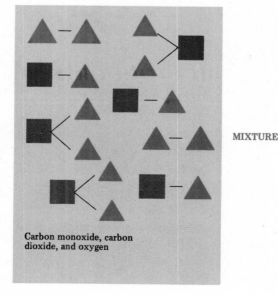

**1-3** *Carbon is an element, as is oxygen, although oxygen is customarily present in the form of two bonded atoms ($O_2$) rather than one. These two elements can form chemical bonds, producing compounds such as carbon monoxide and carbon dioxide. Compounds and elements that are physically near each other may form mixtures without the creation of any new chemical bonds. An example of a mixture can be seen in the gases customarily found emerging from the exhaust pipe of a combustion engine.*

ELEMENT

COMPOUND

MIXTURE

Carbon monoxide, carbon dioxide, and oxygen

When two or more substances, either elements or compounds, are physically very near each other without forming any chemical bond, the result is called a **mixture.** The earth's atmosphere, for example, is a mixture of many kinds of gases. Unlike a compound, a mixture can be separated into its component parts by purely physical means, such as passing the mixture through a strainer or applying a magnet. Mixtures are very important to living systems, for many parts of the cell are mixtures.

Matter can exist in any of three physical states—as a solid, a liquid, or a gas. The physical state of matter varies in response to temperature. At the coldest temperatures, matter will be a solid. With the addition of heat, matter changes from the solid to the liquid state; and if the process of adding heat is continued, the substance will eventually become a gas. Changes in pressure may also affect the physical state of matter; the effect varies with different compounds.

When matter goes from one state to another, no chemical bonds are made or broken. The change in form is due to the difference in speed at which the atoms move. As more heat is applied, the internal kinetic energy of the molecules is increased, and they can move faster and farther apart.

## Atomic structure

The basic substances of which all matter is composed are the elements. But what is the structure of an element? In order to answer this question, scientists hypothesized the existence of **atoms,** the smallest possible particles of matter which cannot be broken down into smaller particles and still retain their characteristic identity. These tiniest bits of matter were named by Greek scientists over 2500 years ago; the term comes from the Greek word *atomos,* meaning indivisible.

Our ideas about the shape and structure of the atom have changed greatly; Figure 1-4 shows

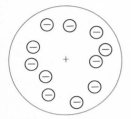

**Thompson's atom (1898)**
minus charges in a
positive sphere

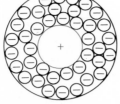

**Rutherford's atom (1911)**
electrons encompass-
ing a nucleus

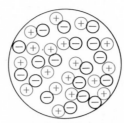

**Lenard's atom (1903)**
paired plus and
minus charges

**1-4** *These sketches trace the historical development of the concept of the atom. Thompson's model (1898), which specified little particles of negative charge embedded in a positive sphere, is jokingly called the raisin pudding model. Rutherford's model (1911), accepted for a time, was invalidated by experiments with light waves that demonstrated the existence of hollow spaces within the atom. Bohr's model was the first to explain satisfactorily all the known evidence regarding atomic structure.*

some of the early conceptions of atoms. The modern picture of the atom is based on a model proposed in 1913 by the Danish physicist, Niels Bohr. Although the existence of atoms is accepted as fact, there is still some controversy over the most accurate model of this particle of matter.

**The atomic nucleus.** The nucleus, the dense center of the atom, consists of a group of small, positively charged particles called **protons** and other small particles that carry no electric charge, called **neutrons.** The number of protons in the atomic nucleus is different for each single element; this number is called the **atomic number** of an element. For example, the atomic number of hydrogen is 1, indicating that it has 1 proton in its nucleus; the atomic number of helium is 2, since it has 2 protons; uranium, the heaviest natural element, has an atomic number of 92.

Every atom of the same element will always have the number of protons that is indicated by its atomic number. However, it is possible for

atoms of the same elements to have different numbers of neutrons in their nuclei. All chlorine atoms have 17 protons, but some have 18 neutrons and others have 20 neutrons. Atoms of the same element that contain different numbers of neutrons are called **isotopes.** Isotopes can be distinguished by their different **mass numbers,** a number derived by adding the total number of neutrons and protons in an atom's nucleus. The number of neutrons in an atom does not seem to affect its chemical properties significantly, for isotopes of a single element all appear to have essentially the same chemical characteristics. Isotopes of elements such as carbon and oxygen are frequently used in biological research, since these atoms can easily be distinguished from others of the same element; an isotope is a labeled atom.

The exact structure of the nucleus is not fully established, but it is now known that other particles, smaller than the neutron or proton, can be found in the nucleus. These particles include

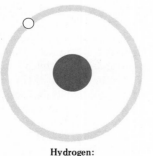

**Hydrogen:**
1 proton
1 electron

**Deuterium:**
1 proton
1 neutron
1 electron

**Tritium:**
1 proton
2 neutrons
1 electron

**1-5** *The two isotopes of hydrogen shown here have chemical and physical properties very similar to those of the hydrogen atom, although in large quantities the isotopes will weigh more. These isotopes can be introduced into the atmosphere or into compounds being used in an experiment, allowing researchers to trace the path of specific hydrogen atoms during chemical reactions within living cells.*

the nucleus is equal to the number of protons in the nucleus of the atom. The attraction of the positive charges of the protons thus cancel out the attraction of the negative charges of the electrons, and the atom is electrically neutral. If, for some reason, the atom gains or loses an electron, it then carries an electrical charge and is known as an **ion.** Ionization occurs frequently, since electrons are particles with very little mass, relatively easy to move.

Since electrons are so small and move so fast, it has been difficult to determine their exact position in relation to the nucleus. The Bohr model of the atom specified electrons moving around the nucleus in concentric circular orbits, rather like the planets moving around the sun. Although this model of electron configuration has been subsequently modified, it is still a useful way to think of the atom, because it helps to explain and predict many kinds of chemical reactions.

According to Bohr's theory, the orbits in which electrons revolve are not located at random; each orbit is at a certain distance from the nucleus. Bohr called these orbits principal energy levels. The orbit nearest the nucleus is the first and lowest energy level. Bohr reasoned that the electrons would stay at that level unless they received additional energy. When an electron at the first level is bombarded by a sufficient number of energy quanta, it is "bumped up" to the second energy level, the next concentric orbit. When an electron loses energy, it falls to a lower energy level. The farther away the electron is from the nucleus, the higher its energy level and the more energy it takes for the protons to hold it in place.

The Bohr model shows seven different energy levels, which are usually designated by the letters K (for the lowest energy level) through Q (for the highest energy level). It was later discovered that there is a limit to the number of electrons that can occupy each energy level. There are never more than 2 electrons at the first

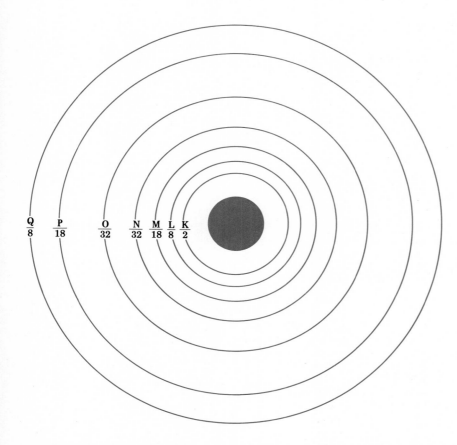

**1-6** *The Bohr model of the atom specifies seven energy levels of electrons circling the central nucleus. Although Bohr envisioned these energy levels as being rather like the orbital paths of planets circling the sun, present-day theory has altered this feature of the model. Modern drawings show the various energy levels in terms of probability clouds, or areas where the probability of finding the electrons is at maximum.*

the muon, the pion, and the neutrino. It is believed that these particles play some important role in holding the nucleus together.

**Electrons.** Moving around the nucleus of the atom are one or more very small particles, carrying a negative electrical charge, called **electrons.** Electrons are very much smaller than the nucleus; it would take at least 1,836 electrons to weigh as much as a single proton or neutron. It is these electrons that largely determine the chemical properties of the atom, and which are directly involved in chemical reactions.

The number of electrons traveling around

level, 8 electrons at the second level, 18 electrons at the third level. Although levels M through P can hold a larger number of electrons, it appears that they are able to achieve a stable condition with only 8 electrons.

Each atom has a characteristic and unique pattern of orbiting electrons, called its **electron configuration.** For example, oxygen has a complete shell of 2 electrons at the first level but only 6 electrons at the second level. Neon has 2 electrons at level K and a full set of 8 electrons at level L. It is the electron configuration of the atom that gives each element its particular chemical properties. Most chemical reactions take place between electrons in the outermost energy level

## 1-2
## Electron energy levels

| Atomic number | Element | 1 | 2 | 3 | 4 | 5 | 6 | 7 |
|---|---|---|---|---|---|---|---|---|
| 1 | Hydrogen | 1 | | | | | | |
| 2 | Helium | 2 | | | | | | |
| 3 | Lithium | 2 | 1 | | | | | |
| 4 | Beryllium | 2 | 2 | | | | | |
| 5 | Boron | 2 | 3 | | | | | |
| 6 | Carbon | 2 | 4 | | | | | |
| 7 | Nitrogen | 2 | 5 | | | | | |
| 8 | Oxygen | 2 | 6 | | | | | |
| 9 | Fluorine | 2 | 7 | | | | | |
| 10 | Neon | 2 | 8 | | | | | |
| 11 | Sodium | 2 | 8 | 1 | | | | |
| 12 | Magnesium | 2 | 8 | 2 | | | | |
| 15 | Phosphorus | 2 | 8 | 5 | | | | |
| 19 | Potassium | 2 | 8 | 8 | 1 | | | |
| 25 | Manganese | 2 | 8 | 13 | 2 | | | |
| 30 | Zinc | 2 | 8 | 18 | 2 | | | |
| 47 | Silver | 2 | 8 | 18 | 18 | 1 | | |
| 53 | Iodine | 2 | 8 | 18 | 18 | 7 | | |
| 79 | Gold | 2 | 8 | 18 | 32 | 18 | 1 | |
| 92 | Uranium | 2 | 8 | 18 | 32 | 21 | 9 | 2 |

of each atom, so the number and arrangement of electrons at that level is important in determining which atoms will react with others, and what the reaction will be. An atom such as neon, with a complete outer shell, will not under ordinary circumstances react with other atoms, whereas oxygen can be expected to react with other atoms that can provide the two additional electrons it needs to make its configuration stable.

Bohr's idea was that electrons moved in regular circular orbits; today this theory has been altered, and scientists refer to a "probability cloud" or an orbital. Both concepts describe an area in which the electrons are most likely to be found, according to the laws of statistical probability. Electrons, which move in random paths as a result of their own kinetic energy, are not necessarily confined to the orbital area; there is a small but statistically measurable possibility that an electron of an atom in this book is somewhere out in the Milky Way. Orbitals describe probabilities, not certainties. It is not certain that all electrons will be on the orbitals shown in Figure 1-6, but that is where the probability of finding them is at its greatest.

### Chemical bonds—the formation of molecules

Two or more atoms may join together to form a *molecule* of a new substance, with chemical properties that differ from those of the atoms from which it is formed. The existence of the molecule depends on chemical bonds between the individual atoms. A chemical bond is the result of electron interaction among atoms; since it involves some exchange of energy, a chemical bond can be measured in terms of energy units. Some chemical bonds contain much more energy than others, as shown in Table 1-3. This fact has important biological implications, since it means that some compounds can store much more energy than others.

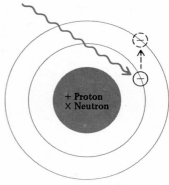

**1-7** When a packet, or quantum, of energy hits an orbiting electron, the electron itself becomes energized, and thus it is enabled to move farther away from the nucleus. If the additional supply of energy gained by the electron is great enough, the electron may move to another energy level or it may even leave the atom altogether, perhaps forming a bond with another nearby atom.

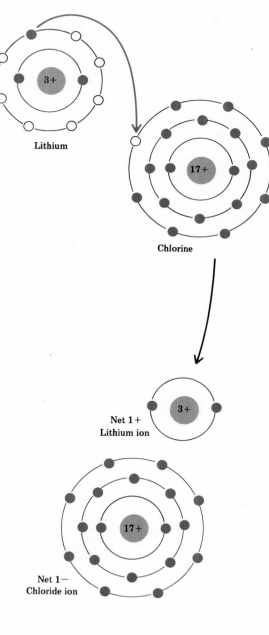

**1-8** *Lithium and chlorine are linked by an ionic bond. Lithium gives up the single electron in its second (L) energy level; this electron completes the third (M) energy level of the chlorine atom. The lithium portion of the molecule has a charge of +1, while the chlorine portion has a charge of −1. The attraction of these opposing charges helps to hold the molecule together.*

**1-3**

**Bond energies (kcal/mole) for common organic elements**

|   | **H** | **C** | **N** | **O** |
|---|-------|-------|-------|-------|
| H | 104   | 99    | 93    | 110   |
| C | 99    | 83    | 70    | 84    |
| N | 93    | 70    | 32    | —     |
| O | 110   | 84    | —     | 33    |

When two atoms come close enough together for their electron clouds or orbitals to overlap, an energy exchange may occur, forming a chemical bond. The exchange enables each of the participating atoms to assume an electron configuration that is more stable than its original configuration. This rearrangement can occur in one of two ways: one atom can give up some of its electrons to the other (ionic bonding); or each atom can share one or more electrons with the other (covalent bonding). Knowing the electron configurations of the atoms involved allows us to predict the kinds of bonds that will be formed.

**Ionic bonds.** In ionic bonding, one atom gives up its outermost electrons to another. This bonding mechanism can be seen in Figure 1-8, showing the formation of a molecule of lithium chloride. The electron configuration of chlorine is 2 electrons in the K level, 8 in the L level, and 7 in the M level. Chlorine therefore needs either to give up 7 electrons or to receive 1, in order to make its configuration stable. Theoretically, it is possible for the chlorine atom to lose its 7 outer electrons, but it would take a great deal of energy to wrench them away, since they are being pulled toward the nucleus of the atom by the positive charge of the 17 protons. Even if the 7 electrons could somehow be removed, the atom would then be so strongly ionized that its configuration would not be stable. The atom would have 17

**21**

protons and 10 electrons, so it would carry a charge, or valence, of +7. This strong charge would pull additional electrons back to its outer shell.

It is much more likely that the chlorine atom will receive one electron from some donor atom, such as lithium, which has an outer shell of a single electron at the L level and therefore needs to lose 1 electron to achieve a stable configuration. The lithium gives up 1 outer electron, so that it now has a stable configuration of 2 electrons at the K level and a relatively weak valence

## 1-4
## An explanation of symbols and signs

**Atomic number**
A number that represents the protons an atom contains. For instance, $_8O$ signifies eight protons within the oxygen nucleus.

**Mass number**
The sum of the number of protons and neutrons present within an element. The symbol $O^{16}$ shows that oxygen contains eight protons and eight neutrons.

**Isotopes**
Atoms of the same element which possess like chemical properties but differing masses. Mass numbers such as $O^{17}$ or $O^{18}$ indicate nine or 10 neutrons within the oxygen nuclei, respectively.

**Energy levels of electrons**
Each orbital, at maximum, can hold the following number of electrons:

2    8    18    32

However, the outermost orbital can function as a complete orbital with only eight electrons.

**Valence**
A measure of the bonding capacity of an atom.

OXYGEN ($_8O$)                          SODIUM ($_{11}NA$)

−2 VALENCE                          +1 VALENCE

If oxygen gains electrons and completes its outer orbital, the electrons (−) will exceed the protons (+) by two. Sodium may lose one electron and have an extra plus charge. Outer orbitals with less than four electrons are usually electron donors and have a plus valence. Outer orbitals with more than four electrons possess a minus valence. Elements share (covalent bonding) electrons if their outer shells contain four electrons, and therefore do not have a plus or minus valence.

**Bonding symbols**

The bonding of elements may be shown in several ways. The single bonding compound, methane (CH₄), is shown below:

Double and triple bonding may also occur between the outer electron orbitals of elements. Benzene (C₄H₆), with a double carbon-to-carbon bond, is represented in the following diagram:

$$
\begin{array}{c}
\text{H} \\
| \\
\text{C} \\
\diagup \ \ \diagdown \\
\text{H--C} \quad \text{C--H} \\
\| \qquad \qquad \\
\text{H--C} \quad \text{C--H} \\
\diagdown \ \ \diagup \\
\text{C} \\
| \\
\text{H}
\end{array}
$$

And, propyne (C₃H₄) has a triple carbon-to-carbon bond:

$$
\begin{array}{c}
\text{H} \\
| \\
\text{H--C--C} \equiv \text{C--H} \\
| \\
\text{H}
\end{array}
$$

of +1. The chlorine atom accepts the electron, giving it a stable outer shell of 8 electrons at the M level and a valence of −1. The attraction between the positively charged lithium atom and the negatively charged chlorine atom holds them together, forming a molecule of lithium chloride.

The process of losing electrons is called **oxidation;** the process of gaining electrons (i.e., reducing the positive charge) is called **reduction.** The lithium atom has been oxidized, and the chlorine atom has been reduced; the two processes are always linked in a single reaction. In living systems, it is generally the hydrogen atom which is oxidized (or contributes an electron) in energy exchanges; oxygen frequently serves as the electron acceptor.

Ionic bonding generally takes place between atoms that need to exchange only one or two electrons. Atoms with fewer than four electrons in their outer shell, such as potassium, sodium, iron, and calcium, tend to form ionic bonds in which they are the electron donors. Atoms such as oxygen, chlorine, and sulfur, with more than four electrons in their outer shell, tend to form ionic bonds in which they are the electron receivers.

Ionic compounds can be classified on the

basis of certain chemical characteristics. Some compounds are **acids** and others are **bases.** A quick chemical test can identify these compounds, for acids will turn litmus (an organic dye) from blue to red, whereas bases turn it from red to blue. Acids and bases can also be distinguished by their taste; acids are sour and bases are bitter.

A more significant way of distinguishing acids and bases is through their chemical reactions. An acid is a substance that can donate protons to other compounds. A base is a substance that can accept protons. In a chemical sense, acids and bases are exact opposites.

Because of their opposing chemical characteristics, acids and bases frequently react with one another. Such a reaction neutralizes both the acid and the base, producing water and a salt. It is a case of two extremes canceling each other out.

Chemists have devised a scale to measure exactly how strong an acid or a base any given compound is. Called the **pH scale,** it runs from a value of 0 to 14; the scale is based on actual calculations of the number of hydrogen ions found when the substance is mixed in water. The strongest acid has a pH of 0; the strongest base has a pH of 14. A substance with a pH of 7 (water, for example) is neutral. Most compounds found in living organisms have a pH between 6 and 8, although there are a few exceptions, such as digestive secretions, which are strongly acidic.

**Covalent bonds.** Another kind of chemical bond is that formed when two atoms share their electrons. This type of bonding is frequently found in elements, such as carbon, with an outer shell of four electrons, or in elements which must gain or lose more than two electrons to attain a stable configuration. An example can be seen in a molecule of methane. To attain stability, the carbon atoms must acquire or lose four more electrons; however, that would leave it strongly ionized, with a charge of either + or − 4. But

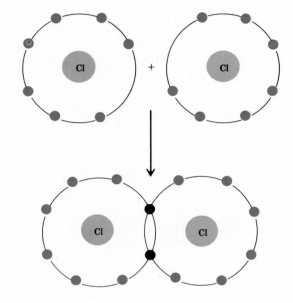

bonding can be accomplished without ionization if the carbon atom shares the single electrons of four atoms of hydrogen. The shared electrons circle both the hydrogen and the carbon atoms.

**Bonds and polarity.** When atoms combine to form a molecule, they establish a definite and predictable geometric relationship, which is largely determined by the electron configuration of the atoms involved. The shape of the molecule then influences the way it will react with other molecules.

Another consequence of the geometric shape of molecules is the uneven distribution of electric charge. The water molecule furnishes a good example. The two hydrogen atoms are not attached at opposite sides of the oxygen atom; rather, they are attached so that they form a bond angle of precisely 104.5°. As a result, the region of the oxygen atom where the bond is formed is slightly positively charged, while the other side of the atom is slightly negatively charged. A water

**1-9** *Two atoms of chlorine may be linked together by covalent bonding. Each atom lacks one electron in its outer shell; by sharing electrons, each can achieve a stable configuration with eight electrons at the M level. Covalent bonding is frequently found in carbon compounds, for each carbon atom can share electrons with as many as four other atoms.*

**1-10** *When two hydrogen atoms and one oxygen atom combine to form a molecule of water, the geometric relationship of the three atoms is always the same. Because chemists know the shape of this molecule, they can predict the way it will combine with other compounds or other molecules of the same compound. The bond angles are also responsible for giving the molecule its polarity; the left side of the molecule pictured here will carry a slight positive charge, while the right side is slightly negative.*

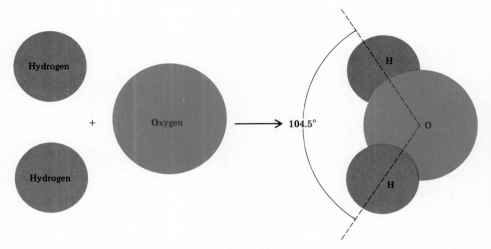

molecule has some degree of **polarization;** that is, it has a weak positive and a weak negative pole.

Polar molecules will always orient themselves in a certain way in regard to other molecules. For example, the negative pole of one water molecule will repel the negative pole of another water molecule and attract the positive pole. Therefore water molecules will always combine in the same predictable way, aligned in a specific pattern. Many organic molecules are polar, and their polarity influences the kind of reactions that can take place in the living cell.

**Weak chemical bonds.** In addition to the strong types of bonds — ionic and covalent bonds — molecules can also be held together by weaker bonds. One such bond is the **hydrogen bond.** This bond is produced by the polar attraction between positively charged hydrogen atoms on one part of a

molecule and negatively charged atoms of oxygen or nitrogen on another molecule. Figure 1-11 shows the hydrogen bonds thus formed between the hydrogen atom of one water molecule and the oxygen atom of the next. These additional bonds give the water molecule great tensile strength.

**High-energy bonds in living systems.** Chemical bonds between atoms may store potential energy, which can be released as kinetic energy when the bond is broken. This is true not only of chemical reactions in the laboratory but also of the chemical reactions that take place in living systems. One problem that all living systems must solve is storing energy so that it can be released quickly when needed, in such a way that the release will not create excess heat which might induce other harmful bond breakages.

In most living systems, energy storage takes

**1-11** *Because of their bond angles, water molecules always line up in a very precise order. Hydrogen bonds will be formed between the positive pole of one molecule and the negative pole of the next. This bonding gives water its great cohesive strength and explains why pure water has such a high degree of surface tension.*

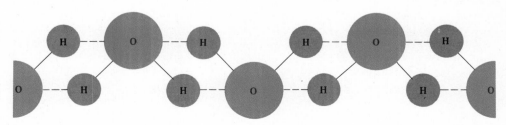

place in the form of high-energy bonds forming a molecule of a compound called adenosine triphosphate, or ATP. The shape of the ATP molecule is shown in Figure 1-12. It consists of a central unit of adenosine to which is attached a tail made up of three phosphate groups. Through an internal rearrangement of the electrons in this molecule, a maximum amount of energy is concentrated in bonding the last two phosphate groups to the molecule; these are the two high-energy bonds. When the organism needs energy to do some kind of work, such as building new molecules or moving molecules from one place to another, the energy locked in the high-energy phosphate bond that holds the last phosphate group can be extracted by breaking the bond. The process produces quick energy that can be used anywhere in the cell, and leaves behind another compound called adenosine diphosphate (ADP) and a free phosphate group. When the organism receives a new supply of energy, it can be stored through another process that bonds the free phosphate group back to the ADP, forming a molecule of ATP. This storage system provides controlled release of energy in amounts roughly equal to that needed for most biochemical reactions; very little energy is wasted.

## Chemical reactions

When one atom or molecule interacts with another, rearranging electrons so as to form new bonds, we call it a chemical reaction. Two substances may join together in a combination reaction, or a complex molecule may undergo a reaction and break down into simpler components. Chemists indicate reactions through the use of a simple rotation system. A combination reaction would be written

$$A + B \rightarrow C,$$

whereas a decomposition reaction would be

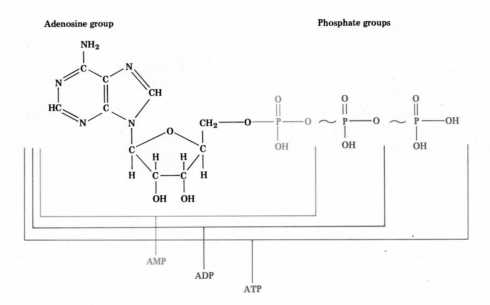

Adenosine group — Phosphate groups

written

$$C \rightarrow A + B.$$

All compounds on the left of the arrow are reactants; those on the right are products.

## Energy exchanges during reactions

Chemical reactions are not random occurrences, but are the result of the conjunction of many different factors. The most important of these factors is the availability of free energy to initiate the reaction.

Atoms and molecules are relatively stable chemical units. The forces that would lead to change, such as the unit's electrical charge or the thermal energy of the subatomic particles, are balanced by forces that hold the unit together, such as the gravitational pull of the nucleus and the amount of energy locked up in bond formation. Only when additional energy is supplied will the unit move from its stable condition to an

**1-12** *The high-energy bonds of the ATP molecule are represented by the wavy lines between the central portion of the molecule and the last two phosphate groups. When the cell needs energy, the bond holding the last phosphate group is broken, yielding free energy, a free phosphate group, and ADP. In an emergency, the second high-energy bond can also be broken; the remaining compound is adenosine monophosphate, or AMP.*

unstable one in which it is capable of entering into a reaction. This additional energy is the energy of activation; it permits molecules to pass "over the hump" or energy barrier that ordinarily prevents reactions.

The energy of activation can come from several different sources. An electrical current is one; chemists frequently induce reactions by passing an electrical current through a chemical mixture. Heat energy can also serve as the energy of activation. A third source is the kinetic energy of the molecules themselves. Most of the time this energy is used in moving the molecule about. But if two molecules bump into each other, they are temporarily prevented from moving, and during that short period their kinetic energy is converted to free energy that can serve as energy of activation. Thus collisions between molecules frequently induce chemical reactions.

Once the energy of activation is supplied, the chemical reaction begins, and during the reaction there will be some exchange of energy among the various compounds involved. The energetics of the reaction can be measured by comparing the energy contained in the various reactant compounds with those in the product compounds. If the total energy content of all the reactants is greater than the total energy content of all the products, then the reaction is **exergonic;** it has released a certain amount of free energy. If the energy content of the products is higher than that of the reactants, the reaction is **endergonic;** it has used free energy. An example of an endergonic reaction is the one involved in producing ATP:

$$ADP + phosphate \rightarrow ATP;$$

energy must be added to make ATP. The reaction

$$ATP \rightarrow ADP + phosphate$$

is exergonic, and provides free energy that can be used by the cell.

It should be noted that in living systems, every endergonic reaction has its complementary exergonic reaction. For every process that stores energy, there is a process that releases it; the two are always linked together.

## Chemical equilibrium

All chemical reactions are theoretically reversible; that is, they can proceed in either direction. When we write a formula like that for the production of lithium chloride

$$(2 \text{ Li} + \text{Cl}_2 \rightarrow 2 \text{ LiCl}),$$

we are expressing the statistical probability that most atoms in the system will be undergoing that combination reaction. But at any given moment, there will also be some molecules in the group that are undergoing the reverse reaction:

$$2 \text{ LiCl} \rightarrow 2 \text{ Li} + \text{Cl}_2.$$

Although, in principle, every reaction is reversible given the right conditions, in reality it often happens that those conditions do not occur. Many reactions are, for all intents and purposes, irreversible. For example, if the reaction releases a large amount of energy, it cannot be reversed unless more energy is added to the system. Chemical reactions may also be considered irreversible if one of the products leaves the site of the reaction—for example, if it is given off as a gas.

But no reaction, no matter how easily reversible, can continue indefinitely. Within a certain period of time, the chemical system will reach a point of equilibrium. The rate of combination, or the forward rate, will be exactly equal to the rate of decomposition, or the reverse rate. Thus the relative proportions of lithium, chlorine, and lithium chloride in the test tube will show no further change. (They will not necessarily be present in equal amounts; the equilibrium point varies, depending on the amount of energy needed to perform the forward and the reverse

Equilibrium

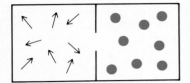 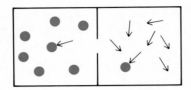 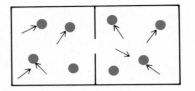

**1-13** *Equilibrium is reached when the rate of the forward reaction combining the two different elements is the same as the rate of the reverse reaction, in which the compound is broken down into its separate elements. As the sketch shows, at equilibrium, there will be a certain number of molecules present, as well as atoms of each element.*

reaction.) Once the equilibrium point has been reached, there will be no net increase in the amount of lithium chloride formed.

All physical systems continually move toward a state of chemical equilibrium. If the system remains closed, equilibrium will eventually be reached; no net forward action can take place. This tendency to move toward equilibrium presents a serious threat to the existence of biological systems. For a living organism, a state of chemical equilibrium is equivalent to death. Biological systems avoid equilibrium by constructing sequences of reactions which remove the product of each reaction from the site immediately. As soon as the combination reaction

$$A + B \rightarrow C$$

takes place, the product of the reaction, C, enters a new reaction,

$$C + D \rightarrow E.$$

This new reaction removes product C from the site of the initial reaction and thereby prevents a reversal of the first reaction, which would eventually cause the system to reach equilibrium. An example of this avoidance of equilibrium can be seen in the sequence of reactions that breaks sugar down into carbon dioxide and water. Over 30 different reactions take place in this process, each one potentially reversible. But because each product is involved in a new reaction, these reversible chemical reactions move along a one-way path, and the living system manages to decompose these organic molecules at a rapid rate.

## The rate of chemical reactions

In any chemical reaction, the exact point of equilibrium is determined by the rate at which the various reactions take place. The rate of a chemical reaction is the amount of reaction that takes place in a given period of time. Reactions may take place at varying rates within the same chemical system. For example, in the reaction

$$2\ H_2 + O_2 \rightarrow 2\ H_2O$$

it is clear that the combination reaction of $H + O$ is occurring at a faster rate than the reverse reaction breaking down $H_2O$. In order to predict the behavior of chemical systems, it is important to understand the factors that influence the rate of chemical reaction.

**The concentration of reacting substances.** The laws of statistical probability tell us that there is more likely to be a collision between hydrogen and oxygen atoms forming a molecule of water when there are many of both atoms in the system than when there are only a few. Generally speaking, the higher the concentration of reacting substances, the greater will be the rate of reaction. For example, it is usually possible to speed up a chemical reaction simply by doubling the amount of reactants in the system; this fact is expressed in the Law of Mass Action. The influence of concentration on the rate of reaction also explains why chemical systems will eventually reach equilibrium. After a while, a large percentage of hydrogen and oxygen atoms will have formed molecules of water; the concentration of hydrogen and oxygen drops significantly, thus lowering the rate at which the combination reaction takes place. Since the combination reaction steadily increases the concentration of water molecules in the system, the rate at which the reverse reaction of decomposition of water occurs will also increase. Therefore the rates of the forward and reverse reactions will eventually become equal, and equilibrium will be reached. Living systems can maintain different concentrations of a substance without reaching equilibrium, dividing them by a membrane that keeps certain molecules from passing through.

**The effect of temperature and pressure.** Under conditions of high temperature, atoms and molecules move very rapidly. The faster they move, the greater is the chance that they will collide with some other particle. Moreover, since each particle possesses more kinetic energy when it moves fast than it does when it moves slowly, the chances that the collision will be successful in altering chemical bonds are also increased when the particles are moving at high speeds. The rate at which a chemical reaction takes place can be speeded up by increasing the temperature. In living systems, this additional heat energy can be dangerous, since it will speed up all reactions and may therefore destroy the balance of long sequences of reactions. Pressure also affects the rate of reactions, but its effect varies so greatly that no general rule can be advanced.

**The effect of catalysts.** Catalysts are chemical substances which speed up the rate of reaction between two other substances but are not in themselves permanently affected by the reaction that takes place. The effect of a catalyst is to lower the amount of energy needed for activation. One example of a catalyst is platinum, which serves to activate and speed up many chemical reactions, such as the combination reaction between hydrogen and oxygen that forms water molecules. Enzymes are catalysts found in biochemical systems.

Catalysis is extremely important in the chemical reactions that take place in living systems. Catalysis has the advantage of being low-heat chemistry, and it can be narrowly controlled by altering the quantity of the catalyst present in the system. Catalysis can be highly specific, affecting only one reaction, without in any way disturbing the other reactions that are going on in the complex system of the cell.

## Inorganic compounds in living systems

Living systems possess the ability to form molecules that are based on the linkage of atoms of carbon; compounds exhibiting this characteristic carbon-to-carbon linkage are called organic compounds. Although chemists have recently learned how to synthesize organic compounds in the laboratory, in nature they generally occur only in association with living (or once-living) organisms.

All other compounds, those with no carbon-to-carbon linkage, are called inorganic compounds. Several inorganic compounds play important roles in living systems.

### Water

About 70 percent of the human body is water. We are virtually walking puddles, and we share this dependence on water with all other living organisms. Obtaining and conserving water is a basic life function.

One of the primary uses of water in the living organism is as a medium in which other molecules can be dispersed. Large molecules can remain suspended and evenly dispersed in water, so that water can be used as a means of transport. By helping to distribute dissolved molecules, water helps promote chemical reactions; it also ionizes many molecules, thus making them chemically active. Water itself plays a role as a reactant in a number of important processes, such as the digestion of proteins and the synthesis of large organic molecules. It is also a product in many biochemical reactions.

Water has several properties which make it particularly useful to living systems. For example, it has the ability to absorb and release heat energy very slowly. That is why you can swim comfortably in a pool or small lake in the evening, even when the air is chilly; since the water releases the stored heat of the sun much more slowly than the air does, it remains warm for a longer time. The water content of the body is a good source of insulation against sudden changes in temperature that might otherwise be fatal to the organism. This insulation is especially important for mammals, because they must maintain their temperature within a fairly narrow range.

Another property of water that makes it useful to living systems is that, due to hydrogen bonding of polar molecules, its molecules cling very tightly to each other and resist separation; it has a high degree of cohesion. This cohesion makes water a very good medium of transportation; where one molecule goes, it tends to pull all the others right along behind it. Without water's great cohesiveness, the Douglas fir tree would not be able to send columns of water 250 or 300 feet straight up in the air to the uppermost needles.

## Carbon dioxide

Carbon dioxide in the atmosphere is the principal source of the carbon from which all organic compounds are made. Green plants take in carbon dioxide, and with the help of light energy from the sun, use it to produce sugar, the basic energy source of all living organisms.

Carbon dioxide, which is generally in the form of a gas, does not readily react with other molecules. In order for most biochemical reac-

tions to take place, carbon dioxide must be dissolved in water, where it forms a new compound, carbonic acid ($H_2CO_3$). The formula for this reaction is written

$$CO_2 + H_2O \rightarrow H_2CO_3.$$

This reaction is easily reversible with only a slight energy cost, so that the carbon dioxide can be released whenever it is needed.

## Oxygen

Oxygen plays an important role in biochemical systems, for it is a highly efficient oxidizing agent. Oxygen removes electrons from organic compounds with which it is combined. It is this removal of electrons, coupled with their eventual transfer to waiting electron acceptors, that is the basis of aerobic cellular respiration.

In the atmosphere, oxygen exists only in the form of a molecule, two atoms joined together ($O_2$). Within living organisms, the $O_2$ may dissolve in liquids, or it may combine with hydrogen to form hydroxyl (OH) ions. Oxygen is found as a component of most organic compounds.

## Organic compounds in living systems

All organic compounds contain the element carbon. Because carbon has a propensity toward covalent bonding, it readily forms long chain molecules by sharing electrons with other carbon atoms. These chains of carbon-to-carbon linkages are not usually in the form of a straight line; when more than two carbons are linked, the chain begins to bend. The chain can turn corners, double back on itself, or close to form rings. Each different shape gives the molecule different chemical properties. Carbon-to-carbon linkages offer an almost infinite variety of complex molecules, each with its own special properties.

## Carbohydrates

The name of this group of organic compounds reveals its components. The basic building unit of carbohydrates is a group of one carbon atom, two hydrogen atoms, and one oxygen atom; in other words, it is carbon plus water. The general formula is expressed as $CH_2O$. This basic unit can be strung together in long chains, or arranged in rings, or twisted in spirals.

In the carbohydrate group, it is quite common to find molecules that have the same molecular formula but very different structural formulas and therefore different chemical and physical properties. Such molecules are called **isomers.** An example of isomeric carbohydrates can be seen in glucose and fructose. They both have the molecular formula $C_6H_{12}O_6$, but their structural formulas are different (see Figure 1-14). Therefore they react somewhat differently and have different properties. For instance, fructose has a much sweeter taste than glucose.

The simplest carbohydrates are the **monosaccharides,** which usually contain no more than six carbon atoms. Among the monosaccharides are the sugars glucose and fructose, ribose, and xylose. Rings of monosaccharides can be joined together by a bond that eliminates two hydrogen atoms and one oxygen atom, to form even more complex molecules. When two monosaccharides are joined, the resulting molecule is called a **disaccharide;** examples are maltose, lactose, and sucrose (common table sugar). There are also **trisaccharides,** with three rings joined together. Beyond three units, the molecules are called **polysaccharides;** these include starch, glycogen, and cellulose. A more general name for any large molecule made up of monomers, repeated groups of atoms in chains or rings, is a **polymer.**

Carbohydrate molecules vary greatly in their size, and therefore also in their solubility and their mobility. Simple sugars, like glucose, dissolve easily and can also easily pass through cell membranes and into a transportation system such as the bloodstream. But large molecules such as starch and glycogen are too big to pass through cell membranes, and they are generally insoluble in water. That is why they are ideal storage units; they do not enter into, or influence, cellular chemical reactions. Much of the metabolic activity of living organisms involves building up large carbohydrate molecules for storage, and tearing these macromolecules down into simpler sugars when they are needed for immediate use.

In living systems, carbohydrates function primarily as a basic energy source. The oxidation of sugar molecules yields end-products of carbon dioxide and water, and a certain amount of free energy that is available to do work. In most

**1-14** *This figure illustrates two common methods of indicating molecular structure. (A) attempts to represent the three-dimensional arrangement of the atoms. (B), with its linear pattern, is less realistic, but it makes the difference between the two isomers immediately obvious. Neither method is more correct than the other; each has its own uses.*

organisms, the fuel is the monosaccharide glucose, and the basic equation is

$$C_6H_{12}O_6 + 6\ O_2 + 6\ H_2O \rightarrow$$
$$6\ CO_2 + 12\ H_2O + Energy.$$

## Lipids

The second major group of organic chemicals is the lipids, the fats and oils. These are also made of the elements of carbon, hydrogen, and oxygen, but in a ratio different from that of the carbohydrates; there is less oxygen and more hydrogen in lipids. They are both a source of energy and a component of living tissue.

Lipids are a rather heterogeneous group of compounds; they show similar chemical properties but are often structurally dissimilar. The typical fat molecule has two distinct parts: an alcohol group, usually glycerol; and a fatty acid. Figure 1-16 shows the structure of a molecule of glycerol. Fatty acids typically contain a carboxyl (COOH) group and a long chain of carbons, each of which may bond on two hydrogens. A fatty acid that contains two hydrogens for each of these carbons is said to be **saturated;** an unsaturated fat has "unfilled" carbons, or carbons that have not formed four bonds.

It is the OH, or alcohol, group of the glycerol to which the COOH group, the acid part of the fatty acid, is attached. In order for the new molecule to be formed, the glycerol must give up a hydrogen, and the fatty acid must give up an OH group; these displaced atoms unite to form a molecule of water. In other words, fat synthesis requires the removal of water. (This type of reaction is called **dehydration synthesis.**) The same thing is also true of carbohydrate synthesis.

Since a fat is formed by removing water, it can be broken down into its component parts by adding water. (Reactions in which the chemical addition of water results in the splitting of molecules are called **hydrolysis** reactions.) The

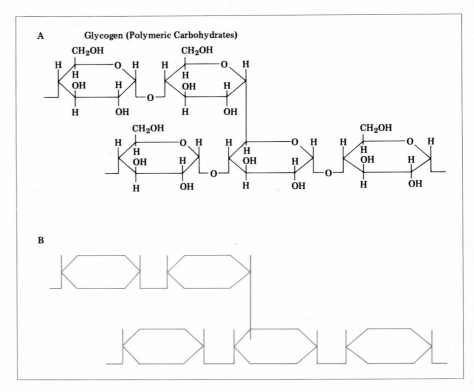

digestion of both lipids and carbohydrates takes place by hydrolysis.

Fats are the most concentrated source of energy available to living systems. The chemical reason that fats contain more energy than carbohydrates can be seen simply by looking at the number of hydrogens in their molecular formulas. For example, the formula for glucose is $C_6H_{12}O_6$; the formula for a common fat (tristearin) is $C_{57}H_{110}O_6$. Since fats have a much higher ratio of hydrogen to oxygen, they can undergo more oxidation, thus releasing more energy than carbohydrates. However, fats cannot be metabolized as efficiently as carbohydrates; it takes about 10 percent more energy to release a given quantity of energy from a fat than it does to release the same quantity of energy from a carbohydrate,

**1-15** *Glycogen is a polymer. It consists of a sequence of monomers (the individual carbon rings) joined together by H—O—H bonds. The representation in (B) is a kind of chemist's shorthand that emphasizes the polymeric structure. Glycogen, a storage molecule, can be easily broken down into its major components (the monosaccharides that are the monomers) when the cell requires additional energy.*

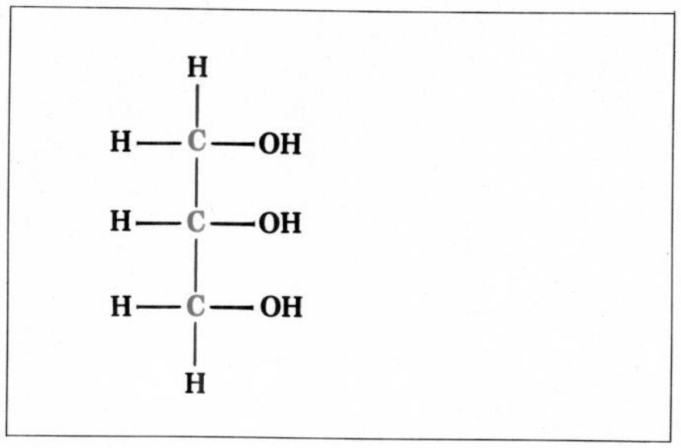

**1-16** *The glycerol portion of a lipid molecule is shown here. A fatty acid can be bonded on at each of the three sites of OH groups. The H will combine with an OH from the fatty acid, forming a molecule of water; the O of the glycerol will bond to a carbon atom of the fatty acid. Complete lipids are those in which fatty acids are attached at all three possible sites.*

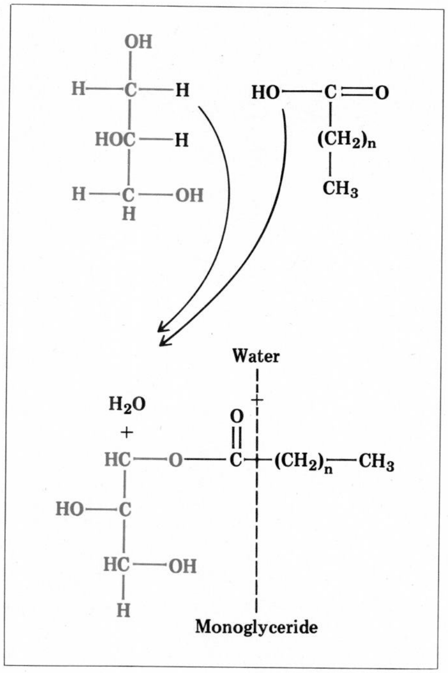

**1-17** *The process by which this lipid is formed is called dehydration synthesis. The glycerol loses an H, and the fatty acid loses an OH group; these combine to form water. The reaction thus produces a lipid (monoglyceride in this case) and a molecule of $H_2O$.*

where they help to absorb certain parts of the light spectrum. Another important subgroup is the steroids, which include the hormones secreted by the sex glands, such as estrogen and testosterone. Steroids also play an important part in cellular membranes, helping to regulate the passage of various molecules from one side of the membrane to the other. The phospholipids, lipid molecules that include a phosphate group, are important components of cell membranes.

## Proteins

Proteins are far more complex molecules than either carbohydrates or lipids. They also play more varied roles in the living organism. Fibrous proteins serve as structural elements; they are the basis of feathers, horns, scales, hair, cartilage, tendons, ligaments, and muscles. Globular proteins are the enzymes and hormones that regulate most of every organism's activities.

The fundamental building block of proteins is the amino acid. Amino acids are made of hydrogen, oxygen, carbon, and nitrogen. Joined to a central carbon atom is an amino group, with the formula $NH_2$, and a carboxyl group, with the formula $COOH$. The carboxyl group gives the molecule its acidic properties. There is room for two other bonds to form with the carbon atom. One bond is formed with a single hydrogen atom, and the other with a group of molecules, the R group. The R group may be as simple as a single hydrogen or it may be a long chain or ring. It usually is a base.

There are 20 different amino acids commonly found in proteins, each with a different number of amino and carboxyl groups. These 20 amino acids join together in long chain molecules to form proteins. It is the different types of amino acids and the sequence in which they are arranged that give proteins their immense variety. Amino acids are often called the alphabet of proteins, because they can be put together to form proteins

such as sugar. That is why sugars are the primary source of energy storage.

The function of lipids is not confined to supplying fuel. For example, one subgroup of lipids is the carotenoids, red and yellow pigments found in plants and also in the eyes of animals,

**33**

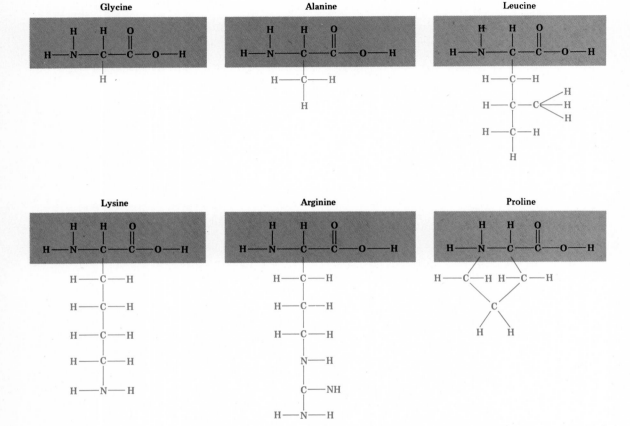

1-18 *This figure shows some sample amino acids. The lower unshaded portion of the molecule is the R group, which may vary considerably from a simple H atom to a complicated carbon ring. The upper portion of the molecule, the amino end, is much the same for all amino acids. Although the possibilities for variations in the R group are almost numberless, there are actually only 20 amino acids commonly found in natural proteins.*

in the same way that letters can be put together to form words. The number of possible combinations is so large it is really incomprehensible. It has been estimated that if someone were to make a mixture containing only one molecule of each of the different kinds of proteins that could exist, the weight of this mixture would be greater than the weight of the known universe! Therefore, the thousands of proteins found in living organisms actually represent only a very narrow selection from the huge range of possibilities.

When amino acids join together to form proteins, they do so by forming a peptide bond, linking the amino end of one molecule with the carboxyl end of the next through dehydration synthesis (the term for the kind of synthesis in which displaced hydrogen and oxygen atoms unite to form water molecules). A group of 2 to 20 amino acids thus bonded together is called a **peptide;** peptides, in turn, join together to form larger groups, called **polypeptides;** and groups of polypeptides form protein molecules. Some proteins, called conjugated proteins, also include atoms of minerals such as iron or sulfur.

The way in which the polypeptide groups join together determines the shape of the protein molecule and therefore its chemical and physical properties. They may join together in one long chain, or the chains may lie side by side, with linkages at intervals between them.

Only in the last 20 years, with the help of new techniques such as crystallization and X-ray diffraction, have chemists begun to learn about the exact three-dimensional structure of proteins. One of the first to be mapped was the protein myoglobin, found in muscles; this work was done in the early 1950s. Gradually the structure of other proteins has also been revealed. We now know that most proteins are coiled or twisted in a variety of ways by the formation of hydrogen bonds between different parts of the molecule.

A shape very often found in protein molecules is the **alpha helix,** produced by a spiral twisting of straight-chain polypeptides. Hydrogen bonds form between adjacent levels of the spiral, to maintain the shape of the molecule. Another common shape is the **beta configuration,** in which chains lying side by side form hydrogen bonds between polypeptides. Because of its complex shape, the protein molecule can react only with a very limited number of other substances, and most proteins are highly specific. Their structure is delicate, however, and the slightest change in a protein's configuration can greatly alter its chemical properties.

Although we have said that hydrogen bonds are relatively weak, the molecule is generally stable because the large numbers of hydrogen bonds reinforce each other. However, the bonds can be destroyed by heat or excess acidity. As the bonds break, the molecule loses its spiral shape. The spiral uncoils in a process that is irreversible, called denaturization. The protein will still have the same molecular formula, but since it has lost its characteristic structure, it can no longer be chemically active in the normal way.

**Enzymes.** A special group of proteins are the enzymes, the biological catalysts. Enzymes speed up the rate of many important chemical reactions in living cells, such as synthesis of large molecules and hydrolysis to release needed energy. They also permit these reactions to take place at body temperature. An enzyme's ability to catalyze such reactions is a direct result of its specific structure and three-dimensional shape.

In order for molecules to be bonded together, the various atoms must come close enough together for their electron clouds to overlap. Sooner or later, this will happen as the result of a collision between randomly moving particles. If the concentration of reactants is high, and the reaction is a simple one, then the laws of statistical probability insure that the reaction will take place; but the bonding of large organic mole-

A

B

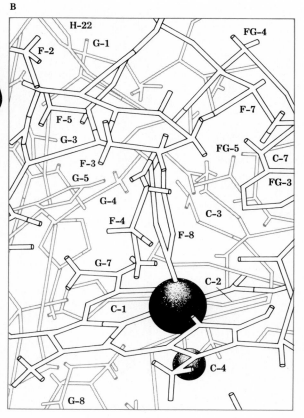

**1-19** *This figure compares two kinds of molecular models. (A) is designed to show the spatial configuration of a molecule of myoglobin, a protein found in muscle. (B), a much more complex model, shows the exact atomic structure of the dark area seen in (A). The round dark ball, present in both models, is an atom of iron.*

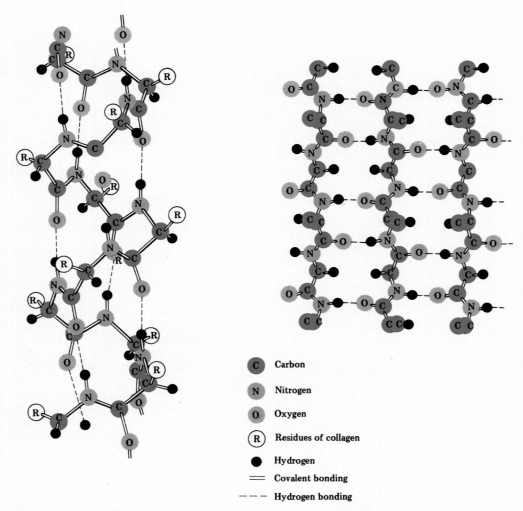

**C** Carbon

**N** Nitrogen

**O** Oxygen

**R** Residues of collagen

● Hydrogen

= Covalent bonding

--- Hydrogen bonding

**1-20** *The alpha configuration of proteins, shown in (A), is a single polypeptide coiled upon itself in a helical fashion. Because of the shapes of amino acids, a protein helix must always coil in this same right-handed direction. (B) shows one version of the beta configuration, caused by hydrogen bonding between polypeptide chains that creates a sheetlike arrangement.*

cules, such as starches, is much less likely to occur unassisted. The concentration of the reactants may be fairly low in any one reaction site; moreover, the peculiar shapes of the bulky molecules mean that many random collisions will not involve the specific section of the molecule that is to be bonded. Thus the rate of reaction would be extremely slow, if it were not for the catalytic action of enzymes.

The basic function of the enzymes is to align reacting molecules in such a way that the reaction can take place. For example, the jagged surface of the enzyme molecule holds the groups of atoms to be bonded (called the **substrate**) in such a way that their electron clouds overlap in exactly the spot where the bond is to be formed. This is shown in Figure 1-21. Enzymes that help split large molecules work in a slightly different way.

**1-21** *The surface configuration of the enzyme molecule causes molecules of the substrate to line up in a certain specific order, thus bringing surfaces to be bonded close together or exposing bonds to be broken. The shaded portion of the enzyme's surface is actually a coenzyme—an ion or small molecule that must fit between enzyme and substrate to produce the correct configuration.*

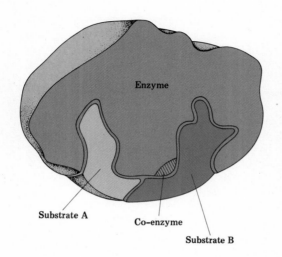

Enzyme

Substrate A

Co-enzyme

Substrate B

A minute difference in the surface configurations of the enzyme and the substrate pushes the substrate slightly out of shape at the location where the bond is to be broken; this slight deformity makes the molecule break apart easily. Many enzymes have associated **coenzymes** (often mineral ions) that must join with the enzyme and substrate in order to produce a perfect fit.

It is significant that for virtually all organic compounds produced by living organisms, there exists an enzyme designed to enable its decomposition. This is not true of man-made organic compounds—DDT, for example—which is one reason that they pose such a serious potential threat to the earth's environment; there is no mechanism that enables them to be decomposed quickly into less harmful substances.

## Nucleic acids

A fourth important group of organic compounds are the nucleic acids. Like proteins, they are very large polymeric molecules. One kind of nucleic acid, deoxyribonucleic acid (DNA), serves to transmit information regarding hereditary traits; another, ribonucleic acid (RNA), serves as a messenger that puts this information into action. Nucleic acids play a crucial role in the synthesis of proteins and the control of the living biochemical system.

The individual units of which nucleic acids are composed are called **nucleotides.** A nucleotide has three major parts: a phosphate group; a 5-carbon sugar; and a nitrogen-containing base. Five bases—adenine, guanine, cytosine, thymine, and uracil—are known, and it is these units that form the "alphabet" of nucleic acids. Like amino acids, they can be arranged in an alternating sequence, which gives them their informational content.

**1-22** *This molecule of adenylic acid is composed of three major parts. There is a base, a 5-carbon monosaccharide. Joined to it are a phosphate, in this case phosphoric acid, and a nitrogen-containing base (adenine). Only five different bases are found in nucleotides, but their sequential arrangement carries all genetic information.*

Phosphate

Base

$O = P - O - CH_2$

Sugar

Nucleotide of
adenine: adenylic acid

# Summary

All matter, whether living or nonliving, is composed of the same basic elements. Therefore the chemical and physical laws that apply to nonliving matter must also apply to living things. One such law is the first law of thermodynamics, which states that energy is convertible but indestructible. The second law of thermodynamics, which has important implications for living systems, states that energy always flows in the direction of increasing entropy, or disorganization. Energy may exist in either a potential or a kinetic state, and it may be changed from one form, such as light energy or electrical energy, into another. Such conversions always entail an increase in entropy.

Matter is made up of a few basic elements; 92 elements occur in nature and at least 12 more have been produced in the laboratory. Each element has its own characteristic atomic structure. The nucleus consists of protons and neutrons and carries a positive electrical charge. Circling the nucleus are a number of electrons, small particles with a negative electrical charge. In an atom, the positive charge of the nucleus equals the negative charge of the electrons. If electrons are lost or gained, the atom acquires an overall charge, or valence, and is called an ion.

When two atoms come close enough together for their electrons to overlap, an energy exchange may take place, forming a chemical bond. If one atom gives up one or more electrons to another atom, it is called ionic bonding; if the two atoms share electrons, it is called covalent bonding. A weak type of bonding, found in many protein molecules and also in chains of water molecules, is the hydrogen bond. Each chemical bond contains a certain amount of energy, the quantity varying with the elements bonded. In living systems, energy to carry on life functions is stored in the high-energy bonds of the ATP molecule.

When atoms or molecules interact with one another, breaking and forming bonds, the process is called a chemical reaction. Initiation of a reaction requires a certain amount of additional energy, known as the energy of activation; this may be heat energy, electrical energy, or the kinetic energy of the atoms and molecules themselves. Chemical reactions that release free energy are exergonic. Reactions that consume energy are endergonic.

All chemical reactions are theoretically reversible, capable of proceeding in either a forward or reverse direction. When the rate of the forward reaction is the same as the rate of the reverse reaction, then an equilibrium is reached and no further net changes will take place. All physical systems continually move toward a state of chemical equilibrium. Biological systems are able to avoid equilibrium through the use of sequences of reactions, which remove the product of each reaction from the site immediately.

Chemical compounds may be classed as organic (those containing carbon-to-carbon linkages) or inorganic. Among the inorganic compounds important to life are water ($H_2O$), carbon dioxide ($CO_2$), and oxygen ($O_2$).

There are four major types of organic compounds. Carbohydrates are made of units consisting of one carbon atom, two hydrogen atoms, and one oxygen atom. The simplest carbohydrates are the monosaccharides, such as glucose and common table sugar. These 6-carbon units can link together in long chains or rings to form polymeric molecules called polysaccharides. Carbohydrates are the basic energy source for all living things. Lipids are a second type of organic compound; they are a heterogeneous group that includes all fats and oils. Lipids, like carbohydrates, are composed of carbon, hydrogen, and oxygen atoms joined in monomers and polymers, but in lipids these three elements are found in different proportions. Lipids are generally utilized as storage molecules.

More complex than either carbohydrates or lipids are the proteins, a third class of organic compound. Proteins are polymeric molecules composed of alternating sequences of 20 different amino acids built up from carbon, hydrogen, oxygen, and nitrogen. Protein molecules form elaborate chains, spirals, and helixes, held in shape by hydrogen bonds between various parts of the molecule. Although their shape makes them highly specific in their reactions, protein molecules are delicate and their shapes can easily be distorted. A special group of proteins are the enzymes, the biological catalysts that speed up the rate of chemical reactions in living organisms.

The fourth important group of organic compounds are the nucleic acids, which serve to transmit information regarding heredity. Nucleic acids are polymers composed of individual units called nucleotides. Like amino acids, they can be arranged in an alternating sequence, giving them an informational content.

## Bibliography

### References

Dyson, F. J. 1971. "Energy in the Universe." *Scientific American.* 224(9): 50–59. Offprint no. 662. Freeman, San Francisco.

Kendrew, J. C. 1961. "The Three-Dimensional Structure of a Protein Molecule." *Scientific American*. 205(12): 96–111. Offprint no. 121. Freeman, San Francisco.

Lambert, J. B. 1970. "The Shapes of Organic Molecules." *Scientific American*. 222(1): 58–75. Offprint no. 131. Freeman, San Francisco.

Lehninger, A. L. 1965. *Bioenergetics*. Benjamin, New York.

## Suggested for further reading

Baker, J. J. W. and G. E. Allen. 1970. *Matter, Energy and Life*. Addison-Wesley, Reading, Mass.
One of the best introductions to this topic. There are separate chapters on matter and energy, the structure of matter and the various organic compounds. Good, clear illustrations and a well-annotated bibliography.

Blum, H. 1962. *Time's Arrow and Evolution*. 2nd ed. Harper Torchbooks, New York.
Blum examines the limitations placed on living systems by the laws of thermodynamics and mechanisms that have evolved to overcome these limitations. An interesting, thought-provoking book.

Bush, G. L. and A. A. Silvidi. 1961. *The Atom: A Simplified Description*. Barnes & Noble, New York.
Written for the introductory student, this book describes atomic structure and includes coverage of energy sublevels.

Loewy, A. G. and P. Siekevitz. 1969. *Cell Structure and Function*. 2nd ed. Holt, Rinehart & Winston, New York. Chapter 2 is a discussion of "Life and the Second Law of Thermodynamics." Chapters 5 through 9 discuss the chemical structure of circular components.

## Related books of interest

Asimov, I. 1960. *The Wellsprings of Life*. Signet (Science Library), New York.
A cogent discussion of the origins and evolution of living matter. Part Three, "Life and the Molecule," traces the history of biochemistry from the eighteenth century through the discovery of the structure of DNA.

Rose, S. 1966, 1968. *The Chemistry of Life*. Penguin Books (Pelican), Baltimore.
A lively, simplified, but demanding account of recent developments in biochemistry. Beginning with a description of basic chemical systems, Rose extends his discussion to systems of intra- and intercellular control.

Snow, C. P. 1958. *The Search*. Scribner, New York.
Lord Snow, who is both a scientist and a novelist, wrote this story of a young
English research scientist who is on the brink of a major breakthrough in his work.
The book does an excellent job of conveying the atmosphere and excitement of
the laboratory.

Watson, J. D. 1968. *The Double Helix*. Atheneum, New York (paperback ed. by
Signet, 1969).
An exciting and even suspenseful narrative, by a participant, of the fundamental
research breakthrough in molecular genetics.

# 2
# The cell: basic unit of life

In every living organism thousands of chemical reactions take place every day. Eating, breathing, walking, turning toward the sun to catch its warmth and light—each of these actions is actually the result of a chain of chemical reactions within the body. This dependence on chemical reactions is true even of the smallest living things, such as tiny bacteria. The chemical reactions that take place in a streptococcus bacterium are just as complex as those that take place in a whale or a redwood tree, but of course the number and range of reactions is greater in the larger organisms.

As we saw in Chapter 1, chemical reactions between two reactants can take place only under fairly specific conditions. The two molecules must come together in a certain way, so that they can exchange or share electrons at one particular point on their surfaces. Moreover, to speed up the reaction, a third molecule (an enzyme) may also be involved, so it is necessary to have all three molecules aligned in one specific order. This order may involve not just one reaction of two molecules and one enzyme, but a whole series of closely related chemical reactions which proceed like an assembly line.

Suppose that all the chemical components of a cell—all the proteins, carbohydrates, lipids, nucleic acids, and inorganic ions—were mixed together in a kind of colloidal soup. The laws of statistical probability tell us that under these conditions of disorder the chances for a successful collision between two reactants are very slight. The cell would die from lack of energy long before the random movement of molecules could be expected to produce even one necessary chemical reaction. It is evident that the chemical components of the cell are not in fact sloshed together in a living Jello, but are very carefully organized. The thousands of interactions on which the cell's existence depends are separated in both space and time. The basis of this organizational separation of chemical systems is the membrane.

**43**

## The importance of membranes

A **membrane** is a partition that both separates and organizes chemical reactions. Membranes in living cells can actually be seen under the electron microscope; they appear as layered structures that define the shape of the cell. But even when biologists were not able to see membranes, they were quite certain that these structures existed. They *must* exist, or living things would not be able to function in the way that they do.

The basic service that the membrane performs is the compartmentalization of chemical substances. A network of membranes within an organism creates a number of small spaces that can be used as individual chemical laboratories. It is really quite like the effect of erecting partitions to divide the floor space belonging to the Research and Development section of Du Pont. If the whole area were just one big open space, with all the scientists milling around together, the confusion would prevent productive use of the laboratory facilities. But if partitions are installed to divide the space into separate offices and laboratories, then each team of scientists has all the facilities and materials it needs for its specialized work close at hand, and there is no need to disturb or interfere with other workers. Moreover, the partitions provide a large increase in the wall surface of the lab area, so there is more space to put up shelves or drawers or cabinets. The partitions greatly enhance the efficiency of the lab.

Membranes do the same thing in living biochemical systems. They partition the space, creating little laboratories where certain reactions can take place independently of other reactions, separating them in both time and space. Each compartment contains its own metabolic pool of chemicals; they are its "team." For example, one compartment of the liver may specialize in the function of converting stored glycogen into glucose, to be used as an immediate fuel source throughout the body. Within that compartment, there would be a supply of glycogen; there would be molecules of water to break the polysaccharide bonds by hydrolysis; there would be specific enzymes to catalyze the hydrolysis. The pH in the compartment would be at the level that favored fast progress of the reaction (a pH of 6 to 8) even though the compartment next to it contained an acid and therefore had a lower pH. And once the glucose was produced, it could be removed from the compartment, so that the reaction would continue its net movement in one direction only.

A second advantage of membranes is that they permit the completion of sequential reactions. A membrane, like any partition, is an essentially linear structure. It runs along in a line—straight, curved, or zigzag—from one point to another. This linear structure is beautifully adapted to organizing linear reaction, or reactions that must proceed in a certain sequence. Such sequential reactions are very common in biological systems, where they serve as a means of avoiding equilibrium and maintaining a one-way direction to the reaction.

A generalized scheme for a sequential reaction can be diagrammed in this way:

$$\xrightarrow{\hspace{4cm}}$$
$$A \rightleftarrows B \rightleftarrows C \rightleftarrows D \rightleftarrows E$$

Substance A represents the raw material that initiates the chain of reactions, and substance E is the final product that will be used in the body. The product of the first reaction in the series will be a reactant in the second reaction, and the product of the second reaction will be a reactant in the third reaction, and so on down the sequence until the final product is reached.

The compartmentalization effect of the membrane means that the reactants needed in this sequential reaction can all be assembled in one metabolic pool. Other substances that might interfere with the reaction, or cause it to veer off in some other direction, can be excluded from the pool.

But suppose that the substances are quite literally in a pool, floating freely about in their little chamber. There is only one chance that they will line up in their correct sequence of ABCDE; there are 119 chances that they will line up in some other pattern, such as ACDBE or EDBAC. What is even more likely is that they will not line up at all. One little clump of BBA will form in one corner, while units of AC float past and bump into similar units of EE. Under these conditions, the sequential reaction could never take place; the odds are only one out of 3,025 chances.

The problem is solved by lining the enzymes up along a membrane, so that they are organized in their proper sequence. The reactants will then automatically be lined up once the first reaction begins. The sequence along the membrane can be diagrammed like this:

$$A \quad B \quad C \quad D \quad E$$
$$\searrow \nearrow \searrow \nearrow \searrow \nearrow \searrow \nearrow$$
$$En_1 \quad En_2 \quad En_3 \quad En_4$$

As each reaction occurred, the product would be moved along the membrane toward the next reactant and the next enzyme. This system of organization would allow the sequence of reactions to take place efficiently and would also help to keep it moving in one direction.

Membranes can also be used to direct the movement of molecules from one compartment to another. For example, in the breakdown of glycogen into glucose, the enzyme that catalyzes the final step in the conversion can be located so that it is straddling the membrane between two compartments. The substrate (glycogen) can hit this enzyme in Compartment A, but the product (glucose) would be formed on the *other* side of the membrane, in Compartment B. This positioning of the enzyme automatically moves the glucose from one compartment to the next. Thus Compartment A can maintain a low concentra-

tion of glucose, so that the glycogen conversion will continue. And the end-product, the glucose, can be ready for another sequence of events in Compartment B.

## Movement across membranes

We have been speaking about membranes as if they were impenetrable partitions that form tightly sealed chambers. But observation tells us that this is not really the case. For example, doctors know that when they give a patient a pill containing some therapeutic drug, the drug will soon be found in many parts of the body—in the bloodstream, in muscles, in organs such as the liver and the kidney—and eventually it will probably also be found in the urine. Since the patient's body contains millions of membranes, creating millions of small chambers for chemical reactions, it is clear that the drug must have passed across many membranes.

This kind of observation tells us that the membranes in living organisms are permeable; that is, they permit the passage of certain substances from one side of the membrane to the other. The movements of molecules across membranes can be explained by the physical laws that govern the behavior of all atoms and molecules.

### Diffusion

The principal basis of transport across membranes is the physical process of **diffusion.** Diffusion is the tendency of molecules to move from an area in which they are densely packed to an area where they are scattered thinly. This tendency should not be attributed to some mysterious urge to head for the wide open spaces. Rather it is the result of the laws of statistical probability when applied to small bodies moving at random. There are more chances for mole-

**45**

cules to move to areas of low concentration (because there is less likelihood of collision) than there are for them to move to areas of high concentration. Therefore, on the average, the net movement will be toward the lower concentration.

Figure 2-1 traces the process of diffusion through the example of burning incense. When a pinch of grape incense is placed in a shallow dish and set on fire, molecules of incense will move out of the dish into the air. At first the incense molecules will be clumped closely together in the immediate vicinity of the dish in which they were burned. But they do not hang motionless in the air. Each molecule is continually darting back and forth, up and down, around and around, as a result of its own thermal energy. The molecules move in random patterns, but when they are bunched together tightly, as in the case of the incense molecules, there are more chances for the molecules to move away from each other than there are for them to move closer together. Some molecules will, of course, move in the direction of the center of the group, but a greater number of molecules will move away from the center. This result is dictated by the statistical laws of probability.

Perhaps the most important factor in the process of diffusion is the possibility of collisions between moving molecules. When one molecule collides with another, it can no longer move straight ahead the way it was going; it will be shunted off in some other direction. A molecule that moves toward the center of the group has a very good chance of colliding with another molecule and thus being diverted off course, so that it no longer moves toward the center. On the other hand, a molecule that moves away from the group has a much smaller chance of being involved in a collision, since it is moving to an area where there are few other molecules. It is exactly this principle that lies behind the fact that more accidents occur on roads with heavy traffic than on roads with very light traffic; the more cars (or particles) that are gathered in one place, the more

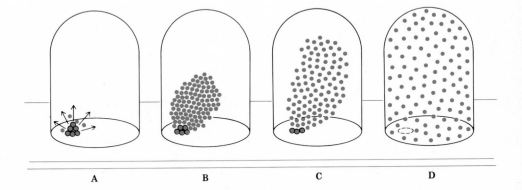

likely it is that they will bump into one another.

Diffusion is generally a very slow process. In the example of the burning incense, the process of diffusion is being supplemented by the action of air currents which are created by the movement of hot air. In a closed system where wind or water currents cannot help to spread molecules out, diffusion may take a very long time. It might take years for the incense molecules to become evenly dispersed throughout the room without the help of currents.

The rate of diffusion of molecules depends on a number of factors. One is the concentration of the substance being diffused—the number of molecules per unit volume. The higher the concentration, the faster diffusion will take place. Another factor is the concentration difference between the area from which the diffusing molecule is coming and the area to which it is going. For example, diffusion will take place much more quickly in a system that has pure water on one side and very sugary water on the other side, than it will in a system with a solution of 20 percent sugar on one side and 30 percent on the other, since the first system has a much higher concentration difference. Raising the temperature also speeds up the rate of diffusion, since it makes all the molecules move faster. A final factor to be considered is that of the electrical

**2-1** *When a pinch of grape incense is placed in a shallow dish and set on fire, molecules of incense will move out of the dish into the air. At first the incense molecules will be clumped closely together in the immediate vicinity of the dish in which they were burned. As they move, some will move in the direction of the center of the group, but a greater number will move away from the center, according to the laws of probability.*

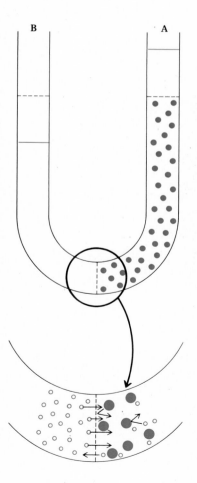

**2-2** *Molecules of water can move through the osmotic membrane, but all other molecules are stopped by its barrier. If pure water is placed on one side of the membrane and a solution of sugar water on the other side, osmosis will take place. The water molecules will move through the membrane in both directions, but the laws of probability dictate that the net movement will be in the direction of the concentration differential.*

charge carried by the molecules. Diffusion takes place faster when the molecules have a high charge, either positive or negative, because the charge will push the molecules away from one another.

## Osmosis and osmotic pressure

When molecules of water diffuse across a membrane, the process is called **osmosis.** Osmosis can be demonstrated in a simple lab experiment that uses an artificial membrane, such as a sheet of cellophane. Molecules of water can move through the membrane, but all other molecules are stopped by its barrier. If pure water is placed on one side of the membrane and a solution of sugar water on the other side, osmosis will take place. The water molecules will move through the membrane in both directions, but the laws of probability dictate that the net movement will be in the direction of the concentration differential. They will move from the side where they are in highest concentration (in pure water) to the side where their concentration is lower (the sugar-water solution). The net movement of water molecules would stop when an equilibrium was reached, with an equal concentration of water molecules on both sides of the membrane.

The artificial membrane displays a characteristic important to biological membranes; it is **differentially permeable.** It permits passage of certain molecules while preventing the passage of others. If the membrane were completely impermeable, no osmosis would take place at all. If it were completely permeable, both water and sugar molecules would pass through. The property of differential permeability allows the membrane to serve as a highly selective barrier.

Any atom or molecule that can be dissolved in water but cannot readily pass through the membrane is osmotically active—that is, its presence lowers the concentration of water on its side of the membrane, so that the long-run

tendency of water will be to move into that area. A calcium ion is an osmotically active particle, and so is a globular protein in the bloodstream. The only molecules which are not osmotically active are those which will not dissolve in water and those which can slip easily through the membrane. Starch is the best-known example of an inactive molecule, since it is insoluble in water. That is why it makes a good storage molecule. Stored starch will not affect the osmotic balance of the organism, whereas stored glucose will.

The more osmotically active particles a solution contains, the greater will be its tendency to take in water by osmosis. Biologists can measure this tendency to take in water and give it a quantitative value. One way this is done is by placing the solution to be measured on one side of a membrane and then applying hydrostatic pressure on the other side. At a certain point, the force of the applied pressure will prevent the movement of water molecules into the solution. That amount of pressure is called the **osmotic pressure** of the solution being measured. Osmotic pressure is generally measured in exactly the same units (bars) as all other kinds of pressure, such as water pressure or blood pressure.

Two solutions with the same osmotic pressure are said to be **isotonic** to one another. When isotonic solutions are separated by a membrane, no net flow of water will take place. Blood, for example, is isotonic to lymph fluid. Blood is **hypertonic** to water, since blood has a much higher osmotic pressure, so there would be a net movement of water into the blood. Blood is **hypotonic** to urine, because the osmotic pressure of urine is greater than the osmotic pressure of blood. When the two fluids are separated by a membrane, there will be a net movement of water out of the blood and into the urine. As we shall see later, in Chapter 8, the balance of osmotic pressure between different parts of the body, and between the organism and its environment, is crucially important to the existence of life.

**47**

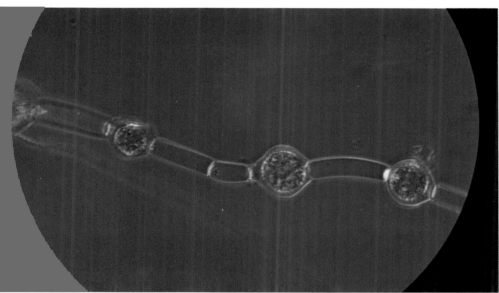

*Oedogonium*

Desmid

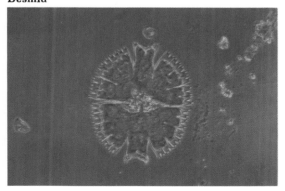

## Acellular organisms

A high degree of differentiation and specialization can be seen in the primitive acellular organisms. *Oedogonium,* a green alga that lives in fresh water and establishes long filamentous colonies, illustrates the high degree of complexity found in the phylum Chlorophyta. The seemingly transparent cells shaped like bricks carry on the processes of photosynthesis and respiration; these cells are purely vegetative. The spherical cells are the egg-forming structures, or oogonia. They produce small nonmotile eggs that store a food supply. Sperm cells are produced by other specialized cells in the colony, called the antheridia. The sperm, adapted for movement, penetrates the oogonium and fertilizes the egg. The zygote develops within the oogonium; when it is mature, the protective thickened cell walls decays, and the zygote is released into the water. It then undergoes two mitotic cell divisions, forming four new cells capable of starting colonies.

Another type of green algae can be seen in the lower picture. Desmids also live in fresh water, and their one-celled bodies are always exactly symmetrical. Their lacy and delicate shapes make them look like little green snowflakes floating freely on the surface of a pond. The trilobed green structures are the chloroplasts. Desmids commonly reproduce by fission, or simple splitting, producing two new organisms.

# Simple multicellular organisms

These pictures give a glimpse of the fascinating variety to be found in the living world. Slime molds are an unusual combination of plant and animal characteristics. Creeping over the soil is the vegetative body, which strongly resembles an amoeba in its changeable shape and lack of cell walls. Yet the rounded fruiting bodies, an aggregation of cells that contains reproductive spores, are similar to those found in bread mold and other fungi.

The green sea anemone is a good example of symbiosis. The anemone, a coelenterate, is covered by a growth of green algae. The anemone benefits from this association because the algae remove its excreted nitrogenous wastes; the algae in turn get a food supply and a safe place to live. Another species of sea anemone lives in close association with a colony of barnacles.

A more extreme example of symbiosis can be seen in the lichen commonly called British soldiers, because of its bright red fruiting bodies. This plant actually consists of two separate organisms, a green algae and a fungus.

The Portuguese man-of-war, like the sea anemone, is a coelenterate, found only in the ocean. The man-of-war is actually a colony of specialized individuals. One forms the balloon, which acts like a sail to move the colony with the wind; other individuals form the dangling sticky tentacles; and other are adapted to carry on the process of digesting food.

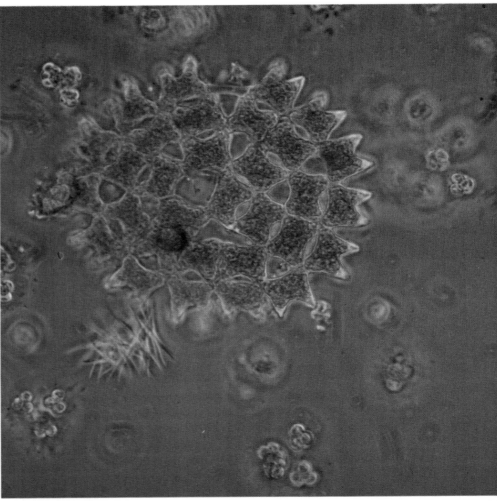

Colonial green algae

**British soldiers**

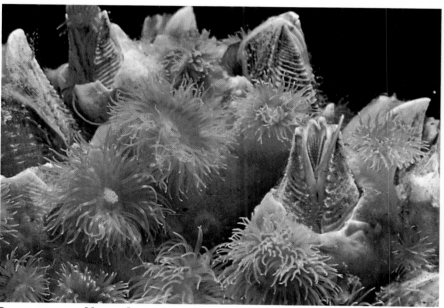

**Sea anemone and barnacles**

**Portuguese man-of-war**

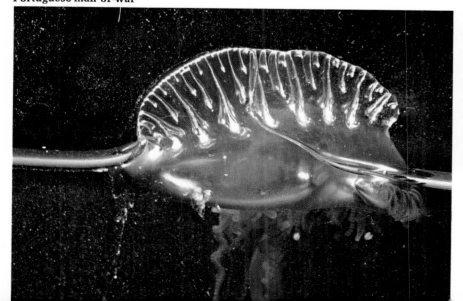

**Green sea anemone**

**Slime mold**

# The plant kingdom

Among the simplest vascular plants are the horsetails. Only one genus of *Equisetum* is now in existence. Shown here are vegetative stems, but the plant can also produce another type of stem with a cone-like reproductive structure. In the stem, cell walls contain silica; a handful of horsetails will scour a dirty pot very efficiently.

At first glance, the cycad may appear to be very like the palm tree, but cycads are primitive gymnosperms, much more closely related to the cypress than the palm. The Mascarene bottle palm shown here is a member of Class Angiospermae, the dominant plants of the present geological era. Palms are monocots, with one seed leaf, scattered vascular bundles, and flower parts in threes.

The Indian wheat, a dicot, grows on the dry open slopes of the Sierra Nevadas. It belongs to the large composite family, which also includes lettuce, artichokes, and asters. The beavertail cactus is another dicot, well adapted for the hot dry desert climate.

The Spanish moss which hangs from the cypress branches is not a moss at all (for that matter, it's not Spanish either). It is a member of the bromeliad family, a near relative of the pineapple. This plant is not a parasite; the cypress furnishes it nothing more than a place to grow. The moss feeds on small drops of moisture and particles of soil that collect in the axils of the cypress branches.

Beavertail cactus

Indian wheat

**Cycad**

**Horsetails**

**Cypress trees**

**Mascarene bottle palm**

# The animal kingdom

Imaginary animals, such as the dragon, griffin, or unicorn, are not nearly as exotic or fantastic as some of the real animals we see here. The green monster that seems to be peering out from the opposite page is simply a larval stage of the spicebush butterfly; its eyes and mouth are no more than colored marks on the skin.

Another nightmarish creature is the horned chameleon, which clutches the branch with its locking feet. This African chameleon belongs to the same order as lizards and snakes. Its eyes can move independently, each looking for prey in a different direction. It eats mostly insects, which it catches with a long sticky tongue. The spotted turtle is a near relative of the chameleon, since both are members of Class Amphibia.

The beautifully luminescent sea slug has unusual feeding habits. It eats coelenterate polyps, but before it actually ingests its prey, it carefully removes all the stinging nematocysts, without exploding a one, and transfers them to its own skin for protection.

The copepod swimming through the water is a female *Cyclops*. The soft masses on either side of her tail are egg cases. Like many other arthropods, Cyclops reproduces large numbers of offspring. This microscopic crustacean is an important link in many food chains; for example, fresh-water copepods constitute the principle food of the spotted turtle.

Porcupine

African horned chameleon

**Spicebush butterfly larva**

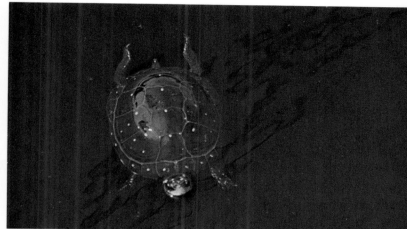

**Spotted turtle**

**Sea slug**

**Female copepod**

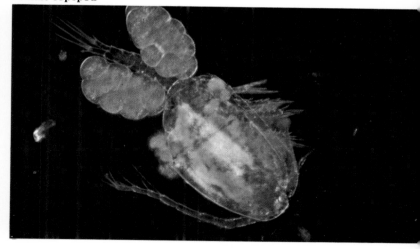

# Variation in frogs

The frogs shown on this page all belong to the same order, Anura, in Class Amphibia. They display marked similarities of anatomy and physiology; they reproduce in exactly the same way. Yet they show amazing variation in their coloration and specific adaptations.

The frogs in the two upper pictures are members of the same genus, but they are adapted to live in very different habitats. The one on the left, *Hyla andersonii*, lives in the pine barrens along the coast of the eastern United States. Its predominantly green color protects it from its enemies and conceals it from its prey. *Hyla ebraccata,* shown in the next photo, lives in Mexico. Its oddly shaped head and its brighter coloration are adaptations to this different environment.

Another Mexican frog is the red-eyed tree frog (*Agalychnis callidryas*). Native to the province of Oaxaca, this beautiful little frog has a green back, yellow belly, and startlingly red eyes. To conceal itself, it has only to perch on a leaf or blade of grass and close its eyes.

The Colombian horned frog (*Ceratophrys calcarata*) lives in South America. This frog, usually no more than three inches high, is surprisingly aggressive. With its strong jaws and sharp pointed teeth, it will attack small rodents, other frogs, and even the ankles of passing pedestrians. Such fierce behavior is very unusual among frogs.

*Hyla andersonii*

*Hyla ebreccata*

*Ceratophrys calcarata*

*Agalychnis callidryas*

**In isotonic
solution**

**Red blood cell**

**In hypotonic
solution**

**In hypertonic
solution**

**2-3** *The importance of osmotic
pressure to living systems is
demonstrated by placing living
cells in isotonic, hypertonic, and
hypotonic solutions. In the isotonic
medium, no net flow of water
takes place. In the solution which
is hypertonic to the cell, water
will follow the osmotic gradient
into the solution, causing dehydration of the cell; in the solution
hypotonic to the cell the reverse is
true: net movement of water across
the cell membrane and into the
cell causes it to expand or even to
burst.*

## Active transport

The movement of water across a membrane by
osmosis is called **passive transport.** The transported molecules are simply responding to a
concentration differential and obeying a physical
law that governs all matter. In living systems, the
net result is that a certain amount of water enters
the cell. No expenditure of energy is required for
this kind of transport.

Biologists know from observation, however,
that certain substances are able to enter the cell
against a concentration differential—they move
from a lower concentration outside the cell to a
higher concentration inside. An example can be
seen in the movement of potassium ions. In
nerves, these ions move into the cell even when
the concentration inside is nearly 30 times as
high as the concentration outside. Experiments
indicate that transportation of substances against
a concentration differential is accomplished
through an expenditure of energy. This is called
**active transport.**

A hypothetical model of the mechanism of
active transport is shown in Figure 2-4. The
molecule entering the cell is coupled with some
carrier molecule within the cell membrane,
through an energy-consuming bonding process.
The carrier is probably some membrane constituent that can easily form and break chemical

bonds; lipid molecules are thought to be the
most likely possibility. The movement of such
carrier molecules across the membrane and into
the interior of the cell can be explained in terms
of diffusion with a **concentration gradient** (a
gradient is a series of concentration differentials,
arranged in a steplike pattern). The concentration
of the carrier molecules that have bonded to
some entering molecule would be high near the
outer edge of the membrane, where the bond is
formed. So the molecule would tend to move to
an area of lower concentration, inside the cell.
There the bond between the carrier and the
transported molecule is broken, probably with
the help of an enzyme lying just inside the cell.
This breaking of the bond, which is another
energy-consuming step, releases the transported
molecule into the cell. The "empty" carrier
molecules, accumulating in high concentration
near the inner edge of the membrane, once again
follow the concentration gradient and diffuse
back to the outer edge, where the process can
begin all over again.

In some cases, bonds with a carrier may form
and break simply as a result of the kinetic energy
of the molecules involved. When no outside
energy source is required, the process is called
**catalyzed transport.** More frequently an outside
energy source is needed. The exact mechanism of
the energy transfer has not yet been traced, but it

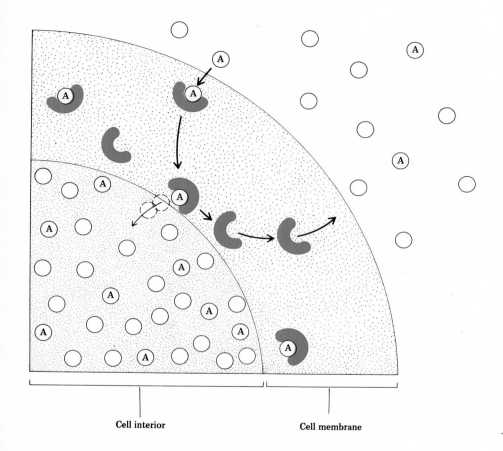

Cell interior

Cell membrane

**2-4** *A hypothetical model of the mechanism of active transport has the molecule entering the cell coupled with some carrier molecule within the cell membrane, through an energy-consuming bonding process. The transported chemical here combines with the carrier molecules (color). The molecule thus formed moves within the cell membrane to a point of lesser concentration at the cell interior, where the transported chemical is released.*

is assumed that ATP plays some important role in the process.

**Active transport of large molecules.** Living cells are able to transport large groups of molecules across the cell membrane by enclosing the molecules in a membranous envelope. When small groups of molecules (especially molecules of liquids) are taken into the cell this way, it is called **pinocytosis.** The group of molecules from the surrounding environment apparently presses against the membrane, which responds by folding in, or **invaginating.** When the molecules are

completely surrounded, the membrane pinches off (vesiculates), forming a sac or **vesicle** which then moves into the interior of the cell. Enzymes can later break down the vesicle and free the transported molecules. When the particle thus transported is a relatively large one—a microorganism, for example—the process is given the different name of **phagocytosis.** In man, white blood cells employ phagocytosis to engulf and digest bacteria. Phagocytosis is sometimes called "cellular eating," in contrast to pinocytosis, which is referred to as "cellular drinking."

## Permeability characteristics of membranes

In the laboratory, biologists have carefully observed the evidence of transport across membranes, obtaining membrane samples from many different living things. By using radioactive isotopes of the transported substance, they can determine exactly how long the substance takes to cross the membrane. Another technique is to use dyes, chemicals whose progress can be traced visually. Since different dyes are composed of molecules of different sizes, it is possible to establish whether or not molecular size and weight affect transport.

The results of such experiments have clearly described the permeability characteristics of membranes in living systems. One of the most important discoveries is that only molecules of a certain size and shape can be readily transported across living membranes. A molecule with a diameter of more than 7 or 8 Angstrom units (an Angstrom unit is 1/10,000,000th of a millimeter) will not be able to cross. A molecule whose configuration is long and thin can pass through the membrane, but does so at a very slow rate. This evidence leads to the hypothesis that such molecules can pass through only when they are aligned in a certain way, probably so that they

**49**

present their thin side to the membrane opening, rather than the long side, which is too big to fit. The smaller the particle, the faster it will move across the membrane. Water molecules can move across very rapidly. Urea, a fragment of an amino acid, can pass through, but its movement is about a million times slower than that of water. Glucose is also small enough to pass, but large molecules of carbohydrate — polysaccharides, for example — cannot penetrate a membrane. In fact, all polymeric molecules are normally unable to cross living membranes.

Another factor which has been found to determine whether or not a given molecule can cross a membrane is its electrical charge. Ions with a positive charge (**cations**) cross a membrane more slowly than would the same size particles with a negative charge (**anions**).

## Structure of the cell membrane

All the permeability characteristics that we have discussed serve as important clues to the nature of the membrane found in living things. On the basis of these characteristics, and with the help of electron microscopy, biologists have been able to construct a hypothetical model of the membrane.

The first modern model of the membrane was proposed by James F. Danielli and Hugh Davson in the mid-1930s, long before it was possible to observe any of the structural details biologists can now see through the electron microscope. According to the Danielli-Davson model, the membrane is a partition built up of several thin sheets firmly bonded together, rather like plywood. The basic structure of the membrane is a double layer of lipid molecules. A well-established property of lipid molecules is that when they are put in a watery medium, they form thin sheets on the liquid's surface. This seems likely to have been the evolutionary path through which membranes first began to form.

## 2-1
## Tables of weights and measures

**Metric system of linear measurement**

| | |
|---|---|
| 10 Angstroms (Å) | = 1 millimicron |
| 1000 millimicrons (m$\mu$) | = 1 micron |
| 1000 microns ($\mu$) | = 1 millimeter |
| 10 millimeters (mm) | = 1 centimeter |

**Metric equivalents — linear**

| | | |
|---|---|---|
| 1 Angstrom | = | 0.1 millimicron |
| | | 0.0001 micron |
| | | 0.0000001 millimeter |
| | | 0.000000004 inch |
| 1 micron | = | 0.001 milimeter |
| | | 0.00003937 |
| 1 millimeter | = | 0.03937 inch |
| 1 centimeter | = | 0.3937 inch |
| 1 meter | = | 39.37 inches |
| | | 1.094 yards |
| 1 inch | = | 2.54 centimeters |
| 1 foot | = | 0.3048 meter |

**Volumetric**

| | | |
|---|---|---|
| 1 milliliter | = | 0.271 fluid dram |
| | | 16.231 minims |
| | | 0.061 cubic inch |
| 1 liter | = | 1.057 liquid quarts |
| | | 0.908 dry quart |
| | | 61.024 cubic inches |
| 1 quart, dry (U.S.) | = | 67.201 cubic inches |
| | | 1.101 liters |
| 1 quart, liquid (U.S.) | = | 57.75 cubic inches |
| | | 0.946 liter |
| 1 pint, liquid | = | 28.875 cubic inches |
| | | 0.473 liter |
| 1 ounce, liquid (U.S.) | = | 1.805 cubic inches |
| | | 29.574 milliliters |

**Temperature**

Centigrade (C) = 5/9 (F − 32°)
Fahrenheit (F) = (9/5 × C) + 32°

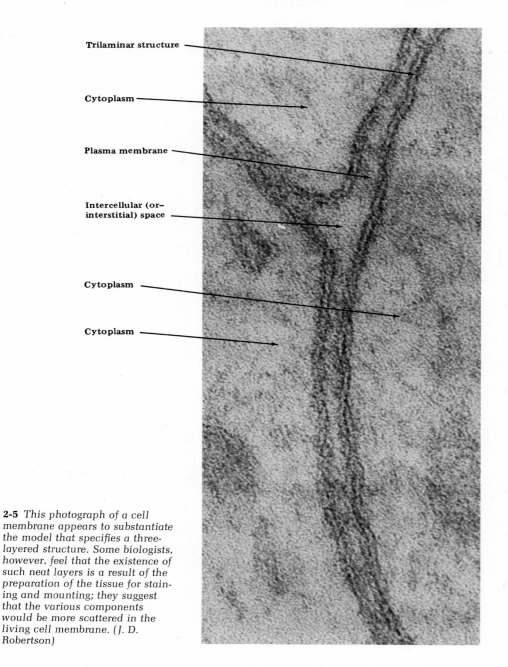

Trilaminar structure

Cytoplasm

Plasma membrane

Intercellular (or–
interstitial) space

Cytoplasm

Cytoplasm

**2-5** *This photograph of a cell
membrane appears to substantiate
the model that specifies a three-
layered structure. Some biologists,
however, feel that the existence of
such neat layers is a result of the
preparation of the tissue for staining and mounting; they suggest
that the various components
would be more scattered in the
living cell membrane. (J. D.
Robertson)*

Since lipid molecules are polar, they always
orient themselves in exactly the same way, with
alternating positive and negative poles. Thus
they form a barrier with highly specific chemical
properties. The presence of this lipid layer helps
to explain the fact that lipid-soluble molecules
generally move across membranes very rapidly;
these molecules probably dissolve in the lipid
layer of the membrane. Chemical analysis has
indicated that some of the lipids in the membrane have attached phosphate groups, and some
researchers believe that the presence of these
phospholipids is closely connected with the
membrane's ability to carry on active transport.
Perhaps the phosphates are used in the synthesis
of ATP, or perhaps the lipids act as temporary
receivers of the phosphate groups that are split
off from ATP. At the moment, both ideas are
mere conjecture, but they are the subject of current research which may produce some sort
of answer.

Associated with the lipid layers is a layer—
or possibly two layers—of protein molecules.
The Danielli-Davson model specified that the
proteins were on either side of the lipid, making a
sort of sandwich. This aspect of the membrane
model is subject to debate, and other relationships have been proposed recently that seem
better to account for new experimental evidence.
Several of these models are shown in Figure 2-6.

However the proteins are actually arranged,
there is no question but that they are present, and
that they cover much of the inside and outside
surface of the membrane. It is the protein molecules that are responsible for the membrane's
ability to expand and contract in response to
changes in osmotic pressure on either side. Chains
of polymeric proteins are really quite elastic,
capable of coiling up and then stretching out into
long chains again. The protein layer is probably
also responsible for giving the outer surface of
the membrane a positive charge, which repels the
movement of positively charged ions and attracts
negatively charged ones.

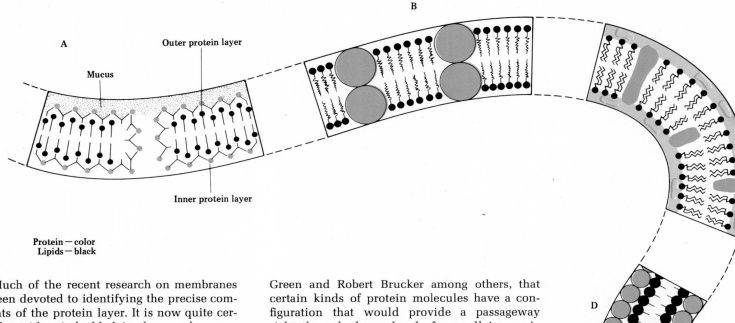

A

Mucus

Outer protein layer

B

Inner protein layer

C

D

Protein — color
Lipids — black

Much of the recent research on membranes has been devoted to identifying the precise components of the protein layer. It is now quite certain that either imbedded in the membrane or attached to it (probably by means of the phospholipids) are many of the enzymes that catalyze biochemical reactions. In our earlier discussion of the way that membranes can organize sequential reactions, we suggested it might work something like this:

$$A \quad B \quad C \quad D \quad E$$
$$En_1 \quad En_2 \quad En_3 \quad En_4$$

Now we can see that $En_1$, $En_2$, $En_3$, and $En_4$ could actually be a part of the membrane, lined up in just that sequence, separated by lipid molecules to give the optimum spacing.

One feature of the membrane model that remains totally unconfirmed is the suggested existence of tiny pores, perhaps no larger than 5–7 Angstrom units (Å) in diameter, which permit the passage of molecules across the membrane. These structures, if they do exist, are too small to be seen in even the most powerful electron microscope, so there is presently no way to prove that the membrane does or does not have pores. It has recently been suggested, by David Green and Robert Brucker among others, that certain kinds of protein molecules have a configuration that would provide a passageway right through the molecule for small inorganic ions such as $Na^+$ or $Ca^{++}$. In other words, the pores might be a built-in feature of the protein constituent of the membrane.

## Membrane systems: the cell

The membrane is the basic structural component of all living systems, but a membrane that is isolated and maintained in an isotonic solution inside a laboratory jar is not really alive, even though it may for a time display some of the characteristics of life. But it cannot grow, it cannot reproduce, it cannot supply its energy needs — in a very short time it cannot even function effectively as a partition. The unique characteristics that separate living things from nonliving matter can be seen only when a series of membranes is organized into a larger system that is called a cell.

The **cell** is the smallest unit that still displays all the characteristics of life. A cell carries on energy exchanges; it receives information about

**2-6** *The drawing shows a cell membrane rendered according to four different theoretical models. All four place proteins and lipids in some kind of double layer, but the precise structure has yet to be determined. The Danielli-Davson model is on the far left; the three other models are more recent.*

**2-7** *The first recorded description of cells was by Robert Hooke (1665). His microexamination of vegetable tissue revealed the generally rectangular structures that were so much like monks' cells of Hooke's day that he gave them that name; cellularity has since then been proven universal in living things. (The Bettmann Archive)*

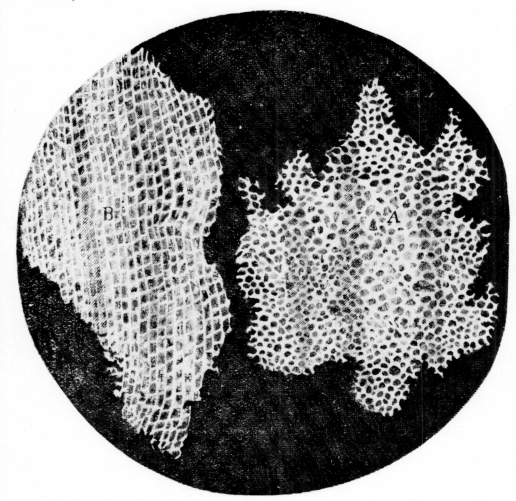

its environment and adapts its activities accordingly; it grows and reproduces itself. Because all known organisms are composed of cells (except for **viruses,** a group of perhaps-living things that never seem to fit into any of the established biological categories or to obey any of the established biological rules), the presence of cellular structure has come to be an acceptable working definition of life—we can say that living things are those which are made out of cells. This defines life in terms of structure, or how it is organized, rather than in terms of process, or what it does.

## The cell theory

Early scientists frequently speculated about the composition of living organisms. It seemed logical to believe that living things, like atoms, were composed of some basic structural unit, even though the unit was too small to be seen. Not until the technological development of strong magnifying lenses came along were biologists able to see what these units looked like. The first person to describe the units (and the one who gave them the name of cells) was the Englishman Robert Hooke, in 1665. Hooke examined a thin slice of cork under an early microscope and found it to be made of many little chambers, uniform in size and shape.

Once the microscope was in common use, a confusing variety of cells was observed, but it was not until 1839 that convincing proof of the universality of cells was offered. In that year, two German scientists proposed the cell theory. M. J. Schleiden was a botanist who had examined all kinds of plant cells; working independently, Theodor Schwann had studied animal cells. The fact that they found evidence of cells in every organism they examined led them to conclude that all living matter is composed of cells. Their cell theory stated that the cell is the basic unit of organization in all living things.

An important addition to the cell theory was proposed in 1858 by Rudolf Virchow, who stated that all living cells must come from other preexisting cells of the same type. To us, this principle of **biogenesis** may sound so obvious that it hardly needs saying, but at that time it seemed a startling idea. Most people, including even Schleiden and Schwann, were quite convinced that living organisms could be generated spontaneously. After all, they had frequently seen molds grow in bread or cheese where no mold had been before, and earthworms crawl out of soil that previously contained no worms. A seventeenth-century chemist, Jean van Helmont, even published a "recipe" for mice: put one dirty shirt in a large bin of wheat germ, let stand for 21 days, and live mice will emerge.

It was Louis Pasteur who proved that Virchow was right. In 1860, the French Academy of Science offered a prize to anyone who could prove once and for all that one side or the other in the biogenesis controversy was correct. Pasteur won the prize with his famous experiment in which he boiled broth in two kinds of flasks—one with a straight neck and one with a bent neck—that allowed air in but trapped all the tiny particles floating in the air. Left standing, the straight-necked flasks were soon filled with microorganisms, whereas the bent-necked flasks remained sterile for months. This was convincing proof that microorganisms were not spontaneously generated in the broth but grew from tiny particles, living cells, that were carried to the broth by the air.

## Studying cells

Since most cells are too small to see with the naked eye, the study of cells began only after the development of the microscope. Leeuwenhoek, who is often credited with the construction of the first microscope form, used an instrument that was little more than a pair of magnifying glasses

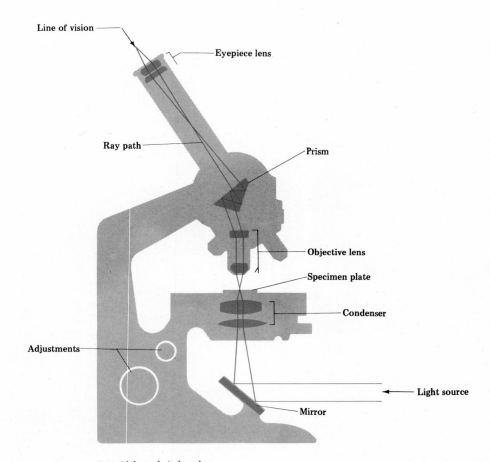

Line of vision

Eyepiece lens

Ray path

Prism

Objective lens

Specimen plate

Condenser

Adjustments

Light source

Mirror

**2-8** *Although it has been surpassed in some respects by more recent inventions, the ordinary light microscope, with a resolving power 500 times that of the naked eye, remains one of the most important laboratory tools in biology.*

screwed onto a wooden paddle. The first lens magnified the image and cast it on the second lens, which further magnified it and then projected the magnified image back to the eye. This basic principle is still utilized in the compound light microscope, but improvements in the techniques of lens grinding have led to the development of instruments much more effective than the one used by Leeuwenhoek. It has also been discovered that the position of the lenses in relation to one another is a very important factor. In Leeuwenhoek's microscope, the distortions of the image in the first lens, caused by the bending of light as it passed through the lens, were magnified and exaggerated by the second lens. In today's compound microscope, the second lens corrects the distortions of the first—in other words, the distortions in the two lenses counteract one another, so that the final image comes out undistorted rather than distorted two times over.

The magnifying power of the light microscope is limited by the nature of light itself. Light consists of waves, and each kind of light has a specific wavelength. Physicists have calculated that the smallest distance between two objects that can be **resolved,** or distinguished as two separate objects, is one-half the wavelength of the light. According to this formula for calculating resolving power, a microscope using visible light cannot resolve structures smaller than 0.0002 mm. (2000 Angstrom units).

The electron microscope is capable of revealing much smaller structures, because it uses beams of electrons rather than visible light for illumination. The electron beams have a much shorter wavelength (which is why we cannot see them), and therefore the limitation on their resolving power is much lower. The resolving power of the electron microscope is about 10 Å.

For some kinds of scientific work, the electron microscope has made the light microscope obsolete. Since it can reveal more detail, the electron microscope is used to study such things as the composition and structure of membranes,

and the way that they are grouped together to form small functional units within the cell. But the electron microscope has a number of serious disadvantages for other kinds of study. Its chief drawback is that it can be used only to look at the corpses of cells; the processes of life elude it. And its resolving power is so great that it can generally focus only on parts of cells, not on the entire unit.

The light microscope, on the other hand, can be used to study living cells. With its help, biologists can watch tiny one-celled plants and animals swimming around in the water that is their home. They can see a cell divide; they can follow the travels of a sperm cell as it moves in search of an egg; they can observe an amoeba eating its food. They can try experiments, such as dyeing the amoeba's food and tracing the path of digestion, or increasing the acidity of the water and measuring the effect of pH on the rate at which cell division takes place. The light microscope can also be used to study the relationships between one part of the cell and another. It can reveal the fact that some structures within the cell migrate from place to place, and that others tend to accumulate in certain regions, and still others invariably align themselves toward each other in exactly the same way.

**Preparing cells for microscopic study.** Suppose that you decide you want to examine the growing tip of a plant root under the light microscope, in the hope of being able to determine the structure of these rapidly dividing cells. How do you go about it? If you just cut off the root tip and place it under the microscope, you will find that what you see is singularly unrevealing. The root cells are a dense colloidal mass; the light cannot penetrate your specimen, so it appears to be no more than a black blob. Even if you had enough light passing through the cells, you would have difficulty in distinguishing one structure from another, since their color is a uniformly dull gray.

The root cells would be much more interesting if they were properly prepared for study under the light microscope. The object of this preparation is to make the cells transparent, so that the light can pass through them and reveal their structure. The process of preparation must do as little damage as possible to the cells, so that cell shapes are not distorted and cell struc-

tures are not destroyed. It is also important that the preparation should not cause any artificial structures, such as crystals or foldings of a membrane, to form. For this reason, the cells should be killed as quickly as possible. The usual method is to drop them in a fixative solution containing alcohol and formaldehyde.

ESSAY **Methods of studying cells**

The three photographs illustrate some of the methods other than microscopy that can be used to study cells. Figure A shows the use of one technique, paper chromatography. This process is used to separate the individual components of mixtures. A drop of the mixture to be resolved is spotted at the top of filter paper, which is then placed into an apparatus that allows a solvent to flow slowly down the paper. The solvent consists of an aqueous phase mixed with an organic phase; since water and organic solvents do not mix (like oil and water), this solvent actually constitutes two distinct liquid phases. The water phase will be trapped in the pores of the paper and can be considered a stationary phase. Each different type of molecule in the spot will be carried along down the paper at a rate proportional to its solubility in the faster-flowing organic phase. Thus, different kinds of molecules can be separated according to very light differences in their chemical properties.

Figure B illustrates the use of tissue culture in studying cells. It is often advantageous to the biologist to grow cells or tissues outside the plant or animal being studied. By supplying the cells with a nutrient medium that provides all the necessary materials usually brought to the tissues by the blood, whole organs can be kept alive for days or weeks. Often cultures are derived from a single cell, and descendants of this one cell can be grown indefinitely in some cases. Such a culture provides a homogeneous or "pure" and reproducible system for the biochemist to study. For example, one might grow a culture of kidney cells and then study the chemical response to some hormone or drug which is added to the medium. Some plant tissues grown in culture for a period of time can be induced to develop an entire new plant. In other types of studies, the development of embryonic tissues can be followed. Under appropriate conditions, cancer cells and normal cells can be distinguished by their behavior in tissue culture. The normal cells growing in a petri dish will stop dividing when a layer one or two cells thick covers the dish, while cancer cells "overgrow" and form a layer several cells thick. Figure B shows a culture of human tissue cells that are being grown to serve as a living host for a strain of virus that is suspected of causing a certain type of cancer.

Autoradiography (Fig. C) is a technique very useful to biologists in pinpointing the location of specific molecules within cells or tissues. The first step is to label the cells by feeding them a radioactive molecule, which will then become incorporated into a specific kind of cellular molecule. For example, in the photo shown here, *Tetrahymona*, a one-celled animal, was fed $^3$H-cytidine, which becomes incorporated into RNA. After the cells have been labeled, they are

A

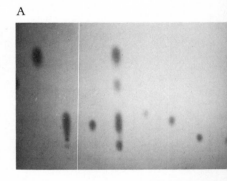

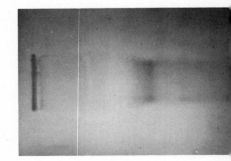

How can the cells be made transparent? The solution is to cut them in very thin slices, using a special tool called a microtome. But as you know if you ever tried to cut thin slices of a tomato, slicing often causes bruising and crushing. The delicate cells might be smashed or torn apart by the knife, rather than sliced cleanly through. In order to make the slices, you must have some way of holding them in a stiff and rigid position while they are cut. This can be done by soaking them in hot wax which stiffens as it cools. After the slices have been cut and placed on glass slides, it will be necessary to remove the wax; this can be done with a few drops of a sol-

treated with a fixative, such as acetic acid, to preserve their structure, then mounted on a glass slide. Next, the slide is coated with either a liquid photographic emulsion or a thin strip of special film. After an appropriate length of incubation in total darkness, the slides are developed by the same process used for the roll of film. In this example, $\beta$-rays emitted in the disintegration of the $^3$H (tritium) molecules are responsible for the blackened regions of the developed "negative." If the slide is viewed under a light or electron microscope, grains which look like black dots will appear over regions where $^3$H-containing molecules are located. If cells are killed and fixed at varying times after feeding the $^3$H precursor, the movement of specific kinds of molecules within a cell or tissue can be followed, using this technique.

B

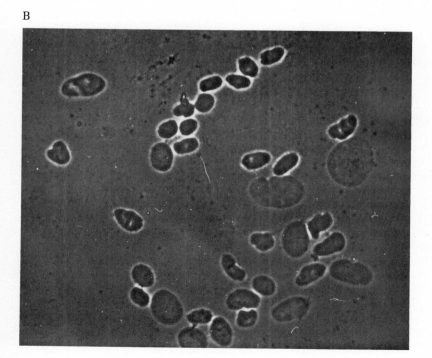

C

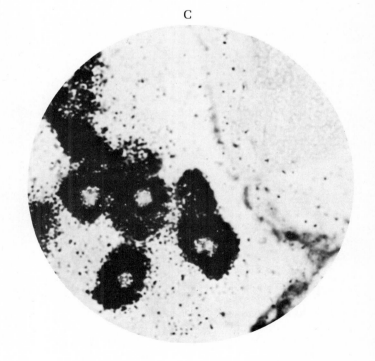

vent. The final step is to protect the slice by glue-ing on a thin glass or plastic cover, to keep the cell from drying out.

This procedure will make the root cells transparent, but under the microscope, it will still be the same uniform gray, so that it is difficult to tell one part of the cell from another. In order to distinguish cell parts, various stains or dyes can be added to the root at any stage of the preparation. For example, if an eosin dye is dropped on the cells, the nucleus will appear dark purple and the rest of the cell will be pink. A stain called Sudan black makes groups of fat molecules show up as black dots. Chromosomes will stand out as blue-black threads if the cells are treated with aceto-orcein. There is an appropriate staining technique for every cell part being studied.

## Cell form and function

Cells exist in such a wide variety of sizes and shapes that it is impossible to speak of a "typical" cell. An ostrich egg is a single cell, and it has a diameter of about 40 centimeters; a pneumonia-causing bacterium, also a single cell, has a diameter of about 0.02 microns. (A micron is 1/1000th of a millimeter; it is commonly abbreviated by the symbol $\mu$.) Differences in shape are just as remarkable. Human nerve cells are usually less than 1 $\mu$ thick but may be as long as three or four feet, so that they look rather like wires. Red blood cells are shaped like poker chips with very thick edges. The amoeba, a single-celled organism, has no determinate shape at all; its shape constantly changes as it moves, feeds, and bumps into other microorganisms.

These variations in size and shape, many of which are caused by the effects of the environment, help adapt cells to perform different functions. The function of the nerve cell, for instance, is to conduct electrochemical impulses, and its long thin shape is ideally suited for this job. The function of an egg cell is to provide food and water for a rapidly growing embryo, and so its large size provides additional storage space. Cells that are closely packed together in a multicellular organism are usually shaped like polyhedrons (solid shapes with many faces), permitting them to fit together without large spaces between them. Single-celled organisms that live in the air are often shaped like perfect spheres, but their shape may change if they land in water. It has been demonstrated that many cells of the human body will assume different shapes if they are removed and grown in a nutrient broth in the laboratory. Only when they are in the chemical and physical environment of the body do they take on their characteristic shapes.

With the exception of some very large egg cells and some very small bacteria, most cells fall within the size range of 0.5 to 40 $\mu$ in diameter. It appears that there is some practical limitation on the size of the cell. One reason for a size limit is the problem of the surface-to-volume ratio. Needed materials such as water, oxygen, and sugar must enter the cell through its surfaces; waste products must leave the same way. As cell size increases, the volume of the cell increases at a much faster rate than the surface, so that it becomes more and more difficult for the limited surface area to carry on the necessary chemical exchanges. This problem of the relationship between surface and volume means that cells simply cannot grow beyond a certain size—the science fiction movies about giant amoebae that devour cities could not possibly occur in real life. Cells whose function requires a rapid rate of chemical activity will typically be very small, with a relatively low surface-to-volume ratio. Less active cells, such as egg cells, can survive with high surface-to-volume ratios and are therefore often much larger cells.

Another reason for size limitation is the problem of internal transport. Larger molecules of carbohydrates, enzymes, and structural pro-

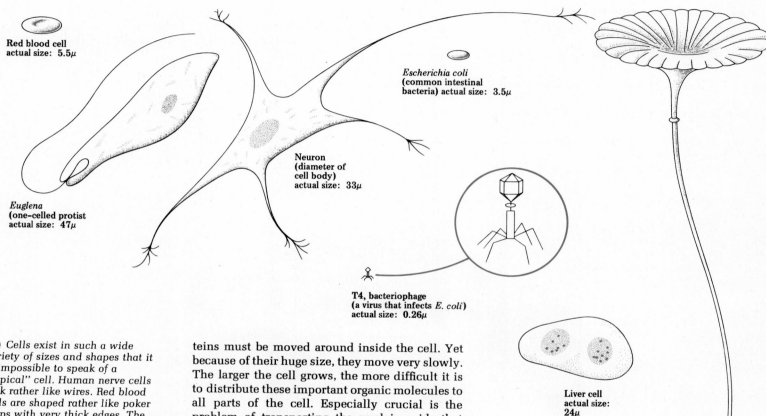

**Red blood cell
actual size: 5.5μ**

*Euglena*
**(one-celled protist
actual size: 47μ**

**Neuron
(diameter of
cell body)
actual size: 33μ**

*Escherichia coli*
**(common intestinal
bacteria) actual size: 3.5μ**

**T4, bacteriophage
(a virus that infects** *E. coli*)
**actual size: 0.26μ**

**Liver cell
actual size:
24μ**

*Acetabularia*
**(giant-celled
green algae)
actual size:
2.3 × 10⁴μ**

**2-9** *Cells exist in such a wide
variety of sizes and shapes that it
is impossible to speak of a
"typical" cell. Human nerve cells
look rather like wires. Red blood
cells are shaped rather like poker
chips with very thick edges. The
amoeba has no determinate shape
at all. The T4 virus, not clearly
a cell or even an organism, is
shown for contrast.*

teins must be moved around inside the cell. Yet
because of their huge size, they move very slowly.
The larger the cell grows, the more difficult it is
to distribute these important organic molecules to
all parts of the cell. Especially crucial is the
problem of transporting the nucleic acids that
control cellular functions to remote parts of the
cell, for there is a limit to the amount of cytoplasm that can be controlled by a single nucleus.

A third problem is that of physical support.
British biologist D'Arcy Thompson has pointed
out that if an elephant's weight were to be
doubled, his leg bones would have to quadruple
in diameter in order to provide support for such
bulk. The same problem occurs on a cellular
level. As cells increase in size, weight, and
volume, they require much more support. The
necessary supporting fibers in turn make the cell
just that much bigger, multiplying all the original
problems of size.

## The structure of the cell

Not long ago, biology students were taught that cells are made of protoplasm, or "living substance"—a concept about as helpful as saying that all cars are made of autoplasm. It is more accurate to say that a cell is a system of membranes forming a functional unit that can carry

on all the metabolic activities associated with life.

The cell is bounded by a continuous membrane that serves both to enclose the contents and to regulate passage of molecules into and out of the cell. Within the cell are a number of specialized membrane systems called **organelles.** Each organelle is specifically adapted for certain functions. According to a theory advanced by J. David Robertson of Harvard, the organelles and the outer membrane have been formed from one single membrane structure, the unit membrane.

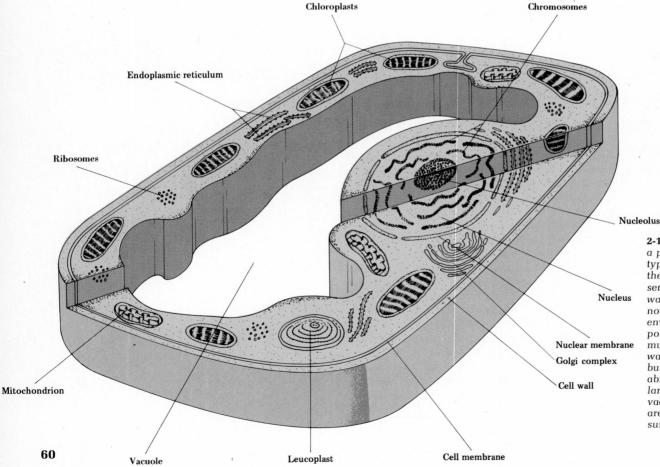

Chloroplasts

Chromosomes

Endoplasmic reticulum

Ribosomes

Mitochondrion

Vacuole

Leucoplast

Cell membrane

Nucleolus

Nucleus

Nuclear membrane

Golgi complex

Cell wall

**2-10** *This generalized drawing of a plant cell shows the organelles typically found in such cells. Note the large central vacuole, which serves as a reservoir for cellular wastes. Unlike animals, plants do not expel their wastes into the environment, since to do so would poison the earth in which they must remain rooted. Instead, wastes are held within the cells but kept isolated by an impermeable membrane. Because of the large size of the waste-containing vacuole, other cellular organelles are crushed up against the outer surfaces of the cell.*

**2-11** *This generalized animal cell can be compared with the plant cell in Figure 2-10. Because it has no rigid cell wall, the animal cell is more spherical and is often an irregular shape. Since the animal cell has no vacuole, other cellular organelles usually appear rather evenly distributed.*

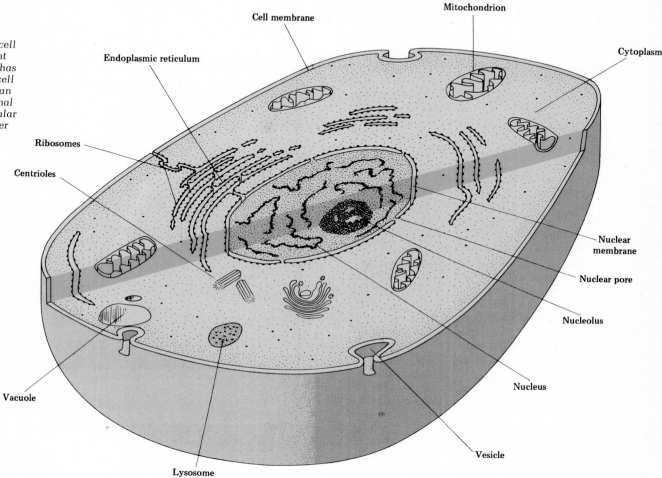

Cell membrane

Mitochondrion

Cytoplasm

Endoplasmic reticulum

Ribosomes

Centrioles

Nuclear membrane

Nuclear pore

Nucleolus

Nucleus

Vacuole

Vesicle

Lysosome

Chemical analysis has cast some doubt on the unit membrane theory, since the specific constituents of the membrane seem to differ according to the function of the organelle. Yet the basic similarity of the membranes in organelles is undeniable, indicating a close relationship and probably a similar evolutionary origin.

In addition to the organelles, the cell contains a large dense structure called the **nucleus,** which serves as the control center of the cell. Both the organelles and the nucleus are embedded in the **cytoplasm,** or cell substance.

Much of the cytoplasm in a cell is a colloidal mixture of protein macromolecules and fat globules dispersed in water. Smaller molecules, such as simple sugars and inorganic ions, are mixed with water to form a true solution. And protein molecules of an intermediate size may mix with water in such a way that they show some of the characteristics of a solution and some of the characteristics of a colloid. The outer portion of cell cytoplasm is usually in a gel state, but farther inside the cell, the cytoplasm becomes more and more fluid.

**61**

## The cell nucleus

The nucleus is a relatively large structure within the cell, and it appears to be denser than the surrounding cytoplasm. It is surrounded by a double thickness of membrane. This membrane exhibits distinct pores that are visible in the electron microscope. Experiments have indicated that in the area of the pores there is a definite regulation of the passage of molecules; many that are clearly small enough to fit through the opening do not pass through. From this data biologists hypothesize that the pore spaces are lined with carefully structured groups of molecules whose electrical charge can attract or repel specific molecules that might enter the cytoplasm.

The nucleus directs most of the metabolic activities of the cell. It performs this function by discharging certain nucleic acids, often called informational molecules, that in turn cause changes in the rates of the synthesis and breaking down of organic compounds. The nucleus also determines the direction in which the cell will specialize or differentiate. It has been demonstrated, for example, that when the nucleus of one type of cell is carefully removed and replaced by the nucleus of another type of cell, the entire cell will undergo changes until it resembles the cell from which its new nucleus was taken.

When the nucleus is stained, it shows certain dark masses, called **chromatin.** These chromatin fibers contain the hereditary material of chromosomes and genes, which control reproduction and pass along from parent to offspring nucleic acids that are actually coded patterns of cell development.

Within the nucleus itself lies a small ovoid structure called the **nucleolus.** There may even be two or more nucleoli in a single nucleus. The nucleolus appears to be the site for synthesis of ribonucleic acids. It also seems to have some special relationship with the chromosomes. Chromosomes are often observed intimately associated with the nucleolus. Most often the attached

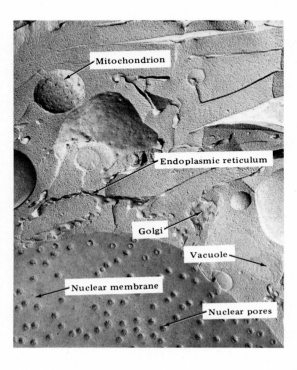

Mitochondrion

Endoplasmic reticulum

Golgi

Vacuole

Nuclear membrane

Nuclear pores

**2-12** *Nuclear pores are clearly visible on the face of the nuclear membrane (large gray structure) in this freeze-etch electron micrograph of a portion of an onion root tip cell. The interior of the nucleus can be seen through the pores. (× 40,000) (Courtesy of D. Branton. 1966. Proc. Nat. Acad. Sci. U.S.A. 55: 1048.)*

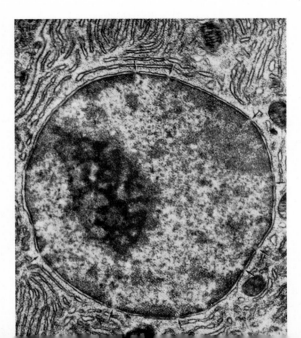

**2-13** *When the nucleus is stained, it shows certain dark masses, called chromatin. These chromatin fibers contain the hereditary material of chromosomes and genes, which control reproduction and pass along from parent to offspring nucleic acids that are actually encoded patterns of cell development. Within the nucleus itself lie one or more nucleoli. They appear to be the site for synthesis of ribonucleic acids. (From A Textbook of Histology by Bloom and Fawcett, W. B. Saunders)*

**2-14** *The endoplasmic reticulum pictured here serves as a kind of canal system through which materials can be transported in and out of the nucleus. It has also been suggested that the membranous structure might serve as a method of separating potential reactants, or reactants and their products. The tiny granular bodies, ribosomes, attached to the surface of the endoplasmic reticulum constitute the site of all protein synthesis in the cell. (From A Textbook of Histology by Bloom and Fawcett, W. B. Saunders)*

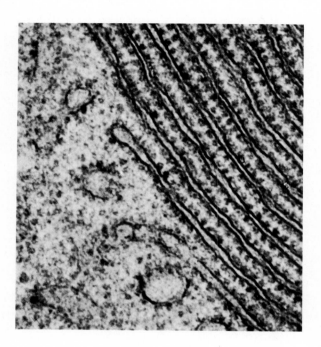

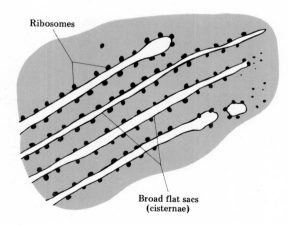

Ribosomes

Broad flat sacs (cisternae)

chromosome is the one that controls sexual development and differentiation, so the nucleolus may play some part in this process.

## The endoplasmic reticulum and ribosomes

The endoplasmic reticulum (ER) is thought to be a supply line to the nucleus. The ER is a network of membranes within the cytoplasm. The membranes of the ER show the same fundamental characteristics found in the cell and nuclear membranes, being made of proteins and lipids. It has been folded many times to form an interconnecting system of channels. In some types of cells, the ER appears to be attached to the nuclear membrane; in other cell types it is attached to the cell membrane, and in some cells, attachment to both membranes has been demonstrated.

It has been suggested that the ER serves as a kind of canal system through which materials can be transported in and out of the nucleus, perhaps even directly from the outer surface of the cell to the nucleus. It has also been suggested that the membranous structure might serve as a method of separating potential reactants, or reactants and their products, during various stages of metabolic activity.

**Ribosomes** may be attached to the surface of the ER. They are tiny granular bodies, about 200 Angstroms in diameter. (ER with ribosomes is called rough ER; if it has no ribosomes, it is called smooth ER.) Ribosomes may also be seen scattered throughout the cytoplasm, unassociated with the ER. They are the site of all protein synthesis in the cell. This has been demonstrated by introducing radioactive amino acids into cells. Within a very short time—15 or 20 minutes—the amino acid, the essential ingredient for protein synthesis, is found concentrated at the ribosome. Later the tagged amino acid is found within a protein molecule.

**63**

## The Golgi apparatus

Electron microscope photographs of the Golgi apparatus show that this structure, too, is made of a folded membrane. Some of the folds produce small flattened sacs or vesicles. These sacs are used to package organic compounds for transport.

The function of the Golgi bodies is still under debate; in fact, when Camillo Golgi first described the structure in 1903, some scientists denied that it even existed—they were convinced it was some distortion of cell structure caused by the process of preparing it for viewing in the microscope. Not all cells contain this structure, but it is especially large and active in cells that specialize in secretion. This association has led to the hypothesis that the Golgi apparatus itself is the site of the synthesis of secretory products. The membranous portion of the apparatus may help separate reactants during the stages of synthesis. The vesicles may serve as bags that transport the products through the cell to the outer membrane, separating them from the rest of the cytoplasm during the trip to prevent harmful reactions.

## Mitochondria

Mitochondria, structures large enough to be visible as tiny dots under a light microscope, are the site of cellular respiration. A mitochondrion is covered by a smooth outer membrane, similar to the membrane found in other cellular structures. Inside is a second membrane, of a slightly different composition. The inner membrane has many folds, called **cristae.** The cristae form interconnected little chambers, each filled with a dense liquid. Like the cell membrane, the membrane covering the mitochondrion can stretch and shrink, and mitochondria have been observed to change size as a cell becomes more or less metabolically active. These changes are probably caused by changes in the permeability characteristics of the surrounding membrane.

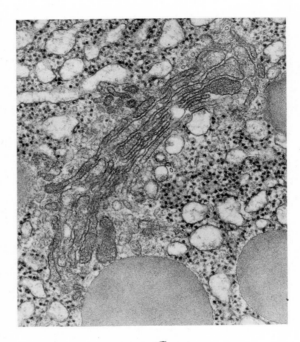

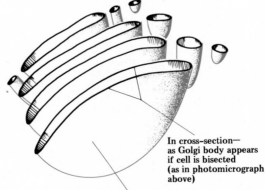

In cross-section—
as Golgi body appears
if cell is bisected
(as in photomicrograph
above)

Lozenge form of body (penetrating interior of cell, i.e., encapsulated by cell)

**2-15** *Not all cells contain a Golgi body, but it is especially large and active in cells that specialize in secretion, such as this cell from a rat's pancreas. This association has led to the hypothesis that the Golgi apparatus itself is the site of the synthesis of secretory products. (Daniel S. Friend)*

**2-16** *The unusually large and numerous mitochondria of cardiac muscle are clearly visible in this tissue sample from the heart of a rat. They permit the tissue to contract repeatedly without fatigue. Note also the striated-muscle pattern of the myofibrils. (× 60,000) (Courtesy of Dr. Bryce L. Munger, Milton S. Hershey Medical Center, Pennsylvania State University, Hershey, Pa.)*

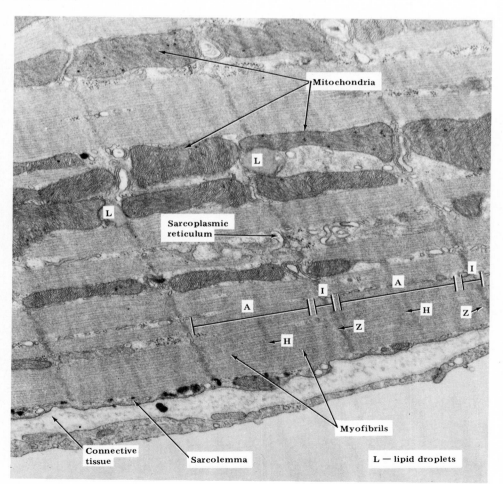

Mitochondria

Sarcoplasmic reticulum

A    I    A    I

I    H    Z

Z

H

Myofibrils

Connective tissue    Sarcolemma    L — lipid droplets

The mitochondria function as the powerhouse of the cell. It is inside their chambers that the greatest portion of energy in simple sugars is released. Mitochondria are usually present in large numbers in the parts of the cell where most work is done, such as the portion of the muscle cell that initiates contraction, and the tail of the sperm that pushes the cell along. This is another example of the great efficiency of cellular organization—providing energy in precisely the location where it is needed.

Small granules can be seen clinging to the folds of the inner membrane; they play a crucial part in the process of respiration, serving as a storage place for needed enzymes and ATP.

Biologists are still debating the possible origin of mitochondria. Since these organelles are composed of a membrane similar to the cell membrane, it is possible that they might have been formed by a folding of that membrane. Reinforcement for this theory lies in the fact that the protein structure of the membrane is particularly well adapted for such folding, as well as in the fact that in some cells many organelles can be seen to be connected to other membranes. A somewhat different hypothesis has recently been summarized by Lynn Margulis. She suggests that the mitochondria might once have been independent organisms that at some long-ago time invaded larger cells and established a permanent host-parasite relationship. (Presumably most cells did not use oxygen in their respiration before that invasion.) According to this theory, the outer membrane of the mitochondrion is formed by the larger cell, as a sort of barrier against the invader, and the inner folded membrane is actually the cell membrane of the small parasite. This hypothesis is given some support by the fact that mitochondria contain nucleic acids that differ in certain marked ways from the nucleic acids found in the cell nucleus, and that they possess the power of reproducing themselves independently of cell division. It has also been shown that when they are carefully extracted from the cell, they seem capable of independent existence for an indefinite period of time.

## Plastids

Plastids are specialized bodies with structures somewhat similar to those of the mitochondria. They have a double outer membrane and a complex inner system of densely folded membranes.

Plastids, which are found in plant cells only, are specialized for the function of synthesis. **Leucoplasts** are the site of synthesis of starch from simpler sugars; they also serve as a storage area if the starch is not needed immediately. **Chromoplasts** are the site of the synthesis of the pigments found in plants; for example, the red pigment in coleus plants is made there.

The most important plastids are the **chloroplasts.** Photosynthesis takes place within the chloroplasts of the green plant. The electron microscope shows that in some places the inner membrane of the chloroplast forms thick stacks called **grana.** Small beadlike structures, called quantasomes, or granules, about 100 Å thick, can be seen between the folds. Molecules of sugar are synthesized along the surfaces of the membrane; the quantasomes probably contain the chlorophyll, needed in the reaction, and many enzymes as well.

Like mitochondria, plastids also contain a kind of nucleic acid not found in the cell nucleus, and they are capable of reproducing themselves. They too might once have been free-living organisms that took up a parasitic way of life.

## Lysosomes

Lysosomes first were seen in 1952, but even before the electron microscope made them visible, their existence was described through the evidence of their biochemical activity. Lysosomes

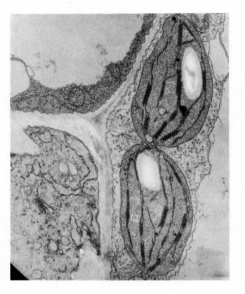

2-17 These tobacco chloroplasts appear to be in the last stages of dividing. Clearly visible in the electron micrograph are the vacuole (to the right of the chloroplasts), the cell wall (left of the chloroplasts), and two starch grains (white structures). (Courtesy of Dr. David Stetler, Dartmouth College)

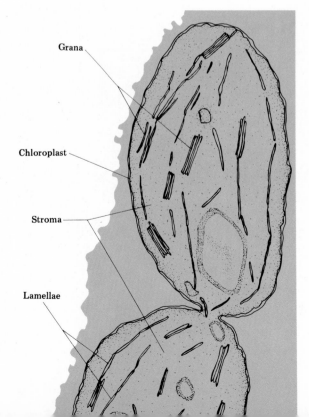

Grana

Chloroplast

Stroma

Lamellae

are small membranous sacs that enclose enzymes used in breaking down cellular materials. Lysosomes store these enzymes until they are needed, thus preventing the disintegration of the living cell.

Under normal conditions the lysosome membrane remains intact and the huge enzyme molecules remain imprisoned and inactive inside their sacs. But when changes in the chemical environment of the cell cause the lysosome membrane to rupture, the enzymes are released and the cell's structure is attacked; the lysosomes play an important role in the death of the cell. This occurs, for example, when tadpoles turn into frogs. The cells in their tails are rich in lysosomes, and at a certain stage of development, the lysosomes burst open and the enzymes break down those cells, freeing the organic compounds they contain for use in building the new cells of the mature frog. One-celled organisms like the amoeba, which get their food by surrounding and engulfing smaller organic particles, appear to use lysosomes in the process of digestion. The lysosome travels to the part of the cell containing the food particle, where the sac bursts and the enzymes digest the food.

Current research is investigating the role of lysosomes in the aging process. Many of the signs of aging are due to the fact that cells are dying faster than they are being replaced. It is possible that this increased rate of cellular death is caused by more frequent ruptures in the lysosome membranes. When we discover the mechanism that causes the membrane to break, we may then learn how to keep it from happening, and thus we might retard the aging process.

## Vacuoles

A vacuole is a pocket of fluid—a solution of organic compounds, inorganic ions and water—enclosed by a membrane called the **tonoplast.** The tonoplast keeps the contents of the vacuole from diffusing out into the cell.

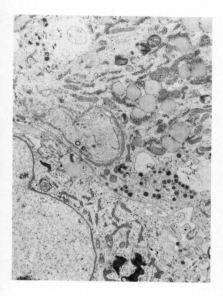

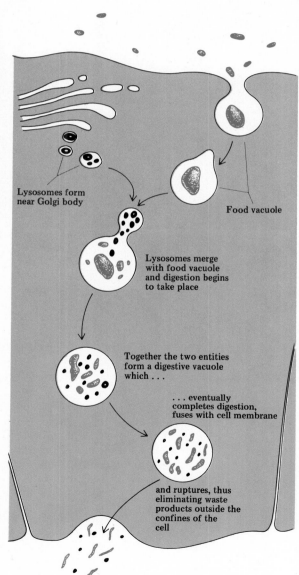

Lysosomes form near Golgi body

Food vacuole

Lysosomes merge with food vacuole and digestion begins to take place

Together the two entities form a digestive vacuole which . . .

. . . eventually completes digestion, fuses with cell membrane

and ruptures, thus eliminating waste products outside the confines of the cell

**2.18** *Lysosomes are small membranous sacs that enclose enzymes used in breaking down cellular materials. Lysosomes store these enzymes until they are needed, thus preventing the disintegration of the living cell. A series of lysosomes can be seen in the lower right hand part of the photo; it is quite clear that their size increases as they move toward the cell membrane.* (Ross—Omikron)

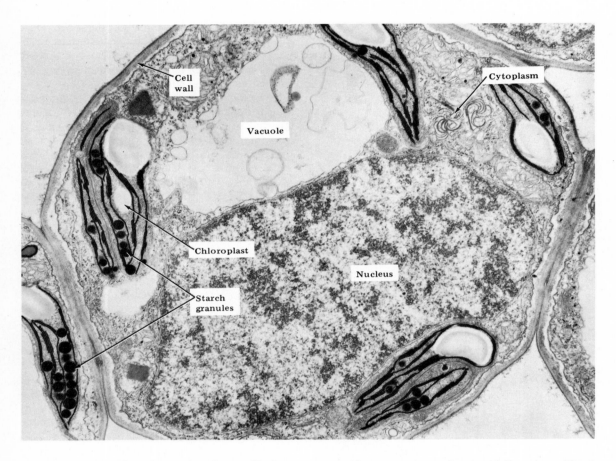

Cell wall

Cytoplasm

Vacuole

Chloroplast

Nucleus

Starch granules

**2-19** *A vacuole is a pocket of fluid, a solution of organic compounds, inorganic ions, and water, enclosed by a membrane called the tonoplast. The tonoplast keeps the contents of the vacuole from diffusing out into the cell. The large vacuole in this cell from a sunflower plant serves to store starch.*

Vacuoles are found in most plant cells, but only a very few kinds of animal cells have them. When the cell is young, it may have several small vacuoles; as the cell matures, the vacuoles move together, forming one large structure. In many plant cells, the vacuole occupies about 90 percent of the volume. The cytoplasm and other organelles are squashed up against the cell membrane by the surface of the tonoplast.

In some unicellular organisms, the vacuole can expand and contract, due to the elasticity of the surrounding membrane, and the contraction pushes water out of the cell. In other cells, the vacuole may serve as the site of digestion of food particles. Lysosomes travel to the vacuole and empty their contents into it. The membrane around the cell prevents the enzymes from leaving the site and breaking down cellular material. In plant cells, the vacuole often contains pigments, such as the one that gives roses their red color. It may also contain such compounds as citric acid (in citrus fruits) and highly concentrated mineral salts, which may be present in such heavy concentration that they crystallize. The vacuole may also serve as a storage place for toxic cellular wastes.

## Centrioles

Centrioles are tiny cylindrical bodies located near the nucleus. They are about 3000–5000 Å in length, and about 1500 Å in diameter. They are found at the outer surface of the nucleus, and are also often associated with the Golgi bodies. One hypothesis is that the centrioles play a role in cellular reproduction, furnishing threadlike structures called **spindles** to support and perhaps guide the chromosomes during the process of division. They are commonly found in animal cells and are very rare in plants.

The electron microscope reveals that centrioles have a very specific organization, no matter where in the cell they appear. If two centrioles are present, they will always be oriented so that one is perpendicular to the other. Each centriole is composed of nine groups of triplet tubules which are evenly spaced around the circumference of the centriole. The centriole is closed at one end, but is apparently capable of growth at the open end. It is assumed that the spindle threads are simply extensions of the groups of tubules. Centrioles are capable of dividing and reproducing. Just before the chromosome begins to divide during cellular reproduction, the centriole divides into two parts, and one moves to each end of the group of chromosomes.

In some kinds of cells, the centrioles divide to form a structure called a **basal body,** or kinetosome. This structure then moves toward the edge of the cell. From it will grow the cilia and flagella.

## Cilia and flagella

Many one-celled organisms have little hairlike structures projecting from the surface of the cell membrane. These adaptations increase the cell's ability to move through its watery environment. Similar structures can be found in specialized cells of multicellular organisms, where they help push large bodies or structures along a canal. An ex-

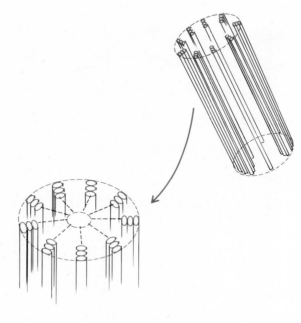

**2-20** *Centrioles are tiny cylindrical bodies located near the nucleus. They are about 3000–5000 Angtrom units in length, and about 1500 in diameter. They are found at the outer surface of the nucleus, and are also often associated with the Golgi bodies. One hypothesis is that the centrioles play a role in cellular reproduction, furnishing threadlike structures called spindles to support and perhaps guide the chromosomes during the process of division. (Reproduced by permission of Dr. E. de Harven, Sloan-Kettering Institute)*

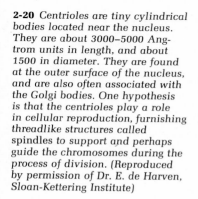

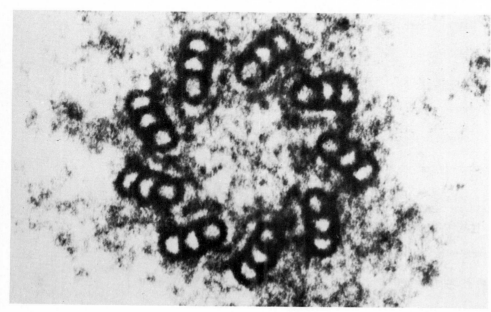

ample can be seen in the cells lining the female Fallopian tubes; the little hairs help push the egg along the tube. If the projection is long and whip-like, it is called a flagellum. If there are many short projections, they are called cilia. Both cilia and flagella grow out of the basal bodies, and like centrioles, they have a very specific structure. Shaped like long cylinders, they contain nine groups of twin tubules arranged around the circumference and two single tubules in the center of the cylinder. The entire structure is covered by a membrane.

## Microtubules

Recent electron microscope studies of cells have revealed many small microtubules running through the cytoplasm. These are tiny hollow tubes, about 230 Å in diameter, and they appear to be quite similar to the tubules found in centrioles, flagella, and cilia. They are abundant in both plant and animal cells, and can often be observed densely clustered just inside the cell membrane.

One function of the microtubules is to act as a support for the cell, in the same way that bones support the human body. Other possible functions are still being investigated. In plant cells, they may play some part in laying down the cell wall. Studies of microtubules in animal tentacles have indicated that they may serve as a transportation device for digestive enzymes which must move out to the trapped foods.

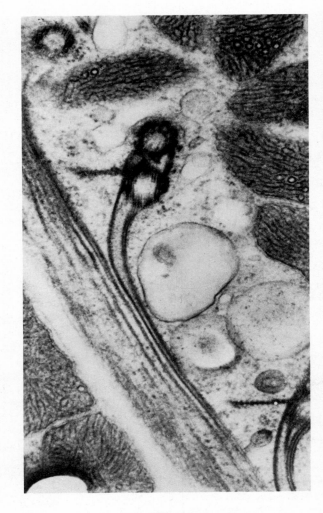

**2-21** *Cilia grow out of the basal bodies, and like centrioles they have a very specific structure. Shaped like long cylinders, they contain nine groups of twin tubules arranged around the circumference, and two single tubules in the center of the cylinder. The entire structure is covered by a membrane. This photograph shows both a cross-sectional and longitudinal view of a cilium. (Dustin Osborn)*

## Cell coatings

Many cells also have a layer of molecules covering the outer protein layer of the cell membrane. This coating often serves to protect the cell. In multicellular plants, it may also provide structural rigidity, just as bones do for animals.

In bacteria, where the function of the cell coating is primarily protective, the coating is a complex structure of simple sugars, amino sugars, amino acids, and lipids that are secreted through the cell membrane. This mucuslike coating provides an additional barrier to the entry of macromolecules from the environment; it also helps the cell conserve water when it is in a dry environment, such as the air.

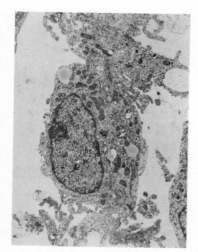

**2-22** *Microtubules are tiny hollow tubes, about 230 Angstroms in diameter, and they appear to be quite similar to the tubules found in centrioles, flagella, and cilia. They are abundant in both plant and animal cells, and often can be observed densely clustered just inside the cell membrane. In this photo, they show up as fine threads running vertically through the cell. (David Soifer—Omikron)*

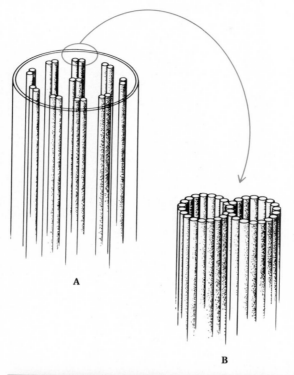

A

B

The most elaborate coatings are the cell walls found in green plants. Plant cell walls are composed of long rigid chains of cellulose, a polymeric molecule. The cellulose is deposited in a pattern that looks woven, creating a thick and rigid structure. The plant cell wall is secreted by the living cell, and to a certain extent it is expandable, allowing for some degree of cellular growth. However, in many parts of the plant, the contents of the cell die after the cell wall has been built.

A young plant cell is surrounded by a **primary wall,** which is relatively thin ($1-3\mu$ thick) and quite elastic. As it matures it lays down a thick inner wall, the **secondary wall,** about $5-10\mu$ thick. Adjacent plant cells share an intercellular substance called the **middle lamella.** It contains a rather gummy carbohydrate called pectin, which binds the cells together (pectin is the compound that gives jelly its characteristic texture). It is believed that the middle lamella may be formed by an extension of the membrane of the Golgi apparatus.

**2-23** *The ultrastructure is clearly shown in this heavy metal shadowed preparation of primary cell wall from a higher plant (Juncus). The composite fibrils on the interior of the wall (top) are perpendicular to the long axis of the cell. Microfibrils on the exterior wall (lower left) are parallel to the long axis. Those in between are intermediate in orientation. (Photo by Prof. P. A. Roelofsen, courtesy of Technische Hogeschool, Delft, Netherlands)*

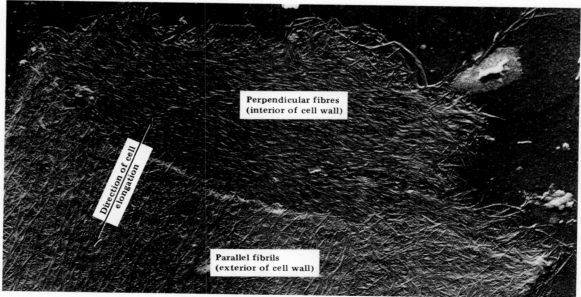

Perpendicular fibres (interior of cell wall)

Direction of cell elongation

Parallel fibrils (exterior of cell wall)

## Cell specialization

No one cell contains all of the structures that have been described here. Some cells are very simply organized. They have a nuclear region which contains chromosomes and necessary nucleic acids, but there is no clearly defined nucleus and no nuclear membrane. They do not possess any distinct organelles, although regions of the cell may still specialize in one particular metabolic activity. These simplest cells are called **procaryotic cells.** More specialized cells—those with a nucleus, nuclear membrane, and organelles such as plastids or mitochondria—are called **eucaryotic cells.**

The primary advantage of all forms of specialization is the increase in efficiency with which life processes can be carried out. For living systems, always struggling to defeat entropy and equilibrium, an increase in efficiency means an increase in survival.

It is thought that specialization began when groups of relatively simple cells began to live together in colonies. Figure 2-26 shows one such colony of green algae cells that live together in a clump; it is called *Pandorina.* The algae cells have actually formed a kind of commune, in which independent individuals share their limited resources for their mutual benefit. This kind of sharing does not create real dependence. If a girl who lives in an organic-food commune decides she cannot stand to look at another bowl of brown rice, she can easily leave the group, and perhaps set up a new commune with different food preferences. In exactly the same way, experiments with *Pandorina* have shown that if one algae cell is isolated from its group, it will immediately begin to divide and form new cells, until another colony of *Pandorina* is created. Each cell in the group is capable of independent existence.

The next stage in the process of specialization can be seen in *Volvox,* which is also a colonial form of green algae. In the *Volvox* colony,

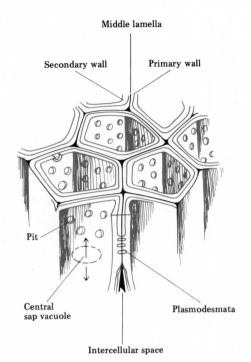

Middle lamella

Secondary wall

Primary wall

Pit

Central
sap vacuole

Plasmodesmata

Intercellular space

**2-24** *Adjacent plant cells share an intercellular substance called the middle lamella, composed primarily of pectin, a viscous carbohydrate. Each cell shows the thin primary wall and the much more heavily thickened secondary wall, made of woven strands of cellulose like those in Figure 2-23.*

**2-25** *Although procaryotic cells contain much of the same metabolic machinery found in eucaryotic cells, procaryotes contain few organelles or areas of specialization. This simplicity is probably less efficient, but it has proven a very effective alternative pattern, for procaryotic organisms, such as bacteria, have been very successful.*

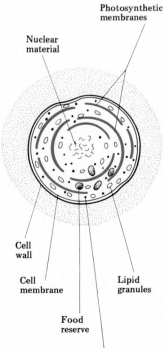

Photosynthetic
membranes

Nuclear
material

Cell
wall

Cell
membrane

Lipid
granules

Food
reserve

Ribosomes

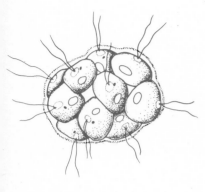

**2-26** *Pandorina is a colonial form of green algae. Individual cells join together in a group, held in a globular shape by a protective gelatinous sheath. Yet despite this degree of cooperation, each cell of Pandorina is capable of carrying on an independent existence.*

**2-27** *The next stage in the process of specialization can be seen in Volvox, a colonial form of green algae. In the Volvox colony, some cells have specialized in the function of reproduction. Cells separated from the colony are able to live alone, but cannot reproduce. Each of the round balls shown in this stained preparation may contain as many as 40,000 cells. (Hugh Spencer from National Audubon Society)*

some cells have specialized in the function of reproduction. Their specialization frees other cells in the colony to work full time at producing food, which increases the entire colony's food supply—a definite advantage for all participating cells. But as the old saying goes, you can't get something for nothing, and that applies just as much to the biological world as it does to the field of economics. The cost of the increase in food supply is the sacrifice of complete independence. Many of the food-producing specialists lose the ability to reproduce. If they are isolated from the colony, they will sooner or later die. In *Volvox*, specialization has reached the point where certain cells are unable to survive independently of the colony.

The general trend of evolution in the biological world has been toward greater specialization, with groups of cells becoming completely interdependent and at the same time highly differentiated. Some cells specialize in chemical activity for the group; others provide protection or serve as a mechanical support; some specialize in conduction of water or electrochemical impulses. A group of specialized cells within a multicellular organization, similar in structure and function, is called a **tissue.** An **organ** is a group of tissues which combine to carry out a single function. Complex multicellular organisms, such as orchids and orangutans, feature groups of organs working together as a system to carry out important life functions.

## Specialized plant cells

Multicellular green plants must solve the same fundamental problems faced by one-celled organisms such as bacteria and algae. The specialized tissues of the poppy function like the specialized cell areas of the bacterium. Let us look briefly at some of the more important kinds of specialized plant cells that group together to form tissues.

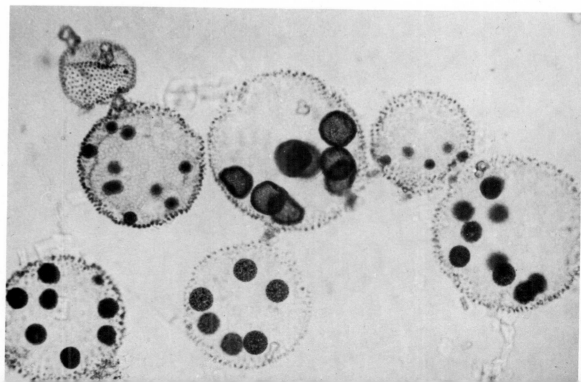

**Epidermis.** The outer cells of the roots, stems, and leaves of green plants are epidermal cells, which function to protect the inner cells from the environment. Epidermal cells are relatively flat, with large vacuoles, and they are shaped like irregular polyhedrons that interlock with no spaces between the cells. Epidermal cells on the stem and leaves often secrete an additional cell coating, a wax **cuticle,** which guards against water loss, invasion by parasitic one-celled organisms, and damage by crushing or puncturing. In many plants, the epidermal cells give rise to projecting hairs or spiny thorns, which also have a protective function. These cells may also contain secretory structures. Certain differential cells in the epidermis, called **guard cells,** regulate tiny openings, or **stomata,** through which excess water can evaporate, and oxygen can be exchanged for carbon dioxide.

**Parenchyma.** Parenchyma is the type of tissue that comprises most of the green plant. Photosynthesis takes place in the parenchyma cells of green plant leaves. These cells characteristically contain plastids that are the site of this important chemical reaction. The most common function of parenchyma is storage. The cells usually have large vacuoles, which can serve as reservoirs to hold needed raw materials, as well as manufactured sugar dissolved in water. In comparison

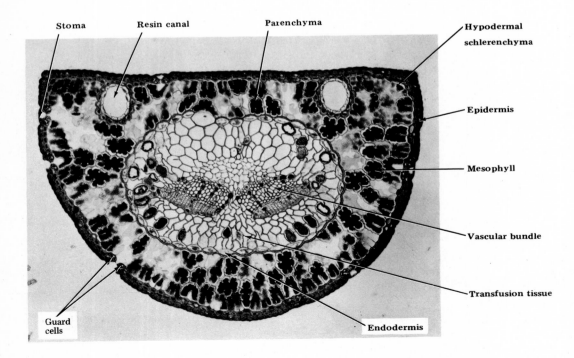

2-28 A cross section of a leaf shows the outer epidermal cells, the wax cuticle secreted by the epidermis, and guard cells which regulate the stomata, tiny openings in the epidermis that permit gas exchange and transpiration. (J. Limbach, Ripon Microslides)

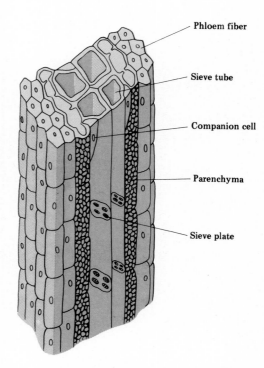

- Phloem fiber
- Sieve tube
- Companion cell
- Parenchyma
- Sieve plate

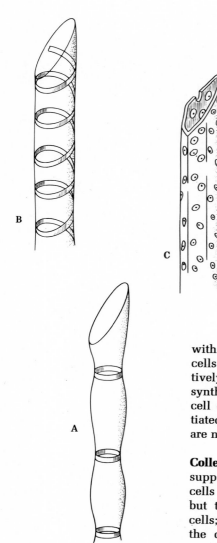
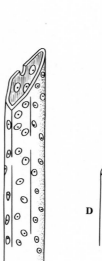
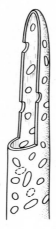

B

C

D

E

A

**2-29** *Individual vessels of various types are shown here. (B) shows how these different vessels may be grouped together within the stem.*

with other kinds of plant tissues, parenchyma cells have fairly thin cell walls. They are relatively unspecialized; in addition to their photosynthetic function, they are frequently capable of cell division, and they can be further differentiated to form more specialized cells as they are needed.

**Collenchyma.** Collenchyma cells specialize in support. Under the microscope, collenchyma cells appear to have a regular rectangular shape, but they have thicker walls than parenchyma cells; the cell walls are often especially thick at the corners. These cells serve to support the parenchyma cells in the leaves, in all young green plants, and in plants that never become very woody.

**Sclerenchyma.** Sclerenchyma cells also specialize in support. There are two basic types:

**fibers,** long thin cells with very thick walls, usually found clustered in bundles (they can be used commercially to make such products as linen and rope); and **stone cells,** irregularly shaped cells that have a granular appearance. Fruit pits are an example of a structure composed largely of stone cells. Sclerenchyma cells are often so thickly walled that they appear to be one solid mass. At maturity the contents of the cell usually die.

**Xylem and phloem.** Xylem cells specialize in conducting water and dissolved minerals from the ground to the leaves, where they will be used in photosynthesis. They are long cylindrical cells that form a continuous transportation network throughout the plant. Once the xylem cell has grown to full size, the contents of the cell die, leaving only the hollow tube of the cell wall. In one type of xylem cell, called a **tracheid,** the

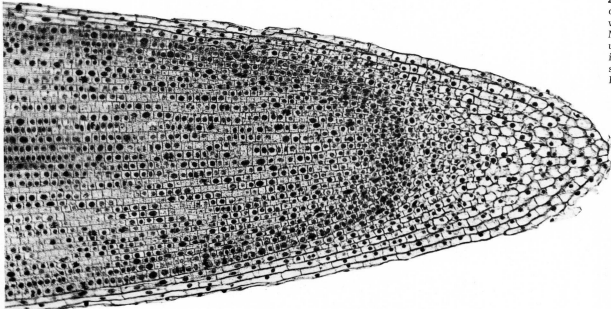

**2-30** *The meristematic cells of an onion have thin cell walls, small vacuoles, and large nuclei. Meristem cells, themselves very undifferentiated simple cells, indirectly give rise to all of the specialized cell types. (Carolina Biological Supply Co.)*

cell walls have small thin spots that permit the passage of water. In another type of xylem cell, the **vessel element,** the end walls are either partially or entirely dissolved. The thick and rigid cell walls also serve to support the plant; the wood of a tree is its densely packed xylem.

Phloem cells specialize in conducting dissolved sugar, or other organic molecules that have been manufactured in the parenchyma cells, to other parts of the plant. Phloem cells are also long cylinders, but unlike the xylem, they remain living cells, since the cell membranes have an important role in conducting large food molecules. One type of phloem cell, the **sieve tube,** loses its specialized organelles, such as the nucleus, mitochondria, and ribosomes, when it reaches maturity. These cells specialize in conduction only; movement of nutrients takes place through porous membranes, at either end of the cell, called **sieve plates.** Other phloem cells,

much narrower than sieve tubes and possessing only the thinnest of cell walls, specialize in another way. They contain nuclei and they develop an unusually large number of the cell organelles that are concerned with metabolic activities. These cells, which lie beside the sieve tubes, are called **companion cells.** It is believed that they perform many metabolic functions for the sieve tubes.

**Meristem.** Plants grow throughout their entire life span, elongating their stems and putting out new leaves and roots. The cells which specialize in this growth function are meristem. Meristem tissue is found only in the growing parts of plants. Meristematic cells typically have thin cell walls, small vacuoles, and large nuclei. They have the ability to divide and reproduce themselves, an ability which most of the other types of specialized cells have lost. Meristem cells, themselves very simple undifferentiated cells, indirectly give rise to all the specialized cell types—xylem, guard cells, parenchyma cells—of the plant.

## Specialized animal cells

Like plant cells, animal cells may also be highly specialized. Since they do not make their own food, as plants do, but obtain food by eating other organisms, animal cells have specialized in a different direction, to meet the specific problems of finding, ingesting, and digesting for their own use, external sources of food.

**Epithelium.** All surfaces of an animal's body are covered by epithelial cells. The outer layer of our skin is composed of epithelial cells, and so are the linings of our digestive tract and our blood vessels, as well as the insides of our nose and mouth. These epithelial cells protect the underlying cells of the body. Epithelial cells are

usually separated from underlying tissue by a fibrous substance called the **basement membrane.**

Epithelial cells are always packed tightly together, with little or no space between their membranes. It has been demonstrated, through experiments using radioactive chemicals and dyes, that the membrane at the edge of the epithelial cell, where it is in contact with other cells, differs from the membrane at the other surfaces of the cell, where it is exposed to the environment. Many compounds that can pass freely back and forth between cells through the membrane at the edge cannot pass through the other membranes of the cell. This difference in permeability is probably due to some difference in the chemical structure of the membrane.

Epithelial cells may be differentiated in shape and function. They may give rise to projecting cilia, and they may also secrete protective mucus. **Squamous** epithelial cells are quite flat; they are usually found in the very outermost

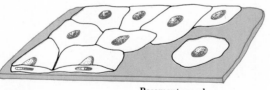

Squamous       Basement membrane

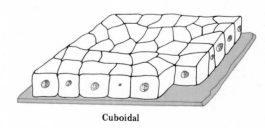

Cuboidal

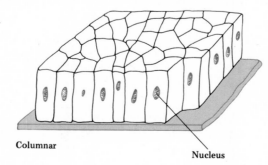

Columnar

Nucleus

**2-31** *Several different types of epithelial cells are shown here. The flat squamous cells are usually found in the outermost layer of the epithelium. Columnar and cuboidal cells may be ciliated, and they often perform a secretory function. In humans, epithelial tissue covers the outer surface of the skin, lines the digestive tract, and covers all organs, such as the liver and kidneys.*

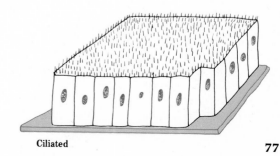

Ciliated

77

layer. **Cuboidal** cells, shaped rather like little cubes, and **columnar** cells, with a bricklike shape, may be found under the squamous cells of the skin, or forming the covering of internal organs. Many of these cells specialize in secretory functions. In groups they may form a gland, such as the sweat gland.

**Connective cells.** Connective cells serve to support and protect the animal's body and to connect the various parts into one functioning organism that can carry on chemical exchanges. There are several different types of connective cells that go to make up connective tissue.

1. Fibrous connective cells—These cells serve to anchor internal organs in place, to form the elastic outer walls of blood vessels, to bind nerves and muscles together. Tendons and ligaments are both made of fibrous connective cells. The cells, called **fibroblasts,** secrete proteins, such as collagen and elastin, that form long connective fibers. Figure 2-32 shows a highly magnified collagen fiber; it is a chain of protein molecules held together by hydrogen bonds. About one-third of the protein in the human body is collagen.

2. Cartilage—Cartilage is a kind of flexible support; it is found in the nose and ears, in the discs between the bones of the spine, and at the ends of the ribs. It is made up of widely spaced spherical cells embedded in a gelatinous matrix secreted by the cells.

3. Bone—Bone cells, or **osteoblasts,** secrete an additional cellular substance that is quite rigid and forms a supporting structure for the body. This secreted matrix consists of collagen and certain inorganic salts of calcium. Each osteoblast is more or less trapped in the matrix it has secreted. At maturity, the cell shape changes to send out long thin projections of cytoplasm, so that the cells can remain in contact with one another and receive food and expel wastes.

4. Fat cells—These connective cells have special-

**Collagen fibres**
**The protein of connective tissue**

ized in the function of storing fats, as a food supply that can be drawn upon if carbohydrate intake halts. Large drops of oil are found in these cells, and the oil often pushes the nucleus and other organelles to the very edge of the cell. When needed, the stored oil molecules pass through the cell membrane and into the bloodstream.

5. Blood cells—The bloodstream consists of a fluid, called **plasma,** in which are dissolved many different organic compounds and inorganic ions. It also contains two types of cells. The red blood cells, or **erythrocytes,** specialize

**2-32** *The most common connective tissue is found in the form of white collagenous fibers, composed, in turn, of fine fibrils of the protein collagen. These fibers (the ones pictured are from human skin) resist stretching and give the body considerable strength. These fibers have been shadowed with chromium for photographic purposes. (Courtesy of Dr. Jerome Gross, Harvard Medical School)*

**2-33** *Each of these samples of bone cell tissue has been prepared and stained in a different way, to reveal various aspects of cellular structure. (From A Textbook of Histology by Bloom and Fawcett, W. B. Saunders)*

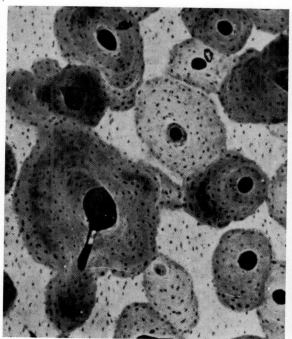 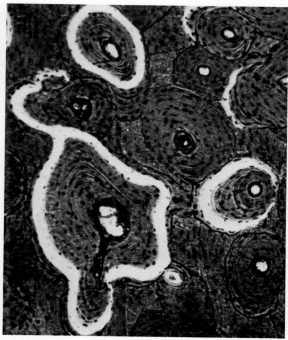

in the function of carrying oxygen molecules. Erythrocytes are filled with the protein hemoglobin, which combines easily with oxygen. Mature red blood cells lose their nuclei and all other organelles; they thus lose their ability to reproduce or to repair themselves, so they slowly "wear out" and must be constantly replaced. The white blood cells are called **leucocytes.** They specialize in destroying alien cells, such as bacteria and viruses, that enter the body. They do this either by manufacturing chemicals (antibodies) that combine with the alien cells to inactivate them, or by engulfing the alien cell and digesting it, in the same way an amoeba obtains its food. White blood cells are capable of independent movement, through the flowing of their cytoplasm, and they often move through the walls of blood vessels to enter an infected or damaged area.

**Muscle.** Muscle cells specialize in contraction. Many kinds of cells possess the power of contraction to a limited degree, but the muscle cell is differentiated so as to increase that power. There are three kinds of muscle cells. One is the **smooth,** or involuntary muscle. These cells are found in the intestinal tract and in the lining of blood vessels. Smooth muscle cells (also called visceral muscle cells) are not as long as other muscle cells, and the protein fibers they contain are much thinner and more delicate. The outer membrane of the smooth muscle cell is characteristically folded to form tiny pits; the role of these pits is still unknown.

A second type of muscle cell is the **striated** muscle cell, found in the muscles that move the arms, legs, and other appendages. Striated muscle (also called skeletal muscle) is often referred to as voluntary muscle, because it can be moved inten-

**79**

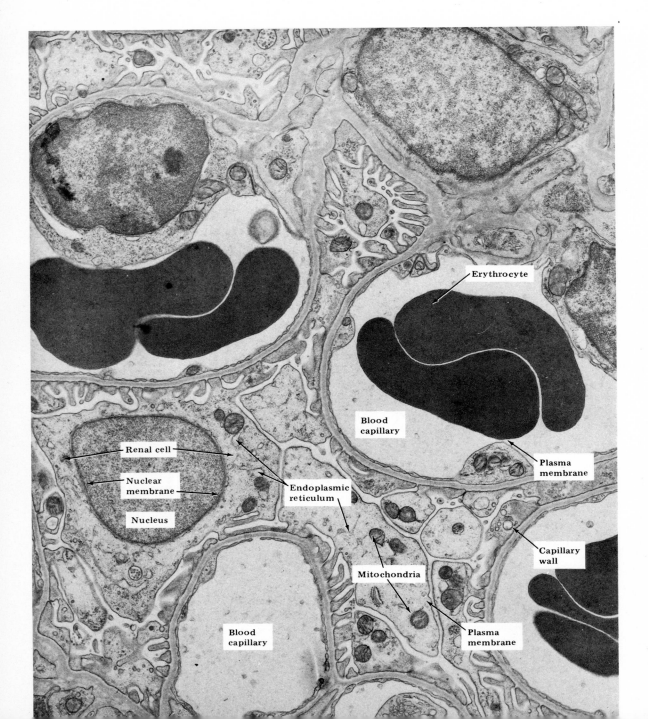

Erythrocyte

Blood
capillary

Plasma
membrane

Renal cell

Nuclear
membrane

Endoplasmic
reticulum

Nucleus

Capillary
wall

Mitochondria

Blood
capillary

Plasma
membrane

**2-34** *Red blood cells, erythrocytes, specialize in the function of carrying oxygen molecules. They are filled with the protein hemoglobin, which combines easily with oxygen. The two shown here are compressed together to fit into the narrow confines of a capillary. (Photo courtesy of Keith R. Porter)*

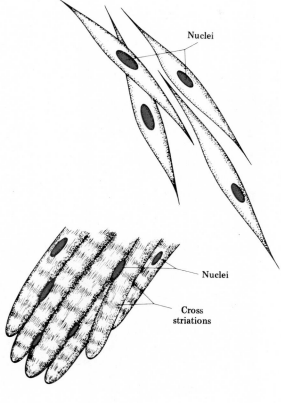

Nuclei

Nuclei

Cross
striations

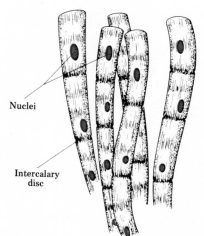

Nuclei

Intercalary
disc

**2-35** *There are three kinds of
muscle cells. One is the smooth
or involuntary muscle. These cells
are found in the intestinal tract
and in the lining of blood vessels.
A second type of muscle cell is the
striated, skeletal or voluntary
muscle cell, found in the muscles
that move the arms, legs, and
other appendages. A third type of
muscle cell is cardiac muscle,
found only in the heart. The fibers
within cardiac muscle cells are
like those in striated muscle, and
their shapes are also similar.
However, the fibers in cardiac
muscles are often branched.*

tionally. A striated muscle cell has a thin shape,
but it may be several centimeters long. Most of the
cell is filled with strands of twisted protein filaments that do the work of contraction. The other
contents of the cell, such as the nucleus and
organelles are pushed to the edge of the cell, just
inside the membrane.

The third type of muscle cell is **cardiac**
muscle, found in the heart. The fibers within
cardiac muscle cells are like those in striated
muscle, and their shapes are also similar. However, the fibers in cardiac muscles are often
branched. Another difference is that cardiac
muscle cells contain an unusually large number
of oversized mitochondria. This adaptation provides a constant supply of energy, so that the
cardiac muscle cell can perform its job of continual contraction. Cardiac muscle also shows
heavy black lines, called **intercalary discs** which
separate individual cells.

**Nerve cells.**  Nerve cells, or **neurons,** specialize
in the ability to conduct electrochemical impulses. Neurons are very peculiarly shaped cells.
There is a small round or oval **cell body,** which
contains the nucleus of the cell and other organelles. Branching out from the cell body, in a
typical muscle-stimulating neuron, is a network
of thin tendrils of cytoplasm called **dendrites;**
they are the receivers of electrochemical impulses.
There is also one long thin branch of cytoplasm
called the **axon,** which may attain a length of
many feet. What is commonly called a nerve is
really a bundle of axons. The axon is the part of
the cell that transmits the impulse, either to the
dendrites of another neuron, or to the muscle
cell that it stimulates. Nerve cells have very
prominent Golgi bodies, which may secrete the
chemicals that convert chemical energy into
electrical energy.

Another specialized type of nerve cell is the
Schwann cell, which lies alongside the axon of
certain kinds of neurons. As the Schwann cell
grows, it wraps itself around and around the

**81**

axon. The Schwann cell's membrane, which lays down sheets of the lipoprotein myelin, thus creates an insulating layer around the conducting axon. An axon that is insulated in this way is said to be **myelinated,** and it has been demonstrated that myelinated axons conduct impulses at a much faster rate than those which lack such insulation.

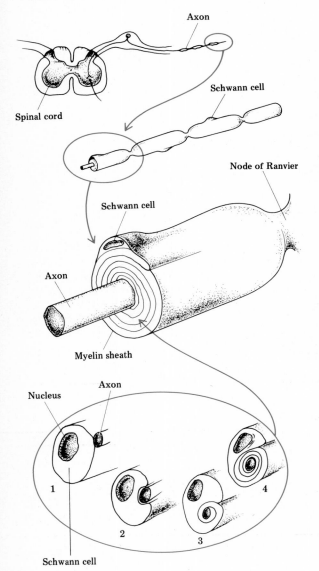

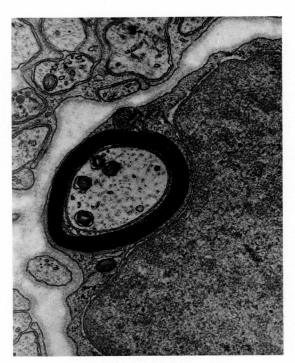

**2-36** *This cross section electron micrograph of a human motor neuron illustrates the development of the fatty myelin sheath. The axon lies within the Schwann cell membrane and next to the Schwann cell nucleus (large gray area). The myelin sheath (dark structure) is seen to be an extension of the Schwann cell membrane, which has coiled around the axon to form a tight, 13-layer spiral. (Courtesy of Johnny Beggs, Barrow Neurological Institute, Phoenix, Ariz.)*

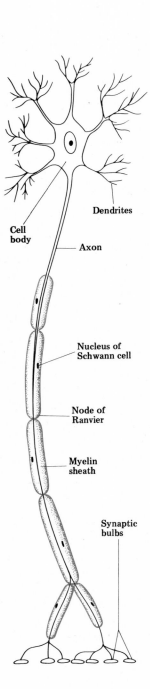

## The multicellular organism

Cell specializations, such as those we have described here, are the consequence of a multicellular form of organization. One-celled organisms, such as bacteria, must perform all life functions within their single cell. In a multicellular organism, specialized tissues or organs can perform some of these functions for the entire body, leaving other cells free to carry on different work.

Is specialization really an advantage? This question is frequently debated by biologists. It can be demonstrated that many multicellular organisms make more efficient use of their food than do one-celled organisms. We know that multicellular animals, who have specialized in mobility, can often escape from environmental problems that destroy one-celled animals. For example, when a forest fire starts, foxes and deer can run away from the fire, while one-celled animals living in the soil are killed by the heat. And man can even take steps to regulate his environment; he can dig trenches and pump water and stop the forest fire.

But for every case in which specialization has proven to be an advantage, we can cite another case in which it has been a fatal disadvantage. The dinosaurs are a prime example. In a time of rapid climatic change, their specialization led to their extinction. We can see the same problem in places that are undergoing rapid change today. All the large fish that once lived in Lake Erie—the bass, whitefish, and pike—are gone, but many kinds of bacteria still thrive in the polluted waters.

Can we really conclude that the highly specialized multicellular human being is a more successful life form than a simple bacterium? In spite of all his technology, man is still unable to live permanently in many places—deep in the sea, on the Antarctic continent—where bacteria can be found in large numbers. Man's efforts to regulate his environment are often unsuccessful, and even when they succeed, in the long run they may do him more harm than good, as some people believe is the case with commonly used herbicides and pesticides. The simple bacterium does not try to influence its environment; instead the cell yields to it. If it is dry, the bacterium encases itself with water-conserving coating and remains dormant until conditions improve. If the rains come, the bacterium simply drifts along on the water. If the environment is poisoned (perhaps by man-made antibiotics), the speed at which bacteria reproduce insures that sooner or later a mutant bacterium will appear that is able to resist the poison, and the species will then quickly reestablish itself. It seems that a great degree of specialization is not necessarily better—it is simply different.

# Summary

The cell is a system of membranes forming a functional unit that can carry on all the metabolic activities associated with life. In order for chemical reactions to occur in the cell, there must be cell membranes, which are layered structures of lipid molecules, probably lined with two layers of protein molecules. The cell membranes define the shape of the cell, and they also compartmentalize the cell's chemical substances; chemical reactions can thus be divided in time and space.

Membranes are permeable by diffusion. The rate and direction of diffusion depend upon the concentration of the substance permeating the membrane and the concentration differential between the area from which the diffusing molecule is coming and the area to which it is going. Also influencing the rate of diffusion of substances into a cell are the entering molecule's size, shape, and electrical charge and the temperature of the cell. Osmosis, known as passive transport, is the diffusion of pure water through a membrane. Active transport involves an energy expenditure because substances are being transported against a concentration differential. Pinocytosis and phagocytosis are employed when larger groups of molecules are transported into a cell.

The cell is the basic unit of organization in all living things, and all living cells must come from preexisting cells of the same type. Cells were not clearly defined until the emergence of the microscope and a proper procedure for their preparation and viewing. Although they vary greatly, cells are limited in size by the necessity for a low surface-to-volume ratio, by problems of internal transport, and by problems of physical support.

Bounded by a continuous membrane, the cell also contains a number of specialized membrane systems called organelles, which, along with the nucleus, are imbedded in cytoplasm.

The nucleus directs most of the cell's metabolic activities by discharging nucleic acids. Within the nucleus are masses of chromatin, containing hereditary materials, and the nucleolus. In the cytoplasm may be any of various structures: endoplasmic reticulum, ribosomes, the Golgi apparatus, vesicles, mitochondria, plastids (in plants), lysosomes, vacuoles, centrioles, and microtubules. Eucaryotic cells are those having the specializations enumerated here; simpler cells are termed procaryotic.

The evolutionary trend is toward specialization and differentiation of cells. In plants, cell specialization results in an epidermis, cuticle, guard cells, parenchyma, collenchyma, sclerenchyma, xylem and phloem, and meristem. Animals specialize cells to form epithelium, connective cells—including cartilage, bone and blood, and fat, muscle, and nerves. In multicellular organisms, specialized tissues or organs perform life functions for the entire body.

# Bibliography

## References

Beams, H. W. and R. G. Kessel. 1968. "The Golgi Apparatus: Structure and Function." *International Review.* 23: 209–276.

De Duve, C. 1963. "The Lysosome." *Scientific American.* 208(5): 64–72. Offprint no. 156. Freeman, San Francisco.

Fox, C. F. 1972. "The Structure of Cell Membranes." *Scientific American.* 226(2): 31–38.

Green, D. E. and R. F. Bruckner. 1972. "The Molecular Principles of Biological Membrane Construction and Function." *Bio-Science.* 23(1): 13–19.

Lehninger, A. L. 1964. *The Mitochondrion.* Benjamin, New York.

Porter, K. R. and M. A. Bonneville. 1968. *Fine Structure of Cells and Tissues.* 3rd ed. Lea and Febiger, Philadelphia.

## Suggested for Further Reading

Burke, J. D. 1970. *Cell Biology.* Williams & Wilkins, Baltimore.
This book is written for the advanced undergraduate student, and incorporates much of the latest research in molecular biology. Chapter 1 is an excellent discussion of the development of microscopic study of cells.

Du Praw, E. J. 1968. *Cell and Molecular Biology.* Academic Press, New York.
The early chapters describe the morphology and analyze the function of cell structures. The importance of the relationship between structure and function is emphasized.

Jensen, W. A. and R. B. Park. 1967. *Cell Ultrastructure.* Wadsworth, Belmont, Calif.
This is an atlas of the cell with full-page electron microscope photographs of each structure and organelle. The brief text introducing each topic is valuable.

Swanson, C. P. 1969. *The Cell.* 3rd ed. Prentice-Hall, Englewood Cliffs, N.J.
A good overview of the cell structures and cell differentiation. Excellent charts and drawings supplement a presentation aimed at the beginning student.

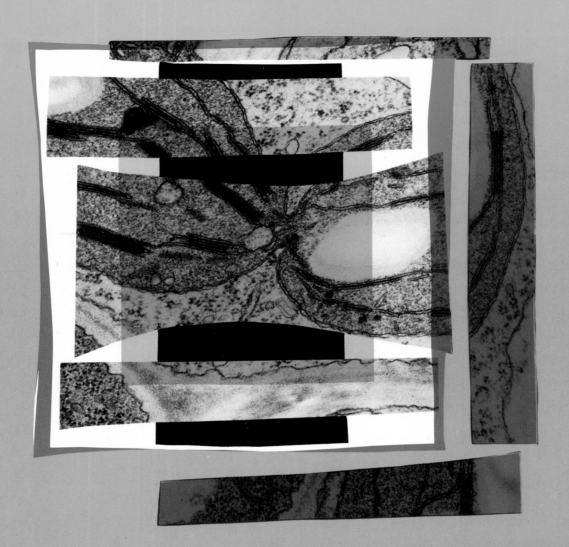

# 3
# Energy and the cell

A living cell is a stable unit. Through the microscope, we can observe a paramecium swimming around in a drop of water, and even if we watch for a long period of time, we will not be able to see any significant change in the organism. Its shape (like the sole of a shoe) will remain the same; its organelles will continue to function in the same way; its cilia will continue their synchronized contractions that row the paramecium through the water in search of food. Even a cell that constantly changes shape, such as a leucocyte, maintains a stable internal organization that appears to be the same from one day to the next.

By observing a cell, we may conclude that living things are immutable. But if we switch the focus of our observation from the cellular level to the molecular level, we will be compelled to come to a very different conclusion. Suppose, for example, that at the beginning of our observation, we were able somehow to label every single molecule in the cell's nucleus. Should we look at the nucleus again a day later, we would find that the nucleus now contains many unlabeled molecules. Moreover, we would find some of those labeled molecules in various other parts of the cell, perhaps even outside the cell in the water. Although the nucleus appears to be a stable organelle, the molecules of which it is made change constantly as they undergo many different chemical reactions.

One way of viewing the change in the paramecium's molecules is in terms of a continual trend toward greater disorganization and chaos which is termed entropy. This tendency toward increasing entropy is not confined to living cells alone; it characterizes the entire universe. Any group of molecules together in a closed system will gradually become more and more disordered until they reach the point of maximum disorganization; this point is the system's equilibrium. An example of this trend toward disorganization in the molecules of the cell can be seen in the process of osmosis. A concentration gradient is a kind of organization; it involves molecules arranged in a specific order. If osmosis proceeds unchecked, the concentration gradient will be replaced by a complete but random dispersal of the molecules. Osmosis is a movement toward greater disorder, or increased entropy.

In a nonliving chemical system, the processes of diffusion and osmosis will eventually bring about a complete equilibrium. All molecules will be evenly dispersed; entropy will be at its highest; and there will be no more free energy to bring about more changes. This is what happens to a cell after it dies. But a living cell continually counteracts the move toward entropy, by utilizing sources of energy outside the system to rebuild order. An example can be seen in the process of active transport. The active transport of sodium ions from inside the cell (where they are in low concentration) to the outside (where they are in high concentration) increases the organization of the cell. It also increases the free energy available to the system, since the concentration differential thus established can be used in the future to move molecules from one side of the membrane to the other. The investment of energy in the process of active transport counteracts the entropic effects of osmosis, and the cell remains a stable unit.

## The role of ATP

The energy used in the process of active transport comes from molecules of ATP. The ATP is found very near the membrane, perhaps even attached to the phospholipid component of the membrane by hydrogen bonds. A molecule of ATP is like a portable battery. It can be taken to any part of the cell where an energy-consuming reaction is taking place, and it will serve as an energy source. It does this by breaking the high-energy bond that holds the last phosphate group onto the ATP molecule. The reaction is catalyzed by an enzyme, called an ATP-ase. Once the bond is

3-1 This schematic drawing shows the flow of energy in the cell. Energy for metabolic activities comes from ATP, which itself is produced by the oxidation of energy-containing food molecules, such as glucose.

broken, the energy formerly locked up in it is free to do any kind of work in the cell.

The reaction that breaks the phosphate bond liberates energy; it is said to be an exergonic reaction. This implies that the reaction responsible for creating the phosphate bond—the **phosphorylation** reaction—required some input of energy. This deduction can be confirmed in the biochemist's laboratory, where the exact amount of energy needed for phosphorylation can be measured. Phosphorylation is an endergonic reaction; it requires energy.

The energy required by the endergonic

**3-2** *Physicists use energy-flow charts to trace the path of energy through its various transformations. Biologists have adapted this concept, and trace the flow of energy through the biosphere. Note that at each transformation, a certain amount of energy is given off as heat and therefore is lost to other organisms.*

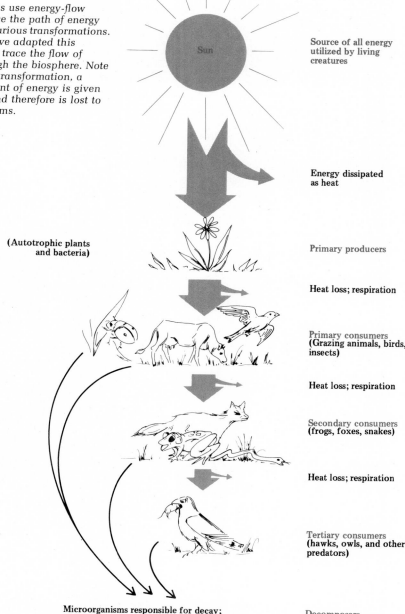

Source of all energy utilized by living creatures

Energy dissipated as heat

(Autotrophic plants and bacteria)

Primary producers

Heat loss; respiration

Primary consumers (Grazing animals, birds, insects)

Heat loss; respiration

Secondary consumers (frogs, foxes, snakes)

Heat loss; respiration

Tertiary consumers (hawks, owls, and other predators)

Microorganisms responsible for decay; ultimately to be reutilized by primary producers.

Decomposers

reaction of phosphorylation comes from outside the cell. Sometimes it is light energy. Phosphorylation that takes place with the aid of absorbed light energy is called **photosynthetic phosphorylation.** Sometimes it is chemical energy, obtained by utilizing the energy stored in sugar and other similar compounds. The compounds are oxidized, and the energy contained in their bonds is used to phosphorylate molecules of ATP. Energy obtained through the oxidation of organic compounds is called **oxidative phosphorylation.**

Photosynthesis is the process through which phosphorylation by means of light energy takes place. Organisms that are capable of carrying on this type of phosphorylation are called **autotrophs.** Photosynthetic phosphorylation can produce molecules of ATP for immediate use in the cell. In fact, the amount of photosynthesis the organism can carry on is usually great enough to produce more ATP than can be used right away. This surplus can be stored for future use by the organism, providing a definite advantage in the struggle for survival. But ATP, although it can be everywhere in the cell to meet energy needs, is not a good storage molecule. Often called biological currency—a kind of money that can be used anywhere in the cell to pay energy debts—ATP is good for spending but not for saving. The molecules are large and bulky, and any significant surplus would take up too much space in a cell. It is as impractical for the cell to store ATP as it would be for Henry Ford to save his fortune in quarters.

The solution to the problem of storage is to use ATP to synthesize small molecules, such as glucose and other monosaccharides, that store energy. These can later be joined together to form insoluble polysaccharides, molecules that can be stored without upsetting the osmotic balance of the cell. They can also be rearranged by combining the atoms in slightly different proportions, to make lipids, which are also good storage units.

At a later date, the energy stored in glucose

can be released through the process of cellular respiration and made available for use in the endergonic reaction of oxidative phosphorylation. Autotrophs, which are mostly the green plants, meet their energy needs through cellular respiration during the night when no light is available, in the winter when the light-absorbing leaves have died, and any time that energy demands are especially high, such as during the process of reproduction. For organisms that cannot carry on photosynthetic phosphorylation (they are called **heterotrophs**), cellular respiration is the sole mechanism for obtaining new supplies of free energy.

We can regard photosynthesis and cellular respiration as being two different versions of the same process. Both are a means for utilizing sources of energy outside the living system of the cell for the work of phosphorylation, thus producing new supplies of potential energy to be used in maintaining the cell's high level of organization. They are the processes by which the cell circumvents the trend toward entropy.

## Cellular respiration

The oxidation of a molecule that stores energy, such as glucose, will always yield exactly the same amount of energy, no matter how the oxidation process is performed. Chemists have set sugar on fire in the laboratory, and they know that the bonds of one mole of glucose contain about 680,000 calories. (A mole of a compound is the number of grams equal to its molecular weight. The molecular weight of glucose is 180, and therefore one mole of glucose is 180 grams.) A mole of glucose will always yield just that number of calories, whether it is oxidized by the flame of a Bunsen burner or by the cells of an organism.

The chief problem faced by living organisms in the process of cellular respiration is how to harness the energy released by oxidation for the work of phosphorylation. In a laboratory the heat energy given off by burning sugar might be caught and utilized. It is relatively easy to design a heat engine in which a gas or liquid absorbs the heat of oxidation and then gives it off slowly to do work as it is pumped away from the combustion chamber to a cooler region. But heat cannot be utilized as an energy source in the cell. One reason is that the cell is of uniform temperature, and heat energy can do work only when it is moved from a warm region to a cool one. Another reason is that the additional heat energy could cause all the molecules of the cell to become excited, breaking existing bonds and forming new ones, thereby destroying the complex organization of organic compounds upon which life depends.

The solution to this problem is to oxidize the sugar in a sequence of small steps, releasing only a few energized electrons at a time. In this way, the oxidation yields chemical energy, which can easily be utilized in the process of phosphorylation. This slow release increases the efficiency of cellular respiration; only a relatively small percentage of the total energy content of the glucose is released as heat energy and consequently lost to the organism.

When we speak of cellular respiration, we generally intend to refer to the oxidation of glucose rather than the other energy-containing compounds that can be used as energy sources. The 6-carbon molecule of glucose is first broken down into two 3-carbon molecules, through the process of glycolysis. The enzymes that catalyze glycolysis are found throughout the cell, so it is assumed that there is no special site for this process. All that is needed are the enzymes and the substrate molecule of glucose; the reaction is anaerobic (it can take place in the absence of oxygen).

Glycolysis releases a small amount of energy in the cell, but it is only the first step in the process of cellular respiration. The 3-carbon

**3-3** *Burning sugar illustrates an alternative mode of releasing energy contained in organic chemical bonds. Here the energy is released suddenly, as heat and light. Biological oxidation, discussed in this chapter, permits controlled release of energy and its use as fuel for the life processes. (Alycia Smith Butler)*

compound produced by the splitting of glucose can then be oxidized by oxygen to release additional energy. If no oxygen is present, fermentation, a less efficient release of energy, may take place instead. Both these processes yield energized electrons which can be used in the oxidative phosphorylation reaction that produces ATP. Fermentation is presumed to take place at various sites within the cell, but oxidation – the process of removing electrons – can occur only within the mitochondria. These organelles contain all the necessary enzymes to catalyze the reactions, the coenzymes that are often necessary to form the complete enzyme complex, and substrates which are involved in the reactions. All of these molecules are probably held in the small granules on the membrane folds of the mitochondrion, and the surfaces of the cristae provide a kind of skeleton around which the necessary chemicals can be organized. It has recently been demonstrated that some of the enzymes involved in oxidation respiration are actually structural components of the folded inner membrane.

Glycolysis and fermentation or oxidation are the final energy-yielding steps in cellular respiration, but there are many other steps that can precede them. For example, before glycolysis can take place, large storage carbohydrate molecules must be broken apart to yield the glucose. This process can occur anywhere within the cell. In a multicellular organism, organs such as the liver may specialize in this phase of cellular respiration. Biologists have been able to trace a large number of different methods by which macromolecules can be split into smaller units for respiration and later recombined; these are called metabolic pathways. The great variety of metabolic pathways insures that the cell (or the organism) will never be without a supply of molecules for respiration. If no carbohydrate molecules are available, lipids can be used instead. As a last resort, even proteins can be broken down to yield a compound that the cell can oxidize.

## Glycolysis

The first step in cellular respiration is glycolysis, a process which breaks down the glucose molecule in preparation for either fermentation or oxidative respiration. Although glycolysis is the most common metabolic pathway in cellular respiration, it is only one of the ways in which organic compounds can be broken down as a first step in respiration. Other pathways may also be utilized, if this one becomes blocked, or energy needs suddenly increase.

To begin the process of glycolysis, a certain amount of energy is required. This initial energy is often called the **energy of activation,** meaning the energy needed to get a chemical reaction started. In an energy-producing, or exergonic, reaction, the energy of activation can be regarded as a kind of investment which will eventually bring multiplied dividends.

In glycolysis, the energy of activation is supplied by two molecules of ATP. One ATP is used when a phosphate group is attached to the 6-carbon glucose, making a compound called glucose-phosphate. This attachment makes the structure of the molecule unstable and it isomerizes, becoming fructose-phosphate. (Glucose and fructose have the same chemical formula but a different structural arrangement.) This molecule receives another phosphate group, and thus uses up another molecule of ATP. The phosphorylated glucose molecule, called fructose diphosphate, now has 6 carbons and 2 phosphates. As we shall see, the two molecules of ATP used up in the production of fructose diphosphate will soon be repaid, at a good rate of interest.

The fructose diphosphate molecule is also rather unstable, and it virtually falls apart, forming two almost identical molecules, each having 3 carbons and 1 phosphate. This compound is called phosphoglyceraldehyde, or PGAL for short.

Up until this point, where PGAL is produced, there has been no oxidation in glycolysis. The molecules have been rearranged, but no elec-

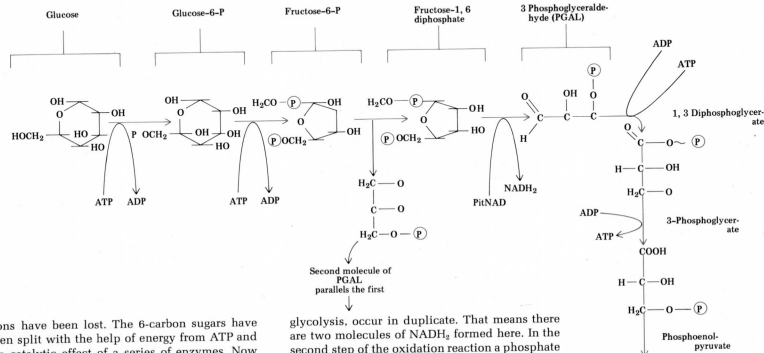

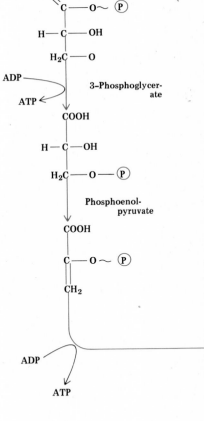

trons have been lost. The 6-carbon sugars have been split with the help of energy from ATP and the catalytic effect of a series of enzymes. Now the first oxidative step begins, releasing the energy tied up in the bonds of PGAL. It is important to remember that although these are oxidation reactions, they do not involve the presence of oxygen but use other electron acceptors. Therefore this is still an anaerobic type of respiration.

In the first oxidative step, a molecule of an electron-accepting compound called NAD (nicotinamide-adenine-dinucleotide) removes two electrons from the molecule of PGAL. The PGAL is oxidized (loses electrons) and the NAD is reduced (gains electrons). Each electron that moves to NAD is followed by a proton, forming an atom of hydrogen (one proton + electron). Therefore the reduced compound is called NADH, or $NADH_2$, depending on the number of electrons it accepts. Since there are actually two PGAL molecules, this step, and all subsequent ones in

glycolysis, occur in duplicate. That means there are two molecules of $NADH_2$ formed here. In the second step of the oxidation reaction a phosphate group is detached from the oxidized PGAL and joined to a waiting ADP. This reaction produces a molecule of ATP.

The loss of the phosphate group changes the oxidized PGAL into phosphoglyceric acid (PGA). The PGA molecule, which has 3 carbons and 1 remaining phosphate, then undergoes a two-step oxidation reaction that removes the last phosphate. As in the previous step, the phosphate joins with ADP and produces a molecule of ATP. The loss of the last phosphate turns PGA into pyruvic acid, a simple 3-carbon molecule.

**Energy yield.** What is the energy yield of glycolysis? The reactions that change one molecule of glucose into two molecules of PGAL consume energy; two molecules of ATP are used in this process. But as PGAL is changed to pyruvic acid, energy is released. As each molecule of PGAL

**3-5** *Electron micrograph of a crista, or fold, projecting into the lumen of mitochondrion. The spherical particles attached to the crista membrane have a very precise protein-lipid structure which permits them to function as the site of ATP-yielding aerobic respiration reactions.* (H. Fernández-Morán, University of Chicago)

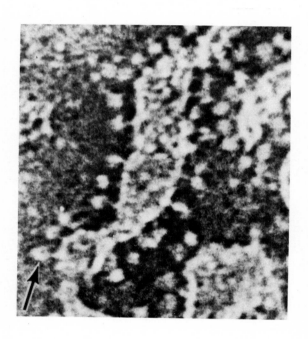

**3-4** *Glycolysis is a long and complex sequence of reactions that ultimately may follow one of three paths: the Krebs cycle, fermentation, or lactic acid anaerobic breakdown. Note that the breakdown of fructose diphosphate produces two molecules of PGAL, each of which will continue along the glycolytic pathway.*

is converted into pyruvic acid, it produces two ATPs and one molecule of $NADH_2$, so the total yield for the original molecule of glucose is four ATPs and two $NADH_2$s. After we subtract the energy that was expended to get the chain of reactions started, we have a net gain of two molecules of ATP and two of $NADH_2$. If we calculate the energy content of these four molecules, and compare that figure with the energy contained in the original glucose molecule, we will find that the products of glycolysis contain less than 5 percent of the total energy available in the glucose. If an organism had to depend on glycolysis alone, it would be making very inefficient use of its food.

## Aerobic respiration

Aerobic respiration is the term used for oxidative respiration using oxygen as the agent of electron removal; the high-energy carbohydrate is "burned" by reaction with oxygen. Aerobic respiration is the most efficient metabolic pathway, the one that extracts the largest amount of energy from the glucose molecule. Aerobic respiration can take place only in the mitochondria. Recent photographs taken through a very powerful electron microscope have revealed that the cristae of mitochondria are covered with regularly-spaced little protrusions that look like lollipops—thin stalks ending in a round knob. There is evidence that this is where the actual electron transfers that produce ATP take place. The initial sequence of reactions that frees the electrons apparently takes place on the inner membrane walls of the mitochondria.

Oxidative respiration begins with the pyruvic acid that was the end-product of glycolysis. As soon as the pyruvic acid enters a mitochondrion, it undergoes a process known as oxidative decarboxylation. Long as it is, this term is a kind of chemical shorthand; it simply means that, with the help of oxygen, atoms of

## Metabolic map

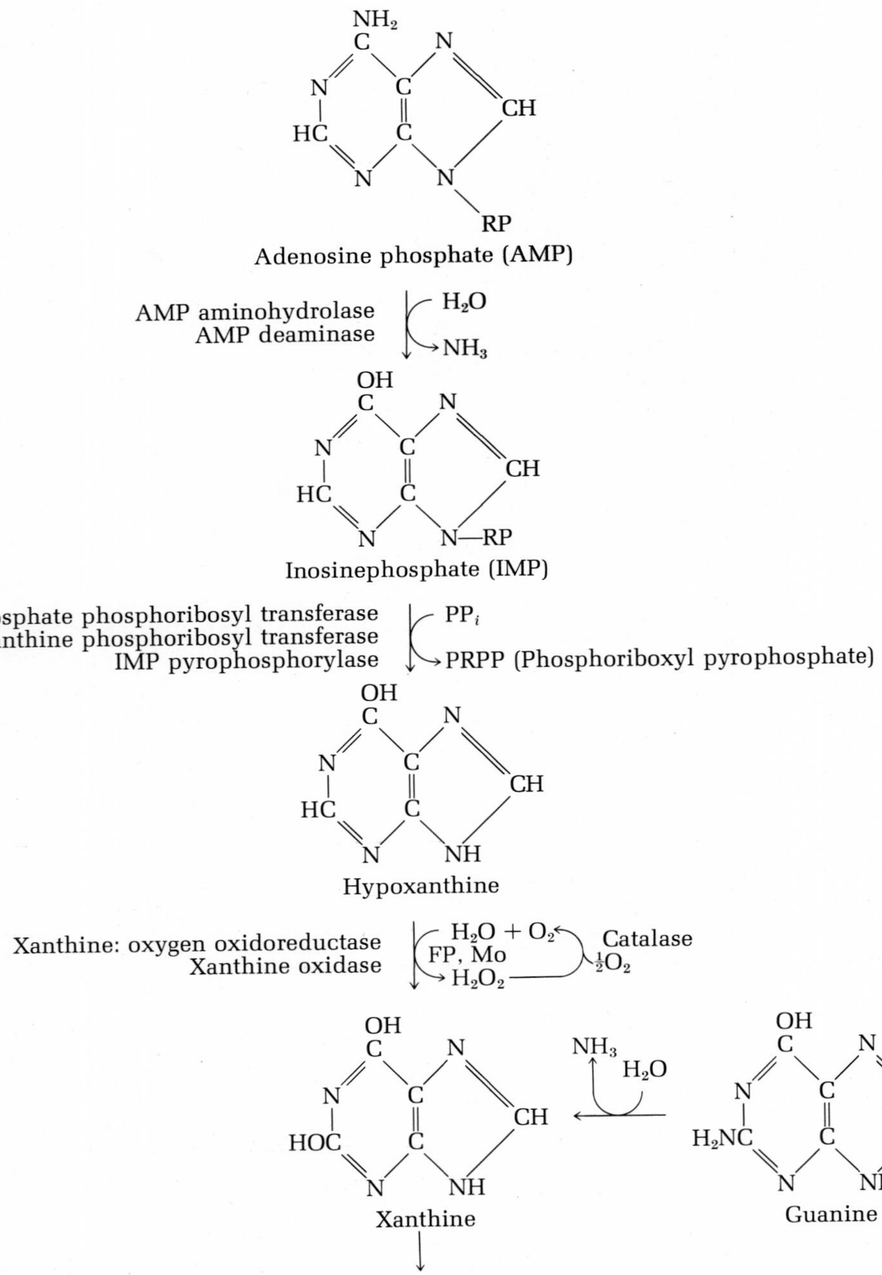

NH$_2$
C
N N
C
HC CH
C
N N
RP
Adenosine phosphate (AMP)

AMP aminohydrolase
AMP deaminase
H$_2$O
→NH$_3$

OH
C
N N
C
HC CH
C
N N—RP
Inosinephosphate (IMP)

IMP: pyrophosphate phosphoribosyl transferase
Hypoxanthine phosphoribosyl transferase
IMP pyrophosphorylase
PP$_i$
→PRPP (Phosphoriboxyl pyrophosphate)

OH
C
N N
C
HC CH
C
N NH
Hypoxanthine

Xanthine: oxygen oxidoreductase
Xanthine oxidase
H$_2$O + O$_2$ Catalase
FP, Mo
H$_2$O$_2$ $\frac{1}{2}$O$_2$

OH
C
N N
C
HOC CH
C
N NH
Xanthine

NH$_3$
H$_2$O

OH
C
N N
C
H$_2$NC CH
C
N NH
Guanine

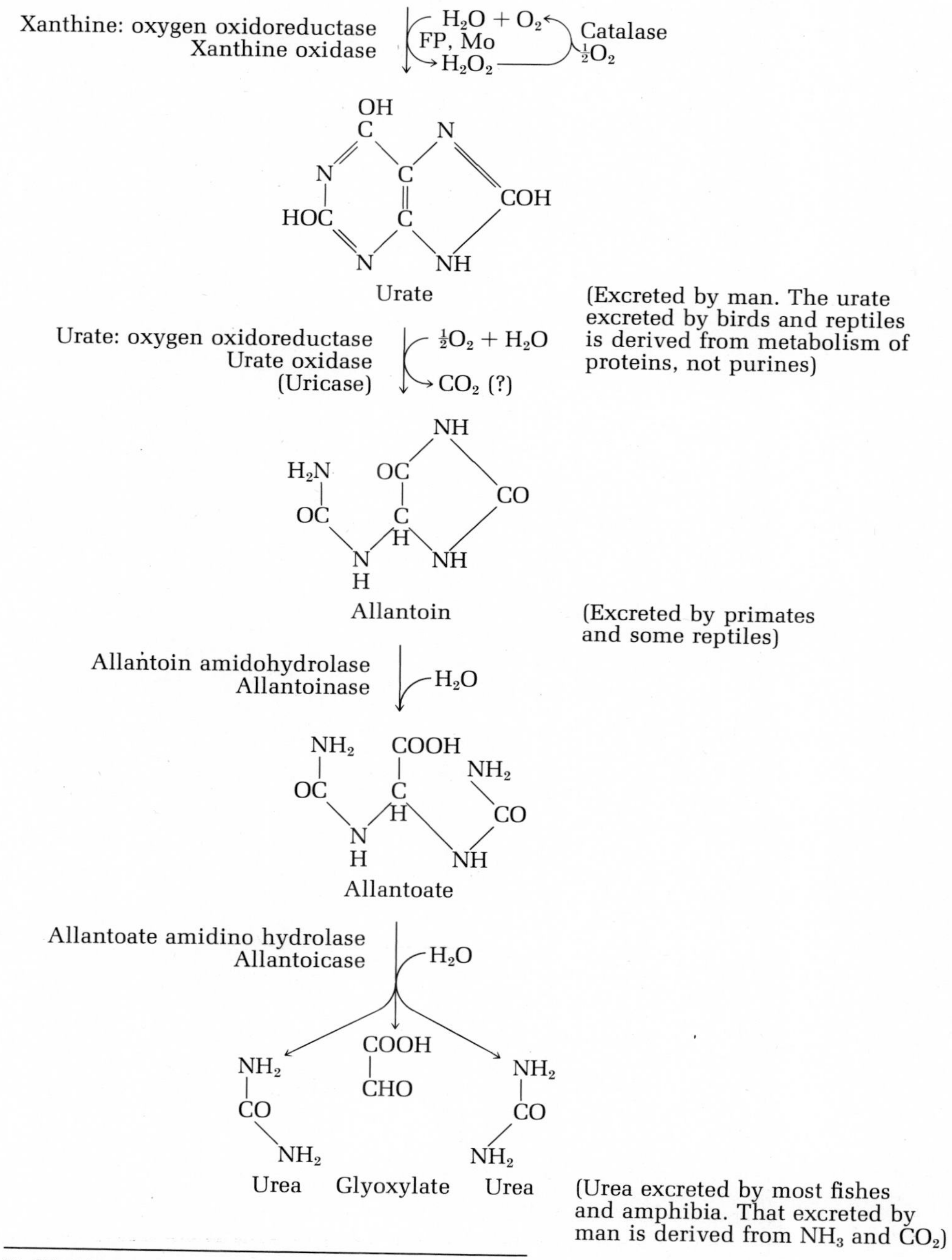

Xanthine: oxygen oxidoreductase
Xanthine oxidase

$H_2O + O_2$
FP, Mo
$H_2O_2$

Catalase
$\frac{1}{2}O_2$

OH
C

N

N

C

COH

HOC

C

N

NH

Urate

(Excreted by man. The urate excreted by birds and reptiles is derived from metabolism of proteins, not purines)

Urate: oxygen oxidoreductase
Urate oxidase
(Uricase)

$\frac{1}{2}O_2 + H_2O$

$CO_2$ (?)

NH

$H_2N$

OC

CO

OC

C
H

N
H

NH

Allantoin

(Excreted by primates and some reptiles)

Allantoin amidohydrolase
Allantoinase

$H_2O$

$NH_2$   COOH

OC

C
H

$NH_2$

CO

N
H

NH

Allantoate

Allantoate amidino hydrolase
Allantoicase

$H_2O$

COOH

CHO

$NH_2$

CO

$NH_2$

$NH_2$

CO

$NH_2$

Urea     Glyoxylate     Urea

(Urea excreted by most fishes and amphibia. That excreted by man is derived from $NH_3$ and $CO_2$)

3-6 *Biochemists use metabolic maps to trace the pathway of chemical reactions in living organisms. One such map is shown here. It follows the breakdown of nitrogenous compounds taken in as food.*

Modified from S. Dagley and D. Nicholson. *An introduction to metabolic pathways.* Oxford: Blackwell Scientific Publication, 1970.

carbon and oxygen are pulled off the molecule. This complex reaction requires the presence of pyruvic acid, oxygen, various cofactors, and at least three different enzyme catalysts.

During the oxidative decarboxylation reaction, one carbon of the 3-carbon pyruvic acid molecule is removed. This carbon atom combines with free oxygen and is given off as $CO_2$. Two hydrogen electrons are also removed; they combine with NAD to form $NADH_2$. The 2-carbon compound remaining, which is actually a form of acetic acid (the acid found in vinegar), is quickly bonded to a nearby molecule called coenzyme A, thus preventing rapid oxidation of this energy-rich compound. If the 2-carbon compound is not bonded to some stable molecule, all its energy will be given off as heat, rather than being converted to usable chemical energy. The end-product of this reaction is called acetyl-CoA.

At this point, the acetyl-CoA usually enters an oxidative cycle, called the Krebs cycle, after its formulator, Hans Krebs, who in 1953 won the Nobel Prize for his work. But in some cases it may instead be shunted into another pathway, in which it condenses to form straight-chain carbon compounds, 14 to 20 carbons long. These compounds are fatty acids, which can later be used in the synthesis of lipid molecules. The acetyl-CoA appears to take this second pathway only when the Krebs cycle is overloaded. If all the molecules of the enzyme that initiates the Krebs cycle are tied up in the cycle, no new acetyl-CoA can enter. In such cases, the acetyl-CoA enters an alternate pathway which will prepare it for storage in the form of lipids.

**The Krebs cycle.** The Krebs cycle is a sequence of chemical reactions that removes a large number of hydrogen electrons from molecules of energy-containing compounds. As the electrons are removed, they are passed to a chain of cytochrome molecules. Cytochromes are specialized protein molecules which incorporate atoms of metallic elements. Cytochromes serve as interim

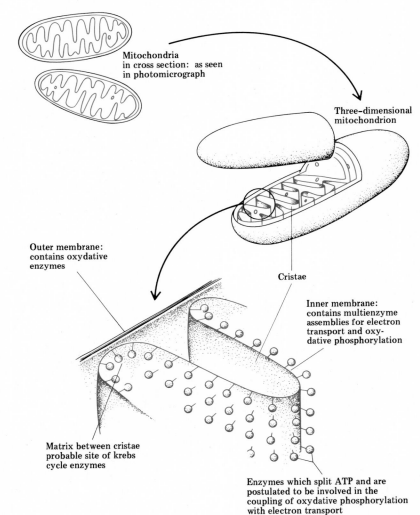

Mitochondria in cross section: as seen in photomicrograph

Three-dimensional mitochondrion

Outer membrane: contains oxydative enzymes

Cristae

Inner membrane: contains multienzyme assemblies for electron transport and oxydative phosphorylation

Matrix between cristae probable site of krebs cycle enzymes

Enzymes which split ATP and are postulated to be involved in the coupling of oxydative phosphorylation with electron transport

**3-7** *The process of aerobic respiration can take place only within the mitochondrion, whose membranous folds provide the organization needed for such a complex sequence of reactions. It is thought that electron transfers take place on the knoblike projections of the membrane, and other steps in the Krebs cycle occur at certain intervals along the cristae.*

electron acceptors. It is the electrons passed along cytochrome chains that provide the energy needed for endergonic reaction that bonds a phosphate group to ADP, producing a molecule of ATP. It has been discovered in laboratory experiments that even when all the necessary substrates and enzymes needed in the oxidative reaction that removes the electrons are present, the reaction will not take place unless ADP and free phosphate groups are also present. When ADP is removed or locked up in another reaction, the Krebs cycle grinds to a halt.

When the 2-carbon acetyl-CoA enters the Krebs cycle, an enzyme joins it to a 4-carbon compound already present in the mitochondrion, oxaloacetic acid. At the same time, a molecule of water hydrolyzes the bond between the acetyl group and coenzyme A, so the coenzyme is freed to reenter the pyruvic acid reaction. It will eventually bring more acetyl-CoA into the Krebs cycle.

When the 2-carbon acetyl is joined to the 4-carbon compound in the mitochondrion, a 6-carbon molecule is produced, called citric acid. (The Krebs cycle is also sometimes called the citric acid cycle.) The citric acid then undergoes isomerization, forming first one, then another 6-carbon compound.

This reaction provides an interesting example of the way living systems keep a reaction moving along a one-way path. The isomerization of the citric acid is caused by an enzyme. Enzymes have the ability to alter the rate of a reaction, but they cannot influence its direction. The enzyme can turn citric acid into isomers, and it can also turn the isomers back into citric acid at the very same rate. Theoretically, then, in a short time this reaction would reach an equilibrium with the reverse reaction going at the same rate as the forward reaction. But this state of equilibrium in the Krebs cycle is never actually reached. As soon as the isomer is formed, it is attacked by a second enzyme which splits off 2 hydrogens. The hydrogens are immediately picked up by NAD. By continually removing the product and involving it in other reactions, the isomerization reaction, which is potentially very easily reversible, is kept moving in one direction. This is an excellent example of the advantage of sequential reactions, which we discussed in Chapter 1.

The citric acid isomer (now minus 2 hydrogens and a portion of its energy content) is rapidly decarboxylated. The atoms of carbon and oxygen which are removed are given off as $CO_2$. The molecule remaining in the Krebs cycle now has only 5 carbons. This 5-carbon molecule undergoes another decarboxylation step, removing 2 more hydrogens and another carbon. The second decarboxylation is accomplished with the help of coenzyme A, and the product is a 4-carbon compound called succinyl CoA.

The next energy-producing step is the removal of the coenzyme A, which has bonded to the 4-carbon molecule during decarboxylation.

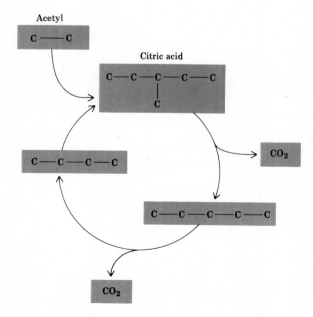

**3-8** *This schematic drawing gives the outlines of the Krebs cycle, in terms of the basic carbon skeletons of the compounds produced by the cycle. This is a convenient kind of chemical shorthand that permits summarization of complex chemical reactions.*

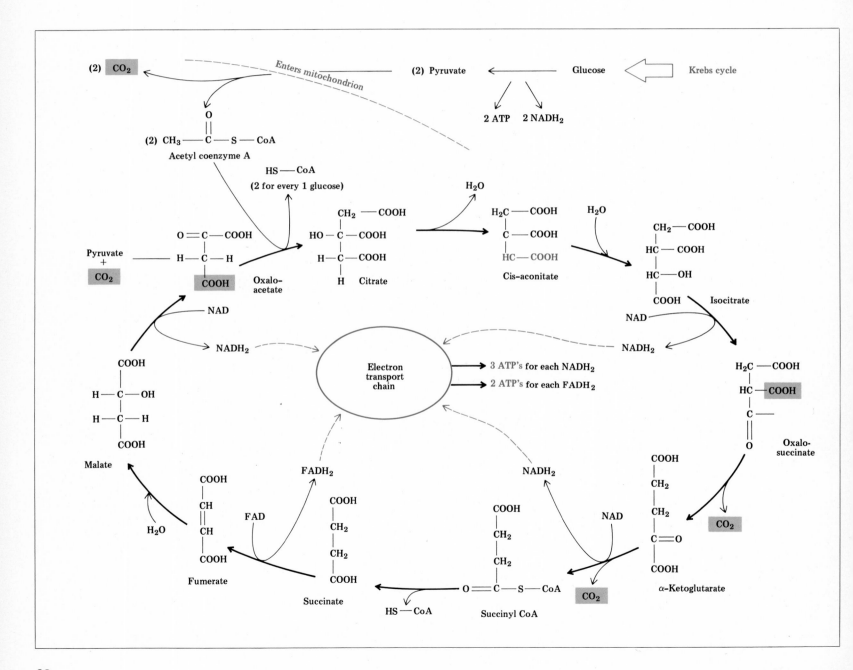

**3-9** *In this expanded version of the Krebs cycle, it is possible to follow not only the carbon skeletons of compounds but also the reactions involving oxygen and hydrogen as well. The carbons are shaded for easy identification.*

When the coenzyme A is removed, more energy-rich electrons come off the molecule. These electrons are eventually used to make ATP; three molecules of ATP are produced for each 2 hydrogens removed in the Krebs cycle.

The 4-carbon compound that remains after all these changes are made has been brought down to a relatively low energy level. It undergoes a sequence of reactions that rearranges the molecule slightly and converts it into oxaloacetic acid, the 4-carbon compound which accepts

acetyl-CoA at the start of the Krebs cycle. With this regeneration, the cycle is ready to begin again.

**Energy yield.** There are four dehydrogenation (hydrogen-removing) steps in the Krebs cycle. Each time hydrogens are removed, the electrons are picked up by a cytochrome chain and used in the production of ATP. For each 2-carbon acetyl compound that enters the Krebs cycle, 12 molecules of ATP are produced. Another 6 ATPs are produced in the reactions converting pyruvic

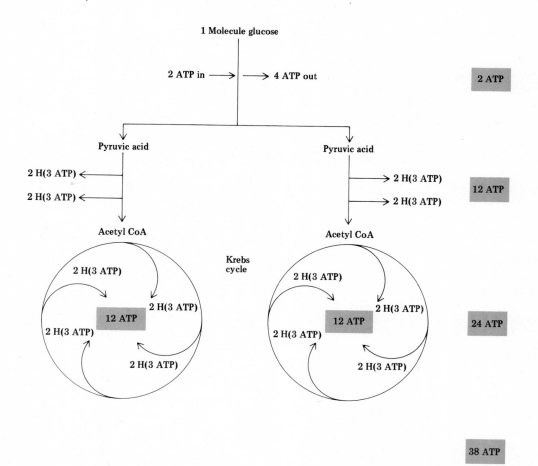

**3-10** *The total energy yield of one molecule of glucose is thought to be 38 molecules of ATP. This means that about 45 percent of the original energy content of the glucose molecule has been harnessed to do chemical work; the other 55 percent is given off as heat or remains locked up in the compounds that are the final products of the Krebs cycle.*

acid into the acetyl compound. Since each original molecule of glucose is split up into two parts, each of which goes through the entire oxidative cycle, we can say that the aerobic respiration of a molecule of glucose yields 36 molecules of ATP ($12 + 6 = 18$; $18 \times 2 = 36$).

The process of glycolysis by which glucose was broken down into pyruvic acid had a net yield of 2 ATPs. When we add these 2 molecules to the 36 produced in aerobic respiration, we get a total energy yield of 38 ATPs from the original molecule of glucose. The energy content of the terminal phosphate bond in a molecule of ATP is estimated to be about 8000 calories per mole. We know that one mole of glucose contains about 680,000 calories. The 38 moles of ATP contain about 304,000 calories. This means that about 45 percent of the available energy in glucose has been harnessed by the process of aerobic respiration; the other 55 percent is released in the form of heat energy, which cannot be used by the organism. This efficiency rating compares very favorably with man-made energy conversion systems, such as the internal combustion engine, which is never more than 30 percent efficient; in fact, in the average automobile engine, it is often as low as 15 percent.

## Fermentation

Organisms that live in an anaerobic environment (one without any atmospheric $O_2$), such as yeast and many bacteria, cannot burn pyruvic acid in a reaction with oxygen, since no oxygen is available. How, then, can they oxidize the compound, releasing the hydrogen electrons? The process of fermentation is one solution to this problem. Although it is much less efficient than aerobic respiration, it does serve to extract some more of the energy locked up in the two molecules of pyruvic acid that were derived from the original glucose. Fermentation simply substitutes compounds other than oxygen as electron acceptors.

In the simplest kind of fermentation, the hydrogen in $NADH_2$ is added directly to the pyruvic acid, with the help of a catalyst. That produces a molecule of lactic acid and a free NAD, which can then return to pick up more hydrogen. This kind of metabolism is seen in some microbes and in skeletal muscle cells when they are under stress. Fermentation in yeast cells is somewhat more complicated. There the pyruvic acid is broken down into carbon dioxide, and then the hydrogen electrons are added, forming ethyl alcohol and $CO_2$, which is given off as a gas. By this process, yeast can convert the sugar in grapes into alcohol, to produce wine. If the wine is fermented in the bottle so that the $CO_2$ cannot escape, then the wine is bubbly, like champagne. The yeast cells in bread dough also produce alcohol and $CO_2$ in respiration. The bubbles in the $CO_2$ cause the bread to rise and make little holes in the dough. (The alcohol evaporates during the baking process.)

Only a few kinds of cells are completely anaerobic, but many cells that ordinarily carry on aerobic respiration have the ability to switch to anaerobic respiration when their oxygen supply is exhausted. This is the case with striated muscle in vertebrates. The rapid contraction of skeletal muscle requires the expenditure of many molecules of ATP. During vigorous exercise, respiration in the cell is speeded up so that more ATPs can be produced. Even though the organism breathes faster and more deeply, it may not be possible for the lungs and circulatory system to supply enough oxygen to carry on such rapid aerobic respiration; therefore the cell will begin to supplement the output of ATP by also carrying on anaerobic respiration. The end-product of this type of respiration is lactic acid. As this compound accumulates in the cells after a period of vigorous exercise, it causes muscle fatigue and sometimes soreness. When new supplies of oxygen become available, the lactic acid is converted back into pyruvic acid, and the pyruvic acid then undergoes aerobic respiration.

The fact that anaerobic respiration is found in so many primitive one-celled organisms has led biologists to hypothesize that the first organisms all depended solely on this form of respiration. It is believed that at the time life first began on this planet, our atmosphere had large amounts of free hydrogen and very little oxygen; it was virtually an anaerobic environment. With so little oxygen available, it seems likely that glycolysis and fermentation were the only practical methods of energy release for many millions of years. It was not until large numbers of green plants had been photosynthesizing for thousands of years that there was enough free oxygen in the air to enable the development of aerobic respiration and a more efficient utilization of food.

## Photosynthesis

The process of photosynthetic phosphorylation is the ultimate source of all energy used in living systems. The process of photosynthesis converts light energy into chemical energy that can be used by the cell. When photosynthesis produces more chemical energy than is needed for immediate use in the cell, the surplus can be stored in the form of carbohydrates or lipids, and later oxidized through cellular respiration.

For at least 2,000 years, men have speculated about the mysterious process through which plants make food. Even today, after several decades of intensive research, using the most modern equipment and techniques, there are still many areas that must be further researched. Although scientists have succeeded in establishing a general outline of the process, and some of the reactions can be specified in exact sequence, many of the details of this process are still unknown. The mechanism by which light energy is changed into chemical energy seems to be one of nature's most closely guarded secrets.

## The historical perspective

The complex interlocking series of reactions in photosynthesis were not, of course, revealed to science all at once. Understanding has come gradually, through a series of experiments conducted over a period of several hundred years. Each of certain key experiments has added another dimension of knowledge, until at last scientists pieced together an accurate outline of the process. It may help in understanding photosynthesis to look at the historical development of its study.

The first recorded explanation of photosynthesis was set down by the Greek philosopher-scientist, Aristotle, in the fourth century B.C. Through a combination of observation and deductive reasoning, Aristotle concluded that plants withdrew food from the soil. According to Aristotle, the equation for photosynthesis would be:

$$soil \longrightarrow food$$

Not until the seventeenth century did anyone attempt to establish proof of Aristotle's equation. The Dutch scientist Jean van Helmont planted a five-pound shoot from a willow tree in a large pot containing exactly 200 pounds of dirt. He allowed the willow to remain in the pot for five years, watering it faithfully all the while. At the end of five years, he weighed the willow, which had gained almost 165 pounds. He then weighed the dirt, which had lost less than 2 ounces. This seemed conclusive proof that the plant material was not made from the dirt. Van Helmont theorized that it must be made from the water. According to van Helmont, the equation for photosynthesis would be:

$$water \longrightarrow food$$

More than a hundred years later, in the late eighteenth century, the English clergyman Joseph Priestley noted that the air might also contribute something to the food-making process. Priestley

**101**

devised an ingenious three-part experiment for testing this hypothesis. In one air-tight glass container he put a green plant; in a second, he put a healthy mouse; and in a third, he put both a plant and a mouse. Within a week, both the isolated plant and the isolated mouse were dead, but the plant and mouse together were still in good condition. From this, and similar experiments, Priestley concluded that breathing exhausted the air, or caused it to become "phlogisticated," whereas plants refreshed, or "dephlogisticated" the air. According to Priestley, the equation for photosynthesis would be:

$$\text{water} + \text{phlogisticated air} \longrightarrow \text{food} + \text{dephlogisticated air}$$

At about the same time Priestley was doing his work, a Dutch physician, Jan Ingen-Housz, was undertaking a set of experiments that added two other important factors to the equation. Ingen-Housz put one group of plants in the light and another group in total darkness, and he found that only the plants that were exposed to light were able to make their own food. He was also able to demonstrate that it was only in the green parts of the plant that food was made and dephlogisticated air was released; in woody stems and roots, neither of these processes occurred. According to Ingen-Housz, the equation for photosynthesis would be:

$$\text{water} + \text{phlogisticated air} \xrightarrow[\text{green plants}]{\text{light}} \text{food} + \text{dephlogisticated air}$$

The last 25 years of the eighteenth century were a time of rapid scientific advancement. During this period, many of the elements were discovered and named. It was established that phlogisticated air was actually carbon dioxide and that dephlogisticated air was oxygen. Working with this new information, the Swiss chemist Nicolas de Saussure was able to measure the rate

at which green plants exposed to light took in carbon dioxide and gave off oxygen. He found that the rate of these two processes was approximately the same, and this finding stimulated his curiosity about the fate of the carbon atom in the carbon dioxide. He therefore planted two sets of bean seedlings—one in the ordinary atmosphere and another in an atmosphere free of $CO_2$. The first group flourished, and added to its total carbon content. The second group was able to survive only a limited time, and there was never any increase in the total carbon content of these plants. Thus de Saussure concluded that the food produced by plants was a carbon-containing compound, and that the carbon came from atmospheric $CO_2$. According to de Saussure, the equation for photosynthesis would be

$$H_2O + CO_2 \xrightarrow[\text{green plants}]{\text{light}} \text{carbon compound} + O_2$$

In the nineteenth century, a German botanist, Julius von Sachs, investigated the nature of the carbon compound formed during photosynthesis. He concluded that it was definitely a starch, and probably was glucose, or $C_6H_{12}O_6$. This adds one more piece to the equation (for the time being, we will take the liberty of leaving it unbalanced), which would now read:

$$H_2O + CO_2 \xrightarrow[\text{green plants}]{\text{light}} C_6H_{12}O_6 + O_2$$

The next major step forward came in 1881, as the result of the work of another German botanist, T. W. Engelmann. Engelmann tried illuminating green algae with a variety of colors of light—red light, blue light, violet light, green light. From these experiments he was able to establish an **action spectrum,** showing the lengths of light waves that are most effective in causing photosynthesis. He found that the peak of the action spectrum came with blue light and red

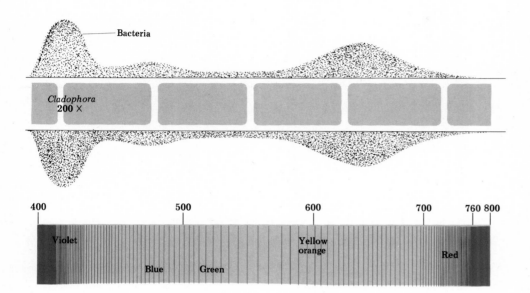

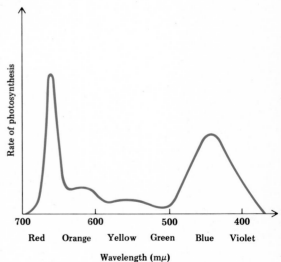

results were evidence that two separate reactions were involved in photosynthesis—a light reaction and a dark reaction.

It seemed logical to suppose that the light reaction was the first in the sequence, and that it involved the initial absorption of light energy and its conversion to chemical energy. The second phase of the process, the dark reaction, would then be the use of this chemical energy to build up molecules of glucose.

One of the results of Blackman's research was an increased interest in the identity of the compound responsible for absorbing the light. The shape of the action spectrum was an important clue. Physiologists knew that the compound must absorb red and blue light, since it is those wavelengths that activate photosynthesis. From

**3-11** *Engelmann's experiment is an interesting example of a bioassay—a test for the presence of certain elements that uses living organisms rather than chemical reactions. Bioassays are frequently used today to test for the presence of certain hormones. They exist in such small quantities that it is difficult to isolate them for chemical tests; but an animal can be injected with blood thought to contain a hormone, and its subsequent reactions will indicate whether or not the hormone is actually present.*

light. So the equation for photosynthesis could be changed to read:

$$H_2O + CO_2 \xrightarrow[\text{green plants}]{\text{blue and red light}} C_6H_{12}O_6 + O_2$$

In 1905 F. F. Blackman published results of some of his experiments regarding the rate of photosynthesis. Blackman's work did not really add any new elements to the equation, but it helped plant physiologists understand something about the sequence of steps involved in the process of photosynthesis, and it stimulated other important research. Blackman's research was designed to test the effects of varying degrees of light intensity and temperature changes on the rate of photosynthesis. His results at first seemed puzzling. He found that when plants were given full light, changes in temperature seemed to have no effect on the rate of photosynthesis; but when plants were in dim light, increasing the temperature served to increase the rate of photosynthesis. Blackman finally concluded that his

**3-12** *An action spectrum can be drawn by correlating the results of experiments like those of Engelmann and other workers. The action spectrum shows the amount of photosynthesis that takes place in each part of the visible light spectrum.*

**103**

this evidence, they decided that the light-absorbing compound was probably green in color. (When we see color, we are seeing the portions of the spectrum that are reflected by specific molecules of matter. For example, the color of an orange is due to the fact that it absorbs all the blue and green lengths of the light spectrum and bounces the orange portion back to our eyes. A green substance absorbs red and blue light, while bouncing back the green.) It did not take long to isolate a green pigment found in plant cells. That pigment was chlorophyll.

Chlorophyll is actually a conjugated protein, containing a molecule of magnesium within one of its chains; it is, in fact, rather similar to the cytochrome pigments used in the electron transfer of cellular respiration. By shining light of different wavelengths on isolated chlorophyll, physiologists were able to establish its **absorption spectrum.** The absorption spectrum of chlorophyll closely corresponds to the action spectrum of photosynthesis, which is good evidence that chlorophyll is the light-absorbing compound involved. With the addition of this evidence, the equation for photosynthesis can once again be revised, to read:

$$H_2O + CO_2 \xrightarrow[\text{chlorophyll}]{\text{red and blue light}} C_6H_{12}O_6 + O_2$$

In 1930, most plant physiologists believed that the light energy was used to split the $CO_2$ molecule. The carbon would then combine with the hydrogen and oxygen of $H_2O$, and the oxygen would be released as a gas. But the work of Cornelius van Niel of Stanford University seemed to disprove that theory. Van Niel had studied photosynthesis in a group of bacteria that can utilize hydrogen sulfide ($H_2S$) in place of water in photosynthesis. Van Niel found that these bacteria released sulfur rather than oxygen. The equation for their photosynthesis would be written:

$$H_2S + CO_2 \longrightarrow C_6H_{12}O_6 + S$$

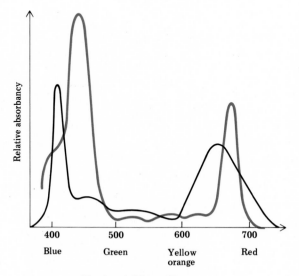

Relative absorbancy

| 400 | 500 | 600 | 700 |
| Blue | Green | Yellow orange | Red |

**Wavelength ($\lambda$) in millimicrons**

**3-13** *By exposing a solution of chlorophyll in liquid to various wavelengths of light, biochemists can determine its absorption spectrum, since the chlorophyll will fluoresce as it absorbs light. A comparison shows a great similarity between this absorption spectrum and the action spectrum, leading to the hypothesis that chlorophyll is the primary photosynthetic pigment.*

This showed that the $CO_2$ was not split, but was combined with hydrogen split from the $H_2S$ to form glucose. Van Niel suggested that the same thing must happen in plants that use water rather than hydrogen sulfide. It is the water, not the carbon dioxide, that is split. Van Niel arrived at his conclusion purely by deductive reasoning, and he was proven right by an experiment performed by a group of California scientists in 1941. Using technology that was a by-product of the research that eventually led to the atom bomb, this group introduced an isotope of oxygen ($O_{18}$) into the water for photosynthesis. It was the tagged oxygen from the water, rather than the unlabeled oxygen in $CO_2$, that was later released as a gas. Van Niel's deduction can therefore be incorporated into the photosynthesis equation, which now reads:

$$H_2O + CO_2 \xrightarrow[\text{chlorophyll}]{\text{red and blue light}} C_6H_{12}O_6 + O_2 \uparrow$$

The last major change in the equation came

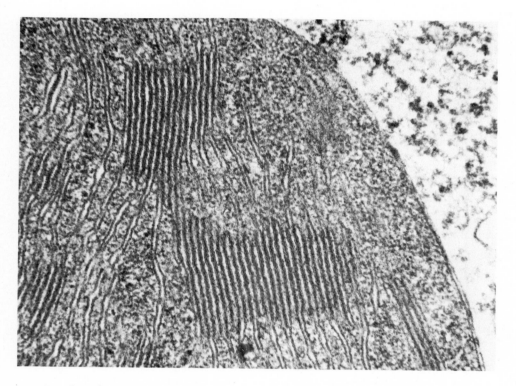

**3-14** *Complete photosynthesis takes place only in the chloroplast. The process of photosynthesis is thought to take place only along the membranous surfaces of the grana. Chlorophyll is a polar molecule, and it may be attracted and held near the membrane by polar molecules in the lipid layer. An alternative theory is that the chlorophyll is itself a part of the protein layer of the membrane. (Omikron)*

a certain amount of water is also produced during photosynthesis. Hill theorized that this water was essentially a waste product, the result of a combination reaction between unused oxygen atoms and unused hydrogen atoms. With the addition of Hill's findings, we can now write the final balanced equation for photosynthesis:

$$6CO_2 + 12H_2O \xrightarrow[\text{chlorophyll}]{\text{blue and red light}} C_6H_{12}O_2 + 6O_2 + 6H_2O$$

## The contemporary view of photosynthesis

In the decades since Hill's death, plant physiologists have continued to add to the knowledge of photosynthetic processes. Although much of this mysterious process has been clarified, some details are still in dispute, as we shall see when we look at a current model of photosynthesis.

Complete photosynthesis takes place only in the chloroplast. The process of photosynthesis is thought to take place along the membranous surfaces of the grana. Chlorophyll is a polar molecule, and it may be attracted and held near the membrane by polar molecules in the lipid layer. An alternative theory is that the chlorophyll is itself a part of the protein layer of the membrane. Enzymes and various substrate molecules are probably also arranged along the membrane in a definite pattern. That pattern is certainly a key to the efficiency of photosynthesis in the living cell. Laboratory experiments that follow "recipes" and mix together the correct ingredients have succeeded in producing a photolytic reaction in which water is split and $O_2$ is given off, but it uses light very inefficiently. Only in the living cell, where all the compounds are neatly arranged in a certain sequence by the presence of a membrane, can the process of photosynthesis take place quickly and efficiently.

The complete photosynthetic reaction takes place in two distinct phases. First there is the

as a result of the work of Robert Hill, a brilliant young scientist who was killed in an automobile accident at the peak of his career. In the 1930s, Hill was studying the light reaction of photosynthesis, and he made a number of important contributions to our knowledge of photosynthesis. For example, Hill established the fact that all necessary components of the light reaction are located in the chloroplasts; he demonstrated that when isolated chloroplasts are exposed to light, they will carry on photosynthesis as far as the release of oxygen (derived from the splitting of water). By putting chloroplasts in a $CO_2$-free atmosphere, where they continued to release $O_2$, Hill also discovered that carbon dioxide was not needed in the light reaction of photosynthesis.

Using a combination of ingenious technology and deductive reasoning, Hill demonstrated that

light, or photo, reaction, in which light energy is converted to chemical energy. The two important products of the light reaction are: ATP, the energy-containing compound that drives most metabolic processes; and free electrons, which can participate in the reduction reactions of photosynthetic phosphorylation, transferring their energy to the reduced compound.

The second phase of photosynthesis, called the dark reaction, utilizes the chemical energy produced by the light reaction to transform a low-energy compound, $CO_2$, into a high-energy carbohydrate. ATP provides the energy to drive the reaction, and the hydrogen electrons serve as the reducing agents.

The complete photosynthetic process involves a chain of at least 30 reactions, probably more. Photosynthesis requires the presence of light, chlorophyll, carbon dioxide, and water. It also requires the presence of molecules of ADP, that can be phosphorylated as light energy is converted to chemical energy. A number of cofactors, chemical compounds that are involved in reactions during photosynthesis but regenerated in their original form by the end of the cycle, must also be present. One important cofactor is nicotinamide-adenine dinucleotide phosphate (biologists have very sensibly given this awkward mouthful the nickname NADP). It is NADP that receives the hydrogen electrons in the light reaction, thus becoming $NADPH_2$. This compound carries the electrons into the dark reaction, giving them up as $CO_2$ is reduced. Also needed in the process of photosynthesis is a series of cytochromes, which can serve as interim acceptors of hydrogen electrons. A number of enzymes are also necessary to catalyze the sequence of reactions.

## The light reaction

It is during the first phase of photosynthesis that the important conversion of light energy into

**3-1**

**Major coenzymes**

| Coenzyme | Symbol | Reactions in which it participates |
|---|---|---|
| Nicotinamide-adenine-dinucleotide | NAD$^+$<br>NADH<br>(reduced form) | glycolysis, fermentation, Krebs cycle |
| Nicotinamide-adenine-dinucleotide Phosphate | NADP$^+$<br>NADPH<br>(reduced form) | photosynthesis, miscellaneous oxidation-reduction reactions |
| Flavin-adenine-dinucleotide | FAD | Krebs cycle, fatty acid oxidation |
| Adenine triphosphate | ATP | glycolysis, muscle contraction, active transport, protein synthesis, many other energy-requiring reactions |
| Coenzyme A | CoA | Krebs cycle, fatty acid oxidation |

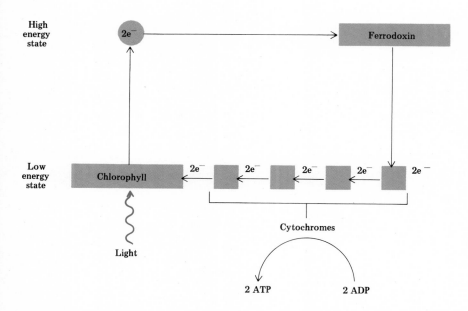

High
energy
state

2e⁻

Ferrodoxin

Low
energy
state

Chlorophyll | 2e⁻ | 2e⁻ | 2e⁻ | 2e⁻ | 2e⁻

Cytochromes

Light

2 ATP          2 ADP

**3-15** *The process of cyclic phosphorylation utilizes the light energy absorbed by chlorophyll molecules in the direct production of ATP. This is accomplished by passing energized electrons along a cytochrome chain, each step easing the electron down the energy gradient.*

and it becomes an excited molecule. This means that at least one of its electrons will move to a higher orbital. The duration of this excited state is very brief. It usually lasts only about $10^{-9}$ second. If the additional energy from the light photons is not utilized within that time period, it will be released by the chlorophyll molecule, in the form of a different, lower energy wavelength of light. This is called **fluorescence.** When chlorophyll is isolated in a test tube and exposed to light, it always fluoresces. This indicates that it is still absorbing light, even outside the living system, but there is no way for the light energy to be used outside the chloroplast, so it is released. It has also been shown that chlorophyll fluoresces inside the living plastid, which means that the process of photosynthesis is not totally efficient. Some absorbed light escapes before it can be used.

After the light energy is absorbed by an electron, the excited electron is "grabbed" by a cytochrome molecule positioned very near the chlorophyll. Cytochromes have the special property of attracting electrons and relinquishing them easily. When the electron leaves the chlorophyll, that compound is oxidized. The cytochrome molecule in turn is reduced.

The electron from the chlorophyll molecule is handed through a chain of cytochrome molecules—probably five different molecules—and with each exchange it gives up some fraction of its energy. One way to visualize this process is to think of the old-fashioned bucket brigade that our ancestors depended on to put out fires. One man would dip a bucket of water out of a pond or trough, and hand it to the man nearest him, who passed it to the next man, and so on down the line until the bucket reached the last man, who threw the water on the fire. Each time the bucket passed from hand to hand, some of the water sloshed out over the rim, so that after every transfer there was a little less water in the bucket. This is just the case with the electron as it is handed down the cytochrome

chemical energy occurs. The work of the American scientist Daniel Arnon in the early 1950s established that there are actually two separate kinds of light reaction that may occur simultaneously within the same chloroplast. One reaction simply manufactures ATP; this is called cyclic phosphorylation. The second light reaction also produces ATP, but it does something else as well. It liberates hydrogen electrons that were bound up in a molecule of water, and joins them to an electron acceptor, producing a reduced compound that holds the electrons securely until they are used in the dark reaction to provide energy for glucose synthesis. This reaction is called noncyclic phosphorylation.

**Cyclic phosphorylation.** The first step in cyclic phosphorylation is the excitation of the chlorophyll molecule by photons of light. When a packet of light (called a photon) hits a molecule of chlorophyll, the molecule passes from its normal state into a state of higher energy. Energy is transferred from the light to the chlorophyll,

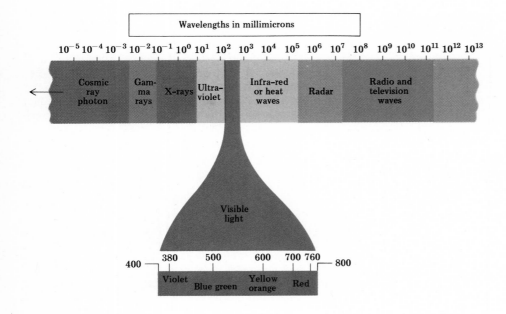

Wavelengths in millimicrons

$10^{-5}$ $10^{-4}$ $10^{-3}$ $10^{-2}$ $10^{-1}$ $10^0$ $10^1$ $10^2$ $10^3$ $10^4$ $10^5$ $10^6$ $10^7$ $10^8$ $10^9$ $10^{10}$ $10^{11}$ $10^{12}$ $10^{13}$

Cosmic ray photon | Gamma rays | X-rays | Ultraviolet | Infra-red or heat waves | Radar | Radio and television waves

Visible light

400 — 380   500   600   700 760 — 800

Violet | Blue green | Yellow orange | Red

**3-16** *This light spectrum specifies the wavelengths of various colors of light. It is interesting to note that photosynthesis utilizes light at both ends of the visible spectrum — blue light with a relatively high-energy content and red light with a relatively low-energy content.*

chain. It loses a little energy each time. But unlike the water that dribbles out of the bucket, the energy that leaves the electron is neither lost nor wasted. It is utilized by the cell to form high-energy phosphate bonds between ADP and a phosphate group, thus synthesizing ATP. At the end of the cycle, the electron moves from the last of the series of cytochromes back to the original chlorophyll molecule, attracted by the charge of that electron-deficient molecule. This explains why the process is called cyclic phosphorylation; the electron has made a complete circle and returned to where it started. The chlorophyll molecule is once more in a normal state, and it is ready to absorb more light, beginning another cycle.

Cyclic phosphorylation produces two molecules of ATP for every electron that goes through the complete cycle. It does not require the presence of water or carbon dioxide, but it must have light and chlorophyll, plus some molecules of

ADP and inorganic phosphate, and the presence of the cytochrome carriers.

**Noncyclic phosphorylation.** The fundamental events in noncyclic phosphorylation are the oxidation (removal of electrons) of water and the reduction (addition of electrons) of NADP. During the reaction, molecules of ATP are also produced. Noncyclic phosphorylation can proceed only in the presence of light.

According to recent theory regarding the events of noncyclic phosphorylation, each of the two phases — the sequence of oxidation reactions and the sequence of reduction reactions — is actually a separate system that takes place in a specialized site of the plastid granum. The system that reduces NADP is powered by the absorption of long wavelengths in the near-red spectrum of light. The system that oxidizes water uses a slightly shorter, and therefore more energetic, wavelength of light.

Plant physiologists explain the absorption of two different wavelengths of light by hypothesizing that several different pigments are involved. At least five different types of light-absorbing molecules in the chlorophyll class have been discovered. The most common, called chlorophyll *a*, is found in all green plants. It is usually accompanied by its close relative, chlorophyll *b*. The two molecules differ only in the structure of a single side chain of atoms.

This slight difference in structure enables these two molecules to absorb different parts of the light spectrum. Chlorophyll *a* absorbs the far-red region, where wavelengths are about 700 m$\mu$. Chlorophyll *b* absorbs the red portion of the spectrum at wavelengths of about 650 m$\mu$. The exact functional relationship of the various chlorophyll compounds is still a subject for research. We do know that photosynthesis cannot take place in green plants unless both chlorophyll *a* and chlorophyll *b* are present; what we

do not know is exactly why they are both needed. Some biologists believe that chlorophyll *b* (and the other forms of chlorophyll as well) acts as a gathering system. The molecules might somehow pass their absorbed energy along to chlorophyll *a*, which would be the only pigment directly involved in electron transfers with other compounds. According to this view, the absorption capabilities of both *a* and *b* are needed in order to accumulate enough energy to activate the electrons of chlorophyll *a*. But other biologists disagree. They think that chlorophyll *a* and chlorophyll *b* are used in separate events during the entire cycle, and that both are involved directly in electron transfers.

Figure 3-18 traces the events of noncyclic phosphorylation in sequence. The first step is the absorption of light by the chlorophyll molecule, causing the excitation of electrons. This absorbed radiant energy may then be used to help split water molecules apart, forming $H^+$, $e^-$ and

**3-17** *The chlorophyll molecule is a conjugated protein, containing an atom of magnesium within its structure. Note the great similarities between chlorophylls* a *and* b; *only a few atoms are different.*

**Phytyl group**

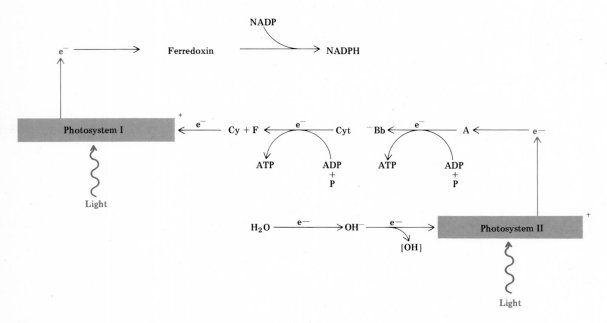

**3-18** *The process of noncyclic phosphorylation also serves to harness the light energy absorbed by chlorophyll, but it is a much more complex process than cyclic phosphorylation. Although in some organisms, both processes occur simultaneously, noncyclic phosphorylation is generally associated with higher, more organized multicellular forms of life, whereas cyclic phosphorylation is chiefly found in procaryotic organisms such as bacteria and blue-green algae. (Modified from Arnon in "Biochemistry of Chloroplasts," T. W. Goodwin, ed. Academic Press, 1967)*

OH groups. The exact sequence of this reaction, and identity of the coenzymes or other compounds that are associated with it, have not yet been determined. We do know that it is an oxidation reaction, in which electrons are taken away from the water molecule. As we saw in Chapter 1, the water molecule is strongly bonded, and ordinary chemical reactions in the laboratory cannot usually break the bonds. Molecular biologists are very curious to find out how this energy-expensive work of breaking the bonds is accomplished during photosynthesis. They suspect that the light energy plays some part in this process.

After the water molecule is split, some of the hydrogen atoms will be used in a later stage of noncyclic phosphorylation; that is, they will be joined to NADP. One of the hydrogen electrons will eventually be passed back to the chlorophyll molecule, which thus regains its missing electron. But the electron that the chlorophyll molecule gets back is not the same one that left the molecule when it acquired additional energy

from the light; it is an electron donated by water. This is why the process is called noncyclic phosphorylation. The oxygen atoms from the water molecule will either be given off as a gas ($O_2$) or will recombine with unused hydrogens to form new molecules of water.

At the same time that water is being oxidized, a complementary reduction reaction is going forward, in which an excited electron from a chlorophyll molecule is handed along a chain of cytochromes. This phase of the reaction is very similar to the process observed during cyclic ATP; this is one reason biologists believe noncyclic ATP evolved as a more complex version of cyclic ATP. We know the identity of most of the cytochrome carriers in noncyclic ATP, and even their chemical and structural formulas have been determined.

In this phase of the process, the excited chlorophyll electron is transferred in a sequence of reactions to at least four cytochromes, each one a slightly less energy-rich compound. In the process of this downhill transfer of energy, one molecule of ATP can be formed. At the end of this chain of cytochrome carriers, the electron has lost a significant amount of the energy it initially absorbed, and by itself it would be unable to bring about a strong reduction reaction. So the electron is passed on to a second molecule of chlorophyll, and there, another photon of light hits the electron to boost it up the energy gradient again. Then the excited electron is passed along another cytochrome chain; again a molecule of ATP is formed in the process. Using the energy of the excited electron, the coenzyme NADP can bond on the hydrogen ions that came from the earlier splitting of water. The product of this reaction is an energy-rich reduced compound, $NADPH_2$. This compound provides a stable and convenient way to transfer the hydrogen to the next stage of photosynthesis, the dark reaction that reduces $CO_2$ to form a carbohydrate. The molecules of ATP produced in both cyclic and noncyclic phosphorylation will also be used in the next stage, as a source of energy to drive the reactions.

## The dark reaction

The second stage of photosynthesis, often called the dark reaction, is the reduction or fixation of carbon dioxide. It uses the stored energy of ATP rather than the direct energy of light quanta. Fixing one molecule of $CO_2$ requires the energy of three molecules of ATP and the reducing power of two molecules of $NADPH_2$.

The dark reaction is so named because it can proceed in the absence of light. It may take place in the parenchymal cells throughout the night, as the plant uses up the ATP and the NADPH that was accumulated through the light reaction during the day. The dark reaction generally goes on during the day as well, occurring simultaneously with the light reaction. Typically, the dark reaction takes place in the chloroplasts, and the carbohydrate molecules thus produced are temporarily stored in the plastid and then later transported to other cells to serve as energy sources. But it has been demonstrated that the dark reaction, which is essentially a storage process, can actually take place in any cell of the plant, and in most animal cells as well, so long as the cell is provided with a supply of ATP and $NADPH_2$. It seems that the membranous structure of the grana in chloroplasts is not a necessary setting for this reaction, although it is definitely needed for the light reaction.

In the dark reaction, carbon dioxide, an inorganic compound, is converted into an organic substance upon combination with a 5-carbon compound. This compound (called ribulose diphosphate) must be manufactured by the cell, using its energy reserves of ATP. Once $CO_2$ is fixed in this new organic compound, which has 6 carbons, the only remaining step is to convert it to a plain 6-carbon sugar such as glucose. This final step utilizes the reducing power of the hydrogens in $NADPH_2$. Any cell that has enough ATP to produce the 5-carbon compound that accepts $CO_2$, and enough $NADPH_2$ to reduce the sugar acid, can produce carbohydrates by fixing $CO_2$.

The dark reaction has been studied intensively, and molecular biologists are fairly certain they have determined the basic sequence of reactions. Much of this work was done by Melvin Calvin and his associates at the University of California in the late 1940s and early 1950s, and so the cycle of events during the dark reaction is sometimes referred to as the Calvin cycle. In his Nobel Prize-winning research, Calvin utilized a radioactive isotope of carbon, carbon-14 (the normal carbon atom has a mass weight of only 12), which could be traced as it entered various reactions. He introduced the carbon into a group of photosynthesizing algae cells, and then at intervals of several seconds he killed some of the cells with hot alcohol and analyzed their contents. By this method, he was able to follow the

**111**

path of the carbon-14 as it was passed from one molecule to another in a chain of reactions. Thus biologists know not only the steps in the process, but also the approximate speed at which each takes place. Calvin found carbon-14 had been incorporated into glucose molecules in less than a minute after its introduction; it is, therefore, safe to assume that the whole sequence of reactions in the Calvin cycle takes place in less than a minute.

The first step in the reduction of carbon dioxide occurs when it is linked to the substrate molecule, ribulose diphosphate. This acceptor molecule must be present for the cycle to begin, and therefore it is not surprising to learn that it

is also one of the products of the cycle, which thus constantly regenerates this necessary substrate.

Most of the reactions in the Calvin cycle, including the initial coupling of $CO_2$ and ribulose diphosphate, are accomplished with the aid of enzymes. In the last few years, research has begun to identify many of these catalysts. It is known that the initial bonding of the carbon dioxide molecule is accomplished with the aid of an exceptionally large enzyme called carboxydismutase. This same enzyme seems to play some part in a number of other reactions in the cycle, probably using a different part of its surface in each of the reactions. This is a good example of

**3-19** *The dark reaction of photosynthesis is the step that is actually synthetic, since it builds up molecules of high-energy glucose from elements with a relatively low-energy content. In many respects, this step is a reversal of the sequence of reactions that breaks down glucose during glycolysis.*

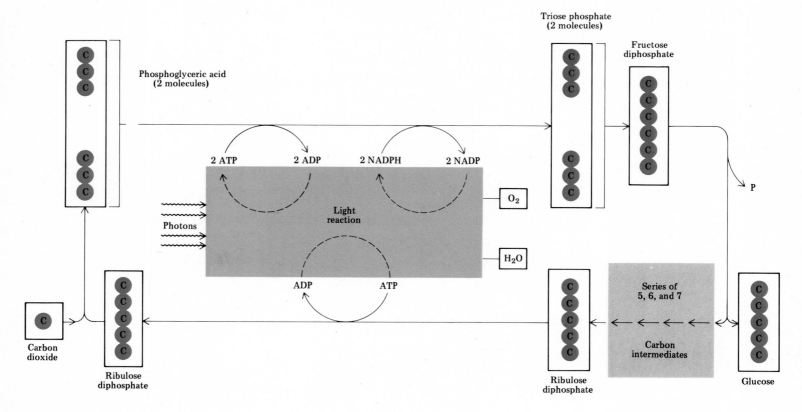

the efficiency typical of living systems: there is only one molecule to be synthesized and stored, rather than several. But it is difficult to imagine the millions of years of evolution—the constant trial and error, the synthesis of untold numbers of enzyme configurations, most of them failures whose inefficiency doomed the organism—that produced such an efficient molecule.

Ribulose diphosphate is a 5-carbon compound, so when it is linked to $CO_2$, a 6-carbon compound is formed. This compound is extremely unstable; it is no sooner formed than it breaks down again, into two molecules of a 3-carbon compound called phosphoglyceric acid, or PGA. The PGA is next phosphorylated by two molecules of ATP and then reduced by two molecules of $NADPH_2$. The result is a 3-carbon compound called phosphoglyceraldehyde (PGAL).

PGAL is an energy-rich compound that can be used in a variety of subsequent reactions. In fact, some biologists regard it as the actual end-product of photosynthesis. PGAL can serve as a direct energy source, like ATP, and it is thought that a number of PGAL molecules are immediately used in other reactions involving by-products of the Calvin cycle. However, many of the molecules of PGAL produced by $CO_2$ fixation undergo a few additional reactions that prepare them for storage. This is how glucose is formed.

When two PGAL molecules are linked together, they form a compound called fructose diphosphate. In a one-step reaction, this can be converted to fructose-6-phosphate. This 6-carbon compound is actually an isomer of a glucose molecule, so a second simple reaction will rearrange the internal structure of the molecule and yield a molecule of glucose. The glucose then can be used as a building block of poly-saccharides, osmotically inactive molecules that can be stored by the cell. Glucose can also be used as the basis for the formation of lipids, which are also storage molecules.

The formation of glucose, although it is an important step, is not the end of the Calvin cycle.

In order for the cycle to be able to start again, a new supply of the 5-carbon $CO_2$ acceptor, ribulose diphosphate, must be produced. The rest of the Calvin cycle is concerned with meeting this need.

The basic problem is to transform the 3-carbon molecules of PGAL into 5-carbon atoms of ribulose. Of course, one way to do this would be to break down the 3-carbon compound into 1-carbon units, and then link these units together in sets of five. But that would involve breaking and forming a great many bonds, calling for a large expenditure of energy. The cell utilizes a more efficient process, with the help of a 6-carbon compound. Fructose-6-phosphate is an interim product in the synthesis of glucose, and it can easily split into two parts, a 4-carbon group and a 2-carbon group. PGAL, with its 3-carbon structure, is also available for use in this regeneration cycle, so the cell has 2-carbon, 3-carbon, and 4-carbon compounds for use as components of the necessary 5-carbon compound. The 2-carbon fragment can join directly with 3-carbon PGAL to derive the necessary 5-carbon compound. An alternative route involves the linkage of PGAL with the 4-carbon fragment; that forms a 7-carbon compound. It is easily split into a 5-carbon group and a 2-carbon group, which can join with PGAL. Since there are so many different reactions that can produce the 5-carbon compound, the cell is assured a steady supply. It is very unusual for any organism to rely on only one metabolic pathway in any important chemical process. Usually there are several alternative pathways, providing an elaborate fail-safe mechanism.

Once the 5-carbon ribulose has been reassembled, it is phosphorylated by a molecule of ATP, the third and last one used in the dark reaction. This phosphorylation yields ribulose diphosphate, the molecule which performs the initial fixation of carbon dioxide, thus starting the cycle all over again.

As the tools of research in molecular biology are developed and improved, we learn more and

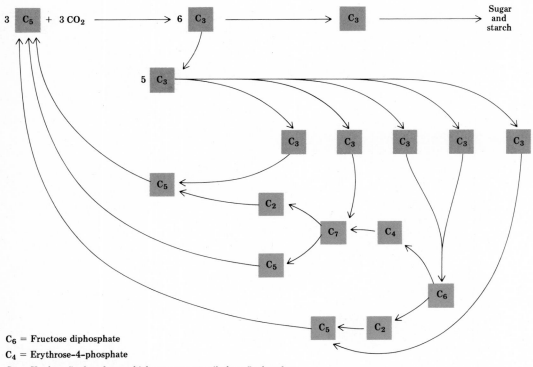

$C_6$ = Fructose diphosphate

$C_4$ = Erythrose–4–phosphate

$C_5$ = Xyulose–5–phosphate, which rearranges to ribulose–5–phosphate

$C_7$ = Sedoheptulose diphosphate

$C_3$ = PGAL

**3-20** *This schematic chart follows the complicated pathways through which the 5-carbon ribulose is regenerated. A large number of different pathways can be utilized, therefore insuring that a supply of this necessary compound will always be available.*

more about the dark reaction. Some of the discoveries have interesting implications. Through radioactive labeling, it has been found that carbon atoms from $CO_2$ quickly show up in a number of amino acids within the chloroplast. One hypothesis is that these amino acids are synthesized through a series of reactions that starts when certain enzymes known to be present in the chloroplast react with PGA, early in the Calvin cycle. Other scientists are now investigating the possibility that the $NADPH_2$ formed during the light reaction is used in other parts of the cell for other reduction reactions. It is known that the syn-

thesis of fatty acids, steroids, and many pigments must be reduction reactions; it may well be that they are formed through the reduction of one of the compounds generated in the carbon fixation cycle.

## Chemosynthesis

Most autotrophs use light as an outside energy source, and water and $CO_2$ as raw materials in their synthetic activities. But neither light nor water is absolutely indispensable for the synthesis of light energy compounds. Several bacterial groups display unusual methods of synthesis. One such group is the red bacteria. These contain a photosynthetic pigment that traps the blue and green light spectrum. They use the light energy to synthesize carbohydrates from the raw materials of carbon dioxide and hydrogen sulfide. The oxidation of hydrogen sulfide provides the electrons needed for the reduction of carbon dioxide.

An even more unusual type of bacteria is the chemosynthetic group. They can synthesize high-energy organic compounds without the aid of light. Instead they utilize the energy released from the oxidation of low-energy inorganic compounds. One group oxidizes ammonia ($NH_3$). Another group oxidizes nitrite ($NO_2$), changing it to nitrate ($NO_3$). Although it is only a small specialized group of organisms that can carry out chemosynthesis, the process is quite important to the overall balance of the biosphere. It is the chemosynthetic bacteria that can obtain nitrogen from the atmosphere and convert it into a form that can be used to build organic compounds, thus supplying much of the nitrogen needed for protein in all organisms.

Such "abnormal" methods of synthesizing energy-rich compounds furnish valuable clues about the way living systems could have evolved on our planet. If carbon dioxide and water were not always freely available, there are other compounds that might have served as substitutes. Through the study of chemosynthetic bacteria, we have learned that a number of different compounds can be utilized as electron donors, and we also know that life can exist even where there is no water and no atmospheric oxygen. This raises interesting questions about the possibility of life on other planets—a form of life that utilizes other compounds, perhaps even other elements, yet still carries on the basic chemical reactions that constitute life. Astronomers have already identified a number of compounds in deep space—water, ammonia, formaldehyde ($H_2CO$),

and a complex molecule called magnesium tetrabenzporpine that is very similar to chlorophyll. Any of these might be the basis of another type of life system.

# Summary

In this chapter, we have examined the ever-changing; ongoing processes that occur within the cell. The cell must work to counteract entropy, the trend toward disorganization and chaos, or equilibrium, which is manifest in osmosis, will occur. To fight entropy, the cell must utilize sources of energy from the outside environment to rebuild order.

One of the major energy sources is ATP. It releases energy when the bond that holds the last phosphate group onto the molecule is broken. This reaction is exergonic, releasing energy for the cell's use. However, before the phosphate bond can be broken, it must first be formed. An endergonic reaction is responsible for creating the phosphate bond, and it requires energy.

Specific phosphorylation reactions that facilitate both the exergonic and the endergonic reactions include photosynthetic and oxidative phosphorylation. The two aforementioned reactions use absorbed light energy, and energy stored in sugar or similar compounds, respectively. Autotrophs carry on both photosynthetic and oxidative phosphorylation; in heterotrophs the sole mechanism for obtaining new supplies of energy is oxidative phosphorylation. For this reason, photosynthesis and cellular respiration are viewed as two different versions of the same process. Both processes utilize outside energy sources, and both are used by the cell to circumvent entropy.

The final energy-yielding steps in cellular respiration are glycolysis, fermentation, and oxidation. Glycolysis is a process which breaks down the glucose molecule in preparation for either fermentation or oxidative respiration. It occurs in the mitochondria.

Fermentation follows the same procedure as glycolysis except that oxygen is not present, and it may occur on various sites in the cell. Called anaerobic respiration, this type of cellular respiration is found in many primitive, one-celled organisms. It has been suggested that perhaps the first organisms all depended on an oxygenless type of respiration.

Oxidative respiration is the most efficient metabolic pathway. It extracts the largest amount of energy from the glucose molecule. Aerobic respiration, in conjunction with the introductory stage of glycolysis, yields a total of 38 ATPs, and harnesses 45 percent of the available energy in glucose.

Photosynthetic phosphorylation converts light energy into chemical energy that can be used by the cell. In the light reaction, light energy is converted to produce ATP and free electrons. Cyclic phosphorylation produces only molecules of ATP, whereas noncyclic phosphorylation both produces ATP and liberates hydrogen electrons to be later used in the dark reaction to provide energy for glucose synthesis.

Noncyclic phosphorylation can proceed only in the presence of light, and it requires the presence of chlorophyll *a*, the most common type of light-absorbing molecule in the chlorophyll classes.

The dark reaction involves the reduction or fixation of carbon dioxide. It uses the stored energy of ATP rather than direct light quanta. This reaction can proceed in the absence of light, and occurs simultaneously with the light reaction, too.

Chemosynthesis is a process whereby high-energy organic compounds are synthesized without the aid of light. Instead, the energy released from the oxidation of low-energy inorganic compounds is utilized. There is speculation that chemosynthesis may give us a clue as to how living systems could have evolved on our planet. Other compounds may have served as substitutes when carbon dioxide and water were not freely available. These questions, in turn, raise the question of the possibility of life on other planets.

# Bibliography

### References

Bogarad, L. 1966. "Photosynthesis." *In* W. A. Jensen, *Plant Biology Today*. 2nd ed. Wadsworth, Belmont, Calif.

Brachet, J. 1961. "The Living Cell." *Scientific American*. 205(9): 50–61. Offprint no. 90. Freeman, San Francisco.

Calvin, M. 1962. "The Path of Carbon in Photosynthesis." *Science*. 135(3): 879–889.

Holter, J. 1961. "How Things Get into Cells." *Scientific American*. 205(9): 167–180. Offprint no. 96. Freeman, San Francisco.

Korn, E. D. 1969. "Cell Membranes: Structure and Synthesis." *American Review of Biochemistry*. 38: 263–288.

Robertson, J. D. 1962. "The Membrane of the Living Cell." *Scientific American*. 206(4): 64–82. Offprint no. 151. Freeman, San Francisco.

Solomon, A. K. 1960. "Pores in the Cell Membrane." *Scientific American*. 203(12): 146–156. Offprint no. 76. Freeman, San Francisco.

### Suggested for further reading

Giese, A. C. 1968. *Cell Physiology*. 3rd ed. Saunders, Philadelphia.
A classic in the field, Giese's book provides comprehensive coverage of the subject in language the first-year student is able to understand.

Hoste, R., ed. 1971. *Modern Biology*. Penguin, Baltimore.
Part 1 is a varied and interesting collection of articles on intracellular processes. The material in this book is also useful in connection with Chapter 2 of the present text.

Jensen, W. A. 1970. *The Plant Cell*. 2nd ed. Wadsworth, Belmont, Calif.
An up-to-date review of all structures found in plant cells, explaining both their form and function.

McElroy, W. D. 1971. *Cell Physiology and Biochemistry*. 3rd ed. Prentice-Hall, Englewood Cliffs, N.J.
A good summary treatment of basic cell processes. See especially Chapters 5 through 9 dealing with energy release and transformation.

# 2
# Functioning
# of the
# organism

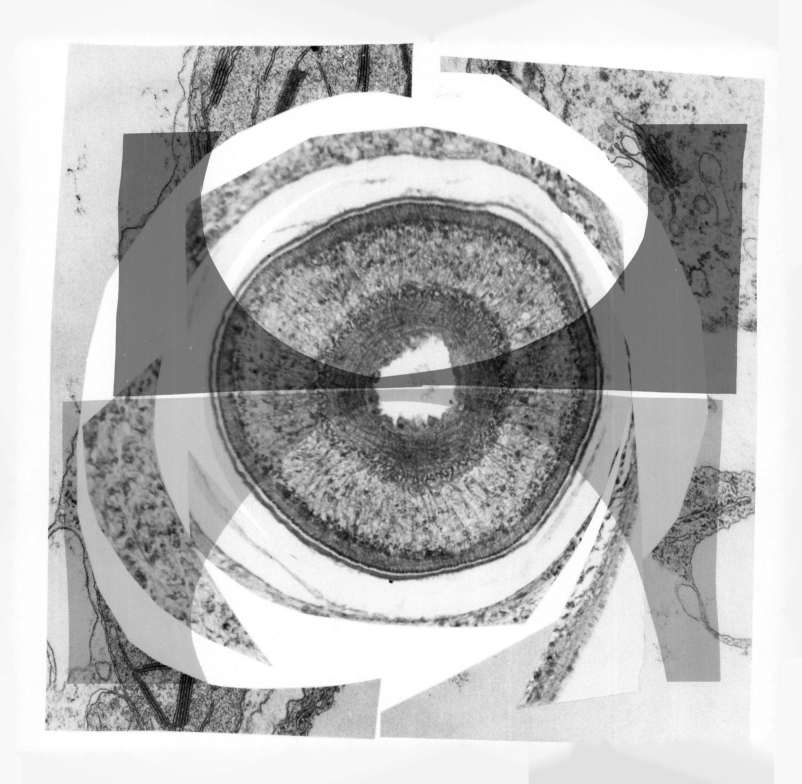

# 4

# The organism: patterns of organization

Biologists believe that the first living organisms were simple procaryotic cells. These cells had almost no specializations—no nucleus, no plastids, no formed organelles of any type. They were little more than membrane systems, enclosing a colloid in which was dispersed a number of chemicals needed in the reactions characteristic of life. These first cells had the ability to transduce energy—to change it from one form to another—and so by utilizing outside sources of energy, they were able to maintain their internal organization and stay alive.

With time, cells began to show some degree of specialization. Organelles such as the endoplasmic reticulum and the Golgi apparatus slowly evolved, perhaps as a result of an infolding of the cell membrane. Many biologists think that mitochondria and chloroplasts first developed as separate organisms and then later invaded larger cells to live inside them in a symbiotic relationship.

The eucaryotic cells, with their regions of specialization, led to the next evolutionary advance, the development of interdependence among groups of cells. Here we see the origin of multicellular organisms. At this point, we also see the emergence of two very different patterns of multicellular organization, each adapted to take advantage of a particular type of environment and position in the ecological system.

One pattern is the plant pattern. On the cellular level, this pattern includes the existence of a thick cell wall and an autotrophic mode of nutrition, with chloroplasts in most, if not all, of the cells. On the organismic level, the plant pattern includes a complete lack of mobility, a lifelong process of growth, and a total reliance on chemical means for the control and integration of all life functions.

The animal pattern on the cellular level includes the presence of centrioles and the absence of any cellular covering except on the outer layer of cells exposed to the environment. Animals have specialized in a heterotrophic

**123**

mode of nutrition and have therefore developed in the direction of increasing mobility, to aid them in their search for food. This increased mobility entails a need for rapid gathering and processing of detailed information about the environment. Because of this need, there is a trend in animals toward replacing mechanisms of chemical control with the faster system of nervous control and integration.

These four basic patterns of organization—procaryotic cells, eucaryotic single-celled or colonial organisms, plants, and animals—have been adapted in a wide variety of ways. There are, in fact, millions of different *species*, or distinct groups of individuals with similar structural and functional characteristics. It is the diversity of living things that most often catches our attention, but biologists are also interested in revealing the unity that lies behind this diversity.

On the basis of structural and functional characteristics, biologists who specialize in the field of taxonomy can separate organisms into groups such as (in ascending order of generality) the species, genus, family, order, class, and phylum. Each organism can then be given an "official" or scientific name, consisting of its genus and species. The giant panda, for example, is *Ailuropoda melanoleuca;* the common amoeba is *Amoeba proteus.* Scientific names often describe the species' outstanding characteristics. Proteus was a Greek god with the disconcerting habit of changing his form and identity in the midst of conversations, and thus the name refers to the amoeba's ability to change its body shape. The panda's name comes from four Greek words meaning "cat-footed black-and-white," a graphic description of the animal.

The four major taxonomic divisions, or **kingdoms,** correspond to the four patterns of organization we have mentioned here. Kingdom Monera contains the procaryotic cells; Kingdom Protista contains eucaryotic one-celled or colonial organisms; the other two kingdoms are the plants (Plantae) and animals (Animalia).

Each kingdom is divided into smaller groups called phyla. Grouping organisms into these divisions helps us to see not only the structural and functional similarities behind apparent diversity, but also to see the evolutionary relationships that account for these similarities. Since these phylum designations will be referred to again and again throughout the remainder of the text, a brief description of each of the important phyla may prove useful. Further details on phylum characteristics, and information about the subgroups into which each phylum is divided, are contained in the Appendix.

## Kingdom Monera

All members of the Kingdom Monera are procaryotic. They lack both a nuclear membrane and recognizable chromosomes; they have none of the organelles generally found in eucaryotic cells. If pigments, such as chlorophyll, are present, they are not confined to plastids but are dispersed throughout the cytoplasm. Since monerans have no nuclei, new cells arise by a simple splitting of the parent cell, a process called **fission.** Species may be either unicellular or colonial.

## Phylum Schizophyta

The Schizophyta includes the tiniest and most primitive of the procaryotes, the bacteria. While a few, such as the sulfur bacteria (*Beggiatoa*), are long, hairlike filaments, all others come in one of three shapes: **cocci** (spherical), **bacilli** (rod-shaped), and **spirilla** (spirals). The smallest of the cocci may be a tenth of a micron, while some of the larger bacilli may be over 100 microns in length.

Most bacteria are **saprophytic,** absorbing organic nutrients from their surroundings. The

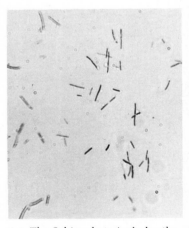

**4-1** *The Schizophyta includes the tiniest and most primitive of the procaryotes, the bacteria. This photo (taken at a magnification of 500X) shows two types. The long rods are bacilli, and the small circles are cocci. The picture was taken from a laboratory culture grown on an agar medium. (Russ Kinne—Photo Researchers)*

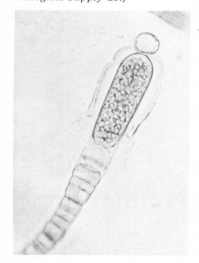

**4-2** *Blue-green algae constitute the second major group within the Kingdom Monera. Gleocapsa, shown here, is a filamentous form of blue-green algae. It is a colonial alga, but there is little differentiation of the individual procaryotic cells, and only asexual reproduction is known to take place. (Courtesy Carolina Biological Supply Co.)*

saprophytic bacteria are vital to the economy of the living world, since they bring about decay and decomposition of dead organisms, thereby reducing them to simpler compounds. In doing so, they perform two important functions: one, the disintegration of unsightly carcasses (imagine having to pick your way to class or work over the bodies of buffaloes or brontosauruses); and two, the recycling of structural components of cells, such as nitrogenous compounds. Although some saprophytic bacteria inhabit the bodies of living organisms, most of these are harmless. Some, such as those that aid digestive processes, are even beneficial. But a number of species are parasitic, and many of these, such as the tuberculosis bacillus, are pathogenic, or disease-causing organisms. Some bacteria may cause human death without actually invading body cells. Botulism food poisoning, for example, is caused by the release of potent toxins by *Clostridium botulinum.*

A number of bacteria have long been useful to man in food-making (sauerkraut, vinegar, or yogurt), tanning of leather, and curing of plant fibers. Bacteria have also become indispensable in the treatment of sewage.

Some bacteria are autotrophic. Perhaps the most important of these are the chemosynthetic soil bacteria. These play a key role in the nitrogen cycle. Some, such as *Nitrosomonas*, oxidize ammonia into nitrites. Others convert nitrites into nitrates. From these oxidation reactions, the bacteria obtain energy to manufacture their own organic food, while making available nitrogenous compounds needed by green plants. A handful of genera, such as *Chromatium*, has chlorophyll, enabling them to carry on photosynthesis.

Bacteria are everywhere—in soil, air, ocean or fresh water, in food or living organisms. Under certain conditions, some bacilli form spores. These are formed from portions of cytoplasmic and nuclear material which become enclosed in tough coats. Metabolic activities of spores are greatly reduced. In this state they are highly re-

sistant to prolonged boiling, freezing, or desiccation. When growth conditions are favorable, the spore germinates and gives rise to a new cell. In 1947, Tatum and Lederberg discovered that certain bacteria exchange genetic material, in a sexual reproduction process called **conjugation.**

## Phylum Cyanophyta

This relatively small group contains the blue-green algae. The most ancient fossil evidence of life on this planet is of blue-green algae estimated to be two billion years old. Blue-green algae are so named because of the presence of accessory pigments (phycocyanin and phycoerythrin) which combine with chlorophyll to give a bluish-green color. Some are unicellular while others occur as loosely organized colonial forms. Colonies may take the form of spherical clumps of cells; or flat square sheets, such as *Merismopedia*, one cell in thickness; or filaments, such as *Nostoc*. Many species of blue-green algae live in standing bodies of fresh water. There are also marine species, and some grow in moist soil or on tree trunks. In addition to cellular fission, colonies may reproduce by breaking up, a process known as **fragmentation.** Sexual reproduction is unknown in this group.

Some of the more intriguing blue-green algae may be found actively growing in hot springs, like those in Yellowstone National Park, where temperatures fluctuate between 85° and 90°C. When we recall that most enzyme systems are inactivated at temperatures of 50° to 60°C, this feat must be viewed as truly remarkable.

While most plants store their food in the form of starch, blue-green algae produce a storage compound called cyanophytan starch, which closely resembles glycogen of animals. Surrounding the cells of blue-green algae is a gelatinous covering or sheath, which helps to protect them against desiccation. In *Merismopedia* this jelly serves as a matrix binding the cells together.

**125**

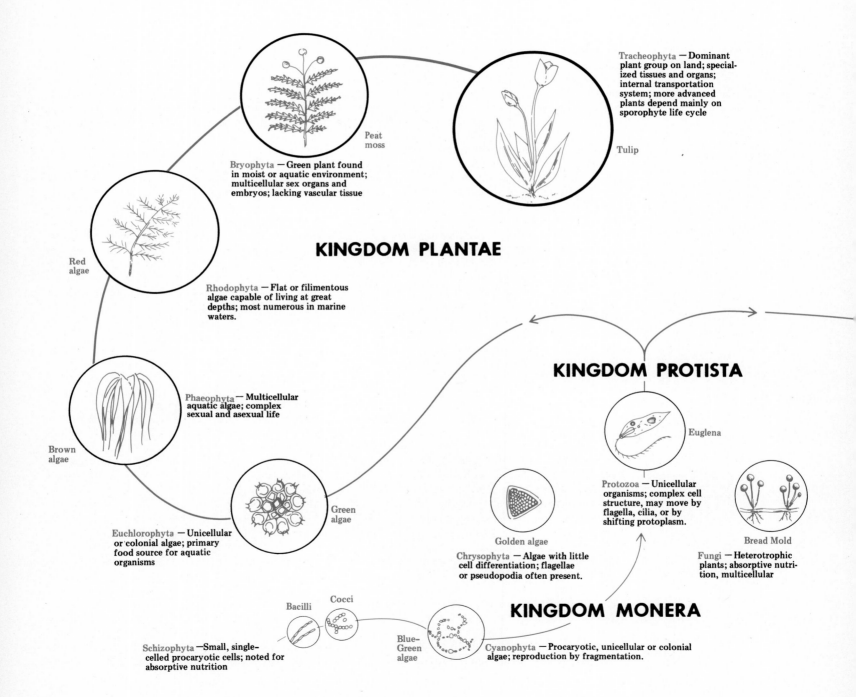

Tracheophyta — Dominant plant group on land; specialized tissues and organs; internal transportation system; more advanced plants depend mainly on sporophyte life cycle

Tulip

Peat moss

Bryophyta — Green plant found in moist or aquatic environment; multicellular sex organs and embryos; lacking vascular tissue

**KINGDOM PLANTAE**

Red algae

Rhodophyta — Flat or filimentous algae capable of living at great depths; most numerous in marine waters.

**KINGDOM PROTISTA**

Euglena

Phaeophyta — Multicellular aquatic algae; complex sexual and asexual life

Brown algae

Protozoa — Unicellular organisms; complex cell structure, may move by flagella, cilia, or by shifting protoplasm.

Green algae

Golden algae

Bread Mold

Euchlorophyta — Unicellular or colonial algae; primary food source for aquatic organisms

Chrysophyta — Algae with little cell differentiation; flagellae or pseudopodia often present.

Fungi — Heterotrophic plants; absorptive nutrition, multicellular

Bacilli

Cocci

**KINGDOM MONERA**

Blue-Green algae

Schizophyta — Small, single-celled procaryotic cells; noted for absorptive nutrition

Cyanophyta — Procaryotic, unicellular or colonial algae; reproduction by fragmentation.

**4-3** *The characteristics of the major phyla, and their relationship to one another, are shown in this chart. The simplest organisms are found in the Kingdom Monera, located at the lower left-hand corner of the chart. The Protista are mostly one-celled or colonial forms. Complex multicellular organisms are found in the Kingdoms Plantae and Animalia.*

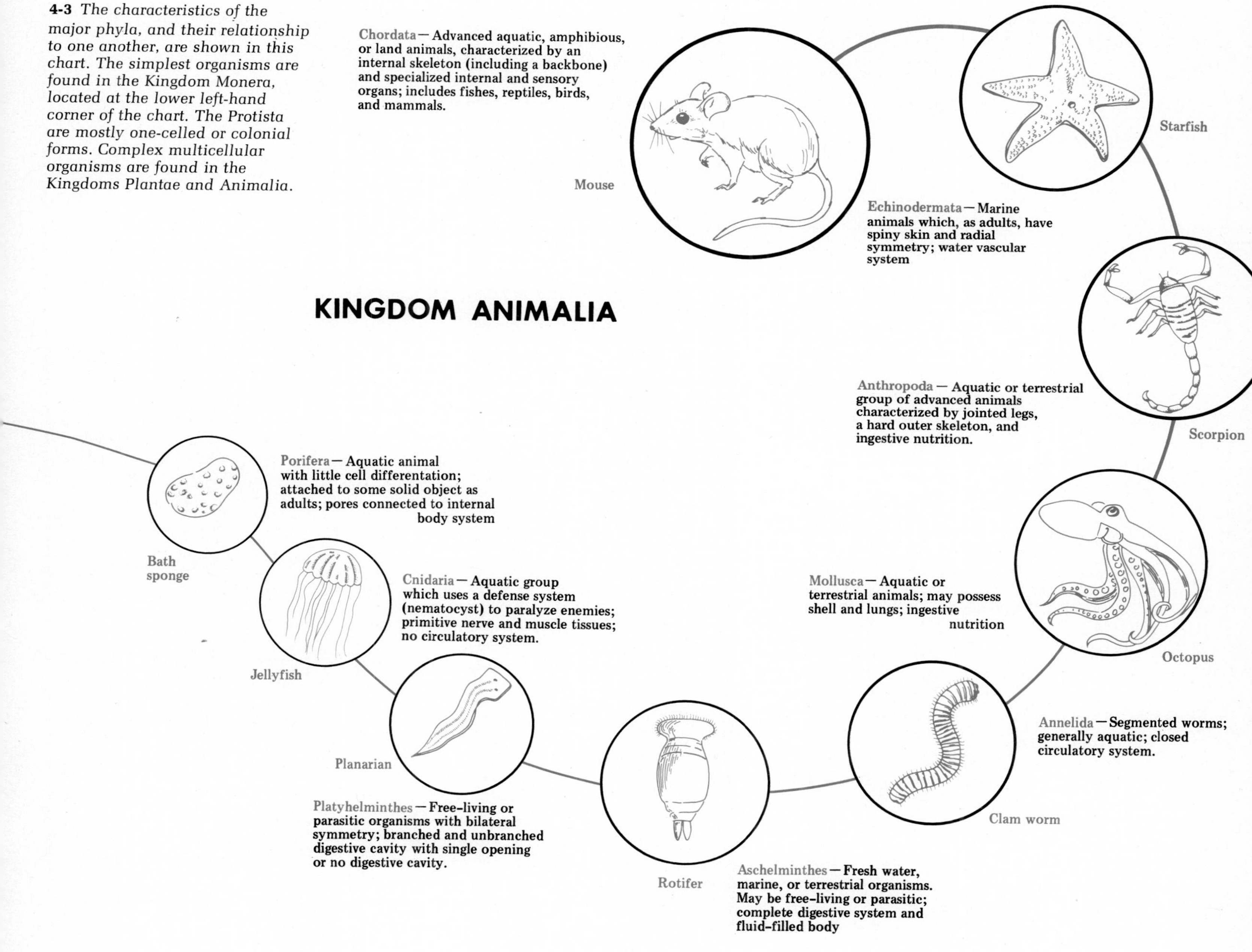

Chordata — Advanced aquatic, amphibious, or land animals, characterized by an internal skeleton (including a backbone) and specialized internal and sensory organs; includes fishes, reptiles, birds, and mammals.

Mouse

Starfish

Echinodermata — Marine animals which, as adults, have spiny skin and radial symmetry; water vascular system

## KINGDOM ANIMALIA

Anthropoda — Aquatic or terrestrial group of advanced animals characterized by jointed legs, a hard outer skeleton, and ingestive nutrition.

Scorpion

Porifera — Aquatic animal with little cell differentiation; attached to some solid object as adults; pores connected to internal body system

Bath sponge

Cnidaria — Aquatic group which uses a defense system (nematocyst) to paralyze enemies; primitive nerve and muscle tissues; no circulatory system.

Jellyfish

Mollusca — Aquatic or terrestrial animals; may possess shell and lungs; ingestive nutrition

Octopus

Planarian

Platyhelminthes — Free-living or parasitic organisms with bilateral symmetry; branched and unbranched digestive cavity with single opening or no digestive cavity.

Rotifer

Aschelminthes — Fresh water, marine, or terrestrial organisms. May be free-living or parasitic; complete digestive system and fluid-filled body

Annelida — Segmented worms; generally aquatic; closed circulatory system.

Clam worm

Although blue-green algae do not have flagella, some filamentous forms, such as *Oscillatoria*, display a slow oscillating or gliding movement.

## Kingdom Protista

Protistans are eucaryotic organisms that are primarily unicellular or colonial. Some may be multicellular or multinucleate, but none has true tissue differentiation. Their methods of nutrition vary. Some are photosynthetic, but others feed on small solid particles, and dinoflagellates combine both types of nutrition. Within this kingdom may be found the euglenoid organisms, the golden algae, slime molds, protozoa, and fungi.

## Phylum Chrysophyta (golden algae)

The golden-brown sheen so typical of the algae in this group is the result of a combination of the pigments carotene, xanthophyll, and chloro-

ESSAY **Viruses**

The viruses represent life in its simplest known form, and may be said to stand at the boundary between living and nonliving matter. They have a definite structure and the ability, under specific conditions, to replicate that structure. But a virus particle, or virion, cannot perform the functions of life except by parasitizing the living cell of a host. The virion consists of a molecule of nucleic acid, comprising between 10 and 100 genes, enclosed in a jacket of protein which both protects the nucleic acid and provides a mechanism for infecting the host. It is the host that provides the virus with the mechanism and materials for carrying out the instructions encoded in its nucleic acid.

Three hypotheses have been offered in explanation of the extremely simple nature of viruses. One hypothesis is that the viruses represent fragments of genetic material detached from their original substrates in cellular organisms; this might explain their highly specific parasitism. A second is that modern viruses are a modified version of free-living acellular ancestors, which have become parasitic only after the disappearance of the primordial "organic soup." The third hypothesis is that viruses arose from cellular ancestors through the gradual loss of extranuclear function, as a result of extreme specialization for parasitism. Interestingly, viral infection is limited to a relatively few phyla. Most bacteria, angiosperms, vertebrates, and arthropods serve as viral hosts, but not protozoa, the lower plants, or most other animal groups.

The T4 bacteriophage (Figs. B and C) has been intensively studied and will serve to illustrate the mechanism of viruses as a group, although it is an unusually complex viral structure. The nucleic acid (DNA) is contained in the head or capsid. The phage infects a bacterium by attaching itself to a specific site on the bacterium's cell membrane by means of the tail fibers and end plate; the tail sheath contracts to force the DNA from the head through a hollow core and into the host cell (Fig. D-1). The bacterial DNA is immediately degraded, and the cell metabolism comes under the control of the viral DNA. The cell's synthesizing apparatus now is used for the synthesis of enzymes (Fig. D-2) for the replication of viral DNA and the synthesis of viral structural proteins (Fig. D-3). As viral DNA forms into viral heads, the component protein parts coalesce around them,

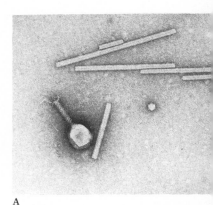

A

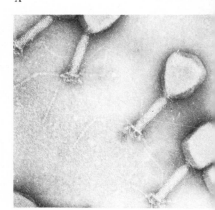

E

phyll. An obvious relationship to protozoa is indicated by the presence in some of flagella or pseudopodia.

Most numerous of the chrysophytes, and of paramount importance to the global ecosystem, are the diatoms. Diatoms are the principal photosynthetic food source for all animal life in the oceans, and probably account for over half of the earth's atmospheric oxygen. These tiny, exquisite forms come in two basic patterns, wheel-like and cigar-shaped. Their walls, which fit together like two halves of a pillbox, are composed of pectin,

upon which is deposited a glasslike or siliceous material. Movement of the slender diatoms is believed to be accomplished by the friction produced by the flowing cytoplasm against the water passing along a groove in the cell wall. Buoyancy is imparted to floating forms by oil droplets.

## Phylum Fungi

The fungi lack chlorophyll and are entirely heterotrophic. While a few are parasitic, most

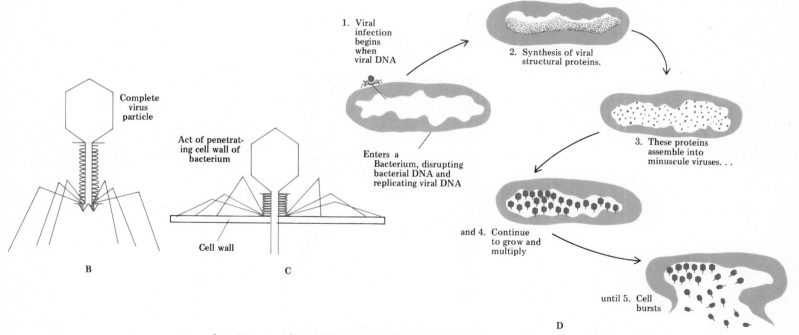

1. Viral infection begins when viral DNA

2. Synthesis of viral structural proteins.

Complete virus particle

Act of penetrating cell wall of bacterium

Cell wall

Enters a Bacterium, disrupting bacterial DNA and replicating viral DNA

3. These proteins assemble into minuscule viruses. . .

and 4. Continue to grow and multiply

until 5. Cell bursts

B

C

D

forming complete new virions (Fig. D-4). This process continues for about 25 minutes, until the lysis, or dissolution, of the host cell releases about 200 new virus particles, each capable of infecting a new host (Fig. D-5). Figure E shows ultra-thin sections of *E. coli* at various stages of infection by the phage T4. Note especially the almost immediate disappearance of bacterial nuclei (white areas in top photo).

**4-4** *Most numerous of the chrysophytes, and of paramount importance to the global ecosystem, are the diatoms. Diatoms are the principal photosynthetic food source for all animal life in the oceans, and probably account for over half of the earth's atmospheric oxygen. This photo shows why diatoms are sometimes referred to as "living pillboxes." (Richard Stevens—Omikron)*

Some cells of a slime mold show protozoan characteristics, while other cells look and behave like true molds. It is difficult to understand how these two different types of cells coordinate their activities; even more difficult is the question of how such widely differing cells arise within the same organism.

True fungi are separated into three major classes on the basis of their spore-forming structures. The alga-like fungi form coenocytic mycelia that look like black powdery growths. These include the familiar bread mold and some parasitic molds. The sac fungi derive their name from the saclike container in which the spores are produced. Included in this class are the economically important yeast used in baking, the morels, and truffles, the delight of gourmets. To most people, the term fungi is synonymous with mushrooms and toadstools. These actually belong to the club fungi, which produce spores on a

feature saprophytic nutrition and absorb dissolved food from the organic remains of decaying plants or decomposing animals. Like the bacteria, therefore, they play an extremely valuable part in the world of the living by breaking down the remains of the dead. A few fungi are unicellular, but typically they are multicellular. In multicellular forms, the plant body, called the **mycelium,** is composed of filaments. These may have no walls separating nuclei (coenocytic condition) as in the bread mold, or the nuclei may be separated by cross walls (septate condition) as in the mushroom.

Two major subdivisions of the fungi are the slime molds and the true fungi. Slime molds have puzzled and intrigued biologists for many years.

**4-5** *Mushrooms are the most familiar of the fungi. The visible (and edible) part of this mushroom is actually the fruiting body; as it matures, it will bear spores on the underside of the umbrella-shaped cap. (Photo by Hugh Spencer from National Audubon Society)*

basidium, or club-shaped filament. Some highly poisonous mushrooms resemble closely the edible species. There is no foolproof method for distinguishing them; even experts make mistakes, in some instances the last they ever make. Other club fungi are the bracket fungi, the puffballs, and the smuts and rusts which are parasites of flowering plants.

## Phylum Protozoa

Protozoa are unicellular organisms. They may occur as single units or as colonies. Some are so tiny they can be seen only with the highest magnification of the light microscope. Others are large enough to be seen with the naked eye. All are heterotrophic, either ingesting particulate food, or absorbing dissolved material.

Protozoa are separated into subgroups primarily on the basis of their method of locomotion. The Sarcodina move by means of pseudopodia; Ciliophora, as their name suggests, use cilia; and the Zoomastigina use long flagella. The remaining species, all parasitic, are placed within the Sporozoa. These often have complex life cycles in which immature stages may have pseudopodia or flagella, and adult stages are nonmotile.

Although the fresh-water amoeba is the most familiar of the Sarcodina, a diversity of form and habit may be found in this group. The famed White Cliffs of Dover represent the accumulated skeletal remains of millions of years of foraminiferans, marine amoebae which secrete a hard shell reinforced with calcium carbonate, or lime. Undoubtedly the most beautiful of the Sarcodina are the radiolaria, floating marine amoebae with siliceous skeletons. Another group, the heliozoa, are spherical amoebae with needlelike pseudopodia; these are found mostly in fresh water. A parasitic relative of the common amoeba is *Entamoeba*, which causes dysentery in humans.

The close relationship between the Sarcodina and the Zoomastigina is illustrated by

such forms as *Mastigamoeba*, which is amoeboid, but also has a flagellum. While some of the Zoomastigina are free-living, the majority live in symbiotic relationships with higher animals. Thus, *Trichonympha* lives in the gut of termites in a mutualistic relationship benefiting both partners. *Trypanosoma* is a parasitic flagellate that causes African sleeping sickness.

Like the Zoomastigina, the Ciliophora are characterized by a fixed shape, often maintained by a living outer covering called the **pellicle.** Ciliophora shows the greatest complexity of cellular organization of all protozoa. Most are free-living aquatic forms, but some are found in commensal or parasitic association with higher animals. Suctoria, which lack cilia, may be either free-living or parasitic.

The Sporozoa are so called because they form spores. All members of this group are parasitic. Some, such as *Eimeria*, cause diseases in domestic animals. Malaria in human beings is caused by *Plasmodium*, a sporozoan which is transmitted by mosquitoes and attacks red blood cells.

## Kingdom Plantae

All members of the plant kingdom are eucaryotic. They have cell walls and vacuoles, and their photosynthetic pigments are confined to plastids. While most are autotrophic, some, such as the Indian pipe, have lost their chlorophyll and are saprophytic. The typical storage compound is starch. Cell division is by **mitosis** (a type of cellular reproduction in which the chromosomes divide, each one duplicating an exact copy of itself). Both sexual and asexual reproduction are common. These two types of reproduction are frequently combined within a life cycle in which sexual, or gamete-producing stages **(gametophytes)** alternate with asexual, or spore-producing stages **(sporophytes).** The advantage of this alternation of generations is that the spores are resistant to extreme environmental condi-

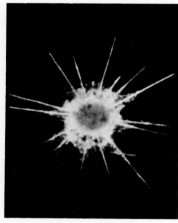

**4-6** *The most beautiful of the Sarcodina are the radiolarians, floating amoebae with spiky skeletons made of silica. These radially symmetrical organisms are an important part of the marine food chain, for many animals have adaptively acquired the means of digesting or crushing the protective spines. (Courtesy of the American Museum of Natural History)*

**131**

tions and can safely remain in their inactive state for long periods of time. This enables the plant to wait out droughts or cold spells. They cannot escape bad weather by moving away from it, as animals do; instead they escape by a transformation that makes them practically immune to the dangers of the weather.

## Phylum Euchlorophyta (green algae)

The green algae, which typically have a grassy green or yellowish green color, are believed to be ancestral to all other groups of plants. They may be unicellular or colonial. *Spirogyra* is a well-known filamentous colony with spirally shaped chloroplasts. Most green algae are aquatic, occurring in both marine and fresh-water habitats. Round bodies in the chloroplasts, the pyrenoids, are believed to be centers of protein around which starch layers accumulate. Many single-celled species swim by means of flagella, and resemble closely some of the Zoomastigina. *Chlamydomonas* is a unicellular flagellated species abundant in fresh water or in damp soil. *Volvox* is a spherical colony often found in fresh-water ponds. Sea lettuce, or *Ulva*, grows as a broad flat blade on rock surfaces within the intertidal zone.

Green algae are most important in their role of primary food source for animals in aquatic communities. In standing water or lakes, green algae (as well as blue-green algae) may undergo a sudden surge or explosion in numbers. This "bloom" usually follows an aeration of the water or an influx of mineral nutrients.

## Phylum Phaeophyta (brown algae)

Brown algae have an accessory pigment called fucoxanthin, which in combination with chlorophyll imparts an olive green or brown color to the plant. This primarily marine group is best

**4-7** *This green alga has acquired the common nickname "merman's shaving brush" because of its brushlike appearance. It is a colonial form, occurring only in a marine habitat. Like all other chlorophytes, it contains chlorophyll and photosynthesizes its own food. (A. W. Ambler from National Audubon Society)*

represented in cooler coastal waters. Brown algae are commonly known as rockweeds and kelps. They usually vary from a few inches to three or four feet in length, but some giant kelps of the Pacific coast of North America may be over 300 feet long. The plant body is a multicellular flat blade, or **thallus,** often provided with gas-filled sacs or bladders to keep it afloat. At the base there is usually a specialized holdfast for attachment to a rock or other firm surface. The life cycle of most brown algae is fairly complex, and both sporophyte and gametophyte generations are well represented.

*Sargassum* is an abundant floating brown alga of the Sargasso Sea. *Laminaria*, a kelp, is a source of algin, used as a thickener in making ice cream.

## Phylum Rhodophyta (red algae)

The distinctive red color of these algae are caused primarily by the accessory pigment, phycoerythrin, which masks the green of the chlorophyll. In some species, phycocyanin is also present. What is the advantage to these algae in being red? Their color enables them to live at greater depths than other algae, perhaps as deep as 200 feet. Phycoerythrin, with its red color, absorbs green light, which is the only wavelength that can penetrate far into the water.

Red algae attain their greatest profusion of species in warm marine waters, although they also occur in abundance in cooler waters. A few species may be found in fresh water.

The multicellular thallus or plant body of red algae may be a flat blade or filamentous. The filamentous species include some of the most delicate and exquisite forms of the algae.

## Phylum Bryophyta (mosses and liverworts)

Bryophytes are multicellular green plants that have made the transition from water to land. They lack vascular tissue, have multicellular sex organs and embryos, and an abbreviated sporophyte generation. The latter lacks chlorophyll and lives essentially as a parasite on the dominant gametophyte.

Liverworts are flat, ribbonlike plants that grow on the surface of rocks or on soil. They may occur in fresh water or in very moist, shady woodlands. Small upright structures, somewhat re-

**4-9** *This close-up view of a colonial red alga indicates the degree of specialization and differentiation which can sometimes be found in this group. The life cycle of this* Polysiphonia *plant is very complex, with large multicellular haploid and diploid generations. (Russ Kinne—Photo Researchers)*

sembling miniature palm trees, bear the sex organs. When released, the sperms swim through a film of water, attracted by chemicals emanating from the female sex organ. The fertilized egg grows into a sporophyte which remains attached to the parent gametophyte. At maturity, the **sporangium,** or spore sac, of the sporophyte plant ruptures and releases spores that are carried away by the wind. If the spore alights in a favorable area, it will germinate and grow into a new thallus.

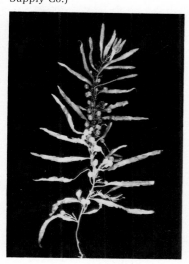

**4-8** *A typical marine brown alga is this Fucus. It exhibits considerable tissue differentiation and structure: note the rootlike holdfast, the stemlike stipe, the leaflike blades and the typical gas-filled sacs. The brown algae are complex plants similar in many respects to the vascular plants, an example of convergent evolution. (Courtesy of Carolina Biological Supply Co.)*

**133**

Mosses generally grow in drier areas than liverworts. The life cycle of a moss includes an intermediate stage called a **protonema,** which closely resembles a filamentous green alga. Buds produced at intervals on the protonema grow into the vertical green gametophyte plants. The moss sporophyte that emerges from the top of the female gametophyte resembles a reddish wire with a capsule on the end. Male gametophytes are generally taller than females, and, as a result, splashing raindrops splatter the female gametophytes with thousands of sperm cells.

Sphagnum moss, a bryophyte typical of acid bogs, grows at the surface of the water. When it dies, it sinks to the bottom and a new layer of moss takes its place. Since decay occurs very slowly in acid waters, the layers gradually become compacted and form peat, which can be dried and used as a cheap source of fuel.

## Vascular plants

All remaining phyla of plants may be referred to collectively as vascular plants because of the presence in the sporophyte generation of specialized conducting or vascular tissue, the xylem and phloem. For this reason, most traditional classification schemes incorporated vascular plants in a single phylum, the Tracheophyta. Although there is now a tendency to abandon this name as a phylum designation, and to elevate each of the major subgroups to phylum status, the concept of a tracheophyte group helps to underline some unifying affinities of the members.

We have seen that in bryophytes, the dominant generation is the gametophyte. In vascular plants, the situation is reversed. It is the sporophyte that is larger and longer-lived. If we were to arrange the groups of vascular plants in a graded series from primitive to advanced, we would see that the gametophyte gradually becomes tinier, more dependent on the sporophyte and shorter in life span. Thus, in flowering plants,

the gametophyte has shrunk to an enlarged multinucleate cell within the flower.

The gametophyte generation culminates in the production of gametes (egg and sperm cells). Even in the lower vascular plants, the gametes require water for fertilization. We may therefore regard the gametophyte generation as being more water-dependent than the sporophyte generation. The sporophyte give rise to air-carried spores and may thus be considered more terrestrially adapted than the gametophyte. With the introduction in conifers of nonswimming sperm, the adaptation to a terrestrial existence became more complete.

Class Filicinae includes the ferns, plants most often found in shady, moist areas. The fern gametophyte is fairly large, flat, and photosynthetic. Fern sporophytes have true roots, stems, and leaves. The stem is a horizontal, underground **rhizome,** and the leaves are often subdivided into leaflets known as **pinnae.**

Seeds characterize both the conifers and flowering plants. In the Class Gymnospermae, two kinds of cones are produced. The larger of the two contains the ovules that will mature into seeds; the smaller one produces pollen grains. Since the seeds remain exposed on the surface of the woody bracts of the cone, this group of plants has traditionally been called the **gymnosperms** (naked seeds). Seeds and pollen are dispersed by wind. It is within this class that we find some of the tallest trees—Douglas fir and redwood—and also the oldest—bristlecone pines, which have been known to live 4,600 years.

Flowering plants are placed within the Class Angiospermae. In all flowering plants, the seeds develop within a special covering called the ovary; for this reason, the plants are called **angiosperms** (enclosed seeds). It is the ovary that ripens into the fruit. While conifers are wind-pollinated, flowering plants are typically dependent upon insects for pollination. It is not surprising, therefore, that flowers have evolved brilliantly colored petals, aroma, and nectar to

Diploid
sporophyte
stage

Haploid
gametophyte
stage

**4-10** *The lower "leafy" plants are the haploid gametophyte generation. Growing out of these plants are the sporophyte stages, wirelike reddish filaments bearing at the top a distal capsule, or sporangium. Asexual reproduction occurs when the spores are released; some will develop into protonemas, which in turn give rise to the mature moss gametophyte.*

attract insects. Some orchids have even heightened their appeal to the wasp species that pollinate them through the evolution of flowers that mimic the appearance of another wasp.

The Subclass Dicotyledoneae is so called because in these plants two seed leaves are first to emerge from the germinating seed. Dicots have their floral parts in fours or fives, or multiples of these. Herbaceous dicots live only one season and do not grow appreciably in diameter. However, the woody annuals (trees and shrubs) increase their stem diameter because of the presence of an embryonic tissue called **cambium,** which produces new xylem and phloem. Examples of dicots include daisies, peas, and all hardwood trees.

The second Subclass of flowering plants, the Monocotyledoneae, has only a single seed leaf. Flower parts are in threes or multiples of three. Secondary growth of the stem does not occur, since there is no cambium, and vascular bundles are scattered throughout the stem. Examples of monocots include lilies, orchids, grasses, and palm trees.

4-12 *The dicot shown here in the early stages of germination has two seed leaves. When the stem grows taller and the leaves appear, these seed leaves will wither and drop off. The second class of flowering plants, the monocots, has only a single seed leaf. The purpose of the seed leaves is to provide food (through photosynthesis) for the young plant in its early stages of growth. (Alycia Smith Butler)*

## Kingdom Animalia

While plants are primarily autotrophic, animals are entirely heterotrophic. That distinction is probably the best single way to separate plants from animals, although like all general rules, it has its exceptions.

The animal kingdom is striking in the enormous number and diversity of species. Although the exact number is not known, it is well over a million and a quarter species. Counting species is made difficult by the fact that many animals can be found in more than one form, depending on the stage of life cycle. The milkweed butterfly, for example, spends part of its existence as a caterpillar, part as chrysalis, and part as adult. Much more complicated is the life cycle of a parasitic flatworm, such as a liver fluke.

4-11 *The photo shows the growth of the sporophyte generation of fiddlehead ferns. As the plant develops, the coiled frond will unroll. The gametophyte generation of this plant is an independent organism, not parasitic on the sporophyte. (Alycia Smith Butler)*

It is only when we classify animals on the basis of similarities and differences of embryonic development and adult body patterns that we emerge with a usable and orderly array, from an apparently bewildering profusion of forms. Modern classification of living things is based primarily on genetic and evolutionary relationships. Recent techniques in protein analysis,

**135**

cytogenetic and electron micrograph studies have added much evidence for or against certain groupings. Whether to place some beast in species A or species B is no longer decided solely by a museum taxonomist pouring over its gray, preserved remains. Today, every conceivable facet of an organism's biology is considered: behavior of both individual and population; chromosome appearance and number; crossbreeding studies; habitat; and serum protein characteristics.

One of the most important distinctions used in classifying members of the animal kingdom is the number of **germ layers,** or layers of tissue present in the embryo. Many of the lower animals have only two germ layers (they are **diploblastic).** There is an **ectoderm,** which gives rise to the outer covering, such as the skin and certain nervous receptors. There is also an **endoderm,** which produces the alimentary system. Animals that are **triploblastic** have a third layer sandwiched in between the other two. The **mesoderm** gives rise to the circulatory and reproductive systems.

## Phylum Porifera

Sponges make up a large group of aquatic animals that are outside the mainstream of animal evolution. Although sponges are multicellular, they are considered loose aggregations of cells that have not achieved true tissue differentiation. One advantage of this low level of specialization is that sponges are able to regenerate lost parts. All sponges are incapable of locomotion and are therefore permanently attached to some surface. Only the immature larval stage, the amphiblastula, swims about; it is propelled by flagellated cells. All sponges are filter feeders. The phylum name, Porifera, refers to the many tiny pores riddling the surface of each sponge. Special collared flagellated cells, the choanocytes, create a current of water which is drawn into the body

through the pores. The water is then expelled through a large chimneylike opening, the osculum.

The simplest sponges are saclike (asconoid) in basic construction, and are supported by hard spicules (little spines) of calcium carbonate. More complex sponges may have an intermediate construction (syconoid) with internal side pockets, or a highly intricate arrangement (leuconoid) of chambers and canals. Very deep sea sponges, such as the Venus' flower basket, have latticelike skeletons of "spun glass" or siliceous material. Bath sponges have a network of proteinaceous spongin fibers. Many sponges, such as the finger sponge, have spongin in addition to siliceous spicules.

## Phylum Cnidaria

The Phylum Cnidaria, also known as the Coelenterata, include the hydroids, jellyfish, and anemones. It is predominantly a marine group, although a few of its members are found in fresh water. All members are diploblastic. They have **radial symmetry,** the kind of symmetry found in a circle. Radial symmetry is a feature associated with limited powers of locomotion or no locomotion at all.

One feature that distinguishes this phylum from all others is a specialized tissue called the **cnidoblast,** the cell which gives the phylum its name. The cnidoblast produces a fluid-filled sac, the **nematocyst.** When the nematocyst is stimulated, it shoots out a long poisonous thread that has a paralyzing effect on the organisms it pierces. Nematocysts may be found anywhere on the cnidarian body, but they are particularly concentrated on tentacles. These stinging structures are used both in defense and to obtain food.

The basic body patterns are found in this phylum: the free-swimming **medusa,** and the attached **hydroid** or polyp. The medusa is hemispherical and has a large amount of mesoglea, a jellylike substance, between the epi-

**4-13** *The term germ layers is used to refer to the layers of tissue present in the embryo. Diploblastic (two-layered) animals contain only ectoderm and endoderm layers; in triploblastic animals, there is also a mesoderm. This is an important evolutionary and morphological distinction.*

**Diploblastic**

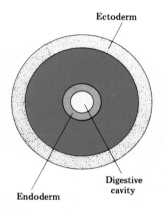

**Triploblastic**

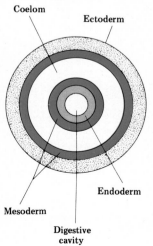

dermis and the gastrodermis. Encircling its edge is a fringe of tentacles. The hydroid is shaped like a hollow cylinder. It is usually attached to some underwater surface by its pedal disc. At the opposite end of the cylinder, it has a mouth surrounded by a circle of long tentacles. Only a thin layer of mesoglea separates the two cell layers.

Hydra is a fresh-water cnidarian that occurs in the hydroid form only. But many salt-water relatives of hydra have a life cycle in which there is an alternation of medusoid and hydroid generations. The Portuguese man-of-war, well known for its powerful sting, is a colony of individuals, each specialized for a different function. One class of cnidarians contains only large jellyfish, such as the moon jelly or lion's mane jellyfish.

Anemones, other members of Phylum Cnidaria, are attractive flower-shaped cnidarians with many rings of tentacles. Corals are colonial hydroid forms which secrete a skeleton, in some

**4-15** *The transparency of these jellyfish is due to the large amount of mesoglea, a jellylike substance, found between the two body layers. This seemingly delicate creature is by no means defenseless; its stinging nematocysts serve both to stun prey and to repel predators. (Dr. Carleton Ray—Photo Researchers)*

**4-14** *This member of Phylum Porifera shows a relatively low level of specialization. As its name suggests, this creature has innumerable pores on the surface of its body; these provide channels through which food can be taken in and wastes expelled. This particular sponge is calcareous; its body is reinforced by deposits of calcium carbonate. (Dr. Carleton Ray—Photo Researchers)*

cases composed of calcareous material (calcium carbonate). Millions of generations of corals have formed the reefs which abound in shallow tropical seas.

## Phylum Platyhelminthes (flatworms)

The flatworms are the first animals to show **bilateral symmetry;** the right and left sides of the

**Bilateral symmetry**

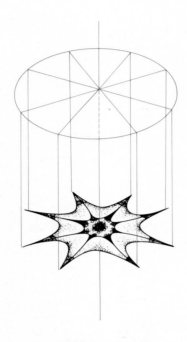

**Radial symmetry**

4-16 *(A) shows a flatworm with bilateral symmetry; (B), a radiolarian, is an example of radial symmetry. Radial symmetry is generally associated with slow locomotion or a sessile condition. Bilateral symmetry, which gives the animal a definite "front end," permits much faster locomotion. It is also associated with a trend toward increasing development of the brain.*

body are mirror images. They are all triploblastic, and there is no body cavity between the body wall and the digestive tract. As in the cnidarians, there is only one opening to the digestive tract. From mesoderm is derived the muscular system, an excretory system in which the smallest functional unit is the **flame cell,** and a reproductive system that is often highly developed.

Since there are no respiratory or blood circulatory systems, the introduction of internal bulk creates problems in gas exchange and food transport. Gas exchange is aided by the extreme flatness of the body, which increases surface area. The problem of food distribution is met in some of the larger flatworms by a highly branched intestinal cavity, a device which reduces the distance food must travel by diffusion to any body region.

Turbellarians are free-living flatworms found in fresh water, in moist terrestrial regions, or in the sea. They may be propelled by beating cilia or by undulations of the body.

Trematodes are parasitic flatworms known as flukes. These usually have sucking organs for attachment, and are lacking in sense organs. The blood fluke, *Schistosoma,* causes serious debilitating diseases among people in tropical areas. In Asia, where fish is often eaten raw, the human liver fluke, *Opisthorchis,* is commonplace.

Tapeworms make up the third major group of flatworms. These make their home in the intestinal tract of other animals. They are well equipped with suckers and hooklets that help them to resist being expelled. The worms are actually colonies of individuals in which the reproductive system is developed at the expense of all other systems. It should not surprise us that a worm frequently bathed in digested food is utterly lacking in a digestive tract of its own. Humans may acquire the enormous fish tapeworm (sometimes over 50 feet in length) by eating improperly cooked fish infested with immature stages of the tapeworm.

## Phylum Aschelminthes

This phylum contains an assortment of wormlike groups, all characterized by a complete digestive tract and a fluid-filled body cavity, called a pseudocoel. It includes the roundworms, hairworms, rotifers, and gastrotrichs.

The nematodes, or roundworms, make up an enormous group which is second only to the insects in numbers of individuals and probably numbers of species as well. Roundworms are cylindrical in cross section, slender, somewhat pointed at either end, generally lacking in appendages, and covered with a noncellular **cuticle.** Cilia are completely lacking.

Many are free-living, occurring in staggering numbers in soil. They are also abundant in fresh and salt water. Some roundworms thrive in the unlikeliest places. In the bygone days of the general store, vinegar would often become clouded by the growth of millions of tiny vinegar eelworms. A relative of the vinegar eelworm was known only in the felt mats set beneath beer steins in Silesia.

Parasitic roundworms abound in both vertebrates and invertebrates. Best studied are the parasitic species that have been mankind's constant companions. These are usually found wherever sanitary practices are poor and the climate is warm and moist.

*Ascaris,* which may attain a length of some 14 inches, is a common intestinal parasite of man and pigs. Female ascarids must rank among the champion egg-layers of the world, with an output of nearly 200,000 per day. Because it has only four chromosomes, *Ascaris* is a popular choice of biologists in the study of genetics and reproduction. Hookworm, a much smaller intestinal parasite, saps the energy and vitality of its human host, leaving him vulnerable to disease. Rotifers are microscopic aquatic creatures named for the circular rows of cilia which resemble two whirling propellers at the mouth end.

**4-17** *Phylum Aschelminthes contains an assortment of wormlike groups, all characterized by a complete digestive tract and a fluid-filled body cavity, called a pseudocoel. It includes the roundworms, hairworms, rotifers and gastrotrichs. Rotifers are microscopic aquatic creatures named for the circular rows of cilia which resemble two whirling propellers at the mouth end. (Photo by Hugh Spencer from National Audubon Society)*

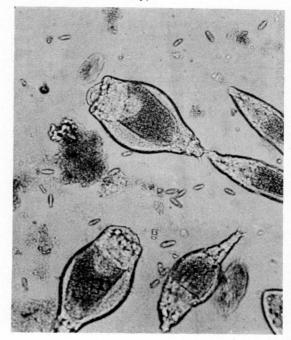

## Phylum Annelida

Segmentation of the body is the key distinguishing feature of annelid worms. Organs and parts are repeated both externally and internally. Annelids have a true **coelom** or body cavity. They also have a closed circulatory system and an excretory system made up of tubules called **nephridia.**

Polychaetes, the most primitive of the three major classes of annelids, are set apart by their paired appendages (parapodia), which are supported by numerous stiff bristles. These are primarily marine worms. Some have a well-developed head and move about actively; others have no head to speak of and lead a sedentary, filter-feeding existence. Sand worms or clam worms are familiar to any salt-water fisherman. Fan worms are among the most beautiful creatures in the group. They use their plumes in filtering food and respiration.

Earthworms are representatives of the second major class, the oligochaetes. This group occurs either in moist soil or fresh water. Oligochaetes

4-18 *Phylum Annelida is a varied group. The earthworm is one familiar member of this phylum; pictured above is a more exotic and beautiful relative, the fan worm. The delicate plumes are used to filter food; they also serve as organs of respiration, operating on the same principle as the gills of a fish. (Des Bartlett —Photo Researchers)*

**4-19** *The snail is a typical example of a mollusk; it belongs to the group called gastropods. The hard shell offers support and protection for the animal's soft body. The coiling of the shell is considered a progressive characteristic. Coiling is accompanied by another, less easily observed characteristic, that of torsion, or the rotation of the viscera at a 180° angle. (Bert Monier—Photo Researchers)*

## Phylum Mollusca

Mollusks are a large group of soft-bodied animals that occur in a variety of habitats. Key features include a flat, muscular foot and a large **mantle** which covers the internal organs. The body is often housed within a calcareous shell that is secreted by the mantle.

Chitons are a primitive group of marine mollusks. They are protected by a covering of eight overlapping plates. A flexible scraping organ called a **radula** is used to rasp food from rock surfaces. Chitons are able to adhere very tightly to the surf-pounded rocks of the intertidal zone.

By far the largest group of mollusks is the Gastropoda, which includes the snails, land slugs, and sea slugs. Many snails are asymmetrical because of the irregular twisting of their shells. In these animals the internal organs also undergo twisting or torsion. Since the shell of snails is secreted as a single unit, these animals are also referred to as univalves. Some gastropods, such as the land slug or sea slug, have lost their shell altogether. Like the chitons, gastropods have a flat creeping foot and a radula. Certain sea slugs have the remarkable ability to transfer nematocysts from the cnidarian hydroids on which they feed to their skin, without exploding them until needed for their own defense.

Scaphopods are a small group of burrowing marine mollusks known as "tooth-shells." They secrete a tubular calcareous shell, once prized by eastern American Indians as a form of money.

Since pelecypods secrete two shells that enclose the internal organs, they are called bivalves. The name Pelecypoda refers to the hatchet-shaped foot. Some common bivalves are the oysters, clams, and mussels. Evolutionarily speaking, it may be said that bivalves have "lost their heads," since they lack obvious sense organs, and the central nervous system is greatly reduced. It is hardly a serious loss, for these animals are sedentary or slow burrowers. The

have no appendages and relatively few bristles in each body segment. Earthworms are economically important because their burrowing activities enrich and aerate the soil. They also serve as food for many birds and small mammals.

The last major group of annelids is the Hirudinea, which includes the leeches. Many of these have sucking organs at the mouth and hind ends. The body appears to be divided into many segments, but actually, it consists of no more than 34 true segments. Many leeches are adapted to sucking blood from larger animals, a habit which has labeled them as parasites. But certain leeches have a varied diet of small animals along with their occasional drink of blood. Even those that subsist solely on blood lead an independent existence between meals.

gills have become important as food-gathering organs, while still functioning in respiration. A constant water current in and out of the body is maintained by cilia covering the gills and mantle.

Cephalopod mollusks are among the most intriguing of animals. This group of marine animals includes the octopus, squid, and cuttlefish. In sharp contrast to pelecypods, cephalopods have a body dominated by a well-developed head, from which projects the foot. As a matter of fact, the class name means "head-foot." The foot is modified into a number of tentacles supplied with powerful suckers. The eye found in the octopus and squid is strikingly similar in structure and function to that of vertebrates. All cephalopods have a radula, a pair of jaws shaped like a parrot's beak for killing prey and tearing food, and a sac for releasing a dark fluid to enable them to elude attackers.

In any contest of speed and gracefulness, cephalopods must surely win top prizes. Shooting water out of their siphons, squids can jet propel themselves faster than many fish. The octopus may clamber leisurely over rocks or use its jet propulsion. The largest of all invertebrates is the giant squid, *Architeuthis*, a titanic 55-foot monster known to tangle occasionally with whales.

## Phylum Arthropoda

Arthropods stand at the apex of invertebrate phyla, both because of their large numbers and their successful invasion of a variety of environments. Only arthropods have paired, jointed appendages and an exoskeleton of **chitin,** a hard material rather like horn. The group includes the crustaceans, insects, spiders, centipedes, and millipedes.

Crustaceans are aquatic arthropods with gills and two pairs of antennae. Their exoskeleton is impregnated with calcium carbonate salts

which make it harder than that of other arthropods. Lobsters, crayfish, beach fleas, and barnacles are among the more familiar crustaceans.

Most numerous of all living things are the insects, of which there are probably well over 850,000 species. The insect body is divided into three regions: head, thorax, and abdomen. Adults have three pairs of walking legs, a single pair of antennae, and a tracheal respiratory system. In most insects there are one or two pairs of wings, although some groups such as the silverfish and springtails are primitively wingless. The compound eye, a feature insects share with crustaceans, is extremely sensitive to even the slightest movements in the visual field.

Insects owe much of their success to their excellent powers of reproduction. Some termite queens, for example, may lay as many as 30,000 eggs each day. Other factors in the success of insects are their small size, flying ability, and resistance to harsh conditions. Furthermore, the enormous array of material that insects as a group can utilize as food is truly impressive. With their sucking or chewing mouthparts they are constantly at work on nectar, wood, tobacco, wool,

**4-20** *This stinging scorpion is an arachnid, closely related to spiders, ticks, and mites. Many evolutionists believe that the scorpion (although not this particular species) was the first air-breathing animal. The scorpion's sting is carried on its tail. (Billy Jones from National Audubon Society)*

cotton, blood, and plants and animals, both living and dead.

While modern man has unleashed a torrent of chemical compounds to control the voracious appetites of insects, he tends to forget that insects are most valuable as pollinators of flowering plants. If pollinating insects were to become extinct, so would most of our fruits and vegetables. Many pesticides hit these insects harder than the pests for which they are intended. Moreover, in several generations the pests can build up resistance to chemicals, making them virtually immune.

Spiders differ from insects in having only two body regions: a cephalothorax and an abdomen. They lack antennae and have four pairs of walking legs. Common to all members of this group of arachnoids is a specialized pair of sucking appendages, the chelicerae.

This group includes many active predators. Spiders secrete silken threads through abdominal glands, weaving webs to snare prey; scorpions make use of a poisonous sting. The horseshoe crab is an aquatic relative of the spider. Centipedes or chilopods are wormlike arthropods with a single pair of legs on each body segment. They, too, are predators, and all are endowed with a pair of poisonous fangs. Millipedes or diplopods are similar in appearance to centipedes, but they have two pairs of appendages per segment. These are mainly scavengers, feeding on decaying plant and animal material.

## Phylum Echinodermata

Although echinoderm adults are radially symmetrical, their ciliated larvae are bilateral in symmetry. The group is also characterized by an endoskeleton of calcareous plates and a water vascular system. This system draws seawater from the outside through a sievelike plate (the madreporite), pumping it through major canals and side branches. Many echinoderms also have small sucker-tipped tube feet which are associated with the water vascular system. The tube feet function primarily in locomotion, but they also serve as respiratory and sensory structures. Echinoderms are entirely marine.

Crinoids are an ancient group of echinoderms; fossil crinoids have been found in rocks at least 500 million years old. Sea lilies are filter-feeding crinoids, some of which occur at ocean depths greater than 12,000 feet. Other crinoids are the feather stars, which use their arms in swimming.

A flattened disclike body and five jointed snaky arms characterize the brittle stars, or ophiuroids. Even slight handling can cause an arm to break off. However, as in most other echinoderms, lost parts are easily and quickly regenerated.

Anyone who has walked along an ocean beach is familiar with the sea stars (often called starfish), or asteroids. These usually have five arms, or rays, although some, such as the sun star, have more. Tiny pincerlike organs called pedicellariae occur in great numbers on the upper surface of sea stars. These have a cleaning function.

Sea cucumbers or holothuroids have a thick, leathery skin with tiny embedded ossicles or

**4-21** *Sea cucumbers, or holothuroids, have a thick, leathery skin with tiny embedded ossicles or little plates that represent the much-reduced skeleton. As their name suggests, these organisms are about the same size (and sometimes even the same color) as a pickle. (Aaron O. Wasserman)*

little bones. These represent the much-reduced skeleton. Organic food is obtained by finely branching tentacles that encircle the mouth. Respiration is enhanced by an internal system, the respiratory tree.

The reason for the name given this phylum (echinoderm means spiny skin) is best demonstrated by the remaining group, the echinoids. Sea urchins have long, movable spines, enabling them to walk on the bottom. In some urchins, the spines are needlelike, break easily, and are quite poisonous. The body is enclosed in a rigid, globular shell. Sand dollars are flattened echinoids with short, fine spines.

## Phylum Chordata

Chordates include all animals having three important characteristics during all or part of their lives: a hollow nerve cord, a notochord, and gill slits derived from the pharynx. The notochord, from which the phylum derives its name, is a supporting rod of gristly material located immediately under the nerve cord.

Two groups of primitive chordate groups are recognized: the urochords or sea squirts, and the cephalochords, or lancelets. Sea squirts are marine filter-feeders that anchor themselves to rocks; they have a pharynx that is used to strain the water. Both notochord and nerve cord are lost in adult sea squirts. In the lancelet (or amphioxus), all three chordate features are well represented throughout adult life. These two groups are often referred to as invertebrates.

The third major group of chordates is the Vertebrata. All vertebrates have a bony or cartilaginous backbone which surrounds the nerve cord, and a cranium which encloses and protects the brain. In addition to the backbone and skull, vertebrates also have red blood cells, a tail (or the remnants of a tail) situated behind the anus, and often specialized sensory organs for sight, hearing, and olfaction.

Vertebrates are sorted into eight classes, one of which, the placoderms, a group of armored fish, is completely extinct. The most primitive of the living vertebrates are the agnathans or jawless fish. These bizarre, elongated animals lack a lower jaw, paired appendages, and scales. Only the lamprey eel and the hagfish represent this class.

Sharks, rays, and skates belong to the chondrichthyans, or cartilaginous fish. These have lower jaws, a skeleton that is originally cartilaginous (but may become hardened by salt deposition), and external gill slits. There are two pairs of appendages, and the skin is roughened by tiny scales. This is primarily a marine group.

The osteichthyans, or bony fish, usually have a bony skeleton. This large group may be found in both fresh and salt water. Some, such as the walking catfish, can remain out of water for prolonged periods. In bony fish the gill slits are covered by a plate, the operculum. Most have a gas-filled swim bladder which increases their buoyancy.

All remaining classes of vertebrates are referred to as tetrapods, or four-footed animals. They are basically terrestrial.

Amphibians have retained a dependence on water or moisture. This class includes the frogs, toads, salamanders, and the caecilians, which have a wormlike appearance and a burrowing way of life. All have a moist or slimy skin that is lacking in scales. The skin assumes major importance as a respiratory organ even when lungs are present. For example, many salamanders remain strictly aquatic throughout their lives, never losing their gills. Some terrestrial salamanders depend entirely on skin respiration, since they are lacking in lungs. Frogs and toads demonstrate their aquatic ancestry by breeding in water, and passing through a gilled tadpole stage.

Reptiles are better adapted to a terrestrial existence than the Amphibia. They have a dry, scaly body and often well-developed teeth and claws. The class includes the primitive tuatara,

4-22 *The owls, members of Class Aves, have a separate order all to themselves. Owls are nocturnal predators, and their eyes show a high degree of adaptation to this way of life. Their feathers are arranged loosely, so as to make very little wind noise during flight; thus, their prey often does not hear their approach. (Alycia Smith Butler)*

lizards, snakes, turtles, and crocodilians. While some reptiles bear live young, most lay a leathery shelled egg.

Birds, or Aves, are separated from other groups because of their feathers. They have a single pair of scale-covered legs, a toothless horny bill, and they maintain a constant body temperature. Their adaptation to flight has resulted in thin, light bones, numerous air sacs, and a well-developed breastbone for the attachment of flight muscles.

All mammals have hair, mammary glands, a muscular diaphragm separating thoracic and abdominal cavities, and in many cases, an external pinna, or ear. With a few exceptions, such as the egg-laying duck-billed platypus, mammals develop their young internally. Among the marsupials, such as the kangaroo, the young emerge from the female's reproductive tract in a highly undeveloped state. They complete their development within a fur-lined pouch or marsupium. All higher mammals are placental. Like the birds, mammals are homeotherms; they maintain a constant body temperature.

# Summary

The Kingdom Monera contains procaryotic cells that are either unicellular or colonial. Since these cells possess no nuclei, the parent cell simply splits in a process known as fission. One representative phylum of this kingdom is the Phylum Cyanophyta, or the blue-green algae.

Organisms within the Kingdom Protista are eucaryotic and primarily unicellular or colonial. Some may be multicellular or multinucleated, but no true tissue differentiation is present.

Several well-known phyla are included within this kingdom. Diatoms of the Phylum Chrysophyta are noted for being the principal photosynthetic food source for all animal life in the ocean. They are assumed to be responsible for over one-half of the earth's oxygen, too. The Phylum Fungi is also mentioned in relation to the saprophytic form of nutrition, whereby dead organic remains are broken down.

The eucaryotic organisms within the Kingdom Plantae are distinguished by several features such as cell walls, vacuoles, and photosynthetic pigments confined to plastids. Most members of this kingdom are autotrophic, but a few exceptions engage in saprophytic nutrition. Cell division is accomplished by mitosis or by a form of alternate sexual/asexual reproduction. This combination of sexual, or gamete-producing stages (gametophytes) alternates with asexual, or spore-producing stages (sporophytes). Plants are thus able to remain inactive during extreme environmental conditions until a suitable growth period comes about.

Several phyla, including brown algae, red algae, mosses, and liverworts, are noted within the Kingdom Plantae. Of special importance is the Phylum Euchlorophyta, or green algae, since they serve as the primary food source for animals in aquatic communities.

All remaining members of the plant kingdom are referred to collectively as vascular plants. This classification is based on the presence in the sporophyte generation of specialized conducting or vascular tissues, the xylem and phloem.

Within the vascular plant phyla, the Class Gymnospermae is characterized by exposed seeds, or gymnosperms, which are wind-carried for pollination. Flowering plants, or members of the Class Angiospermae are noted for having seeds enclosed within a saclike ovary and a reliance on insects for pollination.

Over a million heterotrophic species are included within the Kingdom Animalia. Classification is based on genetic and evolutionary relationships, and germ layers, or the layers of tissue present in the embryo.

Sponges, or the Phylum Porifera, are aquatic and somewhat outside the mainstream of animal evolution. They are unable to move, but can regenerate lost parts.

Hydroids, jellyfish, and anemones are listed within the Phylum Cnidaria. A specialized cell (cnidoblast) produces a fluid-filled sac (nematocyst) that, when stimulated, shoots out a long poisonous thread to paralyze other organisms.

Phylum Platyhelminthes contains flatworms such as turbellarians, trematodes (or flukes), and tapeworms. These animals have bilateral symmetry.

An assortment of wormlike groups falls in the Phylum Aschelminthes. These roundworms, hairworms, and rotifers have a complete digestive tract and a fluid-filled body cavity, or pseudocoel.

Worms with segmented bodies are classified within the Phylum Annelida. The most common, visible representatives are earthworms and leeches.

The Phylum Mollusca embraces a variety of soft-bodied organisms with a large mantle to protect the internal organs. Mollusks thrive in a variety of locations, and are represented in sizes from the snail to the octopus.

Members of the Phylum Arthropoda are at the top of the invertebrate phyla. Crustaceans, insects, spiders, and centipedes may be found in a variety of environments.

Spiny-skinned organisms such as sea lilies, starfish, sea urchins, or sand dollars belong to Phylum Echinodermata, and they inhabit marine waters. The adult species are radially symmetrical, but the larvae are bilaterally symmetrical.

Members of the Phylum Chordata exist in a variety of environments, and have developed a number of adaptations to survive. For instance, the primitive invertebrate groups include sea squirts and lancelets, while the vertebrate group has animals with backbones, cranium, red blood cells, and often specialized sensory organs for sight, hearing, and olfaction. Sharks, rays, and many fish in fresh and salt water fall into this latter category.

Tetrapods, or four-footed terrestrial animals embrace all chordates but the fishes. The amphibians are chordates that are dependent on water or moisture.

Reptiles such as lizards, snakes, and turtles are classified within the Phylum Chordata, as are feathered birds with bodies adapted for flight.

Lastly, organisms that have hair, mammary glands, muscular diaphragms separating the thoracic and abdominal cavities, develop their young internally, and maintain a constant body temperature (homotherms) are classified as mammals. Examples are numerous, and may range from the common rodent to man.

# Bibliography

### References

Alston, R. E. and B. Turner. 1963. *Biochemical Systematics*. Prentice-Hall, Englewood Cliffs, N.J.

Blackwelder, R. E. 1963. *Classification of the Animal Kingdom*. Southern Illinois University Press, Carbondale, Ill.

Blackwelder, R. E. 1967. *Taxonomy: A Text and Reference Book*. Wiley, New York.

Cain, A. J. 1954, 1966. *Animal Species and their Evolution*. Humanities Press, New York.

Cronquist, A. 1968. *Evolution and Classification of Flowering Plants*. Houghton Mifflin, Boston.

Hawkes, J. G., ed. 1971. *Reproductive Biology and Taxonomy of Vascular Plants*. Pergamon Press, New York.

Hennig, W. 1966. *Phylogenetic Systematics*. University of Illinois Press, Urbana, Ill.

Mayr, E. 1942. *Systematics and the Origin of Species*. Dover, New York.

Mayr, E. 1963. *Animal Species and Evolution*. Harvard University Press (Belknap Press), Cambridge, Mass.

Mayr, E. 1969. *Principles of Systematic Zoology*. McGraw-Hill, New York.

Simpson, G. G. 1953. *Life of the Past: An Introduction to Paleontology*. Yale University Press, New Haven.

Simpson, G. G. 1961. *Principles of Animal Taxonomy*. Columbia University Press, New York.

### Suggested for further reading

Briggs, D. and S. M. Walters. 1969. *Plant Variation and Evolution*. McGraw-Hill (World University Library), New York.
An excellent and well-illustrated introduction to the study of plant variation and speciation, demonstrating how these studies have permitted new insight into the processes of plant evolution.

Cain, A. J. 1954, 1966. *Animal Species and Their Evolution*. Humanities Press, New York.
A lucid and stimulating introduction to animal taxonomy and the basic processes of speciation. This book, in conjunction with Briggs and Walters, is also valuable supplementary reading for Chapter 19, "Evolution in Practice."

**Books of related interest**

Brockman, C. F. 1968. *Trees of North America*. Golden Press, New York.
A clear and understandable guide for student or amateur.

Burt, W. H. and R. P. Grossenheider. 1964. *A Field Guide to the Mammals*. Houghton Mifflin, Boston.
A clear, well-illustrated and easy-to-use field guide to the mammals of North America. Attention is drawn to the diagnostic points which differentiate species and races.

Peterson, R. T. 1947. *A Field Guide to the Birds*. Houghton Mifflin, Boston.
A classic field guide for ornithologists and amateurs.

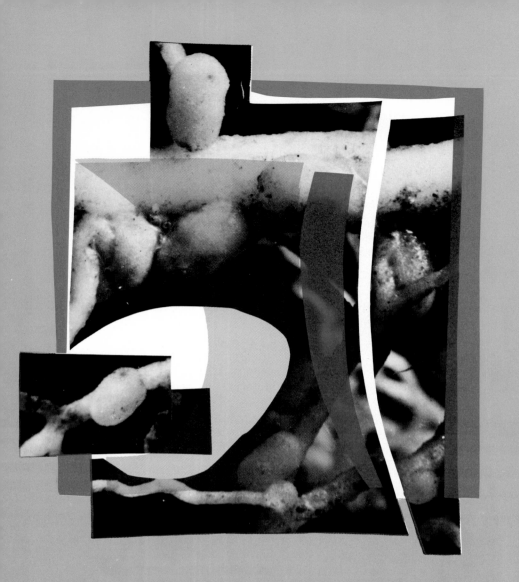

# 5
# Nutrition and digestion

All living things must take in certain raw materials available in their environment. Some substances are needed as energy sources, while others provide the structural materials from which the organism is made. The process of obtaining and utilizing these materials is called nutrition.

All living systems must take in elements and compounds that can be classified as nutrients, but the problems of nutrition vary widely in accordance with the diversity of organisms. Autotrophic organisms need only to obtain inorganic nutrients; they must take in the raw materials of photosynthesis (water and carbon dioxide) and mineral components of the organic compounds they will synthesize. Nutrition in heterotrophic organisms is much more complex. Heterotrophs must obtain the correct quantity of all the organic compounds they are unable to synthesize. This includes not only the energy-rich carbohydrates and fats but also proteins and vitamins. In addition, heterotrophs, too, must take in water and a variety of mineral elements. Another special problem of the heterotrophic way of life is that food does not always come in its most usable form and therefore must be digested. For example, when a cow eats hay, it must consume the entire dried alfalfa plant. Yet only part of that plant consists of immediately available nutrients. While getting those nutrients, the cow has also eaten all of the tough and indigestible cellulose in the plant cell walls (it will eventually be digested by bacteria that live in the cow's stomach and intestine), the waxy cuticle that protects epidermal cells, the pectin that holds cells together. Therefore the cow must be able to break down the plant and free the nutritionally valuable compounds for its own use, discarding the others as waste.

As living organisms have become differentiated, they have specialized in various modes of obtaining and utilizing nutrients. The digestive apparatus of heterotrophs is always adapted to the nutritional mode; by examining an animal anatomy, a biologist can usually predict what it will eat.

**5-1** *Heterotrophic organisms can rarely obtain food in the form of glucose, fatty acids, amino acids and other substances immediately usable by body cells. This browsing hippopotamus illustrates the need for a digestive system to break down bulk food (shoots, roots and leaves in the case of this herbivore) into nutrients that can be absorbed across cell membranes. (Leonard Lee Rue from National Audubon Society)*

## Autotrophic nutrition

Most autotrophs—green plants and chemo-synthetic bacteria—are at the mercy of their environment. They cannot travel around to find the nutrients they need; if the necessary nutrients are not in their immediate vicinity, they will quickly die. Autotrophic organisms can establish themselves only in nutritionally rich areas. Since they require only elements that are present in abundance in every type of environment, auto-

trophs can be found almost everywhere on the planet.

## Required nutrients for autotrophs

The three elements that make up most of the weight of any green plant are carbon, oxygen, and hydrogen, the elements from which most organic compounds are made; these elements account for 94 percent of the dry weight of an average plant. The plant gets its supply of carbon by taking in carbon dioxide present in the air or water; this process will be discussed in detail in Chapter 7, which is devoted to the subject of gas exchange. Plants can obtain the hydrogen and some of the oxygen that they need from water, which is readily available and easily absorbed. The balance of the oxygen they require is absorbed from the air and water in which they live.

In addition to these three elements that are

**5-1**

## Dry weight components of a plant

| Element | % of total dry weight |
|---|---|
| O | 44.4 |
| C | 43.6 |
| H | 6.2 |
| N | 1.46 |
| P | 0.2 |
| K | 0.92 |
| Ca | 0.23 |
| Mg | 0.18 |
| S | 0.17 |
| Fe | 0.08 |
| Si | 1.17 |
| Al | 0.11 |
| Cl | 0.14 |
| Mn | 0.035 |
| Undetermined | 0.93 |

the raw materials of photosynthesis, a green plant requires a variety of other inorganic nutrients. The identity of most of these elements has been known for a long time. More than a hundred years ago, biologists were conducting experiments to determine the minerals essential to plant growth and development. The most widely used research technique has been the use of a water culture. The plant is grown in a glass jar containing distilled water free of all traces of mineral impurities. The surface of the water is protected by a stopper (cork is often used) so that no minerals enter through the air. Then selected elements can be introduced into the solution, and the growth and health of the plant can be observed. If the plant cannot survive, grow, or reproduce in the absence of an element, that mineral can be said to be essential.

The principal problem encountered with this method of research has been the difficulty of obtaining water in which absolutely no mineral ions are dissolved. For many years, chlorine and sodium were considered to be unnecessary elements for plant nutrition, because plants could grow perfectly well in water culture solutions to which these elements had not been added. But recently it was demonstrated that a small number of these elements were present in ionic form in even the most carefully distilled water, and that this small amount was enough to supply the plants' needs.

The elements that are required by green plants are summarized in Table 5-2. Carbon, hydrogen, and oxygen are needed in the largest quantities; most crops require several thousand pounds of these nutrients for each acre planted. Nitrogen, phosphorus, sulfur, potassium, calcium, and magnesium are also required in relatively large amounts — perhaps several hundred pounds per acre. These nine elements are called **macronutrients.** The rest of the elements on the list are required only in very tiny amounts; they are called trace elements or **micronutrients.**

Many of these minerals and their uses are already familiar to us. We have seen that phosphorus, for instance, is a component of the lipid layer of cell membranes, and we also know that it plays an important role in the high-energy bonds of ATP. Our discussion of photosynthesis pointed out the inclusion of magnesium atoms in chlorophyll and iron atoms in cytochrome molecules. By using radioactive isotopes of elements such as iron and manganese, botanists have been able to trace their pathway through the plant; knowing the location in which these elements are used helps us deduce their function.

But molecular biologists have not yet been able to determine the precise manner in which some of these inorganic minerals function within the living cells of the plant. For example, it is known that when plants are deprived of calcium, the permeability characteristics of their membranes change, with consequent damage to the plant cells. But the reason for this change is not clear. One possibility is that the calcium ions may line the spaces of the pores in the membrane. Since calcium ions are positively charged, they may help to attract negatively charged ions (for example, iodine), thus helping to transport them into the cell. Another possibility is that the calcium cations play some part in active transport. It is known that calcium can sometimes serve as an enzyme activator, and also that active transport depends on the presence of catalyzing enzymes. For this reason, some biologists theorize that calcium activates the enzymes, lying just inside the cell membrane, that uncouple transported compounds from their carrier molecules. Proving whether any of these theories is correct is yet to be achieved.

It is quite possible that our list of necessary minerals is still incomplete. Some botanists suspect that green plants also require minute quantities of vanadium, strontium, iodine, and selenium. Many plants contain silicon, and it has been shown that when some of these plants are grown in a silicon-free solution, they become highly susceptible to fungus attack and bacterial

## 5-2

## Plant macronutrients and micronutrients

| Name | Symbol | Role in plant physiology |
|------|--------|--------------------------|
| Macronutrients | | |
| Carbon | C | important constituent of almost every molecule in the plant |
| Hydrogen | H | important constituent of almost every molecule in the plant |
| Oxygen | O | important constituent of the majority of molecules in the plant |
| Nitrogen | N | important constituent of all proteins and nucleic acids, as well as most other types of molecules |
| Phosphorus (usually in form of phosphate) | P | constituent of nucleic acids, proteins, phospholipids, cofactors |
| Sulfur | S | constituent of proteins, stimulates development of root system; important for production of chlorophyll |
| Potassium | K | important for cell division; photosynthesis |
| Calcium | Ca | translocation of food within the plant, development of middle lamellae, regulation of acidity |
| Magnesium | Mg | constituent of chlorophyll, participates in some enzymatic reactions |
| Micronutrients | | |
| Iron | $Fe^{++}$ | necessary for the production of chlorophyll, forms part of respiratory enzymes |
| Copper | $Cu^{++}$ | form necessary parts of various enzymes |
| Zinc | $Zn^{++}$ | |
| Manganese | $Mn^{++}$ | |
| Boron | B | ? |
| Molybdenum | Mo | ? |

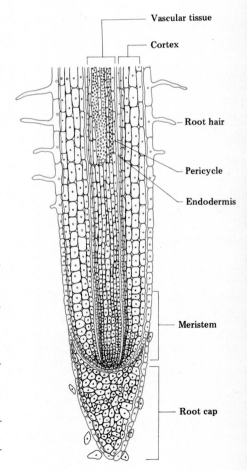

5-2 This longitudinal section of a plant root shows both the zone of cell division, near the tip of the root, and the zone of elongation, where the root hairs develop. The cells of the root cap, with their thick walls, protect the thin-walled meristematic cells, whose division is responsible for the growth of the root.

Vascular tissue

Cortex

Root hair

Pericycle

Endodermis

Meristem

Root cap

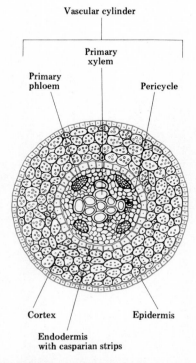

Vascular cylinder

Primary
xylem

Primary
phloem

Pericycle

Cortex

Epidermis

Endodermis
with casparian strips

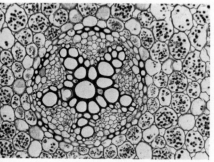

**5-3** *In this cross section of a
Ranunculus (buttercup) root we
can distinguish the major tissues
of the xylem, phloem, pericycle,
endodermis (the thinner-walled
cells of this layer are passage
cells—the thick-walled cells form
the Casparian strip) and cortex,
whose parenchyma cells contain
dark-stained starch grains.
(Courtesy of Carolina Biological
Supply House)*

disease; yet many plants grown without added silicon seem to be perfectly normal and healthy specimens. As new research techniques are developed—for example, better methods of removing all contaminating mineral ions from water and containers—we may see some of these elements added to the list of required nutrients.

## Nutrient absorption in green plants

The first autotrophs were single-celled organisms that lived in the water; they were probably quite similar to the simple blue-green and green algae we know today, the organisms that make up the scum on a stagnant pond. For these organisms, obtaining needed water and mineral nutrients is simply a matter of transporting them from the watery environment across the cell membrane.

As plants continued to evolve, some autotrophs made the move from a water environment to the land. The first land plants lived in the marshy regions around the edges of the oceans and lakes, where nutrients still could be easily procured from the water, but in time, the green plants began to adapt to an inland habitat. They developed specialized organs which enabled them to absorb needed water and mineral nutrients from the soil. The first organs of nutrient procurement were underground extensions of the stem, or **rhizomes;** through evolutionary change these structures eventually developed into true roots.

### The root system

A root is a plant organ, composed of several different kinds of tissue, and specializing in the functions of absorption, storage, and support. When a root is sliced lengthwise (making a longitudinal section) and magnified, it shows a

number of differentiated cells (see Figure 5-2). At the tip of the root is a cup-shaped group of cells, called the **root cap,** that serves to protect the tender growing tip of the root. The friction of the growing root pushing through the soil frequently damages and destroys these cells, but they are replaced rapidly. Just behind the root cap are the meristematic cells, which divide and produce the new cells that are needed for root growth. Above the meristematic tissue, in the very center of the root, can be seen the vascular tissue, the xylem and phloem. The outer layer of relatively small cells is epidermal tissue, and between the epidermis and the vascular tissue is a thick layer of parenchyma cells, called the **cortex;** these cells are often used for the storage of starch granules. Parenchyma cells in roots generally lack chlorophyll, and thus are not green; since they are underground, not exposed to light, they could not in any case make their own food. Like all root cells, they live on carbohydrates transported down through the phloem from the leaves and stems.

If we look at the same root tissue in cross section (Figure 5-3), other evidence of specialization in the root can be observed. Within the central ring of vascular tissue (often called the **stele**), there may often be found a few parenchyma cells. These cells probably serve to help conduct water and dissolved solutes into the vascular tissue. The outer parenchyma cells of the stele form a specialized layer called the **pericycle.** In most plant species, these cells mark the boundary of the vascular tissue.

The innermost layer of the cortex is called the **endodermis.** The end walls of some of these cells are heavily thickened either by a waxy substance (suberin) or by lignin. This thickened layer is called the **Casparian strip.** The function of the Casparian strip is still being debated. At one time, it was thought that its waxy coating helps prevent the seepage of water out of the xylem, but this seems unlikely. The other surfaces of these endodermal cells are quite thin and

**155**

apparently specialize in conduction, so that in spite of the Casparian strip, water could easily seep out of the xylem any time the concentration gradient favored movement in that direction.

An interesting feature of the cortex is revealed in cross section. The parenchyma cells in this layer are not crowded closely together as they are in other parts of the plant, such as the leaves; rather they are loosely packed, with many spaces between the cells. These intercellular spaces are alternate pathways of water conduction, enabling a rapid bulk flow to the xylem.

As roots age, an outer cambium layer called the **cork cambium,** begins to lay down heavily thickened protective cork cells. The growth of these cork cells often destroys many of the cortical cells, and the thick cork covering also prevents the movement of water into the central vascular tissue. At this stage of development, the cork-covered portion of the root has become specialized as a mechanical support, helping to hold the plant upright, and it has lost its function as an organ of absorption.

Most water is absorbed in an area just behind the growing tip of the root. In this region, all epidermal cells absorb water through their membranes. In addition, certain epidermal cells specialize by growing elongated protuberances called **root hairs.** The shape of these cells, with a high surface-to-volume ratio, is especially adapted to absorption. Root hairs vary in length from less than a millimeter to more than a centimeter; they are usually no more than 10 microns thick. Unlike most epidermal cells, they have no waxy cuticle, but they have an especially heavy coating of pectin which serves to bind particles of soil to the root hairs. A root with no root hairs would be a relatively inefficient absorptive organ. It has been estimated that the presence of root hairs increases the absorptive surface of roots by at least 20 times.

The delicate root hair cells have a short life span; their cell walls and membranes are easily torn by friction or pressure, causing immediate

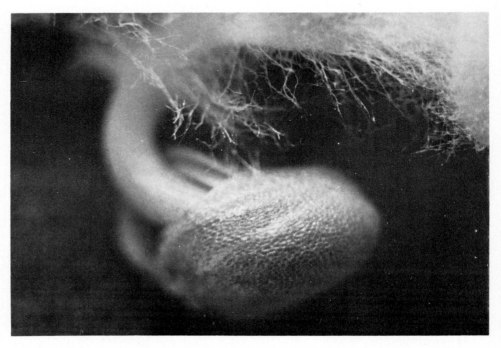

death to the cell. In most plants the root hairs live only a few weeks. But new ones constantly develop, to replace the old ones that have died. In this way the roots abandon areas of the soil that have been depleted of nutrients and begin "eating" in a new fertile location. It has been estimated that a single rye plant contains more than 14 billion root hairs, with a total root area of about 6,500 square feet. Yet this whole system can be contained within two cubic feet of soil.

A variety of different root patterns can be found in green plants. The avocado tree, for example, has a long central root that goes straight down, deep in the soil; this is called a **taproot system.** A rosebush has roots that branch and spread in a wide but shallow pattern; this is called a **fibrous root system.** In begonias, roots can be seen growing out of the leaves (**adventitious roots**); these serve as a supplementary absorptive system. Another kind of adventitious

*5-4 Certain epidermal cells specialize by growing elongated protuberances called root hairs. The shape of these cells, with a high surface-to-volume ratio, is especially adapted to absorption. Root hairs vary in length from less than a millimeter to more than a centimeter; they are usually no more than 10 microns thick. Developing seedlings have so many root hairs they almost appear to be furry. (Alycia Smith Butler)*

**5-5** *Carrots are a good example of a plant with a taproot system; this root serves to store food as well as absorb water. Both corn and maize have prop roots growing from the lower portion of the stem, providing good support in addition to absorption. Adventitious roots are typically found in plants that grow low to the ground in a spreading growth pattern. Fibrous roots may grow either close to the surface or relatively deep, depending on the amount of rainfall and the level of the water table.*

root can be seen in maize, where **prop roots** develop along the stem, providing both additional water absorption and some degree of mechanical support. The extent of root growth and the pattern of root development are also greatly influenced by the environment. When winter wheat is grown in a wet area, its roots will be quite deep, but when the same plant is grown in a dry climate, the root system will be much wider and shallower, since the plant needs to absorb as much surface rainwater as possible.

## Absorption of water

Water is absorbed by the root hairs and by the epidermal cells in the root hair zone of the young root. Much of the movement of water into the root cells can be accounted for in terms of osmosis along a concentration gradient. In the parenchyma cells of the plant leaf, water is used in photosynthesis, and it can also be lost through evaporation when the stomata are open. This water loss

causes an increase in the concentration of sugar and other osmotically active solutes in the leaves. A concentration gradient is set up, moving water out of the xylem and into the leaf cells. This upward pull on the water in the xylem in turn sets up a similar concentration gradient in the roots, with the highest concentration of solutes being found in the cells nearest the xylem vessels.

Water is absorbed by the epidermal cells, because they contain a higher concentration of osmotically active solutes than is found in the soil outside the plant. Once the water has entered the epidermal cells, it continues to move with the concentration gradient, diffusing from the outer root cells, where water is in relatively high concentration, to the inner root cells, where the water is in relatively low concentration. As soon as it moves into the xylem, the water is transported away from the root and up into the stem and leaves, so the gradient is maintained, allowing the process of absorption to be carried on continually.

Water can enter the root by two different pathways of absorption. The water can pass

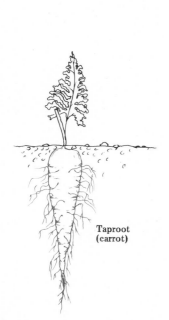

Taproot
(carrot)

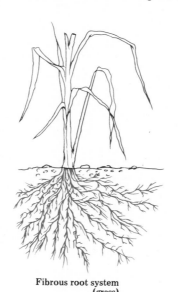

Fibrous root system
(grass)

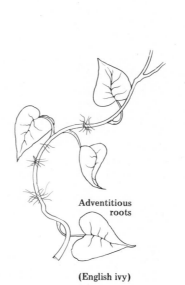

Adventitious
roots

(English ivy)

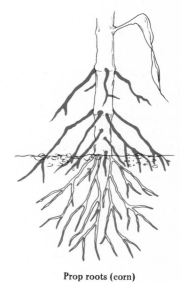

Prop roots (corn)

**157**

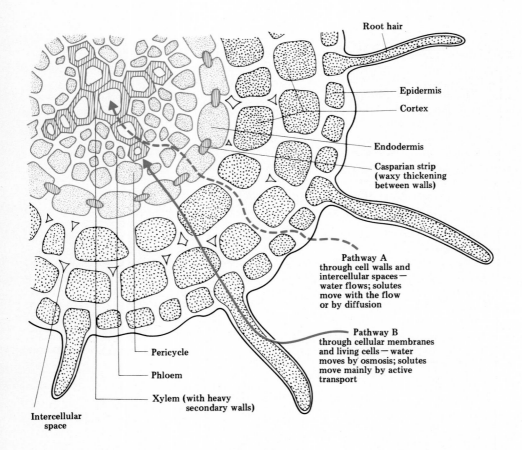

Root hair

Epidermis

Cortex

Endodermis

Casparian strip
(waxy thickening
between walls)

**Pathway A**
through cell walls and
intercellular spaces—
water flows; solutes
move with the flow
or by diffusion

**Pathway B**
through cellular membranes
and living cells—water
moves by osmosis; solutes
move mainly by active
transport

Pericycle

Phloem

Xylem (with heavy
secondary walls)

Intercellular
space

**5-6** *This cross section of a root
shows the different paths by which
water can move into the central
stele. It can move by osmosis
through the parenchyma cells of
the cortex, and it can also move
through the intercellular spaces by
the process of imbibition. Since the
movement of water into the xylem
often occurs at a very rapid rate,
plant physiologists assume that
some mechanism of active
transport is also involved.*

through the cells of the epidermis and cortex, crossing each cell membrane by osmosis. This pathway accounts for much of the water that enters the plant. Water can also go around the cells, through the intercellular spaces. The cellulose fibers in the cell wall can act as a sort of sponge, soaking up the water and helping it pass along the outer edges of the cells. However, when the water reaches the endodermis, it must then move into those cells by osmosis, since the thick Casparian strips prevent further intercellular transport. This final passage through the endodermis guarantees that all absorbed water will be screened by a differentially permeable membrane before it enters the xylem.

In addition to absorption, there is another mechanism at work in plant roots that helps to bring water into the xylem. We have already noted that much of the contents of the living cell exist in colloidal form. Many colloids, including cellular ones, have a marked affinity for water; they are **hydrophilic.** Since hydrophilic colloids attract water, the water does not actually enter the colloidal mixture of particles but simply coats the entire outside surface of the group of molecules. Such water is said to be **adsorbed;** the process of water adsorption by colloids is called **imbibition.** The outer root cells imbibe water from the soil, and this imbibition causes a marked increase in the volume of cell contents. As the volume increases, turgor pressure (the pressure of the cell contents against the cell wall) also increases. This pressure quite literally pushes water out of the full cell and into a neighboring cell with a lower volume of cell contents. In this way, the imbibed water is pushed along a pressure gradient until it finally reaches the xylem.

Absorption of water by osmosis and imbibition may take place even when there is no upward pull on the water in the xylem tissue. It has been demonstrated that if the top of a plant is cut off and its root system is immersed in a sealed container of water, liquid will begin to exude through the cut stem. The amount of liquid that comes out of the stem is roughly equal to that removed from the container. In other words, the roots are acting like a pump, and even with no upward pull on the water in the xylem, enough pressure can develop inside the xylem to force the water through. This phenomenon is sometimes called **root pressure.** It is root pressure that is responsible for the phenomenon of **guttation,** the exuding of occasional droplets of water from the leaves of green plants when water is freely available to the plant.

## Absorption of mineral nutrients

With the exception of carbon, hydrogen, oxygen, and nitrogen (the four most abundant elements in plants), all necessary nutrients enter the plant in the form of ions dissolved in water. Some minerals are brought into the plant by passive transport alone. Each of these ions enters in response to its own individual osmotic concentration gradient. Once they are inside the plant, they may be changed into another form through reactions with other elements. Their conversion helps maintain the gradient. Some ions are found within the vacuoles of root cells; presumably, this imprisonment makes them osmotically inactive. Certain ions, such as potassium, move into the cell against a concentration gradient. This is

an energy-consuming process of active transport. Each nutrient enters at its own rate, regardless of the rate at which water is absorbed.

These nutrients absorbed from the soil were components of the rocks from which the soil was derived, and in most parts of the earth, they are present in abundant quantities. If the cycle of nature is allowed to take its course, no shortage of these minerals will ever occur in the soil. The nutrients that were removed by growing plants will be returned to the soil after the plant dies and decomposes (or after the animal that eats it dies and decomposes). But modern farming methods have created the problem of soil depletion. Crops are picked and shipped far away to be eaten. Instead of being returned to the farmland, the minerals in corncobs and corn husks are mixed in with aluminum cans and plastic bottles and buried in a landfill. The same is true of the livestock that eat the plants; the minerals contained in their bodies become part of the garbage problem rather than a valuable fertilizer for agricultural land. This is why modern large-scale farming must depend so heavily on the use of chemical fertilizers to replace lost nutrients, such as potassium, phosphorus, and nitrogen.

Chemical fertilizer must be used very carefully, or it can cause problems worse than those it is meant to cure. If large amounts of fertilizer are placed next to the roots, then the osmotic concentration will be higher in the soil than it is inside the plant; water will move out of the plant rather than into it, causing the plant to wilt and die. Another danger is the possibility of fertilizer "burn," the poisoning of root cells by high concentrations of salt.

Ever since the end of the nineteenth century, when chemical fertilizer first came into use, some farmers and gardeners have believed that plants grown with their aid were somehow less healthy than plants grown with manure or compost for fertilizer. Today the interest in organic farming is greater than ever, and in many cities, people are willing to pay twice as much for fruits and vegetables that have been organically grown. Yet many are convinced that buyers of organic food are wasting their money. What can the biologists tell us about this controversy?

Most botanists reject the idea that plants grown with organic fertilizer are more nutritious than plants grown with chemical fertilizer. They point out that nothing essential is missing in the chemically fertilized plant; for if some unknown trace element was missing, the plants either would not grow, or would become deformed or diseased. They also point out that many elements are not directly absorbed in the form in which they are applied to the soil; the elements are first changed into various ions by soil bacteria. Therefore the chemical the plant actually absorbs will be the same no matter what the ultimate source was. In short, healthy plants are all alike, and it is most improbable that one could contain a health-giving substance that another one of the same variety lacks.

Although biologists generally reject the claim that organically grown plants are healthier than other plants, many of them do support the organic food movement, for somewhat different reasons. Organic food is always grown without the use of pesticides, such as DDT, or herbicides, such as 2,4-D, and there is general concern over the effects of these chemicals on the natural environment. Moreover, some of the ingredients of the chemical fertilizer itself can be hazardous in large accumulations. The nitrate in chemical fertilizer can be changed to nitrite by intestinal bacteria, and nitrite in the blood combines with hemoglobin so that it can no longer carry oxygen. Recent studies by the Public Health Department have established that the rates of death by nitrite-caused asphyxiation are rising rapidly in agricultural communities, especially among infants, the most susceptible group. The accumulated phosphorus content of fertilizer also threatens the environment. Phosphate ions dissolve in soil water and then drain into creeks and rivers. There the phosphate fertilizes algae, causing a "bloom," or explosion of algae growth. This sets off a chain of events that ends in the consumption of all oxygen in the water and the death of nearly all the organisms in the water.

## The special problem of nitrogen

Nitrogen is not an original component of the soil. The ultimate source of the nitrogen that is used in all plant and animal proteins is the nitrogen gas ($N_2$) in the atmosphere. Green plants are unable to utilize nitrogen gas from the air because it is so stable that it does not readily enter into any chemical reactions. Even in the laboratory, chemists have trouble inducing a molecule of $N_2$ to split apart and react with other elements; they must apply extremes of heat and pressure or use relatively large quantities of an inorganic catalyst, such as titanium, before this inert gas can be made to react.

How, then, is atmospheric nitrogen transformed into a compound that can be used for protein synthesis by green plants? The answer lies in the action of certain bacteria that live in the soil. These bacteria absorb the $N_2$ gas through their cell membranes. Although these bacteria live underground and are not directly exposed to the air, they can obtain the $N_2$ from tiny pockets of air that are interspersed through the soil by the tunneling of earthworms or the action of a plow. Once the $N_2$ molecule is inside the cell, it is split by enzymes and cofactors, causing a reaction between the N atoms and other compounds that produces usable nitrogen compounds such as ammonia ($NH_3$). This may be absorbed directly by plants, or it may be converted by other microorganisms into nitrates ($NO_3^-$) which are then used by plants.

These nitrogen-fixing bacteria actually live in the roots of certain green plants. They invade the root tissue, and the infection stimulates the formation of irregular white knobs, called **nodules,** on the roots. Within these nodules, the bacteria

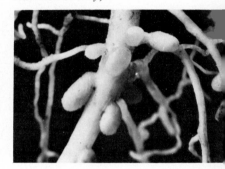

**5-7** *The knobby white lumps on the roots of this legume make the plant look diseased—which in a sense it is. The nodules are caused by an invasion of nitrogen-fixing bacteria into the root tissue. This is one "disease" that is beneficial, since the bacteria supply needed nitrogen to the plant, and to the herbivores that eat the plant. (Hugh Spencer from National Audubon Society)*

perform their work of nitrogen fixation. They supply the host plant with needed nitrogen, and they live by absorbing dissolved sugar contained in the root tissues. This relationship is a form of symbiosis (living together) called **mutualism** in which both parties benefit from the association. Interestingly, when these same bacteria are found in the air or soil, living outside the plant roots, they are no longer able to carry out the fixation of nitrogen.

Peas, beans, alfalfa, clover, and peanuts—the legume family—are examples of plants that can be hosts to nitrogen-fixing bacteria, and it is through these plants that nitrogen enters the life chain. When these plants, and the animals that eat them, decompose, their nitrogen is deposited in the soil or water and provides a source of nitrogen for other growing plants.

A rather peculiar group of plants, the insectivorous green plants, have specialized in another method of obtaining nitrogen. They trap small insects in their leaves, or other differentiated structures, and then secrete digestive enzymes that break down the animal's protein content. Fragments of the protein molecules and other mineral nutrients that are absent in the soil are then absorbed through the epidermal cells of the leaf and used in building plant proteins. Examples of such plants are the Venus' fly trap and the sundew. They are typically found in areas such as bogs, where the nitrogen content of the soil is extremely low.

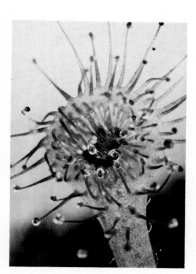

**5-8** *A rather peculiar group of plants, the insectivorous green plants, have specialized in another method of obtaining nitrogen. They trap small insects in their leaves, or other differentiated structures, and then secrete digestive enzymes that break down the animals' protein content. Fragments of the protein molecules and other mineral nutrients that are absent in the soil are then absorbed through the epidermal cells of the leaf and used in building plant proteins. (Courtesy of Carolina Biological Supply House)*

## Heterotrophic nutrition

Heterotrophic cellular organisms evolved long after autotrophic cellular life was well established. The first heterotrophs may have retained a limited ability to synthesize certain foods when they were unable to find other food sources, but as they adapted to their specialized way of life, heterotrophs lost the ability to synthesize their own energy sources from inorganic materials. Many of the more specialized heterotrophs have even lost the ability to synthesize other needed organic compounds, such as amino acids and unsaturated fats. Therefore the nutritional requirements of most heterotrophs are quite extensive.

## Carbohydrates

The basic energy source of most heterotrophic activities is the oxidation of carbohydrates, compounds which they are unable to synthesize from inorganic low-energy components. To provide adequate fuel for daily metabolic activities, about one-half to two-thirds of the total calories an animal eats must be carbohydrates. If the daily intake is insufficient to meet current needs, the animal may draw on stored reserves; if the daily intake is in excess of current needs, then the surplus will be stored. In animals whose methods of obtaining food do not require much movement, such as clams or sea anemones, the excess may be stored in the form of polysaccharides. In mobile animals, such as turtles or jaguars, the excess is usually converted to fat for storage. This is a more compact storage system, since a gram of fat contains twice as much energy as the same weight of carbohydrates.

## Lipids

Although a lipid can be produced by rearranging a carbohydrate and altering the ratio of hydrogen to oxygen, some animals are unable to synthesize certain lipids, even when they eat a carbohydrate-rich diet. For example, rats cannot make linolenic acid, a fatty acid which is a component of several important lipid compounds. Man is unable to synthesize at least three necessary fatty acids—linoleic acid, linolenic acid, and arachidonic acid—and so these are an essential part of his nutritional requirements.

**161**

**5-3**

**Dry weight components of a 70 kg. adult man**

| Components | Number of grams |
| --- | --- |
| Water | 41,400 |
| Fat | 12,600 |
| Protein | 12,600 |
| Carbohydrate | 300 |
| Na | 63 |
| K | 150 |
| Ca | 1,160 |
| Mg | 21 |
| Cl | 85 |
| P | 670 |
| S | 112 |
| Fe | 3 |
| I | 0.014 |

## Proteins

Only plants have the ability to turn inorganic nitrogen into organic compounds; most heterotrophs have lost the ability to synthesize all of the amino acids that constitute the building blocks of needed proteins. Dietary requirements vary from species to species. Rats need to consume 10 different amino acids, but only eight are essential for human nutrition. Cows need none, thanks to a mutualistic relationship they have established with several varieties of bacteria and protozoa. These organisms live in the cows' digestive tracts and produce all 20 of the common amino acids for them.

## Vitamins

The term "vitamin" was coined to describe a group of organic compounds that, in trace amounts, were shown to be essential to the healthy functioning of an organism. Strictly speaking, a vitamin is a substance that must be consumed as part of the diet, but the term is commonly used as a name for all the chemical compounds that have been identified as necessary in human nutrition. Thus ascorbic acid is always called vitamin C, even when we are talking about animals that can synthesize it.

The relationship between nutrition and certain deficiency diseases has been understood for centuries. Physicians in ancient Greece, for example, treated patients who had night blindness (caused by a deficiency of vitamin A in the diet) by feeding them chopped fish liver, which is rich in vitamin A. Of course, the Greeks did not know why this cure worked, but their observations had already established the connection between a state of good health and an adequate diet. In the late nineteenth century, investigation of the role of vitamins in animal nutrition was carried forward on a more scientific basis, using laboratory animals that were fed vitamin-deficient diets. It was discovered that many conditions that had been considered diseases, such as scurvy and beriberi, were actually symptoms of some vitamin deficiency. Current research is investigating the possibility of a link between vitamin deficiency and schizophrenia.

Chemists began the work of isolating vitamins at about the turn of this century, but at that time they were unable to determine their chemical structure; this is why they could not give the compounds any better names than the letters of the alphabet, to indicate the sequence in which they were isolated. The chemical and structural formulas for all the vitamins known to be necessary in human nutrition have been established, and most of these are now routinely synthesized in the laboratory, to be sold as dietary supplements. From sheer force of habit, we continue to use the old, imprecise names, adding numbers to the letters (vitamin $B_1 \ldots _{12}$, for example) as variant forms are discovered.

Table 5-4 lists the vitamins required in the human diet. They are divided into two categories —those which are soluble in water, and those which are soluble in fat. The water-soluble vitamins must be consumed frequently, because there is no mechanism for storing them. The fat-soluble vitamins, such as vitamins A and D, can be stored in droplets of fat in the liver, and so they are not really a daily requirement. There is some clinical evidence that excessive doses of fat-soluble vitamins can accumulate in the tissues through this type of storage and cause harmful effects.

The chart also shows the role each vitamin plays in the function of the organism, where it is known. Vitamin A combines with a protein in the eye to form a pigment that helps catch and use light; it also plays a role in the maintenance of epithelial tissues and in growth of bones and teeth. Vitamins in the B complex are utilized in one complex set of reactions that takes place during cellular respiration: for example, vitamin $B_2$

**5-4**

**Vitamins**

| Symbol | Name | Source | Role | Disease caused by deficiency |
|---|---|---|---|---|
| *Water-soluble* | | | | |
| $B_1$ | thiamine | grain with husks, peas, beans | carboxylations, decarboxylations | beriberi |
| $B_2$ | riboflavin | liver, yeast, wheat germ | dehydrogenation reactions | |
| | niacin | most types of meats | biological oxidations | pellagra |
| | folic acid | ? | synthesis of nucleic acids and amino acids | anemia, growth failure |
| | pantothenic acid | yeast, liver, eggs, milk | forms part of coenzyme A which participates in group-transferring reactions | unknown |
| $B_6$ | pyridoxine pyridoxal pyridoxamine | wheat germ, egg yolks, yeast, liver, kidney | transaminations | convulsions in infants

*(continued)* |

| Symbol | Name | Source | Role | Disease caused by deficiency |
|---|---|---|---|---|
| $B_{12}$ | cobalamin | liver, meats | variety of enzymatic reactions | pernicious anemia |
| C | ascorbic acid | fresh fruits and vegetables | biological oxidation | scurvy |
| *Fat-soluble* A | retinol | green and yellow vegetables | constituent of visual pigments; all tissues require it, but function unknown | cornification of cells, atrophy of some tissues |
| D | calciferol ($D_2$) | fish oil | calcification of bone | rickets |
| E | tocopherol | plant oils, green leafy vegetables | not known | defective fat absorption |
| K | phylloquinone | intestinal flora, fish, grain | not known | failure of blood to clot |
| F | essential fatty acids | plant, animal fats | part of coenzyme, $\alpha$-lipoic acid | |
| Q | ubiquinone | | unknown | |
| H | biotin | beef liver, yeast, peanuts, chocolate | carboxylation reactions ($CO_2$ fixation) | "egg-white injury" syndrome |

is a cofactor that participates in a decarboxylation step of the Krebs cycle. Vitamin D, which in adults can be synthesized in the body, is necessary for the absorption of calcium in the small intestine. Under experimental conditions in the laboratory, a deficiency of vitamin E can cause the destruction of red blood cells in rats.

Although much research has been done on the function of vitamins, a great deal of mystery still surrounds their role in animal physiology. Vitamin E is a good illustration of this problem.

It was discovered in 1922, when H. M. Evans and K. S. Bishop found it to be a necessary part of the diet of laboratory rats. Since then, a variety of claims have been made for this vitamin's effectiveness in humans. It has been said to cure sterility, diabetes, impotence, and severe burns, as well as to retard the process of aging in arteries and skin. None of these claims has ever been proven clinically; on the other hand, neither has it been proven that vitamin E does *not* do these things. The National Research Council says that

Vitamin E is an essential nutrient and lists the recommended daily intake. Yet no case of actual deficiency has ever been reported (except in a very small number of premature infants), and no one has any idea what the symptoms of deficiency might be. Clearly, much more research is needed in this area.

Humans must consume vitamins as part of their daily food intake, but a number of unicellular protozoa and bacteria have retained the ability to synthesize all necessary vitamins. It is common for multicellular animals to depend on these microorganisms as a vitamin source in a symbiotic relationship. For example, intestinal bacteria in man synthesize most of his daily requirement of vitamin K, which plays an important role in blood clotting. Under normal circumstances, therefore, this vitamin is not a necessary part of the daily diet. One harmful side effect of many antibiotic drugs is that they kill these bacteria, creating a vitamin K deficiency, which will last until the bacteria population has a chance to reestablish itself. For this reason, hospital patients are routinely given an injection of vitamin K before surgery, to guard against the possibility of hemorrhage.

Many plants have been shown to contain substances chemically similar to human vitamins. Since these compounds are synthesized by the plant, they cannot really be called plant vitamins—they are not an essential part of plant nutrition. It is assumed that these compounds play an important role in the metabolic processes of plants, as they do in animals, but this has proved difficult to demonstrate.

## Water

Even though animals do not use water as a raw material for the synthesis of organic compounds in the way that plants do, they still need to take in water. Water makes up most of the weight of the animal body—in some animals, it accounts for as much as 95 percent of total weight. The constant loss of water through evaporation and excretion creates a need for replacement.

**5-9** *Water comprises 80 to 95 percent of all living matter. It is the essential environment of all life processes and a factor in many chemical reactions. These drinking kangaroos illustrate the need of terrestrial animals to replenish their water supply periodically, daily in most cases. (Courtesy of E. Slater, Division of Wildlife Research, CSIRO, Canberra, Australia)*

## Inorganic nutrients

Heterotrophs also require specific inorganic components in their diet. These mineral elements, and the role they play in animal metabolism, are listed in Table 5-5. As with plants, the required elements may be categorized as either micronutrients or macronutrients. The intake of these elements must be carefully regulated, since in many cases too much can be just as harmful as not enough. For example, copper, which is needed in an enzyme that catalyzes the formation of hemoglobin, can poison cells if it is present in excess.

Many of the elements required by heterotrophs, such as potassium, phosphorus, magnesium, and zinc, are also required by green plants. These elements will automatically be obtained along with the other nutrient compounds furnished by the green plant, and deficiencies in these elements are rarely seen in animals. Minerals that are not usually found in plants—iodine, for instance—must be obtained as dietary supplements, and deficiencies in these elements are much more common. In parts of the world where iodine is not a normal constituent of the soil or water, such as Switzerland, the large goiters caused by lack of iodine in the diet are seen in an unusually high percentage of the population.

**5-5**
## Mineral requirements of humans

| Name | Symbol | Role in animal physiology |
|---|---|---|
| Calcium | $Ca^{++}$ | muscle contraction, bone formation, membrane permeability |
| Chloride | $Cl^-$ | production of HCl in stomach |
| Cobalt | $Co^{++}$ | forms part of vitamin $B_{12}$ |
| Copper | $Cu^{++}$ | participates in some enzymatic reactions |
| Fluoride | $F^-$ | necessary for dental health |
| Iodine | $I^-$ | thyroid hormone production |
| Iron | $Fe^{+++}$ | production of hemoglobin, myoglobin, cytochromes |
| Magnesium | $Mg^{++}$ | participates in some enzymatic reactions |
| Manganese | $Mn^{++}$ | participates in some enzymatic reactions |
| Phosphate | $PO_4^-$ | production of nucleic acids, phospholipids |
| Potassium | $K^+$ | transmission of action potentials |
| Sodium | $Na^+$ | transmission of action potentials |
| Zinc | $Zn^{++}$ | participates in some enzymatic reactions |

**5-10** *Dung beetles roll a mass of animal feces into a small ball. They will transport it to the nest, where it will be stored for food. These beetles serve as an example of highly specialized behavioral, ecological and nutritional adaptation. (Aaron O. Wasserman)*

## Heterotrophic feeding mechanisms

How do animals obtain their nutrients? The mental picture most of us see when we think of a feeding animal is typified by the ferocious carnivore—a Bengal tiger tearing into the unfortunate goat that is its dinner. But that is only one kind of animal feeding mechanism. The heterotrophic way of life does not necessarily require either great mobility or evident ferocity. Some animals feed by sipping the sugary sap of plants; others peacefully munch grass and leaves; others filter the soil or water and extract tiny microorganisms and food particles. A variety of special organs and adaptations have evolved to meet heterotrophs' needs for obtaining nutrients.

## Mechanisms for feeding on small particles

Some animals are adapted to capture very small food particles, such as bacteria, unicellular algae, the eggs and larva of small invertebrates, or tiny bits of organic material from decomposing plants and animals. Since their food consists of both plant and animal tissues, small-particle feeders are classed as omnivores (eaters of everything). The majority of these small organisms that serve as food are aquatic, and they are most numerous in salt-water environments. Although these food particles are tiny, the animal that feeds on them need not be. Many of the large whales are small-particle feeders.

The pseudopodia (false feet) of the amoeba are one kind of adaptation for feeding on small particles. When the amoeba encounters a tiny bit of food, the pseudopodia surround the particle and engulf it, using phagocytosis to bring it into the cell for digestion. Flagella and cilia are also adaptations for feeding on small particles. The sponge, for example, depends on the action of beating flagella to push food particles into its intercellular spaces, or pores; there special phagocytic cells will digest and absorb the food. Ciliary feeding mechanisms are especially common among larval forms of many animals, and they are also found in some worms, in bivalve mollusks, such as oysters and clams, and sea lilies. The cilia move rhythmically to create a current that sweeps the water and food particles it contains past a special feeding organ. This organ typically secretes a film of mucus that traps the floating particles, and then both mucus and particles are ingested. This kind of feeding is generally a continuous process; the cilia never stop moving.

Some marine animals, such as the sea cucumber, use waving tentacles rather than cilia. The tentacles are coated with mucus, which virtually glues the food particles to the epidermal cells. Periodically, the tentacles deposit the food-filled mucus in a special mouthlike organ, where it is digested.

A more elaborate use of mucus as a feeding mechanism is seen in a number of marine mollusks, such as whelks and limpets. These animals secrete threads of mucus that act exactly like

**5-11** *Phagocytosis, or ingestion by an amoeba of a large nutrient particle (in this case the bacterium Bacillus megaterium), begins when a pseudopod moves toward and around the particle. The amoeba streams forward and finally surrounds the particle in a digestive chamber or vesicle (last picture). (Courtesy of Dr. René Dubos, from* The Unseen World, *Rockefeller University Press, New York 1962)*

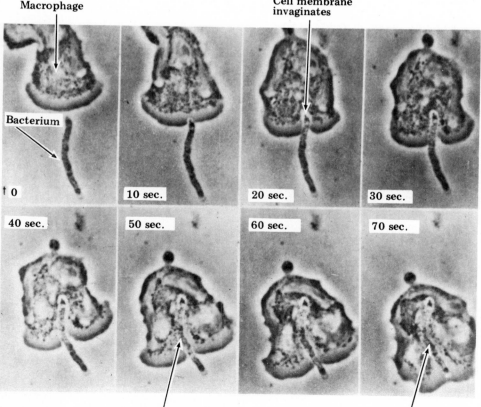

Macrophage

Bacterium

Cell membrane invaginates

0

10 sec.

20 sec.

30 sec.

40 sec.

50 sec.

60 sec.

70 sec.

Bacterium becomes engulfed

Bacterium totally enclosed in a vacuole

**167**

a net hung in the water; all passing microorganisms are entrapped. Then the mucus and the particles can be sucked back into the stomach, where the food is digested. The feeding mechanism of these small-particle eaters emphasizes the point that mobility and ferocity are not necessary accompaniments to the heterotrophic way of life.

Another widespread adaptation for feeding on small aquatic particles can be seen in the filtering systems found in many small invertebrates, as well as in larger fish and a few birds. For example, many of the whales feed by swallowing huge gulps of water and then forcing the liquid back out through a bony strainer (the baleen). The water passes through the baleen, but all small food particles remain in the whale's mouth and are soon swallowed. Flamingos feed in much the same way. Some fish with gills, such as the herring and the basking shark, have modifications of their gills that allow food particles to be filtered out as water passes over them. This efficient system eliminates the need for separate food-gathering organs, allowing an already existing structure to serve two functions.

## Mechanisms for feeding on fluids and soft tissues

Some animals are adapted to feed on the fluids and soft tissues of other organisms, both plant and animal. Aphids, for example, feed by piercing the leaves and stems of plants and sucking out the carbohydrates that are dissoved in water within the parenchyma cells or the vascular tissue. Other insects, such as butterflies and bees, simply suck the nectar and other juices that are secreted by the flowers of green plants. These animals can feed many times without destroying the plant's ability to continue producing food. In fact, some of these insects (bees, for example) have become a necessary part of the plant's life cycle; they carry pollen from one plant to an-

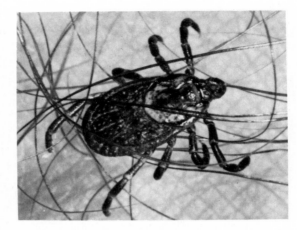

**5-12** *Some animals are adapted to feed on the fluids and soft tissues of other organisms, both plant and animal. This wood tick (Dermacentor variabilis) is one example. This arthropod is related to the spider, as evidenced by the fact it has eight legs. They have no chewing mouth parts; instead they have two modified appendages, called chelicerae, that can pierce the skin and a mouth adapted to sucking blood. (John H. Gerard from National Audubon Society)*

other, serving as an important link in the reproductive process.

Many animals that specialize in fluid feeding are symbionts (animals that participate in a relationship of symbiosis). Although some provide beneficial services, such as vitamin synthesis, to the host animal, many weaken or harm their host, and some even endanger the host's life. This relationship in which one organism is helped and the other is harmed is called **parasitism.** Although some parasites do kill their hosts, such an action poses an obvious threat to the existence of the parasite, since it loses its food supply when the host dies and may die itself unless it finds another host quickly. The most successful groups of parasites are those that have already learned the lesson that you should not kill the goose that lays the golden eggs.

One class of parasites, the **endoparasites,** pass their entire lives within the body of some host animal or plant. These parasites often lack a differentiated feeding or digestive organ; they simply absorb the nutrients that are present in their environment through their cell membranes. **Ectoparasites** live outside the host organism, often on the skin. These include such animals as lice and ticks, which remain on the host for long periods. Some zoologists also classify as ectoparasites mosquitoes and vampire bats, organisms which visit a host for a meal and then depart. More often, these animals are classed as specialized predators.

Animals that feed on fluids and soft tissues typically have highly specialized piercing and sucking feeding organs. They may have needle-like mouth parts, as is seen in the mosquito, or muscular sucking mouths, or razor-sharp jaws like the leech. Another adaptation to this type of feeding is the ability to secrete anticlotting agents into the host's blood at the point of puncture; otherwise the natural clotting action of the blood would rapidly seal up the puncture, shutting off the supply of nutrients. Since the amount of anticlotting agent is small, there is no danger of disabling the host.

## Mechanisms for feeding on large particles

Many animal species feed on large particles of multicellular organisms, sometimes taking in a whole organism at a single gulp. Animals that feed in this way exhibit a variety of specializations that help them seize, masticate, and consume their prey.

Herbivores (plant eaters) do not usually need great mobility to obtain their food, since the grass or seaweed they eat is not likely to run away

from them (they may, however, have to run fast to avoid becoming some other animal's food). All herbivores must have mouth parts that can tear and crush plant tissues, and they must have some means of locomotion from one plant to the next; in marine organisms, this may be accomplished by riding the current. Herbivorous mammals often have broad grinding teeth, to help shred the tough plant cell walls before the food enters the digestive tract.

The typical carnivore (animal eater), on the other hand, relies on its mobility to seize prey. Wolves, owls, and barracuda are examples of swift predators, but even the carnivores that live on dead animals (carrion eaters), such as the hyena and the vulture, must have considerable mobility to enable them to search for their specialized food supply. Carnivores usually have teeth and jaws that are adapted for tearing and biting. Other specialized predatory weapons that have evolved in carnivores include poisonous stings (the jellyfish, for example), or sharp pincers and claws (the lobster), or a long sticky tongue that can be flicked out like a bullwhip (the anteater). Spiders construct nets to trap their prey, and some insects dig holes in the sand and lie in wait at the bottom to seize the smaller animals that fall in.

Few carnivores can be congratulated on the delicacy of their feeding habits. Many large-particle feeders swallow their prey whole, with-

**5-13** *A feeding predator may encounter unexpected defenses in its prey, although this example may be a bit hard to swallow. (By permission of John Hart and Field Enterprises, Inc.)*

out any preliminary chewing; corn snakes, for example, swallow whole mice, and carnivorous sharks swallow chunks of flesh weighing several pounds. A second category of large-particle feeders are animals that break the food into smaller pieces by chewing and also by secreting enzymes in the mouth that speed the chemical breakdown of food; man falls into this group. A third category of animals begins the digestive process outside the organism. For example, spiders inject strong digestive enzymes into their prey and then suck up the dissolved tissues.

### The special case of heterotrophic plants

Not all plants contain chlorophyll; some plants have evolved a heterotrophic way of life. Examples are the molds and fungi such as mushrooms. There are also some heterotrophic flowering plants. These plants may grow in decaying organic tissue, such as a tree stump or an animal carcass. They also can live on the epidermal tissue of either plants or animals. The human habit of preparing food in advance and saving it

for later consumption has provided a whole new category of edibles for heterotrophic plants, and they can live very satisfactorily on such foods as bread, cheese, and chicken soup.

As an adaptation to their parasitic way of life, heterotrophic plants have specialized root-like absorptive structures that can secrete digestive enzymes. In this way, the organic compounds can be digested externally and then absorbed through the cells of the rhizoids. Some heterotrophic vines are able to dissolve their way through the stem tissue of green plants, until their roots enter the host plant's xylem, from which they absorb all needed nutrients.

### The process of digestion

The ultimate use of all nutrients is to serve as an energy source or a structural unit at the cellular level. The carbohydrates, lipids, and proteins that the animal eats must eventually be broken down into smaller and simpler units, so that they can enter all of the body cells, where these com-

5-14 *As this photograph of a snake swallowing a mouse indicates, few carnivores are finicky eaters. The snake's lower jaw is not permanently hinged to its upper one, so its mouth can expand to fit the size of its prey. Snakes can actually eat animals larger than themselves. (Russ Kinne—Photo Researchers)*

pounds will be put to work. Some will be oxidized, to provide energy; others will be reorganized and stored within the cell for future energy needs; and some will be used to create necessary proteins. When a pike swallows a perch, it is obvious that the components of the perch must be changed before they can enter the cells of the pike. Even when the cells of the perch are broken down to yield their component chemical compounds, these macromolecules—large polymeric carbohydrates and proteins—still will be too large to pass through the guardian membranes of the pike cells. The goal of the process of digestion is to break down the particles of ingested food, whatever their original size, into molecules that are small enough to penetrate cell membranes and enter the body cells. In animals that feed on small food particles or fluids, digestion is often a rapid process, and specialized digestive organs and glands are either absent altogether or present only in a very simple form. The most sophisticated adaptations for digestion are found in the animals that feed on large particles, such as the vertebrates.

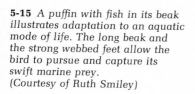

**5-15** *A puffin with fish in its beak illustrates adaptation to an aquatic mode of life. The long beak and the strong webbed feet allow the bird to pursue and capture its swift marine prey.* (*Courtesy of Ruth Smiley*)

## Intracellular digestion

In one-celled animals and some relatively undifferentiated multicellular forms, such as the sponges, all digestion takes place within the cell that is to use the nutrients. The food particle is introduced into the cell through the process of phagocytosis, and it is held in a vacuole, separated from the other contents of the cell by a membrane, for the process of digestion. As soon as the food particle is broken down into fragments small enough to be utilized by the cell, these compounds pass through the membrane surrounding the vacuole, and the nutrients can be put to immediate use within the cell.

**The chemical basis of digestion.** On the cellular level, the digestive process is basically the same wherever it occurs. Complex macromolecules must be broken down into smaller fragments. Digestion is in many ways simply a reversal of synthesis. Most large molecules are constructed through dehydration synthesis, in which two atoms of hydrogen and one atom of oxygen are eliminated from the compound to form a bond. A succession of these bonds link numerous smaller molecules together into a large polymeric molecule. Digestion is accomplished by breaking this sequence of bonds through hydrolysis. When the synthesized molecule is exposed to water in the presence of catalytic enzymes, these new hydrogen and oxygen atoms are added to the molecule, breaking the bonds of the organic compound.

Polysaccharides are broken down by the hydrolysis of H—O—H bonds between individual saccharide groups. Lipid molecules are broken down by the hydrolysis of the H—O—H bonds between the fatty acid and the glycerol portion of the molecule. Proteins are broken down by the hydrolysis of H—O—H peptide bonds. All three processes, diagrammed in Figure 5-18, are basically the same kind of hydrolysis reaction.

**Digestive enzymes.** These hydrolytic reactions

that are the basis of digestion proceed rather slowly; in order to speed them up the catalytic assistance of enzymes is required. The enzymes which accelerate catalytic reactions are typically quite specific. They work on only one type of organic compound, and sometimes they break only one of the bonds. For example, one kind of enzyme breaks the peptide bonds between amino acids of the carboxyl end of the protein chain, but does not attack the same kind of peptide bonds of the amino end of the chain. It is not yet known how these enzymes can be so specific, or what the value of such specificity might be.

Digestive enzymes can be divided into three broad categories, in accordance with the substrate compounds on which they work. The **carbohydrases** specialize in hydrolyzing the bonds of polysaccharide molecules. Examples are amylase, which can break down starch and glycogen, and ptyalin, an enzyme secreted in saliva which begins starch digestion. It is interesting to note that although most herbivores

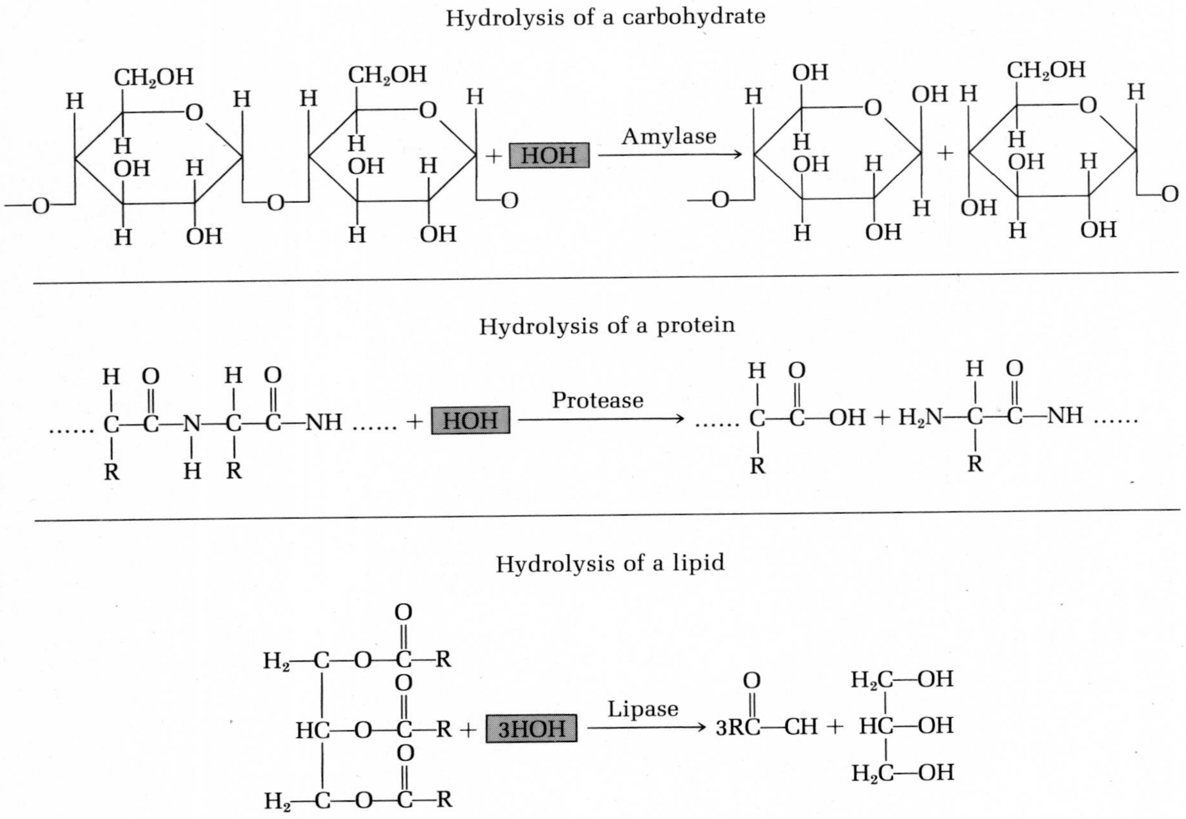

Hydrolysis of a carbohydrate

Hydrolysis of a protein

Hydrolysis of a lipid

**5-16** *Digestion is largely a matter of breaking large polymeric molecules down into their component monomers. Various enzymes attack the bonds linking specific segments of the polymer. This process takes place during the digestion of carbohydrates, lipids, and proteins.*

**5-17** *Herbivores, such as this meadow jumping mouse, do not usually need great mobility to obtain their food, since the grass they eat is not likely to run away from them (they may, however, have to run fast to avoid becoming some other animal's food). Herbivorous mammals often have broad grinding teeth, to help shred the tough plant cell walls before the food enters the digestive tract; rodents like this mouse eating a cherry also have prominent and strong incisors to help in nibbling their food. (Photo by Karl H. Maslowski— Photo Researchers)*

consume large quantities of cellulose, very few of them can secrete cellulases. In many animals, the cellulose is never digested but is expelled with its food value intact; in some vertebrates, intestinal bacteria and protozoa provide this enzyme.

**Lipases** break down the bonds of lipid molecules. Lipases tend to be much less specific than other digestive enzymes. The same lipase can catalyze many different steps in the progressive breakdown of a fat, and it may also catalyze lipid synthesis as well. Therefore if lipid digestion is not to reverse itself, the products of digestion must be removed promptly.

**Proteases** are the enzymes that catalyze the digestion of protein. As a rule, proteases tend to be quite specific, with each step in the sequence

of digestive reactions being catalyzed by a different enzyme. Protein digestion is initiated by endopeptidases, such as pepsin and trypsin in vertebrates. Endopeptidases attack the peptide bonds at the center of the protein molecule, breaking it into smaller polypeptide chains. These chains are then broken down by the action of exopeptases, such as aminopeptidase. Endopeptases attack the ends of the polypeptide chains, breaking off the component amino acids. Some plants contain very powerful proteases that are capable of catalyzing the complete digestion of protein. One example is bromelin, an enzyme found in pineapple juice. In the laboratory, this enzyme has completely digested live roundworms. (Bromelin is completely inactivated by human digestive enzymes, so there is no need to worry about drinking a glass of pineapple juice.)

All digestive enzymes offer a potential threat to the cell or organism that secretes them. A molecule of protease breaks down the protein in food particles; it can just as easily break down the protein component of the cell membrane, or the cristae of the mitochondria. Several different mechanisms have evolved to safeguard the organism from its own digestive enzymes. One is the lysosome, which seals off the enzyme from the rest of the cell until it is needed; at that time the lysosome membrane ruptures and the enzyme is liberated, usually passing into the cell vacuole, where digestion takes place inside another protective membrane. The principal component of mucus, called mucin, also acts as a protection against digestive enzymes.

Another way in which the organism protects itself against the possibility of digesting its own body is by secreting enzymes in an inactive form. Most enzymes, especially the proteases, must be activated or unmasked before digestion can take place. In some cases, the activator is an inorganic ion. For example, salivary amylase must be activated by chloride ions; these ions are normally present in the mouth but not in the cells in which

**173**

the enzyme is synthesized; thus the enzyme cannot be activated at the site of synthesis, but only at the site of digestion. Sometimes the presence of a quantity of the active form of the enzyme also helps to speed up the process of activation. This is the case with the protease pepsin. Pepsin is actually secreted as an inert compound, pepsinogen, but in the stomach certain atoms are removed from pepsinogen, turning it into pepsin. This process takes place much more quickly when the stomach already contains a certain amount of active pepsin.

The activation of enzymes frequently depends on a specific pH. The activation of pepsin, for example, requires a certain degree of acidity in the surrounding medium. It appears that pH is an important influence on the speed at which most enzymes can work; the specific pH required for optimum performance varies from one enzyme to another. In general, the early stages of digestion require an acidic medium, while subsequent stages must take place in an alkaline chemical environment. The variability of required pH creates problems for one-celled organisms. Within the tiny confines of the cell, it is impossible to separate an acid area and an alkaline area. The only way that a unicellular organism can separate these two processes is by time, rather than space. It must begin the process of digestion in an acid medium and then modify the medium, so that the second phase of digestion, that requiring an alkaline medium, can take place.

Temperature also affects the activity level of digestive enzymes. At very low temperatures, enzymes are not active catalysts; at very high temperatures, they, like all proteins, will coagulate, causing irreversible damage to their molecular structure.

The mechanism that sets off the secretion of digestive enzymes has been the subject of extensive study. The presence of food will stimulate this secretion; Pavlov's famous experiments with his conditioned dogs showed that for some animals the expectation of food is also a stimulus to secretion. Emotional conditions, such as rage or sorrow, can also serve to stimulate or inhibit the secretion of digestive enzymes in man. Experiments with laboratory animals suggest that the mechanism governing enzyme secretion is capable of adaptation over a period of time, when there is a change in diet and feeding habits. For example, rats that are fed a very high protein diet slowly develop the ability to secrete larger amounts of proteases, while at the same time their ability to secrete lipases and carbohydrases is diminished. In another experiment, an enterprising physiologist measured the salivary secretions of five Bushmen from the Kalahari Desert who were confined to the city for three months to serve as court witnesses. At home, the Bushmen are almost entirely carnivorous; their diet is high in protein and very low in starch. In the city they were living on a high-carbohydrate diet. At the end of the three-month period, their saliva contained four times as much of the starch-digesting enzyme amylase as it had when they first arrived. Such adaptability would, of course, be a valuable survival characteristic in the wild, where the availability of foods can change rapidly in response to conditions of climate, the effects of disease, and the activities of other members of the food chain.

## Extracellular digestion

Most multicellular organisms carry on the bulk of digestive activity outside the cell, usually in a specialized digestive tract, which is actually a cavity lined with epithelial cells. The adaptation of extracellular feeding has a number of distinct advantages. One is that it allows the ingestion of larger particles; an organism that carries on only intracellular digestion is restricted by lack of space to small-particle feeding. The development of extracellular digestion thus has opened up a

**5-18** *An incomplete digestive tract is one with only a single opening; food particles enter and waste products leave through this same opening. Most animals, with this type of digestive tract are small-particle feeders that eat almost continually.*

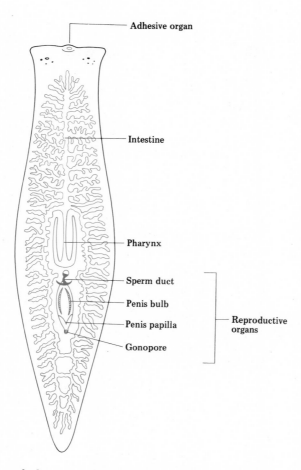

Adhesive organ

Intestine

Pharynx

Sperm duct

Penis bulb

Penis papilla

Gonopore

Reproductive organs

and alkaline phases of digestion can be separated in space, thus allowing them to take place simultaneously. This greatly cuts down the length of time that is needed to complete the entire digestive process.

**The digestive tract.** The simplest kind of a digestive tract is a body cavity with a single opening through which food can enter and wastes can be expelled. The epithelial cells lining the cavity secrete digestive enzymes, and they also absorb organic molecules after they have been adequately broken down. Any bits of food that cannot be digested will eventually be expelled through the same opening. This simple design, called an incomplete digestive tract, is found in many kinds of worms, especially flatworms, and also in coelenterates (the hydra, for instance).

A complete digestive tract, which is found in most other multicellular animals, has an opening at either end of the digestive cavity—a mouth where foods are taken in and an anus where wastes are expelled. A complete digestive tract is therefore a one-way channel. It has the advantage of permitting digestion to take place in a succession of stages. The digestive tract may be modified and differentiated, as various portions are adapted to perform certain steps in the digestive process.

The mouth, which can be considered the first portion of the digestive tract, is specialized for receiving food. It contains teeth for grinding and shredding, and a tongue to help force food down the tract. The mouth also contains the salivary glands, which secrete substances acting to aid swallowing and digestion. The basic secretion is a kind of mucus, which makes chewed particles cohere in a mass; the mucus also helps lubricate the food so that it will slide down the digestive tract easily. In addition to the mucus, salivary glands of some species also secrete a poison, to kill or incapacitate the prey. Other animals secrete acids, which serve the double purpose of killing the prey and activating initial digestive

whole new group of food resources. It also eliminates the need for the continuous feeding that is characteristic of small-particle feeders. The development of a specialized digestive tract allows mechanical as well as chemical digestion; food can be broken down through the squeezing, churning, or grating action of the tract. Another advantage of extracellular digestion is that different cells can specialize in the secretion of the various digestive enzymes, so that it is no longer necessary for every cell to produce every enzyme. Moreover, with extracellular digestion the acid

**175**

enzymes. In man, the salivary glands also secrete amylase, a digestive enzyme that begins to break down starch, such as glycogen and amylose, a starch found in plants.

The second portion of the digestive tract specializes in conduction and storage. Conduction is performed by a simple tube—it is called the **esophagus** in vertebrates. Regular waves of contraction help push food down the esophagus in a process called **peristalsis.** In many animals (but not man) the function of storage is performed by the **crop,** a small pouch in the lower portion of the esophagus. The presence of the crop allows animals to eat more than they can digest at one time; the excess is stored and later released into the stomach. Leeches, for example, have very large crops that can store several times the animal's weight in blood, releasing it later, a little at a time, as the digestive system is able to handle it. This storage organ frees the leech from the need for constant feeding, and also allows it to make the most of its relatively infrequent feeding opportunities.

In certain animals, some of the steps in the process of digestion may take place in the crop. In bees, for example, it is within the crop that the nectar they have eaten is turned into honey. In grain-eating birds, such as chickens and geese, some preliminary digestion of starch may also occur in the crop; the tough outer husks of the seeds are softened in preparation for the later stages of digestion.

The third major region of the digestive tract includes the stomach and the intestine; most of the work of digestion takes place here, with the small intestine being the most active region of all. The stomach continues the process of mechanical breakdown of food that was begun in the mouth. In many insect species, this breakdown is performed by hard toothlike structures within the stomach that actually grind food up; the gizzard in birds is adapted for the same purpose. In man, mechanical breakdown of food **(trituration)** is accomplished by contractions of the muscular

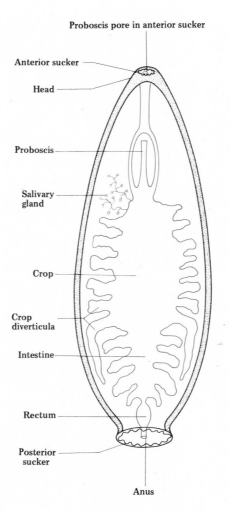

Proboscis pore in anterior sucker

Anterior sucker

Head

Proboscis

Salivary gland

Crop

Crop diverticula

Intestine

Rectum

Posterior sucker

Anus

**5-19** *A complete digestive tract has an opening at either end, so that digestion becomes a directional process. This permits both temporal and spatial separation of the different phases of digestion. The leech, a member of Phylum Annelida, is one of the simplest animals to have evolved a complete digestive tract.*

**5-20** *Although we tend to think of insects as herbivores that munch their way through the leaves of our trees and flowers, there are many insects adapted to a predacious way of life. This wasp is a good example, for its sting can kill or paralyze relatively large prey. (From Insects by Ross Hutchins. Copyright 1966 by Ross Hutchins. Published by Prentice-Hall, Inc., Englewood Cliffs, N.J.)*

wall of the stomach; these peristaltic contractions produce a churning action, similar to the action of a washing machine, that breaks up small particles.

In most vertebrates, the stomach is a simple tube with a valve at either end. The cardiac sphincter at the opening of the stomach allows food to enter and prevents regurgitation; the pyloric sphincter controls the passage of **chyme** — homogenous mixture of mechanically broken down food and digestive enzymes — from the stomach to the intestine.

Some herbivorous vertebrates have more elaborate stomach arrangements, including the division into a number of separate chambers. Cows, deer, and sheep — the group called the ruminants — all have this differentiated type of stomach. It allows them to eat quickly, with almost no chewing of their food. Later the food can be regurgitated in the form of a tightly packed cud, and carefully chewed so that digestion can continue. The value of this adaptation is that it shortens the time that the animal spends actually grazing. This gives the animal a wider feeding range, since it can spend some of its time walking to more distant feeding locations. More importantly, the shortened grazing time contributes to the animal's safety; it is most vulnerable to a carnivorous predator's attack when it is standing exposed in a grassy field, concentrating on eating.

In all vertebrates, some processes of chemical digestion take place in the stomach. Gastric secretions have been quite thoroughly studied in laboratory animals (especially dogs) that have undergone surgery to create stomach pouches or direct openings called fistulas through which chemical reactions can be observed and tested. The two main secretions of the stomach are hydrochloric acid (HCl), which creates the proper acid condition within the stomach, and pepsin, which begins to break down peptide bonds. In young animals, a third secretion, the milk-digesting enzyme rennin, is also present. Initially, this secretion of gastric juice is triggered by nerves in the cells lining the stomach; this is followed by the secretion of a hormone called gastrin, produced by cells in the pyloric wall. This hormone enters the bloodstream and reaches other parts of the stomach, where it stimulates the secretion of large quantities of HCl.

Digestive secretions, especially the strong

**5-21** *The grazing okapi shown here is related to the antelope and also to the zebra, as betrayed by the markings on its hindquarters. The okapi is a fast runner, well adapted to life on the open grasslands of Africa. Its chief predators are the African cats, such as the lion and cheetah. (Jen and Des Bartlett — Photo Researchers)*

hydrochloric acid, pose a threat to the delicate epithelial tissues lining the stomach. There are two defensive barriers that protect the stomach from digesting itself. One is a coating of lipid molecules that covers the outer surface of the tightly packed columnar epithelial cells. Since the digestive secretions are not fat-soluble, they cannot penetrate this lipid barrier. The second protective mechanism is the constant replacement of the epithelial cells. The surface lining of the stomach is replaced every three days, allowing quick repair of any damaged area.

Food is passed from the stomach to the small intestine, where much of the work of digestion takes place. The small intestine consists of three parts — a duodenum, a jejunum, and an ileum. Digestion in the small intestine is both mechanical, through continued peristalsis, and chemical, through the action of enzymes and various reactants secreted by the intestine and other digestive organs.

As soon as the pyloric sphincter opens and permits the passage of the chyme into the duodenum, the chyme's acidity triggers the secretions of hormones by specialized glandular cells in the intestinal wall. Once in the bloodstream, these hormones stimulate the production of a variety of enzymes and digestive juices. Within the intestine itself, at least three different carbohydrases are secreted when the hormone secretin is present in the blood. Two enzymes that break polypeptides down into free amino acids are also secreted in the intestine.

In this phase of digestion, secretory activity is not confined to the intestine alone; the **pancreas,** a large gland located between the stomach and the intestines, plays an important role as well. The pancreas is stimulated by secretin, and it also responds to the presence of another hormone, called pancreozymin. Unlike secretin, which is released by the acidity of the chyme, pancreozymin is released in response to the presence of the products of the first digestive steps. Among the enzymes secreted by the pancreas in response to secretin and pancreozymin are: lipase, which breaks the fatty acid groups off lipid molecules; amylase, which breaks down starch; glucosidase, which breaks down maltose and sucrose, yielding glucose and an alcohol group; and trypsin and chymotrypsin, which split proteins into smaller polypeptides.

The **gall bladder** also aids in this phase of digestion. It is essentially a storage organ, holding the bile salts that are continually produced by the liver, and a quantity of water as well. Secretin in the blood causes the gall bladder to increase the volume of water it stores, but there is no effect on the production of bile salts. The gall bladder also responds to cholecystokinin, a hormone released into the bloodstream by the presence of acidic chyme in the duodenum. Cholecystokinin stimulates the contraction of the gall

## 5-6
## Digestive enzymes

| Where secreted | Enzyme | Control | Role |
|---|---|---|---|
| Mouth | $\alpha$-amylase | stimulated by epinephrine, presence in mouth of foreign material, or condition-reflexes such as smell of food | catalyzes hydrolysis of starch |
| Stomach | pepsin | presence of food in stomach stimulates secretion, mediated by the hormone, gastrin | cleaves internal bonds of proteins (endopeptidase) |
| Pancreas | chymotrypsin | secreted into duodenum whenever large polypeptides are present in it; secretion of enzymes is enhanced by stimulation of the vagus nerve | endopeptidase |
| | trypsin | | endopeptidase |
| | carboxypep-tidases | | exopeptidase (i.e., cleaves amino acids one at a time for the —COOH terminus |
| | steapsin (lipase) | | fat hydrolysis |
| | $\alpha$-amylase | | hydrolysis of starch |
| | deoxyribo-nuclease | | hydrolysis of DNA |
| | ribonuclease | | hydrolysis of RNA |

| Intestine | aminopepti-dases tripeptidases dipeptidases | stimulated by presence of material in the intestine | peptidases |
|---|---|---|---|
| | lipase | | hydrolysis of lipid |
| | nucleases | | hydrolysis of nucleic acids |
| | phosphase | | cleaves phosphate group |
| | maltase | | hydrolyzes maltose |
| | sucrase | | hydrolyzes sucrose |
| | lactase | | hydrolyzes lactose |

bladder, thus emptying bile (bile salts dissolved in water) into the intestine. Bile is not an enzyme, but the bile salts are important to the digestive process. They emulsify fats, thereby speeding hydrolytic breakdown and enabling absorption of these molecules which are not readily soluble.

An important component of all digestive secretions in the intestine is a large quantity of dissolved bicarbonate ions ($HCO_3^-$). These react with the acid chyme to neutralize it (a reaction that produces $CO_2$ gas) and gradually, the pH of the solution of food and enzymes changes from acid to alkaline. This change is a necessity for subsequent steps of digestion and absorption.

Throughout the process of digestion, a great deal of water is poured into the stomach and intestine. Water is needed in the digestive process of hydrolysis; it is also needed to create a liquid solution, rather than a solid mass. A liquid is much easier to move and distribute throughout the tract, and of course the digested nutrients must be in solution to enter body cells. The total production of digestive juices in humans requires more than 16 pints of water a day. Fortunately, most of this water can be reabsorbed and therefore is not lost to the body.

The long digestive tract of vertebrates permits the sequential organization of the processes of digestion. As the food moves from one portion of the digestive tract to another, a step-by-step process of digestion can take place, so that the various phases of digestion are separated in both time and space. Peristaltic contractions are the mechanism that serves to move food through the tract in the right direction and at the proper pace. Most of the digestive ailments that plague all vertebrates are the result of some failure in the sequential organization. For example, it has been demonstrated that one of the main causes of ulcers in humans is the occasional backward flow of chyme from the small intestine back to the stomach; this backward flow is often due to the effect of tension on the muscles responsible for peristalsis. The result of this reverse flow is that bile salts are introduced to the stomach, where they emulsify the lipid layer protecting the epithelium and thus permit the movement of hydrochloric acid into the cells lining the stomach.

**The process of nutrient absorption.** Within the small intestine, the process of nutrient absorption also takes place. The arrangement of the

**179**

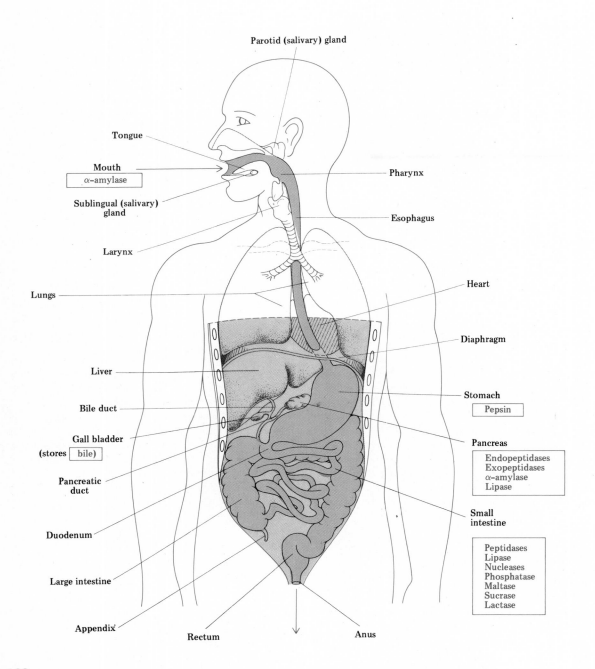

Parotid (salivary) gland

Tongue

Mouth
α-amylase

Sublingual (salivary) gland

Larynx

Lungs

Liver

Bile duct

Gall bladder

(stores bile)

Pancreatic duct

Duodenum

Large intestine

Appendix

Rectum

Pharynx

Esophagus

Heart

Diaphragm

Stomach
Pepsin

Pancreas
Endopeptidases
Exopeptidases
α-amylase
Lipase

Small intestine
Peptidases
Lipase
Nucleases
Phosphatase
Maltase
Sucrase
Lactase

Anus

**5-22** *This schematic drawing shows the location of each phase of digestion, the organs involved, and the digestive enzymes secreted there. Although illustrations of the digestive tract must show it as a sort of tube, it should be remembered that the tract is actually an inner space rather than a structure.*

**5-23** *One of the main causes of ulcers in humans is the occasional backward flow of chyme from the small intestine to the stomach; this is often due to the effect of tension on the muscles responsible for peristalsis. A group of altered surface epithelial cells is seen here in a sample of human stomach from a patient with "stress" gastric bleeding. (Courtesy of Dr. J. M. Riddle)*

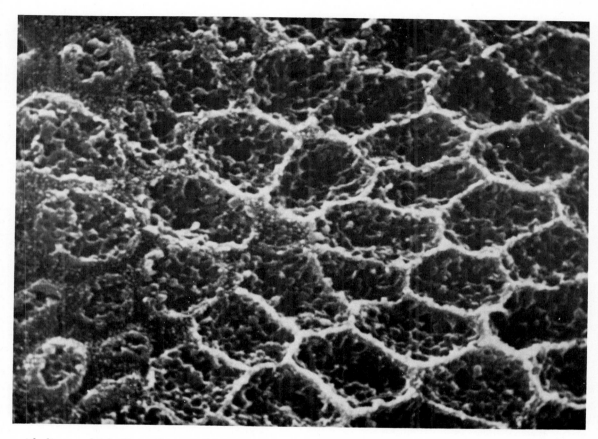

epithelium which lines the small intestine is adapted to give the largest possible absorptive surface. The epithelial tissue is arranged in little fingerlike projections called **villi;** each villus is about 1 mm long, and is a self-contained absorbing unit. The villus is covered with columnar epithelial cells; many of these cells have little projections called **microvilli,** which also increase the absorptive surface. In shape and function, microvilli are much like root hairs. Under the epithelium there is a dense network of tiny capillaries. In the center of the villus lies a larger lymphatic vessel called the **lacteal.** Throughout the villus are located muscle fibers that can con-

tract to change the position and shape of the villus.

An unusual feature of the absorptive epithelial cells covering the villi is that they are replaced very rapidly; here we see another similarity to root hairs. New cells are continually produced by cell division at the base of the villi. The growth of these new cells pushes all the old ones up, and the cells at the projecting tip of the villus are pushed into the intestinal cavity, where they are digested. By using phosphorus isotopes as tracers, it has been found that about 250 grams of epithelial cells are digested every day, and the life span of the cells from time of division at the base until the time they are discarded into the intestine is only about two and one-half days. These cells are rich in carbohydrases and proteases, and their release into the intestine provides additional enzymes for the process of digestion.

Dissolved nutrients enter the epithelial cells through passive diffusion. It seems likely that an active transport mechanism is involved as well, since in many cases absorption appears to take place against a concentration gradient. All nutrients that are water-soluble enter the capillaries and then flow into the **hepatic portal vein,** which connects the digestive tract directly with the liver. This organ regulates the distribution of nutrients, converting them into various storage molecules when an excess is present. If there is an excess of carbohydrates, glucose is converted into glycogen, which can then form the basis of lipid molecules for compact storage. Excess proteins are also broken down (deaminated). The nitrogenous fragments are converted into urea and then passed to the kidney for excretion in the urine; the residue can be converted to glucose and used as fuel.

Fat-soluble nutrients enter the lacteal in the villus and are then carried into the veins that return blood to the heart. The products of lipid digestion thus bypass the liver for a time; they will make an almost complete circuit through the bloodstream before they finally enter the liver.

The nutrients are gradually absorbed by the villi as food passes through the intestine; the final steps of digestion are often assisted by bacteria that live in the intestine and secrete needed enzymes. Near the end of the large intestine, the process of reabsorbing water begins. A great deal of water is used during digestion, to keep the food in a fluid medium, for easy transport and conduction. As a conservation measure, much of this water is reabsorbed in the lower colon. Only enough water is retained in the intestine to facilitate the passage of waste materials, or feces, through the rest of the digestive tract. The feces

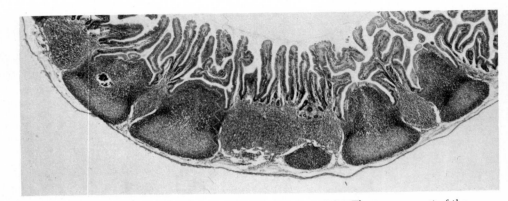

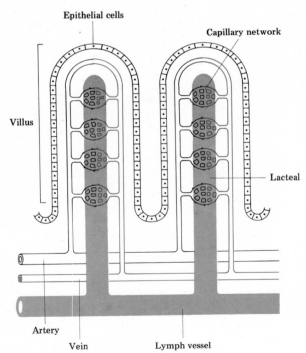

5-24 *The arrangement of the epithelium which lines the small intestine is adapted to give the largest possible surface for nutrient absorption. The epithelial tissue is arranged in little fingerlike projections about 1 mm. long, called villi. Each villus in turn is covered with smaller microvilli; they are too small to be seen at the magnification of this photo. (from Fraser et al., 1969)*

may be temporarily stored in the rectum, a small chamber at the end of the colon, until they are expelled through the anus, the second opening of the digestive tract.

## Summary

All living systems must take in nutrients, but as organisms vary widely, so do their modes of nutrition. Autotrophic organisms need only inorganic nutrients to survive. Heterotrophs need organic food plus water and minerals.

Carbon, oxygen, and hydrogen are the main elements utilized by green plants in the process of photosynthesis. Beginning in a water environment, autotrophs evolved and moved to the land, and began to adapt to an inland habitat. The first organs of nutrient procurement were rhizomes; later these developed into true roots. There is a variety of root patterns found in green plants, depending on the environment.

Movement of water into root cells and throughout the plant can be accounted for in terms of osmosis along a concentration gradient, or imbibition by adsorption. All nutrients except carbon, hydrogen, oxygen, and nitrogen enter the plant in the form of ions dissolved in water.

The basic energy source of most heterotrophic activities is the carbohydrate. Lipids, proteins, vitamins, water, and mineral elements are also required. Omnivores are animals, usually aquatic, which eat small particles of any nature. Herbivores eat only plants, and carnivores eat only meat.

Autotrophs manufacture their own food; digestion is a special problem of heterotrophs. The ultimate use of all nutrients is to serve as an energy source or a structural unit at the cellular level; thus digestion is the breaking down of particles of ingested food into molecules small enough to penetrate cell membranes. It is accomplished by hydrolysis of the bonds linking macromolecules into micromolecules. Enzymes are required as catalytic agents in the process.

Many worms and coelenterates have a simple, or incomplete, digestive tract, consisting of a body cavity with a single opening, through which food enters and wastes are expelled. A complete digestive tract has two separate openings, and permits digestion to take place in a succession of stages. There are three main areas of digestion:

1. The mouth receives food; there food is ground and shredded by the teeth, and the salivary glands secrete substances acting to aid digestion and swallowing.

2. The esophagus conducts food downward by peristalsis; in some animals, a pouch off the esophagus, the crop, serves as a storage organ.

3. The stomach and intestine is where most digestion takes place. Trituration mechanically breaks the food down in the stomach, and secretions there (hydrochloric acid and pepsin) chemically digest it. Both mechanical and chemical digestion take place in the intestine. The pancreas secretes enzymes into the small intestine, and the gall bladder aids digestion there by releasing bile. Bicarbonate ions are an important component of all intestinal digestion, because they gradually change the acid chyme to an alkaline state, necessary for subsequent steps in digestion.

Absorption takes place in the small intestine through epithelial tissue called villi by passive diffusion. Nutrients all eventually enter the liver which regulates their distribution. Water is reabsorbed in the large intestine, and the remaining wastes are expelled as feces.

## Bibliography

### References

Barrington, E. J. W. 1962. "Digestive Enzymes." *Biochemistry.* 7: 1–65.

Beaton, G. H. and E. W. McHenry, eds. 1964–1966. *Nutrition: A Comprehensive Treatise.* Academic Press, New York, 3 vols.

Davenport, H. W. 1972. "Why the Stomach Does Not Digest Itself." *Scientific American.* 226(1): 86–96.

Gregory, R. A. 1962. *Secretory Mechanisms of the Gastro-Intestinal Tract.* Arnold, London.

Robinson, C. H. 1967. *Proudfit-Robinson's Normal and Therapeutic Nutrition.* 13th ed. Macmillan, New York.

### Suggested for further reading

Guyton, A. C. 1969. *Function of the Human Body.* 3rd ed. Saunders, Philadelphia. Illustrations within the book are excellent. The accompanying text adds detailed explanations as to the digestive organs.

Jennings, J. B. 1965. *Feeding, Digestion and Assimilation in Animals.* Pergamon Press, New York. A full-length book on nutrition. Many interesting examples of food-getting devices are discussed. Comprehensive overall treatment of this subject.

Larimer, J. 1968. *Introduction to Animal Physiology.* Brown, Dubuque, Iowa. A useful general physiology book. Thorough treatment of digestion in ruminants is included in the chapter on nutrition. An outside aid for first-year students.

Ray, R. M. 1963. *The Living Plant.* Holt, Rinehart & Winston, New York.
One of the best short treatments of plant function. Popular, easy-to-find outside text for beginning students.

**Related books of interest**

Davis, A. 1970. *Let's Eat Right to Keep Fit.* Rev. ed. Harcourt Brace Jovanovich, New York.
One of the most popular of recent guides to nutrition. Some of Miss Davis' conclusions may be disputed, but her arguments are grounded in experimental data and are always stimulating.

Lappe, F. M. 1971. *Diet for a Small Planet.* Ballantine, New York.
A guide, written in familiar language, to the economics of protein nutrition. Includes chapters on protein theory, recipes, and tables of food value.

Nyoiti, S. and W. Dufty. 1965. *You Are All Sanpaku.* Award Books, New York.
A witty, easy-to-read text on macrobiotics. The summaries at the end of each chapter are useful for review. Overall, a very readable book.

Pauling, L. 1970. *Vitamin C and the Common Cold.* Freeman, San Francisco.
A controversial study of the relationship between the common cold, general good health, and the use of vitamin C. While not basing his views on the traditional research techniques, Dr. Pauling does present a forceful case.

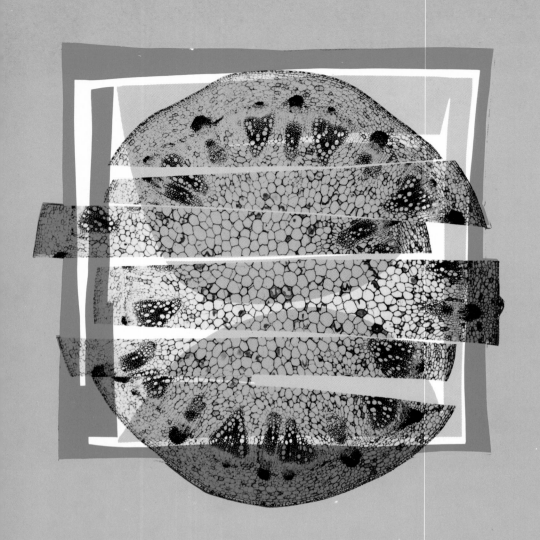

# 6

# Internal transport

All cells are faced with a kind of housekeeping problem, one that should be familiar to anyone who has ever struggled to keep a house or apartment reasonably neat and well organized. Metabolic activities require that raw materials enter the cell and that waste products be expelled. Various storage molecules, such as polysaccharides and lipids, may have to be moved from one location to another. Enzymes must travel from the site where they were synthesized to the place where they will be used to catalyze metabolic reactions.

These problems of internal transport occur even within small one-celled organisms, but multicellular plants and animals have a double problem. They must transport materials within the many cells of which they are composed, and also must develop some mechanism for transporting substances from one cell to another. In this chapter, we will look at some of the solutions to both of these problems.

## Intracellular transport

The cell, with its distinct organelles and precise organization, has a variety of mechanisms to facilitate transport from one part of the cell to another. Simple diffusion accounts for some intracellular transport; the distances to be covered are so short that this rather slow method of travel works perfectly well on the cellular level. Many biologists believe that the canal-like system of the endoplasmic reticulum also functions as a transportation network. This may be the way that large molecules (nucleic acids, for example), which would diffuse very slowly because of their size can be moved quickly from one part of the cell to another. It also seems certain that some mechanism of active transport is at work on the cellular level. A demonstration of this hypothesis can be seen in the fact that radioactive amino acid isotopes are found at the ribosomes within

15 minutes of being introduced into a culture medium. Diffusion alone could not account for such rapid and selective movement of these relatively large molecules, so active transport must be involved.

It has also been observed, by watching living cells through the light microscope, that many organelles are capable of moving through the cytoplasm to relocate in another part of the cell. For example, chloroplasts move around in parenchymal cells, apparently in response to changes in the direction and intensity of absorbed light. This helps the chloroplasts to orient themselves so that they receive the optimum amount of light, and at the same time it helps solve the problem of transport. As the chloroplasts move around, they can supply various parts of the cell with the glucose and ATP they produce, thus shortening the distance over which these compounds have to be transported. The same mobility is characteristic of mitochondria; they have frequently been seen to move toward a part of the cell in which heavy energy demands are being made, such as the contracting fibers of a muscle cell. This movement of the mitochondria eliminates the need for a separate mechanism to transport molecules of ATP, produced by cellular respiration in these organelles, to the site of energy-consuming reactions. Even vacuoles are able to move about within the cell, distributing digested or stored nutrients to all the parts of the cell and picking up metabolic wastes.

Another aid to intracellular transport is the fact that the cytoplasm itself is often in motion. In an amoeba, it can be clearly seen to flow from one portion of the cell to another, as the cell moves through its environment in the characteristic amoeboid movement; this motion is called **cytoplasmic streaming.** In many plant cells, the cytoplasm seems to flow around and around the cell, between the large central vacuole and the cell membrane. This regular flow (it may move either clockwise or counterclockwise) is called **cyclosis.** Cyclosis and cytoplasmic stream-

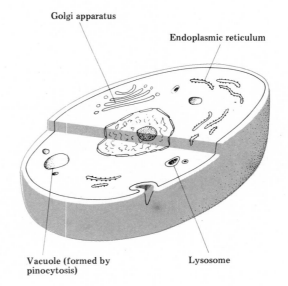

Golgi apparatus

Endoplasmic reticulum

Vacuole (formed by pinocytosis)

Lysosome

**6-1** *On the cellular level, there are many adaptations for the function of transport. The ER helps transport large molecules; lysosomes and Golgi body vesicles also serve a transport function. There is evidence that microtubules, too, may under certain circumstances act as transport pipelines.*

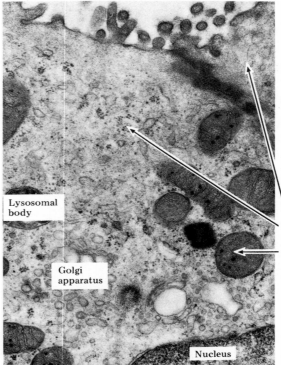

Lysosomal body

Golgi apparatus

Nucleus

Dismosome

Ribosomes

Mitochondrion

**6-2** *The process of pinocytosis is a form of intracellular transport, as well as a means of moving large particles across the cell membrane. Some active transport mechanism is probably involved in the movement of the pinocytatic vesicle through the cytoplasm. (Electron micrograph provided by Professor Lawrence Herman, Department of Pathology, State University of New York, Downstate Medical Center, Brooklyn, New York)*

ing help to distribute organic molecules throughout the entire cell and thus serve to supplement other mechanisms of intracellular transport.

## Transport in multicellular plants

In aquatic plants, the mechanisms of transport utilize the watery medium in which the plants live. Multicellular algae, for instance, absorb all the water and nutrients they need from the sea or pond where they are growing, and they expel their wastes directly into the environment. Their transport system is essentially the same as that of the one-celled organism. The shape of the aquatic plant is well suited to this method of transport. *Pandorina*, for example, is a ball of cells one layer thick; every cell is exposed to the environment along much of its surface. Larger algal forms, such as kelp and other seaweeds, have very broad, flat leaflike blades and slender stemlike connections between the blades. This design gives them a very favorable surface-to-volume relationship, so that no cell is too far from the external environment. Most of these forms have very few specialized organs. Nearly every cell is capable of making its own food, and most cells can still carry on cell division to reproduce new cells. This lack of specialization means that there is little need for extensive communication between cells, and few products that need to be transported from one cell to another; each cell can supply its own needs through its direct contact with the environment. It has, however, been demonstrated that these marine algae, which grow to a length of 50 to 60 feet, can transport throughout the length of the plant certain hormones that control and stimulate metabolic activities. Known mechanisms of transport alone are too slow to account for this movement, so botanists have concluded that there must be some active mass transport mechanism at work. Whatever it is, it has so far eluded all study.

## Transport in vascular plants

The move from an aquatic to a terrestrial way of life brought with it new problems of transport. The limp elongated shape that gives algae such a high surface-to-volume ratio could not be transferred to land; without the support of water, algae become the crumpled heaps of vegetation we often see lying on the beach. Large land plants must have rigid structural tissues to act as a support; they must have underground roots to hold them firmly in the soil; cells that are directly exposed to the environment must have thick protective coatings to prevent excessive water loss. In short, the cells of the land plant must specialize.

Cell specialization brings with it an increased need for intercellular communication. One group of cells, the root hairs, specializes in absorbing water and nutrients; these substances must be moved to the other cells of the plants. Another group, the parenchymal cells of the leaf, specializes in producing the sugar that fuels all energy-consuming reactions; the sugar must be moved to other cells that cannot carry on photosynthesis. Certain cells specialize in the work of cellular growth and sexual reproduction; these cells must be supplied with large amounts of sugar, water, and inorganic ions. The activities of all these cells must be coordinated; the chemical messengers that specialize in coordination must travel to all parts of the plant.

In large organisms with many specialized cells, some rapid method of bulk transport is required. The processes of passive and active transport that move substances within cells, and between one cell and the environment or its neighbor, cannot move a large enough number of molecules at a fast enough speed to serve as the basis of intercellular transport over long distances. In vascular plants (and in vertebrate animals) the mechanism that has evolved is a hydraulic system that moves a liquid through some kind of pipe. Sugar, inorganic ions, and chemical messengers such as hormones, can be carried as solute particles in the fluid. In plants, this mechanism cannot be considered a circulatory system, since the same fluid does not move through all the vessels. It is simply a transport system.

**Vascular tissue: the pipes in the system.** In plants, the pipes of the hydraulic transport system are formed by connected cells of xylem and phloem. The xylem carries only water and inorganic ions, and movement in this network of cells is very rapid. The phloem carries dissolved molecules of sugar and amino acids; the fluid is more viscous (thicker and more syrupy) and moves at a slower speed.

In some plant stems, the xylem cells and the phloem cells appear to be scattered at random. But in others, they may occur in intricate patterns, so that a cross section of the stem shows stars or concentric circles made of vascular cells. In fact, the arrangement of xylem and phloem in the stem is one of the characteristics used to classify and identify various plant groups. Some of the most common patterns are shown in Figure 6-3.

It stands to reason that the most extensive vascular systems would be found in plants that live the longest and grow to be the tallest. A daisy or a sweet pea, with its slender stem and its short life of one summer, does not need an elaborate set of conducting vessels; a few tubes of xylem and a few strands of phloem provide all the transportation required. But a hickory tree is a different case altogether. The tree may well outlive the man who planted it, and it will grow to be 40 or 50 feet tall. If it is not to blow down, it must also grow wider at its base as it grows taller, and it must have an extensive root system to anchor it in the ground. This growth creates transport problems. Water must travel a very great distance to get from the root to the leaves, and food must travel the same distance in the other direction.

Since the tree lives for a long time, it also faces the same problems of aging that plague us all. Elderly xylem cells (which are nothing more than spaces enclosed by a heavy cell wall) gradually fill up with resins and gums and waste products of metabolic activity, clogging the line of water conduction. Therefore new transport cells must be created; this is the function of the tissues that specialize in growth and replacement. If you look at the trunk of an old tree that has been cut down, you can see how this happens. The center of the trunk is usually very dark, and it may be slightly sticky to the touch; this is the area of old xylem, called the **heartwood,** and it is no longer active in transport. Surrounding it are lighter rings of new xylem cells; this area is called the **sapwood,** and it performs all the water conduction for the tree.

Many different kinds of xylem cells may be found in the sapwood, depending on the age and species of the tree. Vessels are elongated cells whose end walls are perforated or completely absent. Tracheids are long skinny cigar-shaped conduction cells with thin spots, or pits, where the cell wall is folded back or pulled away from the primary wall. These cells lie alongside each other, and water readily flows through the pits from one cell to another.

It seems logical to suppose that the xylem cells with open ends would be more efficient conduction tubes than those with pitted walls, since the walls would seem to slow down the movement of water. Interestingly enough, it is the network of tracheids, rather than the open pipeline of vessels, that is found in the very tallest trees. This fact would seem to indicate that the process of water conduction is more complicated than the simple flow of water through a tube, but it is not known where the advantage of the tracheids lies.

Phloem cells make up the vascular tissue concerned with the conduction of dissolved sugar and amino acids. Phloem sieve cells are highly specialized to perform this single task of

A

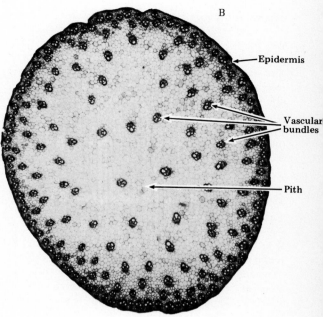

B

Epidermis

Vascular
bundles

Pith

C

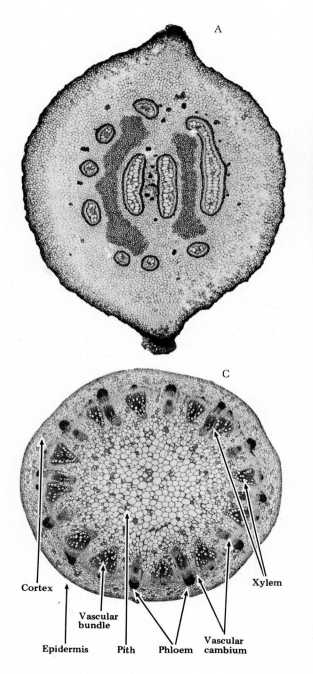

Cortex

Epidermis  Pith  Phloem  Vascular
Vascular            cambium
bundle                      Xylem

**6-3** *A typical monocot (B) has the
vascular tissue scattered in self-
contained bundles throughout
the root. In a dicot (C) the vascular
tissue is located within a central
stele. A more rudimentary
arrangement of xylem and phloem
can be seen in the fern rhizome (A).
(E. J. Kohl—Lakeside Biological
Products, Carolina Biological
Supply Co.)*

conduction. They have no secondary cell wall,
and they lack most of the organelles found in
other plant cells; they have no nucleus, no clearly
defined vacuole, and very few mitochondria or
plastids. Another unusual feature of phloem
cells that may have some effect on their conducting ability is that the cytoplasm is not a true
colloid but is actually a proteinaceous solution.
Most of the necessary metabolic activities are
performed for the sieve tubes by neighboring

companion cells, which have nuclei and are so crammed with other organelles that they look like the white elephant bin in a junk shop. Companion cells provide the sieve tubes with directive nucleic acids, synthesized molecules of protein, and the ATP needed for energy-consuming chemical reactions. In spite of their peculiar adaptations, the phloem sieve tubes, unlike xylem cells, are living units, and if they are killed, they can no longer perform their function of conduction.

## Hydraulics of water transport in vascular plants

It is easy to demonstrate the fact that water does move up the xylem at a fairly rapid rate. A simple experiment can make this fact graphic: take a stalk of celery and cut off the root end with a sharp knife while the celery is in a glass containing water colored with red dye. The red liquid can be observed as it travels up the celery ribs. These are actually bundles of vascular tissues. This movement of water and dissolved solutes is called **translocation.**

The puzzling question is *how* the water is actually moved along the tube, in opposition to the pull of gravity. An engineer can outline the dimensions of the hydraulic problem involved. In a typical Douglas fir tree, for example, which may well reach a height of more than 400 feet, the water that is absorbed by the lowest root hairs may have to travel 450 feet straight up to reach the topmost needles of the tree. In order to raise water 450 feet in the air, a pressure of at least 210 pounds per square inch is required (this is about 13 times the pressure of air at sea level). But the water in a tree is not moving in a vacuum; it is moving through a very narrow pipe. This creates an additional problem of friction, which makes it harder to move the water. And the faster the water must move through the xylem pipe, the greater the friction will be.

When we consider that trees which live in an arid climate, such as palm trees, may pump over a hundred gallons of water a day through their xylem, we realize that water moves through the xylem relatively quickly. Based on a calculation of the average speed at which water must be replaced in the needles of a Douglas fir, engineers have come up with an estimation of the pressure that would be required to move water to the top of the tree — about 420 pounds per square inch.

How does the plant manage to create the necessary pressure? The solution engineers use when they design a system to move water uphill is a pump, but we know that plants do not contain any such structure. Another possible solution is to apply the pressure by squeezing the side of the conducting tube. (This is how animals move food and liquids into their digestive tracts.) But xylem cells have rigid and unyielding walls, so their shape cannot change.

Over the years, plant physiologists have investigated and rejected a number of hypothetical answers to the question of how water transport works in vascular plants. At one time, the mystery of water transport was explained on the basis of "vital movement"—that is, some force or pressure exerted by living cells, using chemical reactions as an energy source. This rather vague theory was disproved once and for all by a determined German botanist, Eduard Strasberger, who liked to do his experiments on a grand scale. He cut down a 70-foot oak tree and propped it up in a huge vat of picric acid, a toxic chemical that killed all the cells as it moved up the xylem. After three days, Strasberger moved the tree trunk out of the picric acid and into a vat of water; he found that water still moved up the xylem cells to the top of the tree, just as efficiently as it had when the tree was alive. But this movement could not in any way be due to the activities of living cells at the base of the trunk, since presumably they had all succumbed to the picric acid.

Another early theory was that water moved up the xylem by capillary action. We know that

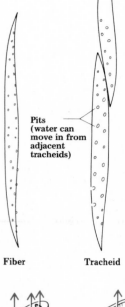

Pits
(water can
move in from
adjacent
tracheids)

Fiber          Tracheid

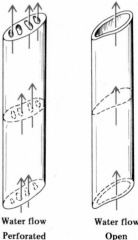

Water flow          Water flow
Perforated          Open
vessel end          vessel end

**6-4** *Although some kinds of xylem vessels are open at either end, permitting the free flow of water, in many cases water must move through thin spots, or pits, in the secondary wall of the vessel. It is believed that the movement of water through the xylem is largely a passive transport mechanism. When water and dissolved solutes travel through the phloem, they must diffuse from cell to cell, probably with the aid of some active transport mechanism.*

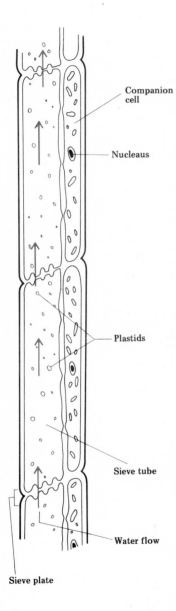

Companion cell

Nucleaus

Plastids

Sieve tube

Water flow

Sieve plate

water molecules cling tenaciously not only to each other but also to any surface with which they come in contact. Water molecules can be observed to creep up the sides of a drinking glass, for example, sticking in a thin layer to the surface of the glass as they move upward. This capillary action is even more pronounced in tubes of very small diameter (in small test tubes, for instance), and therefore could be expected to be quite strong in the narrow xylem cells. Yet experiments have proved that capillarity alone could not possibly move water fast enough or far enough to account for the translocation that occurs in vascular plants.

When the phenomenon of root pressure was discovered, botanists hypothesized that it might be responsible for translocation. But measurement proved this mechanism, too, to be inadequate; it cannot move enough water fast enough to account for the rates of bulk flow observed in most plants. Although the action of root pressure is a supplementary translocation mechanism, and may in part account for the initial establishment of a column of xylem, it is not the solution to the problem of water transport.

**The tension-cohesion theory of water movement.** The best explanation concerning translocation in vascular plants that has been advanced so far is often called the tension-cohesion theory. Its main outlines were contained in a work published by H. H. Dixon in 1895; since then many modifications have been made to account for the results of subsequent research.

The basis for Dixon's theory lies in the unique cohesive properties of water. As we saw in Chapter 1, water molecules are polar, and the negative pole of one molecule attracts the positive pole of the adjacent molecule quite strongly. Moreover, hydrogen bonds may form between water molecules, providing another link that holds them tightly together. Water molecules are so closely bonded to one another that a column of water has enormous tensile strength; it acts just as if it were

**6-5** *The Douglas fir dominates many of our western forests. This beautiful tree may grow to be more than 400 feet high, posing an enormous problem of internal transport of water from roots to needlelike leaves. (Verna R. Johnston from National Audubon Society)*

made of metal wire. It has been estimated that a column of pure water in a tube the width of xylem can withstand a pull of more than 5000 pounds per square inch without breaking. Of course, the liquid in xylem is not pure water. It contains dissolved mineral ions that weaken the cohesion of the bonds between water molecules, as is shown in the diagram in Figure 6-6. Given the constituents of xylem fluid, and the width of xylem cells, botanists calculate that a column of

**193**

xylem fluid could be pulled up a distance of 6500 feet without breaking. This is more than 10 times the height of even the tallest tree.

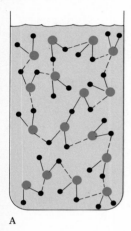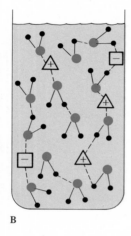

A       B

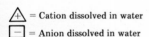 = Cation dissolved in water

▢ = Anion dissolved in water

**6-6** *In a column of pure water, the molecules are bonded together by hydrogen bonds, as shown in (A); this gives the column great tensile strength. Usually the water in the xylem contains dissolved mineral ions, and these do, to some extent, weaken the column of water, because they prevent the formation of all possible hydrogen bonds.*

Another characteristic chemical property of water is its great **adhesion.** It sticks to every surface it touches, spreading out in a thin film that may be only one or two molecules thick. This adhesion (which is the cause of the capillary action of water creeping up a tube) means that when a column of water is drawn through the xylem, it will not pull away from the sides of the vessels and tracheids. It will remain firmly attached to the sides of the xylem cells throughout its entire

length. Thus the column will not elongate and snap in two as it is moved up, but its width will remain constant.

These physical properties of water explain how a pull on the molecules at the top of the xylem will result in moving the molecules up, one after another, in an unbroken column. But what force exerts the pull at the top?

It is believed that the pull is due to **transpiration,** or the movement of water as it evaporates through the stomata, or openings along the surface of the leaf. As the molecule of water at the outer surface of the stoma moves into the drier air, its movement pulls along the molecule just behind it, and in a chain reaction, every molecule in the entire xylem conduction system moves up one "step." This is the scientific version of the Indian rope trick; the column is pulled up and up until it eventually vanishes into thin air!

It is possible to provide visual evidence of this tension-cohesion theory of water movement by showing that changes in the rate at which water evaporates will alter the speed of water conduction in the plant. According to the theory, a greater rate of evaporation ought to lead to a faster rate of conduction. A variation on the celery experiment mentioned earlier will prove that this is indeed the case. Prepare three stalks of celery in three glasses: put the first stalk in a humid greenhouse, the second stalk in an ordinary room, and the third stalk in front of a fan. If you compare the stalks at the end of 15 or 20 minutes, you will find that the first stalk, in humid conditions with low evaporation, shows very little rise of the colored water in the xylem. In the third stalk, where evaporation has been very rapid, because of the wind of the fan, the water will have risen two or three times as high. This demonstrates the direct link between evaporation and water transport.

Another kind of confirmation of the theory can be seen in the fact that translocation takes place at a greatly decreased rate in the wintertime. This coincides with the period when

**6-7** *This drawing presents a visual summary of the mechanism of water transport in plants. Water enters the root hairs, moving either by osmosis or imbibition into the xylem. Once in the xylem, it forms a strong column that can be pulled up from the top, in response to concentration and pressure gradients. The gradients are established when water from the leaves evaporates, creating a water deficit there; the deficit is met by water pulled up and out of the xylem.*

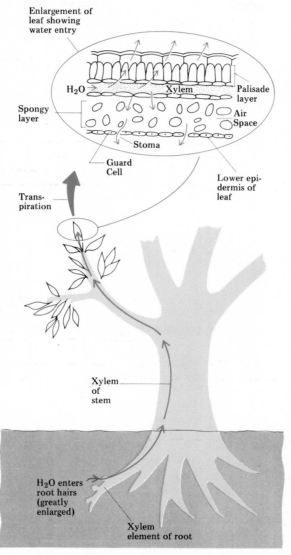

Enlargement of leaf showing water entry

$H_2O$
Xylem
Palisade layer

Spongy layer
Air Space

Stoma

Guard Cell

Lower epi-dermis of leaf

Trans-piration

Xylem of stem

$H_2O$ enters root hairs (greatly enlarged)

Xylem element of root

in plants. But it is clear that many factors in this process are not yet understood. For example, plant physiologists have attempted to build a functional model of a tree, constructed along the outlines indicated by the tension-cohesion theory. (Model building is a common scientific technique that often helps research workers understand the form or the function of the structure they are investigating. Biologists may build models of organic molecules, of hearts, and of the photosynthetic factory found in the leaves.) The results of this model building have been somewhat embarrassing; the man-made model proved greatly inferior to the real tree, being unable to lift the water more than 10 or 20 feet above the ground and transporting it only very slowly. There must be many parts of this conduction mechanism which have not yet been defined.

## Internal water redistribution

Experiments with a variety of plants, such as tomatoes, cotton, and lemon trees, indicate that water is constantly being redistributed inside the plant in order to meet the most pressing water needs. The growing tip of plant stems and roots, where water is needed to form new cells, will continue to be supplied with water even when older parts of the plant are experiencing significant water loss. Water is also supplied to developing leaves and fruits, where it is needed as the basis of new cytoplasm. On the other hand, once a fruit has grown to full size, water may be withdrawn from it to meet other needs. Careful measurements of lemon fruits that are fully grown but not yet ripe disclose that during the height of photosynthetic activity, water is often withdrawn from the fruits and they shrink slightly; the lost water is usually replaced at night. In other words, the water stored in the fruit for future use by developing seedlings can be drawn on by the plant to meet more immediate water needs. The

transpiration is at its minimum rate, because of the absence of leaves and the cool temperatures; water can evaporate only through the epidermal surfaces of the stem. When there is a reduction in transpiration, there is an accompanying reduction in translocation.

The tension-cohesion theory provides a general answer to the puzzle of water conduction

**195**

effect of the overall osmotic gradient is to move water from areas where it is in greatest supply into the areas where it is scarcest. In most cases, this also means that the water will go where it is needed most.

A special problem in water relations arises when water is scarce. When the amount of water in the soil drops below a certain level, it can no longer be absorbed by the root hairs; the osmotic gradient is such that there is more water inside the plant than outside in the ground. For a time, the plant will continue to lose water through evaporation, even though no replacement supply is coming in. This leads to wilting in the plant.

Botanists distinguish between two kinds of wilting. Temporary wilting may occur on very hot days when the plant is losing water through evaporation faster than it can replace it through its water transport mechanism. Temporary wilting usually disappears overnight, when new supplies of water can be distributed to the plant cells. Permanent wilting occurs when the water in the soil is depleted, and it will not disappear unless the soil is given more water. For most plants, a sustained period of permanent wilting leads to death.

The first sign of wilting is the characteristic droop of leaves as turgor pressure drops in the parenchyma cells. As the leaf cells wilt, the stomata are automatically closed, thus preventing further water loss by evaporation. If the wilt continues, there will be marked internal redistribution of existing supplies of water. Water will flow out of leaf cells and into the meristem cells at the growing tip. Water will also move into the epidermal cells of the root, especially the root hairs, since these cells are in direct contact with the dry soil and thus face the steepest osmotic gradient. This redistribution serves the important function of keeping these absorptive cells healthy and ready to go to work as soon as there is a new supply of water.

If the wilt is prolonged, the contents of most cells begin to shrink, and the cell membranes pull away from the cell walls and become wrinkled and folded. The water loss of all cells increases the tension at the top of the xylem column. It has been estimated that in some desert plants, the pressure exerted on the uppermost xylem cells may go as high as 1470 pounds per square inch. The great tensile strength of water will enable it to withstand relatively great pressure, but if the tension becomes too strong, the column will finally snap. Air moves in to fill the xylem at the point of breakage, and usually that dooms the plant. Even if it obtains new supplies of water in the roots, there is no mechanism that can pull up the column of water below the break; the conduction system does not work by pumping the water up from below but by pulling it up from above.

## Transport of nutrients through the phloem

Certain mineral nutrients may be transported as dissolved ions in the water that flows through the xylem, but most nutrients needed by plant cells are transported through the phloem. In order to determine the composition of the sap that moves through phloem tissue, botanists have drafted aphids (insects that live on plant sap) as unwitting helpers. Aphids are apparently more delicate surgeons than people, since they are able to insert their feeding tubes directly into the phloem without disturbing or damaging any nearby tissues. When botanists try to do the same thing with slender needles, they find that they are puncturing many other cells as well, and the contents of those cells gets mixed up with the sap in the phloem, so that it is impossible to tell what compounds came from which part of the plant. For this reason, botanists encourage aphids to begin breakfasting on a succulent stem, and as soon as the feeding tube is in place, they neatly slice off the rest of the aphid. By examining the liquid that flows out of the feeding tube, they

**6-8** *This coleus plant is in a temporary wilt, of the type that often occurs on hot days when the rates of transpiration are high. If the plant is watered soon, it will make a full recovery. (Alycia Smith Butler)*

**6-9** *Aphids are apparently more delicate surgeons than people, since they are able to insert their feeding tubes directly into the phloem without disturbing or damaging any nearby tissues. When botanists try to do the same thing with slender needles they find that they are puncturing many other cells as well, and the contents of those cells get mixed up with the sap in the phloem, so that it is impossible to tell what compounds came from which part of the plant. (Photo by Alexander B. Klots)*

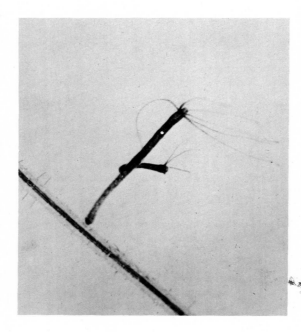

**6-10** *A carnivorous hydra exemplifies the simple multicellular organism with minimal internal transport needs. Prey, captured by the tentacles with their stinging nematocysts, is taken into the central gastrovascular cavity. There extracellular digestion breaks the food down into smaller particles, which are later digested by phagocytic gastrodermal cells.*

have determined that phloem sap is a 10–25 percent sugar solution, the principal sugar being sucrose, which is one molecule of glucose joined to one molecule of fructose. Small amounts of amino acids, and dissolved mineral ions, have also been found in the phloem.

The mechanism of conduction in the phloem is not well understood. By using radioactive tracers, it has been discovered that the rate of flow varies in different substances, and in some cases, it may be as high as 100 centimeters per hour (over a yard); this rate is so high that it seems to indicate some mechanism of bulk transport at work. Most botanists believe that a concentration gradient is involved in mass flow in the phloem. A concentration gradient may exist between the region of high concentration, where recently synthesized sugar from the leaves enters the phloem, and the region of low concentration, where sugar is being used, or converted to osmotically inactive (insoluble) forms for storage in specialized root cells. The most frequent direction of flow is downward, from the leaves to the roots, but it often happens that the flow is reversed, and sugar moves upward, from the leaves to the growing tip, where large supplies of sugar are needed to fuel the activities of growth and reproduction.

The fact that the rate of conduction varies in different solutes has led some plant physiologists to hypothesize that a process of active transport, involving the expenditure of energy, is also utilized. Carrier molecules may be located at the membrane of the sieve plates; these molecules could speed the transport of many solute particles across the membrane.

## Transport in invertebrate animals

One-celled animals, like their plant counterparts, can interact with their environment through the same active and passive processes that serve them as a means of intracellular transport. A protozoan, for example, can absorb nutrients and expel wastes directly into the environment through its outer cell membrane; no other mechanism of transport is needed. Simple multicellular animals, such as the sponge, the hydra, and the flatworm, have a relatively small number of cells, and few of their cells are very specialized. Therefore these organisms, too, can rely on the relatively slow mechanisms of intracellular transport for the small amount of intercellular communication that they must perform.

In many of these lower invertebrates, the central body cavity that serves as an incomplete digestive tract also serves as a means of transport. It is often called the **gastrovascular cavity,** to signify its double function. The sides of the cavity are typically lined with cilia or flagella, whose waving movement causes the fluid in the cavity to circulate throughout the entire space, distributing the nutrient particles to all parts of the organism. Frequently, the gastrovascular cavity is branched, increasing its internal surface area to allow for maximum contact with body cells.

Larger animals, whose cells exhibit a greater degree of specialization, must have some system of bulk transport, linking the individual cells with one another and with the environment. These animals have found the same solution as did higher plants—a system of hydraulic transport. A fluid, which contains dissolved in it the needed nutrients, a supply of oxygen, and hormones, can travel at a rapid speed to the various parts of the body; the fluid can also carry waste products and accumulated $CO_2$ away from the cells, to the organs that specialize in their excretion or expulsion.

When we think of a circulatory system, we usually think of the extensive network of vessels and the strongly pumping heart found in mammals, but many invertebrates have effective circulatory systems that are much simpler in plan. An example of a rudimentary circulatory system with closed vessels to transport blood can be

seen in the common marine worm Nereis, a member of the same phylum as the earthworm. Nereis has two main blood vessels—a **dorsal** vessel that runs all along the upper side and a **ventral** vessel that runs the length of the lower side. The two vessels are connected by a looped tubule in each segment of the worm. Nereis has no heart; instead, waves of contractions (just like the peristaltic contractions in the digestive tract) push the blood along the dorsal vessel and through the loops. The ventral vessel acts as a passive collecting system that eventually returns the blood to the dorsal vessel. This type of circulation is called a **closed system,** since the blood always remains within the conducting vessels. In Nereis, blood does not really circulate; it ebbs and flows in a tidal current created by the waves of contraction. This slow and somewhat uncertain method of transport is adequate for the needs of the worm, since it is small and relatively unspecialized. The famous Greek anatomist Galen (he lived in Rome in the second century A.D.) thought that circulation in humans worked on this same ebb-and-flow principle, but we know now that such a system could never work in so large a creature.

The earthworm has much the same type of circulatory system that is found in Nereis, except that in the earthworm, the loops (called **aortic arches**) help regulate blood pressure and serve as an accessory pumping mechanism. These arches are rudimentary hearts.

Arthropods (a phylum which includes lobsters, insects, spiders, and other groups) have evolved another type of circulatory system, in which the blood is not so much pumped as it is stirred. Arthropods do have a heart, which contracts to force blood into a few large vessels that distribute it to the various parts of the body. But once the blood leaves these vessels, it just seeps through the major body cavity, or **hemocoel,** moving as a result of the contraction of body muscles when the animal moves, and also as a result of pressure gradients between different body parts. Eventually it collects in small spaces

**6-11** *The clam worm, Nereis virens, has a circulatory system typical of annelids, with their segmented bodies. The photo clearly shows the head with its distinct appendages, and the segmental structure of the body. Although this worm looks as if it has legs, they are fleshy paddles supported by bristles; it is the alternate contraction of circular and longitudinal muscles that propels it along the sandy bottom. (Gordon S. Smith from National Audubon Society)*

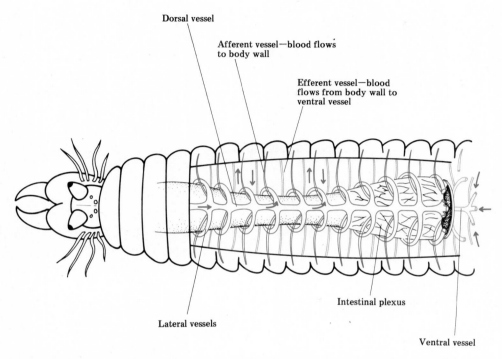

Dorsal vessel

Afferent vessel—blood flows
to body wall

Efferent vessel—blood
flows from body wall to
ventral vessel

Intestinal plexus

Lateral vessels

Ventral vessel

**6-12** *Unlike its close relative, the earthworm, Nereis is aquatic rather than terrestrial. Nereis has a closed circulatory system, with two large blood vessels running its length. There is no heart. Blood moves along the vessels by means of waves of contractions in the vessel walls. Contractions of the body muscles that change the shape of the segments and permit movement also help distribute the blood.*

called **sinuses;** from there, following another pressure gradient, it slowly moves back toward the respiratory and excretory organs and then on to the heart to start another cycle. Even in the vessels, the blood does not necessarily move in one single direction; it may move forward and backward and then forward again. The arthropod system is often called an **open circulatory system,** because there are only a few enclosed vessels to carry the blood. The open system seems to work perfectly well for the arthropods, most of which have relatively small bodies and low metabolic rates. Insects, the most successful branch of the arthropod family, often have high energy demands, but they can survive in spite of their slow circulatory systems, because they do not depend on their blood as a mechanism of oxygen transport. In arthropods, these two systems are separate; in all vertebrates, they are combined.

## Transport in vertebrates

Vertebrates have a relatively large size; they have active metabolisms; they typically exhibit a high degree of cell specialization. To meet their needs for internal transport and intercellular communication, they have developed a very efficient closed circulatory system that pumps blood under pressure in a one-way path, directly to the organs where it is needed. The efficiency of this system allows it to serve also as a means of gas transport, thus eliminating the need for a separate respiratory transport system.

The vertebrate circulatory system performs a number of varied transport functions. Among these are:

1. Transport of oxygen from respiratory organs to all body cells and the simultaneous transport of carbon dioxide in the opposite direction
2. Transport of nutrients and inorganic ions from the digestive tract to all body cells
3. Transport of waste products from all cells to a specialized excretory organ (the kidney)
4. Transport of metabolic products, such as amino acids, intermediate products of cellular respiration (pyruvic acid, for example), and synthesized storage molecules from the organs in which they are formed to the organs in which they are used or stored
5. Transport of regulatory chemicals, such as hormones, from control centers to all parts of the body
6. Transport of heat energy from innermost parts of the body to the surface, so that the body can maintain an even temperature.

The medium of transport is the blood, a connective tissue with various kinds of cells that specialize in different functions. The red blood cells, or **erythrocytes,** are the most numerous cellular component of blood. They specialize in

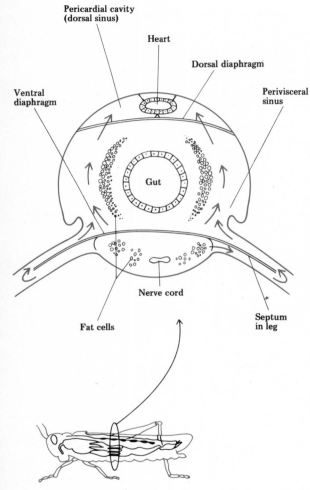

**6-13** *Insects have an open circulatory system. Although there are a few large vessels to carry blood away from the heart, blood travels through much of the body by diffusion. Contractions of skeletal muscles help to push the blood through the body. Eventually it drains into spaces called sinuses and then, following a pressure gradient, moves back to the heart.*

the transport of oxygen and carbon dioxide. Their thin flat shape gives them a large surface area for their size; this surface is used to exchange molecules of oxygen for molecules of carbon dioxide at a very rapid rate. They contain the pigment **hemoglobin,** a conjugated protein that is rather similar to the cytochrome pigments in electron transfer chains. Hemoglobin also has the special property of being able to form bonds easily, and of relinquishing bonded molecules readily. The exact mechanism of exchange will be studied in Chapter 7, dealing with gas exchange.

White blood cells, or **leukocytes,** are part of the body's defense against infection. They surround and engulf foreign cells, such as bacteria and viruses, and they also consume and break down dead or damaged body cells. They are found heavily concentrated at the site of an infection, having moved there by slipping through the walls of the blood vessel. They may travel either in the bloodstream or in the lymph-conducting system, and they are also capable of independent locomotion, using amoeboid movement. Another function of the leukocytes is the secretion of antibodies, which either poison or incapacitate invading cells.

A third type of structure or formed element found in the blood is the **platelet,** which appears really to be only a fragment of a cell. The principal function of the platelet is to initiate clotting when a blood vessel has been damaged or cut. When a platelet comes into contact with the surface of a damaged blood vessel, the platelet begins to disintegrate, and in the process it releases an enzyme called thromboplastin. This enzyme, in combination with the calcium ions dissolved in the liquid of the bloodstream, acts on a globular protein found in the blood, called prothrombin, to unmask (activate) it. Prothrombin is synthesized in the liver, and vitamin K is a necessary cofactor for its synthesis; that explains the connection between vitamin K and blood clotting.

The unmasking of prothrombin is very similar to the unmasking of digestive enzymes that takes

**6-14** *Red blood cells, or erythrocytes, are the most numerous cellular component of blood. They specialize in the transport of oxygen and carbon dioxide. White blood cells, or leukocytes, are part of the body's defense against infection. They surround and engulf foreign cells, and they also consume and break down dead or damaged body cells. A third type of structure or formed element found in the blood is the platelet, which appears to be only a fragment of a cell and is important in initiating clotting when a blood vessel has been damaged or cut. (Dr. G. Schoefl)*

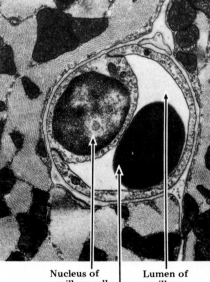

Nucleus of
capillary cell

Lumen of
capillary

Red blood
cell

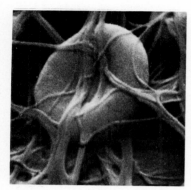

**6-15** *Fibrin is the end-product of the thromboplastin-prothrombin-thrombin-fibrinogen reaction chain. When a clot forms, blood cells like the erythrocyte in the picture are enmeshed in the fibrin threads, which then contract to squeeze serum out of the hardening clot. (Courtesy of E. Bernstein and E. Kairinen. Published in Science (cover photograph) 173, no. 3999. Copyright 1971, American Institute for the Advancement of Science.)*

## 6-1
## Composition of blood

| Component | Amount |
|---|---|
| Amino Acids | 35–65 mg./100 ml. |
| Carbohydrates | 210–320 mg./100 ml. |
| Lipids | 385–675 mg./100 ml. |
| Organic Acids | 12–27 mg./100 ml. |
| Urea | 25–40 mg./100 ml. |
| Proteins | Total = 7.3% |
|    albumins | 4.5% |
|    fibrinogen | 0.3% |
|    globulins | 2.5% |
| anions { Bicarbonate ($HCO_3^-$) | 24–30 mEq./l. |
| Chlorine ($Cl^-$) | 100–110 mEq./l. |
| Iodine ($I^-$) | $6.3 \times 10^{-4}$ — $11.8 \times 10^{-4}$ mEq./l. |
| Phosphate ($PO_4^=$) | 1.6–2.7 mEq./l. |
| Sulfate ($SO_4^=$) | 0.7–1.5 mEq./l. |
| cations { Calcium ($Ca^{++}$) | 4.5–5.6 mEq./l. |
| Copper ($Cu^+$ or $Cu^{++}$) | 8–16 mg./100 ml. |
| Iron ($Fe^{++}$ or $Fe^{+++}$) | 50–180 mg./100 ml. |
| Magnesium ($Mg^{++}$) | 1.6–2.2 mEq./l. |
| Potassium ($K^+$) | 3.8–5.4 mEq./l. |
| Sodium ($Na^+$) | 132–150 mEq./l. |

$$1 \text{ mEq./l.} = \frac{\text{mg/l} \times \text{valence}}{\text{formula wt.}}$$

Source: White, et al. 1968. *Principles of Biochemistry*, 4th ed. McGraw-Hill, New York. pp. 706–708.

place in the stomach and intestines with the addition of acids or mineral ions or other organic compounds. Once prothrombin is turned into its active form, thrombin, it begins to affect another blood protein, fibrinogen, by causing it to gel. The gel form is called fibrin, because it coagulates in long thin fibers. These fibers catch and hold blood cells, just as fish are caught in a net. The fibers weave together and seal the damaged portion of the blood vessel in a temporary repair, rather like the patch on a bicycle tire. The clot will eventually dissolve, but in the meantime, the cells will be able to repair the damage.

These three types of blood cells comprise slightly less than half of the volume of blood in vertebrates. The other half is a nonliving matrix, called the **plasma.** The average adult has about 2½ liters of plasma (and a little over 5 liters, or 10 pints, of blood). The composition of plasma varies somewhat, depending on such factors as age, health, and general physical condition of the individual. The major constituent of plasma is water, which acts as a solvent for all the other plasma components. Plasma contains a larger number of dissolved sodium cations and chloride anions. Other major constituents are bicarbonate ions ($HCO_3^-$), which are the product of chemical changes that take place in the process of gas exchange, and a number of globular proteins. These blood proteins serve a number of important functions. For example, they provide the bloodstream with a relatively high osmotic pressure, thus maintaining the correct amount of water in the blood. The most important protein in this respect is albumin. If the amount of albumin in the blood drops for some reason, then the blood will not perform efficiently its task of collecting water from tissues of the body, and the tissues will become waterlogged and swollen, a condition known as edema. If too much albumin is present, the high osmotic pressure of the blood will cause dehydration of body tissues. Albumin also plays a role in transport, since it bonds and carries fatty acids, calcium ions, and one of the components of bile. Other blood proteins may transport hormones, fat-soluble vitamins, and metallic ions. Blood proteins also function to transport antibodies, and to buffer acid-base reactions. Fibrinogen, the substance that plays such an important role in clotting, is a blood protein. If it is removed from the plasma, the resulting fluid is called **serum;** serum is simply plasma that cannot clot.

## Anatomy of the mammalian circulatory system

In vertebrates, blood is carried through the body in a closed set of pipes, pushed along by the pumping action of the heart. Blood flows from the heart into large vessels, the arterial system, and then into an extensive network of very small vessels, or capillaries. After passing through the capillary system, blood is collected in the venous system for the return trip to the heart. Since one of the functions of blood involves gas exchange, blood returning to the heart must be sent to the respiratory organs, where the carbon dioxide is removed and oxygen is substituted. The re-

oxygenated blood then returns to the heart and starts another trip through the body.

**The heart.** Mammals have a heart with four separate chambers. In effect, they have two separate pumping systems — one that pumps blood to the lungs, under rather low pressure, and one that pumps blood to the rest of the body, under high pressure.

Figure 6-17 shows the general anatomy of the human heart. The two upper chambers are called the auricles, or the **atria** (the singular form is *atrium,* which means hall in Latin). The right atrium receives blood that is returning through the veins from body tissues; this blood is largely deoxygenated. The left atrium receives the blood that returns from the lungs through the pulmonary

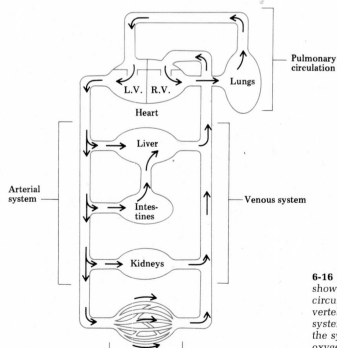

6-16 *This schematic drawing shows the general direction of circulation in the higher vertebrates. This closed circulatory system is really a double one — the systemic circulation of oxygenated blood to the body and the pulmonary circulation of blood to the lungs to be reoxygenated.*

**6-17** *Like all mammals, the human has a four-chambered heart. One-half of the heart (one atrium and one ventricle) carries on systemic circulation, while the other half carries on pulmonary circulation. The septum or wall between the chambers assures that oxygenated blood will remain separate from deoxygenated blood. The valves between the atria and the ventricles prevent backward flow during the heart's contractions.*

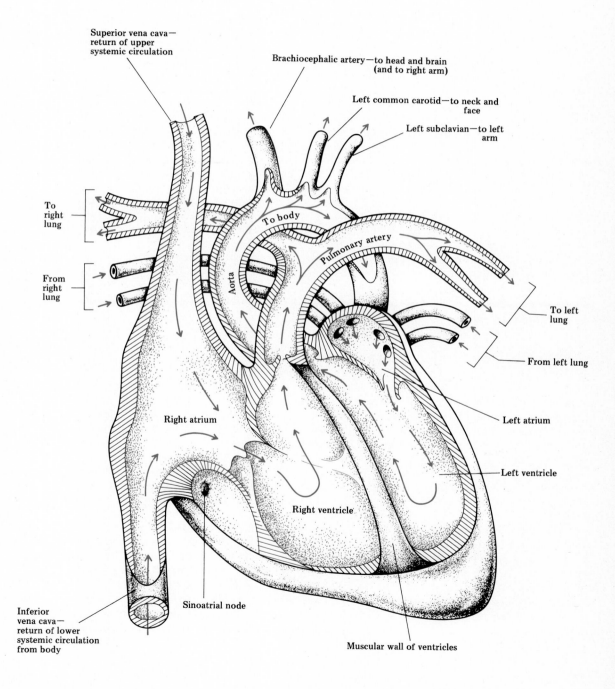

Superior vena cava—
return of upper
systemic circulation

Brachiocephalic artery—to head and brain
(and to right arm)

Left common carotid—to neck and
face

Left subclavian—to left
arm

To
right
lung

From
right
lung

To body

Pulmonary artery

Aorta

To left
lung

From left lung

Left atrium

Right atrium

Left ventricle

Right ventricle

Inferior
vena cava—
return of lower
systemic circulation
from body

Sinoatrial node

Muscular wall of ventricles

veins. The atria, each of which can hold about five ounces of blood, are separated by a thick dividing wall, the **septum,** which assures that there will be no flow of blood between these two chambers.

Below the two atria are two **ventricles,** which do the actual pumping work of the heart. Each ventricle is surrounded by many layers of circular and spiral muscles, making a very thick wall (the walls of the atria are much thinner). When this wall of muscle contracts, it literally wrings every last drop of blood — about 120 milliliters, or four ounces — out of the ventricle and propels it into the arteries. The left ventricle, which receives blood from the storage reservoir of the left atrium, pumps the freshly oxygenated blood into the main body artery, the **aorta.** The right ventricle pumps blood into the pulmonary artery, sending

it to the lungs for new supplies of oxygen. The left ventricle has a much thicker wall than the right ventricle. It is thus able to pump blood under more pressure, which helps it travel further. However, the amount of blood that leaves both ventricles is the same.

The contraction of the heart starts in a thin strip of tissue called the **sinoatrial (S-A) node,** located at the top of the right atrium near the spot where the superior **vena cava,** the vein returning blood from the upper body, enters the atrium. The tissue of the S-A node has a peculiar property; it can contract like muscle tissue, and it can conduct electrochemical impulses like nerve tissues. The S-A node is often called the heart's pacemaker, since it determines the rhythmic beat with which the heart contracts. When a wave of contraction begins in the S-A node, it spreads

ESSAY **Evolution of the vertebrate heart**

The heart found in the first vertebrates was merely a straight tube divided into four chambers: the sinus venosus, which received the systemic veins; the thin-walled atrium for storage; the muscular pumping ventricle; and the conus arteriosus which conducted blood away from the heart. In the fully developed vertebrate heart, found in birds and mammals, there are also four chambers, but they do not correspond to the original four. Instead they represent further divisions of the atrium and the ventricle.

In fish, the primitive tubular heart has evolved into an "s" shape, a curvature which positions the atrium opposite the ventricle. Deoxygenated blood is received in the sinus venosus. The ventricle then pumps the blood first to the gills to be aerated and then on through the rest of the body.

As vertebrates began to develop lungs, they were also faced with a need to keep oxygenated blood returning from the lungs separated from the deoxygenated blood that enters the heart through the systemic veins. The amphibian heart shows the first steps toward a separation of the two types of blood; it has a septum dividing the atrium into right and left halves. Blood from the lungs returns to the left atrium rather than to the sinus venosus. However, oxygenated and deoxygenated blood can still mix together in the undivided ventricle.

In reptiles, a partial division of the ventricle takes place. The sinus venosus is reduced in size and has moved further to the front. The conus is reduced and is subdivided, forming large vessels to conduct the blood.

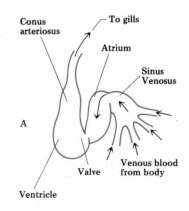

throughout both atria, which contract rather gently and force blood into the two ventricles. These contractions of the atria are not really necessary in order to fill the ventricles. The primary mechanism for filling the ventricles is a strong suction set up after contraction has emptied the chambers. Experiments performed on animals with diseased or damaged hearts have shown that the ventricles can fill even when the atria are incapable of contraction. But the mild contractions of the atria increase the efficiency of the heart by insuring that the ventricles are completely filled; they also serve as a back-up system in case the ventricles lose some of their contractile ability and are thus no longer able to fill up on their own.

After the mild contraction of the atria, which pushes some blood into the ventricles, the valves between each atrium and ventricle close, and the ventricles begin their powerful contraction, called **systole,** which pumps blood out of the heart and into the arteries. After the period of contraction, there is a period of relaxation, called **diastole.** Both these periods can be detected in the electrocardiogram shown in Figure 6-18. The relaxation of the tightly contracted cardiac muscles during diastole sucks blood into the ventricles again, in preparation for the next systole. In man, the heart contracts at an average rate of 70 contractions per minute; during a game of touch football, the players' hearts may well contract more than 200 times a minute, with each stroke pumping about 100 milliliters of blood.

**The arteries.** All the large blood vessels leading away from the heart are called arteries. The term

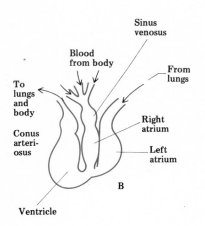

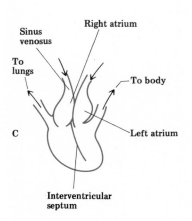

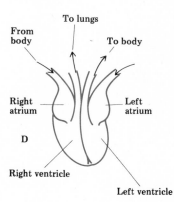

In birds and mammals, the ventricle is completely divided into left and right chambers. The sinus venosus is reduced to a node in the wall of the right atrium, where the systemic veins empty. The conus arteriosus is split into a pulmonary artery connected to the right ventricle, and a systemic aorta connected to the left ventricle. This type of heart completely separates the oxygenated from the deoxygenated blood, thereby achieving the greatest efficiency of heart function.

**artery** is sometimes inaccurately used to mean a vessel that carries oxygenated blood, but actually that is true only of the body arteries, such as the aorta; the pulmonary artery carries blood that has a high $CO_2$ content and very little $O_2$.

Arteries have thick elastic walls, which expand each time systole pushes a new supply of blood into the vessel; most of the energy of ventricular contraction goes into the expansion of these vessel walls. Then during diastole, the arteries gradually contract, increasing the pressure on the fluid in the vessel. In this way, the arteries help to maintain a regular pressure throughout the circulatory system. It sometimes happens that aging interferes with the ability of the arteries to contract, and the arteries gradually turn into rigid pipes rather than expandable tubes. This can cause dangerously high blood pressure during systole.

The large arteries branch off into thinner **arterioles,** which are usually not larger than 0.1 mm in diameter. Arterioles, too, have muscular walls, and by dilating and constricting, they can open and close small valvelike sphincters, thus either permitting or prohibiting the flow of blood into a particular vessel. By this mechanism, which is under the control of the nervous system, they are able to control the amount of blood that enters various areas of the body. This ability to direct blood to specific organs or areas of the body is one of the great advantages of the vertebrate circulatory system. For example, during digestion, when blood is needed to transport nutrients and to supply oxygen needed for the release of metabolic energy, the arterioles leading to the digestive organs dilate, opening the sphincters, and thus allow more blood to flow in that direction. At the same time, arterioles leading toward the head, arms, and legs may constrict slightly, decreasing the volume of blood that flows to those areas during the digestion period. This explains why people so often feel sleepy and mentally sluggish after a heavy meal; after Thanksgiving dinner, the flow of blood to the brain decreases slightly,

while more blood flows into the digestive organs. Because of this system of directing the blood flow, there is no need for the heart to work more rapidly during digestion, so this hard-working organ is spared an additional strain.

**The capillaries.** The smallest vessels in the circulatory system are the capillaries. They are usually about 8–10 microns in diameter, which makes them only slightly larger than the red blood cells they contain. The walls of these vessels are just one cell thick, and they contain no muscles. Circulation in the capillaries, largely because of the blood pressure, is a steady flow rather than a rhythmic pulse. It is estimated that the human body contains a network of capillaries about 50,000 miles long, although not all the capillaries are open at one time; usually about half of them are collapsed, because of the action of the sphincters at the nearest arteriole.

It is the capillaries that actually supply the

**6-18** *This sample electrocardiogram shows both systole and diastole of a normal heart. Any irregularities in the electrocardiogram, such as missed or delayed beats and fibrillation (a condition in which the contraction of cardiac muscle fibers is not coordinated to produce one strong squeeze but a number of small contractive impulses), may be a warning of heart disease.*

Systole | Diastole | Systole | Diastole

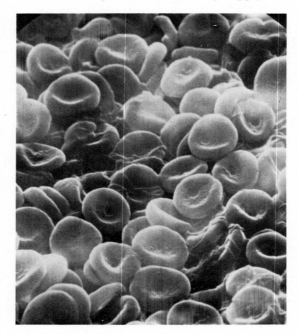

**6-19** *The diameter of most capillaries is only slightly greater than the diameter of the red blood cells that must pass through them. Therefore the erythrocytes usually squeeze together, giving a characteristic clumped appearance. (Dr. Thomas L. Hayes)*

body tissues with needed raw materials and nutrients and remove waste products. The network of tiny vessels throughout the body insures that no cell is more than 50 microns away from a capillary, and simple diffusion along a concentration gradient will suffice to move both liquids and solutes for that short distance.

The capillary wall is actually a differentially permeable membrane. It allows the free flow of water and small molecules, such as sugars and amino acids. However, any particle with a molecular weight of 70,000 or more cannot pass through the membrane and remains trapped within the vessel. Thus the proteins in the bloodstream remain in the capillaries, providing a high osmotic pressure within the vessel.

As blood enters the capillaries, it is still under pressure as a result of the pumping action of the heart—that is, it has a high **hydrostatic pressure.** The hydrostatic pressure, which tends to drive fluid and small molecules out, is greater

than the osmotic pressure of the proteins in the blood, and so the net flow is outward. Water, dissolved ions, and small organic molecules leave the bloodstream and diffuse into nearby cells, moving along a concentration gradient. But as the blood travels through the tiny capillaries, it slowly loses much of its original pressure. At some point, the blood pressure will be equal to the osmotic pressure, and there will be no net flow either into or out of the vessels, although there will, of course, continue to be some random movement of molecules back and forth across the membrane. Eventually, the osmotic pressure exerted by the proteins within the capillaries will be greater than the blood pressure, and fluids and small molecules will begin moving with the pressure gradient back into the capillaries. In this way, exchanges of molecules and ions can be made between body cells and the bloodstream that serves as their transport mechanism. Since the blood flows through the capillaries relatively slowly, there is ample time for these exchanges to take place.

**The veins.** The veins are the vessels that carry blood back toward the heart. The smallest veins, those which pick up blood from the capillaries, are called **venules.** In structure, veins are rather like arteries, but they have much thinner walls and fewer muscle fibers. When the two vessels are compared in cross section, the veins seem much flabbier. They are easily pushed out of shape by the contraction of nearby skeletal muscles or by the accumulation of blood. At any given moment, there is usually more blood in the veins than there is in the arteries; the veins are in some sense a reservoir for blood. Yet the blood pressure in veins is much lower than the pressure in arteries.

One of the chief problems of the venous system is maintaining circulation, especially since in many parts of the body, veins must carry blood upward, against the pull of gravity. For this reason, veins in the legs have valves, to assure that

**6-20** *When blood enters the narrow, thin-walled capillary, its hydrostatic pressure is relatively high, and so liquid and small molecules are pushed out of the vessel into surrounding body cells. This loss lowers the hydrostatic pressure and raises the osmotic pressure within the capillary; at some point (approximately midway through the vessel), fluids and small dissolved solute particles will begin to move from nearby cells into the bloodstream.*

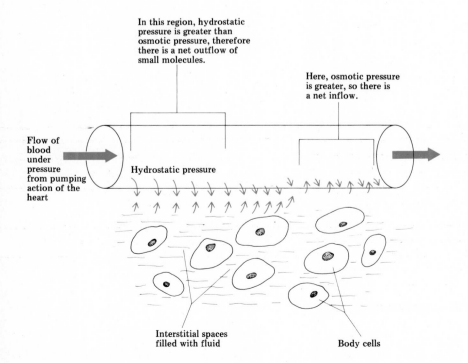

In this region, hydrostatic pressure is greater than osmotic pressure, therefore there is a net outflow of small molecules.

Here, osmotic pressure is greater, so there is a net inflow.

Flow of blood under pressure from pumping action of the heart

Hydrostatic pressure

Interstitial spaces filled with fluid

Body cells

blood can flow only in one direction. Contractions of the leg muscles also help to push the blood along. That is why standing on your feet is much more tiring than walking; there are not enough muscular contractions to help the blood move along in the veins. Venous circulation is also helped by the negative pressure in the heart during diastole. This creates a kind of pull (rather like the pull that draws water up through the xylem) and helps to draw blood up along a pressure gradient back to the heart.

**The lymphatic system.**  The lymphatic system works to supplement some of the functions of the circulatory system. It has been observed that the net outflow in the capillaries is usually greater than the net inflow, which means that some of the plasma fluid is left behind in the body tissues. This fluid will be collected by the small vessels of the lymph system.

Lymph vessels are rather similar to capillaries in their structure, but their walls have different permeability characteristics, being easily penetrated by most protein molecules. The lymph system therefore acts to collect protein molecules or fragments to be returned to the blood. The small lymph vessels drain into collecting points called **lymph nodes,** which contain large numbers of white blood cells. These digest bacteria and cellular debris, breaking them down into their component organic compounds, which can then be carried in the lymph fluid. Other important activities that take place within the lymph nodes are the production of lymphocytes, defense cells that function much like the white blood cells in the bloodstream; the filtration of the lymph fluid; and the distribution of fat molecules, which are absorbed by the lymphatic system in the intestine. The lymph nodes drain into larger lymph veins, whose circulation depends entirely on the movement of skeletal muscles and the action of one-way valves. The large lymph veins eventually empty into one of the main body veins, the vena cava, shortly before

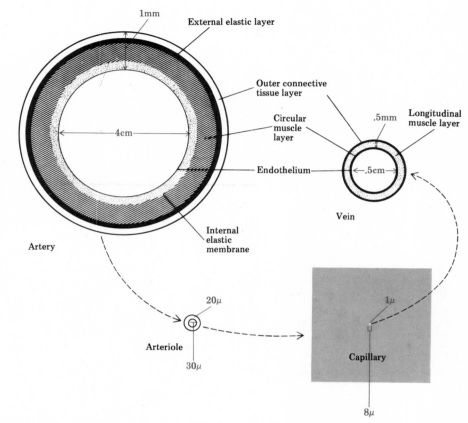

that vein empties into the heart, thus returning the lymph fluid and all the molecules dissolved in it to the bloodstream.

# The hydraulics of mammalian circulation

The rate of blood flow sets limits on the size and metabolic activity of a multicellular organism. If an organism consistently produces metabolic wastes at a rate higher than the rate at which they can be absorbed into the bloodstream in the capillaries, that animal is slowly poisoning itself in an elaborate self-destruct mechanism. Like all other life functions, the rate of blood flow de-

**6-21** *The largest of all vessels are the veins. They are also the most elastic vessels, so they can be stretched to be even wider. This permits them to serve as a kind of reservoir for blood. Capillaries are the smallest vessels; if we draw them to the same scale as the veins, they are only dots on the page. Comparison of the width of the walls of various vessels makes it clear why the arteriole is so important in regulating blood pressure. Its muscular wall is nearly the same diameter as its inner space, or lumen.*

pends on a number of physical factors; blood flow in the circulatory system is governed by the same physical laws that determine the rate of flow in a gas pipeline or sewer system.

Physicists tell us that fluid will not flow in a vessel unless there is some pressure gradient involved. The steeper the gradient, the faster the flow will be. The heart acts as a pump to generate the initial pressure. Thereafter, the system relies on various types of resistance to bring about the decrease in pressure that will allow flow between two points. The three major factors controlling resistance in the circulatory system are the length of the blood vessels, the width of the blood vessels, and the viscosity of the blood. All three of these factors can be utilized to control the rate of blood flow in mammals.

Figure 6-23 shows that the greatest single pressure drop in the circulatory system occurs when blood moves into the arterioles; these are

the vessels that provide the greatest degree of resistance. If we look at a cross section of a typical arteriole, we can understand why. The wall of the arteriole is almost as thick as its opening; the diameter of the opening averages 30$\mu$, while the arteriole wall is usually about 20$\mu$ thick. This thick wall is composed principally of smooth muscle cells, which can be contracted and expanded in response to specific nervous and chemical control mechanisms.

The principal way in which the arterioles can alter the resistance of the vessels is by changing the total cross-sectional width through which the blood flows. Figure 6-24 shows how this is accomplished. Resistance will be greatest when the sphincters at points A and B are closed, because then the total volume of the blood flow must pass through point C, an opening 25$\mu$ wide. To decrease resistance, the sphincter at point A could be opened. Now the blood would

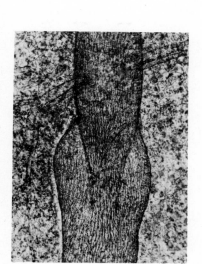

**6-22** *Lymph vessels have much the same structure as veins. This photo shows clearly the valve in a lymph vessel, that prevents the backflow of the fluid. Such valves are especially important in maintaining circular flow in humans, who persist in defying gravity and standing upright on two feet. (Carolina Biological Supply Company)*

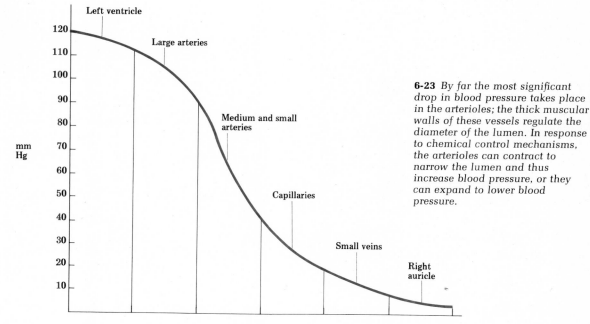

**6-23** *By far the most significant drop in blood pressure takes place in the arterioles; the thick muscular walls of these vessels regulate the diameter of the lumen. In response to chemical control mechanisms, the arterioles can contract to narrow the lumen and thus increase blood pressure, or they can expand to lower blood pressure.*

flow through both points A and C; the total cross-sectional width available is the 25μ at point C plus the 25μ at point A, or 50μ.

It would, of course, be harmful to body tissues if a substantial portion of the capillary system were to be shut down for an extended time period. But, as we have already pointed out, the total length of the capillaries is enormous, and reducing the flow through some of these capillaries provides a means of increasing total volume of flow during periods of stress and exertion. That is why at least half the total length of the capillary system is not in use at any given moment.

A second way to change the resistance of the vessels is to change the actual width of the individual vessels. All blood vessels, except capillaries and venules, can change in diameter as the result of the elasticity of their walls. This elasticity comes in part from the presence of connective tissue that has secreted fibers of **elastin,** a compound that can stretch and contract in response to pressure, without the expenditure of biochemical energy. The major arteries are composed of a large percentage of elastin fibers, so that they automatically and effortlessly expand when they receive a heavy flow and contract as the flow decreases. For this reason, arteries do not contribute much to the total resistance of the circulatory system.

A third factor influencing resistance is the viscosity of blood. The viscosity of plasma is usually about 1.8 times greater than that of water; however, as body temperature drops, the viscosity increases; circulation in a bear during winter dormancy is working against much greater resistance than the circulation in that same bear in the summertime. The actual viscosity of blood depends on the percentage of total blood volume that is represented by red blood cells. This quantity is called the **hematocrit.** Viscosity increases as the hematocrit increases. Human blood, with a hematocrit of about 45, is quite viscous.

There is no mechanism for voluntary changes in the viscosity of blood, but since blood is not a homogeneous solution but a mixture of cells, colloids, and true solutions, it exhibits spontaneous changes from time to time. For example, it has been noted that blood becomes less viscous when it flows through tubes of very small diameter. This can be demonstrated in glass tubing in the laboratory as well as in isolated small blood vessels in animal tissues. No really satisfactory explanation has yet been given for this phenomenon, but it is assumed to involve some reorientation of the hematocrit, perhaps causing each blood cell to be encased in a lubricating bath of plasma; in tubes of narrow diameter, such as the capillaries, it would be impossible for two red blood cells to line up side by side.

Variations in viscosity can also be seen in different portions of a large vessel, if it is viewed in cross section. For some reason as yet undiscovered, the endothelial tissue lining the vessel walls exerts a strong attraction on the plasma component of the blood. Immediately adjacent to the vessel wall there is always a thin layer of plasma that never moves at all; it is held to the wall by a strong adhesive force. Next to that stationary layer is a plasma layer that moves only very slowly, apparently also attracted to the vessel wall. So there is an interesting effect of varying viscosity in different parts of the vessel, and there is also some noticeable variation in the rates of flow. It seems that the fastest flow is directly in the center of the vessel. Rate of flow is slowed at the outer rim of the vessel not only because of the attraction between the wall and the plasma but also because of the effects of friction.

In many mammals, especially man, blood flows down an energy gradient as well as down a pressure gradient, and this additional variable must be taken into account when calculating flow. The energy gradient of blood circulated to most of the body tissues—systemic circulation—is quite similar to the energy gradient described in Chapter 1 in the example of the falling textbook. As blood moves from the level of the heart

**6-24** *(A) shows the minimum total cross-sectional width in a hypothetical set of vessels. This situation, with the valves at points A and B closed, would give the highest possible hydrostatic pressure. To lower the pressure, the total cross-sectional width must be increased; this is easily accomplished by opening the valve at point A. The width is now doubled, and blood pressure will decrease proportionately.*

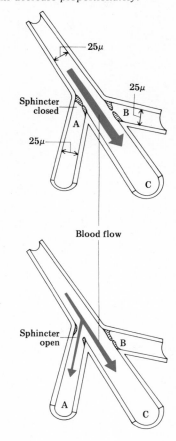

down the body, it is using up stored potential energy. By the time it reaches the toes, all potential energy has been used, and the return trip actually incurs an energy deficit; to meet it, energy must be supplied (this comes from the initial contraction of the heart muscle which utilized the stored chemical energy of ATP). So the blood flows along an energy gradient from the positive potential energy level of the heart to the negative potential energy level as the blood makes its way back up the veins.

## Venous return

Blood circulating to body tissues flows along a pressure gradient and an energy gradient. By the time blood enters the venules, it has a very low pressure and it is entering the negative energy portion of the gradient. Yet even though the pressure is lower in the veins than in the capillaries, the rate of flow is higher. What causes the flow to increase, and the blood to return to the heart?

A number of different mechanisms are at work in venous return. We have already mentioned the role played by the contractions of skeletal muscles and the help given by a system of valves in the legs. Also important is the contractile ability of the veins themselves. Veins are not passive tubes; they can be constricted to increase pressure. Contraction may take place in response to a chemical secreted by the adrenal glands and circulated through the blood, called **norepinephrine.** This hormone affects the muscle cells in the walls of all blood vessels, causing contraction; it also seems to increase the rate and force of contraction in cardiac muscles. When norepinephrine is present in the bloodstream, the rate of venous return is speeded up.

Another mechanism which facilitates the return of blood to the heart is the so-called thoracoabdominal pump. This refers to the effect of pressure differentials inside the body cavities that contain the lungs and stomach. The pressure inside these cavities is usually low, perhaps even negative; the exact pressure varies in response to breathing. During inhalation, the pressure inside the thoracic (lung) cavity drops; the layer of tissue that divides the thoracic cavity from the abdominal cavity—the diaphragm—is pushed down as the lungs expand, and pressure within the abdominal cavity increases. Thus the blood is pushed from the abdominal veins, where pressure is high, up toward the thoracic veins where the pressure is lower. Each breath acts as a sort of pump, and the more frequently and deeply a person breathes, the more efficiently the pump will work to push blood back up to the heart.

Another important factor in venous return is the negative pressure within the heart at the end of systole. As diastole begins, the pressure within the ventricle is much lower than the pressure within the atrium, and therefore blood moves rapidly in to fill up the ventricle. This lowers the pressure in the atrium, creating a pressure gradient that moves blood from the veins into the atrium. Of course, blood does not have the cohesion of water, and so this tension pressure does not pull up a whole column of fluid, as it does in plant xylem. But it does help to move blood from the inferior vena cava, where pressure is very low, into the heart.

## Control of cardiovascular functions

It is important that active mammals be able to adjust the rate and volume of blood flow in response to changing conditions both within the body and within the environment. For example, when metabolic activity is at a low rate—during sleep or hibernation—then the rate of blood flow can be lowered. During bouts of strenuous exercise, such as a cheetah's chase after an impala, the rate of flow must be increased. Aquatic mammals, such as seals, otters, and diving ducks, must be able to adapt their circulation to the conditions of increased pressure under the water.

Condors and the natives of Peru must be able to increase the rate of blood flow to compensate for the lack of oxygen at higher elevations.

One way to control the rate of flow is through the changing of vessel diameter, especially in the arterioles. In cold weather, for example, the arterioles constrict to cut down on blood flow in the capillaries of fingers and toes, thus limiting the loss of heat energy from tiny vessels close to the skin. Much the same redistribution of flow can be found in divers and in animals that suddenly move to a high altitude. As we noted earlier, constriction and dilation of the arterioles occurs in response to both natural and chemical control mechanisms.

Another way to control the rate of blood flow is through changes in the speed of the heart contractions and in the volume of blood that is pumped in each contraction. In this connection, it is useful to look more closely at the mechanisms that control the rate of cardiac muscle contraction.

The vertebrate heart is a self-stimulating organ. An isolated heart can beat for hours, even days, as long as it is maintained in an isotonic nutrient solution under a carefully controlled temperature; no external electrochemical stimulation is needed. The heart contracts in response to the wave of excitation that comes from the sinoatrial node and spreads to the atrioventricular node. However, the rate at which the S-A node initiates contraction can be regulated by the action of two important nerves. The **vagus nerve** acts to slow down the heartbeat. The **accelerator nerve** acts to speed it up.

It is easy to demonstrate in the laboratory that both the vagus and the accelerator nerves transmit their impulse messages to the heart by releasing chemical compounds that affect the rate at which the S-A node contracts. The apparatus for such a demonstration is shown in Figure 6-25. Two vertebrate hearts (from a frog or a mouse) are dissected out; in one of the hearts the vagus nerve is also carefully preserved. An isotonic salt solution is funneled into the heart with the vagus nerve; the second heart is positioned below the first, with a funnel to catch the salt solution pumped out through the severed artery. When the vagus nerve of the upper heart is stimulated, that heart slows down. Within a short time, the second heart, which has no connection with the first except through the salt solution, also slows down. Its slowing can be caused only by a chemical transmitted in the salt solution from the first heart to the second.

The compound released by the vagus nerve is **acetylcholine;** it is released not only by nerves of the heart but by other body nerves as well. The compound which accelerates the heartbeat is **adrenalin.** It is a compound similar to norepinephrine, and it is released by the accelerator nerve. Both these substances are attacked by enzymes shortly after they are released. This prevents them from accumulating and destroying the heart's regulatory system.

**6-25** *In the upper heart, the vagus nerve has been left intact; the lower heart has no external means of regulation. An isotonic salt solution is poured into the first heart, which contracts to squeeze the solution out through the artery, where it is funneled into the second heart. When the vagus nerve of the upper heart is stimulated, that heart slows down. Soon the lower heart slows down also. Since the only connection between the two is the salt solution, the reaction of the lower heart can be due only to a chemical secreted into that solution by the upper heart.*

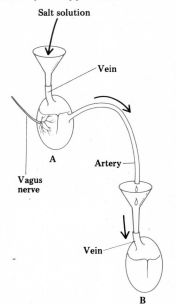

## Summary

The metabolic activities of cells require that raw materials enter the cell and waste products be expelled, and that storage molecules and enzymes move from one location to another. Multicellular organisms have the compounded problem of having to transport materials within the many cells of which they are composed.

Diffusion accounts for much intracellular transport. The endoplasmic reticulum and membrane-enclosed vesicles help move large molecules through the cytoplasm. Cytoplasmic streaming and cyclosis, as well as the movement of plastids, mitochondria, and vacuoles, help to distribute organic molecules. Active transport processes are probably also at work.

Unlike aquatic plants, which utilize their watery medium for transport, the cells of land plants have evolved a hydraulic system of transport, formed by the xylem and phloem. Xylem cells transport water and inorganic ions; phloem cells conduct dissolved sugar, amino acids, and hormones.

Translocation—the movement of water—in vascular plants can best be explained by the tension-cohesion theory and by the evaporation which occurs at the leaves' stomata.

Water is constantly being redistributed inside the plant in order to meet the most pressing water needs due to a pressure gradient. Nutrients are conducted through the phloem mostly along a concentration gradient, but there is also some active transport involving the expenditure of energy.

One-celled animals and simple multicellular animals with gastrovascular cavities rely upon active and passive intracellular transport. Larger animals have developed a circulatory system. *Nereis* has a very simple system relying on tidal current. Earthworms utilize the same system with the addition of a pressure regulator, the aortic arches. Arthropods have a heart, but the blood seeps through the body and relies upon muscular contraction for transport; this is termed an open circulatory system.

Vertebrates have developed an efficient closed circulatory system which serves also as a transport system for many of the body's products. Red blood cells containing hemoglobin are most numerous, and they transport oxygen and carbon dioxide. White blood cells are the body's defense against infection. Platelets initiate clotting when a blood vessel has been damaged by releasing thromboplastin. Plasma constitutes the remaining half volume of the blood. Serum is plasma that cannot clot because of the lack of fibrinogen.

The heart consists of two atria and two ventricles. Blood is oxygenated during its course through the heart by being shuttled via the pulmonary artery through the capillaries of the lungs.

Blood flows from the heart through the aorta to large vessels of the arterial system, through the capillaries, and is collected in the venous system to return to the heart through the venae cavae. It is the capillaries that actually supply the body tissues with materials and remove wastes.

The lymphatic system supplements some of the functions of the circulatory system by digesting bacteria and cellular debris.

The length and width of the blood vessels and the viscosity of the blood control blood pressure and rate of flow. In addition, norepinephrine, secreted by the adrenal gland, and the thoracoabdominal pump help blood move through the less elastic venous system. The speed and volume of the heart contractions are affected by the sinoatrial node, which, in turn, is affected by the release of acetylcholine and adrenalin by the vagus and accelerator nerves.

## Bibliography

### References

Crafts, A. S. and C. E. Crisp. 1971. *Phloem Transport in Plants*. Freeman, San Francisco.

Johansen, K. and A. W. Martin. 1965. "Comparative Aspects of Cardiovascular Function in Vertebrates." In W. F. Hamilton, ed. American Physiological Society's *Handbook of Physiology*. vol. 3. sec. 2. Williams & Wilkins, Baltimore.

Solomon, A. K. 1971. "The State of Water in Red Cells." *Scientific American*. 224(2): 88–96. Offprint no. 1213. Freeman, San Francisco.

Wiggers, C. J. 1957. "The Heart." *Scientific American*. 196(5): 74–87. Offprint no. 62. Freeman, San Francisco.

### Suggested for further reading

Galston, A. W. 1964. *The Life of the Green Plant*. 2nd ed. Prentice-Hall, Englewood Cliffs, N.J.
An up-to-date account of the important life processes of the green plant: metabolism, growth, differentiation, morphogenesis. Stress on the significance of green plants in the economy of nature with clear electron micrographs.

Gordon, M. S. et al. 1968. *Animal Function: Principles and Adaptations.* Macmillan, New York.
Zoological, comparative, and evolutionary approach to the adaptive functioning of animals in relation to the survival of whole organisms. The diverse physiological adaptations found in animals are illustrated.

Graubard, M. A. 1964. *Circulation and Respiration.* Harcourt Brace Jovanovich, New York.
Thorough treatment of the circulatory system. Especially helpful drawings.

Stewart, F. C. 1964. *Plants at Work.* Addison-Wesley, Reading, Mass.
Edition of a series of treatises on plant physiology. Chapter 10 thoroughly covers the plant's respiratory activities.

Wilmoth, J. H. 1967. *Biology of Invertebrata.* Prentice-Hall, Englewood Cliffs, N.J.
Systems and processes concerned with locomotion, neural development, respiration, excretion, food-getting, and reproduction receive extensive coverage. The usual phylum classification with interesting side views of invertebrate adaptations.

## Related books of interest

Hawthorne, P. 1968. *Transplanted Heart.* Rand McNally, Chicago.
Dramatic biographical sketch of Dr. Christiaan Barnard, his patient, Louis Washkansky, and the background of the first heart transplant. A carefully documented and fascinating view of both men.
Nolen, W. A. 1970. *The Making of a Surgeon.* Random House, New York (paperback ed. by Pocket Books, 1972).
Dr. Nolen describes his medical training, with special emphasis on his initiation into the mysteries of heart surgery. He gives a clear outline of the medical aspects of many physiological processes described in Part 2 of this text.

Vroman, L. 1967. *Blood.* Doubleday, Garden City, N.Y. (paperback ed. by Natural History Press, 1971).
A very intelligible and lively introduction to the biochemistry and function of the bloodstream.

# 7
# Gas exchange and respiration

All living things obtain the energy they need through the oxidation of energy-rich compounds. In all but a few organisms, the oxidizing agent is, as the name suggests, oxygen. The cell's demand for oxygen is steady and pressing. So far, no cell has evolved that is capable of storing oxygen, so it must be supplied as it is used. Very active cells, such as nerve cells in the brain or cardiac muscle cells in the heart, use oxygen so fast that they will die in a few minutes if their supply of oxygen is cut off.

In the process of fuel oxidation, carbon dioxide is created. Most organisms can tolerate the presence of a fairly high concentration of this substance, but when the concentration exceeds these limits, harmful effects occur. Since carbon dioxide and water form an acid (carbonic acid), an excess of carbon dioxide will lower the pH of the cell; this change in the cell's environment affects most of the chemical reactions that take place there, either slowing them down or preventing them altogether. To avoid this fatal interference with its chemistry, the cell must get rid of the $CO_2$ produced in cellular respiration.

This need to take in $O_2$ and get rid of $CO_2$ is common to all plants and animals that carry on aerobic cellular respiration. In green plants, there is an additional complication of their gas exchange requirements. They must take in a certain amount of $CO_2$ to use as a raw material in photosynthesis, and they must release the $O_2$ that is a by-product of the photosynthetic breakdown of water and $CO_2$.

Gas exchange is actually a transport problem that is closely related to the other transport needs we discussed in the previous chapter. In one-celled organisms, gases can be easily transported across the cell membrane, so long as the gases are in solution and the membrane is moist. Both the $O_2$ and the $CO_2$ molecules are small enough to slip through the differentially permeable membrane at a rapid rate. Like osmosis, gas exchange takes place along the concentration gradient, so $CO_2$ will move from the cell cytoplasm, where it is in a relatively high concentration as the result of metabolic activity, out of the cell where concentration is usually much lower. The reverse is true of oxygen—there is likely to be more outside the cell, in the atmosphere or water, than there is inside the cell, where it is being used in oxidative reactions. So it will move in the opposite direction, into the cell.

Multicellular algae, aquatic plants with a high surface-to-volume ratio, can also rely on simple diffusion as a mechanism of gas exchange, as can small aquatic animals with relatively low energy demands. But larger and more active organisms need some specialized means of exchanging carbon dioxide and oxygen. As organisms increase in size, their surface-to-volume ratio usually declines, so that diffusion can no longer serve as the sole mechanism of gas exchange. Moreover, these larger organisms often have some outer protective layer—a shell, feathers, sharp thorns—that prevents any significant amount of gas exchange. And the plants and animals that have evolved a terrestrial way of life have another serious problem. Since they are constantly exposed to the dry air, the membrane surfaces across which gas exchange with the environment must take place will dry out. These organisms must somehow devise a way to expose membranes to the air without allowing them to dry out—a problem to tax the ingenuity of the cleverest engineer. We shall look at some of the solutions that have been developed to meet these problems.

## Gas exchange in vascular plants

Gas exchange in vascular plants is carried on primarily by the leaves. The anatomy of roots and stems, both of which have thick coverings of protective cork cells, as well as many layers of dead cells called bark, makes gas exchange almost impossible, since the gases have to move through thousands of heavily lignified cells and also penetrate the waxy cuticle. In the root hair zone of the roots, there may be some gas exchange, as the gases in the soil dissolve in water and enter the plant along with other soluble elements. Along the stem, structures called **lenticels** provide a means of limited exchange for gases.

The leaves have developed specialized openings, or **stomata,** to facilitate the process of gas exchange. Stomata are found scattered throughout the epidermal surface of the leaves of all vascular plants; to conserve water, they are usually more numerous on the underside of the leaf than they are on the side directly exposed to the sun and the wind. The number of stomata on a leaf varies enormously from one plant to another. Plants that are adapted to survive in hot dry climates may have as few as 10 stomata per square millimeter of epidermis, while some plants that live in the tropical rain forests have almost 1300. On the average, the stomata account for about one percent of the epidermal surface.

**7-1** In cross section, lenticels look almost like an injury, rather as if the outer layers of the stem had exploded. This disruption in the regular pattern of closely packed cells provides a channel for gas exchange between the stem and the atmosphere. In deciduous plants, lenticels are the only means of gas exchange during the winter season when the plant has lost its leaves.

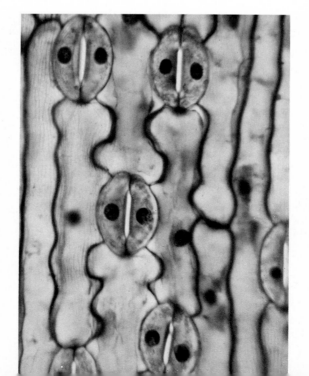

**7-2** This transverse section of a lily leaf's epidermis shows the structure of the stomata and the relationship of the guard cells to the other epidermal cells. The dark dots are the nuclei. Since this particular plant is adapted to a temperate climate, there are a relatively large number of stomata. (S. Cudia—Omikron)

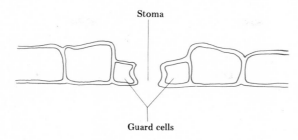

Stoma

Guard cells

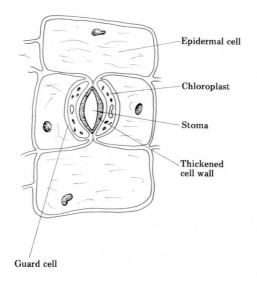

*7-3 The kidney-shaped guard cells, with their heavily reinforced inner walls, border the opening or stoma. Guard cells differ from other epidermal cells in having chloroplasts; photosynthesis helps them accumulate the high concentration of solutes that causes the guard cells to take in water, thus distending the cells and creating the opening.*

Epidermal cell

Chloroplast

Stoma

Thickened cell wall

Guard cell

*7-4 A typical leaf in cross section shows numerous intercellular spaces that can act as channels for gas exchange. At cool temperatures and high humidity, the spaces are often filled with water; when it is hot and dry, the water vaporizes. The actual exchange of gases released from the leaf cells for gases in the atmosphere takes place according to the concentration gradients of each.*

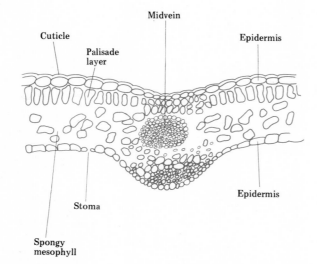

Midvein

Cuticle

Palisade layer

Epidermis

Epidermis

Stoma

Spongy mesophyll

Figure 7-3 shows the anatomy of a typical stoma. (Since the term stoma was taken from the Greek word for mouth, the plural is formed in the Greek way, by adding -*ta*, rather than in the usual English way of adding -*s*.) The stoma is simply an intercellular space that is formed by the changing shapes of the **guard cells** bordering the opening; there is a pair of guard cells for each stoma. These guard cells are typically long and rather narrow; they look something like a kidney or a pork sausage after it has been cooked. The portion of the guard cell wall that borders the stoma is heavily reinforced with cellulose fibers laid down in a pattern of overlapping ridges, and it is much thicker than the normal cell wall. On the other surfaces of the guard cell, the secondary wall is often either abnormally thin or completely missing in some spots, to facilitate bending. Unlike other epidermal cells, guard cells contain chloroplasts and are capable of synthesizing carbohydrates. The whole unit of the stoma and the guard cells that form it is called the **stomatal apparatus.**

The relationship between the stomatal apparatus and the other tissues of the leaf is shown in Figure 7-4. The parenchyma tissue is rather loosely packed, especially in the lower half of the leaf, so that a connecting series of intercellular spaces is formed.

On a cellular level, the process of gas exchange takes place across the membranes of the individual cells in the leaf. The intercellular spaces are full of a mixture of $CO_2$ and $O_2$, each of which will show a net movement in response to individual concentration gradients. Some of the gas may actually be dissolved in water, and in this way it can be transported to other parts of the plant. In the areas of the leaf that are directly exposed to the heat of the sun, the water in the intercellular spaces usually evaporates, turning to vapor. As long as the concentration of vapor stays fairly high, the membranes of leaf cells will remain moist enough for gas exchange to take place.

Gases that accumulate in the intercellular

**219**

spaces, along with the water vapor found there, will move by diffusion to the stomata, and if they are open, the gas and vapor will diffuse into the atmosphere. At the same time, gases that are in high concentration outside the leaf may move into the stomata. During the daylight hours, a green plant usually takes in $CO_2$, a raw material of photosynthesis, and gives off $O_2$; during the hours of darkness, when the plant is converting the sugar it has produced into starch — an energy-consuming process — it usually takes in the $O_2$ needed for cellular respiration and gives off $CO_2$. These are net movements; there is always some exchange in both directions.

This mechanism of gas exchange, which allows water vapor to escape while gas exchange is going on, creates a real dilemma for the green plant. The loss of water vapor, called **transpiration,** accounts for about 95 percent of the water loss of plants. The extensive water loss due to transpiration leads to a constant need for water replacement, and if there is not enough water available in the soil, the plant may wilt and die. The problem facing the plant, then, is to permit necessary gas exchange while trying to control the simultaneous water loss.

One approach to solving this problem is to regulate the spaces through which gas exchange occurs. If the stomata were rigid openings, like the mouth of a sewer pipe, then transpiration would proceed unchecked, no matter how dangerous the resulting water loss was to the well-being of the plant. Since the rate of transpiration increases rapidly with relatively small rises in air temperature and wind velocity, on a hot summer day most plants would wilt beyond the point of no return if the openings were rigid. By regulating the openings, the plant can reach a much safer compromise. When it is very hot and the plant is losing a great deal of water, the stomata can be closed, preventing further water loss at the acceptable sacrifice of a few hours of gas exchange activity. When the plant has been able to obtain additional supplies of water and

the danger of wilting is passed, then the stomata will open again and permit continued gas exchange.

## The mechanism of stomatal regulation

In the early morning, when all the cells of the leaf are turgid because of the water taken up during the night, the guard cells, in their relaxed position, are closed. No stomata are evident, and the inner thickened walls of the guard cells are pressed against each other by the turgor pressure of the other epidermal cells.

As the plant begins its daylight activities, there soon appears a significant difference in turgor pressure between the guard cells and the other epidermal cells. Through a mechanism that is not yet clearly understood, the guard cells build up a concentration of osmotically active solutes that is much higher than the concentration in neighboring cells. The higher osmotic pressure of the guard cells causes water to move into them, increasing their turgor pressure. As the guard cells gain turgor, the pressure of their contents against the cell wall increases. Because of the peculiarities of the guard cell walls, this increased pressure will not be felt equally in all parts of the cell. The thickened cell wall that borders the stoma will resist the pressure, while the thin wall on the other cell surfaces will easily yield to the pressure and start to bulge out. The effect is a distortion of the shape of the cell; there will be a bulge in the thin walls and a bend in the thick wall. You can see much the same effect if you take a long thin balloon and reinforce one surface with heavy tape. When you blow up the balloon, it will no longer look like a cylinder, but will instead be bent at the middle.

It is the bulging of the turgid guard cells that creates a stoma. Since the cells are not cemented together for most of their length, they can easily bend away from each other. As the guard cells press outward, they push against neighboring

**7-5** *At high magnification, the waxy secretion covering these epidermal cells takes on the appearance of an eerie landscape. The function of the covering is to prevent excessive transpiration; this is an adaptation to life in a hot dry climate. (Dr. B. E. Juniper, Oxford University Botany School)*

cells, many of which are specially adapted **support cells**. These cells have rather thin **walls**, and they have at least one surface that **faces** the stomatal opening rather than lying against another cell. Support cells act as elastic buffers between the distorted guard cells and the other epidermal cells. Without them, the bulging guard cells might meet with so much resistance from the relatively unyielding epidermal cells that they would be forced back into their original shape, closing the stoma. The fact that guard cells must push open against the resistance of other turgid cells may mean that it is an energy-consuming process.

The guard cells close the stoma when they lose their high turgor and return to their relaxed state. The walls of the adjacent guard cells gradually come to rest against each other again, when the turgor pressure is normal or somewhat low, and no more gases or water vapor can escape except by slow diffusion through the cuticle of epidermal cells. It is thought that the closing of the stoma is a passive process, due largely to the evaporation of water.

The opening and closing of the stomata in green plants is a phenomenon that has been observed since the early part of the nineteenth century. Biologists know a great deal about the way they open and close, but the question of why they do this is still troublesome. Most researchers in the field feel that there are probably a number of different factors at work causing the variations in turgor pressure, and that some of these mechanisms are still unknown.

The increase in turgor pressure is undoubtedly due to an increase in the number of osmotically active solutes within the cell; this increase, in turn, gives the guard cells a greater osmotic pressure than that in any of the surrounding cells and thus causes water to flow into the guard cells. That conclusion simply pushes the question back one step: why does the osmotic pressure increase? Why do guard cells have a higher concentration of sugar than other cells? Several

different answers have been advanced in response to these questions.

The first hypothesis centered around the presence of chloroplasts in the guard cells; this theory was first suggested in 1856 by the German botanist von Mohl (the same scientist who named the contents of the cell "protoplasm"). It is reasonable to assume that photosynthesis in the guard cells helps increase the concentration of osmotically active carbohydrates, but with the help of sophisticated measuring devices, it has been proved that the action of photosynthesis alone is not enough to account for stomatal opening. By calculating the rate of carbon dioxide fixation and the total volume of the guard cells in onion leaves, M. Shaw and G. A. Maclachlan estimated that the effects of photosynthesis could increase osmotic pressure by 0.34 bar (a bar is a unit physicists use for measuring pressure) in three hours. Yet J. Sayre showed that it takes an increase of at least six bar to open the stomata of onion leaves. That means it would take 60 hours of steady photosynthesis to accumulate enough carbohydrates to open the stomata; yet they actually open within an hour of receiving sunlight. It is evident that photosynthesis accounts for only a fraction of the necessary increase in osmotic pressure.

Subsequent investigation focused on a different aspect of photosynthesis. In the early 1930s, it was suggested that there was some connection between the low level of carbon dioxide found within an actively photosynthesizing leaf and the opening of the stomata. At first, this idea was nothing more than an inspired guess, but it was soon backed up by a great deal of experimental evidence. A link between low $CO_2$ concentration in the atmosphere and the opening of stomata has also been conclusively demonstrated, and researchers have learned not to go around breathing on their plants, lest the experiment be ruined by the $CO_2$ they exhale. It has been observed that in an atmosphere low in $CO_2$, stomata remain partially open even at night,

**7-6** *The structure and function of the stoma is clarified in this scanning electron micrograph. It is the bulging of the turgid guard cells that creates the hill-like stoma. Since the cells are not cemented together for most of their length, they can easily bend away from each other. (from* The Plant in Relation to Water *by R. O. Knight)*

**221**

when there could be no light effects at work.

Even though we know that $CO_2$ can affect the stomata, the causative relationship is not understood. One hypothesis is that a low level of $CO_2$ in the cell shifts the pH from acid to alkaline, causing a change in the rate at which certain metabolic processes occur. Another hypothesis links $CO_2$ to the rate at which starch can be converted into sugar.

It has recently been discovered that stomata open much faster in blue light than in other wave lengths. Meidner and Mansfield, two English botanists, conclude that this mechanism operates independently of other factors, such as the $CO_2$ concentration. They suggest that this blue-light mechanism helps keep the stomata open on cloudy days, when the relatively low rate of photosynthesis might otherwise allow the stomata to close. Since it has been established that guard cell chloroplasts contain substantial amounts of a yellowish photosynthetic pigment, research currently in progress is attempting to discover whether there might be some effect of blue light on these pigments that provides energy for the opening of the stomata. Light itself also might alter the permeability characteristics of the cell membrane.

In 1908, it was noticed that in guard cells, starch is converted to sugar in the daytime; in other leaf cells, this usually happens only at night, when the supply of sugar is running low. The starch-into-sugar hypothesis was quite popular for a number of years, until someone calculated that the maximum rate of conversion was much too slow to account for the increase in osmotic pressure in guard cells. The theory fell into disrepute, and plant physiologists tended to look politely incredulous whenever it was mentioned. Recently, however, this idea has been revived, and eyebrows no longer rise when it enters the conversation. The current version of the hypothesis is that the action of $CO_2$ removal and the illumination by blue light in some way speed up the rate of conversion.

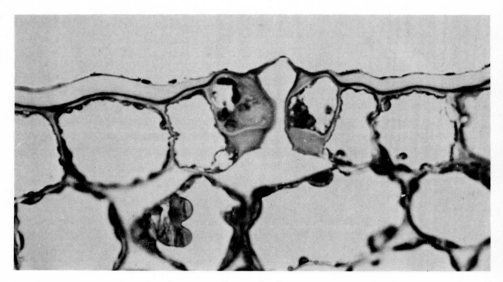

Discovering the mechanism by which stomata open is rather like putting together a jigsaw puzzle. The effect of light is one piece of the puzzle; the effect of $CO_2$ is another; photosynthesis is a third. Somehow, all these pieces of known experimental data have to be combined in one all-inclusive explanation that will give us the whole picture of this amazingly flexible mechanism that allows plants to regulate the opening so precisely, achieving a perfect balance between the demands of gas exchange and the need to conserve water under a wide variety of environmental conditions.

*7-7 As this photo shows, guard cells contain large and prominent chloroplasts. These chloroplasts manufacture sugar, which increases the osmotic concentration of the cell and thus causes water to flow into it. But rates of photosynthesis are inadequate to account for the amount of sugar found in guard cells, so some active transport mechanism is probably also involved. (Omikron)*

## Gas exchange in invertebrates

Among invertebrates, a wide variety of solutions to the problem of gas exchange can be observed. Most aquatic invertebrates, as well as all endoparasites, rely on their outer membranous covering as the major organ of gas exchange. These organisms always have a small diameter, and many are shaped like ribbons. They have a very

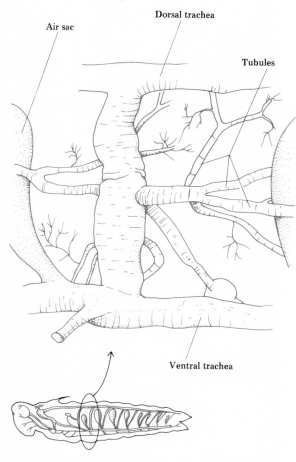

Air sac

Dorsal trachea

Tubules

Ventral trachea

**7-8** *Gas exchange in insects takes place in the tracheal system. Contraction of skeletal muscles, especially during flight, helps force atmospheric gases into the tracheae; the locking of the spiracles, or valvelike covers of the tracheae, traps the air and raises its pressure. On the basis of pressure gradients, it diffuses through the body. Eventually it will move with the concentration gradients of $O_2$ and $CO_2$, leaving the body.*

high surface-to-volume ratio, so the outer membrane is able to serve as the sole respiratory surface. The principal difficulty arises in the absorption of oxygen, which has relatively low solubility in water. Carbon dioxide is 30 times more soluble; therefore its transport presents much less of a problem. In some invertebrates, such as the nematode worms, the oxygen-absorbing capacity of the skin is supplemented by an oxygen transport system. The fluid that fills the major body cavity contains the respiratory pigment **hemoglobin,** which is capable of bonding $O_2$ molecules; in some invertebrates, another

respiratory pigment, **hemocyanin,** is the major oxygen carrier. This system of transport is much faster than simple diffusion, and it also compensates for oxygen's low degree of solubility. In starfish and sea cucumbers, water is pumped through canals or tubules within the body that increase the surface area over which diffusion of gases can take place.

The arthropods, with their heavy external skeletons made of **chitin** (a hard substance rather like the horn of an antelope or big-horn sheep), developed special respiratory organs, since their skin is not in direct contact with the environment. In aquatic arthropods, such as the lobster, specialized folds of skin, or gills, protrude from the joints of the walking legs. These structures function as oxygen-absorbing surfaces; the action of little bristles, or setae, on the gills creates a current that constantly sweeps new water past the surface.

Land-dwelling arthropods, like the insects, have a different respiratory system that is adapted to the problems of obtaining $O_2$ from air rather than water. The body of the insect contains a number of thin tubules, called **tracheae,** which carry air from the atmosphere into the internal tissues; gases can move by diffusion from the tracheae to the individual cells and back again, following the concentration gradient. Contraction of body muscles helps to push the air into and out of the tubules. During flight, for example, most of the tracheal air is changed. The openings of the tracheal system to the atmosphere are covered by a set of valvelike **spiracles.** The spiracles act to lock the openings; the pressure of the volume of air trapped in the tubules causes it to diffuse into body tissues. The spiracles also function to conserve water by preventing evaporation through the tubule opening. The tracheal system of gas exchange functions fairly efficiently for animals as small as insects, but because of the relatively slow rate of diffusion from the tubules into the cells, this method would be inadequate for animals of larger size.

## Organs of gas exchange in vertebrate animals

Vertebrate animals have specialized respiratory systems that allow rapid gas exchange. Oxygen is absorbed in the respiratory organs and then passed to the blood, where it is bonded by respiratory pigments. The combination of these two systems of exchange and transport allows vertebrates to carry on a high rate of cellular respiration.

All vertebrates can be divided into two major groups on the basis of the respiratory medium they utilize—air or water. In most respects, the animals that must obtain their oxygen from water are at a distinct disadvantage compared with those that obtain oxygen from air, and it is among the air breathers that we find the highest metabolic rates. The primary problem for aquatic animals is the low oxygen content of the water. Even the most thoroughly aerated water contains no more than 10 milliliters (mls.) of oxygen per liter of water; air often contains up to 130 mls. Water is also 100 times more viscous and 1,000 times denser than air, so that it takes much more work to push water through or into the respiratory organs. In terrestrial vertebrates, the work of **ventilation,** or the pumping of the respiratory medium over the epithelial tissues that absorb oxygen, uses about 1 to 2 percent of total absorbed oxygen, whereas in aquatic vertebrates, the work of ventilation may require as much as 25 percent of the absorbed oxygen. This represents a heavy energy drain for the water-dwelling animal. In order to survive, it has had to develop a system of oxygen uptake that utilizes available oxygen more efficiently. Fish can absorb up to 80 percent of the oxygen present in water, whereas mammals absorb only about 25 percent of the oxygen in air.

Aquatic vertebrates do have a few advantages over their terrestrial relatives. The water in which

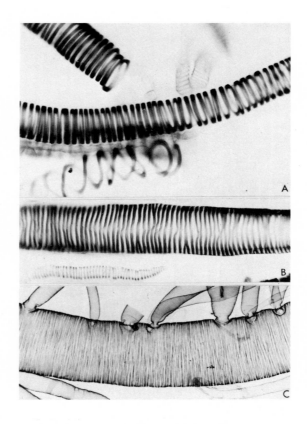

7-9 *A comparison brings out the functional analogy between two different kinds of reinforced conductive tissue, plant conductive tissue and the trachea of an insect. Although they have no evolutionary relationship, the demands of their function has led to some similarities in structure. (Esaw Plants, Viruses and Insects [Plate 2.] Harvard University Press)*

they live solves the problem of keeping epithelial membranes moist enough for gas exchange to take place; there is no danger of drying out. The water also helps to support the respiratory organs.

Because of the difference in the two respiratory media, we would expect to find different respiratory organs developing in aquatic and terrestrial vertebrates. Since aquatic animals are not threatened by drying out, their respiratory organs can be directly exposed to the medium. Because of the buoyant support of water, the surfaces can be delicate and fragile. Terrestrial vertebrates, on the other hand, cannot expose their respiratory organs directly to the environment, and lacking the support of water, the organs must have some kind of noncollapsible surface. These requirements obviously cut down on the efficiency of oxygen uptake, but this loss is more than compensated for by the advantages of air as a respiratory medium.

## Gills

The respiratory organ of aquatic vertebrates is the **gill.** A gill is often defined as a respiratory organ that is turned outward; it is a projection, rather than a cavity like the lung. In most fish, the projecting gill is supported by a series of bony arches. The absorbing surfaces of the gills are actually epithelia that have been folded to form **secondary lamelae.** A cross section of a secondary lamella shows that it is made of two thin epithelial layers attached to a basement membrane. Between the two membranes are tiny blood capillaries and specialized cells called **pillar cells,** which seem to serve a support function, holding the two layers apart. Pillar cells may also have a limited contractile function, helping to move the folds of the gills to create water currents. The structure of the secondary lamellae is

really rather similar to the structure of intestinal villi, projections which also are specialized for the work of absorption.

The work of ventilation is performed by muscular contractions that pump water across the gill surfaces. The water moves in through the mouth, over the gills, and is pushed out through the opening of the gill slits. Two pumping mechanisms seem to be at work moving the water. There is a pressure pump activated when the fish swallows a mouthful of water; there is also a suction pump that is activated by the expulsion of water out of the gill slits. The one-way current thus established cuts down on the amount of energy needed to move the water, since it takes advantage of the inertia of the moving current. Some fish also rely on the effect of their forward swimming motion to force currents of water into the mouth and over the gills. The mackerel, for instance, will suffocate if it is forced to remain stationary in the water.

As the water passes over the secondary lamellae, the oxygen content is absorbed by the epithelial cells and then passes on into the bloodstream for transportation to the rest of the body. One reason for the very high rate of oxygen absorption found in gill breathers is that the flow of blood within the gills moves in a direction opposite to the flow of water. This means that water entering the gill, with its high oxygen content, meets blood that is already partly saturated with oxygen. As the water loses its oxygen content, it comes in contact with low-oxygen blood entering the gills. The advantage of this arrangement, called the **counter-current principle,** is that a more or less constant gradient is maintained between the two liquids, so that the rate at which oxygen moves into the blood remains constant for the entire period of contact.

Consider what would happen if the water flow went the other way, running parallel to the blood flow. When the oxygen-loaded water entered the gills, it would meet oxygen-depleted blood. At that point, the gradient would be very

**7-10** *This schematic drawing of a gill gives some idea of the structure of that delicate respiratory organ. Since it is made up of folded epithelial tissue, it depends on the buoyancy of the water for support. When a fish is taken out of water, the folds of the gill collapse against one another and the poor fish suffocates.*

Filaments

Water flow

Secondary lamellae

Arteries

Gill arch

225

high, and rapid movement of oxygen into the blood would take place. But as the water moved through the gills and lost some of its oxygen, it would begin to encounter blood that was already partly saturated with oxygen. The gradient between the high-oxygen blood and the low-oxygen water would be very low, so movement of oxygen into the blood would be quite slow. In an ingenious laboratory experiment that used mechanical pumps to reverse the water flow over a fish's gills, it was found that under these experimental conditions the fish extracted only about 10 percent of the available oxygen in the water. Under normal conditions, with a counter-current flow, the fish could absorb about 70–80 percent of the oxygen in the water.

The delicate gill tissues are protected in most fish by a bony plate, called the **operculum;** the covering is a set of specialized skin flaps in cartilagenous fish, such as sharks and rays. The covering over the gill opening helps protect against mechanical injury to the gills and also helps to prevent the entrance of foreign particles. This is especially important in bottom feeders, such as sole, because their feeding activities stir up the sand and sludge, and these particles could easily clog the gills. Some fish—carp, for example—have developed the ability to reverse the flow of water over the gills for a brief period. This serves as a cleaning mechanism, rather like coughing in humans.

The structure of the gill is beautifully adapted for use in an aquatic environment. But when a fish is removed from water, the fragile surfaces of the secondary lamellae, lacking the buoyant support of water, collapse against each other and prevent the circulation of air between them. Even though the fish responds to this crisis by slowing its respiratory rate and initiating anaerobic respiration, it soon suffocates.

Certain fish have developed structures to support the gills on land, and they can survive in an air environment for some time, until the drying out of gill tissue prevents further gas ex-

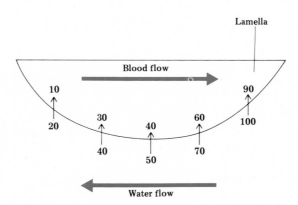

change. Other aquatic vertebrates, such as the eel, with its narrow scaleless body, can rely on gas exchange through the skin to supplement the action of their gills in the air. This enables the eel to move overland from one body of water to another. Cutaneous (skin) gas exchange also may be utilized in the water during times of high oxygen consumption. Some aquatic species have other accessory respiratory organs. The swim bladder, a sac filled with air whose principal function is to give buoyancy to heavy-boned

**7-11** *Counter-current exchange takes place in the gills, for the direction in which water passes over the gills is opposite to the direction in which blood travels through them. Counter-current exchange serves to maintain a more or less constant gradient in the $O_2$ concentration of blood and water, which means that the $O_2$ exchange takes place at the same rate throughout the entire gill.*

**7-12** *This close-up view of the sand shark, found along the coasts of the Atlantic Ocean and the Mediterranean Sea, shows the flaps of tough skin covering the gills. In bony fish similar protection is provided by the operculum. (Arthur Ambler from National Audubon Society)*

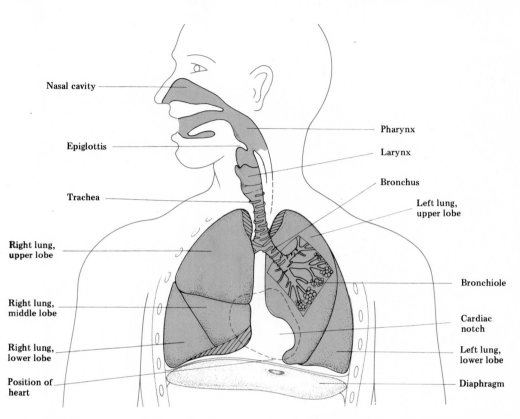

**7-13** *The human respiratory system is generally thought of as having two major divisions—the upper respiratory tract, including the nasal passages and larynx; and the lower respiratory tract, or the bronchial tubes and lungs. The actual gas exchange takes place on the folds of epithelium lining the lung cavity.*

Nasal cavity

Epiglottis

Trachea

Right lung, upper lobe

Right lung, middle lobe

Right lung, lower lobe

Position of heart

Pharynx

Larynx

Bronchus

Left lung, upper lobe

Bronchiole

Cardiac notch

Left lung, lower lobe

Diaphragm

fish, can sometimes serve as a respiratory reservoir; this adaptation can be seen in goldfish and catfish. In some fish, the epithelium lining the digestive tract can carry on supplementary gas exchange, and the fish simply swallows a mouthful of air. The lungfish, as its name suggests, has rudimentary lungs for air breathing. When their pond dries out, these fish go into **estivation,** a state in which metabolism and respiration are reduced. Oxygen consumption may drop to 10 percent of the level of consumption during activity. To supply this small need, the lungfish can breathe air through their lungs while they wait for the pond to refill.

## Lungs

A lung is a respiratory organ specialized for use in an air medium. Any respiratory organ which is turned in, forming a hollow cavity, can be defined as a lung. Internalization of respiratory organs is necessary for animals who live in a terrestrial environment; if the membranes over which gas exchange takes place were to be exposed to the atmosphere they would quickly dry out and cease to function.

The lungs are actually sacs contained in the chest cavity. They are thought to have evolved in early bony fish as an outgrowth of the *anterior* (front) end of the digestive tract. The outer walls of these sacs are made of interlocking fibers of elastin. The inner walls are covered with epithelial tissue that has been folded many times to increase its total surface area; the human lungs contain a surface area of over 70 square meters (about 230 square feet) for gas exchange. The epithelial folds form little saclike compartments (about 150 million of them in an average human lung) called **alveoli.** Each alveolus is about $100-200\mu$ in diameter. Epithelial cells cover the outside; below them are a basement membrane

**227**

and sometimes a few connective cells. Surrounding each aveolus is a capillary network into which absorbed oxygen can diffuse. The surface of the alveoli epithelia is covered by a very thin lipid-protein layer; this coating helps to distribute molecules that come in contact with it evenly over the surface of the alveoli.

The alveoli are bathed by a film of water less than $1\mu$ thick that enables gas exchange to take place. Oxygen moves with the concentration gradient out of the air and into the blood; carbon dioxide moves the other way. The surface of mammalian lungs is so large that once air is inhaled, all gas exchange takes place in less than one second; however, as we have noted earlier, only about 25 percent of available oxygen is absorbed.

Ventilation in air-breathing vertebrates is **tidal;** that is, the respiratory medium flows in and out of the organs in two different directions. Tidal ventilation requires a greater expenditure of energy than does a one-way current, since the flow of the medium must be actively reversed. Tidal ventilation is a luxury air breathers can afford, because air is so much lighter and less viscous than water; pushing it along does not require much energy under any circumstances. The advantage of tidal ventilation is that it cuts down the number of specialized structures needed within the respiratory system, since the same passageway can serve both to bring air in and conduct it out.

There are two different methods of ventilation in air breathers. In amphibians and reptiles, air is forced into the lungs by a positive pressure. In the lizard, for example, air enters the mouth and throat and then all external openings are closed. The pressure of this mouthful of air creates a pressure gradient which forces the air from the throat into the lungs. In frogs, the pressure can be increased by the squeezing action of a muscle in the mouth.

The mammalian system of ventilation depends on negative pressure. At the beginning of inspiration (the inhaling of air) a number of

muscles begin to contract. The diaphragm, a sheet of muscle that separates the lung cavity from the abdominal cavity, contracts and moves downward. Its movement expands the total area of the chest cavity. The action of the diaphragm is supplemented by other muscular contractions that pull the ribs outward and cause increased expansion of the chest cavity. Because of this expansion, the pressure within the chest cavity falls, so that it is lower than the atmospheric pressure outside the body. This pressure gradient causes air to flow into the lungs.

As the air fills the lungs, their elastic walls expand. This movement of the walls outward is analogous to the movement of a book when you lift it to your shoulder; it stores potential energy. As long as air continues to flow into the lungs, more potential energy will be stored in the walls of the lung. When the air flow stops (this occurs when the pressure inside the lungs is no longer lower than the pressure of the atmosphere), the stored potential energy begins to be converted into kinetic energy. It is this kinetic energy that squeezes the lungs shut and forces air back out. Exhalation is therefore largely a passive process.

The normal rate of ventilation in man is 12–14 breaths per minute. The total capacity of the lungs in the average adult is about 5½ liters (11½ pints) of air. However, since there is no way to wring every last molecule of gas out of the lungs, the total breathing capacity, or the amount of air that can be exhaled after the lungs have been filled to capacity—the **vital capacity** of the lungs—is about four liters for the average person; athletes may be able to inhale another liter or so. The difference between total capacity and vital capacity is called the **residual volume;** it is the air that cannot be squeezed out of the lungs. The residual volume is about 1500 ccs. In normal breathing, the amount of air inhaled is far below the vital capacity. On the average, it is about 500–600 ccs. (half a liter). Not all of this amount actually reaches the lungs; some of it, perhaps 150 ccs., fills the so-called dead space of the passageway

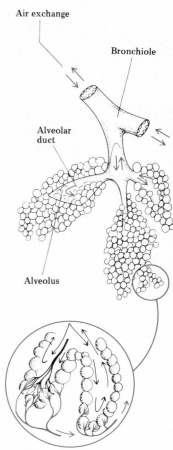

**7-14** *The arrows indicate the direction of air flow inside the alveolus. Once $O_2$ is absorbed by the epithelial tissues, it moves into the bloodstream via the surrounding network of capillaries. The oxygenated blood then returns through the pulmonary veins to the heart, where it is pumped out into the systemic arteries.*

Air exchange

Bronchiole

Alveolar duct

Alveolus

Gas exchange between blood and air

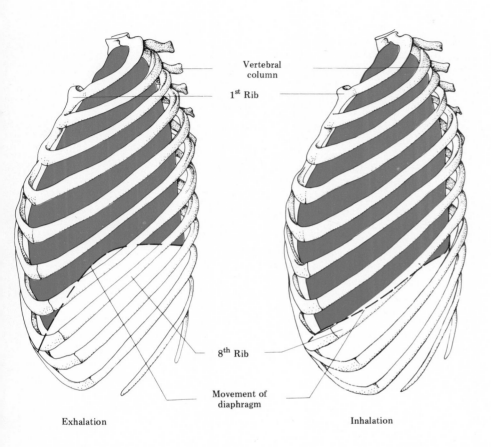

Vertebral
column

1<sup>st</sup> Rib

8<sup>th</sup> Rib

Movement of
diaphragm

Exhalation

Inhalation

**7-15** *Two different mechanisms are at work in helping fill the lung during inspiration. One is a movement of the ribs and the muscles surrounding the rib cage; these muscles literally spread apart the chest cavity, allowing the expansion of the lungs. The second mechanism, which operates most of the time in males but only infrequently in females, is the contraction and consequent downward pull of the diaphragm.*

the respiratory system to provide oxygen is equal to any demand that might be made. The major factor limiting metabolic rate is the relatively slow speed of internal transport once the oxygen has been absorbed into the bloodstream.

Since the rate of diffusion of all molecules is greatly influenced by temperature, and since there may be seasonal variations of as much as 40°C in the temperature of the air, it is important that the respiratory system contain some means of mediating air temperature. This is one of the functions of the upper respiratory tract. When air is inhaled through the nose or mouth, it passes over mucous membranes. These membranes provide a large surface area, serviced by a rich network of blood capillaries, that is ideally suited for the work of heat transfer. If the air inhaled is very cold, heat energy from the blood is transferred to the air; if the inhaled air is hot, its heat energy passes into the blood (heat energy, too, moves along a gradient from high to low). This heat energy transfer mechanism is so efficient that when laboratory animals were exposed to air heated to 260°C and cooled to −37°C, the air in both instances was at the level of body temperature even before it entered the lower portion of the **trachea,** the main passage to the lungs.

The upper respiratory system also helps to protect the lungs against the inhalation of foreign particles. The hairs lining the nostrils trap large particles as soon as they are inhaled. In addition, the complicated arrangement of the nasal bones acts as a kind of baffle that prevents the penetration of large particles; in fact, no particle larger in diameter than 10 $\mu$ will be able to enter the trachea.

The trachea, bronchi, and bronchioles, the passageways that conduct air to the lungs, have another form of defense against foreign particles. They are covered with a sticky mucus, which catches the particles just as if they were flies landing on flypaper. Then the particle and the mucous coating are gradually moved upward toward the mouth by the action of the cilia lining

into the lungs. No oxygen exchange takes place in this passageway.

The difference between the normal volume of air that moves in and out of the lungs with each breath is called the **tidal volume,** and the vital capacity indicates that the respiratory organs are capable of a great increase in the rate of gas exchange when metabolic needs suddenly increase. Additional muscles may come into play, forcing a greater extension of the chest cavity and therefore enabling the inspiration of a greater volume of air. Muscles along the outer walls of the lungs may also contract during exhalation, providing more thorough emptying of the lungs. The rate of ventilation may also speed up. The ability of

the passageways. The cilia beat rhythmically to create a one-way current, and the particles in their mucous coating are slowly but inexorably moved up out of the respiratory system. They will finally be removed by a cough or a sneeze or a clearing of the throat.

**7-16** *If the lung defenses fail, particles may remain in the lungs, causing the growth of non-absorptive fibrous tissues around them and thereby causing a decrease in the total absorptive capacity of the lungs. This cross-section of a damaged lung gives some idea of the changes that take place in the lung under such conditions. (Omikron)*

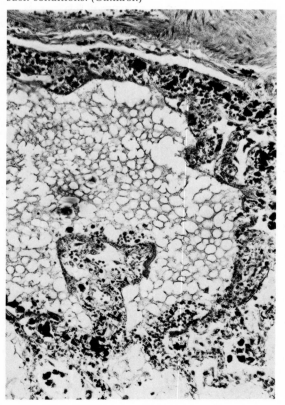

Particles smaller than two microns may escape this protective system and enter the lung; this is especially likely if the particles are encased in fluid coatings. Most of these particles will be attacked by white blood cells, and they will either be digested or carried to the lymph node to be placed in cold storage. However, if all of these defenses fail, then the particles will remain in the lung, causing the growth of nonabsorptive fibrous tissue around them and thereby decreasing the total absorptive capacity of the lungs. People who live in heavily congested cities, miners who breathe in coal dust, cigarette smokers, and construction workers who handle asbestos in any form, are all particularly likely to develop a large number of these fibrous areas in their lungs, to the extent that their ability to absorb oxygen is seriously impaired.

## Transport of respiratory gases

In vertebrates, both oxygen and carbon dioxide are transported in the bloodstream back and forth between individual body cells and the specialized respiratory organ that provides the surface for gas exchange with the respiratory medium. The transport capacity of the plasma is greatly increased by the presence of red blood cells, which contain the respiratory pigment, hemoglobin. It is found in many invertebrates and all vertebrates, except for a few species of antarctic fish with very sluggish metabolisms.

### Oxygen transport

Some of the oxygen absorbed in the lung or gill does dissolve in the liquid portion of the blood. Because of the relatively low solubility of $O_2$, this method of carrying oxygen is not very efficient. G. M. Hughes estimates that if man had to depend on plasma alone as a carrier of oxygen, he would

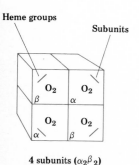

Heme groups

Subunits

O₂ β

O₂ α

O₂ α

O₂ β

**4 subunits ($\alpha_2\beta_2$) = Hemoglobin**

**7-17** *The hemoglobin molecule can be thought of as being rather like a square box (although, of course, its shape is not really that regular). In each quadrant of this square there is a space, or site, where the bonding of an $O_2$ molecule can take place. When all four sites are filled, the molecule becomes oxyhemoglobin.*

**7-18** *This graph shows contrasting $O_2$ dissociation curves for the blood of (A) a trout at 22°C, and (B) a sheep at 38°C. The trout's blood can be seen to absorb $O_2$ more readily. Christian Bohr discovered that an increase in $CO_2$ in the blood will cause the curve to shift to the right, in the so-called Bohr effect.*

either have to have 30 times as much blood, or else circulate the blood he has 30 times faster, in order to meet his oxygen needs.

Most oxygen is transported by being bonded to a molecule of hemoglobin. Hemoglobin is a conjugated protein, containing four atoms of iron. Each iron atom is contained in a heme (iron-containing) group that is covalently bonded to a large carrier portion of the molecule, a globular protein. The iron atom has the ability to form a total of six covalent bonds. Four of those bonds are formed with nitrogen atoms that link the iron to the rest of the heme group; one more bond is formed with the carrier part of the molecule. That leaves one bonding space unfilled, and it is that space which the oxygen molecule will occupy.

Each hemoglobin molecule can hold four molecules of oxygen; when all four are attached, it is said to be fully saturated and it is given the name of **oxyhemoglobin.** Not all hemoglobin molecules will become fully saturated in the lung or gill. Saturation depends on a number of factors. One is temperature; the colder it is, the higher the saturation. Another is pH; under acid

conditions, the saturation will increase. The influence of pH on rate of oxygen uptake has some interesting implications. When the blood has a high carbon dioxide level, blood pH will drop as carbon dioxide combines with water to form carbonic acid. This increased acidity will cause more oxygen molecules to bond to the hemoglobin and thereby increase the rate of oxygen uptake. This is one of the self-regulating mechanisms that keeps the respiratory system closely adjusted to the body's need for gas exchange.

The most important determinant of saturation is the concentration gradient between the oxygen already in the blood and the oxygen in the respiratory medium. The same set of factors determines the rate at which oxygen held in the hemoglobin molecules will be yielded in body tissues. The lower the oxygen pressure in the tissues, the more quickly and completely the oxygen will be given up.

By exposing blood kept at constant temperature and constant pH to oxygen at varying concentrations, a curve can be drawn showing the propensity, or "willingness," of the hemoglobin to load and unload its oxygen molecules. This is called an **oxygen dissociation curve.** Once the basic curve is defined, then temperature and pH can be varied and their effects on loading and unloading measured. Christian Bohr (father of physicist Niels Bohr, who proposed the modern model of the atom) discovered that an increase in $CO_2$ causes more $O_2$ to be given up, shifting the curve to the right; this displacement is called the Bohr effect, in his honor.

Oxygen dissociation curves are different for the various vertebrate groups. The curves also show some correlation with the environment to which the animal is adapted. In aquatic vertebrates, for example, the curve is much to the left of the curve for terrestrial relatives, meaning that the blood of a fish binds $O_2$ more readily than the blood of a mammal. This reflects the need for increased efficiency of uptake when the respiratory medium is low in oxygen content. Animals

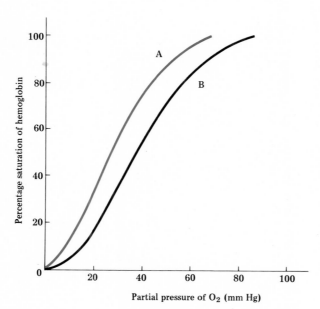

Partial pressure of $O_2$ (mm Hg)

**231**

that move rapidly and frequently have curves much to the left of the curves for slow movers. The hummingbird, for example, binds $O_2$ more efficiently than any other bird. Diving mammals and animals whose native territory is located at high altitudes also have curves far to the left. And fetal blood has a curve far to the left of maternal blood; this allows the fetus to absorb oxygen from its mother's blood.

Several experiments have been carried out to find out whether an individual animal's oxygen dissociation curve will change when it is introduced into a new environment. The answer seems to be—no, it does not. When an animal adapted for low-altitude living is moved to a high-altitude environment, its oxygen dissociation curve does not change, no matter how long it remains in that environment. What does happen is that the hematocrit rises; red blood cells constitute a higher percentage of the blood volume. It appears that the oxygen deficiency the animal experiences when first moving to high altitudes stimulates the release of a hormone produced by the kidney that serves to speed up the production of red blood cells. With more red blood cells, the oxygen-transporting capacity of the blood increases, even though the oxygen dissociation curve remains exactly the same.

It might seem rather inefficient for hemoglobin to be contained within the red blood cells. Loading and unloading of oxygen would certainly take place more quickly if the hemoglobin molecules were set loose in the bloodstream, rather than being crowded together and trapped inside a cell membrane. The red blood cell is actually one of those compromises, rather like the stomata, that allows one function to proceed with only slightly diminished efficiency while guarding against potentially disastrous side effects. If the hemoglobin were to be carried directly in the bloodstream, two major problems would arise. The blood would be much more viscous, and therefore the task of circulation would be more difficult. Moreover, the osmotic pressure of the blood would increase enormously, so that in the capillaries the blood would practically suck the body cells dry.

## Carbon dioxide transport

It is convenient for writers of biology textbooks to divide the subject of gas exchange into two phases—transport of oxygen and transport of carbon dioxide—so that the steps in each process can be described independently. But this division is not found in the actual process of gas exchange. Both molecules are transported in the same place (the bloodstream) at the same time. The molecules are quite literally exchanged, or to be more precise, they displace one another. As the Bohr effect demonstrates, it is the presence of $CO_2$ that causes $O_2$ to be given up by the hemoglobin. In the absence of $CO_2$, the hemoglobin molecule might make a whole circuit through the circulatory system without ever yielding its firmly bonded passenger, $O_2$. The converse is also true, as we shall soon discover; it is the presence of $O_2$ in the lungs that causes $CO_2$ to leave the bloodstream. Transport of $O_2$ and transport of $CO_2$ are completely interdependent processes.

Carbon dioxide is more readily soluble than oxygen; yet only about five percent of the total $CO_2$ content of the blood is carried in the form of a physical solution. Most of it combines with other compounds, both in the plasma and in the red blood cells. These reactions take place relatively quickly and are easily reversible, allowing the $CO_2$ to be unloaded in the lungs and passed back out of the body.

When absorbed $CO_2$ meets the $H_2O$ in the plasma, a combination reaction occurs, forming $H_2CO_3$, or carbonic acid. This product easily falls apart to form bicarbonate ions ($HCO_3^-$) and hydrogen ions ($H^+$). It is in the form of bicarbonate ions that most $CO_2$ travels through the bloodstream. These ions are not just idle passengers; while in the bloodstream they constitute the

**7-19** *This schematic drawing shows the various mechanisms of oxygen transport in the body. The circle within the area marked "Blood" represents a red blood cell, which is the site of most oxygen transport.*

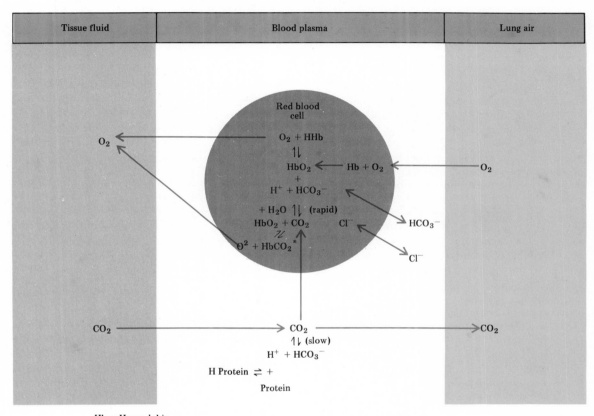

| Tissue fluid | Blood plasma | Lung air |
|---|---|---|

Red blood cell

$O_2 + HHb$

$\updownarrow$

$HbO_2 \leftarrow Hb + O_2$

$+$

$H^+ + HCO_3^-$

$+ H_2O \updownarrow$ (rapid)

$HbO_2 + CO_2 \qquad Cl^-$

$O^2 + HbCO_2^*$

$O_2$

$HCO_3^-$

$Cl^-$

$O_2$

$CO_2 \longrightarrow CO_2 \longrightarrow CO_2$

$\updownarrow$ (slow)

$H^+ + HCO_3^-$

H Protein $\rightleftharpoons$ +

Protein

Hb = Hemoglobin
$HbO_2$ = Oxyhemoglobin
$HbCO_2$ = Carboxyhemoglobin

$\rightleftharpoons$ Denotes—Reversible reaction
$\longrightarrow$ Denotes—Movement of compound

alkali reserve of the body, an important defense mechanism against the excessive production of acids in various body organs. The bicarbonate ions help to maintain the constant acid-base balance that is needed for essential nervous and chemical regulatory reactions.

The free hydrogen ions represent a potential threat to the organism, since they raise the pH of the blood, decreasing the body's alkaline reserve; if there are enough of them, they might finally turn blood into an acid. To prevent this, they must be buffered. When dissociation of carbonic acid takes place in the plasma, the hydrogen ions are buffered by some of the plasma proteins.

The formation of carbonic acid, and its subsequent dissociation into bicarbonate and hydrogen ions, take place much too slowly for this pathway to account for the rate of $CO_2$ uptake in the capillaries. A clue to the mechanism that helps speed the reaction was discovered in 1914, when the Danish scientist Christian Christiansen noticed that $CO_2$ was absorbed faster when the

**233**

$O_2$ content of the blood was low. This observation led to the hypothesis that hemoglobin plays some role in $CO_2$ transport. Later experimental evidence has confirmed this hypothesis. Once hemoglobin molecules unload their $O_2$, they can act as a buffer for the hydrogen ions that are released by the dissociation of carbonic acid.

Most of the $CO_2$ that enters the blood passes into the erythrocytes. Inside the cell, the combination reaction of $CO_2$ and $H_2O$ that produces carbonic acid is speeded up by the action of an enzyme, carbonic anhydrase. Then as the carbonic acid comes apart to produce hydrogen ions, these are buffered by the hemoglobin, thus stabilizing the pH of the cell. The $HCO_3^-$ ions then move with their concentration gradient and cross the cell membrane into the blood plasma. This mass exodus of anions could cause a harmful change in the electrical balance of the cell. However, as they move out, the resulting shift of the cell contents toward a positive charge attracts other anions present in the plasma. The most numerous are the chloride ($Cl^-$) ions, which can easily penetrate the cell membrane. Thus, as the bicarbonate ions move out, the chloride ions move in. This is called the **chloride shift.**

Recently it has been discovered that another $CO_2$-absorbing reaction also takes place within the red blood cell. Amino groups of the hemoglobin molecule can interact with $CO_2$ molecules in solution to form a carbamino compound. About 20 percent of the $CO_2$ content of the blood is transported in this form.

In the lungs, $CO_2$ is unloaded as oxygen begins to move into the red blood cell. The $O_2$ bonds onto the hemoglobin, and the strongly acidic nature of the oxyhemoglobin causes $H^+$ ions to be released. The sequence of reactions that took place when the $CO_2$ entered the blood is reversed; the liberated $H^+$ ions combine with $HCO_3^-$ ions to form carbonic acid, and the acid breaks apart into molecules of $H_2O$ and $CO_2$. Carbonic anhydrase again serves as the catalyst. The resulting concentration of $CO_2$ in the red

blood cell sets up a concentration gradient that causes $CO_2$ to move out into the lungs.

## The control of respiration

The rate of respiration and the tidal volume must be continuously regulated, in order to provide just that amount of gas exchange needed by the organism in its current state of activity. If gas exchange is insufficient, the lack of $O_2$ and the buildup of $CO_2$ may harm the animal; if gas exchange is excessive, the animal is subjected to needless strain and depletion of energy resources.

For centuries, animal physiologists have been searching for the control mechanism that regulates the rate of respiration in response to the changing needs of the body. The first successful experiment to isolate one of the respiratory control centers was performed by César Legallois in 1811. Working with rabbits, he carefully removed small parts of the brain. Legallois found that when he removed small slices—a few millimeters thick—from the lower medulla, respiration ceased. We now know that this small part of the brain establishes the basic respiratory rhythm.

The precise manner in which the medulla receives the information that is needed to coordinate respiratory activities is still rather unclear. It seems to function quite independently. For example, it has been shown that the medulla continues to establish a regular rhythm of breathing even when all nervous connections between it and other portions of the brain have been severed. It has also been shown that regular respiration will persist for some time after the nerves leading from respiratory organs to the brain are severed, thus denying the medulla any specific information from that area.

The fact that the medulla can, under emergency conditions, continue to regulate respiratory activity in the absence of any information input from either the higher sensory centers of the brain

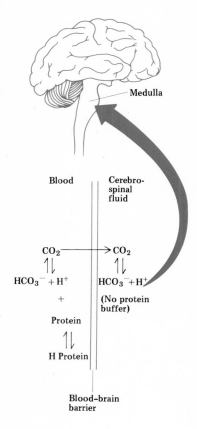

**7-20** *Chemoreceptors located on the medulla itself are an important part of the feedback mechanism that maintains a constant level of $O_2$ in the body and especially the bloodstream. These chemoreceptors are extremely sensitive, responding to very small changes in the concentration of hydrogen ions in the bloodstream.*

Medulla

Blood | Cerebrospinal fluid

$CO_2 \longrightarrow CO_2$

$HCO_3^- + H^+$ | $HCO_3^- + H^+$

$+$ | (No protein buffer)

Protein

H Protein

Blood–brain barrier

or the respiratory organs themselves does not mean that it customarily operates without this input. Under normal circumstances, a wide variety of stimuli can send messages to the medulla and alter the respiratory rate. Pain, heat, cold, the stimulation of certain sights and smells—these have all been found to alter the rate of respiration. In order to account for all known evidence, biologists postulate a control center in the medulla that is capable of completely independent regulation on occasions when informational input is absent, but that usually acts in response to many different kinds of input from the nervous system.

## The role of chemoreceptors

In the latter part of the nineteenth century, the German physiologist Eduard Pflüger demonstrated that both low oxygen and high carbon dioxide pressures in the respiratory medium would stimulate increased respiratory activity. But it was not until the 1930s that another European physiologist, Corneille Heymans, was able to isolate the mechanism by which these changes in gas pressure were translated into changes in respiratory action. Heymans' investigation centered on the **carotid bodies,** small spherical bodies of specialized tissue located in the branches of the carotid artery (the main artery that supplies blood to the head).

Heymans was able to show that the carotid bodies were sensitive to chemical changes; they act as **chemoreceptors.** Any increase in $CO_2$ or decrease in $O_2$ is immediately registered in the carotid body, and in a similar patch of tissue located in the aorta (the **aortic bodies**). The chemoreceptors then send messages through the network of nerve fibers to which they are connected. These messages are received and correlated in the medulla, which then establishes a new respiratory rate to compensate for the problem. Recent experimental evidence indicates that the oxygen pressure in the blood decreases, the

carotid and aortic bodies become even more delicately sensitive, responding to even slighter changes in the pressure of either of the two gases. These chemoreceptors are also able to respond to changes in blood pressure, although this sensitivity seems limited to reduced or low pressure; an increase in blood pressure elicits no response.

In addition to the chemoreceptors in blood vessels, chemoreceptive areas in the brain have also been found. These receptors are on the medulla itself, right on the surface where they are exposed to the cerebrospinal fluid that bathes the brain. They actually respond to changes in the number of $H^+$ ions, rather than changes in $CO_2$ pressure. When the amount of $CO_2$ in the blood vessels near the surface of the brain increases, it diffuses out into the cerebrospinal fluid, where it is immediately changed to bicarbonate and hydrogen ions. Since the fluid has a very low protein content, the $H^+$ ions cannot be adequately buffered and their concentration increases. It is this increase that stimulates the chemoreceptors. In this area, they are somewhat more sensitive to small changes in concentration than are the receptors in the carotid bodies, and since they are located so close to the control center, adjustment can be made very quickly.

The sensitivity of respiratory activity to the level of $CO_2$ in the blood is not the same for all animals. Diving mammals, for example, do not respond to small increases in $CO_2$; it is this adaptation that enables them to hold their breath under water for long periods. And high-altitude dwellers are less sensitive to low $O_2$ than their relatives who are adapted to live at sea level.

## The role of the vagus nerve

The vagus nerve, the same nerve that slows down the action of the heart, also plays a part in regulating the rate of respiration. When the vagus nerve is stimulated, it acts to inhibit inspiration; the filling of the lungs stops. It is the action of the vagus nerve that accounts for the relatively small tidal volume of the lungs during periods of low activity. By shortening the length of time devoted to inspiration (which, of course, means that the period of expiration will also be shortened), the number of breaths per minute can be increased. The inhibiting action of the vagus nerve provides more ventilation with less work.

The effect of the vagus nerve is probably a reflex reaction, coordinated through the medulla. But in certain cases, the reflex can be overridden. In times of violent exercise or oxygen deficiency in the respiratory medium, the automatic inhibition by the vagus nerve stops, and the tidal volume increases. This probably occurs as the result of messages sent to the medulla from other sensory centers.

## Respiration and metabolic rate

The rate of oxygen intake is carefully regulated to coincide with the oxygen requirements of the organism. Of course, these requirements will change as the animal alters its pattern of activity. A rabbit that is trying to escape from a pursuing fox will suddenly need much more oxygen to fuel its frantic muscular activity than it needs when it spends a cozy evening at home in its burrow. In both cases, the rate of ventilation and tidal volume (and the circulatory rate of flow as well) will be adjusted to meet the change in oxygen demand.

Bursts of activity and periods of sluggish idleness are both only limited variations in the basic oxygen demand that is established by the metabolic rate of the organism. This is the rate at which it must consume oxygen in order to fuel the necessary life activities it carries on—obtaining and digesting food, maintaining circulation of the blood, ventilating the respiratory organs, excreting wastes and excess fluids, producing the eggs and sperm needed for reproduction.

The rate at which an animal uses energy under normal, nonstressful conditions can be measured in several ways. The most practical way to measure this rate (the rate of **basal metabolism)** is to measure the amount of oxygen used by the organism. This can be done for small animals with a very simple apparatus. The animal is placed in a closed container that has an apparatus for measuring air pressure inside the container and a hypodermic needle that can inject oxygen inside the container; it also contains granules of soda lime which absorb $CO_2$. As the animal uses $O_2$, the air pressure in the container will drop. The rate at which the $O_2$ is being used can be determined by the amount that must be injected to keep the air pressure constant. In order to compensate for the effects of various body sizes on oxygen needs, the results of this measurement should be expressed in terms of quantity of oxygen consumed per unit of body weight, over some set time period.

Table 7-1 shows the basic rate of oxygen consumption in a number of different animals. As the table shows, there is a wide range of metabolic rates. Part of this difference is due to the level of organization found in the animal. The more highly organized it is—the more extreme the differentiation and specialization of cells and tissues—the greater its energy needs will be. Another important factor is the pattern of activity demanded by the animal's methods of feeding. For example, a fish like the eel, which generally lies in wait for its prey and pounces on it as it swims by, has a lower metabolic rate than the trout, which chases its prey vigorously. The great difference in oxygen consumption between a resting hawk and a hawk that is flying in pursuit of another bird gives us some idea of just how energy-expensive the search for food can be. Interestingly, a study that correlated oxygen consumption with distance traveled (per unit of body weight) showed that flying is usually the cheapest way to go. A mouse, for example, spends twice as much energy in walking as a pigeon does

**7-1**

**Oxygen capacity of blood at sea level and high altitudes**

| Animal | Altitude (feet) | Red cell count (million cells/mm.[3] | O$_2$ capacity of blood (vol %) |
|---|---|---|---|
| Man | sea level | 5.0 | 21 |
|  | 17,600 | 7.37 | 30 |
| Sheep | sea level | 10.5 | 15.9 |
|  | 15,420 | 12.05 | 18.9 |
| Rabbit | sea level | 4.55 | 35 |
|  | 17,500 | 7.00 | 57 |

Source: C. L. Prosser and F. A. Brown, Jr. 1961. *Comparative Animal Physiology*. 2nd ed. Saunders, Philadelphia.

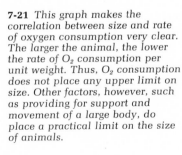

**7-21** *This graph makes the correlation between size and rate of oxygen consumption very clear. The larger the animal, the lower the rate of O$_2$ consumption per unit weight. Thus, O$_2$ consumption does not place any upper limit on size. Other factors, however, such as providing for support and movement of a large body, do place a practical limit on the size of animals.*

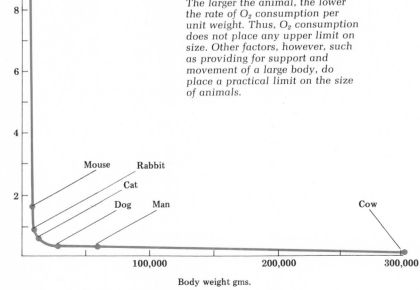

in flying. Swimming is the most energy-consuming way to travel, since the animal has to push its way through a dense and viscous medium.

When we compare metabolic rates on the basis of size (see Figure 7-21), we find that it is the smallest animals that have the highest metabolic rates; in fact, this rule of thumb can also be extended to the plant kingdom, for the same situation is found there. Small animals have to spend nearly all of their lives either looking for food or eating it. Presumably, a mammal smaller than a shrew simply could not exist; it would literally starve to death even though it spent its whole life eating. Energy use seems to become more efficient as size increases, so that very large animals—elephants, for example—require a much smaller energy intake per pound than do little animals, such as mice and chipmunks.

## Oxygen needs and body temperature

The single most important determinant of the rate of oxygen consumption is the animal's body temperature. In order to understand the various effects of temperature on oxygen consumption, we need first to divide animals into two distinct groups, based on the relationships of their body temperature to the **ambient temperature,** or the temperature of the air or water in which they live. The animals which are classed as **poikilotherms** (a group that includes all invertebrates, fish, amphibians, and reptiles) tend to take on a body temperature similar to the ambient temperature. The **homeotherms,** among which are the birds and the mammals, maintain a constant body temperature in spite of changes in the ambient temperature.

The development of homeothermy, which came relatively late in the evolutionary history of animals, greatly extended the range of environments in which these animals could live and thrive. Since the rate of all chemical reactions is influenced by temperature, extreme temperatures present problems for living systems. At low

**237**

temperatures, vital reactions are slowed down, so that the organism can barely function; at high temperatures, the metabolism is speeded up so fast that obtaining adequate supplies of energy becomes almost impossible. Poikilotherms have no defense against extremes of heat and cold; all they can do is go someplace where the climate is more congenial, or try to wait out the bad time. Because their internal temperature is constant, homeotherms can remain in the extreme climate, and what is more, they can also remain active. The arctic fox can function more or less normally in temperatures as low as −40°C; the antelope ground squirrel can scamper over the desert hunting for food at temperatures of +40°C.

But the relative independence of climate that the homeotherms have gained carries a high price tag. About 80 to 90 percent of their total energy output must be used in maintaining a constant body temperature. Homeothermy greatly increases the animal's energy needs, thereby creating a different kind of problem.

In poikilothermic animals, the consumption of oxygen is directly related to ambient temperature. As the temperature rises, so does their oxygen consumption; for each 10°C increase in temperature, their rate of oxygen consumption doubles. The only way they can avoid this increased energy demand is by adapting their behavior so that they avoid the heat outside. Burrowing into the ground, resting quietly in the shade, or migrating to another climate are all examples of behavioral adaptations that help the animal avoid situations where the need for oxygen consumption would be very high and the other dangers of heat, such as water loss in terrestrial animals and protein coagulation at temperatures above 45°C, would be threatening.

Poikilothermic animals have the advantage of decreasing oxygen consumption as the ambient temperature drops, but there are dangers for them in temperatures that drop below certain critical points. Nearby all of the physiological processes of the body — gas exchange, digestion and absorp-

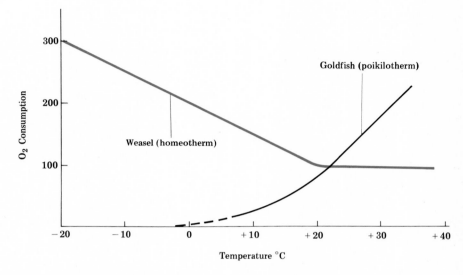

tion, synthesis and secretion of enzymes and hormones — are sensitive to temperature, and if the animal gets too cold it may be unable to carry on some of these processes fast enough to survive.

Although the body temperature of poikilotherms is more or less matched to the ambient temperature, it has been noticed that certain temperature-regulating mechanisms are sometimes put to work. Many marine lizards, for example, are able to mediate the changes in body temperature they could be expected to experience when they move from the rock where they have been basking in the sun into the sea to feed. To keep their body temperatures from dropping too far too fast, they minimize the cooling effects of the water by constricting vessels that carry blood to the surface of the skin and by slowing down the rate of heartbeat. When they come out of the water, they speed up the rates both of circulation and of respiration so that heating up occurs rather rapidly. It has also been demonstrated that poikilotherms can use voluntary muscle contractions to increase their metabolic rate and therefore raise their body temperature. This is the method a python uses to keep its eggs warm while they

**7-22** *In poikilotherms, the rate of oxygen consumption is directly related to the ambient temperature; as the graph shows, each increase of 10°C doubles the rate of O₂ consumption. The curve for homeotherms is quite different, showing a high rate of O₂ consumption at low temperatures. The homeothermic rate of O₂ consumption drops off as ambient temperature approaches the level of body temperature these animals maintain.*

are hatching; this unusual behavior pattern causes oxygen consumption to go up when the temperature decreases, in contrast to the general trend in poikilotherms.

The oxygen consumption curve for homeotherms is quite different from that for poikilotherms. At cold temperatures, the rate of oxygen consumption is very high, and it decreases as the ambient temperature increases. At about 25°C, the oxygen consumption levels off, remaining steady until nearly 40°C, when it begins to increase once more. This curve reflects the effort a homeotherm must exert in order to maintain a steady body temperature in the face of ambient extremes.

Body heat is produced by the burning of energy fuels; the higher the rate of oxidation, the higher the body temperature. That is why oxygen consumption increases so rapidly as the ambient temperature drops. Seals and walruses, for example, produce so much body heat that they are able to stay quite warm in the cold arctic air and the even colder arctic water. In fact, these animals produce so much body heat that when fur seals have to undertake the added exertion of walking any distance, they often die of heat prostration — in a place where the air is never warmer than 10°C.

Increased heat production is not the only homeothermic adaptation to the cold. It is also

**7-23** *These harp seals exemplify the mammals of the far north, able to survive the severe arctic winter through the production of great body heat. The thick fur of the pup reflects its greater inactivity and, therefore, greater need for insulation. (Fred Bruemmer)*

important to reduce heat loss through the surface. This can be done with insulating layers of fat, fur coats, and constriction of the vessels carrying blood to the surface. Animals, like the walrus and the pig, that have no fur often use their skin as insulation. By restricting supplies of blood to the skin they allow its temperature to fall, and this layer of cool skin forms an excellent barrier against body heat loss. Behavioral adaptations, such as hibernation, may also help conserve heat.

Very high ambient temperatures create problems for homeotherms, since the combination of the animal's own body heat plus the heat it absorbs from the environment can easily prove lethal. The temperature-regulating system of homeotherms is much more efficient at heating than it is at cooling, and very few animals can tolerate high ambient temperatures for any length of time. The most common solution to life in a hot climate is behavioral adaptations that allow the animal to avoid the heat. Desert homeotherms tend to use the behavioral solutions found in poikilotherms. They are active at night and sleep by day; they live in underground burrows; the typical posture of a desert homeotherm allows much of its body to stand in its own shadow. Moreover, many of these animals have adapted poikilothermic responses. They can tolerate limited increases in body temperature during the heat of the day. This adaptive poikilothermy is actually characteristic of many homeotherms that live in extreme climates, or climates that are subject to rapid temperature changes.

Cooling by evaporation is a common physiological solution to the problem of overheating. Horses and men sweat, and the evaporation of that moisture from the skin provides some cooling action. Dogs and cats pant, increasing their rate of ventilation, and breathe through open mouths. This evaporates the water on the surface of the respiratory tract and is a very efficient cooling mechanism. It reduces body heat, whereas sweating only reduces the heat on the surface.

But cooling by evaporation is a solution not open to most of the homeotherms that live in really hot climates, since water is likely to be scarce there, and the animal's need for water conservation may be even more pressing than its need for cooling. That is why physiological solutions must be supplemented by behavioral solutions.

## Summary

The energy to sustain life is produced by the oxidation of fuel. The agent of respiration is oxygen; the product is carbon dioxide. Both plants and animals are concerned with the problem of transporting these gases.

In one-celled organisms, as well as in small aquatic plants and animals, gas is transported across the cell membrane along a concentration gradient. Larger organisms must employ a more complicated means of respiration.

Gas exchange in vascular plants is carried on primarily through the stomata of the leaves. Because of transpiration—water loss during the exchange of gases—the opening and closing of the stomata must be regulated. This occurs through changes in the turgor pressure of the guard cells, due to the effects of light, carbon dioxide, or photosynthesis.

Gas exchange in aquatic invertebrates and endoparasites takes place in the organism's outer membrane. Arthropods have gills or tracheae as respiratory organs. Vertebrates have specialized respiratory systems that absorb oxygen and pass it to the blood. The amount of free oxygen present in water is lower than in air, and ventilation of the respiratory organs is more energy-expensive for aquatic animals. Aquatic animals have therefore evolved a more efficient system of oxygen uptake.

The gill is a projecting respiratory organ supported by a series of body arches and consisting of secondary lamellae which are the absorbing surfaces. Water is pumped over the gill's surface—always in a direction in opposition to that of the blood flow within the gill (the counter-current principle), assuring a high rate of oxygen absorption.

Lungs are organs contained in the chest cavity, consisting of millions of saclike compartments, called alveoli, which absorb oxygen. The rate of ventilation depends upon metabolic needs and can vary greatly. Mucous membranes in the upper respiratory tract mediate air temperature and protect the lungs against the inhalation of foreign particles.

The transport of oxygen between individual body cells and the specialized respiratory organ of all vertebrates is facilitated by hemoglobin in the red blood cells. The propensity of hemoglobin for oxygen is determined by temperature, pH, and the concentration gradient. Carbon dioxide travels through the bloodstream mostly in the form of bicarbonate ions, and some travels in carbamino compounds.

The medulla is the part of the brain that coordinates respiratory activities, receiving messages from carotid bodies, which act as chemoreceptors, as well as from chemoreceptors in the medulla itself. The vagus nerve directly regulates the rate of respiration, usually as a result of messages from the medulla.

The rates of oxygen consumption, or metabolic rates, vary among animals. The single most important determinant of the rate of oxygen consumption is the

animal's body temperature. In both homeotherms and poikilotherms, problems
of regulating body temperatures are solved by physiological adaptations as well as
by behavioral ones. Behavioral adaptations include such things as burrowing to
avoid heat or cold, being most active in the night rather than the day, and adopting
postures that minimize the surface presented to the environment.

# Bibliography

## References

Comroe, J. H., Jr. 1966. "The Lung." *Scientific American.* 214(2): 56–68. Offprint
no. 1034. Freeman, San Francisco.

Craighead, J. J. and F. C. Craighead. 1971. "Satellite Monitoring of Black Bear."
*BioScience.* 21(12): 1206–1212.

Hock, R. J. 1970. "The Physiology of High Altitude." *Scientific American.* 222(2):
3–13. Offprint no. 1168. Freeman, San Francisco.

Meidner, H. and T. A. Mansfield. 1968. *Physiology of Stomata.* McGraw-Hill,
New York.

Schmidt-Nielsen, K. 1971. "How Birds Breathe." *Scientific American.* 225(6):
72–82. Offprint no. 1238. Freeman, San Francisco.

Van Dyne, G. M. 1966. "Ecosystems, Systems Ecology and Systems Ecologists."
*Oak Ridge National Laboratory Report.* 3957: 1–31.

Wallace, R. K. and H. Benson. 1972. "The Physiology of Meditation." *Scientific
American.* 226(2): 84–92.

## Suggested for further reading

Andrewatha, H. G. 1963. *An Introduction to the Study of Animal Populations.*
University of Chicago Press, Chicago.
A rigorous and direct study of environmental influence on the survival and
multiplication of animal populations. Describes in detail and assesses the results of
20 case studies designed to measure the responses of animals to components of their
environment, including temperature and respiratory medium.

Hughes, G. M. 1965. *Comparative Physiology of Vertebrate Respiration.* Harvard
University Press, Cambridge, Mass.
Concise study of the structure and function of the vertebrate respiratory
systems. Recent, diverse material.

Ramsay, J. A. 1968. *Physiological Approach to the Lower Animals.* 2nd ed.
Cambridge University Press, New York.
A scholarly study that starts from general biochemical principles and considers the
relationship to the physiology of animals according to their size or mode
of life.

### Related books of interest

Bowen, C. D. 1970. *Family Portrait.* Little, Brown, Boston.
In this book, the well-known biographer turns to the study of her own family,
including her brother, Dr. Philip Drinker, the inventor of the iron lung. Chapter 14
is an insightful and sympathetic view of the problems of scientific research.

Farb, P. 1959. *Living Earth.* Harper & Row, New York.
A fascinating nontechnical description of life within the soil of the forests, the
grasslands, and the deserts. The behavioral adaptations that regulate temperature
are discussed in detail for each type of habitat.

# 8

# Homeo-stasis

The continuity of life depends on a closely interconnected series of chemical reactions. As we have seen, many of these reactions can take place only in a fairly narrow range of temperature and pH. Moreover, the organic molecules involved in these reactions, such as enzymes, vitamins, and hormones, are very delicate. Changes in the temperature or in the chemical environment can alter these molecules so that they are no longer able to play their part in biochemical reactions.

These living chemical systems, so delicately balanced, are continually exposed to sudden and drastic changes in the temperature and in chemical composition of the environment. The hot sun beats down on them in the summertime, the cold snow covers them in the winter. A summer shower falls, and the osmotic concentration of a pond which billions of tiny creatures call home may drop by percentage points within an hour. Men apply fertilizers, herbicides, and pesticides, thereby introducing new elements into the chemical composition of soil and water; a month of heavy rains causes extensive leaching and the composition changes again.

It is evident that the survival of living organisms depends on their ability to maintain some sort of constant internal state in the face of so much external change. The term used for this process, **homeostasis,** was coined in 1939 by a Harvard physiologist, Walter B. Cannon. In his book, *The Wisdom of the Body,* Cannon said:

> The coordinated physiological processes which maintain most of the steady states in the organism are so complex and so peculiar to living things...that I have suggested a special designation for these states, homeostasis. The word does not imply something set and immobile, a stagnation. It means a condition—a condition which may vary, but is relatively constant.

Many of the physiological processes we have already discussed in this book are related to the problem of maintaining a steady state. The regulation of the body's temperature is a kind of homeostasis; the adjustments of the rate of circulation and respiration to the body's need for oxygen are also homeostatic. The opening and closing of leaf stomata in response to changing needs for gas exchange and dangers of excessive water loss is an elaborate mechanism for homeostasis in plants. Even on the cellular level we have examples of homeostasis—the plant's use of the vacuole to hold surplus carbohydrates inside a membrane so that they will not upset the osmotic balance of the cell, the chloride shift that maintains the electrolyte balance of the red blood cell.

As these examples indicate, the concept of homeostasis can be applied on a number of different levels. Individual cells display homeostasis, maintaining some constancy of their internal environment. Organisms as a whole also display homeostasis mechanisms, often very complicated ones. Because of the existence of organismic homeostasis, individual cells of a multicellular organism may lose many of their homeostatic properties. For example, if a protozoan is placed in a hypotonic solution, it can excrete the excess water that moves into the cell by means of the contractile vacuole. A red blood cell in the same solution will continue to take in water until it bursts. The highly specialized blood cell can survive only in the carefully regulated environment of the bloodstream; it has no ability to adapt to environmental change.

The concept of homeostasis can even be applied to whole populations. In many enclosed animal populations—the fish in an aquarium tank, for instance—it appears that there is some self-regulating mechanism that limits the popula-

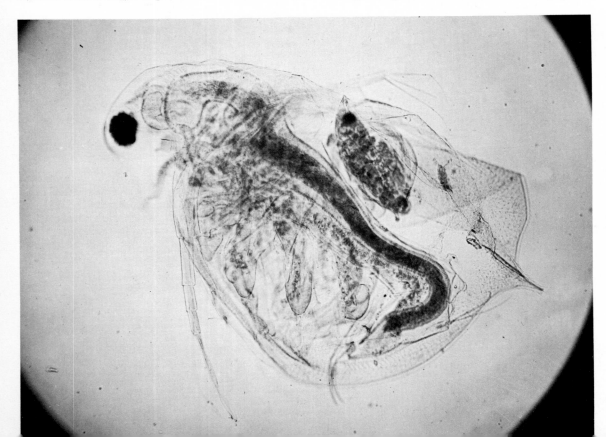

8-1 This tiny, almost transparent water flea faces many homeostatic problems because of the constant variation in its environment. The water flea is a freshwater crustacean, found in ponds and ditches. The osmotic concentration of its environment fluctuates drastically. On a hot dry day, when the evaporation rate is very high, the osmotic concentration of pond water may soar; when it rains, the concentration drops very suddenly. It has been observed that water fleas show seasonal variation in form, a trait that may be a homeostatic adaption to the change in environmental conditions.
(Jeanne White from National Audubon Society)

**8-2** *The schematic drawing indicates the major means of water loss and gain in man. Like all organisms, man must balance his water loss and intake within narrow limits.*

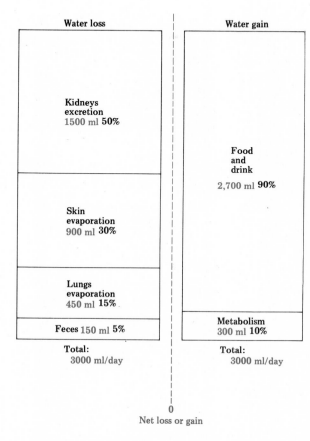

Water loss

Kidneys
excretion
1500 ml **50%**

Skin
evaporation
900 ml **30%**

Lungs
evaporation
450 ml **15%**

Feces 150 ml **5%**

Total:
3000 ml/day

Water gain

Food
and
drink

2,700 ml **90%**

Metabolism
300 ml **10%**

Total:
3000 ml/day

0
Net loss or gain

peach about 85 percent, and a jellyfish about 95 percent. For all organisms, therefore, the problem of maintaining a steady concentration of water is a critical one. This holds true for both plants and animals, whether they are aquatic or terrestrial.

An organism may gain water by eating it in food (perhaps eating a peach), by drinking it, or by absorbing it through the membranes of outer surface cells (plant root hairs, for example). In addition, the organism may manufacture water through the metabolic breakdown of its food. Fat metabolism produces quite a bit of water, so that some desert animals can go for months without drinking water, as long as they eat a high-fat diet. This is true of gerbils, for example, and it is one reason they make good pets. As long as they are fed fatty seeds, it does not matter if their young owner forgets to fill up the water dish.

An organism may lose water by osmosis or by evaporation from the surface cells (and the respiratory tract in vertebrates); it loses water in the urine and feces it produces; it may also lose a small amount of water in metabolic activities, such as photosynthesis, that splits the water molecule.

If the organism is to survive, its water gains must exactly balance its water losses. Too little water and the organism dehydrates and dies. Too much water and its body fluids become so diluted that cells may burst, and respiratory and circulatory mechanisms can no longer provide enough food and oxygen.

### Water balance in marine organisms

It might seem as if aquatic organisms should have no problem of water balance, since so much water is freely available to them. Yet in many cases their problems of water balance are just as severe as those of terrestrial animals. Differences between the osmotic pressure of the organism and that of its environment may cause water to be drawn out of the organism or to flood into it. In all aquatic

tion to the number of individuals that can be adequately supported by the available supplies of food and oxygen. This suggests a direction for certain kinds of ecological research; perhaps it is possible to determine why and how this limitation works and to apply it to the problems of human population growth.

### Water balance

Water is the medium of life on our planet. In humans, about 65 percent of the body weight is water. A chicken is about 75 percent water, a

**247**

forms of life, the problem of water balance revolves around the important issue of osmotic pressure.

Life began in the sea, and the more primitive forms of life are well adapted to the problems of maintaining water balance in that osmotic environment. Marine algae and the lower marine invertebrates have about the same osmotic concentration inside their body cells as is found in the seawater in which they live. In fact, their body fluids are essentially continuous with the sea. They are called **osmoconformers,** indicating that their internal osmotic pressure conforms to that of their environment. They will not lose water to the sea, and they will not gain it, except during certain periods such as cellular or reproductive growth, when the high osmotic concentration of the new cells causes them to take in water as part of their development process. These organisms tend to remain in the same general area for their entire life spans, so they do not need mechanisms to allow them to adapt to differing environments. Their simple system of maintaining water balance works beautifully for their way of life, so long as there is no sudden drastic change in the osmotic concentration of the water in which they live—and that kind of cataclysm is, after all, relatively infrequent in such large bodies of water.

Farther up the evolutionary scale, osmoconformity begins to pose a problem. Many cells which have specialized in one certain direction—nerve cells, for example—lose their ability to adapt to osmotic change; they are completely dependent on a constant internal environment. So many of the complex physiological processes which sustain life depend on specific gradients of concentration in body fluids and cellular fluids, that any osmotic change might cause these systems to fail. Moreover, some of the higher marine invertebrates, such as crabs and snails, have evolved a way of life with more mobility, as a consequence of their active search for food; and this mobility exposes them to a changing environment. Therefore these higher animals have devel-

8-3 *The horseshoe crab has evolved a relatively mobile way of life; it swims (although rather slowly and clumsily) and walks along the shore in search of worms and small molluscs to eat. Since it often feeds in pools near the shore, where evaporation has greatly raised the osmotic concentration of the water, it is faced with a variety of homeostatic problems. It has adapted by evolving mechanisms of osmo-regulation.*
*(Gordon S. Smith from National Audubon Society)*

oped systems of regulating their internal higher osmotic concentration so that it is independent of the osmotic concentration of the environment, and thus they can resist the effects of osmotic change. This power of osmoregulation allows them to move from salt water to the brackish (moderately salty) water of marshes along the shore. Some, such as the Chinese mitten crab, can even live in fresh water for long periods.

One way to adapt to the change in osmotic concentration is simply to shut it out. This is the solution of the oysters and clams that live in tidal pools where the salt concentration fluctuates with the coming and going of the tides. They just close their shells and wait until the osmotic concentration of the water is once again equal to the concentration of their own body fluids. Another way to adapt is to tolerate some decrease in the concentration of body fluids. This is possible in animals that have fairly simple physiological systems, without extensive reliance on concentration gradients as a means of exchange and transport.

A third, more complex, method of osmoregulation is to maintain a constant osmotic concentration in the blood no matter how the concentration of the surrounding water changes. This requires some form of active transport that can move the osmotically active solutes (sodium chloride, for the most part) against the concentration gradient. In the shore crab (Carcinus), this active transport takes place in the gills. When the osmotic concentration of the environment drops for some reason, the gills begin to absorb salt in addition to water. In other animals, the skin may play some role in absorbing the necessary solutes.

This regulatory system solves one problem by allowing the crab to maintain a steady osmotic concentration in its body fluids, but at the same time it also creates a new problem. Since the body of the animal now has a higher osmotic concentration than the surrounding medium, there will be a net movement of water into the animal. So there must also be some mechanism for excreting the excess water. This is the function of the *kidney* (in the crab, it is called the green gland, but this is just a kidney in a green disguise).

The essential feature of the kidney is that it allows the animal to excrete a fluid (urine) that has a different osmotic concentration from body fluids, such as blood or lymph. Urine that is isotonic to blood serves only as a means of eliminating nitrogenous wastes that might poison the organism. Isotonic urine does not help to maintain a water balance, since it removes water and inorganic ions in exactly the same proportion as they are found in the blood. Thus the excretion of the isotonic urine does not change the composition of body fluids. Regulation of water balance requires a kidney that can form urine hypotonic or hypertonic to blood. All kidneys are capable of forming hypotonic urine, or urine that is more dilute than blood. This serves to eliminate excess water in body fluids. The mammalian kidney also possesses the special capability of forming urine that is hypertonic to blood. This concentrating kidney allows water conservation; it is, in fact, the primary evolutionary adaptation that allows mammals to live in a terrestrial environment.

**Marine fish.** The first bony fish were freshwater animals, for it was only much later in their evolutionary history that they developed the mechanisms enabling them to move into the sea. Like all vertebrates, fish maintain a rather low salt concentration in their body fluids—only about one-third that of seawater. The structure of their skin, with its tough coating of scales, creates a very good barrier against diffusion, but the surface of the gills is specialized to allow diffusion

**8-4** *When viewed head on, a lobster looks almost frighteningly carnivorous. It walks along the sandy bottom, searching for food which it seizes with its pincerlike chelipeds. Excretion of the nitrogenous waste products from its high protein diet takes place through the "green glands," small excretory organs located just under the pair of antennules. These organs also serve an osmoregulatory function. (A. W. Ambler from National Audubon Society)*

to take place as a part of gas exchange. So marine fish face a constant problem of dehydration.

## 8-1
## Ratio of osmotic pressure of urine to plasma

| Animal | O.P. urine/O.P. plasma |
| --- | --- |
| Fresh-water fish | 0.14 |
| Earthworm | 0.21 |
| Frog | 0.34 |
| Salt-water fish | 0.85 |
| Man | 3.1 |
| Seal | 6.0 |
| Camel | 6.9 |
| Kangaroo rat | 15.5 |

What if the fish try to compensate for their water losses by drinking more seawater? Excreting the salt they swallow would take more water for the formation of urine than was taken in through drinking, since the fish kidney is not capable of forming urine that is more concentrated than body fluids. Moreover, marine fish cannot, for the most part, solve their problem by taking in additional water with their food, since their diet generally consists of marine algae and invertebrates; swallowing the body fluids of these animals is about like drinking seawater.

The solution of marine bony fish to the problem of water balance has been the evolution of a mechanism that allows them to drink the seawater and excrete the salts through some means other than the kidneys. One adaptation that has minimized the need for salt excretion is found in the digestive tract. The absorptive cells lining the tract are relatively impermeable to all salts with a valence of more than one, such as calcium ($Ca^{++}$) and magnesium ($Mg^{++}$). These ions generally pass right through the digestive tract and

out the other end. The rate of absorption is so low that the intake just meets physiological needs for these ions, and there is no excess to be excreted. The second adaptation helps excrete ions with a valence of one, such as $Na^+$ and $Cl^-$, which are absorbed through the digestive tract. These ions enter the bloodstream and circulate through the body to the gills. There, a process of active transport pumps these ions out into the water, against the concentration gradient. Like all active transport, this is an energy-consuming process, and it requires the presence of a number of carrier molecules and activating enzymes. This extra-renal (outside the kidney) pathway of excretion allows fish to maintain a steady internal osmotic concentration that is much lower than the concentration of the environment. The kidney, of course, continues to function as a means of excreting nitrogenous wastes.

Cartilagenous fish, such as sharks and rays, have solved the problem of osmoregulation in an entirely different way. They maintain the same low salt concentrations in their blood as do all vertebrates, but in addition they have a high concentration of urea, which for some reason they are able to tolerate without harmful effects. This urea raises the osmotic concentration of their blood, making it approximately equal to that of the seawater and thereby eliminating the danger of dehydration.

**Marine birds.** Birds that live on or near the ocean, such as the gulls and penguins, have water balance problems very similar to those of completely aquatic vertebrates. Of course, birds do not have the specialized diffusion surface of the gills, the source of most water loss in fish; in addition, their feathers help to protect them from water loss through the skin. But since they are active homeotherms, their metabolic rate is very high. They must ventilate rapidly, and so they incur water loss by evaporation in the respiratory tract. They also lose water through excretion. Birds do conserve some water by excreting uric

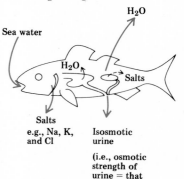

8-5 *Marine fish are protected against water loss through their skin by its coating of scales and mucus, but of course, water can easily diffuse out through the surface of their gills. Their kidneys are not capable of excreting concentrated urine; instead they excrete excess salt extrarenally, pumping the ions out through the gills.*

**8-6** *Marine birds do not experience any significant cutaneous water loss, but they do lose water through evaporation in the respiratory tract. To conserve water, they excrete a semisolid paste containing crystals of uric acid rather than a liquid urine. A modified nasal gland serves as an extrarenal pathway for salt excretion.*

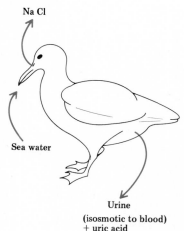

Na Cl

Sea water

Urine
(isosmotic to blood)
+ uric acid

acid rather than urea. This allows the bird to excrete a semisolid paste, containing crystals of the acid, rather than a liquid urine. But since these birds live on a high-protein diet of fish, they produce a relatively large volume of the nitrogenous wastes of protein breakdown. And, in spite of their conservation measures, they register a marked water loss due to excretion.

To balance their water losses, marine birds must have some source of water gain. If they drink seawater, they are faced with the problem of excreting the salts, and they, too, have a kidney which is incapable of forming hypertonic urine. If they eat marine invertebrates, they are consuming about the same quantity of salts as is found in the ocean. For this reason, the diet of the albatross and the penguin often consists largely of

fish (vertebrates), which have already done part of the work of eliminating the salt; their body tissues contain much less salt than does the seawater. Yet even on an exclusive diet of fish, marine birds still have to take in some additional water to balance water losses.

One solution is simply to fly to a source of fresh water, a course which marine birds that live along the shores often adopt. But if no fresh water is within reasonable traveling distance, the birds must drink seawater and then excrete the extra salt. Since their kidneys are not capable of handling this load, they, like the fish, have had to develop some extrarenal pathway for salt excretion. A gland of the upper respiratory tract, inside the **nares** (nostrils), has been modified for this purpose. The salt gland contains secretory

**8-7** *The fulmar shown here is a marine bird that lives along the Atlantic coast; it is related to the gull. In marine birds like these, there is a modification of a nasal gland that permits it to serve as an extrarenal excretory pathway. Excess salt can be excreted through this adaptation. (C. G. Hampton from National Audubon Society)*

tubules adjacent to the blood capillaries that service the area. As the blood travels past, the sodium chloride (NaCl) moves along a concentration gradient from the blood into the tubules, and the effectiveness of this diffusion is increased by the utilization of the counter-current principle. The solution that leaves the tubules, often containing as much as five percent NaCl, passes out the external nostrils, drips down the beak, and plops into the sea. Much the same kind of extrarenal salt excretion is found in marine reptiles, such as the sea turtles. In this group the modified gland opens into the eye rather than the nares, and the salt flows out in large teardrops. It is this physiological peculiarity that has led to the myth that sea turtles are sentimental types, easily given to crying over their sorrows.

**Marine mammals.**   The water balance problems of marine mammals, such as the seal and the whale, are essentially the same as those of marine fish, but mammals have no gills through which to lose water and pump out salts. However, since they are homeotherms, and thus have a high body temperature, they do lose water through evaporation in the upper respiratory tract. One way mammals solve their problem is through careful conservation of water. Their skins are impermeable to water, so they do not have an osmotic loss. Their feces are practically dry and their urine is highly concentrated. Even the milk of nursing mothers is designed to save water, having a high concentration of fat (30 to 40 percent fat content).

Marine mammals have only two sources of water gain. One is through their food, and most of them live on marine fish that have a relatively low salt content. The other source of water is through fat metabolism; as fats are broken down, they yield molecules of water. These two sources of water intake are enough to balance the relatively small water losses of marine mammals. They are assisted in maintaining their water balance by a very efficient kidney that can con-

**8-8** *Marine birds and sea turtles, which may have to drink seawater, must maintain a homeostatic balance with relatively inefficient kidneys. They compensate with excretory glands in the head. The mock turtle from* Alice in Wonderland, *by Lewis Carroll, illustrates the popular myth that these animals have been seen to cry.*

centrate urea wastes in a small volume of urine. For instance, baleen whales, which feed on small plankton, have concentrating kidneys that are able to deal with the high salt content of their diet.

All mammals have a concentrating kidney that can form urine hypertonic to the blood, but only in baleen whales and a few desert dwellers do we find a kidney that can produce urine 15 or 20 times as concentrated as blood. In man, for

**8-9** *Like all mammals, the sea otters nurse their young for the first months of life. The production of milk poses a problem for marine mammals, who must conserve water if they are to survive in their hypertonic environment. The solution has been to produce a small quantity of very rich milk.* (Karl W. Kenyon)

**8-10** *Marine mammals gain water from their food and from fat metabolism. They are able to conserve water because they, like all other mammals, possess a concentrating kidney. Other adaptations which help to cut down water losses are a thick skin or furry outer covering, the explusion of dry feces, and a preference for a diet consisting largely of marine vertebrates with a low salt content.*

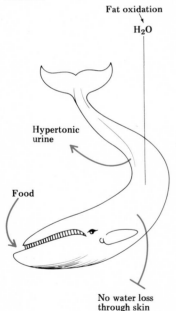

Fat oxidation

$H_2O$

Hypertonic urine

Food

No water loss through skin

example, the urine-plasma osmotic ratio is 4.2. People who are cast away on a raft in the middle of the ocean face a terrible dilemma. If they drink salt water or eat marine plants and invertebrates, the water they need to excrete the salt will be greater than the water they have taken in, causing a net water loss. If they eat marine fish, the water they need to excrete the urea which is the product of protein metabolism will also be greater than the water they have taken in; they will still have a net water loss, even though it is not so large as the loss from drinking salt water. Their best course of action is to eat and drink nothing and hope that help arrives soon.

## Water balance in fresh-water organisms

Animals that live in fresh water have to struggle constantly against the danger of becoming water-logged. The osmotic concentration of their body fluids is significantly higher than the concentration of the water in which they live. This means that there will be a net movement of water into the animal, causing cells to burst. The animal must develop mechanisms that thwart this natural move toward osmotic equilibrium. The problem is made more complex by the fact that inorganic ions tend to diffuse out of the body and into the surrounding water, which has a lower salt concentration.

The problem can be minimized if the surface through which water gain and salt loss can take place is limited. Fresh-water fish, for example, have skins that are covered with scales and a thick mucus; this coating prevents cutaneous osmotic exchange. But the gill surfaces, an area at least twice as large as the skin, still permit exchange. They are what biologists call "windows" that allow osmotic and thermal exchanges to take place.

Behavior patterns also can minimize the water balance problems. Fresh-water animals rarely drink, since swallowing water only adds

to their excretory burden. And they tend to prefer salty foods.

Yet in spite of all these efforts, the fact remains that fresh-water fish take in too much water and constantly lose salt. The ultimate solution to their problem of osmoregulation must come from the kidney. In fresh-water animals, it is capable of producing a large quantity of highly diluted urine, hypotonic to the blood, thus "bailing out" the excess water. In these fish, there is usually no bladder, allowing the urine to be excreted as fast as it is formed. Fresh-water amphibians, such as toads and frogs, do have bladders, which permit them to conserve water and reabsorb some salts when they are in a dry environment.

In addition to their kidney modifications, fresh-water animals must have some means of compensating for the loss of salt through diffusion and through urine formation. The gills of fish and the skins of amphibians contain a salt pump, which actively transports salt from the water into the blood, thus maintaining a steady internal osmotic concentration. A simple laboratory experiment can demonstrate this salt pump in action. When a low-salt solution is placed on one side of a container and a high-salt solution is placed on the other, and they are separated by a piece of skin from the back of a toad or frog, it can be seen that the skin will transport salt against the concentration gradient for hours. The exact mechanism of this salt pump, which can transport both $Na^+$ and $Cl^-$ ions, has not yet been explained, but physiologists are sure that it is controlled at the cellular level and that it is an energy-consuming reaction.

An interesting question, currently receiving much attention from research workers, is the problem of water balance in eggs, embryos, and larvae of aquatic animals. These immature forms are subject to the same stresses as are the adults — in fact, since in many cases the eggs and larvae float at the surface of the water, where the concentration changes constantly in response to rates of evaporation and rainfall, they are subject to even more severe stresses than adults. Yet these immature forms have no working kidneys, no working gills. How can they survive? At the moment, there is not enough data on this subject to permit even an educated guess.

**Adaptation to differing osmotic environments.**
Most aquatic animals can tolerate only small changes in the osmotic pressure of their environment. Severe environmental changes cause a breakdown in their regulatory systems and subsequent death. But there are certain animals that are able to move back and forth from fresh-water to marine environments. Examples are the salmon, which is born, breeds, and dies in fresh water but spends most of its adult life in the ocean; the crab-eating frog, that feeds in a marine environment but spends the rests of its day in fresh water; and the various ducks, herons, and gulls that are equally successful living on the ocean or a river.

A few of these species are osmoconformers; they simply adapt to the alterations in the concentration of their body fluids. This solution can work only for relatively unspecialized animals. In more complex animals, physiological processes are too specific to permit such changes in osmotic pressure. Most of the animals that can survive drastic change in their osmotic environment are good osmoregulators, able to maintain a steady internal environment despite the changes in their ambient environment. The salt gland of birds is an excellent example of a very flexible osmoregulatory device. When the bird is living in a marine environment, the salt gland functions efficiently to eliminate large quantities of ingested salt. When the bird moves to a fresh-water environment, the salt gland ceases to function, secreting only a few drops of mucus that is isotonic to the body fluids.

In many fish that move back and forth between fresh and salt water, the salt pumps in the gills are the means by which they adapt to sud-

den changes in the salinity of their environment. In the sea, the gill pumps work to eliminate salt; in fresh water, the same pumps work in the opposite direction, pulling salt in. In cartilaginous fish, adaptation to changing osmotic environments is due to changes in the urea content of the blood. The basking sharks that live in fresh water, for example, simply reduce the amount of urea in the blood. If they move to the sea, urea can be added until the blood's osmotic pressure is equal to that of the environment. This same adaptation is found in a number of frogs and toads as well. Many animals also show changes in the concentration of the urine when they are moved from one type of environment to another. This indicates that there is some modification of kidney function involved in their adaptation. Physiologists believe that this urine change is only temporary, a means of handling the stress of drastic change. Other adaptations are more effective, and have a lower energy cost, in the long run.

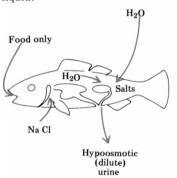

**8-11** *Fresh-water fish must develop mechanisms to prevent becoming waterlogged, since their body fluids have a higher concentration than the water in which they live. These fish do not drink, since the last thing they need is additional water. Their kidneys can excrete very dilute urine, ridding their bodies of excess liquid.*

**8-12** *The squid lays large numbers of eggs, contained in translucent podlike cases. Here the tiny embryos develop, bursting their case when they are developed enough to survive on their own. The developing embryos are subjected to the same problems of osmotic balance as adult forms are, yet they have no known mechanism of osmoregulation. The means by which they survive remains a scientific puzzle. (Russ Kinne—Photo Researchers)*

## Water balance in terrestrial animals

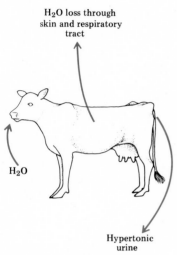

**8-13** *Vertebrates are the only animal group other than insects that has been able to adapt completely to a terrestrial way of life. Outer coverings of scales, fur, and feathers help prevent cutaneous water loss, and a concentrating kidney permits additional water conservation.*

**8-14** *The kangaroo rat is well adapted to its hot dry desert environment. Like many other desert inhabitants, it is primarily nocturnal; its relatively large eyes enable it to see in dim light. Although rodents typically serve as scavengers, the dryness of the desert prevents decay, so there is very little to scavenge. The kangaroo rat has adjusted its feeding habits accordingly, and lives largely on a diet of seeds. (E. R. Kalmbach, Interior—Sport Fisheries and Wildlife)*

The evolutionary move from water to land, which greatly increased the amount of available oxygen and thus permitted the development of an energy-expensive way of life, had its price—the threat of dehydration. Water conservation is a more critical function in terrestrial organisms than in aquatic ones. We have already seen, in Chapter 6, that the conservation of water is a primary concern in all land plants.

The only animal phyla really to have successfully adapted to life on land are the arthropods and the vertebrates. Other seemingly terrestrial animals, such as earthworms and most snails, actually cannot survive unless they live in a damp environment. A worm that is exposed to the air will soon dry out and die. But although desiccation is always a threat to the survival of land animals, most of the arthropods and vertebrates have adapted admirably to their airy environment.

The principal adaptation of the arthropods to life on land is their hard protective exoskeleton, which prevents water loss through evaporation. They also conserve water by excreting almost dry urine and feces.

Vertebrate adaptations to land life show a wide variety of solutions to the problem of water balance. Scales, feathers, and fur help to cut down the rate of evaporation from the skin. Birds and mammals possess a concentrating kidney that is able to excrete urine with a higher osmotic concentration than body fluids, thus saving water. The lower intestinal tract reabsorbs most of the water from feces before they are eliminated. All these measures help cut down the total water losses of vertebrates.

In many vertebrate groups, these conservation measures prove so successful that the animals never have to drink. The quantity of water they take in through their food and produce in the metabolic processes of digestion is great enough to balance their water loss. Gorillas, for example, never drink. Amazingly, neither do desert rats. The kangaroo rat, a native of the American desert, can live in that hot dry climate and eat nothing but seeds and nuts; yet it never takes a drink of water! It gets along on the water content of the seeds (about 5 to 10 percent, even in the driest-looking seeds) and the metabolic water it produces in digestion. The rate also conserves water through behavioral adaptations, such as burrowing underground in the day and hunting for food at night. It has powerful kidneys that produce highly concentrated urine in a very small amount of water; kangaroo rats have such effective kidneys that they can even drink seawater without any harmful effects. The rats have no sweat glands, relying strictly on behavioral adaptations for cooling, thus saving precious water. And they have a special modification of the nasal passages of their respiratory tract which causes the warm damp air coming up from their lungs to condense inside the nose, where the water can be reabsorbed.

Not all vertebrates are such efficient water conservers, for most of them do drink frequently. In reptiles and amphibians, there is such a high rate of water loss through the skin when the animal is out of water, that it must take in water,

Labels on figure 8-13: H₂O loss through skin and respiratory tract; H₂O; Hypertonic urine

**255**

either through drinking or by absorption through the skin, to replace the loss. In birds and mammals, the chief cause of water loss is evaporation through the respiratory tract, due to the high rates of oxygen consumption in homeotherms and the accompanying need for rapid and frequent ventilation. Another reason for water loss in these homeotherms is the need for evaporative cooling, which is a very water-expensive process. A man who is working hard in the hot sun may sweat more than four liters per hour; this is about equal to the volume of water in the blood. This water loss must be replaced fairly quickly, for in man a loss of more than 12 percent of body water is fatal without prompt medical attention. Sweating also causes a significant loss of salt, and this, too, must be replaced.

Contrary to rumor, the camel also must drink to maintain its water balance. The camel does not store water in its hump, nor in its stomach. It survives in the hot dry climate of which it is a native because it is a good conservationist. The camel sweats very little. He will allow his temperature to rise several degrees above normal before he begins the process of evaporative cooling, and during the cool of the night he lets his body temperature drop several degrees, so that he has a low starting point the next morning. The most amazing feature of the camel is his ability to withstand severe dehydration. Man is in serious trouble when he loses 10 percent of his body water, but the camel can lose as much as 40 percent before it begins to experience symptoms of dehydration. Camels have been seen to drink up

**8-2**

**Mechanisms of osmoregulation — summary**

| Animal | Osmotic pressure of environment relative to blood | Mechanism |
|---|---|---|
| Fresh-water fish | Hypoosmotic | Hypoosmotic urine<br>uptake of NaCl in gills<br>does not drink $H_2O$ |
| Marine fish | Hyperosmotic | Isosmotic urine<br>excretes salts through gills<br>drinks seawater |
| Marine birds | Hyperosmotic | Isosmotic urine<br>excretes NaCl through nasal gland<br>drinks seawater |
| Marine mammals | Hyperosmotic | No water loss through skin<br>hypertonic urine<br>obtains water from oxidation of fat |
| Terrestrial vertebrate | (Air) | Regulates water loss through skin<br>hypertonic urine<br>drinks water |

to a third of their body weight in water in just 10 minutes when they finally arrive at a watering place. It would be interesting to know how a camel is able to tolerate such water losses, but so far no physiological explanation has been advanced.

## Electrolyte balance

When we compare the composition of blood with the composition of the extracellular fluid that surrounds cells in most kinds of tissues (see Table 8-3), we find that they are quite similar except in regard to the proportion of the various ions each fluid contains. The fluid found within cells shows yet a third ionic composition, unlike any kind of extracellular fluid. It is logical to assume that this notable difference is not accidental but in some way functional. It is believed that the major function of these differences in ionic composition is the regulation of osmotic pressure.

**8-3**

**The composition of body fluids***

| | $Na^+$ | $K^+$ | $Ca^{++}$ | $Cl^-$ | Protein |
|---|---|---|---|---|---|
| Blood | 142.0 | 4.0 | 5.0 | 102.0 | 17.0 |
| Extracellular fluid | 145.0 | 4.1 | 3.5 | 115.7 | 0.0 |
| Intracellular fluid (muscle) | 12.0 | 150.0 | 4.0 | 40.0 | 54.0 |

*Concentrations expressed in milliequivalents per liter (mEq./l.)

The osmotic concentration of any solution is determined by the number of osmotically active solute particles it contains. Body fluids derive their osmotic concentration by the presence of

a certain number of large organic molecules, such as globular proteins and sugars, and from a larger number of inorganic ions in solution. These ions make excellent osmoregulatory devices. They are small molecules. Through a process of active transport, they can be moved across cell membranes. They are always available, since they are taken in with both food and water, and they can be obtained without the work of synthesis. And most organisms can tolerate relatively large fluctuations in their numbers.

An example of the way that ions can be used to regulate osmotic pressure can be seen in the sodium pump found in most body cells. The table shows that although a large concentration of sodium ions is found in extracellular fluids, there is almost no sodium inside the cell. We know that the sodium ion is small enough to pass through the cell membrane along with the water in which it is dissolved, even though its positive charge may slow down its progress somewhat (due to repulsion by the positively charged membrane). If sodium ions are able to enter the cell, why are more of them not found inside? The answer is that as soon as they get inside, a process of active transport shuttles them right back out again. This process takes place so promptly and efficiently that researchers have called it the **sodium pump**. It is presumed that some carrier molecule with an affinity for $Na^+$ ions is lying in wait just inside the membrane, picking up the ions for transport in the other direction as soon as they enter.

The net effect of the sodium pump is to lower the osmotic concentration inside the cell and increase the osmotic concentration in the extracellular fluid just outside the cell membrane. This tends to equalize the pressure on both sides of the membrane, thus preventing any additional inflow of water into the cell.

In multicellular organisms, the sodium pump is used in a number of physiological processes that involve the moving of water from one part of the body to another. For example, in the intestine there is a sodium pump that moves the $Na^+$ ions from the digestive tract into the absorptive cells and then into the bloodstream. This movement of the sodium ions brings water right along with it, thus helping to conserve the water that is used in the process of digestion. And as we shall shortly see, there is a sodium pump at work in the mammalian kidney, concentrating the urine by moving the water out of it.

Another example of ionic balance on the cellular level is the chloride shift that takes place in red blood cells. (This is described in Chapter 7, p. 234.) This shift allows bicarbonate ions to be transported inside the erythrocyte without upsetting the electrical balance of the cell. Similar electrical shifts, involving sodium and potassium ions, are responsible for the ability of many cells to transmit and react to electrochemical impulses. If electrons are shifted to create an imbalance of anions and cations, then an electrical charge develops.

On the organismic level, the function of maintaining ionic balance is largely handled by the kidney. The excretion of osmotically significant ions, such as $Na^+$, $K^+$, $H^+$, $HCO_3^-$, and $NH_4^+$, can be varied in response to problems of ionic balance. By this method, the osmotic pressure of the extracellular fluids is maintained at a more or less steady level, even though the organism continually gains and loses ions through its daily activities.

## Acid-base balance

Many physiological processes, such as gas exchange, ventilation, and circulation, are quite sensitive to the pH of the surrounding environment. Human blood, for example, requires a pH of 7.4; even a small change from this normal level will interfere with the functioning of the osmotic exchange between blood and body cells. In fact, most tissues of most organisms require a pH no lower than 6 and no higher than 8. There are a few exceptions to this rule. Certain acid-loving bacteria can exist in a pH much lower than 6, and during digestion, the pH of the stomach may drop to 1 or 2. But in most cases, a sudden drop in pH will harm the organism and interfere with its functioning.

The organism needs to maintain a steady pH, but many metabolic activities lead to the release of strongly acidic or basic substances into the bloodstream and other body fluids. For example, in the normal person, cellular oxidation produces 10 to 20 moles of carbonic acid every day. The metabolic breakdown of food produces pyruvic acid. Anaerobic respiration in muscles during periods of strenuous exercise produces lactic acid. Yet in most people, the production of acidic metabolic wastes does not cause any corresponding change in the blood pH; it rarely changes by more than 0.1 pH unit. This means that there must be a very efficient system of maintaining a steady acid-base balance.

The principal method of maintaining a stable acid-base balance in most organisms is the use of a **buffer system.** Buffer systems resist or cushion changes in pH when a strong acid or base is added to them. A good analogy is the shock absorber system of a car. When you drive over a bad and bumpy road, the wheels of your car bounce up and down as they hit the potholes. But much of the motion of the wheels is absorbed by the strong springs in the shock absorbers, so that there is very little bounce felt by the passengers in the car. In just the same way, buffer systems in the body cushion the organism against sudden additions of strong acids and bases to the body fluids.

The major buffer system in vertebrates is composed of carbonic acid ($H_2CO_3$) and bicarbonate ions ($HCO_3^-$). The bicarbonate ions are **proton acceptors.** When a chemical reaction within the body adds a strong acid to the blood, the bicarbonate ions accept the extra protons of the $H^+$ ions forming the relatively weak carbonic acid; thus the strong acid has been buffered, and there is no significant change in the pH of the blood. If OH⁻ ions are added to the blood, the carbonic acid part of the buffer system acts as a **proton donor.** The acid dissociates into bicarbonate ions and free hydrogen ions, and the hydrogens can combine with OH⁻ to form water.

The carbonic acid-bicarbonate ion buffer system can be seen in red blood cells and in blood plasma. Other buffer systems are also found in the body. Like the bicarbonate system, they each consist of a weak acid and the salt of that acid, which are produced in an easily reversible combination reaction. In blood plasma, there is a buffer system involving plasma proteins, and in red blood cells there is a buffer system that utilizes hemoglobin.

The action of these buffer systems is supplemented by the action of the lungs. Through central control mechanisms, the rate and depth of ventilation is increased when the concentration of $H^+$ ions in the blood increases and causes a shift toward greater acidity. The response of increased ventilation removes more $CO_2$, which is the product of the breakdown of carbonic acid. The faster breakdown of this acid, in turn, permits a faster rate of its formation, through the combination of hydrogen and bicarbonate ions. Thus, increased rate of ventilation allows faster buffering of hydrogen ions.

The kidney also plays a role in maintaining a steady acid-base balance. During the production of urine, it can excrete certain ions and reabsorb others; it can even substitute one ion for another. The pH of urine may range from 4.5 to 8.2 in response to the need to maintain a steady internal pH. If body fluids become too acid, then the kidney begins to excrete more and more $H^+$ ions. If body fluids begin to rise in pH, then the kidney will substitute sodium ions for the hydrogen ions, increasing the acidity of body fluids and lowering their pH back to the normal range.

Under normal conditions, these mechanisms can maintain a steady balance of acids and bases. But under some conditions, the body's ability to buffer an excess of acids or bases may be inadequate to the load. For example, certain lung diseases can impair the lung's ability to release $CO_2$. The increase in $CO_2$ in the blood will interfere with the buffering system, and the acidity of body fluids increases, a condition known as acidosis. Diabetes can also cause acidosis, since it causes metabolic irregularities that result in a high production of acidic substances during food breakdown. The problem of an increase in the pH of body fluids, or alkalosis, can be caused by eating an excess of alkaline salts (for example, taking repeated doses of baking soda to cure indigestion), or by the increased breathing rate found in high fevers. Usually, the kidneys can correct acidosis or alkalosis by altering the rate at which various ions are excreted. If the condition is not corrected, the change in pH of body fluids can cause changes in the rate at which necessary chemical reactions occur, with harmful effects on the health of the organism.

## Glucose balance

Maintaining a steady level of glucose in body fluids is a serious problem of homeostasis found among large-particle feeders, a group that includes man along with many other vertebrates. Large-particle feeders do not eat continuously, as do some small-particle feeders, but rather at spaced intervals. As each meal is digested, the end-products of that digestion—simple sugars, proteins, and fats—enter the bloodstream. The presence of these osmotically active molecules will affect the osmotic concentration of the blood. If they were allowed to circulate freely, there would be an enormous increase in the number of these molecules in the blood and therefore in osmotic concentration after every meal, with a subsequent sharp drop in concentration after the meal is entirely digested and no new supply of molecules is added to the blood. This fluctuation in osmotic concentration would interfere with the osmotic exchanges that must take place between the blood and body cells. Moreover, since many important cells—brain cells, for example—do not have the ability to store glucose, they would be unable to meet their energy needs as the concentration of blood sugar dropped between meals.

The important function of maintaining a steady level of glucose in the blood is performed by the liver. Absorbed sugar passes from the intestine, through the hepatic portal vein, directly into the liver. That organ, acting in response to a variety of chemical and neural control mechanisms, carefully regulates the amount of glucose that it passes on into the main vein leading back to the heart. Only enough glucose to supply immediate body needs is allowed to enter the blood. If internal sensory mechanisms report that there is an excess of glucose in the blood, the liver converts the glucose into glycogen, a storage molecule that is held in the liver. Once the storage "warehouses" are full, additional glucose will be changed into fat molecules, which are released into the blood to travel to specialized connective cells (fat cells) for storage.

After digestion of a meal is completed, and no more glucose is being absorbed in the intestine, the liver begins to compensate for the drop in blood sugar. Stored glycogen is reconverted back into glucose and released into the bloodstream so that a constant level of blood sugar is maintained.

The liver can store only enough glycogen to maintain a steady glucose level for 24 hours. Once that glycogen is all used up, the liver will begin to break down fats to produce the necessary quantity of glucose. It can also deaminate proteins, removing the nitrogen groups, and then convert the residue into glucose. If there is an excess of fats or proteins in the bloodstream, these molecules will be used first. The stored fat will be used next. As a final desperate measure, some of the structural proteins of the body will be used for fuel.

In addition to maintaining a steady level of blood sugar, the liver performs other homeostatic functions. It breaks down, or detoxifies, many substances absorbed during digestion that may be harmful to the organism; for example, it is the liver that must detoxify alcohol and barbiturates. The liver is also the organ that maintains the steady composition of the blood. It manufactures the plasma proteins, such as fibrinogen and globulin, adding new supplies when the old ones have been used in various reactions throughout the body. Another function of the liver is destroying old red blood cells. These cells have a short life span of less than four months, due to the wear and tear of circulation (and especially the friction in the narrow capillaries, which are just about the same diameter as the erythrocytes). Red blood cells are replaced at the rate of 2.5 million every second, and the liver must break down the old cells at the same rate. Once the cell is broken

**8-15** *This rainbow trout which is about to make a meal of a minnow exemplifies the homeostatic problems of large-particle feeders. While the minnow is being digested, large quantities of nutrients may enter the bloodstream, greatly increasing its osmotic concentration. In the intervals between feedings, the concentration could drop severely. It is the role of the liver to even out these extremes and maintain a stable glucose concentration in the blood. (Omikron)*

down into its basic structural components, these can be saved for use as fuel or building materials for new cells. Another homeostatic function of the liver is the removal of lactic acid from the blood. This acid, the product of anaerobic respiration in muscles during periods of strenuous exercise, can be toxic to certain body cells; it can also upset the acid-base balance of the body. The lactic acid goes to the liver, where it is resynthesized into glycogen, which can then reenter the cycle of cellular respiration. The liver also stores certain vitamins and micronutrients, releasing them into the bloodstream as they are needed.

The liver is an important homeostatic organ, but it cannot function alone. For example, it performs the job of deamination, but it dumps the waste nitrogenous material into the bloodstream. The same is true of the products of detoxification. They enter the bloodstream and might circulate forever, building up higher and higher concentrations, if it were not for the excretory functions of the kidney.

## The role of the kidney in homeostasis

The great importance of the kidney in homeostasis has already been pointed out many times in this chapter. It is through the excretory activities of the kidney that osmoregulation is achieved, that an electrolyte balance is maintained, that the pH level of body fluids is kept constant, and that the products of metabolic breakdown can be eliminated from the body. Some idea of the importance of the kidney is indicated by the fact that it receives about 25 percent of the total blood supply in the human body.

The three principal functions of the kidney are: **filtration** of body fluids, separating small molecules and ions that are likely candidates for excretion from the large protein molecules that should be retained; **reabsorption,** in which some

water and selected small molecules or ions, such as glucose or sodium chloride, are actively or passively transported back into the blood plasma; and **excretion** of the urine which is the final product of the kidney's activity. In mammals, a fourth function is also found, that of **concentration,** or increasing the osmotic pressure of the urine so that it is hypertonic to the blood.

## Rudimentary excretory organs

In evolutionary terms, the kidney would be called the most advanced model of the excretory organ, but it is certainly not the only model that has ever been introduced. Excretory mechanisms can be found even in single-celled organisms. *Vorticella,* a protozoan first described by Leeuwenhoek when he observed it under his homemade microscope, has a contractile vacuole which expels excess water and some nitrogenous wastes. Free-living flatworms (*Turbellaria*) have excretory mechanisms called flame bulbs, which are hollow tubules lined with cilia. The motion of the cilia sweeps floating waste particles into the tubule and out through the excretory opening, or pore.

The development of the first true excretory organ accompanied the development of the circulatory system. When body fluids could no longer diffuse directly into the environment, a specialized organ of excretion was necessary. We can see a rudimentary type of excretory organ, or **nephridium,** in the leech and the earthworm; it is shown in Figure 8-16. The nephridium is a network of tiny tubules contained within a ciliated sac (the nephrostome) which is actually a specialized part of the central body cavity. The tubules collect some wastes directly from the body cavity. They also collect wastes which filter through the capillaries located adjacent to the tubules. The collected wastes are passed to a small storage bladder, where some reabsorption may take place, and then excreted through a pore-like opening.

## Respiratory adaptations

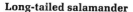

**Mangrove trees**    **Sphinx caterpillar**
**Rat-tailed maggot**    **Long-tailed salamander**

All organisms must carry on some kind of gas exchange, and they have developed a number of specialized adaptations for this process. An interesting example can be seen in the air roots of the mangrove tree. This tree grows along the sea coasts, where the salt water prevents gas exchange in the buried roots. The projecting air roots, 10 to 15 inches long, solve the problem.

Another illustration of a mechanism for gas exchange can be seen in the larva of the Sphinx butterfly, which gets its name from the pose of the larva on the leaf, with an upraised questioning head. Like all insects, the Sphinx has a tracheal respiratory system. The black dots visible down the side of the caterpillar are the spiracles that cover the opening of the tracheal tubes.

The long-tailed salamander shown here is a resident of the temperate deciduous forests. Salamanders show a diverse range of respiratory adaptations. All larval forms have gills, and in some instances the gills may persist into adult life; on the other hand, many salamanders have well-developed lungs. But the long-tailed salamander relies solely on cutaneous methods of gas exchange.

The rat-tailed maggot is actually a larval form of a hover-fly (Order Diptera). Growing from the posterior end of the larva is a long breathing tube connected to the tracheal system. The tube acts as a snorkel, permitting the larva to live in stagnant ponds with a low oxygen content.

# Pollination and seed dispersal

With the move to land, higher plants could
no longer rely on water to carry the sperm to
the egg. This led to the development of the
pollen grain, and accompanying specializations
to increase the chances of pollination. The
photos on the left-hand page show some of these
adaptations. The bright yellow goldenrod flowers
attract bees; in the process of filling their
own pollen baskets, the bees inadvertantly
carry pollen from one flower to another. The
orchid also relies on insects for pollination,
and some even mimic the appearance of female
insects, to attract males into the flower. The
lodgepole pine and the white birch trees are
both adapted for wind pollination. The success
of this adaptation can be judged by the great
cloud of pollen coming from the birch catkin.

Wind may also serve as a method of seed
dispersal, as shown in the dandelion on the
right-hand page. Milkweed seeds have the same
kind of feathery parachutes as the dandelion.
As the milkweed pod matures, it weakens along
the seams, until one day it bursts, flinging
the seeds into the air.

Many flowering plants rely on animals to
scatter their seeds. Burdock seeds, for example,
have a prickly outer covering that catches in
the fur of passing animals. Nutlike fruits
attract the attention of squirrels, who carry
them long distances and then often helpfully
bury them as well. Succulent berries are eaten,
but the indigestible seeds will later be expelled.

**Lodgepole pine pollen cones**
**Green orchid**

**White birch catkin**
**Bee on goldenrod**

**Burdock seeds in bobcat fur**

**Bursting milkweed pod**

**Dandelion**

**Raccoon and wild grapes**

**Red-bellied squirrel**

# Crypsis

A highly specialized kind of adaptation to the environment can be seen in these examples of crypsis, or hiding, that enable animals to escape the attention of would-be predators. The cryptic effect may be due to form, or color, or a combination of both.

A typical example of cryptic coloration can be seen in the dappling of the fawn that blends so beautifully into its shaded leafy environment. The summer plumage of the ptarmigan makes it almost invisible on the lichen-covered rocks; the white winter plumage will make it equally inconspicuous in the snow. *Agalychnis* the tree frog, is superbly camouflaged atop a blade of grass.

Some of the most amusing kinds of crypsis are found among the insects. The saw-toothed elm caterpillar strikingly resembles the serrated edge of the leaf on which it feeds. The thorn insect features a bizarre protrusion of its exoskeleton that makes it look like a thorn when it rests on the surface of a stem or twig. The adult *Cilix glaucata* mimics the color and shape of bird excrement (the insect is on the right of the picture). The lacewing larva feeds on aphids, and then adds insult to injury by masking itself with the skins of its victims so it can creep up on their unsuspecting relatives.

For some of the geometrid moths, crypsis is a way of life. The geometrid larva looks exactly like a twig; the adult moth appears to be only a piece of bark on a tree.

Lacewing larva eating aphids

Ptarmigan

*Agalychnis*

*Cilix glaucata*

**White-tailed deer**

**Thorn insect**

**Geometrid moth**

**Geometrid larva**

**Saw-toothed elm caterpillar**

**Leaf hopper**

# Aposemasis

Another type of protective adaptation is
signalled by aposemasis, or warning coloration.
Aposemasis, indicating an organism genuinely
protected by such defensive devices as a sting,
a nasty taste, or a toxic content, is highly
developed among the insects.

A common example of aposemasis can be seen
in the milkweed leaf beetle. In both the larval
and adult stages, this insect feeds on the
milkweed plant, which contains a toxic substance
similar to digitalis; thanks to their diet, the
insects are also poisonous. The gaudy red and
black coloration of both stages serves to help
"remind" predators that they are dangerous.

The Monarch butterfly also feeds on the
milkweed plant and is thus genuinely protected.
The Viceroy is an edible relative that escapes
predation because its coloration is similar to
that of the Monarch; this is called mimicry.

Many aposematically colored insects are
really quite beautiful. Look, for example, at
the little leaf hopper at the top of this page,
with the green racing stripes on its streamlined
red body. Or notice the elegant design formed
by the black and yellow markings of the current
moth, a European species.

*Catocala concumbens* illustrates both crypsis
and aposemasis. With its tan front wings folded
back, it is hidden against the bark; when it is
disturbed, it flashes its pink underwing, thus
startling predators. A more comical warning
display can be seen in the puss moth larva.

**Monarch butterfly**
**Viceroy butterfly**

**Currant moth**

*Catocala concumbens*

**Milkweed leaf beetle (larva)**

**Puss moth larva**

**Milkweed leaf beetle (adult)**

# Eyes

The eye serves as an excellent example of the way form and function can be adapted to meet the demands of different environments and ways of life. The huge eyes of the saw-whet owl give him a very broad field of vision, so that he is able to watch the movements of the small rodents on which he preys while remaining quite motionless, giving no warning of his presence. As the photo shows, the owl's pupils are each regulated separately, allowing one eye to admit a different amount of light than the other.

The eye of the caiman, a relative of the alligator found in Central and South America, features a vertical pupil. The caiman is a nocturnal predator, and the development of a vertical pupil seems to be closely identified with this way of life. As the strong sharp teeth imply, the caiman is a very successful hunter.

The two photos on the right show the eyes of the male (top) and female horsefly. The structure of these insect eyes, with their countless number of individual ommatidia, is quite evident. What the fly sees is a mosaic assembled from the information received by each separate visual unit.

In addition to serving as organs of vision, the eyes of the horsefly are adapted to serve a second function. The eyes of the female differ from those of the male in both form and color; this sexual dimorphism is a recognition device that permits members of the opposite sex to find each other when they are ready to mate.

**Saw-whet owl**

**Male horsefly**

**Caiman**

**Female horsefly**

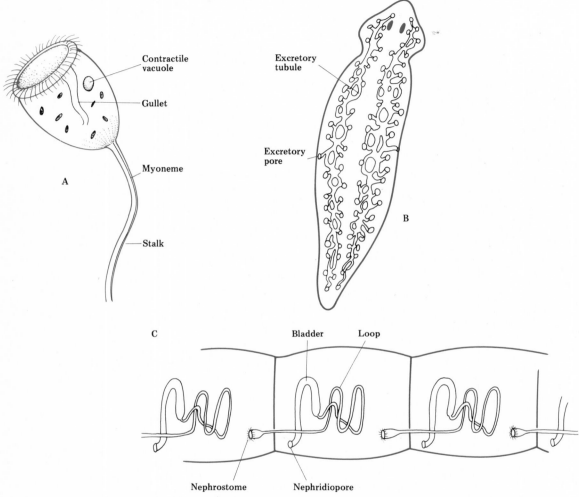

Contractile
vacuole

Gullet

Myoneme

A

Stalk

Excretory
tubule

Excretory
pore

B

C

Bladder    Loop

Nephrostome        Nephridiopore

**8-16** *Among the protists and lower animals, a number of types of rudimentary excretory organs can be observed. Vorticella (A) has a contractile vacuole to expel excess water. Turbellaria (B) excretes water and nitrogenous wastes through excretory pores connected to flame bulbs, where wastes are gathered. The earthworm (C) has many nephridia each of which is composed of a single looped tubule.*

The vertebrate kidney could be described as a nephridium with a great many optional extras. It is interesting to note, however, that in some groups, evolution has taken the course of abandoning the nephridium rather than improving it. For example, the insects have evolved a different kind of excretory system, to accompany their unusual open circulatory system. The excretory organs of insects are the **Malpighian tubules.** (These are named after the Italian scientist Marcello Malpighi, who did pioneering work in studying the anatomy of kidneys in the seven-teenth century.) They are little tubules that branch off the digestive tract between the stomach and the intestine. The tubules are adjacent to the body spaces into which blood drains, and the blood filters from these sinuses into the tubules. Reabsorption takes place as the wastes move through the tubules. The tubules eventually produce uric acid, slightly diluted by water and certain ions. The uric acid passes into the intestine for additional reabsorption of water and is finally expelled as a dry paste along with fecal wastes through the anus.

**8-17** *The excretory organs of insects are the Malpighian tubules. Blood that collects in sinuses is filtered into the tubules, where selective reabsorption adjusts the volume and composition of the wastes, which are excreted in the form of uric acid through the anus.*

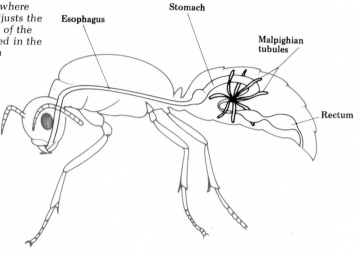

**8-18** *The fundamental unit of both structure and function in the kidney is the nephron. It consists of a spherical body (Bowman's capsule) and a long looped tubule. Filtration takes place in the capsule, and the tubule is the site of reabsorption, concentration, and final adjustments in volume and composition of urine.*

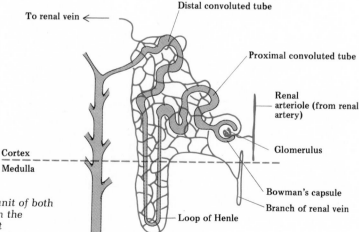

## The mammalian kidney

The structure of the mammalian kidney is basic to its function. The fundamental unit of both structure and its function is the **nephron** (see Figure 8-18). In man, each kidney contains more than a million nephrons. The urine produced in these nephrons drains into central collecting tubes and then moves out of the kidney through the ureter into the urinary bladder. The normal daily volume of urine produced fluctuates from 800 to 2000 milliliters, depending on what the individual has been eating and drinking.

Each nephron consists of two principal parts: a spherical body called **Bowman's capsule,** in which filtration takes place; and a thin convoluted tubule with a loop in the central portion, in which reabsorption and concentration take place. Bowman's capsule is actually a cup-shaped receptacle which holds the **glomerulus,** a bulb-like network of blood capillaries that branch off from the renal artery. The fluid filtered from the blood, called the **filtrate,** moves from Bowman's capsule into the first section of the tubule, or the **proximal tubule.** Here organic substances and required ions are reabsorbed, as is some portion of the water in the filtrate. The fluid then passes into a long, thin U-shaped portion of the tubule, which is called the **loop of Henle,** after the German anatomist who first observed the structure. The final convoluted portion of the tubule is called the **distal tubule,** and this is where the final adjustments in the solute and water content of urine are made.

When we look at a kidney that has been cut in half (see Figure 8-19) we see that the whole organ is very carefully organized. The central core of the kidney—the inner part of the bend—has many large collecting tubules to drain the urine as it is formed. Around this central core the nephrons are neatly layered. All nephrons

**263**

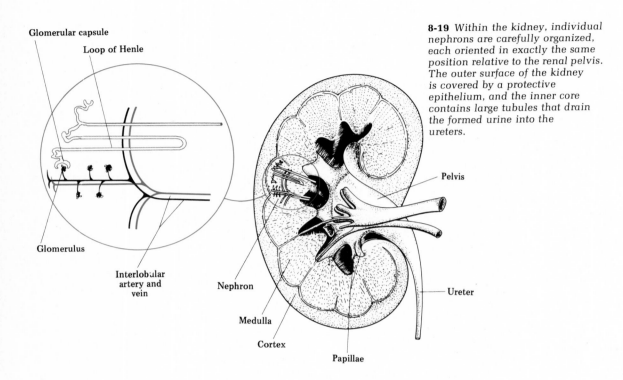

Glomerular capsule

Loop of Henle

Glomerulus

Interlobular
artery and
vein

Nephron

Medulla

Cortex

Papillae

Pelvis

Ureter

**8-19** *Within the kidney, individual nephrons are carefully organized, each oriented in exactly the same position relative to the renal pelvis. The outer surface of the kidney is covered by a protective epithelium, and the inner core contains large tubules that drain the formed urine into the ureters.*

are oriented in the same direction, with the Bowman's capsules and proximal and distal tubules at the outer surface of the kidney and the loops of Henle extending in toward the core. The extreme outer layer of the kidney contains the blood vessels that feed the capillaries, nerves, and endocrine cells to secrete regulatory hormones.

**Filtration.** Biologists long ago realized that the kidney was some kind of selective filter, but the mechanisms of filtration and selection were not understood. In fact, early workers in the field often wrote and spoke of the kidneys as if it were a mysterious little black box, capable of bringing about magical changes. They knew that blood entered the box at the top and that urine, a fluid of entirely different composition, left it at the bottom. But what happened inside?

The first step in the transformation is the filtration of the blood that enters the kidney. The process takes place in Bowman's capsule, as blood enters the glomerulus. Filtration takes place here in exactly the same way as it does in the capillaries during the circulation of the blood. The pressure of the blood in the tiny vessels of the glomerulus is greater than the osmotic pressure of the surrounding tissues. This differential causes many small molecules to leave the blood vessels and move into the tissues of Bowman's capsule. The filtrate is composed of water, with a number of substances dissolved in it—sugars, vitamins, urea, inorganic ions, and amino acids. There are no proteins in the filtrate, since these molecules are too large to pass through the vessel walls.

Only about one-fourth of the total volume of

blood goes to the kidney in each complete trip through the circulatory system, and it is estimated that only about half of that amount is actually filtered each trip. It therefore takes seven or eight trips before the entire volume of blood has been filtered.

Thanks to some ingenious experiments, biologists have been able to measure the rate at which filtration occurs. The glomerular filtration rate (often shortened to GFR) in man is about 125 milliliters per minute. The rate of filtration is about 100 times faster than the rate of urine formation; the kidney produces only about one to two milliliters of urine per minute. This difference indicates that a great deal of the volume of the filtrate must eventually be reabsorbed.

The process of filtration in Bowman's capsule stops when loss of fluid from the capillaries lowers blood pressure so that it is equal to osmotic pressure outside the vessel. To a certain extent, the body can regulate the amount of filtration that takes place. By constricting the arterioles that lead away from the kidney, the pressure inside the glomerular vessels is raised and more filtration will take place. If the arterioles that lead to the kidney are constricted, cutting down on the volume of blood that enters the kidney, filtration will be reduced. This mechanism allows the kidney to adapt to changes in the need for filtration.

**Reabsorption.** Filtrate collected from Bowman's capsule, and from the proximal tubule near the capsule, is about isotonic to the blood. Of course, the composition of these two fluids is somewhat different, since the filtrate does not include any proteins, but their osmotic pressure is equal. As the filtrate passes through the proximal tubule, it undergoes many changes in both concentration and composition. By the time it passes out of this tubule, the fluid has diminished in volume by about 75 percent. Moreover, the dissolved molecules of sugar and amino acids, as well as many inorganic ions, have been removed.

Since the concentration of amino acids and glucose is relatively high at the entrance of the proximal tubule, and practically zero at the end of the tubule, it is clear that the removal of these molecules is taking place against a concentration gradient. It is a process of active transport that is conducted by the cells that line the tubule. At the moment, biologists do not really understand the mechanism of this active transport, nor do they know how the cells differentiate between these molecules and others, left in the filtrate.

Investigations of active transport in the proximal tubule now being performed focus on the movement of sodium ions out of the filtrate. These ions must move not only against a chemical concentration gradient, but also against an electrical gradient, since they travel from an area of low positive charge into an area of higher positive charge. The work of Arthur K. Solomon and his colleagues in the early 1960s demonstrated that the sodium ion is the only inorganic ion actively transported in the proximal tubule; all other ions move in response to changing concentration gradients. The presence of large numbers of mitochondria along the outer surface of tubule cells is an indication of the high energy requirements of the process of sodium transport, but the exact mechanism is still unknown. It is assumed to be similar to the action of sodium pumps in other cells.

The water that leaves the filtrate in the proximal tubule probably does so by a process of passive transport. As sodium ions and valuable molecules of glucose are pumped out of the tubule, the osmotic concentration of the filtrate drops, so that it is more dilute than the fluid in surrounding cells. Water then moves along the concentration gradient, out of the filtrate and into the tubule cells. Other inorganic ions, such as potassium, are transported along with the water in which they are dissolved.

**Concentration.** In mammals, the important function of water conservation is largely performed by the kidneys. The kidneys of mammals (and also of birds) have the ability to concentrate wastes in a very small volume of water and to produce urine that is hypertonic to the blood. Concentration takes place in the loops of Henle, structures found only in mammalian and avian (bird) kidneys.

When the filtrate enters the loops of Henle, it has been reduced in volume, due to the active transport of sodium and the passive transport of water out of the tubule. But the fluid is still isotonic to the blood. Within the loops, great changes in concentration occur. At the curve of the loop, osmotic concentration may be seven or eight times greater than the concentration of the fluid entering the loop and yet when the fluid leaves the loop, it is hypotonic to (more dilute than) blood. The structure of the loop makes these great changes possible.

The process of concentration in the loops utilizes the principle of **counter-current exchange** through the two arms of the loop. The structure of the loop, the properties of the cell membranes of the cells which border the loop, and the spatial arrangement of the loop in relation to surrounding tissues are all important factors in the process of concentration that take place there.

Fluid that enters the descending arm of the loop is isotonic to blood. But as it travels down the loop, it moves through regions of increasing osmotic concentration in the surrounding cells. The cells that line this arm of the loop are permeable both to water and to dissolved ions, so the fluid within the tubules adjusts to conform to the osmotic concentration of surrounding areas. Water moves out of the fluid, while sodium moves in. Through this process of passive transport, the concentration of the fluid increases greatly, while the volume decreases. It appears that the osmotic concentration of the surrounding cells is graded in a series of small steps. So the tubule fluids are not exposed to a sudden and drastic change in concentration, but to a gentle and continuous change.

**8-20** *Within the loops of Henle, counter-current exchange concentrates the urine. When the filtered fluid enters the loop, it is approximately isotonic to the blood. The fluid that leaves the loop and moves into the distal tubule is markedly hypertonic to all other body fluids. The numbers in this illustration are hypothetical, serving only to give some indication of relative concentration values.*

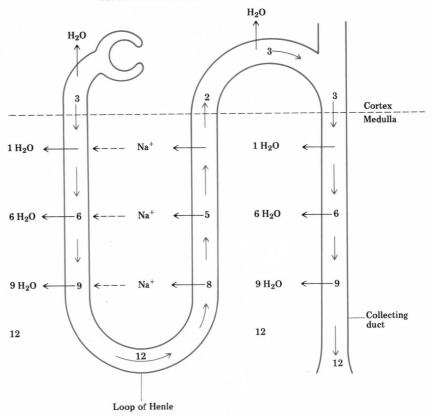

Loop of Henle

The maximum concentration of the tubule fluid is found at the turn of the loop. As the fluid begins to move back up through the ascending arm of the loop, it encounters the very same gradient it met on the way down, but this time it is traveling the other way. In the ascending arm, the fluid moves from an area where the surrounding concentration is relatively high toward an area where it is relatively low. In a mechanical system arranged this way in the laboratory, the result would be that water begins to move back into the tubule, to equalize the concentration inside with the concentration outside. But that, of course, would defeat the concentrating function of the kidney, since the goal is to remove as much water as possible. The movement of water back into the tubule is prevented in the living kidney by the membranes of the cells along the

tubules. The membranes there are relatively impermeable to water.

The impermeability of the membranes prevents the inflow of water back into the tubule, so that the volume of the fluid remains constant rather than increasing. The chief change seen in the fluid as it moves back up the ascending arm is a change in the concentration of the sodium ions. By a process of active transport, these ions are pumped out of the fluid against the concentration gradient. Since the external concentration is continually dropping, the differential between the inside and the outside concentration is never too large. Yet by the time the fluid leaves the loop and enters the distal tubule, its concentration is actually lower than that of blood plasma. This dilution of concentration is accomplished without any addition of water; it is due strictly to the removal of sodium ions.

The sodium ions which are removed from the ascending arm of the loop diffuse across the supporting tissue between the two arms of the loop and then enter the fluid in the descending arm. Thus these ions are constantly being recycled. They move into the descending arm, and across the tissue in between. The effectiveness of this sodium shuttle, and the counter-current exchange that takes place in the two arms of the loop are both dependent on the hairpin turn of the structure.

**Final adjustments in composition and concentration.** The fluid that enters the distal tubule is slightly hypotonic to blood and to the surrounding cells, as a result of the concentration that takes place in the loops of Henle. So as the fluid moves through the tubule, water (traveling with the concentration gradient) moves out of the tubule and into the surrounding cells. This further reduces the volume of urine. Active secretion may also take place in the distal tubule, transporting substances from surrounding cells into the tubule fluid. Potassium ions are often injected into the tubules at this point, when the

**8-21** *The physiological adaptations of the camel permit him to thrive in a most unpromising habitat; not only can he conserve water, but he can also survive severe water loss. This photo shows other adaptations for desert life—the long eyelashes that protect his eyes from the stinging sand and the flattened nostrils that prevent the loss of water vapor at the same time that they keep out the sand. (Gordon S. Smith from National Audubon Society)*

maintenance of ionic balance requires their excretion. It has been observed that penicillin, if it is present in the blood, is also transported directly into the forming urine at this point.

Final adjustments in the volume of urine are made in the collecting tubes, as the urine drains from the nephrons into the ureter. The cells that line this tube can vary their permeability to water, and they can also initiate the active transport of sodium ions out of the tube (water will follow the sodium through the process of passive transport). This variable permeability allows the organism to adjust the volume of urine to compensate for various problems, such as excess salt intake, or loss of blood through hemorrhaging.

The final adjustments in the volume of the urine are largely under the control of two hormones. One, called **antidiuretic hormone** (ADH), affects the rate of water reabsorption in the collecting ducts. The presence of ADH in the bloodstream makes the membranes of the cells lining the collecting duct more readily permeable to water, thus allowing more of it to flow out of the tube and into the surrounding cells and lowering the final volume of urine produced. ADH is secreted in a part of the brain, in response to messages from chemoreceptors there that can measure the osmotic concentration of the blood. If the blood becomes too concentrated, then the production of ADH is increased; if it becomes too dilute, then the production of ADH will be reduced or halted. The volume receptors located in the walls of the atria and major arteries, which can measure the volume of blood that flows through the blood vessels, serve as a feedback mechanism not only for the circulatory and respiratory system but also for the excretory system. When the volume of blood is low, the production of ADH is increased; when the volume is high, ADH production slacks off.

The second hormone that regulates the final adjustment in volume and composition of the urine is **aldosterone.** The presence of aldosterone increases the rate at which sodium ions are reab-

sorbed in the distal tubule. The secretion of this hormone is governed by the action of sensors located in certain cells that line the blood vessels of the kidney. When the volume of blood in the kidney increases, the sensors respond to the stretching of the vessels by secreting a chemical messenger that brings about a change in the rate of aldosterone secretion and therefore a change in the final volume of the urine.

## Summary

The survival of living organisms depends upon their ability to maintain a constant state in the face of external changes. One-celled as well as complex organisms and whole populations must display homeostatic mechanisms, but often specialized individual cells lose their homeostatic properties within a multicellular organism.

One critical homeostatic problem is maintaining a steady concentration of water. Osmoconformers are marine algae and lower marine invertebrates whose osmotic concentration conforms to that of the surrounding water environment. Higher animals have had to evolve osmoregulators to allow them to travel through differing osmotic environments. Some marine forms adapt by closing a shell to shut out the differing salt concentrations in the environment; others have systems so simple that they can tolerate some change. A more complex method is to transport the osmotically active solutes against the concentration gradient. Gills or skin can perform this function. Marine bony fish solve the problem of water balance by a mechanism that allows them to drink seawater and excrete salts through the digestive tract. Marine birds and reptiles also employ specialized extrarenal salt excretion to eliminate excess salt present in their environments and diets.

Fresh-water organisms utilize the kidney, which allows them to excrete urine that is hypotonic to the blood, thus "bailing out" excess water. There is also a sodium pump in the gills of fish and skins of amphibians. Terrestrial animals are threatened by dehydration, and accordingly, have adapted conservation measures to maintain a steady water balance. One of these is the sodium pump, which lowers the osmotic concentration inside the cell and increases the osmotic concentration in the extracellular fluid; this permits the excretion of highly concentrated urine.

A second critical homeostatic process is maintaining an acid-base balance, usually by means of a buffer system which resists pH changes. Buffer systems consist of a weak acid and a salt of that acid produced in an easily reversible combination reaction. The actions of buffer systems are supplemented by an increased rate of ventilation. The kidneys also play a role in stabilizing the acid-base balance.

Maintaining a steady level of glucose in body fluids is a homeostatic problem in animals, including man, which eat at spaced intervals. That function, as well as others, is performed by the liver, which maintains a constant level of blood sugar in the bloodstream.

Excretory organs have evolved from the contractile vacuole and flame bulbs, through the nephridium and Malpighian tubules, to the mammalian kidney. It is through the excretory activities of the kidney that osmoregulation is achieved, that an electrolyte balance is maintained, that the pH level of body fluids is kept constant, and that the products of metabolic breakdown can be eliminated. The fundamental unit of the kidney is the nephron, and it is within the nephron that urine is produced through four processes: filtration of blood; reabsorption of dissolved molecules and ions; concentration of wastes in Henle's loop; and final adjustments in urine concentration and composition, under the control of the two hormones, ADH and aldosterone.

## Bibliography

### References

Cannon, Walter B., 1929. "Organization for Physiological Homeostasis," *Physiological Reviews.* 9: 399–431

Heinrich, Bernd and George A. Bartholomew, 1972. "Temperature Control in Flying Moths." *Scientific American.* 226(6): 70–77.

Krebs, C. J. 1963. "Lemming Cycle at Baker Lake, Canada, during 1959–62." *Science* 140: 674–676.

Pitts, R. F. 1968. *Physiology of the Kidney and Body Fluids.* 2nd ed. Year Book Medical Publishers, Chicago.

Potts, W. T. W. and G. Parry. 1963. *Osmotic and Ionic Regulation in Animals.* Pergamon Press, New York.

Schmidt-Neilsen, K. 1964. *Desert Animals: Physiological Problems of Heat and Water.* Oxford University Press, New York.

Solomon, A. K. 1962. "Pumps in the Living Cell." *Scientific American.* 207(8): 100–108. Offprint no. 131. Freeman, San Francisco.

Whelan, W. J. 1968. *Control of Glycogen Metabolism.* Academic Press, New York.

### Suggested for further reading

Hoar, W. S. 1966. *General and Comparative Physiology.* Prentice-Hall, Englewood Cliffs, N.J.
A basic physiology textbook that assumes the reader has only a minimum knowledge of either physics or chemistry. A good reference for all chapters in Part 2.

Smith, H. W. 1961. *From Fish to Philosopher.* Doubleday Anchor Books, Garden City, N.Y.
The eminent kidney physiologist relates kidney structure and function to evolution. Through examples, he explains how the kidney functions on different levels and how it facilitates the adaptation of organisms to highly specialized environments.

Winton, F. R. and L. E. Bayliss. 1962. *Human Physiology.* 5th ed. Little, Brown, Boston.
A classic guide to the subject, written in a highly readable style. It is not only a reference, but a book to be read.


**Related books of interest**

Longmore, D. 1968. *Spare-Part Surgery: The Surgical Practice of the Future.* Doubleday Science Series. Doubleday, Garden City, N.Y.
Futuristic view of organ (especially kidney) transplants. The author contends that, given the increase in knowledge about genetics and tissue growth techniques, there is no foreseeable limit to what will be grown or made in the way of spare parts.

Tinbergen, N. 1967. *The Herring Gull's World.* Anchor, New York.
A report on researches into the social life of the herring gull with interesting deductions about the brain and nervous system of the gull. Its adaptation to a marine environment is discussed in detail.

# 9
# Movement

The capacity for independent movement characterizes living things. Of course, even the non-living universe is in constant motion, but purposeful, directed movement is the special prerogative of life.

Movement is one way an organism can respond to its external environment, and it is also a mechanism for maintaining some constancy of the internal environment. Through their movements, animals are able to find and consume their food, and plants are able to turn the broad surfaces of their leaves to face the sun. In most organisms, movement is an essential part of nutrition, of reproduction, of homeostasis — of all vital life processes.

Movement is also communication. William Condon, a behavioral scientist who works at the Western Psychiatric Institute in Pittsburgh, has studied the way that people communicate through movement. His analysis of films showing people talking to one another reveals that as they talk, they also move in unison. They nod their heads at the same time; they wave their hands together; they cross and uncross their legs and swing their feet in perfect synchrony, just as if they were responding to an off-stage conductor giving signals for each movement. But of course there is no conductor. The talkers are merely responding to one another's kinetic signals. Each perceives the other's movement and then almost instantly initiates responsive movements of his own. It is a game of follow-the-leader which takes place without any conscious awareness on the part of the players.

Condon believes that these movements are an important part of the total communication each person sends and receives. This conclusion should come as no surprise to biologists. Recent research regarding the activity of the beehive, for example, has shown that bees also communicate through motion. Scout bees can tell the rest of the hive about the location of nectar-bearing flowers by means of a dance. The directions of the dancing bee's movements tell the other bees which way to fly, the speed at which the dance is performed tells them the distance they must travel, and the enthusiasm of the dancer indicates the amount of nectar available.

The speed of the response Condon observed should also be familiar to biologists. The adjustments which allow the talkers to move almost simultaneously take place quite slowly in comparison to the rate of similar adjustments found in many other animals. Consider, for example, the speed at which a lynx must perceive and react to the erratic bounding of the hare it is chasing. From a distance, it looks as if both animals are running along an invisibly marked race-course, an illusion produced by the incredible speed at which the lynx can change course as it sees the hare do so.

Movement is really only the last event in a long sequence of actions. It is the visible outcome of a number of internal processes, involving the receiving and processing of many kinds of information by the organism (these processes are the subject of the next two chapters). The ability to receive and react to information about the environment is characteristic of all living things. Both plants and animals, no matter how simple or complex, show this ability. In nature, the various aspects of this process are combined into one efficient feedback system of continual action

and reaction. Biology textbooks, alas, cannot imitate this remarkable unity, but must instead divide the integrated process into separate little subjects, each allotted its proper share of pages.

## Movement of cells

When we think of movement in living things, we are likely to focus on the contraction of muscle that is found in multicellular animals. But muscles are only the specialization of a fully developed property of movement found within even the simplest of single cells. An easy way to observe the ability of one-celled organisms to move and change shape in response to changes in the environment is to watch a living amoeba under the microscope. The amoeba is a relatively large protozoan (it takes only 20 of them to cover the head of a pin), and it feeds on smaller one-celled animals and plants. Amoebae live on the bottom of muddy ponds, and they continually search the vegetation for suitable prey. Movement is initiated when one surface of the cell begins to bulge out, forming a projection (a pseudopod) that can be used like a foot. The pseudopod adheres to the surface and provides a point of contact. The rest of the cell can then either be pulled

**9-1** *An example of the relationship between movement and social organization can be seen in the precise patterning of these migratory Canada geese. Their spacing is just as measured as that of an Air Force flying team; even more amazing is the relationship of their wing beats. The leader is one stroke ahead of the second in line, who in turn is one stroke ahead of the third. (Alycia Smith Butler)*

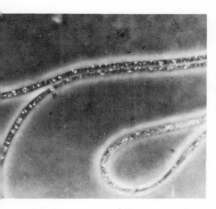

**9-2** *The property of movement is not limited to multicellular animals; even acellular organisms are capable of significant movement. For example, these gliding bacteria (Beggiatoa) can move relatively long distances across their substrate. (Omikron)*

toward the pseudopod or pushed away from it. The amoeba is no speed demon—it moves at the rate of about half an inch per hour. Its very slowness can be an advantage; the gradualness of its attack prevents its victims from perceiving the approaching threat.

The movement we see in the amoeba is an exaggerated version of a kind of movement found in all cells. Both single-celled organisms and the individual cells of a larger organism are able to change shape. They bend, stretch, and even travel from one place to another. A group of biologists at University College London have found that when single cells of body tissues are isolated and maintained in a jellylike culture medium, the cells can be seen to travel over the bottom of the culture dish. They begin this movement by fluttering the forward edge of the cell membrane up and down. As the membrane flutters, it gradually extends itself, and finally it anchors to the bottom of the dish in its new elongated shape. The rest of the cell is then pulled forward by that "foothold." The entire movement is really very similar to that of the amoeba.

It is believed that the microtubules play an important part in the movement of cells. The microtubules function just as bones do in a large multicellular organism. They are a source of stiffening support that helps the cell maintain its basic shape. For example, in cells like those that travel over the bottom of the culture dish, there are microtubules along the side of the cell but not at the undulating forward edge, or front. The presence of these skeletonlike rods restricts contraction to a single plane. This is why the cell succeeds in moving along in a single direction, and how it is able to maintain its general shape as it expands and contracts.

Microtubules act as the skeleton which holds the cell relatively rigid in spite of the changes of expansion and contraction. But what actually causes the movement? What acts as the muscle? In cells that travel or change shape, the electron microscope reveals the frequent presence

of bundles of tiny **microfilaments.** These microfilaments are generally no more than 40 or 50 Å in diameter, much smaller than the microtubules, which are about 250 Å wide. Bundles of microfilaments can be seen lying parallel to the surface of the cell membrane in the tip of the nerve axon, in the undulating edge of the migratory cell, and in the embryonic cells that will bend to form ducts or glands. Chemical studies of microfilaments found in slime molds or certain large protozoa indicate that the filaments are made of a protein that is very similar to **actin,** one of the two contractile proteins found in the muscles of higher animals. The hypothesis is that these microfilaments perform the function of contraction on the cellular level. Microfilament contractions move the cell, bend it into new shapes during the process of development, and even serve to heal tears in the surface of the cell membrane.

Current research is focusing on the question of how the contraction of microfilaments is initiated and controlled. Here too, a definite parallel emerges with the action of muscle fibers in higher animals. It has been found that injections of calcium ions will cause the surface of many cell membranes to contract around the injection site. As we shall see later, it has been established that calcium plays an important part in triggering the contraction of vertebrate muscle. Further research may establish whether or not it does the same thing for cellular microfilaments.

## Movement of cilia and flagella

Another kind of movement on the cellular level can be seen in the rhythmic movement of cilia and flagella. In some tiny organisms, this motion helps to propel the creature through the water. In other cases, it may help to move the water past the organism.

Both cilia and flagella are organelles large enough to be visible under low power in a microscope, but the movement of these structures is so rapid—cilia beat at the rate of about 20 strokes per second—that it cannot be studied through simple observation of ciliated cells. The movie camera has proved useful here. Biologists can take pictures of moving cilia and then run the films at slow motion, analyzing each frame. This type of analysis has shown that cilia employ two distinct strokes in their movement. The first stroke is the **power stroke.** The cilium is held relatively rigid, bending only in one place near the base of the organelle. The second stroke is the **recovery stroke,** which moves the cilium back into position to start another power stroke. In the recovery stroke, the cilium bends throughout its entire length; this stroke moves the cilium through the water with the minimum of resistance, thus insuring that it is not a propulsive stroke. The direction of these two strokes is readily reversible, allowing movement in either of two directions.

The movement of the flagellum is similar to that of the cilium, except that the flagellum can also employ another type of stroke. In addition to the whiplike stroke that moves the cilium, the flagellum has been observed to coil like a corkscrew. This rotary movement helps drive the organism through the water.

A basic biological principle is that structure always reflects function. One of the most striking features of the cilium is its precise and symmetrical structure. This has led to much curiosity about the way that its structure is related to its function. J. R. G. Bradfield, an English biologist, has advanced an interesting theory about this relationship. He suggests that the outer ring of nine filaments makes up the contractile element of the structure. In the power stroke, five of the filaments contract near the base of the stalk. In the recovery stroke, the other four filaments gradually contract throughout their length to return the stalk to its original position. According to this theory, the central two filaments are conducting tubules that carry electrochemical im-

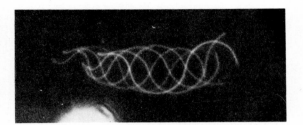

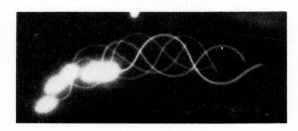

pulses from the basal body to coordinate the sequence of contractions with those of neighboring cilia, so that the correct wavelike rhythm will be produced. Partial confirmation of Bradfield's theory has come from the recent discovery that the enzymes that initiate the ATP-ADP conversion that releases energy for contractile work are found only in the nine outer filaments and not in the center two. Lacking these necessary enzymes, the center filaments probably also lack the ability to contract.

Bradfield's theory so far has neither been proved nor disproved, even though experimental evidence seems to confirm certain elements of the theory. It seems likely that no convincing proof one way or the other will be obtained in the near future, for the tools and techniques presently available are not refined enough for the delicate experiments that are necessary. In the absence of irrefutable evidence for or against the conclusion, how do scientists decide whether or not to accept such a theory as a working hypothesis?

Scientists test new theories with the tools of logic. They check to see that the theory does not

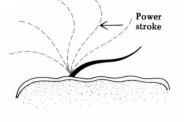

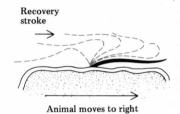

**9-3** *This figure shows two types of mobile organelles. (A) photographs of the sperm flagellum of a sea urchin; it reveals the spiral twisting motion of the flagellum. (B) is a drawing of the sequential events in the power stroke and recovery stroke of a cilium. Biologists suspect that the mobility of these organelles is the result of their specialized 9 + 2 structure. (With permission from C. J. Brokaw)*

contradict any of the known facts of the situation. They check to see if it includes *all* the known facts; a theory that leaves some facts unaccounted for, or explains them as merely coincidence, is not likely to win widespread acceptance, because experience has taught biologists that few of nature's arrangements are coincidental.

They also check to see whether the theory is the simplest possible explanation that can be given to cover all the facts. Most of us use this same logical test in our everyday lives. Suppose, for example, that you are expecting a check in the mail and it does not arrive on time. It is possible, and it does not contradict any of the known facts, that your check is late because the mailman has been kidnapped by a group of extremists and is being held (along with his mailbag) for ransom; it is possible that your check is late because a spy removed it from your mailbox; it is also possible that, out of the 200 million letters processed that day by the United States Post Office, yours was the one that happened to fall through a crack in the floor to be forever lost. But a simpler explanation, and the one that you will surely adopt as a working hypothesis, is that the letter was not mailed in time, or processed quickly enough, to arrive on the day you were expecting it. Scientists choose an explanation of ciliary function in just the same way that you choose an explanation for the lateness of your check; the simplest explanation is also the most likely one.

When we apply these tests of logic to Bradfield's explanation of the movement of cilia, we find only one inadequacy. The theory does not tell us why the contractile filaments should be found in nine pairs. Why are they not in a single, rather than a double, bundle? What is the advantage of having them separated into two structures? Biologists like to say that nature is a conservative architect, meaning that a structure is never any more complicated than its function demands. According to this assumption, the filaments would not be found paired unless there was some function that could not be performed by a single structure. Bradfield's theory does not explain what this function might be. Therefore, although it is probably right in its general outlines, it must be classed as an incomplete explanation of the known facts, according to the logic of science.

## Movement in plants

Many forms of algae are flagellated, and they can move through the water as easily and rapidly as any protozoan. Specialized reproductive cells of higher plants can also move from place to place. The sperm cell of a fern, for example, can swim right up the film of water that lines the female reproductive organ, attracted by a chemical stimulus released by the egg inside.

Most plants, however, are firmly rooted in the ground, and neither they nor their cells can travel about. For this reason, we do not generally think of them when we think of movement in living things. Yet plants do move. They move their leaves, their stems, their roots, their flower petals, and their seed pods, sometimes with startling speed and force.

There are two basic types of plant movement. One is the **tropism,** which is a movement induced by an external stimulus in which the direction of the movement is directly related to the direction of the stimulus. Tropisms are generally growth movements; they take place relatively slowly, and once they have occurred, they are irreversible. The other type of movement is **nastic movement,** in which the movement response is quite independent of the direction from which the stimulus originated. Nastic movements are typically due to changes in turgor; they take place very quickly, and can be reversed. Common stimuli for both tropisms and nastic movements are changes in the degree of light or temperature, contact with unyielding surfaces, and the pull of gravity.

**277**

Tropisms are classified according to the stimulus which causes the movement. For example, the movement which causes leaves to arrange themselves so that each one exposes the maximum possible surface to the sun, and no leaves are overlapping, is called **phototropism.** The movement by which plant roots always grow down toward the center of the earth, no matter how the seed may happen to be placed on the ground, is called a **geotropism.** The growth movement through which the tendrils of a grapevine twine themselves around a supporting lattice is called a **thigmotropism** (the Greek word *thigma* means touch). The growth movement of plant roots toward moisture is an example of a **hydrotropism.** In each of these cases, we see movement either directly toward or away from the source of the stimulus.

The best-known nastic movements are those in plants which appear to go to sleep at night, tucking their leaves together as if tucking themselves comfortably in. The leaves of the clover plant, for example, which during the day are held out horizontally, fold up in the evening, so that they hang vertically, parallel to the stem of the plant. In some plants, the leaf may even fold over along its central rib, closing like an oyster. These sleep movements are both photonastic—a response to changes in light—and thermonastic—a response to changes in temperature. It is thought that these sleep movements help to reduce the amount of transpiration that takes place overnight, conserving water supplies.

The most striking example of nastic movement can be seen in the delicate sensitive plant, a relative of the mimosa tree. The sensitive plant has a compound leaf, one with many little leaflets arranged along a central stalk. If you touch just one of these leaflets with your finger, every leaflet on that stalk will snap shut along the central rib, and the stalk will bend so that it points down toward the ground. The little plant seems to shrink from the touch of a gross finger. It takes at least 20 minutes for the plant to recover and once

**9-4** *An interesting example of a tropism can be seen in the leaf mosaic in this picture. As a result of phototropism, each leaf bends on its petiole so as to expose the largest possible surface to the light, avoiding the shade cast by its neighbors. This tropic response insures that the maximum amount of photosynthesis will take place in the plant.* (F. B. Grunzweig—Photo Researchers)

**9-5** Mimosa pudica, *the sensitive plant described in the text, is shown here before and after the leaflets have folded together along the midrib. Not only touch, but also darkness, heat and certain chemicals can cause this rapid response.* (Courtesy of J. W. Kimball, Biology, Addison-Wesley, Reading, Mass., 1965)

**9-6** *The leaves of the sensitive plant, Mimosa, droop rapidly in response to touch. This reaction is brought about by changes in the turgor of specialized cells in the pulvinus, located at the base of the leaf. Movement of water from the cells on one side of the pulvinus to the cells on the other side causes a corresponding change in the angle of the petiole in relation to the stem.*

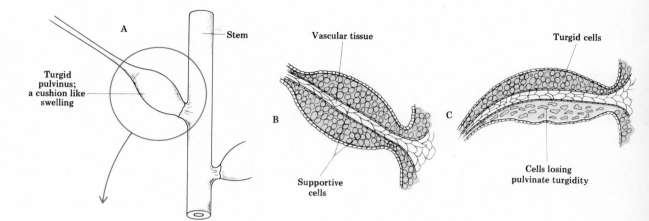

again display its leaves fully open and upright. The function of this exaggerated response is not known.

## The mechanism of plant movements

When an insect lights on the specialized leaf of the Venus' flytrap, the leaf springs shut, imprisoning the insect. This movement is as swift and sure as those of any insect-eating animal, such as a snake or frog. But we know that these animals spring their traps with the aid of muscular contraction, and plants have no muscles. Naturally, plant physiologists have been curious to discover the analogous mechanism in the flytrap.

It appears that the movement of the flytrap is due to sudden changes in the turgor pressure of certain leaf cells. The cells near the upper surface of the leaf lose water; cells near the lower surface gain it. The result of this differential change in turgor pressure is the bending of the entire leaf. The mechanism which causes the movement is very similar to the mechanism we described in Chapter 7 that opens and closes the stomata of leaf cells, except that it happens more suddenly and on a larger scale.

Plant physiologists are unable to explain how water is moved so quickly in the flytrap leaf, but they do know how the movement is controlled. It occurs in response to the presence of a chemical messenger, a hormone called **auxin.** Research on other plants has shown that auxin is involved in many tropisms and nastic movements. We shall discuss this important plant hormone in greater detail in Chapter 11.

Rapid changes in turgor are responsible for many other plant movements. The opening and closing of flowers is due to changes in turgor pressure at the base of the blossom; the movement of the sensitive plant is also due to changes in turgor. The cells in which the change in turgor pressure takes place are specialized cells slightly different in size and shape from other nearby cells. In some plants, such as most of the legumes, the specialized cells are contained in an organ called **pulvinus** (from the Latin word that means cushion). The pulvinus is located at the base of leaves; changes in turgor pressure there can move the leaf stalk, or **petiole,** up and down, opening the leaves in the day and closing them at night.

Most of the rapid and forceful movements of plants are due to differential changes in turgor. The force these changes are capable of exerting can be seen in the seedpod of the dwarf mistletoe. When the seeds are ripe, the pod undergoes changes in turgor pressure that cause it suddenly to act like a catapult, hurling the tiny seeds (about an eighth of an inch long) as far as 50 feet away from the plant!

The second known mechanism of plant movement is differential cell growth. This is the way, for example, that plant roots move in the direction of greatest gravity. The cells which are exposed to the greatest gravitational pull grow relatively slowly; cells with the least gravitational pull grow rapidly. The result of this differential growth is a bend in the plant root that causes it to point downward. The mechanism by which the plants perceive and respond to gravity is still completely unknown. At one time, it was thought to be due to the presence of large starch grains. The theory was that these grains were too heavy to be moved by cyclosis (the circular movement of cytoplasm around the plant cell) and therefore settled at the bottom of the cell, acting as a gravity indicator. However, it has been shown that even when the starch grains are all carefully removed, the geotropism is still displayed. It is evident, therefore, that some other mechanism is at work.

Differential growth is responsible for many phototropic movements, and probably for some of the sleep movements that have been observed as well. Differential growth may also be utilized to supplement the action of turgor changes. For example, in the Venus' flytrap, the initial closing of the trap is due to changes in turgor pressure, but this movement is closely followed by differential growth of cells in the upper and lower surface of the leaf. It is the differential growth that allows the trap to remain shut for the days or weeks it may take for complete digestion of the insect. The trap is then reopened by reversing the rates of growth, with the upper side of the leaf growing faster than the lower side.

In general, differential growth is a slower and more permanent movement mechanism than change in turgor. However, it has been shown that this growth can sometimes take place at a phenomenal rate. When the tendril of a pumpkin plant comes in contact with a surface that offers

**9-7** *This series of photographs follows the sequence of tendril growth in a squash plant. Such growth can be amazingly rapid, because it does not actually involve any cell division or synthesis. It is the result of simple expansion of already existing cells. (Dr. Paul Weatherwax, Indiana University)*

support, it immediately begins a process of differential growth that results in the formation of a coil around the support. This growth has been timed in the laboratory, and the formation of a complete coil has been known to take less than one minute.

It is important to note that this cell growth is not the result of the creation of new cells by cell division, for that process would take much longer. The growth is due to the elongation of tiny cells that had already been created by the division of meristem tissue. Under the influence of a chemical control (which is probably auxin), these cells simply absorb large quantities of water and stretch out. The cell cytoplasm, previously a dense colloid, becomes more dilute; the hydrostatic pressure caused by this increase in the volume of cell contents causes the coiled proteins in the cell membrane to stretch also. Thus this sudden elongation can be accomplished without the need for time-consuming synthesis of new materials. The existing structural elements are simply expanded by the addition of water, just as a dry sponge can be expanded by pouring water over it.

## Movement in animals

Animals have developed a more mobile way of life than plants. Therefore they are capable of faster and more powerful movement. The mechanism of amoeboid movement, or the actions of cilia and flagella, which propel one-celled animals through the water in search of food, cannot be utilized by larger, multicellular animals, because neither of these methods of movement can provide enough force to move a heavy weight. Larger animals have therefore evolved in a different direction. They have developed the contractile property that is present in the single cell into a specialized form of cell, the muscle cell. Muscle cells are capable of contracting with

great force, so that they can serve as a means of rapid propulsion even for a whale that weighs more than 100 tons.

The contraction of muscle develops the force necessary for movement, but movement will not take place unless that force can somehow be transmitted and applied to the surrounding air, land, or water. If the animal is to move forward, the force of the muscle contraction must be used to push backward against the surroundings. For example, we move forward by pressing our feet against the ground; a turtle swims forward by pushing its front legs against the water; when a pigeon flies, the broad flat surfaces of its wings push against the air.

How can the force of muscle contraction be applied as a backward pressure against the surroundings? When man constructs a machine for movement, he generally uses the wheel for this purpose. But a wheel can work only when it has no fixed connection with the rest of the machine, and such a system is obviously impractical for living organisms, which must be completely interconnected by nerves and blood vessels for transport and communication between cells. Instead of wheels most organisms use rods and levers to transmit force. These rods and levers comprise the animal's skeleton.

## The hydrostatic skeleton

We usually think of the skeleton as something hard, like the bones of a horse or the chitinous coverings of a cricket. But if we think of a skeleton in terms of a device to transmit forces from one part of an animal's body to another, then we would have to say that animals like leeches and worms, without a single bone in their bodies, also have a skeleton. Their skeleton is **hydrostatic;** it is made up of body fluids.

We can see the way a hydrostatic skeleton works to transmit forces if we watch the movement of an inchworm. The inchworm has two sets of muscles. One set—the longitudinal muscles—runs down the length of the body; the other set—the circular muscles—encircles the body. Movement begins when the inchworm contracts the circular muscles. This contraction exerts force upon the fluid within the coelom, or body cavity, pushing the fluid through the length of the body. This movement of the fluid stretches the longitudinal muscles and extends the body of the inchworm. When the body is fully extended, a little leglike structure (proleg) at the front of the worm is attached to the ground. Then the longitudinal muscles contract, once again pushing the coelomic fluid, so that the inchworm's body becomes shorter and wider, due to the extension of the circular muscles. When the body is as short as possible, a second proleg, located at the hind end of the inchworm, is attached to the ground, and the sequence of movements can begin all over again.

A similar mechanism of movement by hydrostatic skeleton is seen in many annelid worms, including the earthworm, and in some mollusks as well. In the earthworm, with its longer body and segmental pattern, the two phases of movement may be going on simultaneously in different parts of the body. For example, one segment of the body may be contracting the circular muscles while an adjacent segment is contracting the longitudinal muscles. The body contracts in continuous waves that press the many little bristlelike **setae** against the dirt, enabling a forward movement. This system allows an even faster rate of movement than is found in the inchworm.

This system of muscles exerting their force against a hydrostatic skeleton displays all the elements basic to the movement systems of higher animals. The coelomic fluid transmits the force of muscular contractions just as do the bones of mammals. The proleg, or the seta, acts as the anchor point where the backward push can be applied to the ground; it performs the same function as do our feet. And the longitudinal muscles

and circular muscles work as an **antagonistic pair.** Each set of muscles exerts an effect opposite to that of the other. This same principle of antagonistic pairing is utilized to even greater effect in fast-moving birds and mammals.

## The exoskeleton

In mollusks and arthropods, we find a second type of skeleton, the **exoskeleton.** In these animals, the force of muscular contraction is transmitted by a hard outer covering, such as a shell or cuticle. These coverings look more like plates than the rods we generally think of as comprising a skeleton, but they serve the same functional purpose.

The exoskeleton is hinged to make a bending movement possible, and an antagonistic pair of muscles is attached at each hinge. One muscle contracts to bend the shell at the hinge, and the other muscle contracts to straighten it out again. In oysters and other bivalves, the mechanism is somewhat different. A thick, tough ligament holds the shell open in its normal position; the muscle snaps the shell closed when danger approaches.

For many animals with exoskeletons, movement is rather a slow and laborious process, as moving those hinged shells is hard work. But it would be wrong to conclude that the exoskeletal system is inefficient, or incapable of rapid movement. The fastest movements recorded are those of certain flying insects. A Finnish physiologist has measured the rate at which a number of different insects beat their wings in a complete cycle (one stroke up and one stroke down). Butterfly wings move at the rate of 5 cycles per second; beetles at the rate of 150 cycles per second; the mosquito at the rate of 587 cycles per second; and a tiny midge has been clocked at the incredible rate of 1046 cycles per second.

These insects display a number of anatomical and physiological adaptations that help them

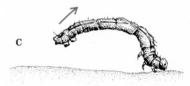

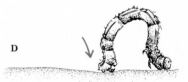

**9-8** *The inchworm begins its movement by contracting the circular set of body muscles, pushing the coelomic fluid the length of the body to extend it in a looping motion. The proleg at the front of the body then attaches the inchworm to the ground, and the longitudinal muscles contract, shortening the body. A second proleg is attached to the ground and the sequence of movements can begin all over again.*

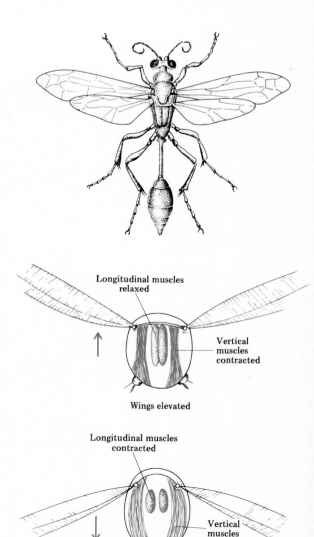

**9-9** *A tough ligament holds the shell of an oyster open in its normal position. When danger approaches, the muscle attached near either end of the hinge contracts, snapping the shell shut.*

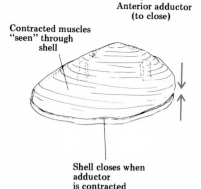

Ligament; holds shell open when adductor muscles are relaxed

Posterior adductor muscle

Posterior retractor

Anterior retractor (for foot withdrawal)

Anterior adductor (to close)

Contracted muscles "seen" through shell

Shell closes when adductor is contracted

move at such rapid speeds. Careful observation of their exoskeleton shows that around the joints where wings and legs are attached, the chitin is paper-thin and as flexible as a spring. It differs markedly from the chitin of other parts of the exoskeleton, such as the upper back, where the chitin has been reinforced by a protein that is hardened by a sort of tanning process. Another adaptation can be seen in the respiratory system of these insects. They have a tracheal system of gas exchange, with tiny tubes that carry $O_2$ into the body and $CO_2$ out of it. The smallest branches of these tubes are called tracheoles. It has been found that in many insects with rapid wing beats, the tracheoles run parallel to the muscles used in flight and are, in fact, actually sheathed inside the external membrane of the muscle. Thus the oxygen needed for respiration in these hard-working muscle cells travels the minimum distance—probably only about five microns. This greatly increases the rate at which gas exchange can take place.

In insects with rapid wingbeats, the muscles are not attached to the wings at all. Instead, they are attached to the central body segment, the **thorax,** in an antagonistic arrangement. The outer bundle of muscles contracts vertically, pulling the upper thorax, where the wings are attached, downward; this compression of the thorax raises the wings. The opposing inner bundle of muscles contracts in the other direction, pushing the thorax out again and lowering the wings. This system is one of great mechanical efficiency, for the distance moved by the tips of the wings is many times greater than the change in length of the muscles in the thorax.

The most interesting physiological adaptation found in these insects is that their flight muscles seem to be capable of self-stimulation. The stretch of one set of muscles (caused by the contraction of the antagonistic muscles) seems itself to stimulate another period of contraction, without the intervention of any nervous stimulation. Thus the muscles expand and contract in a

Longitudinal muscles relaxed

Vertical muscles contracted

Wings elevated

Longitudinal muscles contracted

Vertical muscles relaxed

Wings down

**9-10** *The asynchronous muscle of insects with extremely rapid wing-beats is attached not to the wings themselves but to the thorax. When the outer muscles contract, the roof of the thorax is pulled downward causing the wings to move up; an opposing set of muscles pulls in the sides of the thorax, forcing the roof upward and the wings down.*

series of oscillating vibrations, rather like the vibrations in a taut rubber band when it is plucked. Research in this field has now shifted to the somewhat perplexing question of how this insect version of the perpetual motion machine ever stops. It is thought that some enzyme must temporarily destroy the muscle's responsiveness, allowing it to stop vibrating so that the insect can land.

## The endoskeleton

There are several disadvantages to the exoskeletal system. One is that it places a size limitation on the organism. The larger the animal, the heavier the exoskeleton would have to be in order to support it, and therefore the more difficult it would be to move the exoskeleton. Although aquatic arthropods, such as the lobster, can grow fairly large, since the water will help support the weight of the shell, a large terrestrial arthropod is an impossibility. Another disadvantage is that growth requires periodic molting of the old exoskeleton, which becomes a confining prison as the organism gets larger. During the time it takes to grow a new exoskeleton, the animal is nearly helpless and quite vulnerable to any predator. Moreover, the need to synthesize an entirely new shell creates a significant metabolic drain on the animal. Vertebrates, which are generally larger than arthropods, and which lack the arthropods' ability to reproduce rapidly and in huge numbers, have developed a third type of skeleton, the **endoskeleton,** in which these disadvantages have been avoided. Endoskeletons are also found in echinoderms, such as the starfish.

The vertebrate endoskeleton is composed of bone and cartilage (for descriptions of these cells, see Chapter 2, p. 78). Muscles are attached by tendons to the bones where they are hinged or jointed, allowing flexible movement at those points. The bones in the vertebrate skeleton are arranged so that they act as levers, which are

moved by the contraction of muscle. The muscles of the endoskeletons are found in antagonistic pairs. One set contracts to bend the limb at the hinge and the other set contracts to straighten it out.

Figure 9-11 diagrams the mechanical principles involved in the bending at the elbow. The hinge is found between the upper arm bone, the humerus, and the two bones of the lower arm, the radius and ulna. The humerus acts as the stationary member; the elbow joint is the fulcrum; and the lower arm is the lever. The muscle that contracts to pull the lower arm up is the biceps. The fixed end of this muscle, usually called the **origin,** is attached to the humerus, and the end that moves, the **insertion,** is attached to the forearm bones to the right of the fulcrum. As the

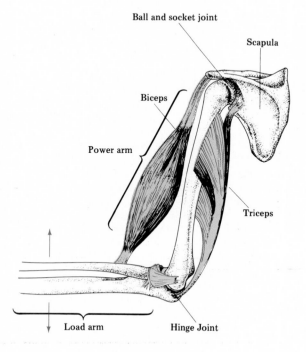

Ball and socket joint

Scapula

Biceps

Power arm

Triceps

Load arm

Hinge Joint

**9-11** *The hinge of the elbow is between the humerus on one side and the radius and ulna on the other. An antagonistic pair of muscles, the biceps and triceps, cause bending or flexure at the hinge. This diagram shows the relative lengths of the power arm and load arm of the lever system found in the arm.*

**9-12** *Mr. Universe demonstrates the extent to which muscular tissue is responsive to nutrition and exercise. Note the highly developed latissimus dorsi and triceps muscles, which are inconspicuous in most people. (Photo by Leo Stern. Used by permission of Bill Pearl.)*

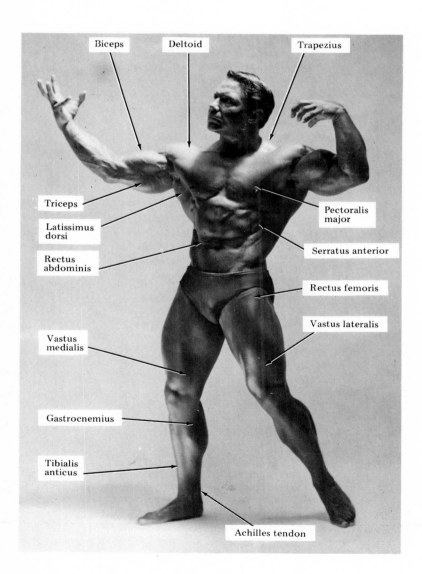

Biceps    Deltoid    Trapezius

Triceps

Latissimus
dorsi

Rectus
abdominis

Pectoralis
major

Serratus anterior

Rectus femoris

Vastus lateralis

Vastus
medialis

Gastrocnemius

Tibialis
anticus

Achilles tendon

biceps contracts, it raises the forearm lever. The antagonistic muscle is the triceps. The origin of this muscle is also found in the humerus. Its insertion is found on the forearm to the left of the fulcrum. When the triceps contracts, it is, in effect, pulling on the other end of the lever, thus lowering the forearm. Note that these two muscles work in opposition. When one is contracted, the other is automatically stretched.

A well-known fact of engineering is that the most efficient lever — the one that obtains the greatest magnification of the initial force — is one in which the distance between the source of the power and the fulcrum (the power arm) is very long, and the distance between the fulcrum and the weight that must be moved (the load arm) is very short. Yet when we study the mechanics of the vertebrate endoskeleton, we find that in most cases, the power arm is short and the load arm is long. In other words, vertebrate muscles and bones are arranged so that they work at a considerable mechanical disadvantage. Interestingly enough, this mechanical disadvantage actually permits muscles to work with maximum speed and efficiency. As we shall soon see, muscles work best when they are heavily loaded. They contract very slowly but develop enormous tension.

In addition to transmitting the force of muscular contraction, the vertebrate endoskeleton has two other important functions. One is to support the weight of the other tissues of the animal's body. The cylindrical shape of the bones is ideally suited to this purpose, since it offers the greatest support while taking up the least space. The second function of the endoskeleton is to protect vulnerable internal organs. It is the **axial skeleton,** or the portion composed of the skull, vertebral column, and rib cage, that specializes in protection. The bones of the **appendicular skeleton,** which includes such appendages as arms and legs, wings, and fins, are specialized primarily for movement with a secondary function of support.

**285**

## The physiology of skeletal muscle

The fact that muscular contraction provides the force for movement is obvious even to an untrained observer. Anyone can see that muscles change shape as people work, and anyone can feel the tension of their contraction. But the question of how muscles contract, and what controls their contraction is not so easily answered.

The first concrete answers to questions about muscle physiology came as an accident. Physiologist Malcolm S. Gordon comments:

> The process that triggers muscle contraction was a subject for long and diverse speculations up until the end of the eighteenth century. It was not until that time, however, that enough had been established about electricity to permit scientists to have in their laboratories machines that let them produce sparks at will. The Italian scientist Galvani had such machines in 1792, and frequently used them. Galvani, among other things, was also a professor of anatomy and often had frogs dissected in his laboratory. One day, apparently quite by accident, Galvani had a freshly dissected pair of frog legs lying on a table near one of the spark machines. An associate apparently touched a nerve on one of the legs with the tip of his metal scalpel at the same time another associate was operating the machine. The muscles of the legs immediately contracted violently. Galvani and his associates repeated the process many times, with the same result. Galvani's interpretation of this was that animal cells contain the same kind of electricity as produced by his machine and that perturbation of this electricity triggered activity in the muscle. We know now that this is essentially what does happen in the initial stages of muscle activation. Thus a technological advance in

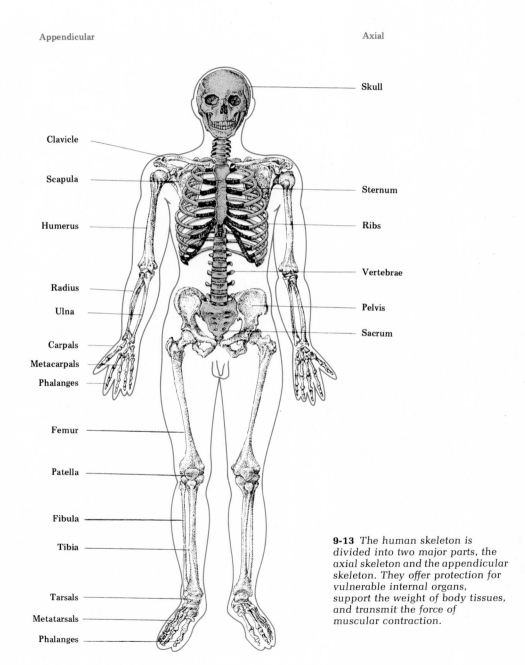

9-13 *The human skeleton is divided into two major parts, the axial skeleton and the appendicular skeleton. They offer protection for vulnerable internal organs, support the weight of body tissues, and transmit the force of muscular contraction.*

the form of the spark machine combined with chance, in the presence of a good observer, to open up the whole field of muscle excitation and, eventually, electrophysiology.*

Galvani's accident led to other experiments, enabling biologists to establish the fundamental contractile properties of muscle. These properties can easily be demonstrated in the laboratory, using very simple equipment. It is best to use the freshly removed muscle of a poikilothermic animal, such as the frog, for it is able to contract effectively at room temperature. The muscle of a homeotherm is much more delicate and creates difficulties for the experimenter. It must be maintained at the animal's normal body temperature and supplied with oxygen, and even then it will soon lose its contractile ability. Frog muscles can remain functional for several hours although they have been isolated and exposed directly to the air.

When the muscle is stimulated by electrical current, it can be seen to contract. If the muscle is attached to a recording device, such as a kymograph or an oscilloscope, the force of the contraction can be measured. This leads to some interesting results which give us clues to the nature of the chemical and structural events of contraction.

Measurements of the contraction of an entire muscle show that the strength of the response increases as the strength of the electrical stimulus is raised (see Figure 9-15). At very low current, no response is seen at all; then as the current is gradually increased, the force of the muscular contraction also increases, until it finally reaches a maximum response.

If the muscle is teased apart to obtain one single fiber, and that fiber is treated in the same way, the results will be quite different. With a very weak current, no response will be recorded. But as soon as the electricity reaches the point

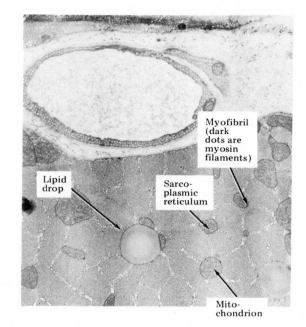

**9-14** *This rather unusual view of a muscle is a transverse section — in other words cut across the grain. Each of the myofibrils is surrounded by membranes of sarcoplasmic reticulum, giving the appearance of dotted lines. The dark irregularly shaped structures are mitochondria, and the single light sphere in the center of the photo is a lipid droplet. (David Soifer—Omikron)*

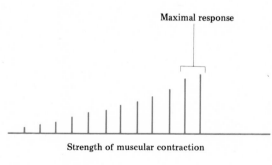

Maximal response

Strength of muscular contraction

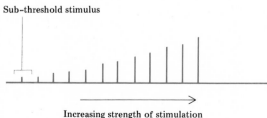

Sub-threshold stimulus

Increasing strength of stimulation

**9-15** *Comparison of these two charts shows the response of a whole muscle to electrical stimulation of varying intensity. The initial subthreshold stimulation produces no response; above threshold value, the strength of the muscular response varies directly with the intensity of the stimulus.*

---

* M. S. Gordon, *et al., Animal Function: Principles and Adaptations* (New York: Macmillan, 1968), p. 112.

where activation begins—the **threshold value**—the response is complete contraction. No matter how much the current is increased beyond the threshold value, there will be no increase in response. This characteristic of muscle fibers is called the **all-or-none response,** and it can be seen in the individual fibers of all vertebrate skeletal muscle.

The results of these two experiments seem contradictory. The individual fibers all exhibit an all-or-none response. Yet the whole muscle, which is simply a bundle of fibers, shows an increase in contractile force as the stimulus is increased. This seems to be a case where the whole is quite unlike the sum of the parts.

We can reconcile this apparent contradiction when we realize that the action of the whole muscle is the cumulative result of the action of individual fibers. Each fiber has an all-or-none response, but the greater the stimulus to the whole muscle, the larger the number of fibers that will be stimulated. If only one fiber is stimulated, the contraction will be very weak. As more of the individual fibers are stimulated, the force of the contraction will increase. It is this cumulative effect of more frequent stimuli, or of stimuli applied to a greater number of fibers, called **summation,** that allows muscles to contract smoothly and to remain contracted for an extended period.

The stimulation of a single fiber produces only one brief contraction, or **twitch.** A simple twitch, which can also be seen in a whole muscle that is given only a brief stimulus, actually has three distinct phases. Immediately after the stimulus is applied, there is a brief period, a tiny fraction (0.003–.004) of a second, in which nothing at all happens; this is the **latent period.** Then the muscle contracts, and this is called the **contraction period.** Then the contraction weakens, during the **relaxation period.** When many fibers are stimulated at once, or the stimulus is re-applied many times, there is a summation of the individual twitches, and a smooth sustained

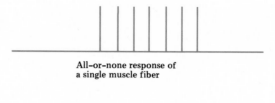

All–or–none response of a single muscle fiber

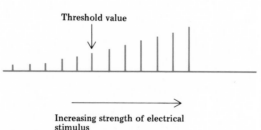

Threshold value

Increasing strength of electrical stimulus

**9-16** *When a single muscle fiber is stimulated by an increasingly strong electrical current, it does not show a corresponding increase in strength of response. Instead it displays an all-or-none response. There is no reaction to subthreshold stimulation and complete contraction when the stimulus reaches threshold level.*

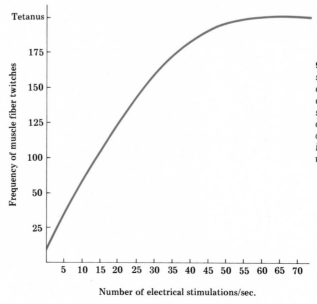

Number of electrical stimulations/sec.

**9-17** *The graph indicates that the smooth sustained muscle contraction called tetanus is actually brought about by the summation of individual contractions of the fibers. The cumulative effect is thus different in nature from the twitches of which it consists.*

**9-18** *This photo of a motor end plate indicates the relationship between the muscle fiber and the neuron that stimulates its contraction. The thick black line is the neuron, and the smaller dots near its end are the actual connections between nerve and muscle cells. The horizontal striations are the individual muscle fibers. (Cameron Thatcher from National Audubon Society)*

than a hundred fibers, allowing small movements for precise control. Where such precision is not important (in the muscles that hold the back erect, for instance), the motor unit may be composed of thousands of muscle fibers.

## The electrical events of contraction

Since it is known that an electrical impulse may act as the stimulus for contraction, early research workers in the field of muscle physiology proceeded on the assumption that muscular contraction is an electrochemical event. Recent research by Christopher Ashley, Graham Hoyle, and others has made the basic outlines of this event quite clear.

Individual muscle fibers have a diameter of between 10 and 100 microns. The fiber is covered by a membrane, the **sarcolemma.** The sarcolemma exhibits strong electrical polarization. The inner surface of the membrane is negatively charged relative to the outer surface; the differential between the two sides of the membrane is about one-tenth of a volt. This differential is maintained through the action of a system of active transport that moves positively charged sodium ions out of the cell. This mechanism is probably similar to the **sodium pump** that operates in the nephrons of the kidney.

When the electrical impulse arrives at the membrane, that structure is suddenly depolarized, probably because the electricity alters the permeability characteristics of the membrane. Positively charged sodium ions flow through the membrane, from the outside to the inside, and it loses its electrical potential. It is this depolarization that initiates contraction.

Electron microscope photographs taken by Keith Porter in 1955 disclosed that the sarcolemma was not the only membrane found in the muscle fiber. Within the fiber itself, there is a folded membrane network called the sarcoplasmic reticulum, which also displays an elec-

contraction, or **tetanus,** is produced. No one fiber contracts throughout the entire period of tetanus. It is the rapid alternation of contraction in these fibers that produces the effect of a sustained muscle contraction. It is really very similar to the effect of the eight-cylinder piston engine, which delivers sustained power through the sequential firing of the individual pistons.

Dissection of the muscle reveals that the electrical impulse that stimulated contraction is delivered to the muscle fibers by a nerve cell, or **motor neuron.** Each motor neuron is connected to a number of muscle fibers; the neuron and fibers together make up a single **motor unit** which fires at one time. In muscles where delicate control is important, such as the muscles that move the eyeball, the motor unit may consist of fewer

trical potential, since it, too, is capable of polarization. This internal membrane network establishes an electrical potential at many points inside the fiber, so that the response to an electrical stimulus can be almost simultaneous throughout the entire fiber.

## The structural events of contraction

Muscular contraction provides another instance of the close relationship between structure and function. When viewed in the electron microscope, a muscle fiber can be seen to be composed of a number of smaller fibers, called **myofibrils.** Each myofibril is about a micron in diameter. These myofibrils can be extracted for study by homogenizing the muscle in an ordinary kitchen blender.

By a combination of chemical analysis and electron microscopy, it has been shown that the myofibrils are themselves composed of even smaller filaments of protein. Two types of filaments are found. **Myosin** filaments are about 160 Å in diameter and about 1½ microns long. **Actin** filaments are shorter and thinner, being about 50 Å wide and about a micron long. The actin and myosin filaments are always arranged in a regular pattern, and it is this pattern that gives skeletal muscle its characteristic striated appearance.

The myosin molecule can actually be photographed. This protein seems to have a small zigzag head and a long straight tail. When the molecules come together to form filaments, the tails line up side by side, and the heads project, as shown in the drawing in Figure 9-20. The myosin filament, then, is not a smooth structure but has a number of projecting zigzag heads.

Knowing the structure of skeletal muscles helps us to understand the mechanism by which they contract. During contraction the muscle shortens, but it can be demonstrated that neither the filaments of myosin nor the filaments of

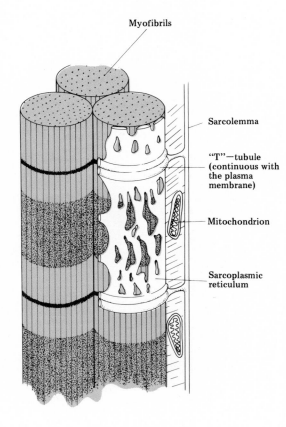

Myofibrils

Sarcolemma

"T"—tubule (continuous with the plasma membrane)

Mitochondrion

Sarcoplasmic reticulum

*9-19 The drawing shows an individual muscle fiber. It is covered by a membrane, the sarcolemma, which can exhibit a high degree of polarization. Within the fiber itself is another membranous network, called the sarcoplasmic reticulum; it, too, can be strongly polarized. The polarization keeps the muscle fiber in a state of readiness to react.*

*9-20 The myosin molecule has a small head with a zigzag shape and a long thin tail. When several are grouped together, the straight tails form a long thin structure from which the zigzag heads seem to protrude. These heads can be altered to a form of ATP-ase, releasing ATP needed to fuel muscle contraction.*

Protruding heads of myosin molecules

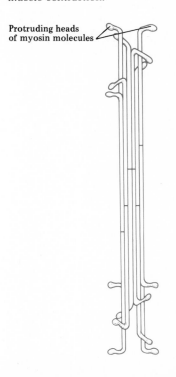

actin actually shorten during contraction. What they do, instead, is slide together. This explains why a contracted muscle becomes both shorter and thicker. The actin and myosin filaments, which in a relaxed muscle are just slightly overlapping at the ends, slide together to create a double thickness of filaments.

The actual coupling of the two fibers occurs at certain sites called cross-bridges. As the filaments slide together, the existing cross-bridges are detached, and new ones are created farther along the filament. It is the cross-bridges that must sustain the full tension of the contraction.

Forming and breaking the cross-bridges is an energy-requiring process. It is hypothesized that at each cross-bridge site one molecule of ATP must be converted to ADP to provide the required

**9-21** *Neither the filaments of myosin nor the filaments of actin actually shorten during contraction. Instead they slide together. Thus a contracted muscle is both shorter and thicker than the same muscle in a relaxed condition.*

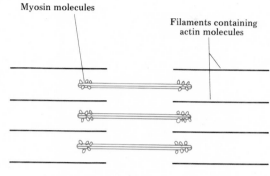

Myosin molecules

Filaments containing actin molecules

Before contraction

**9-22** *The bands seen in this photograph of a muscle reflect its line structure. The thick dark lines are A-bands, corresponding to the length of the myosin filaments. The light central portion of the A-band is called the H-band. The H-band contains myosin filaments alone; in the darker portion of an A-band, the myosin is overlapped by actin. The light areas on either side of the A-band are the I-bands, in which only actin filaments occurs. The thin dark lines in the middle of the I-bands are Z-lines, indicating a separation of functional units. During contraction each unit appears to squeeze together. (David Soifer—Omikron)*

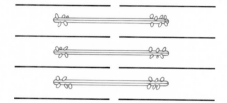

Contracted

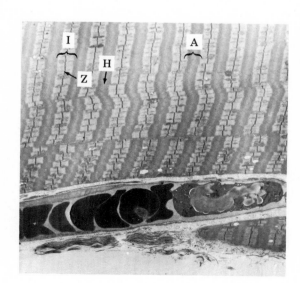

energy for contraction. Early workers in the field of muscle physiology were puzzled by the fact that chemical analysis showed there was very little ATP stored in muscle tissue. However, another phosphate compound, **creatine phosphate,** was found to be abundant in vertebrate muscles. At first it was thought that this compound might be a substitute for ATP, but tests with isolated muscle fibers maintained in a solution of creatine phosphate showed that it was not able to provide energy for muscular contraction. After further experimentation, it was established that the creatine phosphate serves as a storage molecule for the terminal phosphate groups that combine with ADP to form ATP. Thus, as a working muscle uses up its small supply of ATP, the resulting molecules of ADP can be quickly combined with a high-energy phosphate group from the creatine phosphate, permitting the continual resynthesis of ATP.

In order for the ATP molecule to release its energy content, the bond holding the terminal phosphate group must be broken by an enzyme called an ATP-ase. The work of H. E. Huxley established that the protruding zigzag heads of the myosin molecules can be somehow changed or activated so that they serve as an ATP-ase. This points out how well adapted the protein molecule is to the formation of muscle tissue. Protein can serve not only as the structural element of the muscle but also (with slight modifications) as the catalyst for the rapid reactions that must take place there.

Before the myosin molecule heads can function as an ATP-ase they must be activated. Within the last few years it has been conclusively demonstrated that the activation is provided by an influx of calcium ions. These ions are stored in vesiculated sacs along the membrane system within the muscle fiber. When the membrane is depolarized by an electrical stimulus, the electrical change inside the membrane causes the calcium ions to leave their storage sacs and move along an electrical gradient into the filaments, where they activate the myosin and enable it to release energy stored in the bonds of ATP. When the electrical stimulus is stopped and the membrane repolarizes, so that the inside of the membrane is once again negatively charged, the charge attracts the positive calcium ions back into the storage sacs. With the removal of the calcium from the filaments, no more energy can be released, and contraction stops.

The sliding filament mechanism of muscular contraction explains many of the properties that have been observed in muscles. For example, it can be shown that a muscle which contracts rapidly cannot develop as much tension as one that contracts slowly. This is because it takes time to form the large number of cross-bridges needed to sustain a high degree of tension. It also explains why most skeletal muscles have a short power arm. The further a muscle contracts, the more energy it must use in breaking and building cross-bridges. Even though a short power arm presents a mechanical disadvantage, it entails the most economical use of energy.

## Other types of muscle

Research in the field of muscle physiology has concentrated primarily on the skeletal, or striated, muscle of vertebrates. These muscles are convenient to work with, since they are relatively large, easy to remove and to separate into individual fibers, and able to maintain their contractile properties for long periods of time. Moreover, the characteristic striation, with bands of visibly different patterns, proved to be a great help in determining the sequence of events in contractions; the changes in structure in various parts of the muscle could actually be observed and measured.

But vertebrate skeletal muscle is only one out of many different muscle types, each with special properties of its own.

## Invertebrate muscles

The basic mechanism of contraction, utilizing sliding filaments of actin and myosin, seems to be the same in invertebrate muscle as it is in vertebrate muscle. There are, however, significant differences in the functioning of these two different muscle types.

The most important difference lies in the system of innervation, or stimulation by nerves. We have seen one example of this difference in the self-stimulating, or **asynchronous** muscle of insects with rapid wing beats. Another important aspect of the difference is that invertebrate muscles are typically stimulated by at least two nerve cells. In vertebrates, stimulation is always limited to one neuron alone. As shown in Figure 9-25, one invertebrate neuron acts as a fast excitor, stimulating a rapid twitch contraction. The other acts as a slow excitor, stimulating a more sustained contraction in response to a repeated stimulus. In crustaceans, there is also a third nerve cell, which acts as an inhibitor, to stop or slow down muscular contraction.

**9-23** *In spite of the fact that vertebrate muscles are inefficient levers, their total efficiency rating is very high; this is because its short power arm uses relatively little energy to form the cross-bridge that sustains the tension of the contraction. The hovering hummingbird, with its rapid wingbeats, exemplifies the efficiency of muscular contraction. (American Museum of Natural History)*

**9-24** *Although study has centered on vertebrate muscle, especially vertebrate striated muscle, this muscle is only one of many possible different types, each with special properties of its own. Study of the electric organ of the ray shows it to be a modified form of muscle tissue; here the electric potential of the innervation mechanism has become specialized so that the electric charge itself becomes a mechanism of defense and predation. (Schwartz, Pappas, and Bennett—Omikron)*

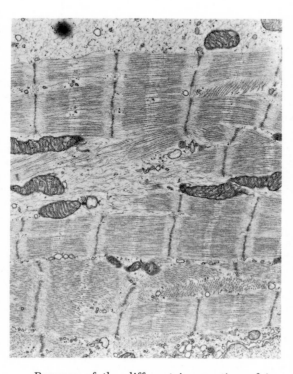

**9-25** *A typical invertebrate (crustacean) muscle has three different kinds of innervation. One nerve fiber acts as a fast excitor. A second acts as a slow excitor. A third acts as an inhibitor, making it an antagonist to the other two.*

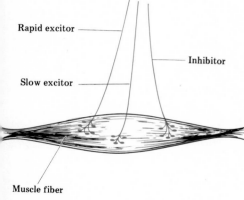

Rapid excitor

Inhibitor

Slow excitor

Muscle fiber

Because of the different innervation of invertebrate muscle, it shows rather different properties. The all-or-none response, so characteristic of vertebrate striated muscle, is not found in invertebrates. If a muscle fiber of a scallop is stimulated more frequently, or more powerfully, the fiber will show a greater response. Another difference can be seen in the length of the period of relaxation. In vertebrates, the relaxation period is very brief, but in many invertebrates it is about a thousand times longer. A long relaxation period would be a fatal handicap to an animal that depends on fast movement, but for less mobile invertebrates it is a real advantage. For example, an oyster or clam is able to keep its shell shut for long periods of time with the minimum expenditure of energy. The shell will remain closed throughout the extended relaxation period without the work of producing a new contraction.

## Vertebrate smooth muscle

In vertebrates, smooth muscles are found in the walls of the intestine and blood vessels, in the female uterus, in the iris of the eye. Smooth muscle differs from striated muscle in both appearance and properties.

Smooth muscle cells are slender, with tapering ends, and generally contain only one nucleus. They are composed of tiny fibers, but there is no regularity in the pattern of the fibers. Although it has been established that smooth muscle contains actin and myosin, it is not known whether the sliding mechanism of contraction is utilized. Some physiologists believe that the lack of patterned organization makes a sliding contraction impossible in smooth muscle.

Innervation, too, is different in smooth muscles. Some cells have no nerve endings on their surface; others have several. Where two nerve endings are present, they seem to come from different parts of the nervous system and to act as antagonists. Some smooth muscle, such as that in the digestive tract, seems to be capable of contracting without the help of any external stimulus.

Smooth muscle contracts more slowly than striated muscle, and the period of relaxation may vary considerably. In fact, it appears that the smooth muscle lining internal organs and vessels has no relaxation phase at all. Contraction of these muscles, called **tonic contractions,** can be maintained for very long periods without any relaxation or fading.

## Vertebrate cardiac muscle

Cardiac muscles show the same striation found in skeletal muscles, but the fibers are organized in a somewhat different way. Rather than occurring in uniform parallel bundles, cardiac muscle fibers are branched so as to form a conducting network. Contraction takes place by the sliding together of filaments of actin and myosin.

Cardiac muscles are not directly innervated by neurons. As we saw in Chapter 6, contraction is initiated by the tissues of the sinoatrial node, specialized cells which are both contractile and conductive. The impulse spreads to other cardiac

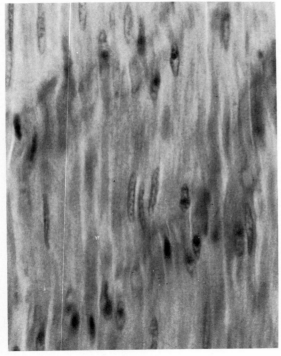

**9-26** *Smooth muscle differs from striated muscle both in appearance and properties. Smooth muscle cells are slender, with tapering ends, and generally contain only one nucleus. They are composed of tiny fibers, but there is no regularity in the pattern of the fibers. Although it has been established that smooth muscle contains actin and myosin, it is not known whether the sliding mechanism of contraction is used. (Permission of Ward's Natural Science Establishment, Inc., Rochester, New York)*

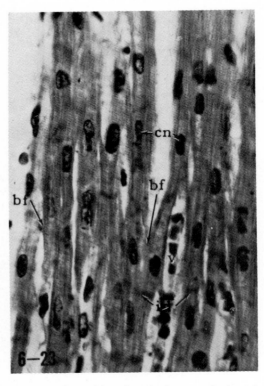

**9-27** *Cardiac muscle is a specialized tissue that is found only in the heart. Cardiac muscles show the same striation found in skeletal muscles, but the fibers are organized in a somewhat different way. Rather than occurring in uniform parallel bundles, cardiac muscle fibers are branched so as to form a conducting network. Contraction takes place by the sliding together of filaments of actin and myosin. (From Piliero and Wischnitza, Atlas of History, Lippincott)*

muscles, conducted by electrical transmission at the intercalary disks, which also serve as a division between individual cells.

The cycle of excitation, contraction, and relaxation in cardiac muscle is very similar to the cycle found in striated muscle, but the entire process takes place more slowly, and the muscle cannot be stimulated again until the entire cycle has passed. This property of cardiac muscle serves to pace the action of the heart and to prevent rapid contractions. This allows the heart time to fill up completely between contractions, making the work of pumping more efficient.

# Summary

Movement is an essential function of living things, for without movement there could be no vital processes. Cellular movement is accomplished through the stiffening function of microtubules and by the contractions of microfilaments which contain a contractile protein similar to actin. Cilia and flagella propel some cells. Two strokes—the power stroke and the recovery stroke—are employed by the cilium. In addition, the flagellum coils in a rotary movement.

There are two types of plant turgor movement: tropism, which is directly related to a stimulus (examples are phototropism and hydrotropism) and usually due to growth movements; and nastic movement, which is generally due to changes in turgor (an example is the movement when leaves of a plant fold up in the evening). Auxin is the hormone that regulates the movement of water within the plant tissues, causing turgor changes.

Plant movement takes place not only through turgor changes but also through differential cell growth, which is slower and more permanent than movement dependent upon turgor changes. Roots that grow in the direction of greatest gravity are an example of this. Differential cell growth also supplements the action of turgor changes.

Most multicellular animals have evolved a specialized muscle cell that works against a skeleton to accomplish movement. A hydrostatic skeleton, consisting of body fluids, is employed by some invertebrates; mollusks and arthropods utilize an exoskeleton (hard outer shell or cuticle) for the same purpose. The movement of invertebrates displays all the elements basic to movement systems of higher animals.

The vertebrate endoskeleton is composed of bone and cartilage. Tendons attach muscles to bones, allowing flexible movement at hinges and joints. The bones act as efficient levers, moved by muscles found in antagonistic pairs. In addition to the function of movement, the appendicular skeleton supports, and the axial skeleton protects, the body organs and tissues.

The whole skeletal muscle contracts in direct relation to electrical stimulus; muscle fibers, however, are characterized by the all-or-none response. The cumulative effect, or summation, of muscle fibers accomplishes the smooth and sustained contraction ("tetanus") of the muscle as a whole.

The twitch of the muscle fiber has three phases: latent, contraction, and relaxation. The motor unit is made up of the motor neuron, which delivers the impulse, and a number of muscle fibers.

Muscular contraction is an electrochemical event. The muscle fiber is covered by the sarcolemma, which normally exhibits strong electrical polarization. The sarcolemma is depolarized by electrical impulse and this initiates contraction. The myofibrils are made up of filaments of two proteins, myosin and actin. During contraction the muscle shortens as a result of the sliding together of myosin and actin filaments; and this is accomplished when ATP is converted to ADP at the cross-bridge site. The activation of the ATP-ase is initiated by an influx of calcium ions.

Invertebrate muscles are similar to those of the vertebrate except in the system of innervation. Vertebrate smooth, or involuntary, muscles are slender and have no regular striation; their system of innervation is different from that of the skeletal muscles. Cardiac muscle is striated; however, the fibers are branched and form a conducting network. Contraction there is initiated by the sinoatrial node and is slower than in skeletal muscles.

# Bibliography

## References

Bourne, G. H. ed. 1960. *The Structure and Function of Muscle.* Vols. 1 and 2. Academic Press, New York.

Currey, J. 1970. *Animal Skeletons.* St. Martin's Press, New York.

Galambos, R. 1962. *Nerves and Muscles.* Doubleday Anchor Books, Garden City, N.Y.

Hoyle, G. 1970. "How Is Muscle Turned On and Off?" *Scientific American.* 222(4): 72–82. Offprint no. 1175. Freeman, San Francisco.

Huxley, H. E. 1969. "The Mechanism of Muscular Contraction." *Science.* 164: 1356–1367.

Kielley, W. W. 1964. "The Biochemistry of Muscle." *Annual Review of Biochemistry.* 33: 403–430.

Margaria, R. 1972. "The Sources of Muscular Energy." *Scientific American.* 226(3): 84–91.

Smith, D. S. 1965. "The Flight Muscles of Insects." *Scientific American.* 210(6): 78–86. Offprint no. 1014. Freeman, San Francisco.

Wessells, N. K. 1971. "How Living Cells Change Shape." *Scientific American.* 225(10): 76–87. Offprint no. 1233. Freeman, San Francisco.

## Suggested for further reading

Asimov, I. 1963, 1964. *The Human Body: Its Structure and Operation*. Signet, New York.
Contains three chapters on the human skeletal and muscular systems. Lucid and very readable, this book is also useful in connection with Chapters 4 through 8 of the present text.

Beadall, J. R. 1969. *Muscles, Molecules and Movement*. Heinemann, London.
A concise explanation of muscle physiology and function. This book starts with the specific impetus to movement, graphically charts its onward progress, and then moves on to the more general responses.

Gray, J. 1953. *How Animals Move*. Cambridge University Press, New York.
A series of six lectures designed to illustrate the more striking features of animal locomotion. No previous biological knowledge is assumed, and the student is shown research problems and the methodology of dealing with them.

Prosser, C. L. and F. A. Brown. 1961. *Comparative Animal Physiology*. 2nd ed. Saunders, Philadelphia.
An excellent general reference book. The chapters on the similarities and dissimilarities among animals and their overall physiology are helpful for the inquiring student.

## Related book of interest

Fast, J. 1970. *Body Language*. Pocket Books, New York.
This popular book suggests ways to "read" the communication being sent by movement when people are in groups. Fast describes the communication content of head, eye, shoulder, hand, and leg movements.

# 10
# Integration: nervous control

The word "nerve" first entered our language as a description of a category of tissue, containing cells with a certain form and function. Over the years, common usage has broadened it to include a number of other meanings: the source or transmitter of vitality, energy, strength or vigor, power and control, spiritual and psychological endurance. By extension, those nerve cells have become identified with the qualities considered to be central to the meaning and mystery of life.

Perhaps it is true that the ability to perceive and respond to variable stimuli in the environment is the essence of life, but it is a case of typical egocentrism for us to believe that the nervous system is the only means of performing this function. Every single cell displays the characteristic of irritability, or the ability to receive information about the environment and to alter behavior and physiology accordingly. Protozoa display irritability; some of them even have specialized organelles that act as sense receptors — spots in the surface of the cell membrane that are sensitive to light, heat, and the chemical composition of the environment. Smaller bacterial cells, which do not appear to have any specialized site of sensory perception, also possess the characteristic of irritability. So do the nerveless cells of the vegetable (a word often used to describe anything considered unfeeling). As their responses reveal, plants have the ability to sense light, heat, presence of water, gravity, and contact with other surfaces.

The complex nervous system of man and other vertebrates, the subject of this chapter, is a specialization of a property present in all living things. Other life forms have developed this property in different ways, while retaining the same function. Botanists do not fully understand how plants receive and transmit information about the environment, but it is obvious that they do. There is some evidence that the plant sensory system is much more complicated than we had imagined. Many people are quite convinced that their house plants grow faster and look healthier when the plants "listen" to certain kinds of music, or to the sound of a soothing and encouraging voice; recent laboratory experiments with the effects of sound waves on plant growth lend a degree of scientific weight to these beliefs.

Perhaps the most significant property of the nerve cell of an animal is its ability to act as a transducer, a term used in physics to refer to any mechanism that alters the form of energy. (One familiar transducer is a television set, which changes radio waves into patterns of light.) The stimuli which the nervous system receives are quite varied—light energy, sound waves, mechanical pressure, heat energy, and the chemical energy of molecules with a strong odor or taste. Sensory neurons can transduce all these stimuli, altering them to produce one uniform electrochemical impulse. Information regarding the source and strength of the original stimuli is coded into the transduced signals. It is this transduction that allows such rapid processing of the signals. It also allows them to be integrated. If the light energy that stimulates sight and the sound waves that stimulate hearing were not transduced to a common energy form, coordinating sight and sound would be as impossible as adding apples and oranges.

The property or process of irritability can be divided into three distinct phases. The first is the receiving of information about both the external and the internal environment of the organism. The second is sending signals from the area of perception to other parts of the organism. The third is the coordination of the information in such a way as to produce the correct response. It is only after these three processes have taken place that some measurable kind of adaptation to the environment, such as the kind of movements we discussed in the previous chapter, will occur.

In vertebrates, highly specialized cells of the nervous system perform all three aspects of the function of irritability. But it should be emphasized that these three steps can also be found in organisms that have no nerve cells and no

10-1 *This potto, an inhabitant of the Kakamega Forest of Western Kenya, has sensory organs that are adapted to the life it leads. A nocturnal animal, its eyes are remarkably sensitive to light, containing a reflective underlayer of guanin crystals, so light can be bounced back and reused. The potto is a primate, with the odd characteristic (visible in this picture) of having a long grooming claw on one finger of each hand, rather than a true nail. (Jen and Des Bartlett—Photo Researchers)*

nervous system. The surface of a single cell can act to receive information about the environment. Signals can be sent along the membranes of the cell by the transport of electrolytes; they can also be sent by the release of chemical messengers rather than by the transmission of nervous impulses. In fact, even in the vertebrate nervous system, chemical messengers are used for communication between individual nerve cells.

## Receiving information

Life is an unending interaction between the complex chemical system inside the protective barrier of the cell membrane and the constantly changing external environment. Under these conditions, survival depends on adequate information about environmental change.

Organisms that are composed of many highly specialized cells need another kind of information. Various groups of cells must exchange information about the rate at which they are functioning and the tasks they are performing—in other words, the right hand must know what the left hand is doing. This is especially necessary if the organism is to maintain a steady internal environment. Sensory perceptors must keep constant surveillance over the various metabolic processes to keep them in balance.

## Information about the external environment

It is easy to see why natural selection would favor the development of information-receiving mechanisms, such as sight, hearing, touch, and pain. As we follow the course of evolution, we see an increasing complexity in these mechanisms. At the same time, there is also an increase in the amount of information received and the precision with which it is defined. For example, the tiny *Chlamydomonas* cell has an eyespot through which it receives information about the presence or absence of light. Man, on the other hand, can perceive the shape, size, color, and movement of an object many yards away. The honeybee can perceive ultraviolet light; the bat can receive sounds of a frequency of about 100,000 cycles per second; the salmon can distinguish the blend of chemicals that distinguishes the smell of the river in which he was born from the smell of all other water pouring into the ocean. Living things show an extraordinary variety of sensory perception mechanisms.

**Vision.** Organisms which remain immobile or move only very slowly through dirt or sand can survive without extensive information about the shape and contour of their environment, or the objects and other creatures found in it. Their photosensitivity is limited to the ability to detect light. Earthworms, for example, can detect the presence of certain light waves, and they immediately shrink back from them. Clams also can detect light; they react to changes in the intensity of light by pulling in their breathing tubes, or siphons. This protective mechanism serves to keep the clams hidden in the silt or sand.

The existence of phototropisms in plants tells us that they, too, can perceive the presence of light. Another light-sensitive phenomenon found in plants is **photoperiodism,** or the regulation of various life processes by the amount and duration of light received. One simple example of photoperiodism is the opening of flowers during the day and the closing during the night. Even the reproductive cycle is photoperiodic. A plant will flower only when it receives a certain amount of light in a certain number of days.

Free-living animals must have more extensive information to guide them as they move through the environment. Consequently they have developed the property of photosensitivity much further, so as to obtain more of the information that light can provide. The organ which specializes in this function is, of course, the eye.

In all animals except arthropods, the eye functions on the principle of the camera, using a lens to form a picture. The eye itself is always shaped like a sphere, although it is not necessarily perfectly round; only in vertebrates is it a fully rounded orb. In the typical eye, there is a small opening through which a limited amount of light can penetrate. The opening is often surrounded by a pigment that screens out all other light rays. This screening permits accurate focusing of a small beam of light, and it also allows the animal to determine the direction from which the light is coming.

The simplest type of focusing arrangement in the eye utilizes the principle of the pinhole camera, as shown in Figure 10-2. Light enters through the tiny aperture and is focused on the retina, or back of the eye, where light-sensitive pigments are located. With this simple eye, the light is always correctly focused, no matter what the distance of the object being viewed. The disadvantage of this type of eye is that it admits such a small amount of light, creating a rather dim picture. This type of eye is found only in animals for whom sight is not a primary mechanism of sensory perception.

A more complex version of a focusing mechanism is found in cephalopod mollusks (the octopus, for example) and vertebrates. This eye uses a lens to focus the light on the retina. Changes in the position or curvature of the lens allow the eye to focus on objects at varying distances. The use of a lens greatly increases the amount of light that can be admitted, and it makes it possible to utilize light coming from many different directions. Nocturnal predators, such as owls, with their large spherical eyes, can pick up light coming anywhere within a 180° field of vision.

Not all vertebrates see with identical acuity; the amount of information they can collect through their eyes varies in accordance with their life styles. Man and a few other primates can see in color, but most mammals see only black and white and shades of gray. A mouse's eye is

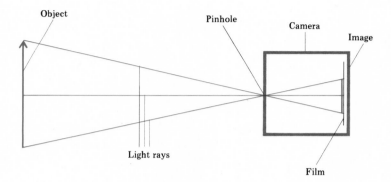

so small that it cannot collect much light; moreover, since the mouse is adapted to a nocturnal way of life, its eyes focus poorly, making its world a blur in which all that can be distinguished is the movement that warns of a possible predator. On the other hand, a hawk circling the top of Yosemite Falls in California, looking for food, would be able to see clearly the shape and size of a mouse on the bank at the bottom of the falls, more than half a mile down.

All photosensitivity in the eye is concentrated in a single layer of visual cells at the back

10-2 *In the pinhole camera, the tiny opening itself acts as the lens, focusing the light on the film at the back of the light-tight box. The image thus formed is a relatively blurry one; a second disadvantage of this kind of camera is that it requires strong bright light. Eyes that work on the pinhole camera principle are found only in aquatic animals; without water to diffuse and reflect light rays, eyes of this type provide only very limited vision.*

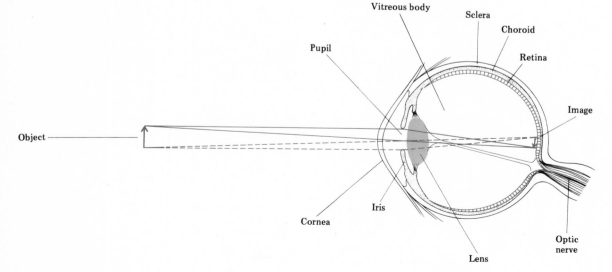

10-3 *The mammalian eye contains a lens, which both collects and focuses light. It works on principles rather like those of the microscope (see Figure 2-8). The dark pigment of the iris helps keep out light coming from other directions, that might blur the image. The pupil of the eye acts like the diaphragm, controlling the amount of light that reaches the lens. The retina is the "film" of the eye, the place where the focused image is projected by the lens.*

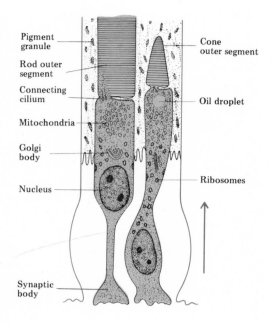

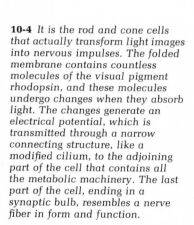

**10-4** *It is the rod and cone cells that actually transform light images into nervous impulses. The folded membrane contains countless molecules of the visual pigment rhodopsin, and these molecules undergo changes when they absorb light. The changes generate an electrical potential, which is transmitted through a narrow connecting structure, like a modified cilium, to the adjoining part of the cell that contains all the metabolic machinery. The last part of the cell, ending in a synaptic bulb, resembles a nerve fiber in form and function.*

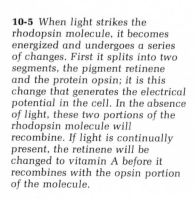

**10-5** *When light strikes the rhodopsin molecule, it becomes energized and undergoes a series of changes. First it splits into two segments, the pigment retinene and the protein opsin; it is this change that generates the electrical potential in the cell. In the absence of light, these two portions of the rhodopsin molecule will recombine. If light is continually present, the retinene will be changed to vitamin A before it recombines with the opsin portion of the molecule.*

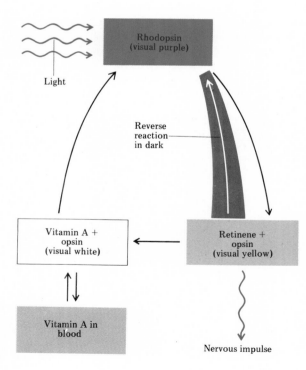

of the retina. There are two types of visual cells in vertebrates. The **rod cells** are highly light-sensitive. They work best in dim light but are not capable of registering color. The retinas of nocturnal animals are composed almost entirely of rods. **Cone cells** require brighter light for their stimulation, and they are capable of forming a much sharper image than can rod cells. An abundance of cones permits color vision. In the human eye, the cones are in the center of the retina and the rods are arranged along the side. This explains why it is easier to see things at night if you look at them out of the corner of your eye rather than head on.

Both the rod and the cone cell are long and slender, with just the sort of shapes that their names imply. The end of the cell that is exposed to the light is composed of a densely folded membrane that, when viewed under the electron microscope, gives the appearance of a stack of disks. Located along the surface of this membranous structure is the photosensitive pigment. In all land vertebrates the main visual pigment is **rhodopsin.** This molecule is composed of a protein, **opsin,** joined to a molecule of vitamin A that has been slightly modified so that it is almost identical to the photosynthetic pigment **carotene,** which is found in plant cells. There is a striking similarity between the pigment (called retinal) and the folded membrane in visual cells, and the pigment and folded membrane found in the chloroplasts of green plants, where another kind of light reaction takes place.

When light is absorbed by rhodopsin, the molecule breaks apart and the opsin portion of the molecule undergoes certain structural changes. These changes are somehow able to generate a corresponding change in the electrical potential of the cell membrane. It is not clear what causes this electrical change. One suggestion is that the rearranged opsin molecule might act as a powerful enzyme that triggers further chemical changes. We have already seen just this kind of mechanism at work in the myosin molecules of muscle fibers. However it happens, the change in electrical potential is relayed from the visual cell to a connecting nerve fiber, and from there it can be transmitted throughout the pathway of the nervous system. The entire sequence of events, from absorption of the light to transmission of the electrical signal, takes place in less than two thousandths of a second. As the signal is transmitted, the rhodopsin is resynthesized, so that the cell is ready to absorb light again.

**The arthropod eye.** The eye of the insect and the crustacean functions on another principle. It is often called a compound eye. Like the vertebrate eye, it is a protruding sphere, but its structure is quite different. The insect eye is made of a large number of independent visual units called **ommatidia.** Each ommatidium is a long tube whose sides are surrounded by a dark pigment that screens out all incidental light. Light enters the exposed end of the tube and penetrates to the bottom, where the photosensitive pigments are arranged. The pigment is quite similar to that found in vertebrates, and it reacts in the same general way to transmit an electrical impulse.

Each ommatidium can receive only the light that is coming in more or less parallel to its long axis. Because of this limitation, the individual ommatidium transmits a picture of only a small part of the environment. But when large numbers of ommatidia are arranged in a spherically shaped group, a composite picture, rather like a mosaic, will be formed that conveys a great deal of information about the surroundings. Moreover, most insect eyes contain the pigments necessary for color vision, adding another dimension of knowledge about the external world. Predatory insects, such as the praying mantis, have the greatest number of ommatidia of all insects, and thus they are able to perceive the most detailed visual information. The insect eye is especially well adapted for the perception of motion.

**Hearing.** It was once believed that the ability to hear was confined to those animals that produced noises; it seemed logical to suppose that only the species able to send communication would find any advantage in receiving it. Certain experimental evidence appeared to support this conclusion. For example, the French scientist Emil

**10-6** *This set of photographs compares the human view of a flower (on the left) with that of the bee (on the right). Because of differences in color perception—the bee can see ultraviolet light and humans cannot—the markings of the flower appear much more distinct to bees than they do to us. (M. W. F. Tweedie from National Audubon Society)*

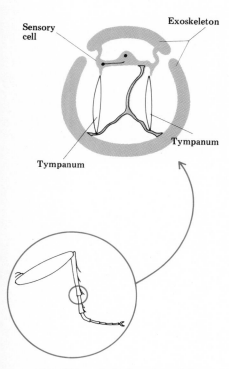

Sensory cell

Exoskeleton

Tympanum

Tympanum

**10-7** *Although the location of the insect's hearing organ may seem a little unusual, his tympanic organ in the leg works on the same principle as the one in the vertebrate ear. Vibrations of sound waves set the tympanum, a thin membrane, in motion, and that movement causes a corresponding movement in a sensory cell located near the membrane. Movement of the sensory cell generates the impulse carried by the neurons to a central coordinating site.*

Du Bois-Reymond submerged a noisemaking machine, consisting of an electromagnet that caused a large steel plate to vibrate, in a pool full of fish. Although he himself could scarcely bear the din, the fish swam about undisturbed.

A series of more carefully designed experiments have refuted the conclusion that fish cannot hear. In fact, it has been shown that some fish even have "perfect pitch," or the ability to recognize any given musical tone. The fish in Du Bois-Reymond's experiment did not respond to the noise of the steel plate because it had no significance for them. They screened it out, just as people who live in large cities often screen out random noises like the honking of horns or the chatter of jackhammers. If the fish learn to associate a certain noise with dinnertime, then they will respond to it immediately. They can recognize it played at different volumes, or even accompanied by many other tones at the same time, but they do not respond when it is played sharp or flat.

In order to hear, the animal must have some mechanism for receiving sound waves, which are, after all, only a form of energy. Theoretically, a hearing organ could be constructed to process any frequency of sound waves, but in fact, the hearing of most animals is restricted to a fairly narrow range of frequencies. Frequencies with a high energy content could easily damage or destroy the hearing organ. Frequencies with a low energy content need amplification, a process that would require a large expenditure of the animal's own energy and therefore would be quite inefficient.

The basis of sound reception in all hearing organs is the **hair cell.** The hair cell has a long thin projection, which is actually a modified cilium. The force of the sound waves, transmitted either through the air or through a liquid, causes the projecting hairs to move, and it is this movement that leads to the generation of an electrochemical impulse. Hair cells are simply an adaptation of ciliated epithelial cells, and no doubt evolved as a specialization of the epi-

thelium. This is true not only of sound receptors; all sensory receptors show some degree of resemblance to ciliated epithelium.

The simplest kind of sound receptors is found in insects. A fine membrane, or **tympanum,** is stretched across the path of sound waves. The waves start the membrane vibrating, and the vibration is picked up and transmitted as an electrical impulse by sensory cells attached to the base of the membrane. The insect's tympanic organ is actually a modification of the tracheal passage through which air enters for gas exchange, and it may be found on either the abdomen or the legs.

The vertebrate ear features a more complicated version of the vibrating membrane. Vertebrates have an outer tympanic membrane (the eardrum) which vibrates independently, with no connection to any sensory cells. The vibration of this structure is transmitted through the bones of the middle ear to the actual organ of hearing, the **cochlea,** located deep inside the skull. The opening to the cochlea, called the oval window, is a much smaller area than the tympanic membrane, so the middle ear acts as a funnel for the sound waves.

When the sound waves enter the cochlea, they are transduced, or converted into vibrations in the viscous liquid that fills the organ. These liquid vibrations are picked up by another membrane stretching across the cochlea, called the **basilar membrane.** This membrane supports a large number of tiny hair cells. The vibrations of the moving liquid cause a corresponding movement in the hair cells. A high frequency sound will travel only a short distance through the liquid, and therefore it causes movement in the hair cells nearest the oval window. A low frequency sound will travel a greater distance, causing movement in the most remote hair cells. This movement of the hair cells generates an electrical impulse that is picked up and transmitted by a neuron. Since the transmitted message tells which hairs are moving, it also con-

**305**

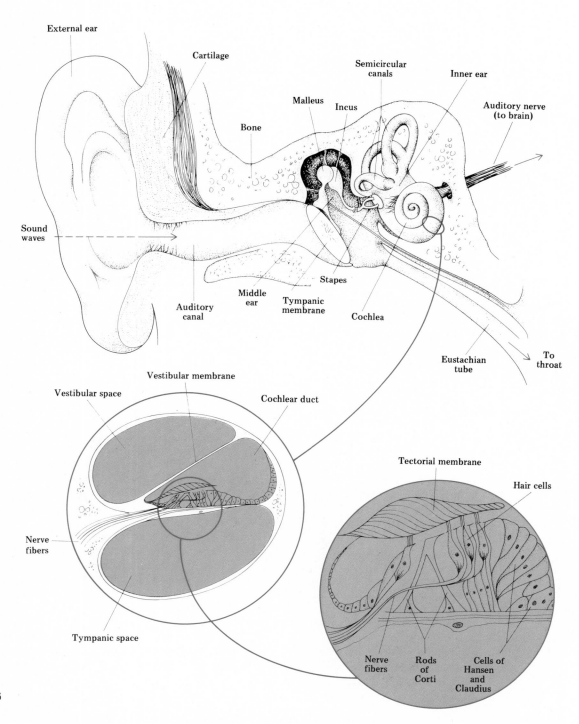

External ear

Cartilage

Semicircular
canals

Inner ear

Malleus

Incus

Bone

Auditory nerve
(to brain)

Sound
waves

Stapes

Middle
ear

Auditory
canal

Tympanic
membrane

Cochlea

Eustachian
tube

To
throat

Vestibular membrane

Vestibular space

Cochlear duct

Tectorial membrane

Hair cells

Nerve
fibers

Tympanic space

Nerve
fibers

Rods
of
Corti

Cells of
Hansen
and
Claudius

**10-8** *The actual organ of hearing, the cochlea, is located deep inside the skull; above it are the semicircular canals, proprioceptors that sense balance. Sound waves enter the outer ear and strike the tympanic membrane, causing it to vibrate. The vibrations are conducted through the middle ear to the cochlea, where a liquid is set in motion by the force of the vibrations. The movement of the liquid causes vibrations in the basilar membrane of the cochlea, and these vibrations are picked up by nearby sensory cells.*

**10-9** *Since bats navigate by echolocation, they must have very sensitive ears. As this photo shows, the ears of most bat species appeared oversized in relation to the rest of its head. Interestingly, the dolphin, which also uses echolocation, has ears that are small and inconspicuous. This reflects the superior conductive properties of water over air. (Karl H. Maslowski from National Audubon Society)*

**10-10** *The raccoon is an example of an animal with very sensitive chemoreceptors. It is primarily a nocturnal hunter and relies heavily on its sense of smell to locate its food. As anyone who has ever camped out in raccoon territory can testify, this animal is very proficient at locating any edible scraps. (Howard Earl Uible —Photo Researchers)*

veys the frequency of the vibration (the pitch of the sound). Thus the vertebrate ear is able to discriminate not only between loud and soft noises but between high and low ones as well.

The range of an animal's hearing depends on the number of the hair cells in the basilar membrane. In the parrot, there are only about 1200 cells, while in humans there are about 25,000, so humans are capable of much more accurate tone discrimination. Humans can hear sounds of a frequency of up to 20,000 cycles per second. Cats can hear about twice that frequency, and porpoises can hear nearly 80,000 cycles per second. Bats have the greatest sensitivity to high tones, being able to hear sounds of almost 100,000 cycles per second. They emit tiny squeaks at very high frequencies, which bounce off objects in their path and back to the bat's ears, serving as an echolocation device like sonar. This mechanism is so accurate that it enables the bat to find a tiny insect in a large cave, even when the cave is full of other bats making sounds that seem almost identical. Bats can even determine textures by sound, being able to tell the difference between a piece of paper and a piece of velvet the same size. For these creatures, being "blind as a bat" is no real disadvantage.

**Smell and taste.** Both smell and taste depend on the presence of specialized epithelial cells, called chemoreceptors. These receptors may be found in a variety of locations. In male moths, they are part of the antennae. Many insects have chemoreceptors on their feet and on the hairs along their legs. Fish have taste buds scattered all over the skin of their heads. In mammals, chemoreceptors are located on the tongue and under a thin film of mucus at the back of the nose. The receptors on the tongue can distinguish only four different tastes: sweet, sour, salty, and bitter. The receptors in the nose can distinguish seven basic scents: minty, ethereal, floral, musky, camphoraceous, putrid, and pungent.

**307**

Chemoreceptors are stimulated by certain molecules or ions. These molecules may travel by air and water currents, or the animal may place the receptors in direct contact with the substance to taste it. We generally use taste to refer to the chemoreception of molecules dissolved in some liquid, and smell to refer to the sensing of molecules which arrive in the air. But this distinction really applies only to land mammals—it is pointless to try to decide whether a fish smells or tastes the water—and even under the best of circumstances it is confusing. We may say that we taste spaghetti, but in fact our tongues can only tell us if it is salty. We appreciate the flavor of garlic and oregano only because the sauce in our mouths releases molecules of these substances for us to smell.

The question of how chemoreceptors actually work has been the subject of much debate, and it has been only in the last decade that a widely accepted hypothesis has been worked out. The theory is that the odorous molecule fits into receptor sites on the membrane of specialized epithelial cells. There are thought to be seven different types of receptor sites for odors, each with its own characteristic shape. Only molecules with a certain configuration can fit into the receptor site; there is a lock-and-key relationship much like that seen in an enzyme and its substrate. It is likely that some chemical reaction takes place between the odorous chemical and molecules at the receptor site, and that it is this reaction that creates the electrical potential to send a message. Some molecules, such as peppermint, can fit only in one receptor, the minty receptor. The almond odor is more complex, for this molecule can fit into three different sites, depending on how it is oriented. It is the combination of the three stimuli that gives almonds their characteristic smell.

**Touch, temperature, and pain.** A number of specialized sensory receptors enable animals to sense contact with other surfaces and the temper-

ature of their surroundings. In most vertebrates, touch receptors are located around the face and on the arms and legs. Insects have touch receptors on their feet, and many fish seem to have them in their fins. These touch receptors may be of several different types. One is the hair cell, a sensory cell that is a specialization of epithelium; it has a thin flagellumlike projection that is highly sensitive to any type of movement, including that caused by tiny currents of air. Another type is the specialized receptor bulb, such as the Pacinian corpuscle. These lie deeper inside the body and register "deep touch" or pressure. These bulbs are made of a nerve ending surrounded by sheets of folded connective tissue arranged like the layers of an onion. Mechanical pressure on the bulb causes it to compress and the layers to slide

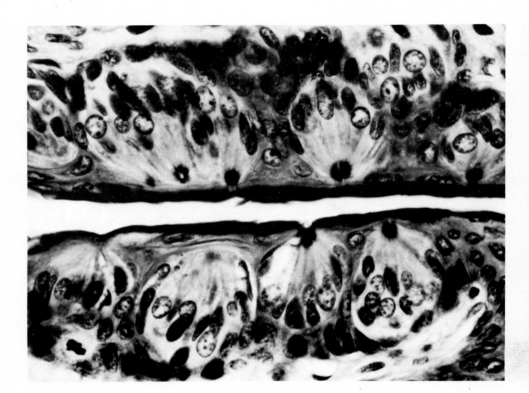

10-11 *This photo of rabbit taste buds illustrates a second type of chemoreception. The sense of taste has survival value, in that it helps animals discriminate between harmless and toxic foods; it gives them an added means of remembering and identifying things they should not eat. (Brian Bracegirdle, from An Atlas of Histology)*

**10-12** *The Pacinian corpuscle is a sense organ that registers pressure or deep touch. As this photo shows, the structure is composed of many layers of folded connective tissue; in the very center of the corpuscle is the ending of a sensory nerve. If pressure is applied, the folded sheets slide against one another and stimulate the nerve inside. (Omikron)*

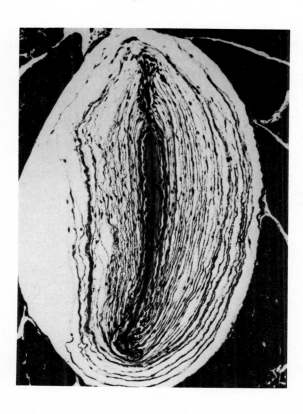

**10-13** *This close-up view of a snake's head shows the pit organs located between the eye and the nostril. These temperature-sensing structures are not found in all snakes but only in the pit vipers, a varied group that includes the rattlesnake and the bushmaster. One advantage of the pit organs is that they help the snake locate warm-blooded prey. (M. W. F. Tweedie from National Audubon Society)*

against one another. This movement stimulates the nerve inside.

Animals in several phyla are able to distinguish very slight changes in the temperature of their surroundings. The finest measurement is found in animals that feed on warm-blooded creatures. Bloodsucking insects respond to changes of as little as 0.5°C.; this accurate information allows them to locate and move toward their homeothermic hosts. Snakes that feed on small mammals have evolved temperature-sensitive organs in their heads, called pit organs. These organs are so efficient that the snakes can detect the presence of a small rodent 40 centimeters away in less than half a second, solely by the difference between the temperature of the rodent and its surroundings. It is known that these organs contain pigment granules that can respond to infrared radiation, but the exact mechanism of the sensory process is not well understood.

An important sensory organ in fish is the **lateral line system.** This system is actually a series of specialized hair cells running down the side of the fish from head to tail. The cells are set inside a canal or a little tunnel opening so that they are not in direct contact with the water. In the laboratory, these cells have been shown to be sensitive to a number of different external stimuli. The lateral line serves as an organ of extended touch; it responds to the currents set up by the movements of other fish and thereby warns of their presence. The lateral line is also sensitive to low-frequency sound waves.

Pain is a special kind of touch sensation about which surprisingly little is really known. At one time, biologists did not even believe pain receptors existed. Pain was thought to be caused only by an overload on the sensory nerves. For example, when a bright light strikes the eyes, so many electrical impulses suddenly bombard the optic nerve that it cannot carry them all, and this overloading causes pain. The same is true of any excessive stimulus.

Recently it has been demonstrated that in addition to the pain of overload, there is another way in which pain messages can be received— through a set of special pain receptors These nerve fibers are found in the epithelium. Unlike the Pacinian corpuscle, they have no specialized covering—they just become thinner and thinner and then disappear. These pain fibers conduct much more slowly than any other nerve fibers, and it appears that the message they carry is not very specific. Many specialists in the field of pain physiology believe that the stimulus for these pain receivers is some chemical released by damaged tissues. Suppose, for example, that you burn your finger on the stove. The cells which are burned are damaged or perhaps even destroyed, and they release a chemical messenger, probably a histamine. This chemical diffuses through the cells in the area, until finally it happens to reach a pain-receiving fiber. The histamine stimulates the fiber, and an electrical message is sent out. This explains why you feel little or no pain immediately after touching the stove but within two or three minutes begin to feel a growing sensation of pain.

## Information about the internal environment

In previous chapters, we have discussed a number of the receptors that serve to keep animals informed of the conditions of the internal environment, which must be kept fairly constant. All homeostatic mechanisms rely on these internal receptors, called **proprioceptors,** to monitor the internal environment and send messages warning of any trend toward fluctuation. Among the examples we have already seen are the stretch receptors in the vessels of the heart that help keep blood pressure constant; the stretch receptors in the blood vessels of the kidney that help keep the volume of blood constant; the chemoreceptors in the carotid body that help keep the pH of the blood constant.

Another important proprioceptor system is the one that gives animals their muscle, or kinesthetic, sense. This sense tells what position the body is in, whether various muscles are contracted or relaxed, and how joints are bent. There are joint receptors, located within the connective tissue of joints, that are sensitive to movement greater than an angle of 10° to 15°; thus they register the bending of the joints. There are sensory receptors located in the tendons and the ligaments that attach muscles to the bones; these measure the movement of the muscles. A more delicate measurement of the activity of muscles is made by specialized muscle fibers. On either end, these fibers contain the same contractile elements found in all other muscle fibers, but in the middle of the fiber, there are thin nerve endings. These are called intrafusal fibers. When the ends of the intrafusal fiber lengthen, as would occur when the muscle relaxes, the nerve fiber is stimulated and sends a message; when the ends shorten, during muscular contraction, the stimulus to the nerve stops and no message is sent. The strength and duration of the impulses from the intrafusal fibers help the animal determine the exact state of muscular contraction in that region.

Another type of proprioceptor system is that which gives an animal its sense of balance. In invertebrates, the **statocyst** serves as the organ of equilibrium. The statocyst is a small particle of sand or calcium carbonate enclosed in a sac filled with fluid and lined with cilia. As the particle moves in response to gravity, the change is registered by the cilia, and an electrochemical impulse is generated.

In vertebrates, the maintenance of equilibrium is the function of a specialized part of the ear, the **semicircular canals** of the inner ear. Each canal is filled with fluid and lined with sensory hair cells. As shown in Figure 10-14, any movement of the head will be registered by these hair cells; this, in turn, causes the transmission of an electrochemical impulse. In most vertebrates,

**10-14** *The three semicircular canals are each set at a different angle, and thus at least one will detect movement in any plane Each canal is filled with a viscous liquid. When the head is moved, the liquid moves, and this movement bends the extended hairs of the sense cells within the canal. These cells then generate an electrochemical impulse causing a message to be transmitted along the nerve fibers.*

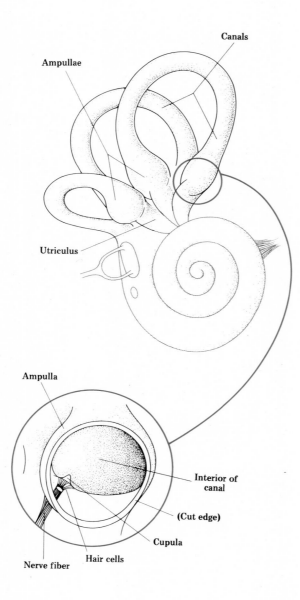

Canals

Ampullae

Utriculus

Ampulla

Interior of canal

(Cut edge)

Cupula

Nerve fiber

Hair cells

there are three of these semicircular canals in each ear, and the three are oriented perpendicularly to one another, so that rotation in any direction can be registered.

## Sending signals

After information about the environment is received by sensory receptors, a signal must be transmitted to the other cells of the body. The transmission of signals is the second phase of nervous activity. It takes place by means of a coded impulse. By a combination of electrical and chemical systems, these impulses can be carried along the nerves to the central control system.

## Membrane potential of the neuron

The basic conducting unit for nervous impulses is the nerve cell, or neuron. This cell consists of a central cell body and a number of elongated projections. The cell body contains the nucleus and the other organelles, such as mitochondria, that are the cell's metabolic machinery. The projections are of two types: very long slender strands, called axons; and shorter thicker branched projections, called dendrites. Dendrites receive impulses from connecting nervous cells, while the axon conducts these impulses on to other cells. In vertebrates, the axon is wrapped in a fatty insulating sheath, made of a lipoprotein (myelin) folded many times. The sheath is interrupted at intervals of one to two millimeters, leaving bare spots called the **nodes of Ranvier.**

It is possible to measure the voltage inside and outside the thin axon of the neuron. This was first done using the giant neuron of the squid, which is often as large as one millimeter in diameter—in fact, much of our current knowledge about the nervous system has come from research using squid neurons. When an electrode is inserted into the axon and another one placed outside the membrane, it can be seen that there is a significant difference in the voltage reading for the two electrodes. The inside of the membrane is always negative to the outside when the nerve is in a resting state. This voltage difference, usually about 60 to 70 millivolts, is called the **membrane potential.**

## The nerve impulse

Early in the twentieth century, physiologists were able to determine many of the properties of nervous impulse action by stimulating neurons through the application of electricity. They found that neurons display the same all-or-none response found in muscles. If the stimulus is great enough to evoke any response at all, the response will be total. An additional increase in the stimulus will not provoke any additional increase in the response. Another characteristic of the neuron is that once an electrical impulse has been evoked, the neuron is incapable of a second immediate response. The period between the first impulse and the time that a second impulse can be generated is called the **refractory period.** If the second stimulus is applied within one millisecond of the first, the neuron will not respond in any way. Within the next several milliseconds, it will respond only to a very strong stimulus. It takes about five to 10 milliseconds for the neuron once again to display its original properties of irritability.

After these two significant characteristics of the neuron's impulse were established, research workers began to look for explanations of this pattern of activity. Measurement of the membrane potential provided an important piece of evidence. It was discovered that during the instant of response, there is a change in the membrane potential. The membrane rapidly depolarizes, and, in fact, for a moment the potential may even

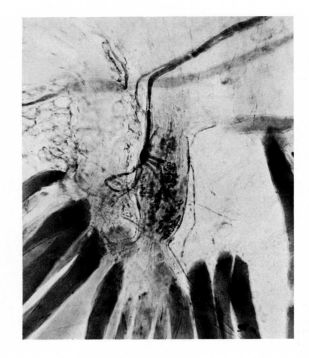

**10-15** *The photo shows an individual neuron. Each neuron displays an all-or-none response like that of a single muscle fiber. If the stimulus applied is great enough to evoke any response at all, the response will be total. Yet an entire nerve does not demonstrate this same property. The effects of selective transmission across the synapse led to the integratory phenomenon of summation. (Omikron)*

be reversed, with the inside of the membrane registering a positive voltage in relation to the outside. This reversal is called the **overshoot.** The entire depolarization sequence, also called the **action potential,** takes place in less than one millisecond.

Figure 10-16 shows the results of the measurement of electrical events before and during a nervous impulse. This graph helps to explain the refractory period of the neuron. There can be no new response until the original membrane potential is reestablished, with the complete repolarization of the membrane.

The membrane potential is created by the special permeability characteristics of the neuron's membrane. In a resting (unstimulated) condition, the membrane is permeable to potassium ions but impermeable to sodium ions. There is usually about 30 times more potassium in the fluid inside the cell than there is in the fluid outside the cell. Potassium ions therefore leak out of the resting cell, moving with a steep concentration gradient. It is this leakage that sets up the membrane potential, since the cell is losing positive charge. Although there is a high concentration of positively charged sodium ions outside the cell, they are prevented from entering by the membrane. In a resting state, the concentration of sodium ions is about 10 times higher outside the membrane than it is inside. Both concentration differentials are maintained by an active transport mechanism, which moves potassium ions into the cell, and sodium ions out. This is often called the **sodium-potassium pump.**

The effect of the initial stimulus is to alter the permeability characteristics of the membrane. The pores of the membrane may change in size or configuration, so as to permit the free passage of sodium ions, or the sodium pump may be turned off. Moving both with a concentration gradient and an electrical gradient, these ions rush into the cell. The influx of the sodium

adds positive charge to the inside while removing it from the outside so that depolarization takes place. In fact, so many sodium ions may move into the cell that the phenomenon of overshoot occurs.

In a very brief period, some mechanism seems to switch off the sudden permeability of the membrane. The change in electrical charge itself might be responsible for this switch. No more sodium ions may enter, and those that are already in cannot leave. No new response is possible until the outward movement of potassium ions once again sets up the membrane potential. The time it takes these ions to diffuse out through the membrane corresponds to the refractory period of nervous response. The number of departing potassium ions will have to be even greater than before to set up the original membrane potential, in order to compensate for the addition of the sodium ions to the fluid inside the cell, increasing its total positive charge. This process can be repeated several more times, creating new action potentials, but eventually the concentration of sodium ions inside the cell will be so great that no outward movement of potassium can compensate for it. In order to maintain the neuron's excitability, the cell must be given time to pump sodium ions back out. The sodium pump mechanism is essential to continued nervous response. It is interesting to note that cyanide's swift and fatal effect is due to the fact that it blocks the action of the sodium pump and thereby causes failure of the entire nervous system.

## Conduction of nervous impulses

The speed at which electrical impulses can be conducted varies, depending on the type and function of the neuron; Table 10-1 lists the conduction rates for several different types. Generally speaking, conduction is fastest in those systems which serve to warn the animal of sud-

**10-16** *This graph shows the effects of three different strengths of electrical current applied to a single nerve cell. Curves 1 and 2 indicate that the current was below threshold level. When a stronger current was applied (Curve 3), an action potential was generated, causing a sharp increase in the electrical potential of the cell. In this experiment, the length of the current was limited. If the current in Curve 2 had been applied for a longer period of time, it is likely that temporal summation would have taken place, giving rise to an action potential.*

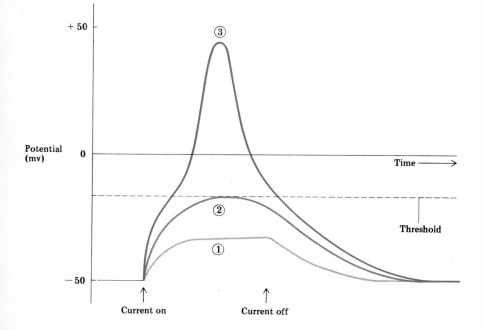

den external changes. For example, messages regarding sights and sounds travel much more rapidly than those regarding muscular contraction or pain. The width of the fiber is an important factor, with the fastest conduction being found in the widest fibers; in fact it has been calculated that the speed of the conduction is equal to the square of the axon's diameter. Another factor is the use of insulation. Myelinated cells conduct much faster than nonmyelinated cells.

## 10-1
## Conduction rate of nerve impulse

| | Rate of conduction (meters/sec.) | |
| | Myelinated | Nonmyelinated |
| --- | --- | --- |
| Mammal | 120 | 1.0 |
| Fish | 35 | 0.2 |
| Frog | 30 | |
| Crab | | 1.5 |

Source: C. H. Best and N. B. Taylor. 1966. *The Physiological Basis of Medical Practice.* 8th ed. Williams & Wilkins, Baltimore.

If we look at the way the action potential actually moves along the membrane, we can see why these two factors are important. The current which is generated by the action potential at one point on the membrane spreads to another adjacent area, causing sudden depolarization of the membrane and an accompanying new action potential, which, in turn, transmits another current to areas farther away. There is, in other words, a wave of action potentials moving along the axon. This means that the strength of the signal does not decrease as it travels. Each new action potential generated has exactly the same strength as the original one, because of the all-or-none nature of the response.

The value of the myelin sheath in increasing the speed of conduction is that it provides precise spacing of the points at which an action potential is generated. At all insulated points, there can be no action potential, since no ions can flow freely through the myelin coating. Action potentials can be generated only at the nodes of Ranvier, so the current leaps from one node to another, rather than traveling slowly and steadily through the whole length of the axon.

## Transmission across the synapse

When a nerve axon is electrically stimulated somewhere along its membrane, the action impulse thus generated will spread out in both directions from the site of stimulation. No matter which way it travels, the impulse is car-

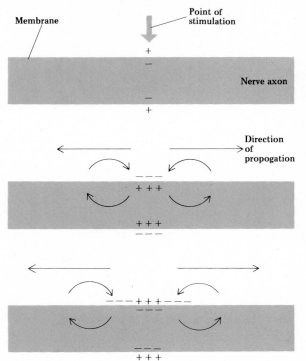

10-17 *All neurons are capable of carrying impulses in either direction along the membrane. This can be demonstrated by applying an electrical stimulus midway along the axon. At the point of stimulation, there will be a reversal of electrical charge, generating an action potential. This potential will then spread in both directions with equal speed. However, the impulse traveling in one direction will reach and cross the synapse, while the impulse traveling in the other direction will be unable to leave the cell.*

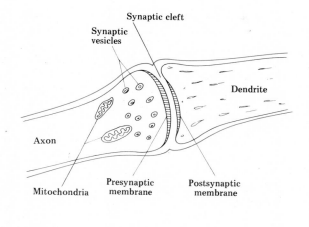

Synaptic cleft

Synaptic vesicles

Dendrite

Axon

Mitochondria

Presynaptic membrane

Postsynaptic membrane

Direction of transmission

**10-18** *The synaptic bulb of the axon is adapted for chemical transmittal. Mitochondria are numerous, providing ample energy for transmission. Synaptic vesicles contain the chemical transmitter, such as acetylcholine. Under the stimulus of an action potential, vesicles release the transmitter into the cleft. Specialized receptor sites on the surface of the dendrite provide a "landing place" for the chemical. The transmitter then creates an action potential in the second neuron, whereupon it is immediately broken down by enzymes. (Leonard Ross—Omikron)*

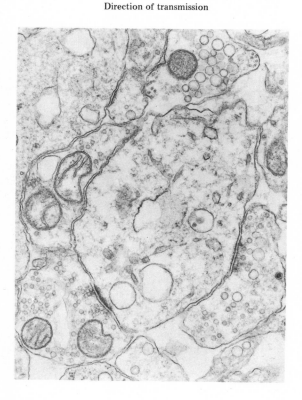

ried with equal efficiency; there is nothing in the structure of the neuron's membrane that would provide direction to the impulse.

There is an obvious disadvantage to a circuit in which impulses can travel two ways. A stimulus would give rise to an impulse that spreads out in all directions, rather than being carried directly to a center of coordination. The nerve net of the hydra is just such a circuit. Since in the hydra there is no directionality to the impulse coming from the site of stimulation, there is a corresponding lack of directionality in the animal's response. For example, if one tentacle is given an electrical shock, the hydra does not just move that tentacle away; it moves its entire body. All of the hydra's responses show this same level of generality.

In order to produce specific responses, such as the response of a dog that moves only one paw away from a damp spot on the ground, impulses from the site of stimulation must travel in one given direction to a site of control. In the vertebrate nervous system, the directionality is provided by the **synapse,** or junction between neurons. Impulses can be transmitted across the synapse in one direction only. Impulses moving the wrong way are not able to travel from one neuron to the next, and the strength of the impulse gradually fades out and dies.

When motor neurons are examined under the electron microscope, it can be clearly seen that the axon of one cell and the dendrite of the next are not actually connected. There is a space of 200 to 500 Angstrom units between the two structures in most synapses. When the existence of this space, or **synaptic cleft,** was discovered, physiologists immediately grew curious about the mechanism of transmission across the cleft. Could the electrical impulse jump this gap?

As recently as 1950, some physiologists maintained that synaptic transmission was electrical. But by the early years of the decade of the fifties, research results pointed unmistakably to the action of some chemical transmitter sub-

**315**

stance. Using electron microscopy to show structure and refined chemical techniques to detect the presence of small amounts of the transmitter substance, scientists were finally able to develop an accurate model of synaptic transmission.

The axon ends in a sort of bulge, called the **synaptic bulb.** Inside this bulb can be seen mitochondria and a number of very small vesicles. When a nerve impulse reaches the synaptic bulb, a few of these vesicles, or packets of chemicals, immediately move to the surface of the cell membrane and empty their contents into the synaptic cleft. The most common chemical transmitter substance is acetylcholine. There are several other transmitters with a similar chemical structure, and it has been found that the amino acid glutamic acid also can act as a transmitter.

It takes about one to two milliseconds for the chemical to diffuse across the cleft to the membrane of the next neuron. The molecules of the chemical fit into specialized receptor sites on the folded surface of the dendrite membrane. The effect of the chemical is to cause a change in the permeability of the membrane, thereby creating an action potential in the neuron. As soon as this action potential is generated, the transmitter chemical is attacked by the enzyme cholinesterase. This enzyme, which is present in high concentrations in the synaptic region of the dendrite, hydrolyzes the transmitter chemical so that it can no longer be effective.

Chemical transmission across the synapse usually serves to excite the adjacent neuron. However, there are certain synapses which are inhibitory; the message that leaves the first neuron prevents the second one from firing. This permits the careful regulation of response to any given stimulus, as an overresponse can be corrected by sending a message along an inhibitory circuit. No one synapse can send both excitatory and inhibitory messages, for each is specialized in only one way. But a nerve fiber may contain both kinds of synapses between individual neurons, so that a fine adjustment is possible.

## Information coding

The primary purpose of the signals transmitted by neurons is to convey information gathered by sensory receptors about the external and internal environment. This raises the question of how such information can be carried by the kind of electrochemical impulses just described.

Much of the necessary information is built into the circuit. For example, there is no need to discriminate between a visual stimulus and an auditory stimulus, because each type of impulse is carried along a different circuit to a different control center. The way in which the neurons are hooked up provides the information about the type of stimulus being received. All that must be carried on any one circuit is the information about the strength of the stimulus.

Since neurons exhibit an all-or-none response, it is not possible to convey information about the strength of the stimulus by grading the response of individual neurons. A neuron cannot vary the strength of its signal, but it can vary the frequency. For this reason sensory information is coded according to the number of impulses sent within a given time period, and the adjacent cell can "read" the code even though it is transmitted by chemical rather than electrical means.

The initial coding depends entirely on the refractory period of the neuron. The absolute refractory period, as we saw, is about one millisecond; under no circumstances can a second impulse be generated within this period. But in the second phase of the refractory period, it is possible to elicit a response if the stimulus is great enough. So the frequency with which impulses are sent can vary with the strength of the stimulus. With a weak stimulus, it might take the full five-millisecond refractory period. With an extremely strong stimulus, impulses might be generated every millisecond.

How does this code survive the synapse, where it must be translated into chemical means? The answer lies in summation of the stimulus.

Let us suppose that each impulse launches one packet of the chemical transmitter. When the impulses are five milliseconds apart, there will be exactly this same interval between the launching of the packets. Since the chemical begins to break down as soon as it arrives in the adjacent neuron, that cell may never accumulate enough of the chemical to provide the threshold stimulus and cause an action potential. But if the packets are sent out at the rate of one every two milliseconds, there will begin to be an accumulation of this chemical in the second neuron, since it is arriving faster than it is breaking down. After two or three packets have been sent, the second neuron has enough chemical transmitter to cause firing. This is an example of summation over time, or **temporal summation.** Summation may also occur in space, when several axons are sending transmitter chemicals to the same neuron. **Spatial summation** is the means by which a single neuron can coordinate the various messages it receives from the several other neurons with which it is in contact.

The process of summation provides integration at the lowest level of the nervous system.

Since the organism is quite literally bombarded by stimuli from its environment—probably thousands of individual stimuli per second—it is important to have some mechanism for selecting and screening the impulses. Very weak stimuli may not prove sufficient even to launch transmitter packets, and so the impulse never leaves the first neuron. Some weak impulses may launch only a few transmitter packets, which decay in the second neuron before they can stimulate it to threshold and propagate an action potential. Variations in the level of the threshold value provide another means of discriminating between strong and weak stimuli. By these means weak stimuli can be suppressed, or selected out.

## Coordination

The third phase of nervous activity, the coordination of the information received in order to produce an appropriate response, is the phase about which we know and understand the least. The new field dealing with the study of information processing started in the 1940s; it is called **cybernetics,** a word that comes from the Greek term for helmsman, the man who guides the ship. The most successful applications of cybernetics so far have been in the area of computers. In spite of a great deal of study and research, the amazingly swift and accurate information processing that is performed by the nervous system is still to a great extent a mystery.

## Reflexes

The simplest process of information sorting and coordinating can be seen in the motor reflex; an example is the knee jerk, that familiar part of the medical examination. Only two neurons are involved in producing the knee jerk. One neuron picks up the original stimulus and passes it to a

**10-19** (A) Shows the frequency of impulses generated by a very intense flash of light; (B) and (C) are weaker flashes. This difference in frequency of generated impulses serves as a kind of code that relays information about the strength of the initial stimulus. By means of temporal and spatial summation, this information can be coordinated by other nerve cells.

A
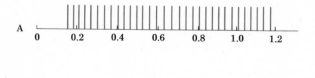
0    0.2    0.4    0.6    0.8    1.0    1.2

B
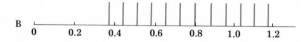
0    0.2    0.4    0.6    0.8    1.0    1.2

C
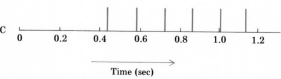
0    0.2    0.4    0.6    0.8    1.0    1.2

Time (sec)

second neuron which directly controls the muscle that causes the knee to jerk. This stimulus-response pathway, linking a receptor and an effector, is called a **reflex arc.** When only two neurons are involved, as is the case with the knee jerk, it is called a monosynaptic reflex arc. The actual coordination of the response takes place in the second neuron. It registers the strength of the excitatory stimulus and responds by generating an action potential. The process is simple, for the excitatory signal is a strong and urgent one, and the neuron does not receive any inhibitory signals at the same time. But the jerk of the knee is really only one small part of the total response to the stimulus. If we look at the whole response, we will begin to appreciate the complexity of the entire nervous system.

Suppose that, as you are walking across the room, you bash your knee on the corner of a low table. The action potential generated by that initial stimulus will do the following things:

1. Stimulate the motor neuron that controls the muscle in the knee that will contract and cause the jerk
2. Send impulses to the spinal cord that inhibit the firing of the neuron that controls the antagonistic muscle in the leg, so that no kind of movement other than the jerk will occur
3. Send impulses to the neurons that control the muscles of the other leg and the trunk, so the body will assume the correct position for supporting all its weight on the uninjured leg
4. Send impulses to the glands that secrete regulatory hormones, such as adrenalin, preparing the circulatory and respiratory systems for a state of emergency in case the injury proves to be serious
5. Send impulses to the higher learning and memory centers of the brain, to enable a conscious understanding of why the painful sensation arose and how it can be avoided in the future.

The transmission and coordination of these various messages may involve millions, some-

times even billions, of neurons. Many of these neurons will also be sending and receiving other informational impulses at the same time. This flow of information must be processed to produce an orderly sequence of responses.

## The autonomic nervous system

The neurons controlling all muscles and organs which are not subject to voluntary control are collectively referred to as the **autonomic nervous system.** These organs and glands are controlled by automatic responses of the nervous system to the information received from proprioceptors.

Under normal conditions, the purpose of the autonomic nervous system is the maintenance of a steady internal environment; all homeostatic mechanisms are a part of this system. In times of emergency, however, the animal may need to do more than simply maintain a steady state; it must be ready to take whatever action is needed. Under these circumstances, the autonomic nervous system functions to prepare the animal for "flight or fight." Blood pressure increases, more blood flows to skeletal muscles, the heart beats faster, the rate and depth of ventilation increase. In some animals, hair may stand up and claws may be unsheathed.

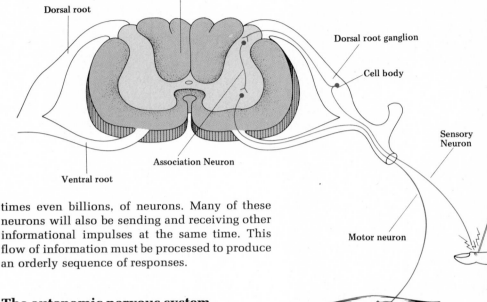

Spinal cord

Dorsal root

Dorsal root ganglion

Cell body

Sensory Neuron

Association Neuron

Ventral root

Motor neuron

**10-20** *The simplest reflex arc contains only two neurons, the sensory neuron that generates the initial impulse and the motor neuron that stimulates the effector. This drawing shows a three-neuron arc which features an intermediary — the association neuron — between the sensory neuron and the motor neuron. This association neuron may be connected to other nerves in the spinal cord, allowing an even greater degree of integration of response.*

**10-21** *An example of a reflex action can be seen in the way an earthworm shrinks away from the light. Sensors located within the outer epidermal tissue respond to the light and they cause stimulation of nerves linked to the muscles of each segment. Although the earthworm shows some degree of cephalization, there is very little integration of such reflex responses. (Alycia Smith Butler)*

**10-22** *Something has angered or alarmed this caracal, so that it responded by preparing for "flight or fight"; this response is under the control of the autonomic nervous system. The animal's ears have moved downward and back; its teeth are bared; its claws are unsheathed; its fur is ruffled; and it growls and spits. All these reactions make it look like a more threatening adversary. (Eric Hosking from National Audubon Society)*

It is the **sympathetic** nervous system that dominates autonomic nervous function in times of stress. Connected to the heart, lungs, adrenal gland, and other important organs and glands, the postganglionic fibers of the sympathetic system release adrenalin or noradrenalin, a chemical that stimulates changes in the rates at which these organs function, thereby readying the animal for stress conditions.

The parasympathetic nervous system predominates in the absence of stress. This part of the autonomic nervous system is connected to the very same glands and organs as the sympathetic, but its effect on them is just the opposite. For example, the sympathetic system increases the rate of heartbeat; the parasympathetic system slows it down. The parasympathetic system releases only acetylcholine as a chemical transmitter in all fibers.

These two parts of the autonomic nervous system can be viewed as an antagonistic pair, operating on the same principles as antagonistic pairs of muscles, or a pairing of excitatory and inhibitory synapses. This kind of paired control mechanism is very common in the biological world, since it permits very fine variations in the degree of response to any stimulus.

## The central nervous system

The central nervous system, or CNS, of vertebrates is dominated by the brain. The spinal cord, which is also a part of the CNS, is largely subordinated to the control of the brain, although the spinal cord is capable of certain kinds of independent action in producing reflexes. The knee jerk, for example, is a spinal reflex.

The trend of evolution seems to have been toward the increasing dominance of the brain over other neurons. In lower invertebrates, such as the hydra, there is no brain at all. Nerve cells radiate throughout the entire body, forming an interconnected nerve net. The nerve net can provide relatively rapid conduction of impulses from receptor sites, but integration in this system is limited to the coordinating abilities of the individual neurons. We can see the next evolutionary step in the flatworms, or Platyhelminthes. With the development of bilateral symmetry and a distinct front end to the animal, there came a corresponding increase in the number of neurons located there and the complexity of their interconnection. In the annelid phylum there is increasing specialization in this area. The earthworm, for example, has a pair of cerebral **ganglia,** or cluster of nerve cells, which provides some degree of coordination of the ganglia found in each segment.

In arthropods and mollusks, we find a highly developed set of ganglia, or rudimentary "brain," which receives most of the sensory input and coordinates much motor activity. However, there are still a number of independent reflex arcs, consisting of sensory cells, two or three neurons, and a muscle. Headless insects are capable of existing for some period of time, carrying on all necessary metabolic functions. The same is true of lower vertebrates. For example, headless frogs can survive for hours, and can exhibit certain kinds of integrated behavior, such as scratching an irritated patch of skin.

In higher vertebrates, the dominance of the brain in the CNS is almost complete. All the important sensory input goes directly to the brain, through the 12 **cranial nerves.** The **spinal nerves** are connected with the `spinal cord rather than directly with the brain, but their input can travel up the cord to the brain for additional processing, as was the case in our example of the informational impulses associated with the knee jerk reflex.

The CNS coordinates the input of the sensory nerves and in turn stimulates some response. In the case of the knee jerk, the response comes from striated muscles that are under voluntary control. The nerves that cause this response are **motor neurons.** In vertebrates, the CNS is also linked

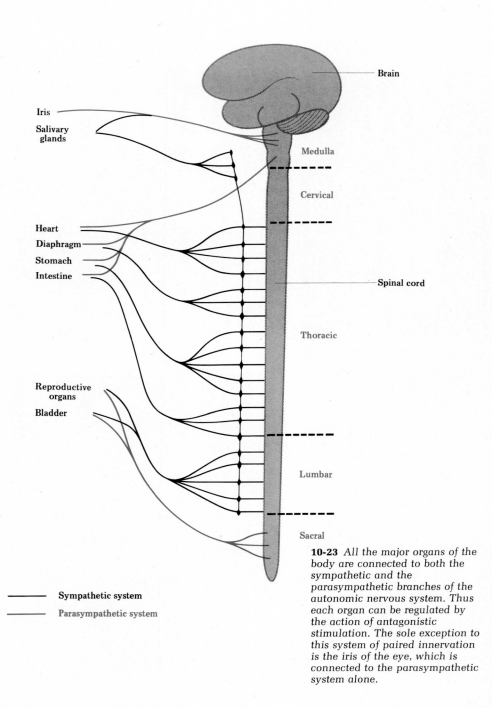

**10-23** *All the major organs of the body are connected to both the sympathetic and the parasympathetic branches of the autonomic nervous system. Thus each organ can be regulated by the action of antagonistic stimulation. The sole exception to this system of paired innervation is the iris of the eye, which is connected to the parasympathetic system alone.*

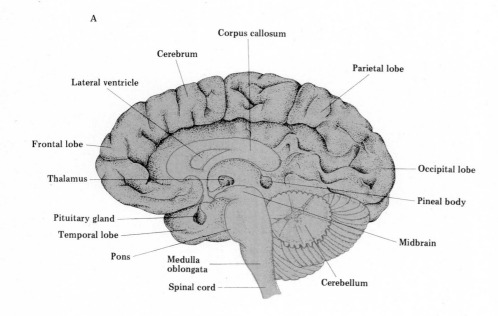

A

Corpus callosum

Cerebrum

Parietal lobe

Lateral ventricle

Frontal lobe

Occipital lobe

Thalamus

Pineal body

Pituitary gland

Temporal lobe

Midbrain

Pons

Medulla
oblongata

Cerebellum

Spinal cord

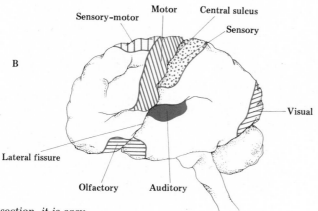

B

Sensory-motor

Motor

Central sulcus

Sensory

Visual

Lateral fissure

Olfactory

Auditory

**10-24** *In cross section, it is easy
to see how the cerebrum has
expanded to cover most of the
brain, seeming to dominate most
of the "old brain." This view also
shows the intimate connection of
the pituitary gland, a part of the
chemical control system, with the
system of nervous control. The
view of the brain's surface maps
the various control areas.*

to the smooth muscles—those that are not under
voluntary control—as well as to glands that
secrete agents of chemical control. Thus the
CNS can stimulate a wide range of responses to
any kind of input.

**The vertebrate brain.** The human brain is a
dense mass of more than 10 billion neurons,
interconnected in complex circuits. Some of
these circuits have been precisely mapped; others
are so dimly known that there are whole areas of
the brain whose function we can only guess at.
The knowledge and experience gained in the
building of computers has furnished valuable
clues to the workings of the brain, but functions
such as learning, judgment, and memory storage
are still largely mysteries.

The complicated human brain is the product
of great evolutionary changes. The primitive
vertebrate brain has three main divisions: hindbrain, midbrain, and forebrain. Each of these
parts has changed in form and function in the
course of vertebrate evolution.

The main part of the hindbrain is the
**medulla,** which is really just the enlarged end
of the spinal cord. Most of the sensory nerves
pass through the medulla, as do the fibers controlling nearly every motor neuron. In birds and
mammals, the coordinating function is supplemented by the **cerebellum,** a large outgrowth of
the medulla. Although a small cerebellum is
present in frogs and fish, its size has been greatly
increased through evolution, providing more
synapses and therefore additional possibilities of
information coordination. The increased size of
the cerebellum in birds and mammals permits
them to move at much faster speeds, because it
makes possible more rapid and more delicate
muscular control.

The midbrain has changed very little in size,
but there has been a significant change in its
function. In lower vertebrates, the midbrain coordinates most complex behavior. It receives
visual input and integrates this with the informa-

**321**

tion concerning functioning of the motor nerves and the autonomic nervous system that is passed along from the hindbrain. In higher vertebrates, the midbrain has lost most of this important integrative function. It still regulates the reflex control of the irises and eyelids, but for the most part, it functions primarily as a relay station, moving information from the hindbrain to the forebrain.

The forebrain has become more and more important in the course of vertebrate evolution. It is divided into a number of different areas. The **thalamus** in higher vertebrates integrates input from all sensory systems and channels information on to more specialized parts of the forebrain and back down to the spinal cord. The **hypothalamus** is an important link between the systems of neurons and chemical control, as it regulates the release of many hormones. It is the center of regulation of water balance and body temperature. It is also the site where feelings of pleasure and pain, hunger and thirst, rage and sexual desire are regulated.

The part of the forebrain that has grown the most through evolution is the **cerebrum.** In mammals, it has become the ultimate control center, and it has expanded so that it covers all the other regions of the brain. In order to obtain more area for the important functions of this part of the brain, its entire surface has been folded and convoluted, producing the characteristic wrinkled appearance of the human brain. In lower mammals, every square millimeter of this surface is concerned with a specific sensory or motor function. In man, however, there are large areas which cannot be assigned a specific sensory or motor function. If these areas of the brain are removed (as they are in the operation called a lobotomy, a once-fashionable treatment for mental patients with violent tendencies) all sensory and motor functions remain unimpaired, and even the functions of learning and memory can still be exercised. These areas seem to be concerned with the kind of responses we label

"personality"; it is here that such characteristics of the human mind as imagination, creativity, and ingenuity seem to originate.

## Summary

The complex nervous system present in man and animals is an expression of a property present in all living things. Every organism possesses some degree of irritability and response mechanism.

However, animal nerve cells have the ability to act as a transducer, or to alter forms of energy. Therefore, they can reduce various stimuli to one uniform, electrochemical impulse, and integrate the response accordingly.

The process is divided into three general phases of irritability. They are (1) the receiving of information, both internally and externally, (2) the sending of signals from the area of perception to other parts of the organism, and (3) coordination.

Information may be received by several external means. One of these, sensitivity to light, is present to some extent in almost every organism, from the flower reacting to light by opening and closing to the human eye which can perceive depth and color.

Another receiving device, hearing, serves to perceive and screen significant noises. The hair cell, an adaptation of ciliated epithelial cells, is the basis of sound reception in all hearing organs. The tympanum is also present within even the simplest forms of sound reception. It is found in insects as a modification of the tracheal passage, and in the vertebrate ear as the eardrum.

Smell and taste depend on specialized epithelial cells, called chemoreceptors, which are found in a variety of areas. These sensations are transmitted by air, water, or direct touch. A varying combination of the four taste and seven smell receptors give the total sensation in both tasting and smelling.

Touch, temperature, and pain receptors are varied. The hair cell, a receptor of the epithelium, is sensitive to almost any type of external movement, whereas the Pacinian corpuscle registers "deep touch" or pressure. Pain is often explained as a response to an overload of stimuli to a receptor.

All homeostatic mechanisms rely on internal receptors, dubbed proprioceptors, to monitor the internal environment and warn of trends toward fluctuation. These systems give animals muscular sense and a sense of balance.

Once information is received, either externally or internally, it is then transmitted, via signals, from the area of perception to other parts of the organisms. The basic conductive unit for nerve impulses is the nerve cell, or neuron, which consists of two types of projections. The dendrite receives impulses from connecting nerve cells, while the axon conducts these impulses on to other nerve cells.

The synapse, or junction between neurons, receives impulses from axons and relays them to the dendrite. This process was at one time thought to be electrical in nature, but it has now been found to be a chemical reaction resulting from the diffusion process within the synaptic bulb of the axon.

Two coordination systems are employed to produce the correct responses after information is received and signals sent. The autonomic nervous system controls involuntary muscles and organs. Under stress, autonomic nervous function is dominated by the sympathetic nervous system; and the parasympathetic nervous system is predominant under normal conditions. The brain-dominated central nervous system coordinates the input of sensory input and stimulates some response.

## Bibliography

### References

Attneave, F. 1971. "Multistability in Perception." *Scientific American.* 225(6): 62–71. Offprint no. 540. Freeman, San Francisco.

Baker, P. F. 1966. "The Nerve Axon." *Scientific American.* 214(3): 74–82. Offprint no. 1038. Freeman, San Francisco.

Bower, T. G. R. 1971. "The Object in the World of the Infant." *Scientific American.* 225(4): 30–38. Offprint no. 539. Freeman, San Francisco.

Case, J. 1966. *Sensory Mechanisms.* Macmillan, New York.

Day, R. H. 1971. *Perception.* 2nd ed. Brown, Dubuque, Iowa.

Eccles, J. C. 1963. *The Physiology of Synapses.* Academic Press, New York.

Heimer, L. 1971. "Pathways in the Brain." *Scientific American.* 225(1): 48–60. Offprint no. 1227. Freeman, San Francisco.

Menaker, M. 1972. "Nonvisual Light Reception." *Scientific American.* 226(3): 22–29.

Pribham, K. H., ed. 1969. *The Brain and Behavior.* vol. 2: *Perception and Action.* Penguin, Baltimore.

Wyburn, G. M. et al. 1964. *Human Senses and Perception.* University of Toronto Press, Toronto.

### Suggestions for further reading

Gordon, M. S. et al. 1968. *Animal Function: Principles and Adaptation.* Macmillan, New York.
Chapters 9 through 11 outline in some detail the role of the nervous system in integrating and coordinating animal behavior.

Granit, R. 1955. *Receptors and Sensory Perception*. Yale University Press, New Haven.
Presents results of electrophysiological research in the special senses. Ranges from primary neural processes to sensory discrimination and integration.

Katz, B. 1966. *Nerve, Muscle and Synapse*. McGraw-Hill, New York.
Rather technical but extremely up-to-date study of neuromuscular structure and function.

Lentz, T. L. 1968. *Primitive Nervous Systems*. Yale University Press, New Haven.
Proposes a hypothesis to explain the origin of the nervous system, with data and illustrations from sponges, hydra and planaria.

Von Buddenbrock, W. 1958. *The Senses*. University of Michigan Press, Ann Arbor.
Popular but precise guide to the eight (not five) senses.

Wilentz, J. S. 1968. *The Senses of Man*. Thomas Y. Crowell, New York.
Comprehensive, well-illustrated and up-to-date summary of the neurophysiology of human perception. Touches on anatomy, physiology, histology, embryology, and evolution.

Woolbridge, D. E. 1963. *The Machinery of the Brain*. McGraw-Hill, New York.
Explores in some detail the structure and function of the brain in man and other animals. Includes recent developments in brain research.

## Books of related interest

Asimov, I. 1963. *The Human Brain: Its Capacities and Functions*. Signet, New York.
An especially clear introduction to the complex interrelationships between the human brain and the endocrine and nervous control systems.

Droscher, V. B. 1971. *The Magic of the Senses: New Discoveries in Animal Perception*. Harper & Row (Colophon), New York.
Profusely illustrated, clear and precise guide to recent research in animal perception.

Piaget, J. 1971. *Biology and Knowledge*. University of Chicago Press, Chicago.
Extensive and fascinating essay by a noted psychologist, originally trained in biology. He attempts to link three forms of cognitive process in man with their bases in biological function.

Ryle, G. 1949. *The Concept of Mind*. Barnes & Noble, New York.
A well-known modern philosopher develops a concept of mind acceptable to both the philosopher and the natural scientist. An important and intriguing book, and one that can be understood by the nonphilosopher.

Von Frisch, K. 1956. *Bees: Their Vision, Chemical Senses and Language*. Cornell University Press, Ithaca, New York.
A classic work in the study of insect perception and communication.

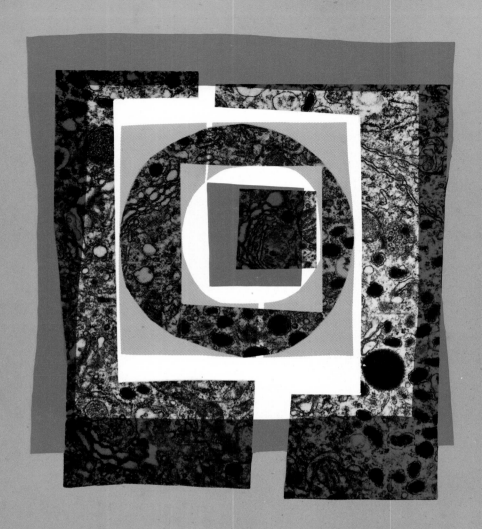

# 11

# Integration: chemical control

Living organisms have a constant and pressing need for communication, as well as a need for integration of response and metabolic functions. The previous chapter showed how, in many organisms, these needs can be met by the nervous system. An even more widespread mechanism for communication and integration is the carefully timed release and transport of chemical compounds that can affect the functioning of various cells or cell parts. All one-celled organisms, all plants, and all animals contain agents of chemical control that regulate important processes. In earlier chapters, we have already discussed a number of examples. In digestion, we saw that the hormone secretin stimulates the secretion of digestive enzymes. The hormone norepinephrine acts to cause contraction of blood vessel walls, thereby increasing blood pressure and hastening the rate of venous return. Regulation of the volume of urine is performed by the antidiuretic hormone. Movement of the leaves in the Venus' flytrap is controlled by a plant hormone, auxin.

In plants and many small invertebrates, chemical control is the major means of integration. Although chemical control is slower than nervous control, it is by itself adequate for the responses of nonmotile organisms. A number of adaptations have helped reduce the time lapse between stimulus and response to a bare minimum. One such adaptation is the extreme sensitivity organisms show to the agents of chemical control. In laboratory experiments, plants responded to hormones that had been diluted to a concentration of one part hormone to a billion parts water. This high degree of sensitivity means that it takes quite literally only a few molecules of a hormone to create a response. Since few molecules are needed, the task of synthesizing them is not very time-consuming.

Another adaptation that speeds up the rate at which chemical control can take place is the presence of hormone components at strategic locations throughout the organism. Hormones and other agents of chemical control cannot be stored in large quantities in the body. They must be produced as they are needed, and they are destroyed by enzymes shortly after their production. Small quantities may be stored in animal cells in the form of secretory granules. However, they need not be synthesized from scratch. The organism can store molecules that are similar in composition and configuration to necessary hormones. These molecules, called **precursors,** can quickly be altered to produce a hormone in a very short time. The alteration is very much like the process of unmasking found in digestive enzymes, and may consist only of adding or removing a side chain from the molecule. Cholesterol is an example of a common hormone precursor in humans.

In animals with a well-developed nervous system, chemical control supplements the swifter action of the nervous system. The separate consideration of chemical and nervous mechanisms is often useful in studying the subject of control and integration. But the separation is in some respects quite misleading since the two systems are really quite closely interconnected in the organism. One example can be seen in the fact that the transmission of nervous impulses across the synapse is performed by chemical agents. It is chemical control that provides the existence of one-way circuits in the nervous system and also permits the integratory processes of temporal and spatial summation. Nervous control and chemical control in vertebrates must be viewed as two aspects of an entire integration system.

In spite of such interconnections, it is possible to make a few general distinctions between nervous control and chemical control. As we have previously noted, nervous control is faster than chemical control. Therefore chemical control alone is never found where response must be immediate. Nervous control is also more specific than chemical control. A nerve impulse stimulates only one small area—one muscle fiber, for instance—whereas a chemical agent may affect

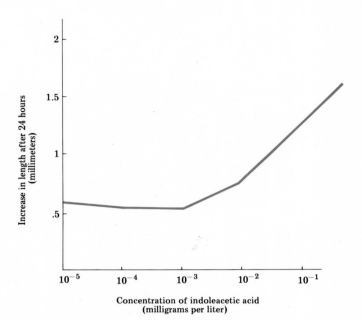

many parts of the body. A well-known example illustrating the broad effects of chemical control is the action of the hormone adrenalin. When it is present in the blood, it causes changes in the rate of heartbeat, rate of ventilation, contraction of both smooth and skeletal muscles, and level of blood sugar. This lack of specificity is in part due to the fact that the agents of chemical control are typically transported in the bulk circulatory system.

Chemical control is frequently used to establish regulatory cycles. The female menstrual cycle is one example of this type of function, and current research in this field indicates that there are probably many other cycles which have not yet been defined. Laboratory animals show definite patterns of daily activity, called **circadian rhythms,** involving regular fluctuations in body temperature, blood pressure, rate of heart beat, and electrolyte balance. There is some evidence that longer cycles, of perhaps two to four weeks, are also present, with periods of optimum health and efficiency alternating with periods of greater vulnerability to stress and strain. These cycles are thought to correspond with fluctuations in the level of certain hormones in the bloodstream. The sex hormones (estrogen in females and testos-

**11-1** *This graph gives some indication of the degree of sensitivity found in plants when they are exposed to very small concentrations of hormones. Even 1/1,000th milligram of auxin per liter of water stimulated rapid growth in plant shoots.*

terone in males) are probably especially significant in this regard.

In an interesting attempt at a practical application of this information about cyclic rhythms, the Japanese government recently collected data regarding the cycles of bus drivers. The data were fed into computers that worked out each man's cycle and scheduled his work assignments to correspond with the optimum points on his cycle. The result was a significant decline in the number of accidents involving buses. Such experiments indicate the possibility of useful new developments in this field of research.

## Agents of chemical control

In the nineteenth century, biologists began to suspect that certain organs of the body secreted chemicals that affected other parts. In the 1840s, the French scientist Berthold demonstrated that castrated roosters, which do not ordinarily develop secondary sex characteristics, would grow combs and wattles when the testes of normal roosters were grafted on. Berthold was able to achieve this effect even when the testes were grafted in unusual places, such as on the back or neck. These results seemed to indicate that the testes produced some chemical that was transported to other parts of the body, producing secondary sex characteristics.

The actual mechanism of chemical coordination, with the isolation of the chemical involved, was first demonstrated in 1902, by the English scientists William Bayliss and E. H. Starling. Their work focused on secretin, a component of digestive fluid. Starling proposed that secretin and similar chemicals be called hormones, a word that comes from the Greek verb that means "to excite."

## Hormones

A hormone is generally defined as a chemical substance produced in one part of the body and carried by bulk circulation to other parts of the organism where it stimulates various kinds of activity. The first hormones to be studied were digestive hormones, but it soon became evident that hormones were also involved in many other regulatory processes, not solely in vertebrates, but in other animals and plants as well.

Early researchers in the field of hormone physiology assumed that since the effect of various hormones was similar in many respects, their chemical composition would be similar also. Subsequent chemical analysis proved this assumption to be erroneous. The group of substances classed as hormones actually includes a wide variety of compounds, differing greatly from one another in their chemical makeup. Some hormones, the steroids, are synthesized from lipids; others, such as secretin and insulin, are modified proteins; auxin and the antidiuretic hormone are closely related to amino acids. Two important plant hormones, gibberellin and cytokinin, are synthesized from nucleotides, the components of nucleic acids such as DNA. In fact, the chemical composition of cytokinin was first discovered through the use of an old sample of herring sperm in an experiment involving plant growth. In the months it had been stored in a cupboard, the sperm sample had begun to deteriorate, and the DNA had broken down to produce a compound almost identical to cytokinin.

**11-2** *The molecule on the left is adenine, one of the nucleotides found in nucleic acids. The molecule on the right is the hormone kinetin, a cytokinin. The hormone has the same basic ring structure as the nucleotide; the only difference is the carbon ring attached to the hormone.*

Adenine
(purine)

Kinetin
(6—Furfuryl amino-
purine)

ESSAY **Photoperiodism**

Variations in the length of the daily light or dark period produce responses ranging from growth, flowering or germination in plants to feeding, hibernation, or migration in animals; this phenomenon is termed photoperiodism. Photoperiodic reactions are initiated by the effect of light on a chromophore, a small, colored, light-sensitive molecule associated with a protein molecule. Absorption of colored light by the chromophore sets in motion a chain of reactions, that results in the photoperiodic behavior. In plants the chromophore is phytochrome, a pigment chemically related to the pigments of human bile and of blue-green algae. Ordinarily present in such low concentrations as to be invisible, isolated phytochrome is a blue or blue-green substance, found in two mutually photoreversible forms; the blue form absorbs red light, the blue-green absorbs far-red light. Figure A shows the effect of light on flowering.

A simple experiment (Fig. B) shows the mechanism of photoperiodism in flowering. A long-night plant (one that flowers only when exposed to the long nights of fall), the cocklebur is stripped of all but one-eighth of one leaf. Phytochrome activation in the leaf fragment initiates development of burrs throughout the plant when it is exposed to the necessary period of darkness. If the dark period is interrupted by a flash of light, flowering is inhibited, but this reaction can be forestalled by immediate exposure to far-red light, which returns phytochrome to its red-absorptive blue form.

To demonstrate the presence of a traveling stimulus or hormone within the plant, (Fig. C), two plants are grafted together near the base. The plant on the left is exposed to long days; this

A

(Photo courtesy Dr. John M. Arthur, Boyce Thompson Institute for Plant Research, Inc.)

exposure inhibits flowering. When the right-hand plant is given at least 8 hours and 40 minutes of darkness, burrs develop in both plants. A hormone evidently crosses the graft to cause burr development in the left-hand plant. In many plants phytochrome also controls growth; when hours of daylight drop below a threshold point, the growing buds go into a dormant state, in which they are protected against the cold of winter.

Photoperiodism in animals embraces such diverse phenomena as migration and sexual activity in birds, circadian activity and metabolic rhythms in rats, mice, cockroaches and crabs, and diapause (or suspended animation) in insects. Codling moth larvae, for example, go into a dormancy period in the fall because release of a brain hormone is suspended in the absence of long light periods.

In vertebrates, the hypothalamus or the pineal body may be involved at an early stage in the release of hormones that activate circadian or circannual rhythms. In some species these brain parts may contain pigments directly sensitive to light that penetrates the skull. Evidence pointing in this direction has been derived from experiments with blinded birds.

The circadian rhythm of birds kept in continuous darkness becomes "free-running"; the active period starts a little later each day but lasts a constant period of time, reflecting the running of an internal circadian "clock." Even a single brief exposure to light suffices to "reset" the free-running rhythm. If the cage is lit even very dimly for a set time every day, the period of activity develops a regular relation with the light cycle, and the bird is said to be entrained.

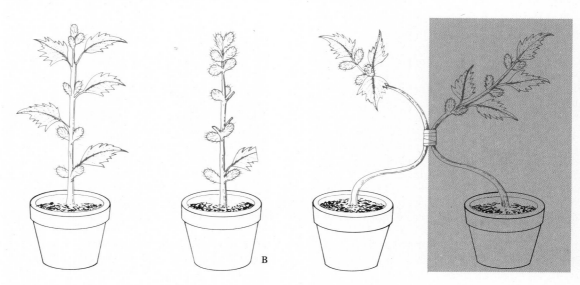

B

C

Thirty or 40 years ago, no one had any trouble deciding what was and was not a hormone. But as time went on, and experimental evidence accumulated, bothersome questions began to arise. Many of these questions focused on the chemical agents of synaptic transmission. These chemicals are produced in one cell and affect another, thus meeting one criterion of a hormone. Moreover, biochemical tests showed that one of these synaptic transmitters, nor-adrenalin, was chemically identical to a hormone secreted into the bloodstream by the adrenal medulla, an important part of the chemical control system. However, the transmitter is secreted by a neuron and functions as an essential part of the nervous system. Should we then call these substances hormones even when they are secreted by neurons? Is a hormone a kind of chemical, or is it a name for chemicals that travel in a certain path, from a large ductless gland **(endocrine gland)** into the bloodstream? It was questions like these that blurred the once clear-cut distinction between nervous and chemical control.

## Prostaglandins

In the last 15 years, a whole group of chemical control agents in animals has been discovered, the **prostaglandins.** These compounds (at least 16 have been identified) seem to act very much like hormones. However, they are not secreted by an endocrine gland but are synthesized by a wide variety of body cells. Prostaglandins are such powerful chemical effectors that they are synthesized in very small traces — perhaps a tenth of a milligram per day of the more important ones — and they are quickly broken down by enzymes almost immediately after their synthesis. Since the body normally contains only very small amounts, isolating and analyzing these compounds presented something of a problem.

Recent technological innovations have permitted the study of the composition of prosta-

glandins. Biochemists were surprised to find that these compounds were actually fatty acids, for no one had previously suspected that fatty acids could play a regulatory role in the metabolism. All prostaglandins are synthesized from common polyunsaturated fatty acids, and synthesis appears to take place in almost every type of tissue.

The study of the physiological effects of common prostaglandins has yielded results that are equally surprising. Prostaglandins can cause relaxation of smooth muscles lining the respiratory and digestive tracts; can lower blood pressure, by causing an increase in the excretion of sodium ions and an accompanying increase in the volume of urine; can constrict blood vessels; can counteract the effects of important metabolic hormones, such as those that stimulate the formation of lipases. This wide variety of effects indicates that prostaglandins must play a major role in regulating many physiological functions.

Recent experiments by a group of Swedish physiologists provide a clue to the way that prostaglandins function. Using isolated organs of mammals, the group found that when they added prostaglandins to the nutrient solution in which the organs were maintained, nervous action was inhibited. The effect of prostaglandin is to decrease the rate at which neurons in the sympathetic nervous system release their chemical transmitter. This evidence suggests that prostaglandins may be part of the inhibitory feedback mechanism of the nervous system.

Much of the current research on prostaglandins is centered on possible medical applications. The potential of these drugs seems quite exciting. They may be used in the treatment of asthma. They can lower blood pressure. There is some evidence that prostaglandins can prevent internal blood clots. They can be used as contraceptives (and also as "morning-after" pills when conception is suspected), and they have fewer side effects than the female hormones now used for contraception. They will induce labor, and they can also be used as a treatment for sterility in both males and females. Although it will take additional laboratory and clinical research, the safe and effective medical use of prostaglandins can be expected within the next 10 or 15 years.

## The mechanism of hormone action

Although scientists have known for a long time that certain chemical substances do act as control and regulatory agents, it has only been in the last decade or so that they began to understand something about the mechanism of their action. The investigation of this problem has focused on three primary areas.

1. Enzyme reactions. It has been demonstrated that certain hormones (adrenalin and noradrenalin, for instance) stimulate the action of important enzymes, thus speeding up chemical reactions and producing larger amounts of the end-products. The presence of adrenalin has been shown to stimulate the activity of an enzyme called glycogen phosphorylase, which splits glycogen and liberates molecules of glucose. Therefore, the end result of the secretion of adrenalin is to make more high-energy compounds available for cellular respiration.

An interesting hypothesis, which has not yet been confirmed, is that hormones may also stimulate the enzymes within mitochondria that catalyze the phosphorylation reactions involved in the production of ATP. It has been demonstrated in laboratory situations that the presence of relatively large amounts of thyroid hormones can make the production of ATP more energy expensive—that is, it requires the burning of more molecules of glucose with more molecules of oxygen to make the same amount of ATP. It has been suggested that this might serve as a mechanism for raising body temperature of homeotherms, for it would produce additional heat as a by-product. This hypothesis remains purely speculative, since it has never been shown

that such a process occurs in a normal organism.

2. Membrane permeability. Practically all vertebrate hormones have been shown to alter the rate at which water and various solutes move across cell membranes. In some cases, the effect of the hormone is quite unspecific. The antidiuretic hormone, for example, seems to increase the permeability of the membrane to water, inorganic ions, amino acids, and sugars. The action of aldosterone, the hormone that brings about the final adjustments in urine volume, seems more narrowly limited to an effect on only two ions, potassium and sodium.

The recent discovery of the important role played by prostaglandins in chemical control has served to intensify the research involving membrane permeability. The cell membrane is composed of a large percentage of phospholipids, and these compounds can supply the structural components of prostaglandins. It is quite possible that the effects of prostaglandins are a result of their ability to alter the structural characteristics of the cell membrane. It is also possible that the arrival of hormones at the membrane of the cell stimulates the synthesis of prostaglandins within.

3. Protein and nucleotide synthesis. It has been demonstrated that hormones can alter the rate at which certain compounds are synthesized within the body. For example, it appears that insulin stimulates the synthesis of the carrier molecules that are involved in the active transport of glucose. Even more important in this area are the findings in regard to RNA synthesis. Both thyroid hormones and the sex hormones are known to speed up the rate of RNA synthesis.

Although the evidence is still not entirely conclusive, it appears that the mechanism by which hormones influence protein synthesis occurs within the gene itself. One piece of evidence is the fact that when the genes are blocked or inactive (this can be caused by the introduction of a powerful antibiotic, such as actinomycin D), sex hormones will have no effect on the cell at all. This evidence leads to the conclusion that hormones must exert their influence by way of the genes, probably by stimulating the genes to synthesize increased amounts of messenger RNA. Even more significant is the fact that certain hormones seem to stimulate genes which are ordinarily repressed by the action of other more dominant genes. For example, when the female hormone is administered to roosters, the effect is to stimulate the activity of female genes, which are present in the rooster but usually are repressed by dominant genes for masculine characteristics. Much to the surprise of the barnyard, the rooster develops the comb and plumage characteristic of the female, and his interest in the hens becomes merely platonic.

## The role of cyclic AMP in hormone action

While studying the effects of hormones on the level of blood sugar in rats, physiologist Earl Sutherland of Duke University observed that the hormone epinephrine served to increase the amount of another substance in the cells of the liver. That substance was **cyclic AMP,** or cyclic-3′5′-adenosine monophosphate, a compound related to ATP. Sutherland's curiosity about the meaning of this unexpected finding diverted him from the object of his original research and caused him to begin investigating the role of cyclic AMP in the cell.

Sutherland finally established that epinephrine stimulated the production in the liver cells of cyclic AMP, which in turn stimulated the synthesis of an enzyme involved in the amount of glucose which enters the bloodstream. In 1971, Sutherland received the Nobel Prize for his work.

As Sutherland's results were published, other physiologists began to do the same kind of research with other hormones. It has now been established that a number of hormones are effec-

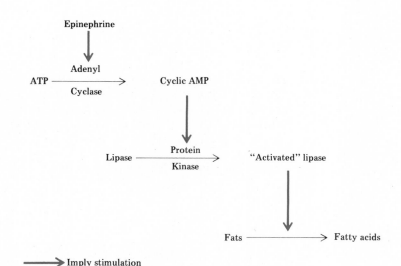

Imply stimulation

tive by means of a "second messenger," or chemical compound that is affected by the hormone and, in turn, causes an effect on cellular activities. The concept of the second messenger suggests that in many cases the hormone never enters the cell, and thus there is no need to explain how a relatively large molecule can penetrate the cell membrane so readily. While still outside the cell, the hormone stimulates an enzyme, adenyl cyclase, that is a component of the membrane; the enzyme breaks down nearby molecules of ATP into cyclic AMP; and the cyclic AMP acts as the second messenger to stimulate the production of various enzymes important in metabolic reactions. It is possible that other compounds, yet to be identified, can also act as secondary messengers, bringing about different effects on cell activity. Prostaglandins for example are suspected at modifying adenyl cyclase activity.

## Chemical control in plants

Plants have evolved a way of life that does not for the most part depend on immediate responses to environmental change. They have not developed a nervous system for rapid conduction of impulse-coded information, but rely almost

**11-3** *This diagram shows the role of cyclic AMP (cAMP) in stimulating the release of fatty acids in adipose tissue. The presence of the hormone epinephrine, secreted by the adrenal medulla, causes ATP molecules to be converted to cAMP. The cAMP in turn activates an enzyme that unmasks the enzyme lipase. The lipase then breaks down stored fats to produce fatty acids.*

exclusively on chemical control instead. The hormones and other agents of control travel through the phloem, along with other solutes, and can thus be conducted to every part of the plant. Like most other hormones, the agents of chemical control in plants are typically not very specific in their action. This lack of specificity, coupled with the fact that the hormones are widely distributed throughout the plant, raises an interesting question: Why does the hormone act only on one target area?

Experiments with plant hormones administered under carefully controlled conditions show that the same hormone can have different effects, depending on the condition of the cell to which it is introduced. For example, the hormone that controls the development of vascular tissue in growing stems will produce many xylem vessels when it is administered to a stem with a low sucrose concentration. When the concentration of sucrose is raised, the same hormone will produce more phloem cells in the same stem. Such evidence leads to the conclusion that there are two important factors in chemical control. One is the presence of a given hormone, and the other is the condition, the chemical environment, of the specific cell on which the hormone has its effect.

## Plant hormones

Because the final effect of plant hormones depends so heavily on the condition of the cells on

### 11-1

### Plant hormones

| Hormone | Chemical type | Site of synthesis | Target tissue | Effect |
|---------|---------------|-------------------|---------------|--------|
| Florigen(s) | never been isolated | leaves | apical meristem (buds) | induces flowering |
| Gibberellin(s) | five-ringed diterpenoids (acidic) | found in high concentrations in seeds, and in young seedlings | seeds, growing stem tissue, buds | stimulates stem elongation, flowering; application of large amounts can cause excessive growth; in experiments, it inhibited bud and root formation |
| Auxin(s) | indole acetic acid; also other related compounds | root tips, stem tips | stems, roots, lateral buds | enlargement of cells in stem, resulting in the stem bending toward light or upwards; also has an effect on root elongation; also inhibits lateral bud formation |
| Cytokinin(s) | 6-amino substituted purines | ? | buds, roots, leaves | stimulates cell division, stimulates differentiation of buds, leaf enlargement, lateral root formation |

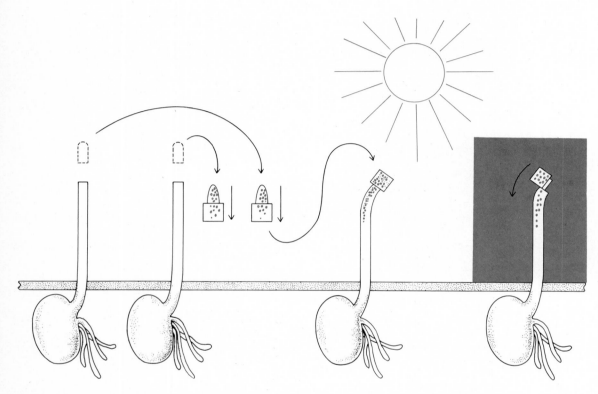

**11-4** *Plant physiologist Fritz Went cut off the growing tips of young seedlings and placed the tips on blocks of absorptive agar. When the blocks were then placed on the cut stem, the stem continued normal elongation. By varying the placement of the hormone-saturated blocks, Went was able to cause differential growth of the stem. The stem bends away from the side on which the block is placed.*

which they act, it is difficult to define the exact effect of each hormone. But the general outlines of hormonal control in plants have been clearly established. Plant hormones generally cause growth responses.

Some of the earliest recorded experiments regarding chemical control in plants were those performed by Charles Darwin. Darwin demonstrated that without the presence of light, characteristic growth movements would not occur, and he suggested that there might be some chemical connection between the light and the pattern of growth. This connection was demonstrated conclusively in 1926 by the plant physiologist, Fritz Went. Since that time, a number of major plant hormones have been identified, and their effects are still being investigated.

**Auxin.** Auxin was the first plant hormone to be identified, as it was the chemical involved in Went's experiments (see Figure 11-4). There are actually a number of different auxins, each of which has been isolated and its chemical structure analyzed. Since they are so similar in both structure and function, this group of compounds is often referred to simply as auxin.

It has been established that most plant tropisms are controlled by differential concentrations of auxin. Auxin is known to stimulate the elongation and differentiation of newly formed cells produced by the meristem (the area of cell division). The effect of auxin seems to be to soften the rigid cell wall, allowing the membrane-enclosed cell contents to expand and thus bringing about a decrease in the cell's turgor pressure.

This decrease permits more water to enter the cell, and the contents expand so that they once more press against the cell wall. This pressure stretches the softened cell wall once again, and the entire process is repeated. In this way the plant cells can grow, with consequent development of stem, roots, and even fruits.

Auxin is apparently produced in the cells at the growing tip of a stem or leaf, and the chemical is conducted from the site of synthesis down through the rest of the plant. It seems that one effect of auxin is to block the action of other hormones, such as the one that causes leaves to drop off. As long as a certain quantity of auxin is moving from a leaf into the stem, the leaf-shedding hormone is repressed. But when the supply of auxin flowing in from the leaf drops below a critical point, as happens during periods of cold temperature and low light such as are characteristics of wintertime, the leaf-shedding hormone is activated, and the leaf falls off.

**11-5** *An eight-week experiment shows the effect of gibberellin on* Pelargonium. *Four biweekly applications of gibberellic acid have stimulated shoot growth in the right-hand plant; the plant at the left received plain water. (Courtesy of P. P. Pirone, New York Botanical Garden)*

CHECK WATER SPRAY

4 APPLIC. GIB.10PPM AFTER 8W

At high concentrations, auxins can be fatal to a plant, perhaps because they completely disrupt control processes. This fact led agricultural botanists to consider the possibilities of auxin's use as a commercial weedkiller. Of course, auxin, like all hormones, is easily broken down by enzymes, since an accumulation of the potent molecules inside the organism would be very harmful. But biochemists found that they could synthesize an **analogue** of auxin. A chemical analogue is a compound that will function in much the same way as the original, entering into many of the same reactions. An analogue usually has the same general structure as the original but may differ in the number and arrangement of side chains. The change makes the molecule more stable and less susceptible to enzyme attack. The auxin analogue in common commercial use is the weedkiller 2,4-D.

**Gibberellin.** Auxin was the first plant hormone to be identified, and for some time, plant physiologists believed it was the sole regulatory agent in plants. But by the 1950s, this idea had to be revised in the light of new evidence. Another group of closely related hormones, the **gibberellins,** was shown to play an important part in plant growth. Although gibberellin had been discovered by a Japanese botanist 30 years earlier, its role in plant development had been overlooked by Western scientists for all those years. When they finally isolated the chemical and began using it experimentally, they observed profound effects.

It has been shown that gibberellins can cause flowering, even when conditions are otherwise unsuitable; they can turn dwarf plant species into plants of normal height; they can stimulate the development of pollen within the flower. Gibberellins are particularly important in the development of seeds. The seed, which consists of an embryo and a food supply, surrounded by a hard coating, remains dormant for a given period of time, probably because of the action of some chemical inhibitor. But certain conditions of temperature, light, and humidity will bring about either the synthesis or the activation of gibberellin, and this hormone counteracts the inhibitors. It stimulates the synthesis of digestive enzymes, which break down the stored food and make nutrients available for the embryo, thus causing germination of the seed.

**Cytokinin.** Cytokinin is a modified nucleotide, and it probably exerts its effect on plants by stimulating—perhaps even altering—cellular RNA. Cytokinin can cause buds to develop; in combination with auxin, it appears to stimulate cell division in the meristem. The continued presence of cytokinin in leaves prevents their turning yellow, or showing other signs of age. In seeds, cytokinin seems to be manufactured by the breakdown of certain stored food, and once it is present in the embryo, it controls the early processes of development.

Auxin and cytokinin often interact. In some cases, such as the growth of leaves, the two hormones seem to act as synergists, with each one boosting the action of the other. In other cases, they counteract one another. For example, as long as auxin is produced in the uppermost growing tip of a stem, the auxin will suppress development in nearby lateral buds along the stem; this is called **apical dominance.** But if the concentration of cytokinin in some lateral bud is increased, apical dominance is overridden, and the lateral buds will develop. Auxin and cytokinin may also combine to produce an effect unlike that produced by either chemical alone. For example, auxin alone in developing stems stimulates cell elongation; cytokinin alone has no effect at all. Together, they cause an increased rate of cell division.

**Abscissic acid.** The effect of abscissic acid was known some years before its chemical composition was identified; it was formerly referred to as the dormancy hormone, in reference to its effects.

**11-6** *The coleus plant (top photo) shows the results of apical dominance, with marked growth of the top shoot and repression of the lateral buds. After the top shoot was pinched off (bottom photo) stopping the downward flow of auxin, the lateral buds began to grow, under the influence of added amounts of cytokinin. (Alycia Smith Butler)*

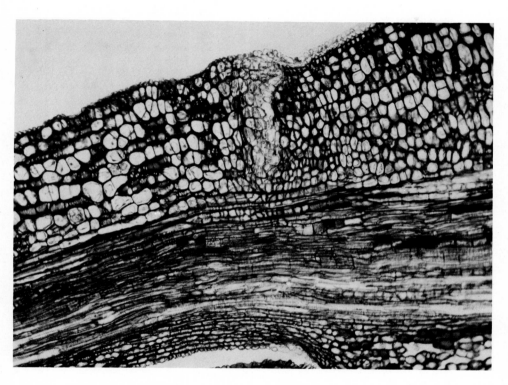

**11-7** *This micrograph of a leaf petiole shows the abscission layer, the point at which the leaf will eventually drop off. When the concentration of abscissic acid increases and the flow of auxin from the leaf decreases, the cells in the abscissic layer die. Their death weakens the petiole at that particular point, and the leaf falls off. (Omikron)*

Abscissic acid causes the dropping of leaves and fruit, and it can, under experimental conditions, stunt the growth of normal plants.

Abscissic acid acts as an antagonist to the other three hormones already mentioned. Its action is analogous to that of the inhibitory neuron in the nervous system. By varying the concentration of abscissic acid and other hormones, a fine control of growth and development can be achieved.

**Other plant hormones.** The existence of at least two other hormones is suspected by their action, although they have not been chemically isolated. One is a hormone that controls flowering, and the other is the so-called wound hormone, a chemical that helps to reduce the traumatic effects of injury and promotes rapid healing.

Another chemical compound, ethylene, is known to act as an agent of control, but botanists are uncertain whether or not to call it a hormone, since it is dispersed by air currents rather than moving through the vascular tissue. Ethylene is a gas, a simple 2-carbon molecule, that is released from flowers, leaves, stems, developing fruits, and the roots of plants. Its presence in the air serves to speed up the process of ripening. Ethylene is so nonspecific that molecules of this compound released by a ripening orange will bring about faster development in bananas stored nearby.

## Chemical control in invertebrates

Chemical control is important in invertebrates, many of which have only poorly developed systems of nervous control. Unfortunately, the practical difficulties of research have hampered the study of hormones in invertebrates. Since they are generally rather small animals, the quantities of potent hormones present in their bodies are so minute that it is almost impossible to collect enough to use in testing and experimentation. For example, in order to collect a very small sample of 750 milligrams (about one-fortieth of an ounce, or barely enough to cover the bottom of a small test tube) of a hormone that controls insect development, researchers had to use over four tons of silkworms!

In both the earthworm and *Nereis*, hormones are known to be responsible for the regeneration of lost segments. The stimulus for the secretion of the hormone seems to be the anterior group of ganglia that is sometimes called a brain but is really more of a neural intersection. These ganglia may also be the site of secretion. Hormones also have been found in the cephalopod mollusks, such as the octopus. These animals have two small glands, one near either eye that produce hormones in response to control by the brain and

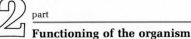

also in response to the absorption of light. The hormones secreted there regulate development of reproductive organs and sexual behavior.

## Hormones in arthropods

The only invertebrate phylum in which chemical control has been extensively studied is the arthropod phylum. A wide variety of developmental changes in arthropods is influenced by hormones. At least six hormones have been identified that are produced in specific body tissues to control local functions, such as laying of eggs and contraction of the heart. There are also two hormones produced by special endocrine glands—the molting hormone and the juvenile hormone, which prevents or delays metamorphosis. Another type of chemical control agent is secreted by a gland inside the mouth parts, or mandibles, of the queen in bee and ant colonies. This chemical is transferred to the workers as they perform their daily task of grooming the queen, and it has the effect of preventing the development of their reproductive organs. Although this substance does not fit into the strict definition of a hormone, because it travels from one individual to another, its effect and mode of action are very similar to that of certain endocrine hormones. It is called an **exohormone.**

**11-2**

**Invertebrate hormones**

| Hormone | Animal(s) | Chemical | Where secreted | Effect | How controlled |
|---|---|---|---|---|---|
| Hormones affecting pigment concentration and dispersion | crustaceans | not known | sinus gland at the base of the eyestalk; tritocerebrum (part of the brain) | stimulation of pigment movement inside the branched chromatophores (cells containing pigment) | release stimulated by light; probably different hormones stimulate concentration or dispersion of the pigments |
| Ecdysone | insects | steroid | prothoracic gland | initiates the molting process, whereby an insect sheds its outer shell-like coat | secretions of the prothoracotropic hormone from cells in the brain promote release of ecdysone |
| Juvenile hormone | insects | lipid (*not* a steroid lipid) | corpus allatum | maintains continuity of the larval form, i.e., suppresses metamorphosis | the corpus allatum is stimulated by neurosecretions |
| Maturation hormone | cephalopod mollusks (octopus) | | optic glands located on the eyestalk | promotes maturation of the ovaries or testes | inhibitory signals from certain nerve fibers prevent secretion from this gland prior to sexual maturation |

| Hormone | Organism | Chemical type | Source | Effect | How controlled |
|---|---|---|---|---|---|
| "Growth" hormone | nereid annelids | | cerebral ganglion | stimulates proliferation of new segments in growing worms and regeneration of lost segments | |
| Sexual development regulating hormone | polychaete annelids | | cerebral ganglion | inhibits premature development of sexual organs; when its level drops, sexual maturity rapidly takes place | the moonlight acts to coordinate the level of this hormone, so that sexual maturity is reached simultaneously by the entire population |
| Queen suppressor substance | bees | 9-oxodec-trans-2-enoic acid (fatty acid) | queen's mouth | suppresses ovarian growth in worker bees | queen feeds it to the worker bees that attend her |
| Estrodiol progesterone | echinoderms | steroids | found in the ovaries | not known | not known |

## 11-3
## Vertebrate hormones

| Hormone(s) | Gland | Chemical type | Target organ(s) | Effect | How controlled |
|---|---|---|---|---|---|
| Thyroxine & triiodothyronine | thyroid | amino acid derived | general | increased rate of metabolism | thyrotropic hormone (TSH) from the anterior pituitary causes secretion of thyroid hormones; TSH release is stimulated by a CNS factor and by low levels of thyroxine in the blood |
| Epinephrine (adrenalin) | adrenal medulla | amino acid derived | liver, muscle, circulatory system | breakdown of glycogen, faster pulse rate | release is stimulated by insulin and is also influenced by the hypothalamus |

| Hormone(s) | Gland | Chemical type | Target organ(s) | Effect | How controlled |
|---|---|---|---|---|---|
| Norepineph-rine (noradrenalin) | adrenal medulla | amino acid derived | circulatory system | constriction of periph-eral blood vessels | release is stimulated by the influence of the hypothalamus or stimu-lation of the sympa-thetic nervous system |
| Aldosterone | adrenal cortex | steroid | general | water-electrolyte balance | released in response to adrenocorticotropin (ACTH) |
| Corticosterone | adrenal cortex | steroid | general | formation of glycogen, amino acid metabolism, water-electrolyte balance | released in response to adrenocorticotropin |
| Estrogens | ovary, placenta | steroid | uterus, mammary glands | maturation and devel-opment of the female sexual organs; cyclic proliferation of the uterine lining | it is synthesized and released in response to gonadotropins (FSH, LH) from the anterior pituitary |
| Progesterone | corpus luteum, placenta | steroid | uterus, mammary glands | development and nor-mal functioning of female sexual organs; maintenance of preg-nancy | secretion is stimulated by prolactin, a hormone released by the anterior pituitary |
| Testosterone | testis | steroid | male sex-ual organs | maturation, normal functioning of male sexual organs; develop-ment of secondary sex characteristics | secretion is stimulated by FSH and LH |
| Insulin | pancreas (beta cells) | polypeptide | general | increases utilization of sugar in blood resulting in lower level of blood sugar; increased syn-thesis of lipids | insulin is secreted whenever the level of blood sugar is high; somatotropin, e.g., raises the level of blood sugar and leads to in-sulin secretion |
| Glucagon | pancreas (alpha cells) | polypep-tide | | breakdown of glycogen, thus raising the blood-sugar level; breakdown of lipid | low blood sugar stimu-lates its release |

| Hormone(s) | Gland | Chemical type | Target organ(s) | Effect | How controlled |
|---|---|---|---|---|---|
| Adrenocorti-cotropin (ACTH) | anterior pituitary | peptide | adrenal cortex | stimulates secretion of adrenocorticosteroids | releases substances into the blood which stimulate release of ACTH |
| Somatotropin (growth hormone) | anterior pituitary | protein | general | stimulates growth | unknown, but may be related to levels of sex hormones |
| Luteinizing hormone (LH) | anterior pituitary | protein | ovary or testis | stimulates production and secretion of sex hormones | secretion is in response to substances released from the hypothalamus; these substances are released in response to circulating sex hormones |
| Follicle-stimulating hormone (FSH) | anterior pituitary | glycoprotein | ovary or testis | stimulates estrogen production, ovulation; stimulates production of sperm | same as for LH |
| Oxytocin | posterior pituitary | oligopeptide | smooth muscle of uterus, mammary glands | initiation of labor, contractions; enhances ejection of milk | suckling causes signals to be sent to the hypothalamus which results in release of oxytocin from the posterior pituitary |
| Vasopressin | posterior pituitary | oligopeptide | arterioles, kidneys | antidiuretic, i.e., reabsorption of water; raises the blood pressure | released in response to CNS signals due to stress or trauma; also released in response to changes in blood pressure; release inhibited by epinephrine |
| Secretin | duodenum | polypeptide | pancreas | stimulates pancreas to secrete digestive enzymes and juices | secretion is stimulated by presence of acid released into the upper duodenum from the stomach |
| Parathyroid hormone | parathyroid glands (4) | polypeptide | general | maintains a relatively high level of $Ca^{++}$ and low level of phosphorus in the blood | calcium concentration of the blood regulates release of this hormone |

**Hormonal control of molting.** One specific example, that of the molting hormone, can serve to illustrate the general mechanism of hormonal control in arthropods. The molting hormone, also called **ecdysone,** was first isolated by the Japanese physiologist S. Fukada in 1940, although its existence had been established in earlier experiments on insect development.

Molting, or the shedding of the exoskeleton, is a necessary phase of growth in all arthropods. A butterfly may molt four to six times during the caterpillar stage of its life, and a lobster often molts seven times in its first summer. Arthropods must escape the confining prison of their outer covering if they are to increase in size.

The actual process of molting is initiated through the secretion of an enzymatic molting fluid by glands in the outer body layer, the epidermis. This fluid digests certain structural components of the covering exoskeleton, thereby making it softer and more stretchable. When a new exoskeleton is fully developed, the old one splits and is discarded (some species eat it, to recover lost nutrients). At first, the new exoskeleton is pliable, and its folds expand as the animal grows. But in a short time, exposure to the air brings about an oxidation reaction that hardens the chitin, and the exoskeleton may be additionally reinforced by calcium salts.

The secretion of the molting fluid is stimulated by ecdysone. The mode of action of ecdysone is not known, but experiments have shown that one dose alone, no matter how large, is not enough to stimulate the complete molting process; the hormone must be continuously present throughout the entire period of the molt. It is suspected that ecdysone may stimulate the formation of the new exoskeleton as well as the secretion of molting fluid.

Ecdysone is secreted in a pair of small endocrine glands, called the prothoracic glands. The actual process of secretion is stimulated by the presence in the bloodstream of a second hormone, called the activation hormone. The activa-

11-8 *The head of a male moth exhibits chemoreceptors which function in the communication between the female and the male. When the female is ready to mate, she gives off a pheromone, an odiferous exohormone. The male's chemoreceptors pick up the scent, and he flies off in her direction. There are reports of pheromone reception taking place at distances of more than a mile. (Photograph by Alexander B. Klots)*

11-9 *The process of metamorphosis in insects is completely controlled by hormone secretions. It is not so much the secretion of a particular hormone that starts the process, but a change in the balance of hormones. (Photograph by Alexander B. Klots)*

tion hormone causes a number of other effects as well, involving the maintenance of water balance, the secretion of digestive enzymes, and the development of the ovaries. This hormone, sometimes called a master hormone, must be present in a certain concentration before the secretion of ecdysone will take place.

The secretion of the activation hormone takes place in a specialized organ near the brain, containing both neurons and secretory cells. The dendrites of the organ are connected by synaptic junctions to the axons of special cells of the brain, which have both a nervous and a secretory function. Impulses transmitted from these brain cells apparently cause the secretion of the activation hormone.

In this hormonal control system, we see a number of features also characteristic of the chemical control system in vertebrates. One such feature is the use of a series of hormones, rather than only one. A single-hormone activation mechanism would certainly provide more rapid responses, but it would also be less precise and more prone to accidental failure. There are a number of conditions, such as injury, disease, or nutritional deficiency, which might accidentally trigger a simple on-off switch involving only one hormone. The use of a series of hormones, some of which must be continually present, guards against such an untimely (and probably fatal) accident. The use of a series of hormones also can increase the specificity of the response. For example, there are a certain number of factors that will stimulate the production of the first hormone. The presence of the first hormone, plus a certain number of other factors, will stimulate the production of the second hormone. The presence of the second hormone, plus a certain number of other factors, will stimulate the secretion of the molting fluid. At each step, the response can become progressively more sensitive to a great variety of internal and external conditions, all of which must be favorable before molting actually occurs.

A second important characteristic of the molting control mechanism is that ultimate control of the process lies in cells of the brain. In lower animals, chemical control often takes place at the level of the individual cells or the tissues. Certain control mechanisms at this level are also found in insects, and they are in fact found in man. For example, any damaged tissue responds by releasing **histamine,** a compound often classed as a hormone, since it stimulates effects on other cells of the body. But there are limitations to the effectiveness of chemical control at this level of biological organization. True integration of response can take place only through a centralized control center which processes all information regarding the internal and external environment. The dominance of the brain over the functions of chemical control is an important evolutionary advance.

## Chemical control in vertebrates

In vertebrates, hormonal control is part of a complex system of integration, and hormones may play many different roles in this system. Some hormones may serve as simple activators of a specific function. An example can be seen in secretin, the hormone studied by Bayliss and Starling. When the pH in the small intestine drops below a certain level (because of the acidity of chyme coming from the stomach), cells in the lining of the intestine begin to release the hormone secretin. Secretin is a relatively large and quite specific hormone. It affects only certain secretory cells in the pancreas which respond by pouring digestive enzymes into the intestine. This process will continue until the pH in the intestine returns to a higher level, stopping the secretion of secretin.

Not many hormones in higher vertebrates act in so direct and simple a fashion. Often they act as antagonistic pairs, with one serving to stimulate a certain function and the other causing its inhibition. An example of such a pairing can be seen in the two hormones insulin and glucagon. When the level of glucose in the blood is high, the hormone insulin is secreted, causing an increase in the rate at which glucose is transported out of the blood and into muscle and fat tissue. When blood sugar is low, the hormone glucagon is secreted, stimulating the conversion of stored glycogen into glucose to be released into the bloodstream. The action of this antagonistic pair of enzymes serves to keep the level of glucose in the blood relatively steady.

A more complex model of hormone action can be seen in hormone pairs with a direct feedback relationship. For example, the adrenocorticotropic hormone (ACTH), when present in the blood, stimulates the secretion of another hormone, cortisol. The cortisol secreted in a second location serves to stimulate and coordinate a number of repair and healing processes; at the same time, its presence in the bloodstream acts as a feedback mechanism that stops the release of ACTH. Without this kind of automatic limitation built into this chemical control system, an endocrine gland might go on secreting its hormone for years in response to one stimulus.

The most complicated control system, and the most advanced in an evolutionary sense, can be seen in the dominance of the central nervous system. The CNS controls the **hypophysis,** often called the "master gland." The hypophysis is an outgrowth of the brain, and is directly linked to the **hypothalamus.** The hypophysis (also called the pituitary) produces hormones that control secretion in other endocrine glands. It is generally the hormones produced in these glands that have a direct effect of stimulating or inhibiting certain metabolic, homeostatic, or developmental processes. At every level, both positive and negative feedback mechanisms help to regulate hormonal secretion. The dominance of the CNS insures that chemical control will be integrated with nervous control, and it also permits chem-

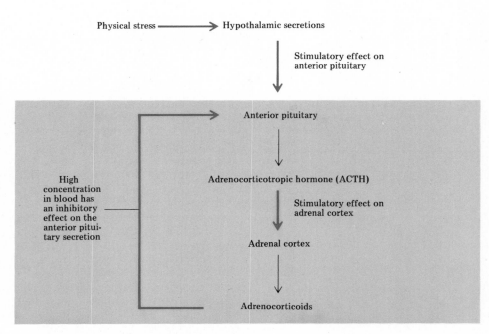

ical response to changes in the external environment. The secretion of adrenalin, in response to the control of the sympathetic nervous system, during flight-or-fight stress situations is one example of chemical response to external conditions, another CNS-dominated hormonal control mechanism can be seen in the regulation of body temperature.

11-10 *This schematic diagram shows a mechanism of hormonal control that contains a negative inhibitory feedback. A high concentration of adrenocorticoids in the bloodstream acts to inhibit the secretory activity of the anterior pituitary. This feedback mechanism monitors the response to the initial stimulus, preventing overreaction.*

## Endocrine organs in mammals

Although some hormones, such as secretin and histamine, can be secreted locally, most hormones are secreted by ductless endocrine glands and are transported in the bloodstream to their target organs.

**Hypophysis.** The most complex of all the endocrine organs is the hypophysis. The hypophysis is divided into three sections, or lobes. The anterior lobe is formed during embryonic growth by a

**11-11** *The hypophysis or pituitary gland is closely connected to the brain, as can be seen in Figure 10-24. The posterior lobe, or pars nervosa, is formed as an outgrowth of the hypothalamus; the anterior lobe is formed during embryonic growth by a specialized outgrowth at the roof of the mouth. (Electron micrograph provided by Professor Lawrence Herman, Department of Pathology, State University of New York, Downstate Medical Center, Brooklyn, New York)*

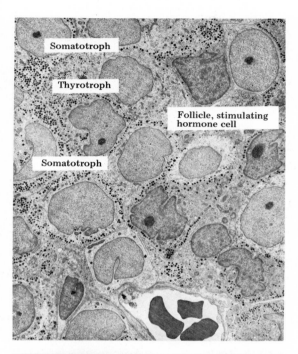

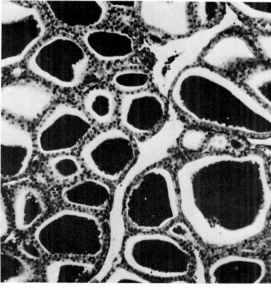

**11-12** *This photo shows a number of spherical follicles in a slide of monkey thyroid tissue. The follicle is the functional unit of the thyroid containing a colloid in which thyroid hormones are dispensed. (Brian Bracegirdle, from An Atlas of Histology)*

specialized outgrowth of the roof of the mouth; the posterior lobe is an outgrowth of an area of the brain, the hypothalamus. These two lobes are connected by a third, or median lobe. A capillary network maintains constant communication by means of the blood circulating from the hypothalamus to the hypophysis. It has been demonstrated that the hypothalamus releases a chemical control agent, called a releasing factor, that stimulates the activity of the hypophysis.

Microscopic examination of sections from the hypophysis reveals that at least six well-differentiated cell types can be found in the anterior lobe. It is assumed that each cell type secretes a different hormone, since there are six different hormones associated with this lobe. Among the hormones secreted in the anterior lobe of the hypophysis are: the growth hormone, which controls bone growth; ACTH, which stimulates the growth of the adrenal glands and activates secretion of their hormones; thyrotropin, a hormone which stimulates the growth and secretion of the thyroid gland; and three sex hormones, which stimulate development of reproductive structures and regulate certain aspects of sexual behavior.

The posterior lobe of the hypophysis is largely concerned with the maintenance of water balance in the organism. The antidiuretic hormone is secreted there, in response to information received from proprioceptors regarding blood pressure and stretching of blood vessel walls. A second hormone secreted in this area is oxytocin, which stimulates contractions of the uterus. Oxytocin is released during childbirth and also during copulation, when the uterine contractions help transport the sperm up to the egg.

**Thyroid gland.** The thyroid is an endocrine gland located in the neck, just below the larynx. In man, the thyroid has fused into one single gland; in other vertebrates, there are two glands, connected by a thin bridge of tissue.

The functional unit of the thyroid gland is the **follicle.** It is a spherical structure formed by a layer of epithelial cells (usually squamous or cuboidal epithelium). The sphere is filled by a colloid, in which are dispersed molecules of thyroid hormones. When the follicle is stimulated by the thyrotropin released by the hypophysis, the thyroid hormones move out of the follicles and are secreted into the bloodstream.

The principal hormone secreted by the thyroid gland is *thyroxine*. This hormone affects a number of different target areas; its ultimate effect is to increase the rate of basal metabolism. This helps to regulate body temperature. If the ambient temperature is low, relatively large amounts of thyroxine will be secreted, causing a higher rate of metabolism and producing more heat. If the ambient temperature is high, very little thyroxine is released, so that the problems of maintaining a low body temperature are not increased by the production of additional body heat.

Thyroxine is a modified amino acid, into which iodine ions have been incorporated. The iodine needed to synthesize this molecule must be taken in as part of the nutritional requirements. A person who eats an iodine-deficient diet will not be able to produce sufficient thyroxine, and will become both mentally and physically sluggish. He will also develop a goiter, which is an enlargement of the thyroid gland. The gland receives constant feedback messages of thyroxine insufficiency, and it responds by adding more cells to produce more of the hormone. But without iodine, the new cells are no more able to produce thyroxine than the old ones were, and so the feedback continues to send the message "make more thyroxine."

**Parathyroid glands.** The four parathyroid glands are found either attached to the surface of, or embedded in, the thyroid. At one time, it was assumed that the parathyroids had evolved as an outgrowth and specialization of the thyroid (hence the name), but closer examination has

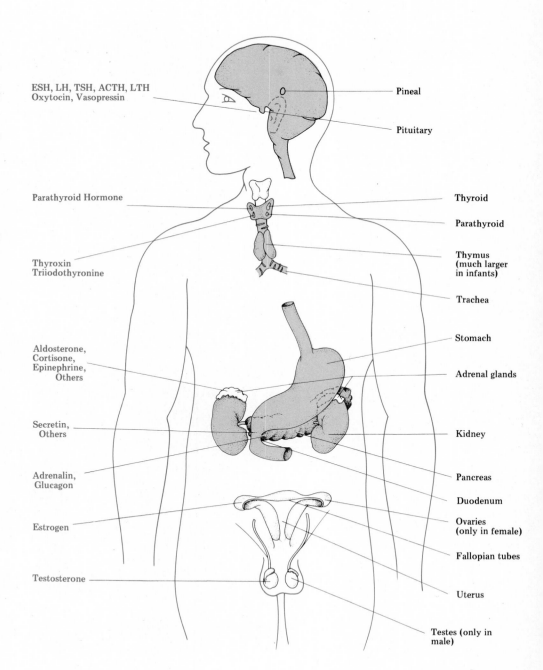

ESH, LH, TSH, ACTH, LTH
Oxytocin, Vasopressin

Pineal

Pituitary

Parathyroid Hormone

Thyroid

Parathyroid

Thyroxin
Triiodothyronine

Thymus
(much larger
in infants)

Trachea

Stomach

Aldosterone,
Cortisone,
Epinephrine,
Others

Adrenal glands

Secretin,
Others

Kidney

Adrenalin,
Glucagon

Pancreas

Duodenum

Ovaries
(only in female)

Estrogen

Fallopian tubes

Testosterone

Uterus

Testes (only in
male)

**11-13** *The location of endocrine glands in man, along with the hormones they secrete, are shown here. The exact secretory function of the pineal in humans has not been determined. In other vertebrates, the pineal regulates a number of photoperiodic cycles, and it is thought to function similarly in humans.*

disproved this theory. Under the microscope, the two glands look quite different.

The parathyroid functions to regulate the level of calcium ions in the blood. Since calcium ions serve as a trigger in muscular contraction, and also play a role in the propagation of nervous impulses, it is important that calcium ions be freely available to these nerve and muscle cells. When the concentration of calcium ions in the blood drops below a certain point, the decrease stimulates the activity of the parathyroids, which release a hormone (a modified polypeptide) that acts on a number of different target organs, raising the level of calcium in the blood. The parathyroid hormone inhibits the excretion of calcium ions, causing more calcium to be

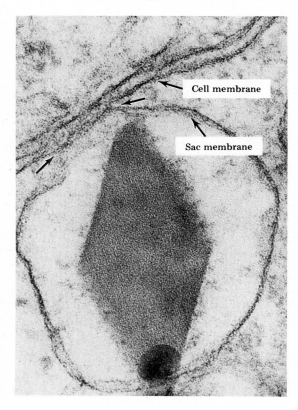

Cell membrane

Sac membrane

**11-14** *This photo shows a complex of secretory cells in the human parathyroid. When this micrograph is compared with the photo in Figure 11-12, of a thyroid follicle, it indicates the great organizational difference in these two closely associated endocrine glands. (Electron micrograph provided by Professor Lawrence Herman, Department of Pathology, State University of New York, Downstate Medical Center, Brooklyn, New York)*

transported out of the digestive tract and out of the kidney tubules into the bloodstream. The parathyroid hormone also causes calcium deposited in bones to be freed and released into the bloodstream.

Recently it has been discovered that the parathyroid hormone is antagonistically paired with another hormone. This hormone is released by the thyroid and serves to decrease the level of calcium in the blood, by causing it to be excreted at a faster rate, and also causing it to be deposited in the bones. Together, this pair of hormones can set upper and lower limits on the level of calcium in the blood.

**Thymus gland.** The thymus gland is found in the upper chest, below the thyroid gland. The thymus is usually embedded in a coating of fat cells and tucked away between the aorta and the lungs, behind the breastbone; its location has made it difficult to observe, even with the help of X-rays.

In children, the thymus gland is generally quite large, but its size gradually decreases, and by maturity it appears atrophied. Surgical removal of the gland in adults seems to have no harmful effects at all. This evidence suggests that the thymus is active only in childhood, and therefore that it probably plays some role in development rather than in regulation.

In the last decade, several investigators have followed the line of research suggested by this hypothesis, with the result that they can now give a fairly detailed account of the mechanism of action of this gland. The chief role of the thymus is establishing the body's immunity to disease. When the thymus is experimentally removed from your laboratory animals, such as mice, they generally survive only a few weeks before their inability to combat foreign bacteria and viruses leads to their death.

The action of the thymus gland can be divided into two stages. At birth, the gland begins to release **lymphocytes,** the disease-fighting cells

**349**

that manufacture antibodies. These lymphocytes are stored, densely packed, in the outer layer of the gland. They can easily be released into the bloodstream and carried throughout the body, concentrating in the other lymphoid organs, the spleen and the lymph nodes. These lymphocytes serve as a temporary line of defense for the newborn animal.

The second stage of thymus activity is the release of a hormone that stimulates both the production of lymphocytes in other organs and the production of antibodies in the lymphocytes. This hormone seems to act rather like the water used to prime a pump. The production of lymphocytes and the manufacture of antibodies cannot take place unless the hormone starts the process, but the continued presence of the hormone is not required.

**Islets of Langerhans.** Scattered among the cells of the pancreas are many small groups of cells, called islets of Langerhans, that secrete hormones. Two different cell types can be seen in the islets. One type secretes the hormone insulin, while the other type secretes its antagonist, glucagon.

The secretion of these two hormones seems to occur in direct response to the level of glucose in the blood circulating through the pancreas. As far as is known, there is no nervous control over this mechanism of hormonal regulation.

**Adrenal glands.** The adrenal glands are always closely associated with the kidneys. In some vertebrates, the adrenals are scattered throughout the kidneys, but in birds and mammals, the two adrenals merely lie at the upper end of the kidneys.

Each adrenal is composed of two different glands that are mixed together in the one structure. In mammals, the outer portion of the adrenal is called the **adrenal cortex;** the inner portion is called the **adrenal medulla.** The two elements have different functions, different structures, and a different evolutionary history.

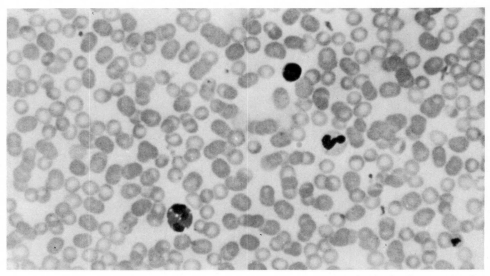

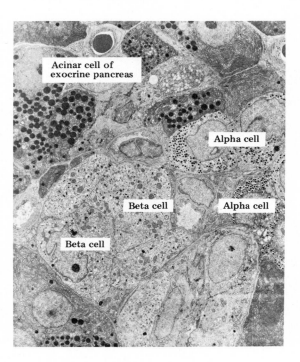

Acinar cell of exocrine pancreas

Alpha cell

Alpha cell

Beta cell

Beta cell

11-15 *At birth, the thymus gland begins to release lymphocytes, the disease-fighting cells that manufacture antibodies. These lymphocytes are stored, densely packed, in the outer layer of the gland. They can easily be released into the bloodstream and carried throughout the body, concentrating in the other lymphoid organs, the spleen and the lymph nodes. These lymphocytes serve as a temporary line of defense for the newborn animal. (Permission of Ward's Natural Science Establishment, Inc., Rochester, New York)*

11-16 *The pancreas is considered an endocrine gland because of the presence scattered through it at groups of cells called islets of Langerhans. This photo shows several types of pancreatic cells. (Electron micrograph provided by Professor Lawrence Herman, Department of Pathology, State University of New York, Downstate Medical Center, Brooklyn, New York)*

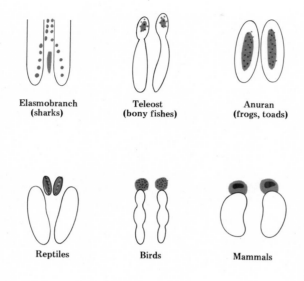

Elasmobranch
(sharks)

Teleost
(bony fishes)

Anuran
(frogs, toads)

Reptiles

Birds

Mammals

Gray = Adrenal cortex

Color = Adrenal medulla

**11-17** *Although the adrenal glands are always located near the kidneys, their exact position varies in the different vertebrate classes. In sharks, bony fishes, and anurans (frogs and toads) the adrenal glands are actually imbedded in the surface of the kidneys. In reptiles, the glands are completely separate; in birds and mammals the adrenals rest on top of the kidneys. There are also significant differences in the separation of the two types of secretory cells. In sharks, for example, they are wholly distinct, whereas in anurans and birds they are intermingled.*

The adrenal cortex apparently evolved from the epithelium covering the main body cavity, or coelom. The adrenal cortex secretes a number of different hormones, regulating a variety of processes. Among these are aldosterone, the hormone that is active in maintaining water and electrolyte balance; cortisol, the hormone that promotes repair and healing of injured tissue; and a hormone that stimulates the development of male sex characteristics. The activity of the adrenal cortex is under the direct control of hormones released by the hypophysis.

The other element of the adrenals, the adrenal medulla, is actually a specialized part of the sympathetic nervous system. Although the cells of the adrenal medulla lack axons and dendrites, they do show certain relationships to more typical nerve cells. They have evolved from neurosecretory cells, and their secretion is under the direct nervous control of fibers from the sympathetic system. The adrenal medulla secretes both adrenalin and noradrenalin in times of emergency fight-or-flight situations.

**Pineal gland.** The pineal is a small rounded lobe on the surface of the thalamus, and it is doubtless derived as an outgrowth of that part of the brain. There is still a great deal of uncertainty about the exact role the pineal plays in regulating metabolism and development. In many vertebrates, the pineal is known to secrete melatonin, a hormone that affects skin pigmentation; it is this hormone, in combination with a paired antagonist, that allows fish and lizards to change color and adapt to their environment's appearance. But it is suspected that the pineal may play an even more important role in regulating sexual development and long-range cycles. Experiments with a number of vertebrates, especially birds, indicate that the pineal may act as a light receptor. It directly absorbs light through the skull, and it also receives chemical or electrochemical messages from the retina when light is absorbed there. It is thought that the pineal may use this information to coordinate photoperiodic activities.

**Gonads.** The gonads—ovaries in females, testes in males—secrete a number of hormones that regulate sexual development and reproductive behavior. The activity of the gonads is controlled by hormones released by the hypophysis. The role of hormones in reproduction and sexual development will be discussed more fully in Chapter 15.

## Summary

Chemical control is the most widespread, and in plants and many small invertebrates, the only mechanism of integration at the organism level. Although slower than nervous control, chemical control is involved in the operation of regulatory cycles, such as circadian rhythms, menstrual cycles, and other cycles of perhaps several weeks' duration.

Hormones, the direct agents of chemical control, include a wide variety of compounds synthesized not only in the ductless, or endocrine, glands, but also—in the case of the recently discovered prostaglandins—in a wide variety of cells. After synthesis, the hormones are carried by bulk circulation to other parts of the organism, where they stimulate various kinds of activity. They are then broken down by enzymes, which prevents their accumulation in large quantity.

The mechanism of hormone action seems to have three aspects: (1) hormones may stimulate enzyme action, as does adrenalin; (2) hormones may affect the permeability of cell membranes; (3) hormones may affect the rate of protein or nucleotide synthesis within the body, possibly within the gene itself. In addition, it is possible that hormone action involves a "second messenger" molecule, cyclic AMP, which mediates the enzyme-producing effect of hormones.

A number of major plant hormones have been identified and studied. Auxin stimulates cell growth and differentiation; it has the further effect of contributing to seasonal cycles by blocking the action of other hormones. Gibberellins can cause flowering and normal growth in dwarf plants, stimulate development of pollen and initiates seed germination. Cytokinin probably functions by stimulating and perhaps even altering cellular RNA. It can cause buds to develop or prevent leaves from aging. It also seems to control early plant development, and frequently interacts with auxin as both synergist and antagonist. Abscissic acid acts as antagonist to auxin, gibberellin, and cytokinin.

Chemical control is also important in invertebrates. In Phylum Arthropoda, at least six specific hormones, including a juvenile hormone and a molting hormone have also been identified, as have certain exohormones in bee and ant colonies. The action of the molting hormones illustrates the general mechanism of hormone control in arthropods, and exhibits, as well, characteristics typical of vertebrate hormone control: first, the use of a series of hormones in increasingly complex and sensitive interaction; second, ultimate control over the hormone system in cells of the brain, an important evolutionary advance.

In vertebrates, hormone control is part of a complex system of integration dominated by the CNS. At least eight hormones are synthesized by the two-lobed hypophysis, the "master gland," which mediates all of the body's complex long-range cycles. Among the other endocrine glands, the thyroid functions in regulating basal metabolism; the parathyroids facilitate nerve and muscle function by regulating the blood calcium level; the thymus is active in early growth, where it establishes the body's lymphocyte system; and the Islets of Langerhans secrete the antagonistic hormones insulin and glucagon, which regulate the level of blood sugar. The adrenals are actually two distinct glands, the cortex and the medulla, in one structure. The cortex secretes a number of hormones, including aldosterone, cortisol, and a male sex hormone; the medulla, actually a part of the sympathetic nervous system, secretes adrenalin and noradrenalin. The gonads function in sexual development and reproduction.

## Bibliography

### References

Barrington, E. J. W. 1963. *An Introduction to General and Comparative Endocrinology*. Oxford University Press, New York.

Burnet, Sir M. 1962. "The Thymus Gland." *Scientific American*. 207 (5): 50–57. Offprint no. 138. Freeman, San Francisco.

Davidson, E. H. 1965. "Hormones and Genes." *Scientific American*. 212 (6): 36–45. Offprint no. 1013. Freeman, San Francisco.

Etkin, W. 1966. "How a Tadpole Becomes a Frog." *Scientific American*. 214 (5): 76–88. Offprint no. 1042. Freeman, San Francisco.

Frye, B. E. 1967. *Hormonal Control in Vertebrates*. Macmillan, New York.

Hamilton. T. H. 1971. "The Vertebrate Endocrine System and Its Regulation." In *Topics in the Study of Life*. Harper & Row, New York.

Pike, J. E. 1971. "Prostaglandins." *Scientific American*. 225 (5): 84–92. Offprint no. 1235. Freeman, San Francisco.

Robison, G. A., R. W. Butcher, and E. W. Sutherland. 1968. "Cyclic AMP." *Annual Review of Biochemistry*. 37: 149–174.

Telfer, W. and D. Kennedy. 1965. *The Biology of Organisms*. Wiley, New York.

Turner, C. D. 1966. *General Endocrinology*. 4th ed. Saunders, Philadelphia.

Whalen, R. E. 1967. *Hormones and Behavior*. Van Nostrand-Reinhold, New York.

Zuckerman, Sir S. 1957. "Hormones." *Scientific American*. 196 (3): 76–87. Offprint no. 1122. Freeman, San Francisco.

**Suggested for further reading**

Galston, A. W. 1964. *The Life of the Green Plant*. Prentice-Hall, Englewood Cliffs, N.J.
Chapters 4 and 5 explore the role of plant hormones in growth, differentiation and morphogenesis.

Galston, A. W. and P. J. Davies. 1970. *Control Mechanisms in Plant Development*. Prentice-Hall, Englewood Cliffs, N.J.
Similar to Galston's *The Life of the Green Plant*, but a broader and more detailed treatment of developmental processes.

Gordon, M. S. et al. 1968. *Animal Function: Principles and Adaptation*. Macmillan, New York.
Chapter 12 is a succinct but detailed introduction to invertebrate and vertebrate endocrinology; with tables and excellent illustrations.

Went, F. W. 1962. "Plant Growth and Plant Hormones." In W. H. Johnson and W. C. Steere, eds. *This Is Life*. Holt, Rinehart & Winston, New York.
A good introduction to the functioning of hormones in vascular plants.

**Books of related interest**

Asimov, I. 1963. *The Human Brain: Its Capacities and Functions*. Signet, New York.
Chapters 1 through 5 describe the endocrine system in man, with separate chapters on pancreatic, thyroid, adrenal and gonadal hormones.

Corner, G. W. 1963. *The Hormones in Human Reproduction*. Atheneum, New York.
A well-written general introduction to research in the endocrinology of reproduction. With many illustrations.

Greene, R. 1970. *Human Hormones*. McGraw-Hill, New York.
Detailed and profusely illustrated to the study of human endocrinology.

Riedman, S. R. 1962. *Our Hormones and How They Work*. Collier Books, New York.
Brief, popular introduction to human endocrinology.

# Reproduction

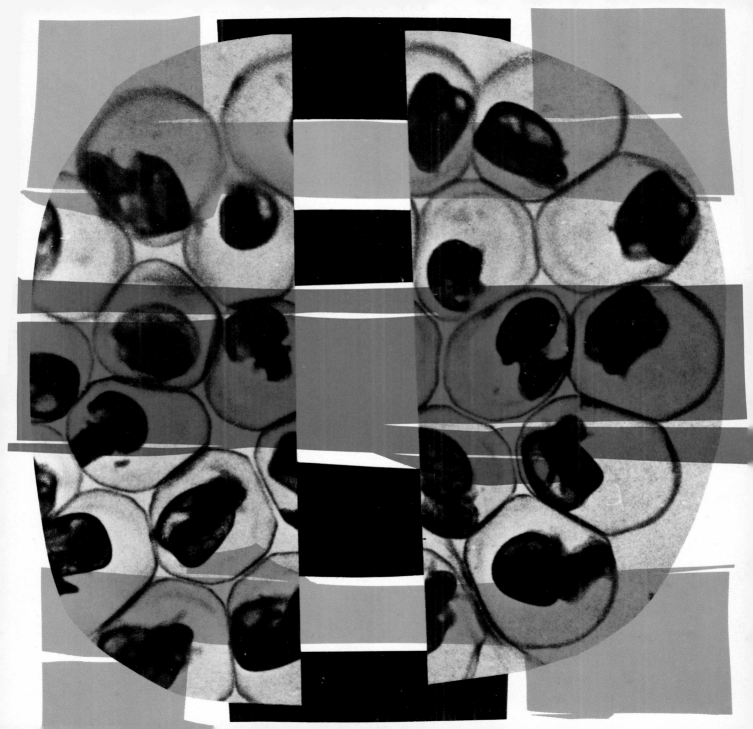

# 12

# Cellular reproduction

The origin of a creature that can carry on metabolic functions, such as respiration and digestion, seems somehow less miraculous than its perpetuation. Since we know that living systems are composed of the same elements as nonliving matter, we can readily accept the premise that a random combination of elements enabled some group of atoms and molecules to transduce energy and carry on a primitive type of metabolism. In fact, given the long stretches of time in which random combinations could take place, it would almost be surprising if no such lucky accident happened.

It is much harder to imagine the random occurrences that led to the ability to reproduce themselves that is found in all organisms. Even the simplest kind of cellular reproduction, such as the division of a one-celled organism into two identical cells, is a complex process. It requires the presence of specialized information-carrying molecules that are capable of self-replication; it involves a long series of reactions that must be precisely coordinated; it can take place only when a number of different environmental conditions are favorable. It seems reasonable to suppose that during the early stages of life on this planet there may have been many chemical systems capable of carrying on certain metabolic functions, but only after millennia of trial and error did a self-reproducing organism appear.

The trend of evolution has been in the direction of greater and greater elaboration of the mechanisms of reproduction. Biologists assume that the earliest living organisms reproduced by simple cell division, an asexual process involving only one parent. At some later date, two-parent or sexual reproduction came into being, bringing with it the interesting possibilities of sharing and recombination of inherited characters. A number of organisms, especially in the plant kingdom, utilize both asexual and sexual reproduction in alternating cycles.

Many organisms seem to live only to reproduce, and much of their behavior and activity is directed toward the end goal of successful reproduction. Some borderline organisms, such as the viruses that live on bacteria, have virtually abandoned all other life functions, becoming nothing more than reproducing systems. A bacteriophage such as the one shown in Figure 12-1 consists only of coiled strands of DNA, the molecules carrying genetic information, within a membranous sheath that acts as a hypodermic needle to inject its contents into the host bacterium. This streamlined virus is incapable of performing any other function beyond reproduction.

Even higher plants and animals devote a considerable portion of their time and energy to the function of reproduction. Elaborate physiological and behavioral adaptations help to produce eggs and sperm, to attract mates, to nourish and protect developing embryos, to safeguard the young. Vertebrates too can seemingly turn into breeding machines. A well-known example is the behavior of the salmon, which stays in the ocean only long enough to grow to physical and sexual maturity. Then these fish hurl themselves upstream in the rivers where they were born, to breed and then die.

In this chapter, we will look at reproduction on the cellular level. Subsequent chapters will focus on the molecular and then the organismic levels of the complex process of reproduction.

## Reproduction of cells

Cellular reproduction is often called cell division, a misleading term if there ever was one. Cell division sounds like a process that could be accomplished by chopping the cell down the center with a knife—presto, two cells instead of one! Nothing could be further from the truth. A cell cannot be divided into two parts and produce two new cells, any more than computer production can be doubled by ripping the wiring diagram

in half and giving one piece to each factory. The wiring diagram, and the cell, must be duplicated rather than divided.

Observation and experimentation have shown that a cell does not necessarily have to be duplicated in every last detail. Although some

**12-1** *Living organisms show a wide variety of reproductive processes. These copulating marine sea slugs (Glaucus atlanticus, an inhabitant of the upper layer of the ocean) reproduce sexually but are hermaphroditic. (William Stephens from Omikron)*

**12-2** *Making its way back from the ocean, the Atlantic salmon bucks the strongest currents and leaps waterfalls to spawn and die in the same mountain brook where it was born. In order to reproduce itself, after wandering the seas the salmon manages to find its way home through a complex network of rivers and streams. It is suggested, though not proven, that the salmon is guided by a highly developed sense of smell or taste. (Mitchell Campbell from National Audubon Society)*

kinds of cellular reproduction—fission, for example—often produce two new cells that are virtual mirror images, successful reproduction can also produce two cells that are morphologically dissimilar. Reproduction by budding, for instance, produces one large and one small yeast cell. Yet in time, the small cell will develop into a duplicate of the larger parent, because the small cell contains informational molecules that dictate such development.

These informational molecules are nucleic acids, called deoxyribonucleic acid, or DNA. As long as the new cell contains a complete set of all the DNA molecules found in the parent cell, it will be able to develop into a cell that is morphologically and physiologically like the parent. The DNA will dictate the formation of other cell

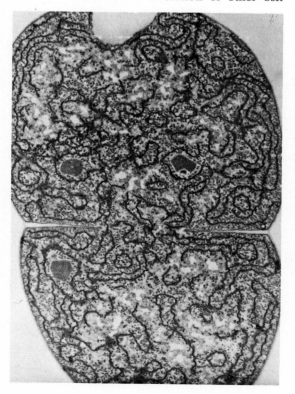

**12-3** *Both bacteria and blue-green algae are procaryotic cells, in which there is no distinct nucleus. As this photo of* Anabaena *shows, the chromatin is confined within a central nuclear area, but it is not bounded by a membrane. (Omikron)*

organelles, such as vacuoles or ribosomes; the addition of a certain amount of cytoplasm to increase the size of the cell; and perhaps the production of a cell wall or a protective mucous coating. It may also dictate the eventual reproduction of yet another new cell.

In recent years, molecular biologists have established the shape of the DNA molecule, and they have discovered the way in which it is able to replicate itself, producing another molecule with exactly the same properties. This aspect of cellular reproduction will be discussed in more detail in Chapter 14.

In certain procaryotic organisms, or procaryotic cells of multicellular organisms, DNA, in the form of thin strands of chromatin (dense, easily stained threads embedded in the cell cytoplasm), is loosely packed in a more or less central location called the nuclear area. In eucaryotic organisms, DNA is arranged in fixed patterns, forming double-stranded structures called **chromosomes.** Each of the strands is called a **chromatid.** The chromosomes are contained within the nucleus and bounded by the nuclear membrane. It is this duplication of the nucleus, or **karyokinesis,** that is the crucial part of cellular reproduction. The division of the cell cytoplasm and separation of the two nuclei by a cell membrane, called **cytokinesis,** may occur at the same time, or it may take place subsequently.

Duplication of nuclear material was first observed in plant cells during the 1870s by Eduard Strasburger (the same botanist who used a 70-foot oak tree to demonstrate the movement of water through the xylem). Later, Walther Flemming observed much the same sequence of events in animal cells. Chromosomes are large enough to show up under the light microscope as fibers within the nucleus. Once coal-based, or aniline, dyes became available to stain these structures so that they would stand out from the rest of the cytoplasm, observation of the chromosomes (the term means "colored bodies," a name given them because they stain so readily) was an easy matter.

**361**

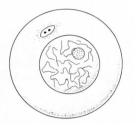

**A Interphase**

**B Early prophase**

**C Middle prophase**

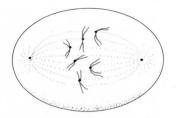

**D Late prophase**

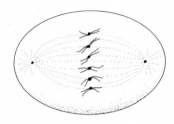

**E Metaphase**

In 1882, Flemming published drawings that showed the sequence of stages in the duplication of chromosomes. The use of the electron microscope has added a number of details to our knowledge of karyokinesis, but the basic outline proposed by Flemming remains unchanged.

The process that Flemming observed and described produced an exact replication of the nucleus of the original cell. If the parent cell had 46 chromosomes, as human cells do, the new cell also had that number. This exact reproduction of the original cell is called **mitosis,** a term coined by Flemming from the Greek word for thread. All asexual reproduction, whether by fission, regeneration, or budding, takes place through mitosis. Mitosis is also the process by which a young multicellular organism, such as an embryo or a baby, grows, and by which mature organisms replace damaged or elderly cells.

The process of sexual reproduction involves the fusion of two cells. If each of these cells were an exact replica of its parents, what would happen? If the human egg had 46 chromosomes and the sperm had 46 as well, then the fertilized egg, or **zygote,** would have 92. When that generation of individuals grew to maturity and mated, their offspring would have twice that number of chromosomes. Within a few generations, there would be a ridiculously large number of chromosomes. Clearly this multiplication of chromosomes does not actually occur. With only rare exceptions, each cell of every new human being has exactly 46 chromosomes.

*12-4 Mitosis in an animal cell begins with interphase (A), when the nucleolus is visible and the chromosomes are dispersed throughout the nucleus. In early prophase (B), the nucleolus begins to disintegrate, the centrioles move apart, and the chromosomes appear as long thin threads. The spindle begins to form at middle prophase (C), and by late prophase (D), the centrioles are nearly at opposite edges of the cell, linked by the nearly complete spindle. The nuclear membrane disappears, and the double chromosomes move toward the equator of the spindle. In metaphase (E), the double-stranded chromosomes become attached by their spindle fibers to the spindle fibrils and become aligned at the equator. In early anaphase (F), the centromeres divide and begin to move toward opposite poles of the spindle. In late anaphase (G), cytokinesis begins as the two new sets of single-stranded chromosomes move close to the opposite spindle poles. Cytokinesis becomes nearly complete in telophase (H); new nuclear membranes appear, the chromosomes becomes less distinct as they become longer and thinner, the centrioles are replicated, and the nucleoli reappear. When cytokinesis is complete (I), the new cells are in interphase.*

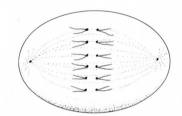

**F Early anaphase**

**G Late anaphase**

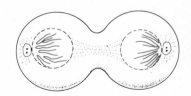

**H Telophase**

**I Interphase**

Early investigators hypothesized that egg and sperm production involves a different process of nuclear division, one in which each new cell contains only half the number of chromosomes in the parent cell. In 1885, this hypothesis was confirmed by the Belgian embryologist Eduard von Beneden. This reducing type of karyokinesis is called **meiosis,** from the Greek word for reduction. Meiosis produces haploid cells, with only one of each chromosome. When the gametes fuse, they produce a diploid cell, the zygote, in which the chromosomes are paired, two of each type being present.

## Mitosis

Cellular reproduction by mitosis is an extended sequence of activities. Evidence of some of these activities can be seen very clearly under a simple light microscope; others have not yet been described even with the aid of the electron microscope.

*12-5 During interphase, the nondividing cell carries out the numerous activities essential to any living, functioning cell, including growth, differentiation, respiration, and protein synthesis. Three distinct periods are distinguished during interphase: post-mitotic growth, synthesis (when DNA is replicated), and a short pre-mitotic period. During interphase, the nucleus and one or more nucleoli are clearly visible. However, chromosomes cannot yet be seen (at least with present techniques) as distinct structures. They appear only as an irregular granular mass of chromatin.*

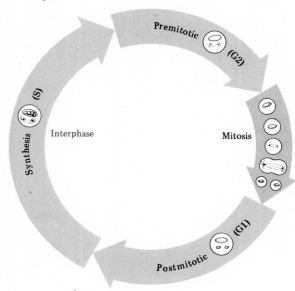

The process of cellular reproduction actually begins shortly after the formation of a new cell. At some stage, the DNA replicates itself. This replication cannot be observed, although evidence that it occurs has been established. If the DNA replication does not take place within 24 to 48 hours after the "birth" of a new cell, it will never be able to reproduce itself. This is the case with many specialized cells of a multicellular organism, such as neurons or red blood cells.

Some time before the process of mitosis begins, each chromosome produces a duplicate of itself; the new chromosome contains a full set of the DNA molecules earlier duplicated. The new chromosome lies very close to the old one, so close that it is usually impossible to tell that the structure is doubled. The process of chromosome duplication has not been accurately observed, and so it is not known exactly how long a period elapses between the doubling of the chromosomes and the observable start of mitosis.

The division of nuclear material into two distinct nuclei takes place in a definite sequence, although not all species go through each one of the steps. The entire process takes place within about an hour in most cells; it appears, however, that the process can be speeded up or slowed down in response to environmental factors such as temperature and the action of certain chemicals. For convenience of description, biologists have divided the process into a number of stages (these, too, were named by Flemming), each of which represents a distinct change in the appearance of the nuclear material.

## Interphase

Before cell division takes place, or after it has been completed, cells are said to be in interphase. Interphase was once called a resting state, but that is misleading terminology. A cell in interphase is likely to be extremely active in carrying on metabolic and growth functions. Moreover,

**363**

it has been shown that the process of duplication that produces a new double-stranded chromosome from the single chromatid also takes place during interphase.

Biologists divide interphase into three distinct periods. Immediately after cell division, there is a first growth (G1) period. The length of the G1 period may vary considerably, and this variation is the cause of significant differences in interphase time among specific cell types. A synthesis (S) period follows G1, during which the synthesis of DNA takes place. The S period in vertebrate cells takes about six to eight hours; a similar S period has been observed in the cells of plant roots. A much shorter S period is found in certain tumor cells, in the cells of a pollen grain, and in many members of the Kingdom Monera. A second growth (G2) period follows synthesis. G2 is usually quite brief, and it does not vary much from cell to cell.

## Prophase

When prophase begins, the nuclear material is scattered throughout the entire nucleus in what appears to be completely unorganized fashion. At one time, it was thought that the chromosomes existed only during cellular reproduction and were absent at other times, but new visual and biochemical evidence demonstrated this assumption to be erroneous. It can be shown that the DNA content of the cell is still present in interphase, and in fact increases during the S period. And electron microscope studies reveal slender structures that are probably the loosely coiled strands of DNA. The chromosomes are still within the nucleus during interphase, but they have become mere elongated threads, too thin to be visible in the light microscope.

The first sign of the onset of prophase is the appearance of the chromosomes in the form of visible threads. These fine threads look as if they are contracting, for they gradually become shorter

and thicker throughout prophase. This contraction probably serves to prevent the tangling and breaking of the chromosome threads during division. The division of the genetic material must be as accident-free as possible, or the daughter cells will differ from the parent cell; if this condition were the rule, there would be no perpetuation of the species.

As the chromosomes condense and become easier to see, the nucleolus of the cell undergoes a change in the opposite direction. Like an aspirin dropped in a glass of water, its outlines slowly become less and less distinct. By the end of prophase, the nucleolus is no longer visible. This seems to be part of the general breakdown of the nucleus that must occur in order for the nucleus to divide in two. Some biologists believe that proteins contained in the nucleolus coat the chromosomes during mitosis, acting as a kind of barrier between the DNA and the cell cytoplasm. Others suggest that the nucleolus is a packet of genetic information regarding the synthesis of RNA, and that it is itself undergoing some process of replication and division. The nucleolus will reappear near the end of the division.

In animal cells, another change can be seen in the cell at the onset of prophase. The centrioles, paired structures lying in the cytoplasm just outside the nucleus which have divided before prophase begins, begin to move farther apart, apparently repelling each other like the two positive poles of a magnet. As the centrioles move

**12-6** *In this slide prepared from the meristematic tissue at the growing tip of an onion root, the individual chromosomes are already quite distinct by late prophase. It may take 30 to 60 minutes for mitosis to advance to this stage. (Permission of Ward's Natural Science Establishment, Inc., Rochester, New York)*

*12-7 By metaphase, the chromosomes are arranged in a more or less orderly sequence on the equatorial or metaphase plate. The spindle fibers are faintly visible in this photo. (Permission of Ward's Natural Science Establishment, Inc., Rochester, New York)*

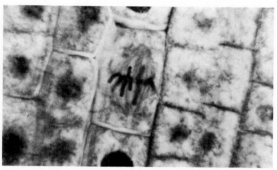

through the cytoplasm, they appear to spin connecting threads. These threads are proteinaceous structures; some that have been examined under the electron microscope have an appearance very similar to that of microtubules.

Many cytologists believe that the centrioles merely organize existing microtubules into a network. A more ingenious model of centriole activity has recently been proposed, in which the structure behaves like a rotary engine. The triplet strands of the centriole, according to this model, are hinged to spokelike microfibrils, enabling the centriole to whirl around like a pinwheel. As it whirls through the cell, it releases RNA molecules from the surface of the triplet strands, and the RNA, through a self-assembly process, becomes a network of microtubules. This model does not contradict any known data, but it must for the present be classed as a speculation, since there is as yet no direct evidence to support it.

The microtubules radiate throughout the cell, forming a network called a **spindle.** Shorter fibers connected to the centriole, or **asters,** also appear. Asters are not found in plant cells, so there is some question about the function of these structures.

Plant cells do not contain centrioles, but during the later stages of prophase they nevertheless form spindles. The mechanism by which this occurs is not known. However, it has been demonstrated that the release of RNA from any source into the cytoplasm will result in the

formation of microtubules, as long as certain proteins that form the walls are available in the cytoplasm. So, in theory, microtubules can be formed by any RNA-containing organelle.

By the end of prophase, very dramatic changes have taken place in the cell. The chromosomes are short and thick. Their doubled structure, consisting of two chromatids, is now quite evident. The two chromatids are joined together at one point by a kind of specialized structure called a **centromere.** The centromere may be centrally located on the chromosome, or it may be near one end, giving the chromosome two "arms" of uneven length. The nucleolus is invisible, and the nuclear membrane is disintegrating. A network of spindles crisscrosses the center of the cell. If centrioles are present, they are located at opposite ends of the nuclear area.

## Metaphase

The metaphase is relatively brief in most types of cells. During this phase, the chromosomes appear to be aligned on the spindle fibers. In slides viewed in the light microscope, the chromosomes, with the two chromatids still tightly attached to one another, can be seen more or less suspended in the center of the spindle, in a region called the **metaphase** or **equatorial plate.**

Recent work with the electron microscope permits a more detailed explanation of the events of metaphase. It is not actually the chromosome that is attached to the spindle fibers, but the centromere, which acts as a kind of handle by which the entire chromosome may be moved. Figure 12-8 shows the structure of the centromere and its relationship to the rest of the chromosome. The centromere (also sometimes called the kinetochore) is really two dense clusters of filaments; one cluster develops on each side of the chromosome, and the filaments appear to grow toward each other, perhaps meeting in the center of the chromosome. In some species, the centro-

Centromere
(primary
constriction)

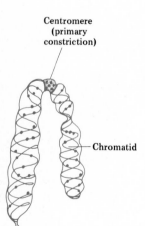

Chromatid

*12-8 By the end of prophase, the chromosome appears as a short, thick rodlike structure with constrictions at the centromere and elsewhere. The centromere serves to unite two identical chromatids and as a point of attachment to the spindle fibers during metaphase.*

**365**

mere is doubled on each side, providing four separate points of attachment.

Most of the fibers that make up the spindle are clearly attached to the centromere, but at all stages of mitosis it can be observed that some fibers go straight through the chromosome. This evidence seems to support a theory held by many biologists, that the chromosome is not really a solid structure. Its clearly defined shape is the result of the coiling of the long thin fibers (probably strands of DNA) seen before the onset of mitosis. If the chromosome is a tightly coiled spiral, there would, of course, be some spaces through which the fibers could penetrate.

By the end of metaphase, the spindle is a tight network; researchers have found that even persistent micromanipulation cannot push the spindle apart or move it to another area of the cell. The chromosomes are aligned in the center of the spindle, and cablelike bundles of fibers are attached to each centromere. The nuclear membrane is invisible; the nucleus appears to be continuous with the cytoplasm.

## Anaphase

The major event of anaphase is the separation of the two chromatids of each chromosome. At the beginning of anaphase, the chromosomes are still aligned in the center of the spindle, but the cleavage between the two chromatids is increasingly evident. By the end of anaphase, the once-paired chromatids have moved to opposite ends of the spindle. In order for this separation to take place, the doubled centromere, which holds the two chromatids together, must first divide.

In most cells, anaphase is the briefest stage of mitosis (see the comparison of time lapse in Table 12-1); in a typical multicellular organism, anaphase takes less than five minutes. A great deal of research has been carried out to determine the mechanism which makes such rapid movement of the chromatids possible. The evi-

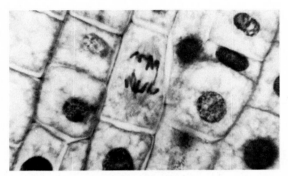

dence does not offer any clear-cut answers to this question, but several theories have been advanced in light of the experimental data obtained.

**Chromosome movement.** The theory that is most widely accepted at the present time is that the fibers of the spindle contract, pulling the attached chromosomes back toward the two poles of the spindle. Visual support for this theory comes from the appearance of the chromosomes during the later stages of anaphase; they do indeed look as if they were being dragged backward by the "handle" of the centromere. Another piece of supportive evidence is the fact that isolated spindle fibers do contract when they are placed in a medium containing ATP; this suggests that the mechanism of chromosome movement is microtubule contraction.

The second theory of chromosome movement, which has some very persuasive adherents, suggests that the chromosomes are pushed rather than pulled apart. According to this theory, a colloidal component of the cell's cytoplasm begins during anaphase to take in water through the process of imbibition. This imbibition may be triggered by the release of some chemical from the chromosomes at late metaphase. The colloid swells, pushing the two chromatids of the chromosome apart. The spindle fibers function as tracks that guide the moving chromatids to the two spindle poles, preventing their dispersal throughout the entire cell. Support for this theory

**12-9** *During anaphase, the chromosomes begin to move along the spindle fibers toward opposite poles of the cell. In most cell types, anaphase is the briefest of the mitotic periods. (Permission of Ward's Natural Science Establishment, Inc., Rochester, New York)*

**12-10** *Two alternative models are offered to explain the migration of the chromosomes along the spindle fibers. In the first, the spindle fiber shortens by contraction. In the second, it shortens by disintegrating at the end closest to the centromere. Pressure within the cytoplasm may help force the chromosome toward the aster.*

A.

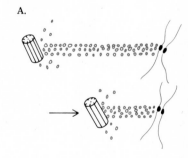

B.

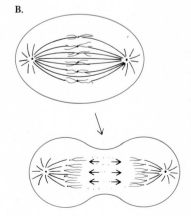

## 12-1
## Frequency of mitosis

| Type of cell | Duration of interphase | Duration of prophase | Duration of metaphase | Duration of anaphase | Duration of telophase |
|---|---|---|---|---|---|
| Onion root | 17 hr. | 88 min. | 1.4 min. | 3 min. | 4.6 min. |
| Human liver cells | 19 hr. | 30–60 min. | 2–6 min. | 3–15 min. | 30–60 min. |
| Pea root (15° C.) | 22.6 hr. | 126 min. | 24 min. | 5 min. | 22 min. |
| Pea root (25° C.) | 14.5 hr. | 54 min. | 14 min. | 3 min. | 11 min. |

lies in the fact that colloids are known to swell through imbibition, as can be seen when a spoonful of gelatin is soaked in water. The bending of the chromosomes into a V shape could be expected to happen if they were being pushed apart at the center, just as it would if they were being pulled. Another support for this theory is a bit of negative evidence. It can be shown that the spindle fibers do not thicken during anaphase, although if they contract by the sliding filament model that characterizes other types of movement, such a thickening would certainly be expected. Supporters of the pushing theory say

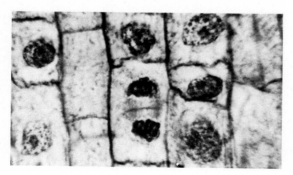

**12-11** By late telophase, the cell wall is beginning to form between the two new daughter cells of this onion root tip. In the succeeding interphase, both growth and synthesis of new DNA will take place. (Permission of Ward's Natural Science Establishment, Inc., Rochester, New York)

this evidence indicates no contraction takes place in the spindle fibers; supporters of the pulling theory say it merely indicates some method of contraction other than sliding filaments is utilized.

At the moment, no conclusive evidence is available to settle this question, but the balance seems to be slowly tipping in favor of the pulling rather than the pushing theory. For example, electron micrographs published by B. R. Brinkley and Elton Stubblefield, in 1970, show that during anaphase, the centromeres are stretched in the direction of movement. Such stretching would occur only if the chromatids were being pulled, not if they were being pushed.

## Telophase

The last phase of mitotic cellular reproduction is telophase. As soon as the chromatids are well separated, they begin to uncoil, appearing longer, thinner, and less distinct. The nucleolus reappears, formed at a specific point on one chromosome; this locus is called the nucleolus organizer.

A nuclear membrane begins to form around each group of single-stranded chromosomes. The spindle fibers fade from view. In animal cells, it is generally at this point that the centrioles replicate, making two pairs in each cell.

Karyokinesis, or nuclear division, is completed when the nuclear membranes are completely re-formed and the chromosomes are thin jumbled strands. In certain instances, the cell may remain multinucleate for a time, postponing the division of the cytoplasm and the formation of a second independent cell. In most cases, however, cytokinesis takes place during the last stages of telophase. In animal cells, cytokinesis begins by a pinching in of the outer cell membrane. This takes place in the same location and along the same plane as the metaphase or equatorial plate. An indistinct membranous division between the two cells can also be observed at this site; it is called the **cleavage furrow.** In plant cells, there is no pinching in at the line of division. Instead, a structure called a **cell plate** becomes visible. It may be formed by remaining spindle fibers; some botanists believe that it is secreted by the endoplasmic reticulum; a third possibility is that it is produced by the coalescence of scattered membranous vesicles. At first, this structure is very delicate, and under the microscope appears to be only a fine line. It becomes more evident through telophase and finally forms a complete division between the cells. When the process of division is complete, the cell plate becomes the middle lamella, the connecting layer between the primary walls of the new cells.

## Control of mitosis

The length of time the cell stays in interphase varies considerably from species to species and from cell type to cell type. The average length of a bacteria generation is 20 minutes; this means that only a small portion of the cell's life cycle is spent in interphase. The meristematic root cells of a mature pea plant spend about 22 hours in interphase, and the slow-growing fibers of the potato plant spend nearly a week in interphase. The cells in human bone marrow, which can produce 10 million red blood cells per second, obviously spend very little time in interphase, whereas other body cells, such as those of striated muscle, may divide only once every two or three days. The rate of mitotic cellular reproduction is very high in a recently fertilized zygote, but it is low in an organism reaching the end of its life span. Cells in the bone marrow, kidney tubules, and stomach epithelium carry on relatively rapid mitosis throughout the organism's entire life; cells in the central nervous system of higher vertebrates stop reproducing by the second or third month after birth.

An understanding of the mechanism that stimulates mitosis and controls its rate would have many valuable medical applications, and therefore extensive research has been carried out in this field. Biologists know that the surface-to-volume ratio is a critical factor in cell division. If, through cell growth, the ratio decreases, so that there is less surface per unit volume than before, the cell will be stimulated to reproduce, for the two new cells together will have the same volume as one single parent cell, and of course they will have much more surface. Yet knowing that changes in this ratio act as a stimulus does not explain the mechanism through which stimulation takes place.

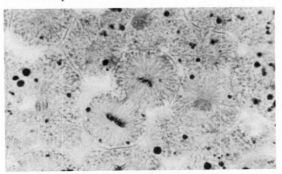

12-12 *The cleavage furrow of this dividing animal cell is clearly visible as the process of pinching in begins to take place during late telophase. Note that, in the surrounding cells, the individual chromosomes are no longer distinct. (Permission of Ward's Natural Science Establishment, Inc., Rochester, New York)*

**12-13** *Extracellular influences on the processes of cell division can be studied by culturing simple organisms, or tissues of multicellular plants and animals, and then introducing various chemical agents into the culture. Shown here is a culture of Aspergillus niger, or black bread mold. (Omikron)*

It can be demonstrated that in plants the so-called "wound hormone" serves to stimulate mitosis. This was demonstrated in a series of experiments by Haberlandt. He worked with potatoes, cutting them in half and timing the rate at which mitosis took place. He found that if he carefully cleaned away the damaged cells along the cut, there was no change in the rate of mitosis, but if the damaged cells were permitted to remain, mitosis occurred more often and more rapidly along the cut edge. Haberlandt was able to produce the same effect of speeding up mitosis by applying a paste of cut-up cells to the surface of undamaged tissue. Apparently, the damaged cells release the wound hormone, and it stimulates mitosis in other cells.

Hormones that stimulate mitosis have also been identified in animals. A hormone called the growth hormone, secreted by the anterior lobe of the hypophysis, stimulates mitosis in tissue cultures and experimental animals. Presumably it is by speeding up mitosis that it produces its effect of increasing body growth. Other hormones secreted by the hypophysis are known to stimu-

  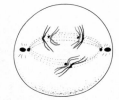 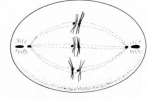 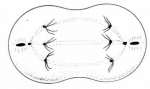 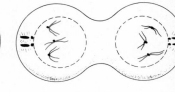

A  Early prophase I　　　　B  Middle prophase I　　　　C  Late prophase I　　　　D  Metaphase I　　　　E  Anaphase I　　　　F  Telophase I

late the growth of the thyroid gland, the adrenal cortex, and the gonads, so these hormones also are presumed to have some ability to stimulate mitosis.

The widespread use of tissue cultures in biological research has revealed that mitotic reproduction, like many other functions of the organism, is probably controlled by a pair of antagonists—one chemical that stimulates mitosis and another that represses it. In almost every case, tissues that are removed from the body and maintained in a nutrient medium grow at a faster rate than they do within the body; it is this peculiarity that makes tissue cultures so useful a method of studying cells. The fact that so many cells grow faster outside of the body than in it leads to the hypothesis that the body secretes some growth inhibitor. Such an inhibitor would obviously be necessary to prevent multicellular organisms from growing beyond their optimal size, so that we do not see 400-pound canaries struggling to get off the ground. This inhibitor is probably absent during the early stages of the zygote's development but is secreted in increasing quantities as the organism reaches maturity. At a certain size level, predetermined by the genetic information inherited by the organism, the effect of the inhibitor prevents net growth.

There must be some antagonist to the inhibitor that works to stimulate mitosis under certain conditions. The wound hormone is one example of such an antagonist. Since animals are also capable of healing wounds, there must be some analogous chemical substance released by

animal tissues. Histamines are a likely candidate. Medical researchers suspect that cancer cells also secrete some chemical that overcomes the effect of the inhibitor, enabling those cells to continue reproducing. Some cell physiologists believe that the accumulation of certain proteins produced in the cell acts to provoke inhibition of mitosis, and therefore the absence of those proteins will overcome the inhibitory effect.

## Meiosis

Meiotic cellular reproduction is found only in organisms that reproduce sexually. The process of meiosis serves to reduce the number of chromosomes in each newly formed cell. Rather than having the diploid (2n) condition, with a pair of each type of chromosome, meiotically reproduced cells have only one of each chromosome type. They are haploid (n). The value of meiosis is that it keeps the chromosome number constant throughout all members of the species.

Meiosis actually consists of two different cell divisions. The first division, consisting of prophase, metaphase, anaphase, and a brief telophase, sorts out the pairs of chromosomes; this is the essential meiotic process. The second division separates each of the pairs by the same steps found in mitosis—a prophase that is essentially a continuation of the telophase of the previous division, metaphase, anaphase, and a concluding telophase.

G  Interkinesis

H  Prophase II

I  Metaphase II

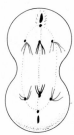 

J  Anaphase II

**12-14** *The stages of meiosis resemble those of mitosis. In early prophase I (A), the replicated chromosomes become visible as long, well-separated strands. In middle prophase I (B), homologous chromosomes pair (synapse) and become shorter and thicker. The double-stranded nature of the chromosomes becomes visible in late prophase I (C), as the nuclear membrane starts to disintegrate. In metaphase I (D), each synaptic pair becomes attached as a unit to a single fibril of the spindle, and the centromeres become aligned at the equator. The double-stranded chromosomes migrate to opposite poles of the spindle in anaphase I (E). In telophase I (F), new haploid nuclei are formed, and the chromosomes, which remain double-stranded, become indistinct. Interkinesis (G) resembles interphase, except that the genetic material is not replicated. The later phases of meiosis (H-L) resemble mitosis, but result in the formation of four haploid germ cells rather than the two diploid cells formed by mitosis.*

Meiosis typically results in the production of four new cells, rather than two as in mitosis. Mitosis involves the doubling of chromosomes followed by one division; diploid cells (2n) produce new diploid cells through the sequence:

$$2n \xrightarrow{\text{interphase}} 4n \xrightarrow{\text{mitosis}} 2n.$$

Meiosis involves the doubling of chromosomes followed by two successive divisions, producing haploid (n) individuals. The sequence is:

$$2n \xrightarrow{\text{interphase}} 4n \xrightarrow[\text{division}]{\text{1st meiotic}} 2n \xrightarrow[\text{division}]{\text{2nd meiotic}} n.$$

### First prophase

The early stages of the first prophase of meiosis are much like those of mitosis. The double-stranded chromosomes begin to shorten and thicken, and as they become denser, they also become easier to see with the microscope. The centrioles begin to migrate to opposite poles of the cell, and the network of spindle fibers once more appears. Although visually it appears that these fibers are created during cell reproduction, since they cannot be seen at any other time, most biologists believe that, like the chromosomes, they are present during interphase. At that stage, they are the same density, and have many of the same staining properties, as the rest of the cell cytoplasm, so that they are not visible. It is quite possible that these fibers, too, contract and thicken during prophase, making them easier to see and to stain during the preparation of cells for microscopic study.

During meiosis an important process called

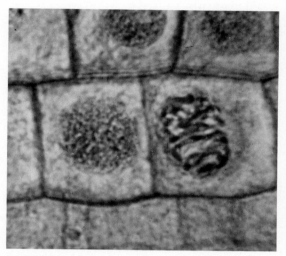

**12-15** *During the early stages of prophase, adjacent chromosomes often become tangled up as they sort out into homologous pairs. It is quite possible for this process of synapsis to result in exchanges of parts between the various chromatids within a bivalent. (Courtesy of Ripon Microslides)*

**synapsis** takes place in the middle stages of the first prophase. In synapsis, chromosomes of the same type come together and lie closely alongside each other. Each pair of similar, or **homologous,** double-stranded chromosomes then acts as a single unit throughout the rest of prophase and metaphase. This unit is called a **bivalent,** or **tetrad,** because it has four complete sets of DNA molecules on different chromatids. The two pairs of chromosomes are not physically joined together, since each has its own centromere. The arms of the homologous chromosomes may intertwine.

In the last stages of prophase, the bivalents can be seen moving to the center of the cell. As in mitosis, the nuclear membrane is fast disappearing, and subsequent events of cell reproduction take place in the cell cytoplasm.

K Telophase II

L Interphase

## First metaphase

During the first metaphase of meiosis, the bivalents line up along a single plane in the center of the cell. As the bivalents line up along the metaphase or equatorial plate, they are attached by their centromeres to the structure of the spindle. Each bivalent is attached by a single fiber to the spindle pole; the centromeres of both chromosomes in the bivalent are attached to the same fiber. The fact that the centromeres are attached by the same spindle fiber does not necessarily mean that the two centromeres are lying side by side. In many cells, it can be seen that the centromeres of homologous chromosomes appear to repel one another. The arms of the chromosomes lie closely together along the metaphase plate, but the centromeres seem to be pointing toward the two poles of the spindle. This alignment insures that the homologous chromosomes making up the bivalent will separate and move toward opposite poles in the next stage of meiosis.

## First anaphase

In the first anaphase, the bivalents move apart, with one chromosome of each homologous pair moving toward each pole of the spindle. At this stage, it is clearly evident that each chromosome is itself double-stranded, consisting of two chromatids. The chromatids, however, do not separate but remain joined at the centromere. Separation occurs only during the second division.

The mechanism of movement during anaphase is probably the same in meiosis as it is in mitosis. The chromosomes assume the same V shape, and the centromere appears to be the initial point of movement.

## First telophase

By the end of the first telophase, the cell is visibly divided into two parts. Each half of the cell contains one of each type of chromosome. Each chromosome, however, is still composed of two sister chromatids. The two daughter cells formed at this stage of meiosis have exactly the same number of chromatids as do the two daughter cells formed by a mitotic cellular reproduction. For example, in humans, in which $n = 23$, each daughter cell contains 46 chromatids. Yet there is a significant difference in the daughter cells formed in the first meiotic division and those formed in mitotic division. In mitosis, the 46 chromatids are scattered throughout the nuclear area, each one independent of all the others. In meiosis, the 46 chromatids occur in paired sets, as the two chromatids of 23 different chromosomes; in meiosis, there has been a separation of homologous chromosomes. This organization makes another immediate division possible.

Except for this important difference in the way the chromatids are arranged, the first telophase in meiosis is much like the telophase of mitosis. In animal cells, a cleavage furrow begins to appear along the metaphase plate. In plant cells, the region of the metaphase plate becomes the division plate. The chromosomes begin to uncoil and therefore become more difficult to see. The nuclear membrane starts to reappear. The centriole at each pole of the spindle replicates itself. The spindle fibers seem to fade. In most cells, some cytokinesis takes place at this stage.

## Interkinesis

In meiosis, there is a brief period of reproductive inactivity following telophase of the first division. In some respects, this stage is analogous to the interphase of cells that have undergone mitosis. In mitosis, however, interphase is the longest stage in the cycle, whereas interkinesis is often the shortest stage of meiosis.

# 13

# Mechanisms of inheritance

That reproducing organisms pass along some kind of hereditary material is self-evident. We really take for granted the remarkable consistency of form and function among individuals belonging to the same species. We often recognize in people, and also in pets, what appears to be a similarity of temperament or personality in parents and offspring.

In an animal like the amoeba, which reproduces simply by splitting in two, it is not surprising that the two new daughter cells are very much like the original parent cell. The mechanism of heredity is more difficult to comprehend in the case of the family cat that usually gives birth to long-haired, tiger-striped, green-eyed kittens that look just like their mother, but in other instances produces a miniature black panther.

People used to explain the phenomenon of consistent inheritance by saying it was "all due to the blood." They thought that offspring received blood from their fathers and blood from their mothers, and the blending of the blood produced a similar blending of physical characteristics. This belief is the basis of expressions like "blood relatives," "blood will tell," and the term "bloodlines" to describe the ancestry of livestock and race horses. This theory that blood was the mechanism of heredity fitted in with a general emphasis on the importance of blood. At that time, all medical treatment was designed to improve the blood, by thinning it out or removing its poisons or giving it increased vitality. Even unhappiness and melancholy were treated with medicines to change the state of the blood. If blood could be responsible for health and happiness, why not for green eyes as well?

The modern scientific explanation of the mechanism of heredity—portions of nucleic acid molecules contained in functional units called genes—is in many ways even harder to believe. The economy with which this process operates is most remarkable. Someone has estimated that if every gene contained in every cell of an adult human body was to be gathered together and placed on the head of a pin, less than one-half of one percent of the surface would be covered. Now we need only divide that quantity of genes on the pin by about 100 billion (the approximate number of cells in the human body) and we will have the total amount of hereditary material that passed from the parents to the child and dictated the development of that human being. It is rather like building the Golden Gate Bridge from plans hidden on a piece of microfilm on the period at the end of this sentence.

In this chapter we will first study the mechanisms by which these tiny bits of information are sorted out and passed along to the next generation, looking at the way that consistency is maintained and variation is introduced. We will save the even more puzzling question of how genes determine development for the next chapter.

## Mendelian patterns of inheritance

By the start of the nineteenth century, the theory of inheritance by blood had been discarded. When male sperm fluid was examined under a good light microscope, it could be seen quite clearly that it did not contain any blood. Nor did it contain a perfectly formed little embryo, as maintained by another early theory of inheritance. The fluid consisted of only a relatively small number of identical looking cells. Biologists therefore concluded that within some, or all, of those cells lay the hereditary material.

While professors of biology were arguing over the identity of the hereditary material, an Austrian monk was engaged in a series of experiments to discover a statistical pattern in the way this material influenced the next generation. Gregor Mendel was a priest and teacher at an Augustinian monastery in Brünn, Austria. He had studied mathematics and natural history at

13-1 *Like parent, like child is a phenomenon that occurs in every species. The markings and features of this kitten are strikingly like those of its mother. Physical and anatomical characteristics are kept fairly constant through the mechanisms of genetic inheritance. (Jeanne White from National Audubon Society)*

the University of Vienna, but even more important to his work was the fact that he was an enthusiastic amateur horticulturist and a person with a great deal of curiosity.

Mendel had noticed that in any plant species, individuals may vary considerably. For example, marigolds may be bright orange or dull rust in color; they may be tall with long stems or dwarf with short stems; they may be hardy in cold climates or they may blacken with the first frost. What Mendel wanted to find out was whether or not these characters were inherited according to some identifiable pattern.

Mendel chose the ordinary garden pea as the subject of his experiment. He had observed a number of simple variations in its appearance that seemed to fall into paired traits. Pods were either yellow or green, full or indented; no intermediate stage was noticeable. Another advantage of the pea was purely anatomical. Its flower is capable of self-pollination, so that pure strains can be produced. Yet if the stamens are removed from the flower, cross-pollination is possible, permitting planned crossing of plants. He selected seven characteristics to study. They were:

1. Color of pods—yellow or green
2. Height of plant—tall or dwarf

13-2 *Gregor Mendel, a modest Austrian monk, first established the chromosomal theory of inheritance based on experiments that he performed on ordinary garden peas. In 1866 he published his results in a classic paper that was promptly forgotten. Not until 1900, after the details of cell division had been worked out, was Mendel's work rediscovered, confirmed, and accepted by the scientific community. (Courtesy of the American Museum of Natural History)*

3. Position of flowers—distributed along the stem or clustered at the top
4. Form of ripe pods—full or indented
5. Color of seed coats—white or gray
6. Color of ripe seeds (peas)—yellow or green
7. Form of ripe seeds—smooth or wrinkled

Mendel's choice of these characters was grounded in his observation of their random occurrence, but he was also just plain lucky. It happened, although he of course did not know it, that each of the seven traits he chose is controlled by genes located on a separate chromosome. Thus, during cell division, each trait was passed along independent of the other six traits. Without this piece of good luck, Mendel's results would probably have been too puzzling to interpret.

## The principle of segregation

It is easiest to understand Mendel's work if we begin, as he did, with a single trait. In his first experiment with peas, Mendel planted yellow seeds that came from plants with yellow-seeded parents and grandparents. He also planted a second group of pure-bred green seeds at the same time. When the plants grew to maturity and produced flowers, he carefully removed the pollen from the yellow-seeded plants and used it to fertilize the green-seeded plants. The yellow-

**13-3** *Mendel worked with the garden pea, because he was familiar with several different sets of phenotypic traits that he thought might be genetically inherited. To notice the difference in such characteristics as pod shape and seed smoothness requires careful and minute observation. (Yeager and Kay—Photo Researchers)*

seeded plants were fertilized by pollen from green-seeded plants. Mendel's assumption in performing this experimental cross was that the trait of seed color was determined by a pair of factors, one for green and one for yellow.

Mendel collected the seeds produced in his first cross and planted them the following season. These plants, which geneticists call the $F_1$, or first filial, generation, were allowed to self-pollinate. After the flowers developed and seeds were produced, Mendel collected and counted them. He found that all seeds were yellow; there were no green seeds at all.

From this evidence, Mendel concluded that the trait of yellow seeds is **dominant** to the trait of green seeds. Would the dominance extend to the next generation? Would all the offspring of any $F_1$ yellow-seeded plant also have the trait for yellow seeds? To answer these questions, Mendel planted the seeds of his $F_1$ generation. These plants, the offspring of self-pollinizing plants, were grown to maturity, and Mendel once again collected and counted their seeds, the $F_2$ generation. Out of a total of 8023 seeds, he found that 6022 were yellow. However, there were also 2001 green seeds. This result indicated that the trait of green seeds had not really been lost in the $F_1$ plants, as some scientists thought. Nor had the trait been diluted, or changed in any way; the green $F_2$ seeds were indistinguishable from their green grandparents. Mendel concluded that the trait of greenness had simply been repressed by the dominant yellow factor.

Mendel obtained much the same results with the other pairs of traits he studied. All individuals in the $F_1$ generation inherited the dominant trait, but in the $F_2$ generation, only about 75 percent of the plants showed the dominant trait. The other 25 percent had the alternate, or **recessive,** trait. Thus he hypothesized that the $F_1$ generation had carried the trait for the recessive characteristic even though it did not show up in the plants, and that the trait was passed along to some of the $F_2$ generation.

**385**

We can diagram Mendel's results, using a specialized table called a Punnett square. The inheritance factors contributed by the female parent are placed on the vertical axis of the table, and the factors contributed by the male parent are placed on the horizontal axis. Let us assume that the parents with which Mendel started were pure-breds carrying only one type of genetic factor. This is the **homozygous** condition. If we use the symbol Y to represent the dominant trait of yellow seeds and the symbol y to represent the recessive trait of green seeds, we can describe the gametes produced by each plant. The yellow plant can produce only Y gametes; the green plant can produce only y gametes. In the cross $Y \times y$, all offspring will be Yy.

Every individual in the $F_1$ generation will possess one hereditary factor for Y and one for y. This is the **heterozygous** condition. Organisms that are heterozygous for any known trait are sometimes called **hybrids;** although the term originally was used for the offspring of a cross between members of different species, its meaning has been extended to cover a cross between any two individuals with significant genetic differences. Since Y is dominant to y, the heterozygous $F_1$ plants will all have yellow seeds, just like their homozygous YY parent. These offspring look like the dominant parent; they have the same **phenotype.** But the hereditary information they carry is different from that carried by their yellow-seeded parent; they have a different **genotype.** This difference in genotype is what causes the change in the appearance of the $F_2$ generation.

The $F_2$ generation when diagrammed on a Punnett square will look like this:

|  | Y | y |
|---|---|---|
| Y | YY (yellow) | Yy (yellow) |
| y | Yy (yellow) | yy (green) |

Since three-fourths of the offspring have a dominant Y factor, they will have the trait of yellow seeds. The other fourth of the offspring, with an inheritance of yy, will have the recessive trait of green seeds. Thus the ratio of phenotypes (physical appearance traits) will be 3 yellow seeds to 1 green seed, or 3:1.

It is important to remember that the Punnett square gives us only the statistical probability that an offspring will have a particular trait. If we breed four Yy hybrid plants, we cannot predict in advance which ones will have yellow seeds and which will have green, nor can we predict that there will necessarily be 3 yellow and 1 green. However, if we take 40,000 plants, we can be certain that about 30,000 will be yellow and about 10,000 will be green. Mendel worked with a sample of just over 8000 pea plants and found that 24.9 percent of the $F_2$ had the recessive trait of green seeds. When his experiment was later repeated by W. Bateson, using a larger sample of 15,806 plants, the tabulation of the $F_2$ generation showed 24.7 percent green seeds. With a large sample, the ratio will always be consistent, but with a small sample—the offspring of one mating, for example—the ratio will be unpredictable.

By looking at the Punnett square for the $Yy \times Yy$ cross, we can see that the genotypic ratio will be different from the phenotypic ratio. One-fourth of the offspring will be homozygous for yellow seeds (YY); two-fourths will be heterozygous (Yy); and one-fourth will be homozygous for green seeds. The genotypic ratio can be expressed as 1:2:1. To confirm this ratio, Mendel carried his experiment a step further, letting the $F_2$ plants self-pollinate. He found that about one-third of the yellow-seeded plants bred true—that is, they could be shown to be homozygous for green seeds because all their offspring also had yellow seeds. But the other two-thirds repeated the ratio found in the $F_2$ generation, with about 3 yellow-seeded plants for every 1 green-seeded plant. This experimental

13-1

**Observed ratios of homozygous recessive phenotypes in peas**

| | Observer | No. yellow | No. green | Green; percent |
|---|---|---|---|---|
| | Mendel | 6,022 | 2,001 | 24.9 |
| | Correns | 1,394 | 453 | 24.5 |
| Seed color | Tschermak | 3,580 | 1,190 | 24.9 |
| | Bateson | 11,903 | 3,903 | 24.7 |
| | Hurst | 1,310 | 445 | 25.4 |
| | Lock | 1,438 | 514 | 26.2 |

| | Observer | No. round | No. wrinkled | Wrinkled; percent |
|---|---|---|---|---|
| | Mendel | 5,474 | 1,850 | 25.2 |
| | Tschermak | 884 | 288 | 24.6 |
| Seed shape | Bateson | 10,793 | 3,542 | 24.8 |
| | Hurst | 1,335 | 420 | 23.9 |
| | Lock | 620 | 197 | 24.1 |

evidence confirms the fact that the $F_2$ generation contained both homozygous and heterozygous yellow-seeded plants, in a ratio of 1:2.

From his studies following the inheritance of a single trait, Mendel concluded that the off-spring of a cross between two distinct varieties possess inheritance factors from both parents. Moreover, he concluded that the two factors do not become joined or merged in the offspring, but can be passed along separately to the next generation. This conclusion is known as the **principle of segregation.**

## The principle of independent assortment

The second phase of Mendel's project was the study of patterns of inheritance involving two traits in the same individual. Mendel knew that if he crossed homozygous yellow-seeded plants with homozygous green-seeded plants (a **mono-**

**hydrid,** or single trait cross), the $F_1$ generation would be all heterozygous yellow-seeded (Yy) plants, and the $F_2$ generation would contain 3 yellow-seeded plants for every 1 green-seeded plant. He also knew that the traits of smooth seeds (S) and wrinkled seeds (s) would be inherited in the same way. The $F_1$ plants would all be heterozygous smooth (Ss) while the $F_2$ would show a ratio of 3 smooth to 1 wrinkled. But what would happen when homozygous smooth yellow plants were crossed with homozygous wrinkled green plants?

Mendel performed this **dihybrid** (two-trait) cross. The $F_1$ generation consisted of all smooth yellow plants, showing that the inheritance of the dominant trait of smoothness did not in any way interfere with the inheritance of the dominant trait of yellow seeds. The $F_1$ was allowed to self-pollinate, and with his usual meticulousness, Mendel tabulated the results. Out of a total of 556 $F_2$ seeds, he found 315 smooth yellow seeds, 108 smooth green, 101 wrinkled yellow, and 32 wrinkled green. If these figures are rounded off to express a ratio, it will be 9:3:3:1.

This result is easily explained. The $F_1$ generation would be heterozygous both for the trait of smoothness and for the trait of yellow seeds; all plants would have the genotype SsYy. Assuming that these factors would segregate independently, the $F_1$ generation could pass along four different assortments of factors: SY, Sy, sY, and sy. A Punnett square with these gametes would look like this:

|  | SY | Sy | sY | sy |
|---|---|---|---|---|
| **SY** | SSYY smooth yellow | SSYy smooth yellow | SsYY smooth yellow | SsYy smooth yellow |
| **Sy** | SSYy smooth yellow | SSyy smooth green | SsYy smooth yellow | Ssyy smooth green |
| **sY** | SsYY smooth yellow | SsYy smooth yellow | ssYY wrinkled yellow | ssYy wrinkled yellow |
| **sy** | SsYy smooth yellow | Ssyy smooth green | ssYy wrinkled yellow | ssyy wrinkled green |

The ratio of phenotypes predicted by the Punnett square would be 9 smooth yellow plants to 3 smooth green plants to 3 wrinkled yellow plants to 1 wrinkled green plant, or 9:3:3:1 — just the ratio that Mendel found in his peas.

Mendel's next step was to try trihybrid crosses involving three different traits. He crossed plants with smooth seeds, yellow seed color, and gray seed coats (all dominant traits) to plants with wrinkled seeds, green seed color, and white seed coats (all recessive traits). The $F_1$ plants all had the phenotype of the dominant parent. The self-pollinated $F_2$ showed the following ratio of phenotypes:

27 smooth yellow gray
 9 smooth yellow white
 9 smooth green gray
 9 wrinkled yellow gray
 3 smooth green white
 3 wrinkled yellow white
 3 wrinkled green gray
 1 wrinkled white

With three traits, there are 64 possible combinations of hereditary factors, and the ratio of the phenotypes will be 27:9:9:9:3:3:3:1. In other words, the chances are 27 out of 64 that an $F_2$ plant will display all three dominant traits, while there is only 1 chance out of 64 that it will display all three recessive traits. There are eight different phenotypes from a cross involving three traits, but the number of different genotypes is even larger (27). In a cross involving five traits, there will be 32 phenotypes and 243 genotypes; with 10 traits, the numbers jump to 1024 phenotypes and 59,049 genotypes; and with 20 traits, there are 3½ billion different genotypes possible. We all like to think we are in some way unique, and the statistics of genetics furnishes support for this cherished belief in our individuality. Since it is estimated that humans have at least 10,000 different gene pairs, the number of possible genotypes is much larger than the total number of people ever born. With the exception of identical siblings, each man's genotype is uniquely his own.

## Genes and chromosomes

Gregor Mendel read a paper containing a summary of his work and findings before the Brünn Society for the Study of Natural Science in 1865; the records of the Society state that there were neither questions nor discussion following his presentation. Like a stone dropped down a well, Mendel's work disappeared from the view of the scientific community without causing so much as a single wave. Mendel himself became disillusioned about the significance of his findings, when he tried to reproduce his results by studying other organisms. He first chose hawkweed, an unfortunate choice because it goes through a stage of asexual reproduction of which he was unaware; the results of the hawkweed study did not confirm Mendel's earlier findings. His second choice was bees, and his results were even more confusing, since he did not know that both fertilized and unfertilized eggs can develop into mature adults.

It was not until 1900 that Mendel's work was rediscovered, when three other scientists, who had searched through the published scientific literature to find experimental data to support their own hypotheses, cited Mendel's findings.

Other researchers, observing the cellular events of mitosis and meiosis, were quick to see that there might be some connection between Mendelian principles of inheritance and the process of cell division. By 1905, W. S. Sutton, an American cytologist, had published a paper in which he stated his belief that "the association of paternal and maternal chromosomes in pairs during the reducing division [meiosis] . . . may constitute the physical basis of the Mendelian law of heredity."

## Chromosome assortment during meiosis

If we review the process of meiosis in light of what we know about Mendelian inheritance, we can see how the assortment of hereditary factors takes place. For convenience of illustration, we will imagine a hypothetical cell that happens to have only two different types of chromosomes. One pair of homologous chromosomes will be labeled A and a; the other pair will be called B and b. During the cell's interphase, these four chromosomes will be intertwined, arranged in no special order. But during prophase, they will be seen to emerge from the background of other nuclear matter. As the chromosomes begin to line up on the metaphase plate, the homologous chromosomes pair up. There is an element of chance in their pairing. The first pair of chromosomes could line up in the sequence Aa, or they could line up in the sequence aA. The same is true of the other homologous pair, which could line up as Bb or as bB. The chromosome arrangement is a completely random occurrence, so it is impossible to predict the sequence in any given instance.

Figure 13-4 shows the results of the first meiotic division. The chromosomes may be lined up as AB and ab, or as Ab and aB; the chances for the occurrence of each arrangement are 1 out of 2. At the second metaphase, the chromosomes will line up on the metaphase plate, and the double

**13-4** *The genetic material, in the form of homologous chromosomes, is assorted during meiosis (color). One strand of each chromosome is shunted at random into each of the daughter cells. This insures that the gamete will carry the haploid number of chromosomes, representing a random sample of the parental genetic material.*

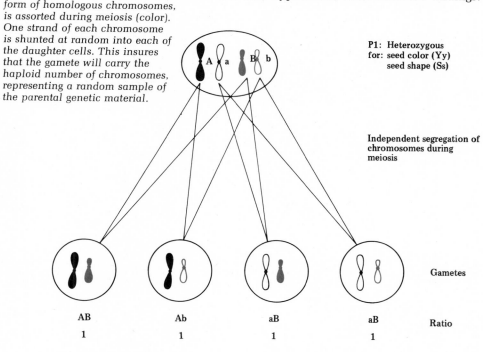

P1: Heterozygous
for: seed color (Yy)
    seed shape (Ss)

Independent segregation of
chromosomes during
meiosis

Gametes

| AB | Ab | aB | aB | Ratio |
|----|----|----|----|-------|
| 1  | 1  | 1  | 1  |       |

strands of each chromosome will separate. From the original mother cell, four different types of gametes are possible: AB, ab, Ab, and aB. The ratio in which these gametes will be produced is 1:1:1:1.

If we compare this hypothetical example with the dihybrid cross performed by Mendel, the similarity is clear. The dominant traits S and Y and recessive s and y will segregate in just the same way. The Punnett square predicts that gametes will be formed in the ratio of 1 SY: 1 Sy: 1 sY: 1 sy. In other words, it can be shown that chromosomes behave in just the same way as the hypothetical bits of genetic information that Mendel thought controlled inheritance of paired traits. Therefore, it is logical to assume the two are identical.

## The sex chromosome

During the early years of the twentieth century, geneticists were carefully observing and recording the process of meiotic cell reproduction in a number of different organisms. They noticed that in certain organisms, such as the fruit fly (*Drosophila melanogaster*), one pair of homol-

ogous chromosomes presents a dissimilar appearance in the male of the species. One chromosome is the usual rod shape, but the other is shaped like a rod with a hook on the end, or a J. In the female, no such dissimilar chromosome pair can be observed.

This observation of a difference in the chromosomes of males and females seemed to fit in with a hypothesis advanced by C. E. McClung, that sex was determined by the presence or absence of a specific genetic factor. The normal female fruit fly has two of the rod-shaped chromosomes, called X chromosomes. The normal male has one X chromosome and one hook-shaped chromosome, called a Y chromosome.

The discovery of the X and Y sex chromosomes explains why the ratio of males to females is approximately 1:1. After a female's primary oöcyte undergoes meiosis, all four of the resulting gametes will have X chromosomes. In the male's primary spermatocyte, there will be random assortment of the chromosomes, so that two of the gametes will have X chromosomes and two will have Y chromosomes. The combination of X and X will produce a female zygote, whereas the combination of X and Y will produce a male. There is a 50-50 chance for either type to occur.

**13-2**

**Sex index and sexual types in Drosophila melanogaster**

| Phenotype | Number of X chromosomes | Number of sets of autosomes (A sets) | Sex index = $\dfrac{\text{number of Xs}}{\text{number of A sets}}$ |
|---|---|---|---|
| Superfemale | 3 | 2 | 1.5 |
| Normal tetraploid female | 4 | 4 | 1.0 |
| Normal diploid female | 2 | 2 | 1.0 |
| Intersex | 2 | 3 | 0.67 |
| Normal male | 1 | 2 | 0.50 |
| Supermale | 1 | 3 | 0.33 |

**13-5** *In most higher organisms, the chromosomal endowments of males and females are different. There are two sorts of sex chromosomes: X chromosomes, bearing many genes, and Y chromosomes, bearing only a few genes. Among many animals, females are designated by XX, while males inherit an X and a Y. The pair of sex chromosomes can be easily distinguished in a male mammal because the Y chromosome is far smaller than the X chromosome, and the pair thus has a rather unbalanced appearance. (Morris Huberland from National Audubon Society)*

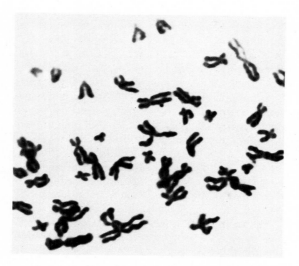

Later studies of sex inheritance in *Drosophila* indicate that genic balance is an important determinant of sex. It is not so much the Y chromosome that determines maleness as it is the balance of sex chromosomes to other chromosomes (called **autosomes**). An individual with two sets of autosomal chromosomes (the diploid number) and two X chromosomes will be a female. The combination of two sets of autosomes and one X produces a male, whether or not the Y chromosome is present. Many aberrant combinations, such as two sets of autosomes and three X chromosomes, or a triploid number of autosomes and one X chromosome have been discovered, and these bear out the importance of the X chromosome/autosome ratio rather than the presence of the Y chromosome. If the ratio is at least one set of autosomes to one X, the individual will be a female. If it goes above that 1:1 ratio, it cannot be female. The ratio of two sets of autosomes to one X will be male. Variants such as 1:5:1 (an individual with 3 n autosomes and 2 X chromosomes) will be an intermediate stage, called intersex.

The XY of type sex determination is found not just in fruit flies but in many other animals as well, including man. Human sex inheritance differs from sex inheritance in fruit flies in one important respect. In humans, it is the presence or absence of the Y chromosome that determines maleness, rather than the ratio of autosomes to X chromosomes.

It is interesting to note that birth certificate records show that slightly more males than females are born every year. The average ratio is about 106 males to 100 females. This is not a chance variation in one particular year or one specific location, but a pattern that is seen every year and everywhere. There are several possible explanations for this imbalance in the birth ratio. One is that female zygotes might be more delicate than males, perhaps not remaining viable long enough for the mother even to be aware that she is pregnant. The explanation which is currently most widely accepted is that the Y-containing sperm cells may have a slight competitive advantage in the race to fertilize the egg. They might be able to move faster or to remain active longer, and thus have a better statistical chance to fuse with the female gamete. If it is true that there is some functional difference between X-containing sperm and Y-containing sperm, that opens the way for an interesting technological development in the future. It might be possible to use a chemical inhibitor that would selectively kill or immobilize either the X or the Y sperm, so that couples could choose the sex of their child.

Not all species have the XY chromosome type of sex determination. Alternate patterns are frequently found, especially among the insects. For instance, in the bug *Protenor*, the females have two X chromosomes and the males have only one; this is the XO sex inheritance pattern. In one kind of wasp, there are at least nine different genetic factors involved in sex determination, and males are produced only when two or more of these factors are homozygous. If the individual is heterozygous for all factors, it will be a female. This indicates that femaleness may be dominant to maleness in this species.

**391**

## Gene location

Terms like "inheritance factor" and "genetic particle," which we have thus far used in describing patterns of heredity, were coined by early investigators. By the second decade of the twentieth century, biologists had agreed to replace these terms with the word **gene.** At the time the word came into general acceptance, it was assumed that a gene was a specific and separate entity, and the chromosome was thought of as being more or less a matrix for the gene. But as additional research data have accumulated, this notion has been revised, and it has become more and more difficult to define a gene.

Today most geneticists agree that the chromosomes are long coiled strands of DNA, which contract to a shorter and thicker shape during mitosis or meiosis. A gene is probably nothing more than a segment of the DNA molecule, a portion that carries an informational content. As a physical entity, a gene is not a separate structure but rather a location, like a dot on a map. As a concept, the term gene still refers to an independently inheritable unit that is passed from parent to offspring. Geneticists use the term **allele** to refer to alternate genes that may be found at the same locus on homologous chromosomes. An example of a pair of alleles would be the Y gene for yellow seeds and the y for green seeds.

**Linked genes.** Early geneticists were perplexed by the relationship between genes and chromosomes. In 1903, when Sutton pointed out that the mechanism of assortment in Mendelian factors was just like that in chromosomes, he suggested that some chromosomes must contain more than one factor. Cytological studies showed that the total number of chromosomes in a haploid set is usually less than 50, even in rather complex animals like man. Yet the number of inheritable traits is considerably larger; therefore some of the factors must be located on the same chromosome.

What happens to the Mendelian principle of independent assortment if some of the genes are on the same chromosome? Can they assort independently, or must they be transmitted as a unit? In the early twentieth century, there were some supporters of each theory. It was suggested that the chromosomes might completely break apart during meiosis, allowing the genes to be independently assorted. Yet there was no cytological evidence that this phenomenon actually took place. The question of linkage was not answered until T. H. Morgan began to collect experimental data on linked inheritance in *Drosophila.*

Morgan and his co-workers kept complete genetic records of countless generations of fruit flies (they breed very fast). They found that certain traits seemed to constitute a linked group that tended to be inherited together. For example, males with white eyes usually also had yellow bodies, miniature wings, and singed bristles. C. B. Bridges found a linkage group consisting of shortened wings, dark body color, small rough eyes, and certain bristles reduced.

After all the data was compiled, *Drosophila* was found to have four definite linkage groups. This corresponds with the number of paired chromosomes; thus it can be assumed that the genes on a chromosome are linked throughout

13-6 Drosophila, the tiny fruit fly, is one of the organisms most extensively studied by geneticists. Inheritance patterns can be examined over many generations because the fruit fly breeds so rapidly. The mechanisms of gene linkage appear more clearly than in most animals because Drosophila has only eight chromosomes. Geneticists have also been able to map the chromosomes of fruit flies in quite a bit of detail. (Grace Thompson from National Audubon Society)

**13-7** *Morgan's experiments demonstrated the sex linkage of the gene for eye color in Drosophila. W is the dominant gene for normal "wild" or red eye color; w, the recessive gene for white eye color. Both alleles are found only on the X chromosome.*

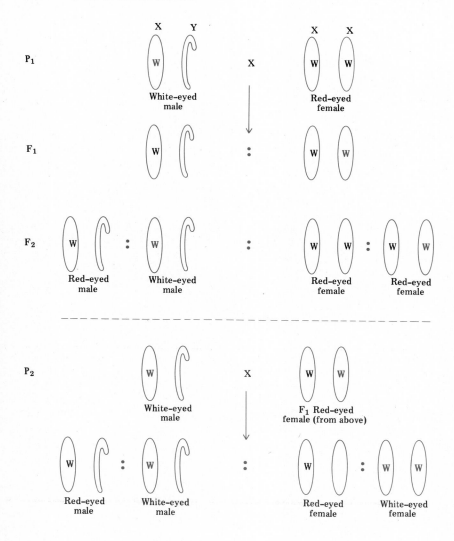

the process of meiosis and are normally inherited as a group. Linked genes are not assorted independently. If Mendel had happened to choose for study any linked traits, his results would have been quite different, and he probably would have been unable to draw any conclusions from his data.

**Sex-linked genes.** Morgan's work with linkage groups in *Drosophila* revealed that there is linkage of sex-determining genes and genes for other nonsexual traits, on the X chromosome. Morgan demonstrated the phenomenon of sex linkage by keeping careful records of the appearance of a mutant gene for white eyes. The phenotype first appeared in a male fruit fly, and Morgan crossed that male with its normal red-eyed sisters. The resulting $F_1$ generation was all red-eyed, indicating that the mutant white gene was recessive to the normal red gene. It was in crossing members of the $F_1$ generation that Morgan made his initial observation of sex linkage. About half of the males in the $F_2$ generation had white eyes, but there were no $F_2$ white-eyed females.

To explain these unusual results, not in keeping with the predicted Mendelian ratio, Morgan hypothesized that the recessive white gene was located on the X chromosome. Since the male has only one X chromosome (which he gets from his mother), he will display the recessive phenotype if he inherits a white gene, no matter what genes are on the Y chromosome he inherits from his father. But the female, who inherits an X chromosome from each parent, will show the white-eyed phenotype only if she gets the recessive mutant gene from both parents. Morgan tested this hypothesis by performing a **back cross;** he mated a heterozygous $F_1$ female to her recessive male parent. The back cross produced both red-eyed and white-eyed females, in a 1:1 ratio, as Morgan had predicted. To further confirm his theory, Morgan crossed the white-eyed females with white-eyed males. All offspring were white-eyed.

## 13-3
## Sex-linked traits in man

| Trait | Inheritance |
| --- | --- |
| Color blindness (red/green) | recessive |
| Congenital deafness | recessive? |
| Muscular dystrophy of pelvic girdle | recessive |
| Spastic paraplegia | not determined |
| Parkinsonism | dominant |
| Hemophilia | recessive |
| Nystagmus — quivering of eyeballs | dominant and recessive |
| Diabetes insipidus | recessive |
| Glucose-6-phosphate dehydrogenase deficiency | not determined |

Evidence of the phenomenon of linkage on human sex chromosomes was first noticed by the writers of the Jewish Talmud, although at that time the significance of the observation was not understood. In carrying out the ritual of circumcision, it was noticed that a small number of boys bled severely, and that such bleeding seemed to be hereditary. The Talmud stated that the sons of all female relatives of a bleeder should not be circumcised, but permitted the circumcision of sons of male relatives. Through observation and recorded experience, the Jews had realized that females carried this trait, even though it was the males that were affected.

Bleeder's disease, or hemophilia, is probably the best understood example of a sex-linked trait. The recessive gene for this trait is found only on the X chromosome, never on the Y. A father with hemophilia therefore cannot transmit the disease to his son, because the son receives only the Y chromosome. However, the X chromosome carrying the gene for hemophilia can be passed from the father to a daughter, since he transmits an X chromosome to her. The daughter then will have one X chromosome with the hemophilia gene and one X chromosome with the allele for normal clotting. The normal gene is dominant, and so the daughter will not display any symptoms of hemophilia. The daughter will produce both normal X and hemophiliac X chromosomes in a 1:1 ratio.

It is difficult to trace the inheritance of recessive genes in the human population. The samples are small, since most people have only two or three children; and no experimental back crosses can be tried. Human geneticists must work by averaging the available records. Only by studying many families in which the recessive trait is inherited can they arrive at some estimate of the genetic ratios involved. If the daughter who carries the trait marries a normal man, statistical averages predict that about half of her daughters will inherit her hemophiliac X chromosome, and they too will be carriers. The other half will inherit her normal X chromosome and will not carry the trait of hemophilia. About half of her sons will inherit the hemophiliac X chromosome and they will be hemophiliacs, for the Y chromosome carries no masking gene; the other half will inherit her normal X chromosome and will be normal. This pattern of inheritance is summarized in Figure 13-8.

If by any chance a carrier female should marry a hemophiliac male, the statistically predictable pattern of inheritance in their sons would be just the same as in the previous example; about half would be normal and half would be hemophiliacs. But there would be a difference in the inheritance of the female children, who will get one hemophiliac X chromosome from their father. Since about 50 percent of the females would get a normal X chromosome from their mother, they would be heterozygous for hemophilia, carrying the disease but not suffering from it. However, the other 50 percent would inherit a hemophiliac X chromosome from their mother as well as from their father, and they would therefore be homozygous for hemo-

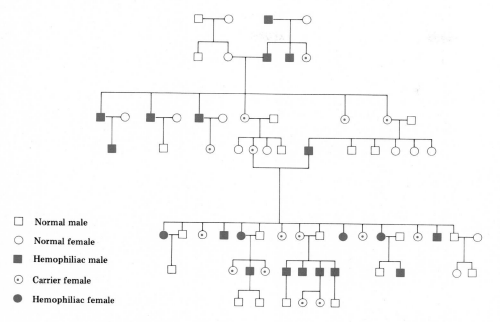

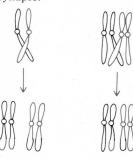

☐  Normal male

○  Normal female

◼  Hemophiliac male

⊙  Carrier female

⬤  Hemophiliac female

**13-8** *This hemophiliac family from Essex, England, was studied between 1886 and 1961. Their family tree traces seven generations of hemophilia to one ancestor. The marriage of a hemophiliac male and a carrier woman gave a high incidence of hemophilia in subsequent generations, including hemophiliac women.*

**13-9** *Crossing over occurs at the four-strand stage, before the first meiotic division when the chromosomes lie together in synapse.*

At 2-strand
stage

At 4-strand
stage

philia. Lacking a dominant normal gene to mask the hemophilia gene, they too would suffer from the disease. The incidence of hemophilia in women is so low that it was once thought that woman could not have this disease. Hemophilia in women is rare because the gene is rare, and so the odds against finding it in both the mother and the father are quite high.

Another example of a sex-linked trait is red-green color deficiency, and it is inherited in just the same way as hemophilia. Women can be carriers, but they are only rarely the victims of the defect. A certain type of muscular dystrophy is also known to be transmitted through the X chromosome. No trait has yet been positively linked to the Y chromosome, although it seems reasonable to suppose that, like the X chromosome, it carries traits other than sexual characteristics. The trait of luxuriant hair growth on the ears is known to be handed down from father to son, but studies have not yet demonstrated that the Y chromosome is responsible for the transmission.

**Mapping chromosomes.** Through studies of traits such as hemophilia and color deficiency, geneticists know that the X chromosome contains at least four genes. That is about as informative as saying that Ohio contains at least four community colleges; the obvious question is, where are they? In recent years, great progress has been made in answering that question and producing a map that shows the location of each gene. The process of gene mapping is based on an analysis of the frequency of crossing over of chromosomes. **Crossing over** refers to the accidental tangling of chromosomes during synapse in the early stages of cell division. One long thin strand of DNA may lie over the top of the other, and they will break when they become entangled. Although the broken strands are quickly linked together again, the chances are good that the homologous chromosomes may have exchanged parts in the process. This is shown in Figure 13-9.

Crossing over happens fairly frequently. Researchers have kept track of the rates of crossing over in the four chromosomes of fruit flies, and they have found that some pairs of traits combine more frequently than others. Chromosome mapping is based on the assumption that this difference in frequency is due to the location of the various genes. The farther apart the genes are on the chromosome, the higher will be the frequency of crossing over. Suppose, for example, that one chromosome contains genes A–Z; allelic genes a–z are located on its homolog. A break during synapse anywhere between genes B and Y will lead to the A + z recombination; there are in fact 25 chances for this recombination to occur. But there is only one site at which breakage can occur to produce an A + b recombination; unless the breakage occurs between those two genes, they will not be recombined. Thus the statistics on crossing over should show that A + z happens about 25 times more frequently than A + b. The amount of crossing over of linked genes depends on the distance between them on the chromosome.

C. B. Bridges and T. M. Olbrycht were able to map three genes on the X chromosome of *Drosophila*, using this method of analyzing cross-over frequencies. They studied three genes: the scute gene (sc), which produces a phenotype with certain bristles missing; the echinus gene (ec), with the phenotype of rough eyes; and the cross-veinless gene (cv), with the phenotype of certain wing veins missing. Out of a total of 2635 flies studied, 80 percent showed no crossing over; all three genes remained linked in the next generation. In 9.1 percent, the genes sc and ec were separated; in 10.9 percent, the genes cv and ec were separated; in 20 percent, the genes sc and cv were separated. Since sc and cv had the highest frequency of separation (caused by crossing over) they were put at the ends of the chromosome. The ec genes were then placed midway between them, since the frequency of crossing over was about the same in either direction. Thus Bridges and Olbrycht established a map that looked like this:

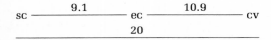

The actual work of mapping chromosomes is not quite that simple, since there are a number of complicating factors. One is that the genes are not spaced out evenly along the chromosome like beads on a necklace. A and B may be bunched together at one end, with a huge space between B and C; this arrangement would permit B+c recombinations to take place more frequently than A+b recombinations. Another complication comes from the fact that chromosomes may cross over more than once; there may be one recombination at A+b and another at M+n. It is also possible that some parts of the chromosome are more fragile than others, breaking more frequently.

Despite these handicaps, geneticists have been able to map the chromosomes of fruit flies in quite a bit of detail. There is great interest in the possibility of mapping human chromosomes, but at the present time little real progress has been made. Humans have a much larger number of chromosomes, and compared to the fruit fly, they are terribly slow at reproducing. The technique of cross-over frequency analysis is not a very promising aid in the study of human genetics, but new biochemical techniques may soon offer an alternate method of study.

## Gene expression

Mendel's work showed that there is a distinction between genetic inheritance and physical appearance, for he found that plants with both YY and Yy genotypes would have yellow seeds. As luck would have it, the traits he chose for study were all controlled by one gene pair, and they were all cases of **complete dominance.** In such cases, the presence of the dominant gene completely masks the presence of the recessive allele, and there are no intermediate stages between the dominant and recessive phenotypes. The seed is either yellow or it is green; the homozygous YY is not yellower than the heterozygous Yy. This type of inheritance assumes full **penetrance,** that is, the individuals carrying the dominant gene will always display the gene's phenotype. It also assumes that the phenotype will always be exactly the same, or that there will be full **expressivity** of the gene.

Not all patterns of inheritance are so simple, and both penetrance and expressivity may be only partial. Some genes are only partially dominant; some dominant genes can be masked by other dominant genes; some traits are controlled by as many as 14 different genes and their alleles. Moreover, the environment can also affect the action of a gene. In the last several decades, most genetic research has focused on these complications in the pattern of inheritance.

**13-10** Drosophila's *four types of chromosomes have been mapped to indicate the position of each gene based on the observed cross-over frequencies. The figures indicate their position in cross-over map units from the zero end of the chromosome. One map unit equals the distance within which crossing over occurs one percent of the time. It is not a measure of absolute distance and cannot be expressed in microns or millimicrons. (Strickberger, M. W., 1968. Genetics, Macmillan)*

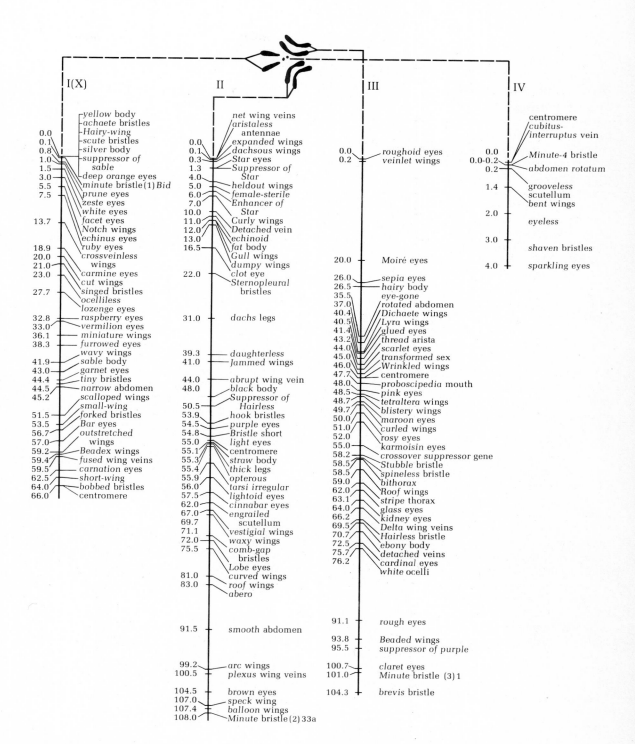

**I(X)**

| | |
|---|---|
| 0.0 | yellow body |
| | achaete bristles |
| 0.1 | Hairy-wing |
| 0.8 | scute bristles |
| 1.0 | silver body |
| 1.5 | suppressor of sable |
| 3.0 | deep orange eyes |
| 5.5 | minute bristle (1) Bid |
| 7.5 | prune eyes |
| | zeste eyes |
| | white eyes |
| 13.7 | facet eyes |
| | Notch wings |
| | echinus eyes |
| 18.9 | ruby eyes |
| 20.0 | crossveinless wings |
| 21.0 | |
| 23.0 | carmine eyes |
| | cut wings |
| 27.7 | singed bristles |
| | ocelliless |
| | lozenge eyes |
| 32.8 | raspberry eyes |
| 33.0 | vermilion eyes |
| 36.1 | miniature wings |
| 38.3 | furrowed eyes |
| | wavy wings |
| 41.9 | sable body |
| 43.0 | garnet eyes |
| 44.4 | tiny bristles |
| 44.5 | narrow abdomen |
| 45.2 | scalloped wings |
| | small-wing |
| 51.5 | forked bristles |
| 53.5 | Bar eyes |
| 56.7 | outstretched wings |
| 57.0 | |
| 59.2 | Beadex wings |
| 59.4 | fused wing veins |
| 59.5 | carnation eyes |
| 62.5 | short-wing |
| 64.0 | bobbed bristles |
| 66.0 | centromere |

**II**

| | |
|---|---|
| | net wing veins |
| | aristaless |
| | antennae |
| 0.0 | expanded wings |
| 0.1 | dachsous wings |
| 0.3 | Star eyes |
| 1.3 | Suppressor of Star |
| 4.0 | heldout wings |
| 5.0 | female-sterile |
| 6.0 | Enhancer of Star |
| 7.0 | |
| 10.0 | Curly wings |
| 11.0 | Detached vein |
| 12.0 | echinoid |
| 13.0 | fat body |
| 16.5 | Gull wings |
| | dumpy wings |
| 22.0 | clot eye |
| | Sternopleural bristles |
| 31.0 | dachs legs |
| 39.3 | daughterless |
| 41.0 | Jammed wings |
| 44.0 | abrupt wing vein |
| 48.0 | black body |
| | Suppressor of Hairless |
| 50.5 | hook bristles |
| 53.9 | purple eyes |
| 54.5 | Bristle short |
| 54.8 | light eyes |
| 55.0 | centromere |
| 55.1 | straw body |
| 55.3 | thick legs |
| 55.4 | opterous |
| 55.9 | tarsi irregular |
| 56.0 | lightoid eyes |
| 57.5 | cinnabar eyes |
| 62.0 | engrailed |
| 67.0 | scutellum |
| 69.7 | vestigial wings |
| 71.1 | waxy wings |
| 72.0 | comb-gap bristles |
| 75.5 | Lobe eyes |
| | curved wings |
| 81.0 | roof wings |
| 83.0 | abero |
| 91.5 | smooth abdomen |
| 99.2 | arc wings |
| 100.5 | plexus wing veins |
| 104.5 | brown eyes |
| 107.0 | speck wing |
| 107.4 | balloon wings |
| 108.0 | Minute bristle (2) 33a |

**III**

| | |
|---|---|
| 0.0 | roughoid eyes |
| 0.2 | veinlet wings |
| 20.0 | Moiré eyes |
| 26.0 | sepia eyes |
| 26.5 | hairy body |
| 35.5 | eye-gone |
| 37.0 | rotated abdomen |
| 40.4 | Dichaete wings |
| 40.5 | Lyra wings |
| 41.4 | glued eyes |
| 43.2 | thread arista |
| 44.0 | scarlet eyes |
| 45.0 | transformed sex |
| 46.0 | Wrinkled wings |
| 47.7 | centromere |
| 48.0 | proboscipedia mouth |
| 48.5 | pink eyes |
| 48.7 | tetraltera wings |
| 49.7 | blistery wings |
| 50.0 | maroon eyes |
| 51.0 | curled wings |
| 52.0 | rosy eyes |
| 55.0 | karmoisin eyes |
| 58.2 | crossover suppressor gene |
| 58.5 | Stubble bristle |
| 58.5 | spineless bristle |
| 59.0 | bithorax |
| 62.0 | Roof wings |
| 63.1 | stripe thorax |
| 64.0 | glass eyes |
| 66.2 | kidney eyes |
| 69.5 | Delta wing veins |
| 70.7 | Hairless bristle |
| 72.5 | ebony body |
| 75.7 | detached veins |
| 76.2 | cardinal eyes |
| | white ocelli |
| 91.1 | rough eyes |
| 93.8 | Beaded wings |
| 95.5 | suppressor of purple |
| 100.7 | claret eyes |
| 101.0 | Minute bristle (3) 1 |
| 104.3 | brevis bristle |

**IV**

| | |
|---|---|
| | centromere |
| | cubitus-interruptus vein |
| 0.0 | Minute-4 bristle |
| 0.0-0.2 | abdomen rotatum |
| 0.2 | |
| 1.4 | grooveless scutellum |
| | bent wings |
| 2.0 | eyeless |
| 3.0 | shaven bristles |
| 4.0 | sparkling eyes |

## Incomplete dominance and codominance

Incomplete dominance occurs when heterozygous offspring show traits that are a combination of two contrasting alleles, in gradations of gene expressivity. An example can be seen in a small flowering plant called a four-o'clock. When homozygous red-flowered plants are crossed with homozygous white-flowered plants, the $F_1$ generation will be neither red nor white, but pink. The red gene has only incomplete dominance over its white allele, and so a blending of the two traits occurs.

Incomplete dominance was once viewed as a rarity, but recent research has indicated that even $F_1$ offspring with the same phenotype as the dominant parent may differ from the parent in important ways at the cellular level. This is true, for example, of one of the traits studied by Mendel, that of seed shape. The appearance of smoothness in seeds is due to the presence of many large globular starch grains, which enables the cell to retain a relatively large amount of water. Wrinkled seeds possess only a few small starch grains, and thus they lose water as they ripen, so that the seeds take on their characteristic withered appearance. Microscopic examination of heterozygous Ss plants has shown that their seeds contain fewer starch grains than do the homozygous SS plants, although there are enough grains to give the seed a plump, smooth appearance. Apparently the gene for smooth seeds is not fully dominant, although the incomplete dominance is not noticeable in the phenotype.

Several cases which were once thought to be examples of incomplete dominance have been reclassified as examples of a form of inheritance called **codominance,** in which characteristics of both phenotypes appear in the heterozygous offspring. An example of this can be seen in the feather color of Andalusian chickens. When a homozygous white-feathered bird is crossed with a homozygous black-feathered bird, the heterozygous offspring will appear to be blue, but a close examination of the feathers shows that each has alternating small areas of black and white coloration; this checkerboard pattern refracts light in a way that makes it appear blue. Neither the black nor the white gene has dominance, and there is no blending of the two traits to form a third new character (gray, for example).

## Multiple alleles

Mendel assumed that inheritable traits came in pairs, and his work seemed to support that assumption. Seeds were either smooth or wrinkled, yellow or green, and genes were either dominant or recessive. More recent research makes Mendel's assumption appear to be an oversimplification.

Certain traits appear to have three or more allelic genes. Of course, only one allelic gene can appear on each chromosome, and so a diploid individual can have only two of the alleles. But those two alleles are not the only possibilities; the individual might breed with another individual who has two other alleles. To predict the outcome of such a mating, we need to know the hierarchy of dominance in the group of multiple alleles.

The fruit fly again furnishes one of the most

**13-11** *The "blue" coloring of Andalusian chickens is actually a checkerboard of white and black areas. As the heterozygous offspring of a homozygous white-feathered and a homozygous black-feathered parent, the "blue" chicken inherits and expresses genes for both colors. This form of inheritance is called codominance. (Alycia Smith Butler)*

| | self–fertilization | cross fertilization | | |
|---|---|---|---|---|
| Parental genome | $S_1 S_2 \times S_1 S_2$ ♀ ♂ | $S_1 S_2 \times S_2 S_3$ ♀ ♂ | $S_1 S_2 \times S_3 S_4$ ♀ ♂ | |
| Pollen development | $S_1$ $S_1$ $S_2$ $S_2$ — $S_1$ $S_2$ | $S_2$ $S_2$ $S_3$ $S_3$ — $S_1$ $S_2$ | $S_3$ $S_3$ $S_4$ $S_4$ — $S_1$ $S_2$ $S_2$ $S_1$ | pollen — style — pollen tube — egg — ovary |
| Progeny | None | $S_1 S_3$   $S_2 S_3$ | $S_1 S_3$   $S_2 S_3$  $S_1 S_4$   $S_2 S_4$ | |

**13-12** *Self-sterility mechanisms have evolved in many plant species otherwise capable of self-fertilization, and apparently function to insure a maximum degree of genetic recombination. Fertilization does not take place unless the pollen and ovum differ in certain genetic factors.*

thoroughly studied examples of inheritance through multiple alleles. Eye color in *Drosophila* appears to be controlled by a large number—perhaps 10 to 12—of multiple alleles. Thus, for example, the genes for eosin, cherry, apricot, wine, and other colored eyes are all allelic. Geneticists are uncertain of the exact number of alleles involved, because the inheritance of eye color is further complicated by the fact that nonallelic genes may also act in combination to alter phenotypes.

Another example of a multiple allele can be seen in a number of plants that are sterile when fertilized by their own pollen, or by pollen from other genotypically identical individuals. Certain varieties of tobacco are self-sterile plants. Fertilization can take place only when the pollen contains at least one allele not contained in the female plant. The value in this adaptation lies in the fact that by preventing continued self-

fertilization, it encourages new combinations of genes and limits the spread of the most common combinations.

The group of genes existing in all the individuals within any particular interbreeding population is called its **gene pool.** A relatively large gene pool is generally an advantage for the species' survival, and thus many genetic mechanisms, such as self-sterility, have arisen to help maintain the size of the gene pool. If every tobacco plant had exactly the same genetic inheritance, they would all be vulnerable to the same environmental threats, such as a change in climate or the introduction of a new parasite. With a larger gene pool, there may be some individuals whose different alleles make them resistant, and thus the species will be saved.

Multiple alleles are also found in human inheritance. The inheritance of blood groups is one example of this pattern of heredity.

There are three alleles for blood groups: type A, type B, and type O. Both A and B are dominant to O, so a genotype of AO will result in an A group phenotype, and BO in a B group phenotype. The phenotype of O group blood can be found only with an OO genotype. Although A and B are both dominant to O, they are codominant when present together. The genotype AB yields a fourth phenotype, in which both A and B are expressed equally. Thus the three alleles yield six different genotypes and four different phenotypes.

Blood is grouped on the basis of its clumping reaction, or agglutination, when in contact with other serum. Agglutination is a natural defense of the body. When foreign materials, or antigens, are introduced into the body, the host produces substances (compounds of protein and sugar) called antibodies, that agglutinate and thus immobilize the foreign antigen.

Blood of type A contains antibodies for both B and AB blood; type B contains antibodies for A and AB blood. Type AB blood has no antibodies, and type O has no antigens. These differences are extremely important considerations in giving blood transfusions. People with AB type blood are considered universal recipients, because they can be given any type of blood successfully. However, their blood can be given only to other AB type recipients, for it will cause agglutination in any of the other three blood types. People with type O blood are universal donors: if there is no time to type the blood group of a person needing a transfusion, it is safe to give him type O blood. The donor-recipient reactions of these four blood groups are summarized in Figure 13-13.

## Nonallelic gene interaction

Implicit in Mendel's work was the assumption that each phenotypic trait, such as color of the flower or shape of the seed, was controlled by one inheritance factor, or gene. But modern geneticists have proven that the one gene-one trait assumption is erroneous. Although Mendel was lucky enough to have chosen traits that do depend on only one gene for their expression, many similar traits are produced by the interaction of two or more pairs of nonallelic genes. For example, if Mendel had happened to work with sweet peas rather than garden peas, he would have found that his $F_2$ generation displayed the recessive flower color in about half of the individuals; the exact ratio would be 9:7, a modification of the 9:3:3:1 pattern. This is a marked difference from the ratio of 3 dominant to 1 recessive that Mendel found in garden peas. The reason for the difference is that in sweet peas, flower color is controlled by 2 pairs of genes, called **complementary** genes. The first gene pair consists of the gene for purple flowers and its recessive allele for white flowers. A second interacting gene pair consists of a gene for the expression of flower color and its recessive allele for colorlessness. If the second gene pair is homozygous for the recessive allele (colorless-

ness), then the flower will be white (colorless) even though the dominant gene for purple flowers is present in the other gene pair.

**Epistasis.** This type of nonallelic gene interaction, in which the effects of one pair of alleles overrides the effect of a second pair of alleles, is called **epistasis.** In sweet peas, the homozygous recessive is epistatic to the effect of the other gene. Epistasis in homozygous recessives is very common; it is found in the coat color of mice and in the feather color of many kinds of birds, as well as the color of the hard exoskeleton of certain beetles.

Another type of epistasis can be seen when the presence of one dominant gene masks the effect of another nonallelic dominant gene. An example can be seen in the inheritance of fruit

**13-13** *Blood group inheritance follows the simple dominance-recessiveness model of genetic theory. Inheritance is by multiple alleles: $I^A$ and $I^B$ are located at the same locus on the chromosomes, and result in blood types A and B respectively. As neither is dominant to the other, the $I^A I^B$ genotype has blood of AB group. Individuals with blood group O have the genotype ii.*

| Alleles of ABO blood group gene | $I^A$    $I^B$    $I^O$ | | O | AB | B | A |
|---|---|---|---|---|---|---|
| Genotype | | Phenotype (Blood group) | | | | |
| $I^A$    $I^A$ <br> $I^A$    $I^O$ | | A | | | | |
| $I^B$    $I^B$ <br> $I^B$    $I^O$ | | B | | | | |
| $I^A$    $I^B$ | | AB | | | | |
| $I^O$    $I^O$ | | O | | | | |

P₁

| YYWW White | × | yyww Green |

Gametes: (YW)     (yw)

F₁

YyWw White

Gametes: (YW)  (Yw)  (yW)  (yw)

F₂

|  | YW | Yw | yW | yw |
|---|---|---|---|---|
| **YW** | YYWW White | YYWw White | YyWW White | YyWw White |
| **Yw** | YYWw White | YYww Yellow | YyWw White | Yyww Yellow |
| **yW** | YyWW White | YyWw White | yyWW White | yyWw White |
| **yw** | YyWw White | Yyww Yellow | yyWw White | yyww Green |

F₂ dihybrid ratios: White ($--W-$) = 12
Yellow ($Y-ww$) = 3
Green ($yyww$) = 1

**13-14** *The white color of this summer squash exemplifies one type of epistasis. The presence of a dominant gene for white has masked the phenotypic expression of another nonallelic dominant gene for yellow. In contrast to dominance, which it superficially resembles, epistasis refers to interaction between nonallelic genes, not to interaction between alleles. (Alycia Smith Butler)*

color in summer squash. In one pair of alleles, the gene for yellow fruit is dominant to the gene for green fruit. We could call these alleles Y (yellow) and y (green). In a second gene pair, we find a gene for white, or colorless, fruit (W) dominant to a gene for colored fruit (w). The presence of the W gene masks the presence of both the Y and the y genes. In the F₂ generation (see Table 13-4), we will find 12 whites to 3 coloreds, or a ratio of 3:1. Only where no W is present will the effect of the second pair of alleles be visible. In those individuals, inheritance will conform to the same 3:1 ratio. So the results of a cross between heterozygous F₁ individuals with a genotype of WwYy will be 12 whites, 3 yellows, and 1 green, a modified 9:3:3:1 ratio.

Interaction between nonallelic gene pairs is often made more complex by incomplete dominance or codominance in one or both gene pairs. In the flour beetle, for example, there is complete dominance in one gene pair for body color and incomplete dominance in the second; moreover, the homozygous recessive is also epistatic to all other combinations. In bean plants, flower color is determined by two pairs of codominant alleles. A test cross between F₁ heterozygous individuals will yield nine different colors of flowers, a phenotype for each genotype.

Perhaps the most complex example of gene interaction yet studied can be seen in the inheritance of coat colors in guinea pigs. Geneticist Sewall Wright and his colleagues have been investigating this problem for more than 50 years, and many of the details of this inheritance process are still unclear. At least seven nonallelic genes are involved, and many of those genes appear to have a number of alleles, so there is an interaction between multiple genes and multiple alleles. Wright's research indicates that each of the seven genes is probably active at a different stage in the formation of the coat pigments. This discovery has important implications for solving the problem of how genes actually affect phenotypes, as we shall see in the next chapter.

**13-4**

## Nonallelic gene interaction

| | | Phenotype | Genotype | Ratio |
|---|---|---|---|---|
| Each gene pair affecting different characters | 1. Complete dominance of both gene pairs. | yellow–round | A – B – | 9/16 |
| | Example: Mendel's peas. Seed color: A–yellow | yellow–wrinkled | A – bb | 3/16 |
| | a –green | green–round | aaB – | 3/16 |
| | Seed shape: B–round b–wrinkled | green–wrinkled | aabb | 1/16 |
| | 2. Complete dominance of one gene, partial or codominance of the other. | polled–roan | A – Bb | 6/16 |
| | | polled–red | A – BB | 3/16 |
| | | polled–white | A – bb | 3/16 |
| | Example: cattle. | horned–roan | aaBb | 2/16 |
| | Horns: A–polled | horned–red | aaBB | 1/16 |
| | a –horned | horned–white | aabb | 1/16 |
| | Hair color: B –red b –white Bb–roan | | | |
| Both gene pairs affecting the same character | 1. Complete dominance: one gene when dominant is epistatic to the other. | white | A – – – | 12/16 |
| | Example: fruit color in summer | yellow | aaB – | 3/16 |
| | squash. | green | aabb | 1/16 |
| | A–white a –colored B –yellow b –green | | | |
| | 2. Complete dominance: interaction between the two dominants produces new phenotype. | disc | A – B – | 9/16 |
| | Example: fruit shape in summer squash. | sphere | A – bb or aaB – | 6/16 |
| | A–sphere a –long B –sphere b –long A + B–disc | long | aabb | 1/16 |

**13-15** *Comb shapes in fowl are described as single, pea, rose, and walnut. Inheritance of comb shapes is governed by the genes P, p, R, and r.*

Rose

Pea

Walnut

Single

**Collaboration.** Not all nonallelic gene interactions involve epistasis. In some cases, two nonallelic genes that affect the same character may interact to produce new phenotypes unlike any ancestral types. This phenomenon can be seen in the gene controlling comb shape in chickens. The pattern of comb inheritance was worked out by Robert Punnett, the geneticist who also gave his name to the Punnett square.

One pair of alleles (R and r) are the dominant rose comb and the recessive nonrose (single) comb; another pair (P and p) are the dominant pea and the recessive nonpea (single) combs. Any $F_2$ genotype that combines both R and P is a new phenotype, the walnut comb. The combinations RR and Rr with pp will produce rose combs; either PP or Pp with rr will produce pea combs. The homozygous recessive rrpp yields the single comb phenotype. As summarized in the Punnett square in Figure 13-16, the ratio will be 9:3:3:1.

## Environment and gene expression

In the early twentieth century, as geneticists began to have real success in working out the complicated patterns of inheritance and cytologists were able to observe the structure of chromosomes within the cell nucleus, biologists developed a theory of heredity in which the importance of the gene was paramount. The nucleus, always referred to as the control center of the cell, was thought to direct all metabolic and reproductive activity, and the cytoplasm was seen as nothing more than a matrix for the nucleus. Like dictators, the genes were presumed to control everything within their sphere of influence, the boundary lines of the cell.

As research techniques grew more sophisticated, new data forced a modification of the omnipotent gene theory. It was shown that cell organelles, such as the plastids, mitochondria, and ribosomes, can exert a strong influence over the activity of the cell; moreover, it was dis-

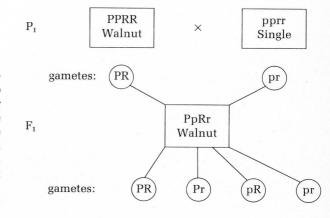

$P_1$   PPRR Walnut   ×   pprr Single

gametes:   PR            pr

$F_1$   PpRr Walnut

gametes:   PR   Pr   pR   pr

| $F_2$ | PR | Pr | pR | pr |
|---|---|---|---|---|
| **PR** | PPRR Walnut | PPRr Walnut | PpRR Walnut | PpRr Walnut |
| **Pr** | PPrR Walnut | PPrr Pea | PprR Walnut | Pprr Pea |
| **pR** | pPRR Walnut | pPRr Walnut | ppRR Rose | ppRr Rose |
| **pr** | pPrR Walnut | pPrr Pea | ppRr Rose | pprr Single |

$F_2$ dihybrid ratios: Walnut (P—R—) 9
Pea (P—rr) 3
Rose (ppR—) 3
Single (pprr) 1

**13-16** *Inheritance of comb types in chickens involves a form of collaborative interaction between nonallelic genes. Here, the allele R is the dominant rose comb and r is the recessive single comb. In a second gene, P is the dominant pea comb and p is the recessive nonpea comb. Collaboration in the genotypes produces a new phenotype, walnut comb, that neither gene pair could produce alone.*

**403**

covered that plastids and mitochondria contain their own DNA and thus constitute a cytoplasmic inheritance factor, in addition to the nuclear inheritance factors of the chromosomes. Recent research has stressed the importance of the cell membrane, acting in direct response to environmental factors, as an independent agent of organization, synthesis, and control. Current concepts of the cell emphasize the interrelatedness of all specialized areas of the cell.

This new concept of the cell has led to research by geneticists into the effects of the environment, both cytoplasmic and extracellular, on gene expression. They have concluded that no gene can be said definitely to produce a particular trait. It would be more accurate to say that a certain genotype, given the right environmental conditions, will influence development toward the formation of a particular trait or character.

**Gene expression and the cellular environment.** After a mitotic division, each cell contains all the genetic information it will ever have. Its future specialization or differentiation, its metabolic adaptations, its aging, and even its eventual death, are all contained within the DNA "program" in its chromosomes. For example, a new cell produced by the meristem in the tip of an onion root begins its life as a small cube-shaped cell. Later it elongates into a bricklike shape, and then it may differentiate to become a xylem or phloem tube, a root hair, or a thick-walled link in the Casparian strip. If it becomes a root hair, it will die within a few days; if it becomes a xylem cell it will die in the process of differentiation, and if it becomes part of the Casparian strip, it may live for years.

But these events in the life cycle of the cell happen sequentially, not all at once, and the sequence differs from cell to cell, even though each cell has exactly the same genes. It is clear that other factors in the cell must exert an influence over the process of gene expression. The exact mechanism of gene expression within the cell, and the role played by other specialized areas of the cell in regulating that expression, will be the subject of the next two chapters. We can summarize the process by saying that certain genotypes influence the development of a cell in a certain direction, resulting in the appearance of observable phenotypic traits. The development of these traits then permits the further expression of the genotype. In other words, development to a certain point provides the conditions in which the gene can express additional effects. Thus the age of a cell, its differentiation, its chemical environment, regulate the sequence of events in which the gene expresses itself, and in some cases, they may even alter the sequence.

**Gene expression and the extracellular environment.** The development of some genetic traits has been linked to the existence of certain environmental conditions. A good example of this phenomenon is the coloration of Siamese cats. The phenotype of the purebred seal point Siamese is light tan fur with darker brown or black fur on the paws, tails, ears, and muzzle. The gene that controls this pattern of coloration is called a Himalayan gene, for this pattern is also found in Himalayan rabbits. The expression of the Himalayan gene depends on a temperature differential; the warmer areas of the body develop light fur while the extremities, which typically have a lower temperature, develop dark fur. (This coloration helps keep extremities from freezing, since the dark fur absorbs more heat.) This pattern can be altered by changing the temperature. For example, if a purebred Siamese cat's tail is shaved and then wrapped in a warm insulating covering, the fur that grows back will be light tan like body fur, rather than the normal dark color. Conversely, light fur can be made to grow back as dark fur if an icepack is applied to a shaved area.

Another important environmental influence on gene expression is the presence of light. All sunbathers know that light can temporarily alter at least one phenotypic trait, that of skin color. Interestingly, it has the same type of effect on a variety of other organisms. The presence of light makes a certain type of corn turn red permanently; the gene for redness is inherited but blue-violet light must be present for it to express itself. Plants will fail to develop chlorophyll if they are kept in the dark, even though they contain the genes for that trait.

Other environmental characteristics that may influence the way a genotype is expressed in phenotypic characteristics include the nutrition of the cell and its chemical environment. A sad demonstration of the power of certain chemicals to alter a phenotype was seen when pregnant women were given thalidomide, causing deformities in their children.

## Heredity or environment?

A topic of much debate over the years, not just among biologists but also psychologists, sociologists, and educators, is the heritability of certain traits, such as intelligence, the tendency to be fat or thin, and the susceptibility to diseases like cancer and arteriosclerosis. How much of an individual's appearance, physiology, and personality is due to heredity and how much to environment?

This question has proved difficult to answer in any direct way; unlike a road map, a genetic code cannot yet be read, its components can only be guessed at by observation over a number of generations. Although it is clear that both heredity and environment are potent influences on an individual's development, the results of recent studies have led to an increased stress on the role of environmental conditions. For example, an interesting study carried out by Eleanor Storres on inheritance in armadillos demonstrated a wide range of variations in phenotypes with the same genotype. Dr. Storres chose the

**13-17** *The beautiful markings of a purebred seal point Siamese cat are actually quite functional. Coloration is controlled by the Himalayan gene, but its expression depends on a temperature differential. The dark fur which develops on colder areas of the body, such as the extremities, helps to keep them from freezing. Cat breeders take note that this pattern can be altered by changing the body temperature of specific areas. (Mary E. Browning from National Audubon Society)*

**405**

armadillo because it always gives birth to four identical quadruplets. Many sets of quadruplets, which had identical genotypes, were carefully weighed, measured, and subjected to biochemical tests. Dr. Storres found a range of 25 to 50 percent variation in such things as heart size, electrolyte balance, and metabolic rates. She concluded that this variation must be due to differences in the environment of each embryo because of its position in the uterus; some embryos are exposed to a richer supply of maternal blood, allowing their genotype to be more completely expressed.

Studies of environment and heredity in humans have focused primarily on twins. Phenotypic similarities are compared in identical twins (those which develop from a single fertilized egg and thus have identical genotypes) and fraternal twins (those which develop from two different eggs fertilized by two different sperm). Where the same trait is found in identical twins but not in fraternal twins, it is assumed that the presence of the trait is due primarily to heredity rather than environment. Some of the findings of these twin studies are shown in Table 13-5. According

to these results, even such personality traits as a tendency to criminality, or the habits of smoking cigarettes and drinking coffee, seem to be genetically influenced. However, geneticists are cautious in drawing conclusions from these studies, because as the experiments with armadillos show, both heredity and environment may be involved in the phenotypes of identical twins. It is quite likely that there are more similarities in the environment of identical twins before birth than there are in the environments of fraternal twins, since identical twins are usually attached to the placenta by the same cord and develop within the same protective membranous covering, whereas fraternal twins are separately packaged. Environmental similarities in identical twins may account for many instances of their phenotypic similarity.

One aspect of the environment versus heredity debate that has recently received a great deal of publicity is the question of hereditary differences in intelligence among races. A statistical study of I.Q. test scores undertaken by Arthur R. Jensen argues that on the average, blacks are less

**13-18** *All armadillos may look alike to you, but biochemical tests have shown that heart size, metabolic rate, and electrolytic balance may vary 25 to 50 percent even in identical quadruplets. Distinct environments are created by the position of the embryo in the uterus and affect expression of the genotype. (Karl H. Maslowski— Photo Researchers)*

## 13-5
## Twin studies

| | Phenotypic trait | Percentage of identical twins sharing trait | Percentage of fraternal twins sharing trait |
|---|---|---|---|
| Traits strongly determined by heredity | Hair color | 89 | 22 |
| | Eye color | 99.6 | 28 |
| | Diabetes mellitus | 84 | 37 |
| | Epilepsy | 72 | 15 |
| | Rickets | 88 | 22 |
| | Feeblemindedness | 94 | 47 |
| | Schizophrenia | 80 | 13 |
| | Criminality | 68 | 28 |
| Traits less strongly determined by heredity | Handedness (left-right) | 79 | 77 |
| | Measles | 95 | 87 |
| | Alcohol drinking | 100 | 86 |
| | Mammary cancer | 6 | 3 |

intelligent than whites, for racial averages show that blacks tend to be 11–15 I.Q. points lower than whites. Jensen's explanation for the difference is a difference in the gene pools of the two races, leading to inherited phenotypic differences in intelligence similar to the differences in skin color and hair texture. Although Jensen's suppositions are not unreasonable, his conclusions are highly debatable. Some scientists question his methods of data analysis, and others point out that standard I.Q. tests contain a cultural bias, since the vocabulary and general knowledge assumed is that of the dominant white culture rather than the black subculture. Jensen himself acknowledges this shortcoming and has suggested that other scientists begin research in this controversial area. One of the chief difficulties of such research is the problem in separating the effects of heredity from the effects of environment. Psychological studies have shown that exposure to certain kinds of mentally stimulating environments can raise a child's I.Q.; biological

and medical studies have shown that nutrition during pregnancy and the first few months of a child's life are critical for the fullest expression of the genotype. Since the average black in America belongs to a lower socioeconomic class than the average white, black parents often do not have the opportunity to provide either the nutrition or the stimulation needed for the development of a high I.Q. The significance of I.Q. scores, and the factors that affect them, are thus open to further study.

## Genetic change

The mechanisms of genetic inheritance serve to keep the anatomical and physiological characteristics of any given species fairly constant; most new individuals look remarkably like their parents and grandparents. But the process is not infallible, and it often happens that in the offspring there is some change of phenotype, caused by a change, or **mutation,** in the hereditary material from the parent. The rates of mutation are really quite high. One well-informed estimate is that about two billion people, or two-thirds of the total human population, carry some kind of genetic mutation. The vast majority of these mutations are insignificant. In some cases they produce no phenotypic change and in others, the change is so slight that it is virtually undetectable; only a small percentage of the mutations cause visibly different phenotypes. Most of these are in some way harmful to the individual. The proportion of harmful genes within a population is called its **genetic load,** and biologists assume that a high genetic load may endanger the existence of a species.

Although most detrimental mutations are **lethal,** causing the death of the mutant phenotype, others, called **semilethals,** permit some proportion of affected individuals to survive. Dominant lethal genes tend to disappear from the gene pool in a short time, since a single dose of the gene kills the individual. Recessive lethals, however, may remain within the gene pool, since individuals who are heterozygous for the trait are viable. An example of a recessive lethal can be seen in a variety of snapdragons, *Antirrhinum majus aurea.* The plant has yellow leaves in the heterozygous condition, due to the dominance of the gene for yellow leaves (Y) over the gene for green leaves (y). When a cross of Yy × Yy is performed, the result is an $F_2$ consisting of yellow and green plants in the unusual ratio of 2:1. When the yellow $F_2$ plants are backcrossed to green plants, they produce some green offspring, proving that they are heterozygous rather than homozygous for yellow. The odd 2:1 ratio can be explained by the fact that homozygous YY plants are inviable; it turns out that they lack the ability to make chlorophyll.

The gene for sickle-cell anemia is an example of a recessive lethal in the human population. In the homozygous condition, this gene causes either death or debility; the penetrance of the gene varies, so that not all affected individuals display the fully lethal phenotype. In the heterozygous condition, however, this gene may be a valuable adaptation, since it is thought to alter the red blood cells just enough to make them resistant to the malarial parasite. Thus sickle-cell anemia, which is found largely in blacks, is really a part of their African heritage.

A third possibility is that the mutation will be advantageous to the individual, helping it to become better adapted to the environment. Although we tend to think of mutations in terms of harmful effects (no doubt because of the monsters so prominently featured in science-fiction books and films) the process of mutation is also responsible for the ability to adapt to the

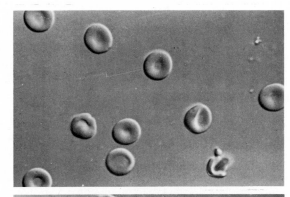

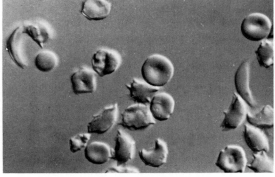

**13-19** *As long as the oxygen content of the blood remains high, sickled red blood cells maintain a normal shape, and their function is unimpaired. But when the oxygen content of the blood drops, the cells display the characteristic sickled shape shown here, and their efficiency at transporting oxygen decreases noticeably. (Courtesy of Dr. Cerami)*

**13-20** *Fruit may be larger as a result of polyploidy. In these photos, taken to scale, the wild raspberry on the left has the normal number of chromosomes. The cultivated raspberries on the right are tetraploids resulting from mutations deliberately induced by chemicals. Polyploidy may inhibit reproduction, affect flavor, and have other less predictable physiological effects. (Alycia Smith Butler)*

environment that characterizes living things and distinguishes them from the nonliving world.

## Change in chromosome number

One source of genetic change is a change in the number of chromosomes. Variations in the number of sets of chromosomes in an organism is called **euploidy,** and it is known to happen rather frequently in plants. It may be the result of fertilization of one egg by two sperms, yielding an organism with 3n chromosomes. It can also be the result of the failure of meiosis, so that gametes are 2n rather than n. Euploid individuals are often larger and hardier than normal individuals, but they are frequently sterile or fail to reproduce euploid individuals. A number of cultivated plants, such as coffee, corn, and alfalfa, are the results of euploid mutations purposely induced by chemicals that inhibit meiosis. New seeds must be obtained for many of these crops every year.

When the change in chromosome number involves only a single chromosome within the set, it is called **aneuploidy.** Aneuploidy is probably the result of an accident during meiotic division. The intertwined homologous pairs fail to separate properly, so that one new cell has two of the same chromosome, while the other cell has none. Aneuploidy is frequently lethal. The individual lacking a chromosome cannot develop properly since it is missing some genetic information; the individual with an extra chromosome usually has some imbalance of genetic factors that is disadvantageous and leads to early death. One known example of the effects of an additional chromosome in humans can be seen in the disability known as mongolian idiocy, or Down's syndrome. This condition, characterized by both mental retardation and physical deformities, is caused by the presence of an extra chromosome added to the pair designated as 21.

Abnormalities in the numbers of X and Y

**409**

chromosomes are less likely to be lethal. Both males and females may have additional X chromosomes. Apparently these extra chromosomes can in some way cancel each other out; although they do cause some mental and physical abnormality, the individual is viable. In the early 1960s, it was found that some males may also have more than one Y chromosome, and that the additional Y may lead to increased aggressiveness in the man's behavior. A number of men convicted of crimes of violence have proved to have this extra Y chromosome, but not all criminals have this abnormality, nor do all those having the extra chromosome exhibit criminal behavior.

## Change in chromosome structure

A second source of genetic change is change in the structure of single chromosomes. The process of crossing over is one example of this type of change. Chromosomes may tangle and break during prophase, recombining parts of homologs. Breakage and recombination can sometimes take place in such a way that one chromosome is missing a piece while the other has several extra genes; this process, called **translocation,** is diagrammed in Figure 13-21. Events of breakage and recombination are also thought to be responsible for the existence of duplications of sections of the chromosome, as when the genes are found in the sequence ABCDEDEFG. Duplication has been found frequently in the chromosomes of the fruit fly, and the double dose of some genes often affects the phenotype, causing a new eye color or wing type.

Chromosomes may also change structure due to **inversions,** or loops formed during meiosis. One of the homologous chromosomes may form a loop during prophase; then if breaking occurs the nearest end of the chromosome will recombine (see Figure 13-22) so that the genes on that part of the chromosome will be present in reverse

Unequal crossover — gene duplication

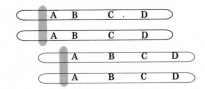

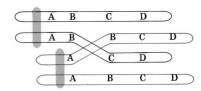

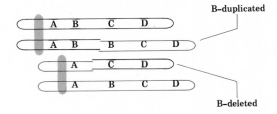

B–duplicated

B–deleted

**13-21** *In translocation, sections of a chromosome are replaced by analogous sections from another. This can be an important means of recombination.*

Chromosomal inversion

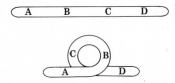

**13-22** *Another important means of recombination is inversion. This occurs when a section of the chromosome becomes looped during synapsis. The result is to reverse the order of a number of genes, possibly affecting their expression and their subsequent sorting during reproduction.*

order. A. H. Sturdevant and C. E. Plunkett, in mapping fruit fly chromosomes, found an example of inversion in the third chromosome. In *Drosophila melanogaster*, the genes are present in the sequence ABCDEF; in its near relative, *D. simulans*, the sequence is ABEDCF.

It sometimes happens in the process of breakage and recombination that one chromosome becomes split into two fragments. Generally, the fragment that bears the centromere will continue to move in the normal way. The other fragment does not separate and does not move from the metaphase plate. It is often incorporated at random in one of the two new nuclei. Some biologists believe that this type of accident accounts for the wide variation in chromosome numbers in living things. Some organisms, such as the fruit fly, have as few as four pairs of chromosomes, whereas parthenogenically reproduced brine shrimp have 160. The supposition is that in the course of evolution, some organisms had many instances of chromosome fragmentation; in some of the fragments, other genes were able to take the place of the centromere, enabling both parts of the fragment to act as an independent chromosome. It is interesting to note in this connection that a larger number of chromosomes does not necessarily mean a larger quantity of genetic information. There are just fewer bits of information on each chromosome.

## Change in gene structure

A third source of genetic change is some accidental change in the fine structure of the gene itself. Such changes may be due to some mistake in the replication of the DNA that occurs naturally without environmental stimulus. Certain chemicals can speed up the rate of genetic mutation by interfering with the chemical reactions through which duplication takes place. One such chemical commonly used in laboratory experiments is the dye proflavin; the antibiotic streptonigrin and the mustard gas used in trench warfare in World War I are also known to cause genetic mutation. Some of the chemicals used as food preservatives are also suspected of this property.

Perhaps the most important source of genetic mutation is irradiation. X-rays, ultraviolet rays, and radioactive rays from radium and cobalt are all known to cause increases in the rate of mutation. Studies of the survivors of the atomic attacks on Hiroshima and Nagasaki have shown that the rate of some kinds of harmful mutations has increased greatly.

Corn and a few other plants have been found to contain special mutator genes on their chromosomes. The presence of these genes enables mutation of other genes for seed color and plant size, permitting some degree of adaptive change. It is possible that other mutator genes will be found as research in this field goes on.

One question that geneticists are trying to answer is why the rates of mutation are different for different genes. Although the accident that causes a mutation is presumably a random process, it can be demonstrated statistically that certain genes mutate much more frequently than others. One possible explanation is that the structure of the DNA molecule at that point is relatively unstable. A more intriguing possibility is that there might be some hereditary factor that serves to control or direct the process of mutation to some extent.

# Summary

It is evident that reproducing organisms pass along some kind of hereditary material, but until Gregor Mendel studied heredity in garden peas, no pattern of transmission had yet emerged. Working with paired traits, such as yellow and green seeds, Mendel established the fact that one trait is dominant to the other. He also determined that plants heterozygous for the dominant trait will have the same phenotype as homozygous dominance, even though they have a different genotype.

Mendel's work led to two important genetic principles. One, the principle of segregation, states that dominant and recessive factors do not merge in the offspring, but can be passed along separately to the next generation. The principle of independent assortment states that the inheritance of one trait will not in any way interfere with the inheritance of another trait.

When Mendel's findings were brought to the attention of the scientific community, other researchers began to realize that Mendel's inheritance factors (or what we now call genes) behaved just as chromosomes do during the process of meiosis. Thus it was established that genes are located on chromosomes.

One type of inheritance which has been the subject of much study is sex determination. In fruit flies and man, the XY pattern of sex determination is found; females have two X chromosomes and males have one X and one Y. In fruit flies, the ratio of X chromosomes to autosomes, or genetic balance, is also important in sex determination. Another pattern of sex determination is the XO type, in which females have two X chromosomes and males only one.

Since there is more than one gene on each chromosome, not all genes actually do assort independently; certain linkage groups have been established. The sex-linked traits such as hemophilia and red-green color deficiency are well-known examples of linkage groups in humans. In Drosophila, four linkage groups have been established, corresponding to the haploid number of chromosomes, and these linkage groups have even been mapped. Mapping is performed by the analysis of the rates of crossing over in chromosomes. The farther apart the genes are on the chromosomes, the higher the frequency of crossing over.

Neo-Mendelian genetics has focused on the study of the penetrance and expressivity of genes. It has been found that not all pairs of alleles have a simple dominant-recessive relationship. In some cases, there is limited expression of both genes, producing a different phenotype for the heterozygous genotype. This is called codominance. In other cases, more than two alleles may be involved, so that there is a hierarchy of dominance. This can be seen in the inheritance of blood groups in humans.

It has also been found out that nonallelic genes may interact with one another. The effect of one pair of alleles may underride the effect of a second pair; this is

**412**

called epistasis. Nonallelic genes may also interact to produce new phenotypes. This pattern of inheritance, through complementary genes, is found in the comb shape of domestic fowls.

Although it was once assumed that the program contained in genetic material was the sole influence on the development of a cell or individual, it has more recently been established that the environment of the genetic material plays an important part in heredity. Chemical and physical changes in the cell's environment can lead to alterations in the individual's phenotype.

Changes in phenotype may also be due to changes in the genetic material itself. One such change may occur in the chromosome number, producing polyploid or aneuploid individuals. Change may also take place in the structure of the chromosome, due to translocations or inversions. A third source of change is some alteration in the fine structure of the gene. Such mutations may take place naturally, through accidents in the replication of DNA. Mutations can also be caused by a number of chemical agents and by irradiation.

## Bibliography

### References

Carlson, E. A. 1966. *The Gene: A Critical History*. Saunders, Philadelphia.

Dobzhansky, T. G. 1967. *Evolution, Genetics, and Man*. Wiley, New York.

Fincham, J. R. S. and P. R. Day. 1965. *Fungal Genetics*. 2nd ed. Davis, Philadelphia.

Rieger, R. 1968. *A Glossary of Genetics and Cytogenetics*, 3rd ed. Springer-Verlag, New York.

Shields, J. 1962. *Monozygotic Twins*. Oxford University Press, New York.

Srb, A. M., R. D. Owen, and R. S. Edgar. 1965. *General Genetics*, 2nd ed. Freeman, San Francisco.

Stahl, F. W. 1964. *The Mechanics of Inheritance*. Prentice-Hall, Englewood Cliffs, N.J.

Sturtevant, A. H. 1965. *A History of Genetics*. Harper & Row, New York.

Wills, C. 1970. "Genetic Load." *Scientific American*. 222(3): 98–107. Offprint no. 1172. Freeman, San Francisco.

## Suggested for further reading

Brink, A., ed. 1967. *Heritage from Mendel*. University of Wisconsin Press, Madison.
A collection of 21 papers by different authors presented at the 1965 symposium of
the Genetics Society of America. Such titles are included as "Mendel and the Gene
Theory," "The Molecular Basis of Genetic Recombination," "Gene Action at the
Level of the Chromosome," and others on such topics as gene action, mutation,
genetic coding, and practical applications of genetics.

Burnet, M. 1971. *Genes, Dreams and Realities*. Basic Books, New York.
This book, written by the 1960 recipient of the Nobel Prize for medicine, first
introduces the concept of what a gene is, tells what genes do, and then goes on
to discuss genetic diseases, applications of molecular biology to curing cancer,
and other topics, with emphasis on social relevance.

Jinks, J. L. 1964. *Extrachromosomal Inheritance*. Prentice-Hall, Englewood Cliffs, N.J.
This book treats the subject of forms of inheritance not directly transmitted via the
DNA. Topics are discussed such as whether mitochondria and chloroplasts are
capable of independent division. Many other examples are given where strict
Mendelian genetics do not seem to explain everything that is observed.

Medvedev, Z. A. 1969. *The Rise and Fall of T. D. Lysenko*. Trans. by I. M. Lerner.
Columbia University Press, New York.
This is a largely political account of the bizarre and unfortunate effects of the
imposition of governmental control on the direction of research in biological
sciences in the Soviet Union between the years 1937–1964. One man's fallacious
theories of genetics were taught and practiced during this time, putting the Soviet
Union, for a time, far behind other countries in the development of the modern
science of molecular genetics.

Paterson, D. 1969. *Applied Genetics*. Doubleday, Garden City, N. Y.
This book, which contains color photographs and many attractive charts and
graphs, includes a history of genetics, a discussion of breeding of better crop
plants, and especially concentrates on applications of genetics toward the
treatment and understanding of at least 800 known genetic diseases.

Stebbins, G. L. 1971. *Chromosomal Evolution in Higher Plants*. Addison-Wesley,
Reading, Mass.
The book discusses the adaptive value of chromosome structure as it is known, as
well as how this structure may play a role in directing genetic recombination and
gene expression. It also discusses the biological consequences and evolutionary
significance of polyploidy, the phenomenon in plants of occurrence of multiples
greater than two of the haploid number of chromosomes.

Strickberger, Monroe W. 1968. *Genetics*. Macmillan, New York.
This is a good basic text that covers all topics discussed in Part III of our text. A
useful reference to the beginning student.

## Books of related interest

Osborn, F. 1968. *The Future of Human Heredity.* Weybright & Talley, New York.
This book is about eugenics, the application of genetic knowledge towards improving the human race. It begins with a reasonably technical introduction to the subject of population genetics and discusses how birth control relates to it. The author optimistically pictures a future world where man will have taken the regulation of reproduction entirely into his own hands.

Scheinfeld, A. 1972. *Heredity in Humans.* Lippincott, Philadelphia.
This well-illustrated book deals with such topics as predicting a child's appearance, from eye color to personality. It also discusses hereditary aspects of criminal behavior, longevity, and Rh babies.

# 14

# The molecular basis of heredity

When Mendel began his work, more than a hundred years ago, he was trying to determine how the fundamental units of heredity operate. He succeeded in establishing certain patterns of inheritance, but he was unable to discover anything about the nature of the units through which the pattern was transmitted and controlled. Real progress in this aspect of genetic study has come only in the last 20 to 25 years.

In the preceding chapter, we formulated two working definitions of a gene. Defined structurally, it was said to be a portion of the DNA molecule that carries all informational content; defined functionally, it was considered the smallest independently inheritable unit. This pair of definitions reflects the two different approaches to the study of molecular genetics, which yielded important new information and, not surprisingly, revised definitions of the structure and function of a gene.

Seeking to understand the chemistry of the gene, some investigators attacked the problem directly, through various kinds of chemical analysis. Others took a different tack; since genes produce morphological effects (wrinkled peas or smooth peas, for example), their conviction was that if they could gain understanding of the steps between gene and effect—i.e., of the workings of genes—then they should be able to gain insight into what a gene is.

Both approaches were to prove rewarding. One might say that they converged—as the arms of a Y do at the apex—in the now-famed Watson-Crick double helix, or spiral staircase, model of DNA in 1953, and in the many significant advances since.

## The nature of gene action

A relationship between genes and body chemistry was postulated only a few years after the rediscovery of Mendelian genetics, but it was nearly

40 years before convincing evidence gathered in support of this hypothesis.

As early as 1902, Sir Archibald Garrod, an English physician, studied a rare human heredity disease, alkaptonuria, in which a major symptom is urine that turns black on exposure to air. The disease had been recognized for hundreds of years, and it was known that a crystalline acid, alkapton, was involved in the urine darkening which was the result of the body's inability to break the substance down by oxidation. In normal individuals, an enzyme catalyzes a reaction by which alkapton is converted into carbon dioxide and water. Garrod suggested that alkaptonuria is inherited according to the typical Mendelian pattern. People who suffer from the disease are homozygous for a recessive gene (probably a mutant) that does not allow production of the enzyme needed to break down alkapton.

Like Mendel, Garrod was far ahead of his time. It was almost three decades before scientists rediscovered Garrod's striking new concept that "inborn errors of metabolism" stemmed from genes that did not permit normal enzymatic reactions to take place. Proof of Garrod's very early insight that genes control enzymes had to wait for the work of G. W. Beadle and E. L. Tatum in the 1940s.

## The one gene–one enzyme concept

Beadle and Tatum, who were to win the Nobel Prize for their achievement, made use of a new experimental approach to the study of genetic control of metabolic reactions. Their approach—based on the idea that it should be possible to discover what genes do by making them defective—has been explained by Beadle in this analogy:

The manufacture of an automobile in a factory is in some respects like the development of an organism. The workmen in the factory are like genes—each has a specific job to do. If one observed the factory only from the outside and in terms of the cars that come out, it would not be easy to determine what each worker does. But if one could replace able workers with defective ones, and then observe what happened to the product, it would be a simple matter to conclude that Jones puts on the radiator grill, Smith adds the carburetor, and so forth. Deducing what genes do by making them defective is analogous and equally simple in principle.

Beadle and Tatum chose to work with a fungus, the red bread mold *Neurospora*. One reason for the choice was that *Neurospora* has a short life cycle (only 10 days) and multiplies profusely by asexual spores; a strain can be multiplied a millionfold within days, without genetic recombination. Most important, *Neurospora* cell nuclei are haploid. This means that any mutations that occur in a gene will not be masked by the action of a normal allele. Any change in the genetic material will bring a corresponding change in the phenotype.

Beadle and Tatum placed individual *Neurospora* spores in a series of test tubes containing a complete chemical growth medium (nitrate, sulfate, phosphate, other inorganic substances, sugar, and biotin, a vitamin of the B group). Then they transferred samples of the fungal growth to test tubes containing a minimal medium. A normal fungus, one that can conduct all metabolic processes, is able to grow in a minimal medium. If a fungus could not grow there, it must have come from a spore with a gene mutation that blocked an essential metabolic step. Growth could take place only in a medium that supplied the substance the fungus had lost the ability to produce for itself. It was possible to determine exactly which substance the mutant could not produce, by placing it in various minimal media to which single supplements were added. For example, one strain of the mold which did not

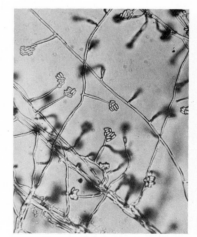

**14-1** *The rapid rate at which* Neurospora *reproduces makes it an excellent subject for genetic study. A useful trait of this fungus is the fact that new daughter cells are contained within a sac and are thereby fixed in the order in which they are produced. (Omikron)*

**14-2** *Pioneering experiments performed by Beadle and Tatum with Neurospora crassa, the red bread mold, led to formulation of the one gene–one enzyme hypothesis. First, through irradiation they induced mutations affecting nutrient requirements of the mold. A mutant unable to grow normally in a minimal medium was found to grow in a medium supplemented with amino acids. It was then established which particular acid the mutant was unable to synthesize. In this experiment, Beadle and Tatum isolated a mutant for methionine. Experiments mutating other single genes indicated that one gene alone directed formation of one particular enzyme.*

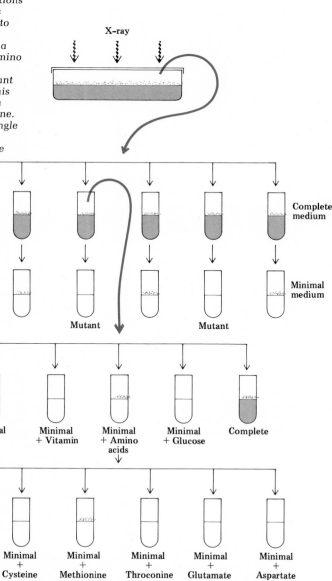

grow on the minimal medium began to show growth when pantothenic acid was added. Thus it was apparent that the mutation could not synthesize pantothenic acid, a necessary vitamin.

Beadle and Tatum proceeded to determine whether inability to produce the vitamin was transmitted as a single unit of inheritance. When they crossed a mold lacking the ability with a normal mold, the resulting spore sacs always contained four spores that produced molds like one parent and four that produced molds like the other parent. In each set of eight daughters, four could make pantothenic acid, and four could not.

By irradiation, Beadle and Tatum mutated genes involved with other specific chemical reactions in *Neurospora*, and in each case they established that single genes are directly concerned with single chemical reactions. Although they still had no idea of the chemical nature of a gene, Beadle and Tatum, on the basis of their experiments, proposed a one gene–one enzyme theory. Each gene, they postulated, has just one major function—to direct the formation of one enzyme and thus to control the chemical reaction made possible by that enzyme.

## Bacterial evidence

Soon after the bread mold experiments, Tatum joined with Joshua Lederberg to extend the one gene–one enzyme concept to bacteria. They chose to work with the ordinary colon bacillus, *Escherichia coli*, and were able to isolate two mutant strains. Each strain had lost the ability to produce two important amino acids. Each mutant could be grown successfully only on a nutrient medium containing the amino acids it was missing. Yet, when both strains were grown together on a medium lacking all four of the amino acids, some of the bacteria flourished. It seemed that these bacteria had somehow regained the ability to produce the amino acids.

**419**

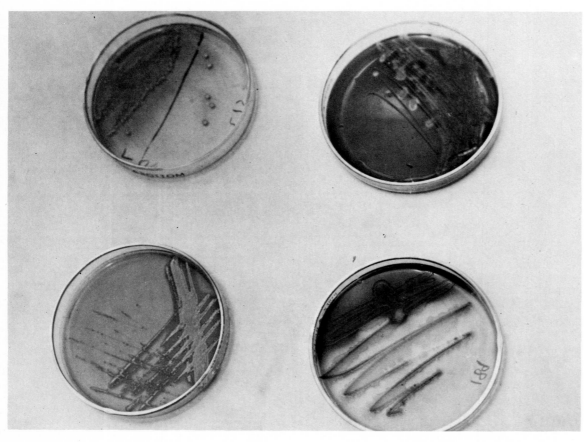

**14-3** *Much of our present understanding of the molecular basis of life has come from studies of bacteria. Many species can be cultivated in the laboratory in various media, and they are highly suitable test organisms in biochemical research. As a group, bacteria are the smallest cells known. However, they always contain both DNA and RNA, many different proteins with enzymatic functions, and the raw material and metabolic machinery for their own reproduction. (Morris Huberland from National Audubon Society)*

It had long been believed that bacterial reproduction is entirely asexual. But the recovery by some organisms of the ability to produce the amino acids suggested that they had inherited and recombined traits from the two mutant strains. Lederberg and Tatum demonstrated that, in fact, sexual reproduction does occur. Bacteria of unlike mating types, they note, adhere side by side, and a tiny conjugation tube forms between them. Genetic material can pass through this tube from one cell to the other. *E. coli* has a single chromosome that passes from donor to recipient cell.

Subsequent experiments established that the chromosome enters the conjugation tube always by the same end, and that a complete transfer may require about an hour. But the transfer is often interrupted, and the cells break apart before it is completed. The recipient cell then retains whatever portion of the chromosome has been transferred, while the remainder is left with the donor cell. Some genes are transferred even during very brief conjugations; others, not unless full-term conjugation is achieved. That would explain why only some, not all, the bacteria regained the ability to produce the amino acids.

Isolating mutant *E. coli* had been tedious work, involving the random hand-picking and testing and retesting of thousands of colonies. But in 1948, Lederberg simplified the process greatly. He made use of the fact that the antibiotic penicillin kills growing bacteria by interfering with production of the bacterial cell wall. When bacteria grow in a medium containing penicillin, the organisms soon reach the point of outgrowing their cell walls, and they literally burst apart.

Therefore, if bacteria are placed in a minimal medium along with penicillin, the normal bacteria will grow rapidly. The growth of mutants however, will be slowed, because of the lack of nutrients they cannot produce for themselves. Before long, the rapidly growing normal bacteria will be killed off by the penicillin. The remaining bacteria will be those that cannot grow—the mutants; they can then be removed, transferred to a complete medium to grow, and then transferred to various incomplete media for study.

The development of this technique meant that many different mutants with varying mutations could be selected and studied. On the basis of analyses of various bacterial as well as fungi mutants, support steadily grew for the one gene–one enzyme theory.

**Bacterial transformation.** In 1928, Fred Griffith, an English bacteriologist, was studying the pneumococcus, one of the agents responsible for human pneumonia. The bacterium is very deadly for mice; within 24 hours after mice are injected with sputum from a patient with pneumococcal pneumonia, they die. The pneumococcus owes its ability to produce serious disease to a thick polysaccharide capsule it forms outside its cell wall. The capsule serves to protect the organism against the usual defense mechanisms of the host animal.

In addition to studying a virulent strain of pneumococcus with a thick capsule, Griffith was experimenting with a nonvirulent strain that had no capsule. He turned up a puzzle. When he injected mice with a mixture of the nonvirulent strain and heat-killed virulent organisms, the mice died. Post-mortem examination showed their bodies to be full of live virulent organisms. Somehow, nonvirulent bacteria had been transformed to virulent; when the transformed organisms were removed from the dead mice and cultured, they went on producing virulent offspring. It seemed that hereditary material from the dead virulent bacteria in the injection mixture had restored to nonvirulent bacteria the ability to form capsules they had earlier lost by gene mutation.

Other investigators, repeating Griffith's work, got the same results, even without using mice. If a culture of mutant, capsuleless pneumococci was grown in a test tube in the presence of heat-killed virulent bacteria, the mutant strain soon regained virulence. At Rockefeller University, James L. Alloway went further and demonstrated that it was not even necessary for the transformation to have whole dead bacteria of the virulent strain present. If mutants were placed in a test tube to which was added only some fluid in which dead virulent organisms had been dissolved, transformation took place.

In 1944, at Rockefeller University, Oswald T. Avery, Colin MacLeod and Maclyn McCarty analyzed the cell-free extract of dead virulent bacteria such as Alloway had used. When they fractionated the material, they found that they could remove proteins, lipids, polysaccharides, and ribonucleic acid. What was left still transformed nonvirulent bacteria into virulent. With still further fractionation, they established that what was left was virtually pure DNA. When they added just a tiny amount of this DNA to a culture of nonvirulent pneumococci—as little as one part in $6 \times 10^8$—transformation took place. Moreover, when they extracted DNA from a culture of virulent bacteria that had descended from a transformed bacterium, that DNA, too, could produce transformation. It appeared that DNA could return to mutant bacteria the ability to

**421**

produce the enzyme required for making the capsule polysaccharide, and that the DNA could replicate itself so that offspring were virulent.

Despite the evidence, the scientific world was not yet convinced that DNA was the stuff of genes. One objection was that the DNA which transformed the mutant bacteria might not have been 100 percent pure; the tiniest bit of protein lingering as a trace contaminant might really have been the hereditary material. Even when, in 1949, Rollin D. Hotchkiss succeeded, through further fractionation, in getting DNA with maximum protein contamination limited to 0.02 percent, there was still doubt.

Another objection was that even if DNA did cause the transformation, it might do so because of some direct chemical effect on formation of the capsule rather than as a carrier of hereditary instructions. Another experiment by Hotchkiss helped overcome the latter objection. He succeeded in finding a penicillin-resistant mutant of pneumococcus. When he extracted its DNA and added the DNA to a culture of penicillin-sensitive organisms, some of them were promptly transformed into penicillin-resistant organisms. It appeared that in addition to carrying instructions for capsule formation, DNA also carried blueprints for the formation of other structures that provide resistance to penicillin.

### Evidence from viruses

Additional support for considering DNA to be the carrier of hereditary information came in 1952 from work showing that bacterial viruses transmit genetic specifications to their offspring in the form of DNA.

At the Carnegie Laboratory of Genetics, at Cold Spring Harbor, New York, Alfred D. Hershey and Martha Chase carried out a series of experiments with a bacteriophage that attacks *E. coli* bacteria. It is well established that viruses consist of a protein shell or overcoat and a core of nucleic acid. Hershey and Chase made use of the fact that DNA contains phosphorus but no sulfur; protein, on the other hand, contains sulfur but no phosphorus. Hershey and Chase grew bacteria in a medium containing a radioactive isotope of phosphorus and another of sulfur. Phages were then introduced, and they promptly took up the sulfur and the phosphorus, incorporating the sulfur into their protein and the phosphorus into their DNA. In the next step, other nonradioactive bacteria were infected with the radioactive phage. After the phage had become attached to the bacterial walls and sufficient time had passed to allow them to inject their hereditary material into the bacteria, the bacteria were agitated in a common food blender and shorn of any particles attached to their walls. Analysis of the particles showed they contained no radioactive phosphorus, only radioactive sulfur; therefore only the phage's protein overcoat had been left outside the bacterial cells. Analysis of the bacteria showed they contained no sulfur, only phosphorus; therefore, only DNA had been injected into them by the phage. It was the DNA alone which carried into the bacterial cells all the genetic instructions needed for the rapid production within the cells of new phage.

### The chemical structure of the gene

As far back as 1869, a Swiss biochemist, Friedrich Miescher, was studying pus cells and decided to use pepsin, the chief enzyme of gastric juice, to make the proteins of the pus cells soluble. Miescher noticed that the nucleus of a cell behaved peculiarly: it became just a bit smaller but otherwise withstood the action of pepsin. It didn't behave like a protein.

Miescher also noted that nuclear material contained phosphorus in addition to carbon, oxygen, hydrogen, and nitrogen found in proteins. He named the material "nuclein," and

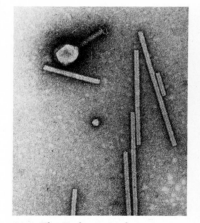

**14-4** *The T4 bacteriophage is a virus that parasitizes the common colon bacillus* Escherichia coli. *Research involving such bacteriophages has revealed many important facts about basic genetic processes and mechanisms.*

**14-5** *Hershey and Chase designed an experiment to determine the chemical composition of genetic material in one type of bacteria-destroying virus. They were able to prove that DNA alone transmitted all the genetic information needed to produce new bacteriophage cells. Their results supported earlier evidence that DNA rather than protein was the basic material of inheritance.*

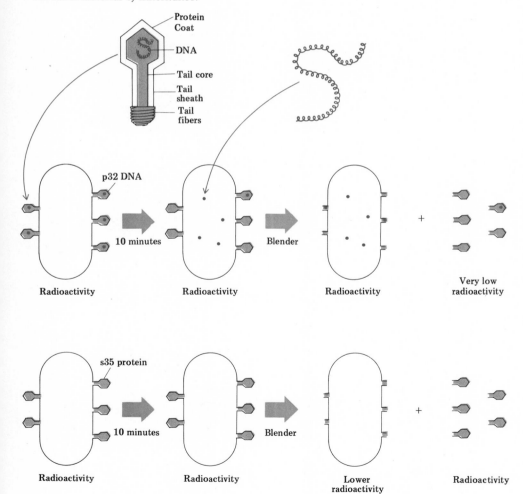

later, because of its acidity, it became known as nucleic acid. Miescher had been working with one of several types of nucleic acid—DNA, or deoxyribonucleic acid—and his was a first, very early clue to the possible chemical nature of the genetic material.

Decades went by before another clue came. In 1914, Robert Feulgen, a German chemist, developed a stain that proved capable of showing up DNA in crimson. Some 10 years later, Feulgen, using the stain, established that DNA is located in the chromosome.

Then, in 1948, the staining technique was taken up simultaneously by two different research teams: A. Boivin and R. Vendrely at the University of Strasbourg, France; and Alfred E. Mirsky and Hans Ris at Rockefeller University. They employed staining as a means of measuring the DNA content in nuclei from cells of different tissues—liver, kidney, nerve, muscle, heart. They found that while there were considerable variations in amounts of many other substances, the amount of DNA was the same in all the nuclei. Both teams also showed that in the nuclei of egg and sperm cells there was only half as much DNA as in cells from all other tissues. This was as it should be if DNA were the stuff of genes, since in meiosis gamete cells receive only half the amount of genetic material distributed to body cells.

But in an essay he wrote two years later, even Mirsky had reservations. He went only as far as to suggest that "if this component (DNA) of the chromosomes is indeed present in constant amount in the different somatic cells of an organism and in one-half this amount in the germ-cells, then it may be said that DNA is a *part* of the gene substance."

The fact was that even in 1950 most biologists, aware that chromosomes of most organisms contain protein as well as nucleic acid, were convinced that protein *must* be the genetic material. It seemed to them that no other material but protein was complex enough to be able to handle so much complex information. DNA, which was thought to be a monotonous macro-molecule, was dismissed as being too simple.

## The structure of DNA

Although many thousands of atoms are present in DNA, the molecule is made up of a sequence of just a few building blocks. Each such block, called a nucleotide, has a phosphate group, a 5-carbon sugar called deoxyribose, and an organic nitrogen-containing base. The base and the phosphate group are bonded to the opposite ends of the sugar.

It was known that DNA has four types of nucleotides, all with identical phosphate groups and sugars, differing only in their nitrogenous bases. The four bases were adenine, guanine, thymine, and cytosine. Adenine and guanine, with double-ring structures, were classified as purines; thymine and cytosine, with single-ring structures, as pyrimidines. The bonding of the DNA molecule, with the sugar of one nucleotide attached to the phosphate group of the next nucleotide, establishes a long chain of alternating sugar and phosphate groups, while the nitrogenous bases form side groups of the chain.

For many years it was thought that DNA was a simple tetranucleotide (composed of four alternating nucleotides). Even in the 1940s, when it became apparent that DNA had a molecular weight much greater than could be explained by the tetranucleotide concept, it was still accepted that the basic repeating unit of the large DNA polymer was the tetranucleotide, with four purine and pyrimidine bases recurring in regular sequence.

DNA did, indeed, seem to be a monotonously uniform macromolecule to which it seemed improbable to assign a genetic role. If such a molecule could carry information, it would probably be no more information than could be conveyed by a "sentence" made up of a single letter.

Basic Structure:

4 Nitrogeneous bases of DNA:

Thymine

Adenine

Cytosine

Guanine

But, by 1948, Erwin Chargaff at Columbia University had begun to adapt paper chromatography—a newly developed technique for separating and quantitatively analyzing chemicals—to nucleic acids. Soon Chargaff was able to show that, contrary to the pattern called for by the

**14-6** *The basic structure of a nucleotide from DNA includes deoxyribose, a 5-carbon sugar, attached to a phosphate group, and a nitrogenous base. The four nitrogenous bases of DNA include two pyrimidines—thymine and cytosine, both single-ring bases; and two purines—adenine and guanine, both double-ring bases. Invariably adenine pairs with thymine, and guanine with cytosine.*

**14-7** *The early conception of DNA pictured a simple tetranucleotide repeated to form a long DNA polymer. The chain was monotonously composed of adenine, thymine, guanine, and cytosine linked in unvarying sequence and quantity.*

**14-8** *The earlier simple tetranucleotide model was revised when it was found that the four nucleotides do not always occur in equal amounts in DNA, but rather in a 1:1 ratio.*

**14-9** *The specificity of the DNA is determined by the particular sequence of bases that occur in any nucleotide. The order of occurrence varies for different DNA molecules.*

simple tetranucleotide view of DNA, the four bases are not always present in equal proportions. When he extracted DNA from calf thymus nuclei, he found that it contained 29.8 percent adenine, 20.4 percent guanine, 20.7 percent cytosine, and 29.1 percent thymine. When he went on to analyze DNA samples from various organisms, he found that the base composition varied with the source—and varied within wide limits. DNA, it seemed, might not be a monotonous polymer at all.

Therefore it might be the sequence of the four bases along the polynucleotide chain which represented the hereditary code; this sequence, in turn, could determine the sequence of the amino acids of the proteins whose synthesis was directed by genes.

Chargaff made another important observation. When he reported his studies of DNA composition, he wrote: "It is, however, noteworthy —whether this is more than accidental, cannot yet be said—that in all deoxyribose nucleic acids examined thus far the molar ratios of total purines to total pyrimidines, and also of adenine to thymine and of guanine to cytosine, were not far from 1." This "equivalence rule"—that no matter how great the variations in composition between types of DNA, the proportion of adenine is always very nearly equal to that of thymine, and the proportion of guanine is always very nearly equal to that of cytosine—was to suggest a critical feature of the structure of DNA.

Another major step toward understanding the DNA structure came when it was learned how to adapt X-ray crystallographic techniques to the study of DNA. Such techniques use the ability of crystals to diffract X-rays in order to get precise data about the three-dimensional arrangement of the atoms in the crystal molecules. As far back as 1912, the techniques had been used successfully to determine the structure of ordinary salt. But X-ray crystallography, while practical for studying relatively simple structures, did not seem useful for larger ones; and proteins were

considered to be much too complicated for crystallography analysis.

Still, for many years, some workers persisted in efforts to apply X-ray crystallography to complex molecules. During this time, Linus Pauling was able to establish the structures of some complicated inorganic silicate molecules. In 1951, he guessed (correctly) that amino acids linked by peptide chains might sometimes tend to assume helical configurations, and he proposed that a configuration he called the alpha helix might be an important element in protein structure. In the alpha helix, all backbone atoms have identical orientations, and the folded polypeptide chain can be viewed as a spiral staircase in which the steps are formed by amino acids. Pauling's alpha helix concept for proteins was soon substantiated.

Meanwhile, however, at King's College, London, a team led by M. H. Wilkins had found a way around the difficulty of applying X-ray crystallography to solutions of DNA, which have a gluelike consistency. Wilkins and his coworkers found a way to draw crystalline fibers of DNA out of the solutions; with the fibers they could get much sharper diffraction patterns than ever before possible. One of the Wilkins' team, Rosalind Franklin, took a remarkably clear picture that showed a precise distance of 3.4 Angstrom units between nucleotides in DNA. It was such information from the King's College analysis that helped fill in some major gaps in knowledge, enabling James Watson and Francis Crick, another team trying to establish the structure of DNA, to work out that structure within a few weeks after many previous false starts.

## The Watson-Crick model

In 1953, a short report appeared in the British science journal, *Nature*. It was written by two young men—James D. Watson, then 24, and Francis Crick, then 36. It was to be quickly recognized as marking a major achievement in the history of biology, one for which Watson and Crick were to receive the 1962 Nobel Prize. It was a report proposing the double-helical, spiral-staircase structure of DNA.

Watson and Crick knew that the DNA molecule had four kinds of nucleotides arranged in long strands, and that for each adenine unit there is a thymine unit and for each cytosine a guanine. It had been proposed that the molecule might have a spiral or helical structure. Watson and Crick, using Tinker-Toy-like pieces, had been attempting to construct a scale model of the DNA molecule. Finally, aided to no small extent by the X-ray diffraction data coming from the laboratory of Maurice Wilkins, they had hit upon a model that was notably consistent with the Wilkins' data.

When the distances between the various dots on the X-ray diffraction photo of DNA were measured, the measurement revealed three major repeating patterns: one of 3.4Å, another of 20Å, and a third of 34Å. Watson and Crick were sure that the 3.4Å pattern was related to the distance between successive nucleotides in the DNA chain, and that the 20Å pattern reflected the width of the chain. About the third pattern of 34Å, they decided to make an assumption. If the chain of nucleotides could be thought of as being coiled in a helix—as if they were wound around an imaginary cylinder—then the 34Å distance could indicate the space between windings or turns of the helix. And, since the 34Å distance would be 10 times the 3.4Å distance between nucleotides, that could mean that a turn of the helix would be 10 nucleotides long.

But there was still a problem. DNA had a known density twice as great as any single chain of nucleotides coiled in such a helix could account for. After considerable arranging and rearranging of their scale model, Watson and Crick hit upon one arrangement that best fitted all the known data. It consisted of two nucleotide chains wound in opposite directions—a double helix.

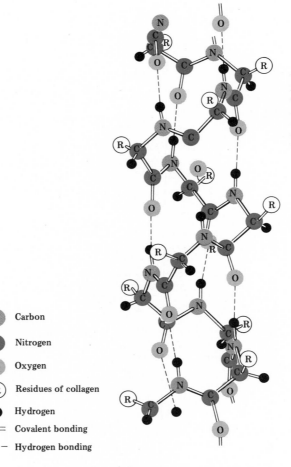

**14-10** *The alpha helix model of many proteins, discussed in Chapter 1, occurs widely in living matter. Knowledge of protein structure was of great help in constructing the double helix theory of DNA structure.*

| | |
|---|---|
| **C** | Carbon |
| **N** | Nitrogen |
| **O** | Oxygen |
| **(R)** | Residues of collagen |
| ● | Hydrogen |
| = | Covalent bonding |
| - - - | Hydrogen bonding |

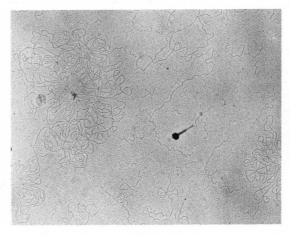

**14-11** *In living organisms, DNA molecules are joined in long thread-like chains. This photo shows the DNA content of the "SPIS phage", a bacteriophage. One can observe the DNA coming out of the tail of one phage. (Phillipe Male, Albert Einstein Medical College)*

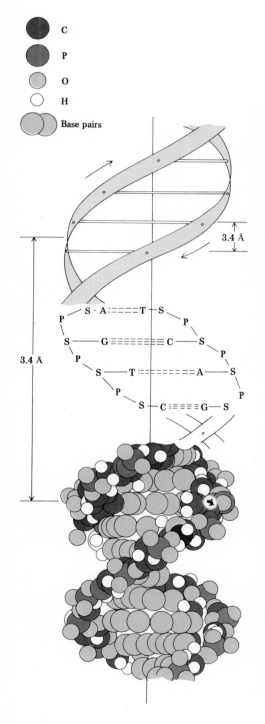

C

P

O

H

Base pairs

3.4 Å

3.4 Å

S — A ═══ T — S

S — G ═══════ C — S

S — T ═══════ A — S

S — C ═══ G — S

The paired chains have opposite polarity so that when one strand is read from bottom to top, it tells the same story as the other strand read from top to bottom. The steps in the spiral staircase were composed of adenine-thymine and cytosine-guanine nucleotide pairs; the base pairs were held together by hydrogen bonds; the sugar-phosphate side chains formed the external supports. The arrangement was such that, if unwound, a DNA molecule would resemble a ladder. Two long chains of sugar and phosphate groups would make up the uprights of the ladder while two nitrogenous bases bonded together by hydrogen bonds would form each cross rung.

**14-12** *The Watson-Crick DNA model shows how the three-dimensional DNA ladder is twisted to form a double helix. Two complementary nucleotide chains are wound in opposite directions. If the chains separate, either can serve as a template to form a new chain of DNA. A complete molecule consists of thousands of turns of the helix and has a molecular weight of several million. (Photo courtesy of the American Museum of Natural History)*

The model also required that neither two purines nor two pyrimidines could form a rung. Two purines, with their double-ring structures, would take up too much room; two pyrimidines, with single-ring structures, would be too short to reach the uprights. A rung therefore had to consist of one purine and one pyrimidine. It also turned out that although adenine and cytosine as a pair and guanine and thymine as a pair would be right in size, there was no way to line them up as pairs so that stabilizing hydrogen bonds could form between them. Therefore, there could be only two possible pairings: adenine with thymine; and guanine with cytosine. That fitted Chargaff's proposition that the ratio between adenine and thymine in a DNA molecule is

**427**

almost always 1, as is the ratio of guanine to cytosine.

Many of the known physical and chemical properties of DNA could be understood in terms of the double-helix model. For example, the structure has a stiffness that explains why DNA keeps a long, fiberlike shape in solution. The behavior of DNA changes in response to changes in pH, which is accounted for by the hydrogen bonds of the bases. The model also suggested how DNA might possibly produce a replica of itself. The model has two parts, one the complement of the other. Either chain then could act as a mold, or template, on which a complementary chain could be synthesized. If the two chains unwind and separate, each could begin to build a new complement onto itself so that where there had been one pair of chains, now there would be two. As Crick has noted: "Moreover, because of the specific pairing of the bases the sequences of the pairs of bases will have been duplicated exactly; in other words, the mold has not only assembled the building blocks but has put them together in the right order."

## Replication of the structure

Replication—biological reproduction at a molecular level—is going on constantly in thousands of body cells at any given moment. During mitosis, each cell receives its full quota of genetic material. How is this achieved? What is the copying mechanism?

In a paper published several weeks after their proposal of the double-helix model, Watson and Crick suggested an answer. The complementary pairing of purine and pyrimidine bases of the two polynucleotide chains of the double helix, they thought, made it possible for DNA molecules to replicate after a simple separation, a kind of "unzipping" through the rupture of the hydrogen bonds between the paired purines and pyrimidines. Each chain could then act as

14-13 *James Watson and Francis Crick were relatively young when they did the work that won them the Nobel Prize. Watson later wrote a best-selling book, recounting the steps that led to their joint discovery. The story of this team's efforts to be first with this new information is as enthralling as a good mystery novel. (United Press International)*

template, or mold, for formation of its own complementary chain. Each purine and pyrimidine base could attract and hold a complementary free nucleotide by means of hydrogen bonds. Then the nucleotides could be bonded to each other to form a new chain. After the replica chains had been structured along the length of each of the parental chains, there would be two new DNA molecules with exactly the same sequences of the four bases as the original DNA double helix. In the next replication, the process would be repeated.

Was the Watson-Crick concept really correct? In 1957, Arthur Kornberg at Washington University in St. Louis devised a method for test tube synthesis of DNA. He combined enzymes extracted from bacteria with nucleotides labeled radioactively with carbon 14, added some primer DNA, plus a little ATP as an energy source. When a culture medium was inoculated with the mixture, new DNA formed. It contained carbon 14, showing that the labeled nucleotides had been incorporated in the DNA. Kornberg established that if primer DNA was not available, there was

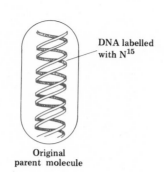

DNA labelled
with N¹⁵

Original
parent molecule

First generation
daughter molecules

Second generation
daughter molecules

**14-14** *Meselson and Stahl experimented with E. coli bacteria produced with N¹⁵, heavy nitrogen and subsequently grown on N¹⁴, lighter nitrogen. They found that each first generation daughter molecule consists of one old and one new, or one heavy and one light strand of DNA. In the second generation, 50 percent were light stranded, and 50 percent intermediate. In the third generation 75 percent were light stranded. Their research supports the Watson-Crick replication model.*

the molecules of primer DNA to be distinguished from those of newly formed DNA. They grew *E. coli* bacteria in a culture medium in which all the available nitrogen was N¹⁵, a slightly heavier isotope of the more common N¹⁴. The DNA of such organisms has a molecular weight about 1 percent greater than DNA containing N¹⁴. After producing a culture of *E. coli* with all-heavy DNA, Meselson and Stahl put a sample of N¹⁵ bacteria in a medium containing only N¹⁴. At intervals, they sampled the bacterial population for DNA density measurements, and when the population had just doubled, they found that virtually all DNA had a density intermediate between heavy and light. Evidently, as the Watson-Crick concept predicted, the newly formed DNA molecules were hybrids containing one old and one newly synthesized strand in each molecule.

If Watson and Crick were correct, then in the next generation, there would be equal numbers of hybrid and all-light molecules. This proved to be the case. A third replication also followed the predicted pattern: three-fourths light and one-fourth heavy molecules. It appeared then that Watson's and Crick's replication concept was correct.

Still other evidence to support the replication concept of Watson and Crick has come from the work of J. Herbert Taylor of Columbia University. Taylor labeled thymidine, a component of the thymine nucleotide, with radioactive tritium. He found that the thymidine is taken up by growing plant root tips and incorporated into the DNA of the chromosomes of active cells. When properly stained, the chromosomes can be seen directly in microscopic sections of the root tips. It is also possible to let the incorporated tritium make a "picture" on radiosensitive photographic film: the electrons emitted by disintegrating tritium atoms expose the film and show clearly where thymidine has been incorporated into DNA. This is the process of autoradiography.

no synthesis; when synthesis did occur, the new DNA was identical with the primer DNA. While the work established that new DNA is a replica of primer DNA, it did not indicate necessarily that replication was as Watson and Crick postulated.

A year later, M. S. Meselson and F. W. Stahl at the California Institute of Technology found evidence to support the Watson-Crick concept, using isotope-labeling techniques that allowed

**429**

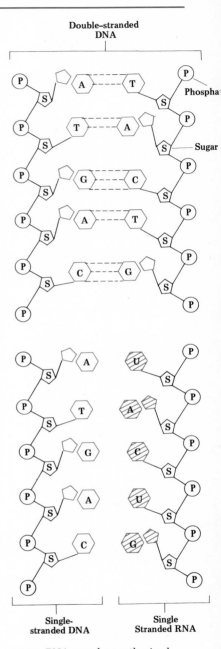

Taylor and his associates applied a solution of labeled thymidine to bean root tips until all the chromosomes were tagged. After one cell division, both daughters of a chromosome showed the tagging, just as did the first generation of hybrid DNA molecules in the Meselson-Stahl experiments. In the next division, the chromosome was separated into labeled and unlabeled daughters just as in the second replication in the Meselson-Stahl work. DNA, it appeared, replicates in dividing plant cells exactly as it does in bacterial cells.

## Protein synthesis and the genetic code

The work summarized in the preceding pages proved that DNA is the hereditary material, and that it has its effect by in some way directing the synthesis of protein within the cell. Yet DNA always remains in the nucleus of the cell, whereas protein synthesis takes place in the cytoplasm. DNA's direction of protein transcription, therefore, could not be direct but involved some intermediary. What was the intermediary?

## RNA

In the late 1930s, investigators had found large amounts of a second nucleic acid, ribonucleic acid (RNA), in the cells of such tissues as the pancreas and liver in vertebrates and the silk gland in silkworms—tissues that produce large amounts of protein. Relatively little RNA appeared in cells such as those of the heart, voluntary muscles, and kidneys which, although physiologically very active, synthesize little protein.

Moreover, microscopic studies had determined that almost all RNA is found in the cytoplasm, in particles which also contain substantial amounts of protein. By the 1950s, thanks

to the nuclear age's discovery of radioactive labeling, it had become evident that the RNA-protein particles, or ribosomes, were the sites of protein synthesis. When, for example, scientists injected a labeled amino acid into a rat and soon afterward extracted ribosomes from tissue samples, they found the labeled amino acid in the ribosomes, indicating that it was being used in the construction of polypeptide chains there.

DNA and RNA are similar compounds with only minor differences. In DNA, the sugar is deoxyribose; in RNA, it is ribose. RNA contains uracil rather than the thymine found in DNA, so the RNA pyrimidine bases consist of cytosine and uracil, whereas those in DNA are cytosine and thymine. DNA usually is double-stranded, but RNA usually is single-stranded.

Despite these differences, DNA can serve as a template for the synthesis of RNA, and the synthesis can proceed in similar fashion to that for new DNA. When the double helical DNA molecule opens up and reveals one of its strands, ribonucleotides (nucleotides containing ribose) orient themselves along the strand so that a uracil ribonucleotide faces each adenine on the DNA, an adenine ribonucleotide faces each thymine, a cytosine opposes each guanine, and a guanine ribonucleotide is opposite each cytosine. In this way, deoxyribonucleotides of DNA establish where ribonucleotides go in the RNA strand, and the ribonucleotides are then bonded to each other to form a new RNA molecule.

According to this model, the base composition of the RNA and DNA from the same organisms should be equal, allowing for substitution of uracil in RNA for thymine in DNA. And scientists have been able to demonstrate this. For example, they have studied E. coli cells before and after infection with bacteriophage and have shown that RNA synthesized in the bacterial cells before infection reflected the base composition of E. coli DNA, while those RNA molecules synthesized after infection had the same base composition as bacteriophage DNA.

**14-15** *RNA may be synthesized on a DNA template despite certain minor differences in their structure. DNA is double-stranded, whereas RNA is single-stranded; in RNA, thymine is replaced by uracil; the sugar in RNA is ribose, while that in DNA is deoxyribose.*

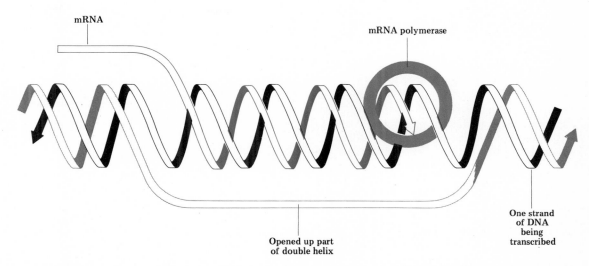

**14-16** *Despite slight structural and chemical differences, RNA synthesis occurs in much the same way as that of DNA. The DNA double helix untwists and opens up to allow transcription of the DNA base sequence into a single-stranded RNA molecule. The sequence of deoxyribonucleotides in the single DNA strand determines the sequence of ribonucleotides in RNA.*

Ribosomal RNA is synthesized on DNA templates in the nucleolus. Nucleolar DNA is thought to be a copy of DNA located at the nucleolar organizer sites of the chromosomes. But it has become apparent that the genetic code is not transmitted from DNA directly to ribosomal RNA. There is an intermediary kind of RNA, called messenger RNA.

**Messenger RNA.** Messenger RNA (mRNA) is one of three forms of RNA now clearly established. The others are ribosomal RNA (rRNA) and transfer RNA (tRNA). Each has a different structure and function. The structural differences are not in the nucleotides but in molecular weight and configuration.

Messenger RNA is lighter in weight than ribosomal RNA, but the configuration of mRNA is still not clear. mRNA is built up and broken down so rapidly that for a time there was some doubt that it existed. But exist it does, and its function is to carry the genetic code from the DNA in the nucleus to the ribosomes where protein synthesis takes place.

After mRNA molecules are formed in the nucleus of a cell, they move into the cytoplasm. Movement probably takes place through the nucleus pore; electron micrographs have shown the molecules of mRNA on both sides of the nuclear membrane, and some molecules have even been observed stuck in the pores. The mRNA then moves to the ribosome. Either the mRNA rides over (or through) the surface at the ribosome, or more likely, it moves along the mRNA strand, "reading" it as it goes.

**Transfer RNA.** Once the message comes, via mRNA, tRNA plays a vital role in the construction of a protein. It is this relatively small molecule which assembles the amino acid building blocks required to make a specific protein.

Each tRNA is constructed of 70 to 80 ribonucleotides, and there is a different tRNA for each of the 20 amino acids. A tRNA becomes attached to its amino acid and, in effect, becomes an address tag, enabling the amino acid to reach its destination in a polypeptide chain. Tagged amino acids collide at random with exposed surfaces on mRNA molecules; when there is a match of the tRNA tag and the mRNA template, the amino acid is fitted into place in the growing polypeptide chain. Some interaction seems to take place between tRNA and ribosomal RNA, but the nature of that interaction still is not clear.

It is worth noting that only very recently a previously unknown class of RNA molecules associated with chromosomes (cRNA) has been discovered. It has been suggested that such cRNA

**431**

may be an artifact resulting from the degradation of tRNA. But David Bonner and his co-workers have recently been able to show that cRNA is not detectably contaminated with either tRNA or rRNA, or the degradation products of either. They argue that cRNA should be considered a distinct class, and that there should be, as there undoubtedly will be, efforts to determine what function it has.

At the moment, then, the picture of protein synthesis appears to be this: the process begins in the nucleus with the transcription of the DNA message onto mRNA. The mRNA carries the message or pattern to the ribosomes; tRNA molecules bring amino acids, the raw materials. In the ribosomes a protein is constructed with the amino acids positioned in the protein chain while ribosomal RNA serves as the translation template. There is an exact correspondence, or **colinearity,** between the linear sequence of DNA nucleotides and the amino acids of the proteins synthesized by that nucleotide sequence. This was shown by S. A. Yanofsky, who mapped the mutation sites inside a gene.

## The code

How can information written in a four-letter language—the four nucleotides of DNA—be translated into a 20-letter language (the 20 amino acids of proteins)? Almost from the moment of announcement of the Watson-Crick model of DNA, scientists were intrigued by the problem.

Working according to the scientific principle that the simplest solutions should be considered first, they started with the possibility of a single-unit code, in which one nucleotide codes for one amino acid. This obviously was inadequate, since a single code could specify only four amino acids. A double-unit code of two-letter DNA words would also fall short, since it allows only four (the number of bases) to the second power (the number of units is two) number of varia-

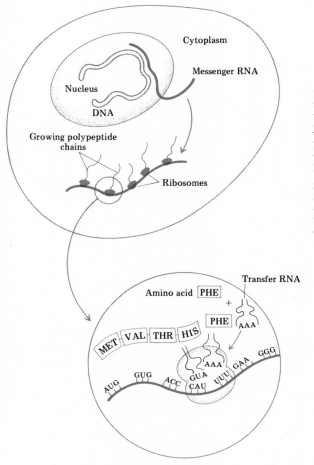

**14-17** *During protein synthesis, the genetic information contained within the genes is translated into form and function. Messenger RNA, complementary to DNA, passes from its place of synthesis in the nucleus into the cytoplasm. There it becomes associated with ribosomes, the sites of protein synthesis. As the ribosomes move along the messenger RNA they translate the genetic information and synthesize polypeptide chains.*

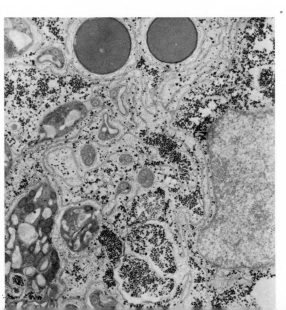

**14-18** *Although ribosomes may be found anywhere within the cell cytoplasm, they are often associated with the membranous folds of the endoplasmic reticulum. In this photo of a Sertoli cell from frog testis, the ribosomes are seen as dots associated with the ER. The large dark globules are drops of glycogen, a stored food supply. (David Soifer—Omikron)*

Reading overlapping codons

G A G . . . Glycine
A G C . . . Serine
G C A . . . Alanine
C A U . . . Histidine
A U G . . . Methionine
U G A . . . Nonsense
G A C . . . Asparagine

Messenger RNA sequence

G A G C A U G A C

No overlap of codons

G A G . . . Glycine

C A U . . . Histidine

G A C . . . Asparagine

Sequential reading of triplets

Met    Gly    His    Asp
– – – A U G   G A G C A U G A C – – – – – – U A A
      ⇧①        ②      ③      ④

Reading always starts at this point

Triplets read sequentially from starting point

Nonsense triplet terminates reading

**14-19** *Alternative ways of reading the genetic code have been proposed. The first overlapping code was proven inadequate because it limits the order in which amino acids could be arranged as a protein (CAU could only be followed by an amino acid that begins with AU, for example). No such limitations actually occur in proteins, however. A non-overlapping code settles this problem but raises the question of how the proper triplets are distinguished. Crick proposes that the triplets are "read" sequentially, starting from a fixed point.*

tions; thus only 16 amino acids could be coded. The next simplest possibility was a triplet code with sequences of three DNA words encoding for individual amino acids. That would provide $4^3$, or 64, code words. That provides more code words than amino acids, but still it is the simplest system that could provide the minimum number of variations.

In 1961, Crick and a team of co-workers produced some experimental evidence to support the triplet code idea. They were able to obtain a quantity of bacteriophages that attacked *E. coli* but differed from normal viruses in the number of nucleotides in a region of DNA. They established that viruses with either one or two extra nucleotides could not function normally. But those with three extras did. Apparently, for normal functioning, the virus had to produce some still-unidentified protein which could not be formed when one or two nucleotides were added to the virus DNA. This suggested that the code

takes a triplet form. While the addition of one or two nucleotides upset the reading of the code, a third extra restored the message. It was true that the added three nucleotides would spoil a message—cause a "misreading"—in the gene area where they were incorporated, but they would not disturb the sequence of the rest of the message.

## The genetic code

In order to determine the triplets, or sequence of three nucleotides, that code for various amino acids, it was necessary to have synthetic molecules of mRNA of known composition. Happily, in 1955, Severo Ochoa at New York University had isolated from bacteria an enzyme that could be used to synthesize RNA molecules of any desired base composition in a test tube.

In 1961, Marshall W. Nirenberg and J. H. Matthaei of the National Institutes of Health discovered that if they substituted a synthetic RNA made entirely of uracil nucleotides for natural mRNA in a laboratory protein-synthesizing system, they got a polypeptide consisting solely of phenylalanine. This was the first indication that synthetic RNA could be a template in protein synthesis; it also immediately indicated that the RNA code word for phenylalanine is UUU, and its DNA complement is AAA—i.e., the DNA sequence AAA, translated by RNA as UUU, directs production of the protein phenylalanine.

Soon, other synthetic RNAs were being studied. There was progress in identifying code words for other amino acids. RNA composed of only adenine nucleotides, for example, directed incorporation of lysine, so the RNA triplet for lysine was found to be AAA and its DNA complement TTT.

Ochoa found that a synthetic RNA beginning with one or two A units followed thereafter only by U units directs production of a polypeptide containing phenylalanine except for one tyro-

**433**

sine. It was already known that a U-A polyribo-nucleotide with twice as many Us as As leads to incorporation of the predicted amount of tyrosine, thus indicating that the code word for tyrosine is UAU. From other similar experiments, the RNA triplet word for cysteine was found to be UGU, for serine UCU or UCC.

In 1964, Nirenberg worked out a valuable technique that helped break more of the code. In separate test tubes, he charged various tRNA molecules with their respective amino acids that had been made radioactive. After incubating the charged tRNAs with ribosomes and known trinucleotides, he placed them on filters that allowed tRNA molecules to pass but held back the ribosomes. If a radioactive amino acid could pass through the filter, the tRNA was not being bound to the ribosome by the particular trinucleotide being studied and so that trinucleotide was not acting as a messenger. But if the amino acid

## 14-1
## The genetic code

| Amino acid | Abbreviation | DNA code | mRNA codons |
|---|---|---|---|
| 1. Alanine | ala | CGA CGG CGT CGC | GCU GCC GCA GCG |
| 2. Arginine | arg | GCA GCT GCC TCT GCG TCC | CGU CGA CGG AGA CGC AGG |
| 3. Asparagine | asn | TTA TTG | AAU AAC |
| 4. Aspartic Acid | asp | CTA CTG | GAU GAC |
| 5. Cysteine | cys | ACA ACG | UGU UGC |
| 6. Glutamine | gln | GTT GTC | CAA CAG |
| 7. Glutamic acid | glu | CTT CTC | GAA GAG |
| 8. Glycine | gly | CCA CCG CCT CCC | GGU GGC GGA GGG |
| 9. Histidine | his | GTA GTG | CAU CAC |
| 10. Isoleucine | ilu | TAA TAG TAT | AUU AUC AUA |
| 11. Leucine | leu | AAT AAC GAA GAG GAT GAC | UUA UUG CUU CUC CUA CUG |
| 12. Lysine | lys | TTT TTC | AAA AAG |
| 13. Methionine | met | TAC | AUG |
| 14. Phenylalanine | phe | AAA AAG | UUU UUC |
| 15. Proline | pro | GGA GGG GGT GGC | CCU CCC CCA CCG |
| 16. Serine | ser | AGA AGG AGT AGC TCA TCG | UCU UCC UCA UCG AGU AGC |
| 17. Threonine | thr | TGA TGG TGT TGC | ACU ACC ACA ACG |
| 18. Tryptophan | try | ACC | UGG |
| 19. Tyrosine | tyr | ATA ATG | UAU UAC |
| 20. Valine | val | CAA CAG CAT CAC | GUU GUC GUA GUG |
| Terminating triplets | | ATT ATA ACT | UAA UAG UGA |

**14-20** *The thistle on the right is an abnormal phenotype, caused by genetic mutation, or an error in the replication of DNA. Such mutations provide the raw material upon which evolution works, through the process of natural selection. (Karl H. Maslowski from National Audubon Society)*

did not pass through the filter, the trinucleotide sequence would have to be the proper code for that particular tRNA and therefore presumably for the particular amino acid.

Thanks to this ingenious technique, the sequence three nucleotides, or **codon,** in all 20 amino acids could be established. It became clear that almost all amino acids have at least two codons and some (arginine and leucin) as many as six. It appears that there are three codons— UAA, UAG, and UGA—that do not code for any amino acid. They are called nonsense codons, or terminators, and if they are included in any mRNA, the message stops right at that point even though the polypeptide chain is incomplete. AUG, a codon for methionine, also appears to be an initiator of messages as well as an amino acid specifier.

**Modifying the Beadle-Tatum concept.** The one gene–one enzyme theory of Beadle and Tatum was brought under question by certain findings of later investigators. In 1956, V. M. Ingram of Cambridge University showed that the abnormal behavior of the red cells in sickle cell anemia is caused by a single change in the complex hemoglobin molecule. That molecule consists of some 600 amino acids arranged in four polypeptide chains, two alpha chains, and two beta chains (see Fig. 7-17, p. 231). Yet the fault that produces sickle cell anemia is the replacement of one amino acid (glutamic acid), by another (valine) in only one type of chain. A mutant gene causes a change in the beta chains, but the alpha chains remain as they were. This strongly suggests (although it does not prove) that synthesis of the alpha chain is controlled by one gene and synthesis of the beta chain by another.

Later research on another hereditary blood disease supported the suggestion contained in Ingram's findings. Thus, it appears that the Beadle-Tatum concept should be modified slightly—to one gene–one polypeptide—to remain valid in terms of all now-known facts.

## Mutations

DNA replicates with great precision yet the process is not infallible. Errors, or mutations, occur. As we saw in the previous chapter, many mutations involve changes in the numbers of chromosomes, or of the sequence of genes on the chromosome. Mutation may also occur in the fine structure of the DNA molecule itself.

Mutations on the gene level, as contrasted with those on the chromosomal level, are called "point mutations." They may involve additions, deletions, or substitutions of nucleotides and can be likened to typographical errors in a printed message.

For example, an original genetic message may look thus:

EEN TAN ANT NAT TEN ATE TNT NAE NAT
 1   2   3   4   5   6   7   8   9

The nine triplet words are instructions for the synthesis of a segment of a protein chain to contains nine amino acids. The message is read left to right. If any one letter is added or subtracted, the reading frame would be so altered that triplets to the right of the error would be changed and there would be a corresponding change in the protein chain.

For example, if the twelfth letter, the T in NAT, were omitted, the rest of the message must still be read in triplets and now, because of the single omission would become:

EEN TAN ANT NAT ENA TET NTN AEN AT
 1   2   3   4   5   6   7   8   9

There would be equally marked changes in the meaning of the message and the assembly of amino acids into the protein chain if a single letter were added.

In another type of gene mutation, called base substitution, one nucleotide replaces another so that, for example, a TNT triplet may be used instead of a TET triplet. Intragenic recombination is another type of mutation, involving the recombination of parts of genes. If one homologous chromosome has a slightly different allele than the other homologous chromosome, and if there should be breaks in the alleles followed by recombinations, one pair or more of the triplets may trade places. Thus the message in the new recombinations would be different.

Some evidence has been found that mistakes in the replication of DNA may sometimes be repaired. For example, for many bacteria, exposure to ultraviolet light is fatal. Yet some organisms survive the photochemical damage to their DNA inflicted by ultraviolet, apparently because of an ability to repair the damage.

A model proposed for the repair mechanism considers that the process is a kind of "cut and patch" one. The hypothesis is that one or several enzymes may constantly move along a bacterial DNA molecule, looking for faults in structure in much the same way as a test trolley seeks out faults in railroad tracks. Encountering an error in DNA, the enzyme cuts the molecule in two locations — ahead of and behind the faulty nucleotide — and removes the faulty one or perhaps a section containing the bad nucleotide and several others on each side of it. With the error-free complementary DNA strand serving as template, a new nucleotide or sequence of nucleotides is pushed and bonded into place.

# Regulation of genes

We know that every somatic cell of a complex multicellular organism contains exactly the same genetic information, and yet the cells specialize in different functions. We also know that many metabolic processes are catalyzed by sequences of enzymes; if all enzymes were present simultaneously, the chain of reactions would be thrown out of cycle. Clearly, cells must be able to turn their genes on and off.

In 1961, François Jacob and Jacques Monod of the Pasteur Institute in Paris proposed a model for gene regulation based on experiments with the metabolism of lactose in *E. coli*. They found that the various enzymes needed for utilizing lactose are produced only when needed. Four genes are involved; mutations in three of the genes can affect the structures of enzymes, but mutation in the fourth gene changes no enzymatic structure but rather affects regulation of enzyme synthesis. A typical mutation of this "i" gene, as it was called, led to continual synthesis of lac enzymes even in the absence of lactose.

## The Jacob-Monod model

It seemed to Jacob and Monod that the product of the i gene must act as a repressor; it could diffuse through a cell and shut off lac genes. The two men went further to predict — and then to find — mutations that could make lac genes insensitive to the repressor, so that there was continual enzyme production even in cells known to contain active repressor. The mutations apparently inactivated some structure (the operator) that otherwise would interact with the repressor. After determining that the same system of gene action holds true for other metabolic activities of bacteria, Jacob and Monod proposed a model to account for it.

According to the model, an operator gene switches on or off the activity of a structural gene (or genes) next to it. Once switched on, the structural genes produce mRNA which then moves to the ribosomes in the cytoplasm to serve as templates for protein synthesis. One operator gene may control a series of nearby structural genes; the unit (operator gene together with the structural genes it controls) is called an **operon.**

Investigators were subsequently able to establish that the structural genes making up a given operon code the production of enzymes with related functions, a kind of family of en-

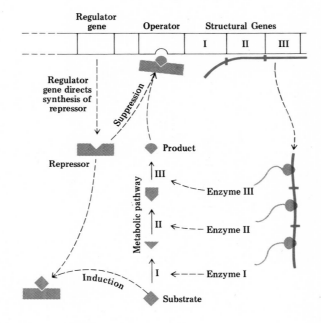

zymes concerned with a single metabolic pathway. In fact, the gene sequence in the operon turns out to be the same as the order in which enzymes are needed in the metabolic process. Therefore, one operator gene is able to control — to turn on or off — not just a single step in a metabolic process but the whole of the process.

**Regulator gene** **Operator** **Structural Genes**

I    II    III

Regulator gene directs synthesis of repressor

*Suppression*

Repressor

Product

*Metabolic pathway*

III

Enzyme III

II

Enzyme II

I

Enzyme I

*Induction*

Substrate

**14-21** *In the speculative Jacobs-Monod model of gene control, a regulator gene produces a repressor that may block the operator gene. With the operator blocked, the structural genes can not produce mRNA. Alternatively, an inducer combines with the repressor and prevents it from blocking the operator. The structural genes then synthesize mRNA. Induction can occur when a substrate or sugar is present. If the sugar is disposed of, the repressor is free to block the operator.*

In the Jacob-Monod model, the operator gene is controlled by a regulator gene which produces or activates a repressor material. This material can combine with certain operator genes to block their activation of structural genes.

It appears that when some starting material or substrate — a sugar, for example — is present and requires the operation of a metabolic pathway, the substrate combines with the repressor substance so the repressor cannot bind to and inhibit the operator gene. It is when the substrate is present and enzyme activity is required, and only then, that the operator gene is left free to turn on the structural genes in the operon. When all the sugar, or other substrate, has been disposed of, leaving none to bind to the repressor, the repressor is free to bind to the operator gene, and operon activity is shut down. This genetic level is known as **induction.**

Some biologists had expected that the repressors, Jacob and Monod identified in bacteria, might turn out to be similar to histone proteins which are found associated with DNA in the chromosomes of higher organisms. Histones are very small proteins with a large excess of positive charges carried on their molecules at neutral pH. But repressors under study thus far are weakly acidic and at neutral pH have an overall negative charge. Moreover, they are large proteins. The lac repressor, for example, has four identical subunits, each with a molecular weight of 38,000. This seems to imply a modification of the Jacob-Monod model when applied to higher organisms. On the other hand, histones may simply be a link in a more complex chain of control found in eucaryotic cells, which have a slightly different type of chromosome than is found in procaryotic cells.

As it now stands, based on the Jacob-Monod model, gene control may be viewed as working in this fashion: While a particular substrate or raw material may nearly always be available in the cytoplasm of a cell, the end-product made from it may be needed only in some

**437**

definite quantity. As the end-product is used up by the organism and stores of it fall to a critical level, the repressor which has been activated by the presence of the product in large quantity now becomes inactivated. The operator gene, then, is no longer repressed and can switch on structural gene activity leading to enzyme production to catalyze the conversion of raw material to more end product. Again, when enough end-product has been produced, its presence in adequate amounts activates the repressor and the production machinery is switched off. This central mechanism is particularly important to the development of an embryo; the regulator gene insures that the turning on of a gene or combination of genes for a particular organ or chemical pathway is regulated at the proper time and in the correct place. Thus, transparent lens cells and not, say, liver cells are formed in the precisely correct site of the embryonic eye at the exactly appropriate time.

## Defining the gene and its function

Once a gene could be defined with relative ease. But many investigations—most notably those of Seymour Benzer, then at Purdue University— have added to the problems of defining the gene. Since there are regulator as well as operator genes, a gene cannot be defined simply as a length of DNA that blueprints an enzyme or an amino acid sequence in a polypeptide chain. Such a definition would fit a structural gene, but a regulator or operator gene would have to be regarded in another light, perhaps as a length of DNA dedicated to regulating the functioning of other lengths of DNA.

Much of Benzer's work on this problem was done with the bacteriophage T4, a virus that infects the colon bacillus. The infective process begins when T4 injects its genetic core, consisting of a long strand of DNA, into the bacillus.

Within 20 minutes, the virus DNA causes manufacture of 100 or more copies of the complete virus particle, consisting of a DNA core and a shell. The bacillus is killed and the virus particles spill out.

The entire DNA molecule in T4 contains about 200,000 base pairs. Benzer was able to explore the fine structure of the A and B genes at a locus of the molecule and was able to show that each of the two genes consists of some hundreds of distinct nucleotide arrangements in linear order.

Should the sequence of hundreds of nucleotides be considered to be the gene because they control one characteristic? Or should they be thought of as hundreds of genes? Although mutation at any one of the nucleotide sites leads to the same defect, it does not follow, Benzer argued, that the entire structure is a single functional unit. For example, growth could require a series of biochemical reactions, each controlled by a different portion of the region. The absence of any one of the steps could be enough to block the final result.

Benzer sought, through experiments with bacteriophages, to determine whether or not such a region could be subdivided into parts that function independently. He made use of the fact that a phage which has lost the ability to reproduce can regain that ability if it is injected into a bacterial cell along with a normal phage. The normal virus supplies an intact copy of genetic structure so it does not matter what defect the mutant had; mutant and normal virus are both able to reproduce.

If the intact structure of the normal phage could be split into two parts and if this were to destroy its activity, the two parts could be regarded as belonging to a single functional unit. Although such an experiment is not feasible, the next best thing is to supply an intact piece A through a mutant having a defect in piece B and an intact piece B through a mutant having a defect in A. If the two pieces A and B can function independently, then there should be activity since each mutant supplies what the other is lacking. But if both pieces must be together to work, the split combination should be inactive.

Benzer found that the two pieces could function independently. He called the functional units **cistrons.**

Through this work and other studies, Benzer established that there are genetic units of various sizes: small units of mutation and recombination, much larger cistrons, and finally a region which includes both cistrons. A unit of mutation, called a muton, could be only one nucleotide long; so could a unit of recombination, called a recon. The codon is bigger, three nucleotides long. The biggest unit is the cistron, the unit of function, which is usually the nucleotide chain that determines a polypeptide chain; it may be as much as 1500 nucleotides long.

Which one should be regarded as the gene? If the gene is defined as the smallest inheritable unit, then it would be the recon and muton. But by this definition the gene would have none of the functional characteristics classically assigned to it. If the gene were defined as the codon, it would be somewhat larger but still without traditional gene characteristics. To consider the cistron to be the gene—and this is implicit in the one gene-one enzyme theory—would be to continue to make the gene the unit of functional activity. While many geneticists do, indeed, continue to regard the gene as the equivalent of the cistron, it is now apparent that it is a large and complex unit and no longer can be thought of as being, as Beadle termed it only a little more than two decades ago, an "irreducible" unit of inheritance.

## RNA-directed DNA

Ever since the Watson-Crick model was formulated, a general supposition has been that genetic information flows in one direction: from DNA to RNA to protein. The model contained no prohibition against a "reverse" flow from RNA to DNA, but there seemed to be no reason for such a flow. To the surprise of many molecular biologists, however, it has recently been shown that reverse flow can take place. The discovery may have important implications for studies of human cancer.

**14-22** *Genetic research in the last several decades has unlocked many of the secrets of cellular replication and control. Now scientists are attempting to put that knowledge to work in research dealing with cancer and viral infections. Shown in this photo is the leukemia virus, cause of an often-fatal blood disease. (Phillipe Male)*

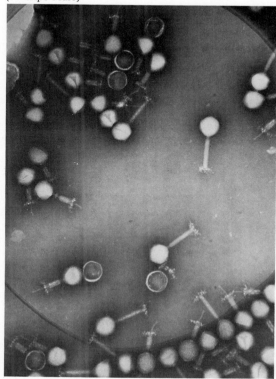

ESSAY **Killer paramecia**

Certain strains of *Paramecium aurelia* are called killers because they carry a nonchromosomally inherited trait causing them to produce paramecin, a toxic substance which destroys defenseless strains of this species called sensitives. Cytological examination of killer paramecia reveals *Kappa particles*, structures approximately 0.2 micron in diameter that contain DNA as well as RNA and protein. Since most paramecia lack *Kappa* particles, they cannot be considered a normal cytoplasmic component. These particles may be symbiotic organisms that originated independently and evolved a unique relationship with certain strains of paramecia.

To determine the genetic basis of this killer trait, killer and sensitive animals are crossed (Figure A). They exchange haploid nuclei across a narrow cytoplasmic bridge. After conjugation, the cells separate and the phenotypes of each can be determined. Since conjugation results in identical heterozygotes, according to expected inheritance patterns both conjugants should be either killers or nonkillers, depending on dominance relationships. Surprisingly, however, the paramecia that were killers remain killers, while those that were sensitives remain sensitives.

After division heterozygous sensitive cells produce only sensitive offspring, but heterozygous killers produce both sensitive and killer progeny. A 1:1 segregation has taken place, clearly indicating that a genetic element does affect expression of the killer trait.

A second experiment reveals that when conjugation between a killer and a sensitive cell is prolonged, the killer remains a killer, but the sensitives are now transformed into killers (Figure B). During prolonged conjugation, the nuclear exchange occurs just as it did during normal conjugation. However, massive cytoplasmic exchange now occurs as well, suggesting that an element in the cytoplasm is responsible for transmission of the killer factor. This cytoplasmic factor is the *Kappa* particle.

Following division, both cells in this second experiment show a 1:1 ratio of killers and sensitives in their offspring. In order for *Kappa* particles to be functional, the animals must possess a specific nuclear gene *K* (Figure C). Genotypically, a killer must be either *KK* or *Kk* since *K* is dominant. A sensitive, however, may be *KK*, *Kk*, or *kk*. Despite their genetic potential, as long as *KK* or *Kk* sensitives contain no *Kappa* particles, they will remain sensitives unless they receive the particles through cytoplasmic transmission. Of course *kk* sensitives will always remain sensitives because they do not possess the gene *K* necessary for maintaining killer particles. The killer trait is thus gene-controlled but not gene-initiated.

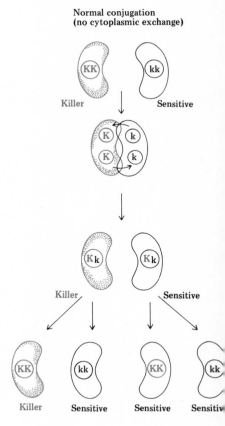

Normal conjugation
(no cytoplasmic exchange)

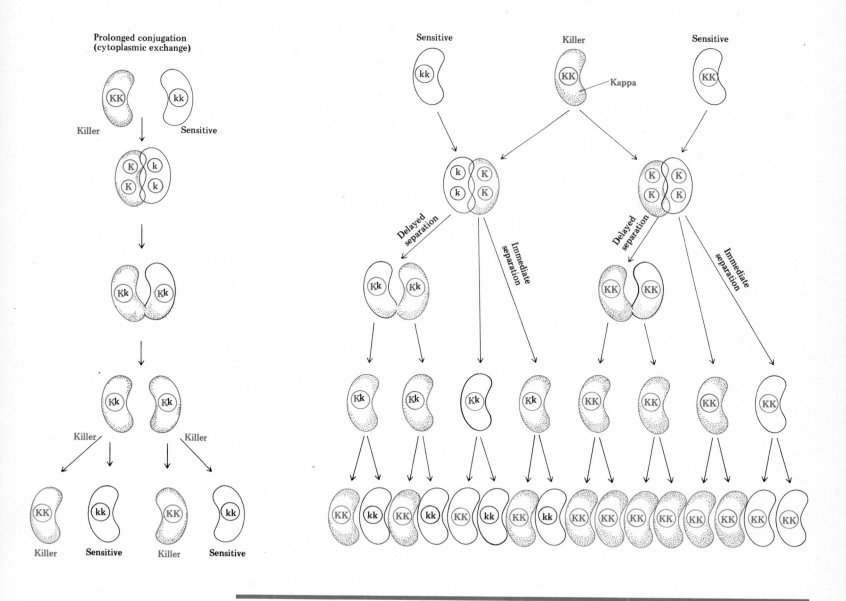

Two broad classes of viruses exist: those whose genome, or complete gene set, consists of DNA, and those with a genome of RNA. When they infect cells, DNA viruses replicate their DNA into new DNA and transmit information from the DNA to RNA and then to protein. Most RNA viruses—which include those that cause poliomyelitis, the common cold, and influenza—replicate their RNA directly into new copies of RNA and into protein; there is no DNA involvement.

But recently, some viruses have been found to use information transfer from RNA to DNA. In one of a series of experiments, Dr. Howard Temin of the University of Wisconsin was able to demonstrate that if a cell is inoculated with a rousvirus, and the synthesis of DNA in the cell is immediately inhibited, the cell is protected against infection. This was an indication that infection requires synthesis of new viral DNA produced on an RNA template. In further work, additional evidence has been developed indicating that with the aid of a recently discovered enzyme, a polymerase, rousvirus can produce DNA on an RNA template.

The DNA polymerase present in RNA tumor viruses may not only explain how such viruses produce cancerous transformations in the cells they infect but also may account for viral latency, a phenomenon in which a virus, after infecting an organism, disappears and then may reappear months or years later. Once an RNA virus has transferred its genetic information to DNA, suggests Temin, it could remain latent in a cell and be replicated by the systems that replicate the cell DNA. After some latent activation, the virus could appear again as infectious particles.

## Molecular genetics and the future

At least 1500 human diseases are already known to be genetically determined and new examples are reported yearly. Many are rare. Phenyl-ketonuria, for example, occurs in about one per 18,000 live births, or about 200 to 300 cases per year in the United States. Others, such as cystic fibrosis of the pancreas, occur about once in every 2500 live births. But, considered together as a group, genetic diseases are a significant medical problem. Although the molecular basis for most of them is not yet understood, a review in 1970 listed 92 for which a genetically determined specific enzyme deficiency has been identified.

To many investigators, it seems likely that in some cases it may be possible to supply a missing enzyme, administering it by new techniques of delivery—possibly including the attachment of the enzyme to resins or its enclosure in microspheres thereby making it resistant to antibody destruction.

There has been much talk—often wild-eyed—of genetic engineering. But some serious investigators believe that some genetic diseases may even be treated at the level of the gene. With growing knowledge of gene structure, possibilities for synthesizing genes to replace damaged ones are increasing. In 1971, the complete synthesis of a gene was achieved—and the technique may be adaptable for synthesizing other genes as their nucleotide sequences become known. Some workers believe that the DNA polymerase found in RNA tumor viruses could be used for gene synthesis since the enzyme can make DNA copies from an RNA template.

An intriguing recent experiment brought results that seem hopeful to some investigators. Galactosemia in humans stems from lack of an enzyme needed to metabolize galactose, a common sugar. *E. coli* bacteria possess the gene for the enzyme, and it was possible to produce a stock of a bacterial virus which had integrated the bacterial gene into its own genome. When defective human cells were treated with the bacterial viruses in a test tube, a few of the human cells regained the ability to synthesize the enzyme, apparently incorporating the bacterial gene and using it for enzyme production.

**14-23** *Another type of virus often used in studies of DNA and RNA is this phage with the catchy name "PBSX." This photo was taken through an electron microscope at a magnification of 110,000; additional magnification is obtained by enlarging the print of the negative. (Phillipe Male)*

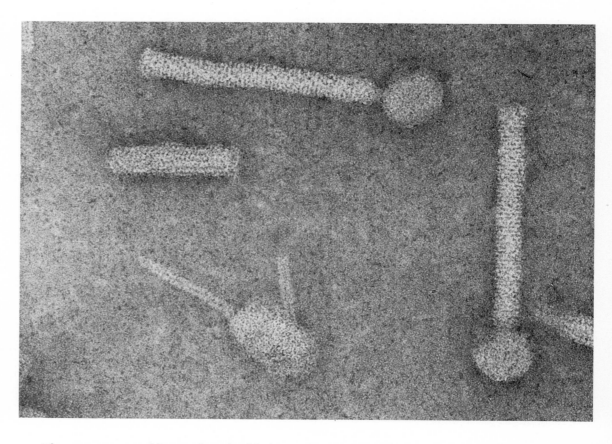

There are many problems to be solved before genetic engineering can become a reality. For one thing, many human genes are active in only a small fraction of body cells. Methods would have to be developed to deliver the new DNA to appropriate target tissues. Another problem lies with regulating the amount of enzyme production. Would it be possible to make certain that only the correct amount of enzyme will be made by newly introduced genes? And still other problems remain—not necessarily insoluble but certainly not to be solved overnight. At this writing, scientists are urging intensified research directed at developing techniques for gene therapy, but they are also cautioning against any rush to attempt gene therapy in human patients until much more is known about gene regulation and genetic recombination in human cells, and until much more, too, is known about both short-range and long-term side effects of gene therapy.

# Summary

The operation of Mendelian genetics rests upon the molecular basis of heredity. This has been investigated in two distinct ways, analytically and physiologically. Garrod's pioneer physiological work with alkaptonuria eventually bore fruit in the "one gene-one enzyme" hypothesis of Beadle and Tatum, which grew out of their work with *Neurospora*. Later work with *Escherichia coli* reinforced this hypothesis. Attention was focused on the hereditary role of deoxyribonucleic acid, or DNA, by experiments with mutant pneumococcus bacteria. Nonvirulent or penicillin-susceptible strains of pneumococcus could be transformed by the presence of DNA from virulent or penicillin-resistant strains. Finally, experiments in which bacteriophages were tagged with radioisotopes of sulfur and phosphorus established that the viral infection was caused by viral DNA.

Staining techniques showed that in any species, all somatic cell nuclei contained a consistent amount of DNA, and that germ cells contained half that amount. This suggested the hereditary role of DNA, although some students felt that the more complex structure of protein made it a more likely candidate for that role. DNA was found to consist of an arrangement of four nucleotides in unknown sequence. While the proportions of nucleotides varied, the ratio of adenine to thymine, and of guanine to cytosine, always approached 1. Meanwhile, X-ray crystallography techniques had established the helical structure of proteins and had thrown some light upon the structure of DNA. This in turn enabled Watson and Crick, in 1953, to propose that DNA consists of a double helix of linked nucleotides, making a complete turn every 10 units.

This structure permits replication of the genetic material: each half of the double helix can separate from the other and, since the chains are complementary in structure, each acts as a template for the replication of its own complementary chain. This model has been amply substantiated and now forms the theoretical basis of molecular biology.

DNA exerts its control over growth and function by directing the synthesis of proteins in the cell. The site of protein synthesis is the ribosome, where proteins are found in association with another, single-stranded molecule, ribonucleic acid, or RNA. The genetic code is transmitted to ribosomal RNA through intermediary messenger RNA, mRNA, which moves from the nucleus through the cytoplasm to the ribosome, which probably moves along the strand of mRNA, "reading" as it goes. Transfer RNA, tRNA, actually assembles the amino acids into a particular protein.

Later research determined that the genetic code is read by the ribosome in "triplets," units of three bases each. Each such triplet, called a codon, codes for a particular amino acid, and the sequence of codons determines the structure of the final polypeptide. The Beadle-Tatum model, modified to read "one gene equals one polypeptide," is accepted today as valid.

This in turn allows a detailed understanding of mutations on the molecular level. These may involve the addition or deletion of a base, substitution of one base for another, or intragenic recombination. Evidence is also appearing of the ability of DNA to repair the "errors" of point mutation.

Regulation of gene action is explained by the Jacob-Monod model, in which special operator genes and repressor substances initiate or halt the activity of the synthesizing genes. The association of operator and structural genes is called an operon, and its activity seems to be initiated by the binding of repressor substance by a substrate chemical in a process called induction.

More recent research has disclosed that there are functional genetic units of different sizes, known variously as mutons, recons, codons, and cistrons. The unit of physiological function is the cistron, a large, complex molecule. This has led to modifications in the definition of the gene.

The complexity of molecular biology is suggested by the discovery that certain tumor viruses, the rousviruses, produce a DNA transcript of their viral RNA, a reversal of the usual pattern.

The findings of molecular genetics are believed to have tremendous consequences for the possibility of genetic engineering, the curing of genetically determined diseases. Although gene therapy is still a remote possibility, several avenues of approach are being actively explored.

## Bibliography

### References

Borek, E. 1955. *The Code of Life.* Columbia University Press, New York.

Cairns, J., G. S. Stent, and J. D. Watson. 1966. *Phage and the Origins of Molecular Biology.* Cold Spring Harbor Laboratory of Quantitative Biology. Cold Spring Harbor, N. Y.

Huberman, J. A. and A. D. Riggs. 1968. "On the Mechanism of DNA Replication in Mammalian Chromosomes." *Journal of Molecular Biology.* 32: 327.

Jacob, F. 1966. "Genetics of the Bacterial Cell." *Science.* 152: 1470–1478.

Ptashne, M. and W. Gilbert. 1970. "Genetic Repressors." *Scientific American*. May. Offprint no. 1179. Freeman, San Francisco. 222(6): 36–44.

Watson, J. D. 1970. *Molecular Biology of the Gene*. 2nd ed. Benjamin, New York.

Woese, C. R. 1967. *The Genetic Code, the Molecular Basis for Genetic Expression*. Harper & Row, New York.

## Suggestions for further reading

Dupraw, E. J. 1970. *DNA and Chromosomes*. Holt, Rinehart & Winston, New York.
Concerns the relationship between chromosome structure and the function of DNA as the genetic material. Covers chemistry and ultrastructure of organelles related to cell division, such as nuclear envelopes, centrioles, spindle fibers. Includes DNA replication, enzymes for nucleic acid synthesis, description of viral and bacterial chromosomes, as well as special cases such as giant polytene chromosomes and lamp-brush chromosomes.

Koller, P. C. 1971. *Chromosomes and Genes*. Norton, New York.
Discusses chromosomal basis of genetic changes, emphasizing how the structure of DNA is a primary factor in these changes. Enumerates several types of chromosomal aberrations that lead to mutations.

Oparin, A. I. *Origin of Life*. Dover, New York.
This classic work, translated from the Russian, presents a believable theory as to how the very first living organisms may have originated in the "primeval soup." He speculates on the chemical constituents of the "soup" and describes what reactions would have had to occur in order for the first organic molecules to spontaneously form. Some of the reactions he predicted have actually been duplicated in chemists' laboratories.

## Books of related interest

Gaito, J. 1971. *DNA Complex and Adaptive Behavior*. Prentice-Hall, Englewood Cliffs, N. J.
Discusses the structure of DNA and other molecules associated with the DNA in chromosomes. Goes into lines of research involving the administration of drugs that affect DNA regulation or activity and thus, hopefully, affecting behavior. Also injection of RNA or brain homogenates.

Gretzer, W. B., ed., 1971. *Readings in Molecular Biology*. Macmillan, New York.
This is a collection of readings taken from the informal "news" section of *Nature* magazine. Current literature is summarized and criticized in areas such as transcription, translation, tRNA, nucleic acid structure, enzymes, and so on.

Koestler, A. 1972. *The Roots of Coincidence*. Random House, New York.
Koestler makes a case for a philosophical view directly opposed to that of Monod; he suggests that life has a purpose beyond mere survival, and that it is somehow more than the sum of its operational processes. Monod sees mutation as a deplorable accident that is sometimes by necessity turned to adaptive use, but Koestler views mutation as evidence of a universal will to change, perhaps even to improve.

Monod, J. 1972. *Chance and Necessity: An Essay on the Natural Philosophy of Modern Biology*. Knopf, New York.
In this book, Monod presents his own philosophic interpretation of the origin of life and its "purpose"—which Monod believes to be nonexistent. The book helps illustrate the nature of the interface between science and philosophy.

# 15

# The reproductive process

All reproduction and growth takes place through the process of cell division. Cytological and karyological studies have shown that only two types of cell division, mitosis and meiosis, ever occur. Yet if we study reproduction in the whole range of living organisms, from algae to eagles, we find an immense variety in the patterns of organismic reproduction. It seems that each living thing has found its own set of variations on the two basic themes of mitosis and meiosis.

All processes of asexual reproduction involve only mitotic cell division. The new cells have exactly the same assortment of genetic information as did the mother cell. This does not necessarily mean that the daughter cells will look just like the mother cell. The process of differentiation (the subject of the following chapter) allows cells with identical genetic constitution to develop in very dissimilar ways. Thus we may find asexually reproduced individuals that look nothing at all like their parent. An illustration of this phenomenon can be seen in many coelenterates; *Obelia* is a good example. A colony of sessile polyps has specialized individuals that undergo mitosis and produce new cells. These cells grow into medusae, a free-living sexually reproducing stage of the organism that looks so different from the polyp it might be another animal altogether. Yet in spite of the difference in the appearance of the parent polyp and its medusa offspring, they have exactly the same genetic constitution.

The process of sexual reproduction includes the fusion of two specialized cells, the gametes. An organism that reproduces sexually must at some point undergo meiotic cell division, reducing the number of chromosomes in the cell; without this reduction-division, the organism's chromosome count would double with each fusion. Much of the variation in life cycles is due to differences in the stage at which meiosis takes place. A typical pattern in the animal kingdom is:

mitosis → meiosis → fusion

**449**

In the plant kingdom, we often see a different pattern:

$$mitosis \rightarrow meiosis \rightarrow mitosis \rightarrow fusion$$

Thus in animals, the haploid stage is generally brief, whereas in plants, large and long-lived haploid organisms are not uncommon. On the whole, the trend of plant evolution seems to have been toward the reduction of the haploid stage. The survival value of this trend is not clear.

## Asexual reproduction

Most people associate asexual reproduction with a relatively low level of development, but that assumption is erroneous. It is true that most acellular organisms reproduce asexually, although many of them also undergo periods of sexual reproduction as well. However, asexual reproduction can be seen in many of the higher plants and animals. Lilies and wasps both commonly undergo forms of asexual reproduction; even more interesting is the fact that laboratory experiments indicate that all organisms are potentially capable of the kind of asexual reproduction called cloning. In this process, a single cell of a highly developed individual—perhaps even a human being—undergoes the process of dedifferentiation, reverting to its embryonic state, and then divides mitotically, just like a fertilized egg, to produce a new, fully formed individual.

## Acellular reproduction at the level of the cell

Acellular organisms, such as bacteria, blue-green algae, and various protists, commonly reproduce asexually, although it is by no means their only method of reproduction. On this level, asexual reproduction may take place by fission or budding.

**Fission.** The simplest kind of asexual reproduction is **fission,** or the splitting of the body of the parent (usually a one-celled organism) into two new equal parts. Each of the new cells is then capable of its own independent existence. Nearly all bacteria reproduce in this manner. Since bacteria are procaryotic cells, with no specialized organelles, the basic genetic information is not contained within a nucleus but is scattered on threads of chromatin throughout the nuclear area. Replication of this information is really the only necessary step, and this kind of simple division poses no particular problems of distribution. Once the genetic material has been replicated, the cell just divides in half by forming a cell wall. This step in the process is shown in the photograph in Figure 15-1.

As a form of reproduction, the chief advantage of fission is its speed. A bacterial cell formed by fission is itself ready to divide and produce two new cells within 15 to 20 minutes. In the space of a single hour, three generations of bacteria can arise; to produce three generations of human beings, it would take about 60 years!

The speed at which this kind of reproduction can occur helps to compensate for its major disadvantage, which is the lack of genetic recombination that produces new types of individuals. The chance for variation among bacteria is about as high as the chance for variation in automobiles on an assembly line. Each bacterium, like each car, is made to a specified pattern. Variation occurs only when a mistake is made.

All organisms are adapted to their environment, which they must in some way exploit to obtain necessary nutrients and raw materials for the synthesis of organic compounds. In the case of bacteria, the immediate environment is also the sole food source, and each species is adapted to obtain the nutrients it needs from its environment. The fact that new individuals are exact carbon copies of the old is not a disadvantage as long as the environment remains stable and the food supply is satisfactory. But if changing

15-1 *The one-celled Paramecium can reproduce asexually through binary fission. Before splitting into two new equal parts, the nucleoprotein material that makes up the chromatin network is replicated. Each daughter cell receives the same kind and amount of chromatin that the mother cell had originally. (Courtesy Carolina Biological Supply Company)*

conditions or the problems of overpopulation require some degree of adaptation to a new environment, there is no mechanism inherent in the reproductive process for producing new types of individuals as experiments in adaptation. New types of individuals, with new adaptive characteristics, will appear only when there is some accidental mutation (change) of genetic material. In the case of bacteria, with their exponential growth rate and their short generations, the chances of producing a better adapted individual from random genetic mutations are fairly good. This fact has been demonstrated repeatedly by the development of bacterial strains resistant to common antibiotics such as penicillin and sulfanilamide. A species with a slower rate of reproduction might become extinct before mutation could produce a well-adapted variant.

**Budding.** A second type of asexual reproduction in acellular organisms is budding, in which a new individual develops as a bud from the larger and older parent. As in all asexual processes, the genetic material of the new individual produced by budding will be exactly the same as that of its parent.

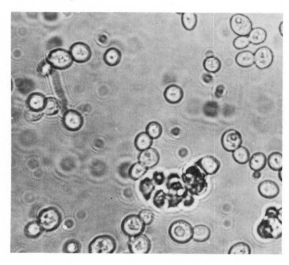

**15-2** *Unlike most Ascomycetes, yeasts are unicellular and reproduce asexually by budding. Here, small outgrowths can be seen developing from the larger mother cells and then pinching off (1500X). Yeasts also have a diploid stage. Two cells come together to form a diploid zygote which then multiplies by budding until a number of cells are present. (Hugh Spencer from National Audubon Society)*

Figure 15-2 shows the process of budding as it takes place in yeast, a microorganism that is usually classed along with the fungi in the Kingdom Protista. The yeast cell is eucaryotic, containing a formed nucleus and several other organelles, such as a large vacuole. When the yeast cell attains a certain size, the nucleus moves to the outer perimeter of the cell and begins to divide. Once the new nucleus, containing an exact replication of the genetic material of the original, is formed, the nucleus and a small amount of the cell cytoplasm are extruded from the cell. The process is very much like that of vesiculation, by which waste products and large molecules leave the cell; in fact, the bud is really a vesicle that happens to contain a nucleus. After the bud is pinched off from the parent cell, the informational content of the nucleus dictates the formation of other cellular organelles and the organization of the cell in a manner just like that of its parent.

## Asexual reproduction on the organismic level

Although we tend to identify asexual reproduction with individual body cells and simple acellular organisms, it is by no means limited to the cellular level. Whole organisms may reproduce asexually; this is true even of vertebrates and higher plants. In most instances, asexual reproduction is used to supplement sexual processes. The plants and animals mentioned in this section are also capable of sexual reproduction.

**Budding.** The hydra is an example of an organism that reproduces by budding. Buds develop along the outer surface of the parent hydra, breaking off to lead an independent existence. Budding can also be seen in certain higher plants among the tracheophytes. Buds form along the margins of the leaves; these eventually drop off,

**451**

and many of them will take root and develop into new plants. An example of a plant that reproduces in this way is *Bryophyllum*, commonly called the air plant.

**Fragmentation.** Fragmentation is an asexual reproductive process in which the parent organism spontaneously splits into two new individuals. It differs from fission in that it involves a colonial or multicellular organism, rather than a single cell.

Fragmentation is a common method of reproduction among flatworms. It is also found in filamentous green algae and in certain members of Phylum Cnidaria, such as the sea anemone. The body of the organism may simply begin a slow process of division and duplication along a central axis. All specialized structures are duplicated. When the new individuals are fully formed, they split apart and each begins to function independently. This type of fragmentation is called **binary fission.** Another type of fragmentation occurs when small pieces of an organism break off to form a new individual; this is common among some cnidarians. Since fragmentation involves the reproduction of a number of different cells, it is considerably slower than the process of fission, or the splitting of single cells.

Fragmentation is found in only a small number of species, most of them aquatic. In terms of number of species or individuals and in terms of distribution, fragmentation would have to be classed as less successful than either fission or sexual reproduction. This is probably due to the low rate of variation inherent in asexual reproduction coupled with a relatively slow reproductive process.

**Regeneration.** Certain plants and animals are able to reproduce by regeneration, a process in which a severed part of the organism grows into a whole new individual. At the same time, the parent organism is also able to regenerate the missing part. The process is closely allied to the growth that takes place after fragmentation.

Perhaps the best known and most dramatic example of regeneration can be seen in the starfish. One of the favorite foods of the starfish is oysters, and since this bivalve is also relished by human appetites, fishermen are much annoyed when they find the starfish getting to the oysters first. They used to try to beat out the competition by catching the starfish and then chopping them apart, casually tossing the remnants of what they took to be a dead starfish back into the ocean. The fishermen were dismayed to hear from biologists that each portion of the starfish (as long as it contains a piece of the central disc) is able to develop into a whole new organism, and that their activities had actually helped the starfish population to multiply.

Certain vertebrates possess a limited regenerative ability, but this should not be classed as a form of reproduction, for it does not produce a new individual but only a new part of the same individual. Examples of this type of regeneration can be seen in some lizards and salamanders, which can grow new limbs and/or tails to replace those lost to predators. This kind of regeneration could be considered an exaggerated form of the healing process of which all vertebrates are capable. Instead of developing scar tissue to close the wound, salamanders grow replacements of the original tissues.

**Formation of spores.** Many plants have a complicated life cycle in which a stage of sexual

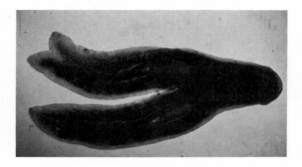

**15-3** *Planaria commonly reproduce through transverse fission. As this photo shows, the entire organism simply divides in half, forming two new planaria. (Courtesy Carolina Biological Supply Company)*

**15-4** *This nine-pointed starfish is regenerating a new arm through the process of mitotic cell division. This kind of reproduction in animals is most often found in radially symmetrical forms. (Hugo H. Schroder from National Audubon Society)*

**15-5** *Here, puffballs use a dead branch as a nutritive substrate. The species of fungi that parasitize seed plants do not produce fruiting bodies but instead produce a mass of spores within the host. The circular elevations are Ascomycetes already broken open for spore dissemination. (John R. Clawson from National Audubon Society)*

reproduction alternates with a stage of asexual reproduction, usually through the formation of spores. A spore is a specialized cell containing a nucleus, a small amount of cytoplasm, and a tough outer covering to protect the cell from the environment. To the casual observer, a spore may seem indistinguishable from a seed, but there is in fact a very significant difference. A seed contains an embryo formed by the union of genetic material from two different parents; it has grown and developed from a fertilized egg. A spore usually consists of an exact replication of the genetic material of one parent alone. The spore can undergo mitotic division and produce a new individual that looks unlike the parent, but this is simply another stage of the life cycle; there has been no change in the genetic material (barring the accident of mutation). Only a seed can develop into a new type of individual that combines traits from both its parents.

Figure 15-5 is a photograph of spores being discharged from a fungus plant; spores are also found in most algae, in mosses and liverworts, in ferns (the little dark dots on the underside of a fern frond hold the spores) and in seed plants as well.

**Vegetative propagation in plants.** In addition to spore formation, budding, and fragmentation, plants can propagate by a variety of other asexual means. Vegetative propagation can take place very rapidly, as evidenced by the spread of the water hyacinth along southern inland waterways. It has been estimated that 10 water hyacinths could, in the course of a single growing season, produce 655,360 new plants. With such a growth rate, these plants have become troublesome weeds that clog the canals and block all navigation. Control of these water weeds cannot be achieved by the conventional herbicides, since they would also kill many plants that are an essential part of the food chain in the canal. Ecologists suggested an interesting solution to this problem; they stocked the canals with manatees, aquatic mammals whose preferred food is the water hyacinth.

The most common type of vegetative reproduction is by means of some underground type of stem. The cattail, for example, produces new plants by rhizomes which grow through the mud. Blueberries also produce new individuals through rhizomes, as do iris, rhubarb, peonies, and canna lilies. Corms or bulbs, such as those of the tulip, daffodil, crocus, and gladiolus, are another variation of vegetative propagation through underground stems.

Another common type of vegetative reproduction involves aerial stems. An example of this type can be seen in the strawberry, which sends out long runners or stolons. When the runner touches the ground, it sprouts adventitious roots and a new plant grows. Plants in the cactus family often propagate by the rooting of broken stems. When an animal brushes up against the plant, a stem breaks off and becomes entangled in the animal's coat, thus being transported some distance away from the parent. When it finally falls to the ground, the stem, through the stimulus of auxin and other hormones, develops roots and becomes a new individual.

Plants may also propagate by the division of roots, as in dahlias, phlox, and sweet potatoes, or by the rooting of a portion of the leaf. African violets, certain species of begonias, and the night-blooming cereus will all develop new plants from a fragment of the leaf.

## Sexual reproduction in plants

Green algae were the first plants to discover sex. The American writer and naturalist Joseph Wood Krutch has pointed out that in so doing, they also invented death. Asexual reproduction by fission is a kind of immortality, since the entire substance of the original individual is contained in

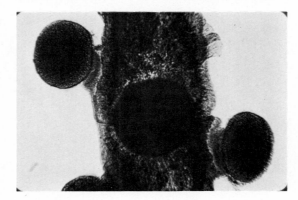

**15-6** *Vegetative reproduction is of great importance in horticulture. The peppermint plant sends out procumbent stems (runners) from which new plants develop. This process is more rapid and easier to manipulate than sexual reproduction. Some plants, like the banana and navel orange, do not produce seeds and therefore must be propagated vegetatively. Vegetative reproduction is responsible for most fruits and ornamental trees grown commercially. (Jerome Wexler from National Audubon Society— Permission of Ward's Natural Science Establishment, Inc., Rochester, New York)*

the two identical daughter cells. Sexual reproduction brings with it the consequence of the death of the individual. Only a single egg or sperm cell, the tiniest fraction of the original individual's being, is passed on to the next generation; the rest of the cells—the totality of the individual—will die some time after reproduction takes place.

In evolutionary terms, individual death is not too high a price to pay for the notable advantages of sexual reproduction. The mingling of genetic inheritance from two individuals permits the creation of a new type of individual. In many cases, the variation is insignificant, a mere rearrangement of viable characteristics. For some individuals, the genetic recombination will produce a disability, a new character that handicaps the individual in the struggle to survive and

reproduce. But in a very few cases, the recombination may produce an individual that is better adapted to survive and reproduce in the surrounding environment. Genetic recombination also brings the possibility of new characters that could prove to be survival traits in a changing environment. Sexual reproduction increases the rate at which the species can adapt to the environment, since it increases the rate of random variations, some small percentage of which may be advantageous.

## The evolution of sex in plants

A very primitive form of sexual reproduction—that is, the combination of genetic inheritance from two parents rather than the duplication of the genetic inheritance of one parent only—can be seen in *Chlamydomonas*, a simple green alga. Although reproduction is generally asexual, through fission, under certain favorable conditions *Chlamydomonas* will form gametes, or sex cells. These gametes have the same number of chromosomes as are found in other *Chlamydomonas* plants, for all individuals in this genus are haploid. The gametes also look like all other *Chlamydomonas* cells, and there is no specialization into eggs and sperm; this plant is **isogamous** (gametes of both sexes are identical). Two of these haploid cells unite to form a diploid zygote. After a brief resting stage, the zygote undergoes meiosis, forming four new cells, each of which develops into a *Chlamydomonas* plant.

A variant of this same primitive type of sexual reproduction can be seen in another green alga, the colonial *Spirogyra*, which is found in the form of long multicellular filaments. Certain of the haploid filaments can produce gametes, and the cells of one filament can then fuse with those of another in a process called **conjugation.** As in *Chlamydomonas*, the gametes are visually indistinguishable from other cells. One important difference is that in *Spirogyra*, there is a slight

**15-7** *All stages of the life cycles of* Chlamydomonas *are shown here, both sexual and asexual. Notice that only the zygote stage is diploid. The first sexually reproducing acellular organisms probably had life cycles resembling that of* Chlamydomonas, *where the diploid and haploid cycles produce the identical mature cells.*

■ Diploid phase of life cycle
   (2n chromosomes)

■ Haploid phase of life cycle
   (n chromosome)

**455**

degree of differentiation in the gametes. There are receiving gametes and supplying gametes, and each filament can produce only one type. Again, meiosis takes place after the fusion of gametes.

Another filamentous green alga, *Oedogonium*, illustrates the next step in the evolution of sex in plants. In this alga, the sex cells are differentiated from other body cells. Moreover, the two haploid cells that unite also differ from one another. One is a relatively large cell, containing more than the normal amount of cytoplasm (the egg); the other is a small and highly motile flagellated cell (the sperm). This specialization of sex cells into eggs and sperm is called **heterogamy.** Heterogamy increases the chances of a successful mating of the two haploid cells. The motile sperm can quickly move to the waiting egg; the bulky egg cell's food supply insures the viability of the zygote.

The final step in the evolution of sex is the development of specialized cells to produce the gametes, rather than having them produced by just any cell in the body. This last step can be seen in *Volvox*, a colonial form that may contain as many as 40,000 cells in a single sphere. Some of these cells specialize to form an **antheridium,** or sperm-producing structure, while others form an **oögonium,** or egg-producing structure. Some colonies may contain both antheridia and oögonia, but others contain only one type of gamete-producing structure and may thus be said to be either male or female plants. An interesting feature of *Volvox* is that if an egg is not fertilized, it may undergo mitotic divisions and produce a new colony all by itself.

## Plant life cycles

Subsequent steps in plant sexual evolution are merely elaborations of the reproductive structures and processes found in *Volvox*; the green algae are the most significant of phyla in terms of the evolution of all higher plants' life cycles and

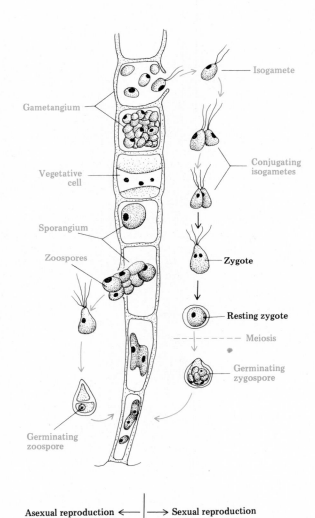

**15-8** *The haploid plant* Ulothrix *reproduces both sexually and asexually, though not simultaneously, as shown here. Asexual reproduction is the more common, and it occurs when certain cells of the filament develop into sporangia, which release zoospores. In sexual reproduction certain cells of the filament become specialized as gametangia, which produce and release isogametes. The zygote produced by the fusion of two such isogametes divides meiotically and releases zoospores.*

Gametangium

Vegetative cell

Sporangium

Zoospores

Germinating zoospore

Isogamete

Conjugating isogametes

Zygote

Resting zygote

Meiosis

Germinating zygospore

Asexual reproduction ⟵ | ⟶ Sexual reproduction

■ Diploid phase of life cycle

■ Haploid phase of life cycle

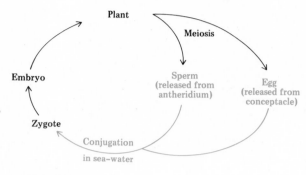

Plant

Meiosis

Embryo

Sperm
(released from
antheridium)

Egg
(released from
conceptacle)

Zygote

Conjugation
in sea–water

■ **Diploid phase**

■ **Haploid phase**

reproductive processes. One basic pattern, or life cycle, can be traced throughout the plant kingdom, in the alternation of generations. A stage of asexual reproduction, the spore-producing or **sporophyte** generation (diploid), alternates with a gamete-producing or **gametophyte** generation (haploid) that features sexual reproduction. This pattern of alternation of generations can be observed in all the plant phyla. In some plants, such as the mosses and ferns, the generations have such a different appearance that early taxonomists were misled into classifying them as completely different plants.

**Phylum Chlorophyta.** Advanced green algae, such as *Volvox*, show clear sexual differentiation, with an alternation of sporophyte and gametophyte generations. The life cycle of a typical chlorophyte, *Ulothrix*, is shown in Figure 15-8. When the two gametes, formed by the specialized sexual structure called the **gametangium,** are released, they join together to form a single diploid cell, the zygote. The zygote, with its new combination of genetic material, then germinates to form several spores which will develop into the filamentous colonies. These may reproduce either asexually through spores or sexually through gametes.

**Phylum Phaeophyta.** The common brown alga, *Fucus*, has specialized gamete-producing structures, and the gametes are heterogamous. *Fucus* is quite unusual among nonvascular plants, in that the only haploid cells in the life cycle are those· of the gametes and a 16-celled gametophyte; all other stages are diploid. This type of life cycle is common among animals and tracheophytes, but most lower plants feature large multicellular haploid generations. In *Ulothrix*, for example, the haploid stage is the most conspicuous and long-lived. The diploid stage, the zygote, is very brief.

Figure 15-10 illustrates the reproductive structures of the *Fucus* plant. When the gametes

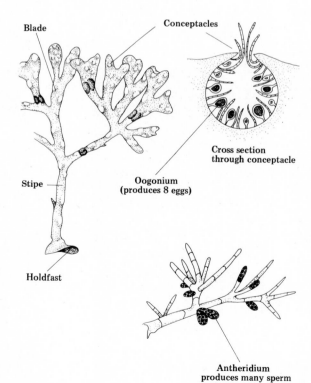

Blade

Conceptacles

Cross section
through conceptacle

Stipe

Oogonium
(produces 8 eggs)

Holdfast

**15-9** Fucus, *also known as rockweed, is a common brown alga along the New England coast. It is formed of a branched thallus attached by a holdfast to the rock on which it grows. Sexual reproduction occurs when reproductive structures called receptacles develop at the tips of fertile thalluses. Openings in the surface lead to cavities (conceptacles) containing the sex organs.*

Antheridium
produces many sperm

**457**

are released into the surrounding water, they
will be carried by currents around the surface
of the floating mass of algae. Since large numbers
of gametes are released, the chances for fertilization to take place are relatively good. *Fucus*
shares its dependence on water as the medium
of fertilization with all other plants except the
tracheophytes.

**15-10** *The reproductive structure
of the common marine* Fucus *plant
includes separate specialized male
and female conceptacles or
cavities. The female gametophyte
(A) contains large haploid gametes
and the male (B) contains a greater
number of small gametes. These
structures discharge numerous
heterogametes into the
surrounding water where
fertilization occurs. (Courtesy
Ward's Natural Science
Establishment)*

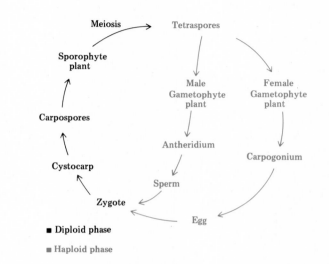

■ Diploid phase

■ Haploid phase

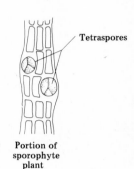

Portion of
sporophyte
plant

Portion of ♂
gametophyte plant

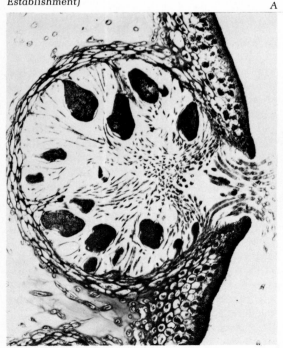

A

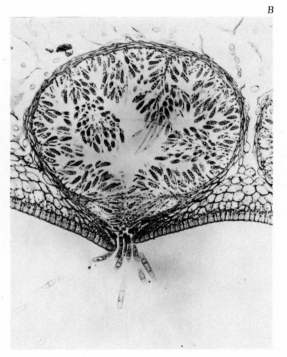

B

Cystocarp

Portion of ♀
gametophyte plant

**15-11** *The complex life cycle of*
Polysiphonia *includes two kinds of
sporophyte and two kinds of
gametophyte. Male and female
gametophytes are much less
conspicuous than the sporophyte
plants, a trend which is prominent
in the life cycles of higher plants.*

**458**

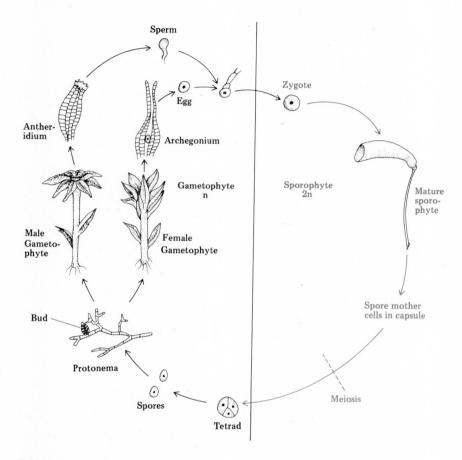

Sperm

Egg

Antheridium

Archegonium

Gametophyte
n

Male
Gametophyte

Female
Gametophyte

Bud

Protonema

Spores

Tetrad

Zygote

Sporophyte
2n

Mature
sporo-
phyte

Spore mother
cells in capsule

Meiosis

■ **Haploid phase**

■ Diploid phase

**15-12** *The young Polytrichum
plant develops from a spore into
the filamentous protonema, from
which the green gametophyte plant
develops. Fertilization takes place
within the archegonium, and gives
rise to the diploid sporophyte,
consisting of a sporangium at the
end of a wirelike stalk.*

**Phylum Rhodophyta.** The life cycles of the red algae are generally quite complex. There is alternation of generations, and both the sporophyte and the gametophyte generations are well-developed multicellular organisms. *Polysiphonia*, one of the most advanced of this group, has four distinct body types. There are two types of sporophyte, the carposporophyte and the tetrasporophyte, each anatomically different but both producing haploid spores. Male and female gametophytes are two separate plants. In many cases, the gametophyte is much smaller than the sporophyte and remains attached to it throughout the life cycle. This trait is exaggerated in the course of evolution, until in higher plants the gametophyte is an almost invisible structure that is completely dependent on the larger sporophyte plant.

**Phylum Bryophyta.** In the bryophytes, we generally find a dominant gametophyte generation. Moss spores give rise to an intermediate stage called a protonema, a filamentous form that grows on the surface of the soil. From this stage will develop the large gametophyte plant. The gametes are produced in specialized structures, and the fertilization of the egg actually takes place within the structure where the egg was produced; in higher plants this is called an **archegonium** rather than an oögonium. The process of fertilization cannot take place without the aid of water; either heavy rain or standing water during the rainy season will suffice. Because of this dependence on water, mosses cannot be considered true land plants, even though much of their life cycle takes place on dry land.

The fertilized zygote develops into a small diploid sporophyte plant that is attached by a footlike structure to the archegonium. At the end of a long stalk, a small sporangium develops. The mature sporophyte has no chlorophyll but is completely dependent on the gametophyte plant to which it remains attached. Once the haploid spores are released, the sporophyte dies. The

**459**

spores will give rise to a new protonema, and the life cycle will continue.

The life cycle of liverworts is very similar to that of mosses, except that there is no protonema stage. Spores are dispersed when the dead sporangium dries and bursts. Special structures within the sporangium, called elaters, twist as they dry out and help forcibly expel the spores. Liverworts can also reproduce by a form of budding. Little **gemma cups** develop on the leaves of the liverwort; when the older part of the plant dies, they can grow into new gametophyte plants.

**Phylum Tracheophyta.** As the tracheophytes have adapted to a terrestrial environment, significant changes have taken place in their life cycle. The two most important trends of change are the dominance of the sporophyte generation and the development of mechanisms of fertilization even in the absence of water.

The ferns, members of Class Filicinae, are an intermediate stage in the adaptation to life on dry land. The large fern plant with which we are most familiar is a sporophyte; the gametophyte is a small flat heart-shaped plant that is entirely independent during its brief life span. Although the sporophyte is a vascular plant, the gametophyte is not, and it can survive only in very damp locations. Fertilization takes place within the gametophyte, as the flagellated sperm cells are released from the antheridium and swim to the female reproductive structure, or archegonium, where the eggs are contained. The zygote immediately develops into a sporophyte plant as soon as fertilization takes place.

In the seed plants (gymnosperms and angiosperms) the final adaptation to terrestrial life has taken place. The sperm cells have no flagella and do not depend on water for their movement. Sperm-containing pollen grains may be dispersed by wind, by birds, by insects, or by forcible discharge from the antheridium. They move into the female gametophyte by growing a

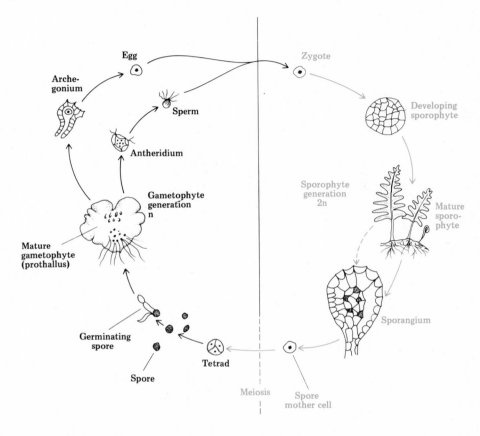

special connective tube. Reliance on such a random method of fertilization means that millions of pollen grains must be produced to insure survival. The fertilized zygote develops into an embryo that goes through a seed stage, with a protective outer coat and an abundant supply of stored food. This permits a waiting period between the time of fertilization and the growth of the new sporophyte; growth takes place only when environmental conditions are optimal.

**15-13** *The life cycle of a typical fern shows further adaptations to terrestrial life. The gametophyte is an inconspicuous heart-shaped nonvascular plant. Flagellated sperm swim through a film of moisture from the antheridium to the archegonium. Upon fertilization, the predominant sporophyte plant develops from the zygote.*

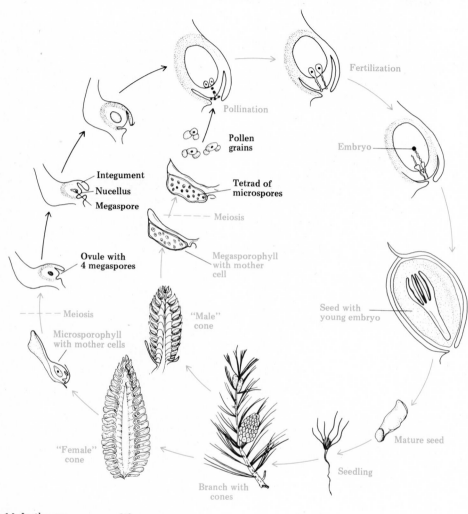

**15-14** *In the gymnosperm life cycle, the diploid sporophyte produces two kinds of cones. The male cones produce and release the wind-borne pollen. Pollen reaching the female, or ovulate, cone enters the ovule and, after a delay of as much as a year or more, fertilization occurs. A seed is formed which later develops into the mature sporophyte, the tree.*

■ **Haploid phase**

■ Diploid phase

## Reproductive structures in seed plants

In all seed plants, the gametophyte is a small structure contained within the sporophyte plant and completely dependent on it. The actual reproductive structures differ in the two main divisions of seed plants, the gymnosperms and the angiosperms.

**Gymnosperms.** In most gymnosperms, the reproductive structures occur in the form of cones called **strobili**. The cones are actually clusters of modified leaves, and the tiny gametophytes develop within ovules upon these leaves, called bracts. Large ovulate cones produce megaspores that develop into female gametophytes; small staminate cones produce microspores that develop into male gametophytes.

Division of the microspores follows the general pattern described in Chapter 12. In a representative example, the pine, the microspore mother cell undergoes meiosis to produce four haploid microspores. The microspores develop little projections, or wings, by a process of maturation, not cell division. Then the microspore divides mitotically to form a large central cell and a smaller vegetative cell. Although the vegetative cell will undergo another mitotic division, both new cells will degenerate. The central cell also divides mitotically, forming a generative nucleus and a tube nucleus. At this point the pollen grain is mature, and it is released from the cone on which it developed.

The megaspore mother cell, which is contained within a structure called an ovule, also undergoes meiotic division, forming four new cells but only a single functional megaspore. This megaspore undergoes a number of mitotic divisions, in a development process that culminates in the formation of two to five archegonia, each containing an egg, and a large vegetative mass of endosperm.

Pollination takes place when the wind blows a grain of pollen onto the bract of an ovulate cone.

**461**

A

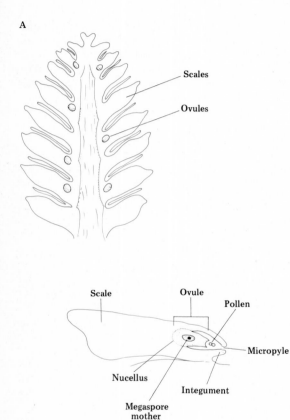

Scales

Ovules

Scale

Ovule

Pollen

Micropyle

Nucellus

Integument

Megaspore
mother
cell

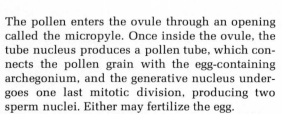

B

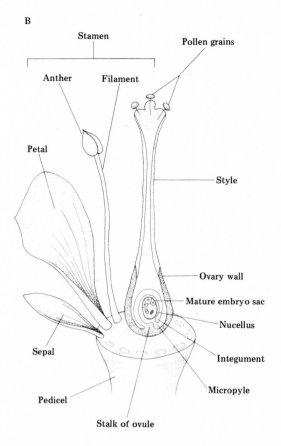

Stamen

Anther    Filament

Pollen grains

Petal

Style

Ovary wall

Mature embryo sac

Nucellus

Sepal

Integument

Pedicel

Micropyle

Stalk of ovule

**15-15** *A comparison of typical flower and cone structures indicates that both are leaves, modified evolutionarily so as to carry out reproductive functions. The flower contains both male and female reproductive organs, while the cone is either staminate (male) or ovulate (female). In both fertilization takes place in the ovule, but the secondary structures of the flower permit pollination by birds and insects, as well as by wind.*

The pollen enters the ovule through an opening called the micropyle. Once inside the ovule, the tube nucleus produces a pollen tube, which connects the pollen grain with the egg-containing archegonium, and the generative nucleus undergoes one last mitotic division, producing two sperm nuclei. Either may fertilize the egg.

The fertilized egg remains inside the ovule while it undergoes the process of growth and development that turns it into a plant embryo. During this process, the stored food of the endosperm is utilized. The outer walls of the ovule harden, turning into the seed coat. When the embryo is completely developed, the entire ovule is released from the plant. When conditions are favorable, it will germinate and form a new sporophyte plant.

**Angiosperms.** In angiosperms, the reproductive structures are flowers, rather than the cones of the gymnosperms. A flower is like a cone in that it consists of modified leaves attached to a stem. An outer circle of **sepals,** or leaves often modified to serve a protective function, forms the **calyx.** Inside that, a circle of petals, often brightly colored, forms the **corolla.** Inside the corolla, a number of **stamens,** which gives rise to pollen grains, form another circle. At the center of the flower is a group of modified leaves that have fused into one single structure, the **carpel,** that will contain the ovules.

The male gametophyte develops within the **anthers,** small sacs borne on the tops of the stamens. As in the gymnosperms, each microspore mother cell undergoes meiosis to form four haploid microspores. The microspore, which represents the first stage of the gametophyte generation, then develops into a pollen grain by a thickening of the cell wall; pollen grains are the best protected of plant cells.

The female gametophyte develops within a specialized structure known as the ovule; it is located inside one carpel. A megaspore mother cell divides meiotically to form four haploid megaspores. Three of these megaspores die and disintegrate. The fourth undergoes three mitotic

**15-16** *In angiosperms, the actual plant embryo, or seed, is covered by some kind of fruit, and the fruit, in addition to providing additional protection for the seed, also serves as a means of dispersal. Angiosperms are thus considered the final step in the process of adaptation to terrestrial life. (Ross Hutchins from Omikron)*

divisions to form eight nuclei. In some angiosperms, three haploid nuclei may fuse to form triploid nuclei. In others, the mature female gametophyte contains six haploid nuclei and one 2n nucleus, the polar nucleus, formed by the fusion of two haploid nuclei.

Pollination takes place when a pollen grain is introduced onto the upper portion of the carpel, the stigma. Prior to this, while the microspore was still within the anther, germination and development has proceeded to a considerable extent. The nucleus of the pollen grain has divided mitotically to form a larger tube cell and a smaller generative cell. The generative cell divides again to form two sperms. It is at this stage that the pollen grain or male gametophyte is transferred to the stigma. As in the gymnosperms, the tube cell controls the growth of the pollen tube which will transport the sperm cells to the female gametophyte within the ovule.

In angiosperms, a double fertilization takes place. One sperm fuses with one of the haploid nuclei nearest the micropyle, the egg; the zygote thus formed will be diploid (2n). The second sperm unites with the 2n polar nucleus. The result is a 3n nucleus. In some angiosperms, the pattern is varied, and the endosperm is 5n. Like many euploidal cells, the endosperm grows quickly and vigorously, developing the endosperm or food supply for the embryo.

The zygote undergoes a series of divisions as it develops into a plant embryo, using the stored energy of the endosperm. At the same time the outer layers of the ovule also undergo development, hardening into a protective coat. The ovary, a reproductive structure that contains all the individual ovules (now called seeds), develops into a fruit, usually by the thickening or expansion of the ovary wall.

The two significant evolutionary advances of the angiosperms are the development of flowers and fruits. Flowers aid in the process of pollination. Their bright colors and sweet smells attract birds and insects, which then carry the pollen from one flower to another, insuring a mixed genetic inheritance. Some angiosperms continue to rely on the wind for pollination, and in these, the calyx and corolla are either very small or absent altogether; but the flower permits adaptation to a number of other modes of pollination. Some of these are illustrated in the color pages of Portfolio 2.

The development of fruits helps in the dispersal of seeds. The fruits can be carried off by small mammals and birds; squirrels, for example, obligingly bury nuts in ideal germinating conditions. The fruits may be eaten and the seeds expelled with other food wastes in the feces, at a location far distant from the site of the seed-producing plant. Some fruits (burrs, for instance) are designed for the purpose of hitching a ride in some animal's fur, while others, like the dandelion seed, can be propelled by the wind for long distances.

## Sexual reproduction in animals

Although some lower animals can reproduce by budding or by regeneration, in the animal kingdom sexual reproduction can be considered the rule and asexual reproduction the exception. The mobility on which animals have specialized makes it relatively easy for them to locate and mate with members of the opposite sex, to deposit eggs and give birth to young in a favorable environment. For animals, an excessive reliance on sexual reproduction is not so risky as it is for plants, those immobile subjects of the environment.

Many animals have complicated life cycles, but except in the case of lower animals, such as jellyfish and other coelenterates, these life cycles are not alternations of generations but simply stages in the individual's development. These cycles have very little reproduction significance and, therefore, will not be discussed here.

Another important difference between plant and animal reproduction is that in animals, reproductive structures are generally a permanent part of the anatomy, whereas in plants these structures (flowers and cones, for example) appear only at certain seasons. The permanent presence of reproductive structures, however, does not necessarily mean that breeding can take place at any time. In many species, both physiological and behavioral mechanisms serve to limit breeding to a specific season.

## Patterns of reproduction in animals

The fundamental outline of sexual reproduction in animals is the same in all phyla. Specialized reproductive organs produce haploid gametes, which are differentiated morphologically and/or physiologically into eggs and sperm; that is, animals are heterogamous. The fusion of these gametes produces a zygote that grows into a new individual. Despite this basic similarity of reproduction in most animals, adaptation to different environments and ways of life has led to a number of varying patterns of reproduction.

**Hermaphroditism.** Hermaphrodites are animals that develop the specialized reproductive structures of both sexes, rather than only one. This condition is found in a number of lower animals, of which the earthworm is a well-known example.

Although to us hermaphroditism might seem something of a social embarrassment, it is a very successful solution to the reproductive problems of endoparasites, such as tapeworms and liver flukes. In a majority of cases, only a single individual happens to enter the host's body, where it must remain in order to obtain nourishment. Under such circumstances, sexual reproduction between individuals of different sexes is impossible. The tapeworm's solution is to fertilize itself; although this limits the possibilities of genetic recombination, it does serve to perpetuate

**15-17** *Hermaphroditism is particularly common among slow-moving, sluggish animals. Mutual fertilization, however, occurs more frequently than self-fertilization among earthworms. Here, the sperm produced by one night crawler is passed through a pore to the second worm who holds it in a seminal receptacle. Note the clitellum, the thick glandular region near the anterior end of the body. This region secretes a sticky substance that glues the worms together during copulation. (Alvin E. Staffan from National Audubon Society)*

the species. When other individuals are available as mates, the tapeworms engage in mutual fertilization. Each fertilizes the other's eggs.

Mutual fertilization is the rule among many hermaphrodites, and again the earthworm furnishes a familiar example. The worms copulate by lying together with their heads pointing in opposite directions. They are virtually glued together by a sticky secretion of the clitellum, a specialized region of several segments near the anterior end of the body. Sperm produced by one earthworm is passed through a pore to the second worm, who holds it in a kind of storage tank called a seminal receptacle. Some time after copulation, the clitellum produces a slippery albuminous fluid that hardens into a cocoon and slides over the worm's body as it moves. The eggs are deposited in the cocoon as it passes the fe-

male pores on segment 14; the sperm is deposited as it passes the receptacles on segments 9 and 10. The cocoon then slips over the worm's head and remains behind him/her on the ground. The openings shrivel up and a little capsule is formed, where the eggs will develop into tiny worms.

There are some hermaphrodites, such as the oyster, in which this fertilization cannot take place simultaneously. Although oysters produce both eggs and sperm, they do not produce both types of gametes at the same time. This prevents self-fertilization from ever taking place.

**Parthenogenesis.** Parthenogenesis is a process by which unfertilized eggs can grow into new individuals, without the need for fusion with sperm. Parthenogenesis in nature is found mostly in arthropods. In this phylum, there are a few species that consist only of females and reproduce by parthenogenesis alone. In most instances, parthenogenesis supplements regular fertilization. There may be some alternation of generations—one fertilized generation followed by one parthenogenetic generation—or, as in honeybees, a single group of eggs laid by the queen may contain both fertilized and unfertilized eggs, to develop into workers and drones respectively.

Laboratory experiments have indicated that the eggs of some species of amphibians that do not normally reproduce by parthenogenesis can be stimulated to develop more or less normally without fertilization. Among the factors that will induce such development are changes in the pH of the water in which the animals live, changes in the ambient temperature, and mechanical manipulation of the egg. Pinpricks are especially effective in inducing parthenogenesis. The fact that sudden dramatic environmental changes can induce parthenogenesis suggests the hypothesis that it is a kind of fail-safe device of the reproductive system.

**External fertilization.** In external fertilization, both the eggs and the sperm leave the body of the parent individuals before fertilization takes place. This type of fertilization is confined to aquatic animals, because it requires water as a medium for carrying the sperm to the eggs. Since actual copulation does not take place, these animals lack accessory sex organs and have only rudimentary secondary sex characteristics. A simple duct that carries gametes from the site of production to the exterior is all that is needed.

An example of external fertilization in marine animals can be seen in *Obelia*, a hydrozoan belonging to Phylum Coelenterata. One type of medusa produces eggs, the other type produces sperm. Both kinds of gametes are released into the water in great quantities; their random distribution with the water currents enables the chance meeting of eggs and sperm, which fuse to form a zygote.

In many aquatic vertebrates, fertilization is also external, but behavioral mechanisms have eliminated much of the element of chance so prevalent in *Obelia's* mating process. The rituals of courtship insure that male and female will be close together when gametes are released. One or both may make a nest which serves the durable purpose of providing a target for gamete deposition and a protection for the developing eggs. The Siamese fighting fish is an extreme example of this kind of behavior. The male courts the female and then induces her to lay her eggs right under a nest he has built, of spittle and air-bubbles, on the surface of the water. As the eggs slowly sink through the water, the male fertilizes them and then swims after them, catching them gently in his mouth. He coats the eggs with aerated spittle and then blows them into the nest, which is safely out of the line of vision of most predators.

**Internal fertilization.** Internal fertilization is an evolutionary advance that is generally associated with the move to a terrestrial environment. Without water as a medium of dispersal of the eggs and sperm, the chances for successful external fertilization are practically zero. Internal fertilization, in which the male sex organ deposits the

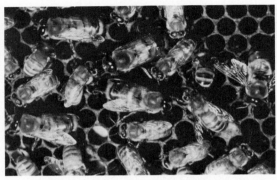

sperm directly inside the female reproductive tract, insures the meeting of eggs and sperm.

The evolution of a mechanism of internal fertilization has brought with it a number of other significant developments. In most species that practice internal fertilization, the sexes are distinctly differentiated in physical appearance as well as physiological characteristics. There are differences in coloration, size, patterns of plumage or body markings, length of feathers, horns, and tusks. This kind of divergence is called **sexual dimorphism.** Behavioral adaptations also help differentiate the two sexes. For example, male song birds, such as the warbler or nightingale, sing, but females do not; male grouse drum and dance while females watch. Such

15-18 *A queen bee generally mates once, and then uses the stored sperm to fertilize some of her eggs throughout the rest of her reproductive career. Fertilized eggs will develop into female worker bees; the unfertilized eggs will develop into male drones. Workers are generally noticeably larger and sturdier than drones. (Carolina Biological Supply Company)*

15-19 *The sea catfish, this one seen in the Gulf of Mexico near the Florida coast, is typical of the fish, which reproduce through external fertilization. Usually a complex and delicately timed mating ritual insures that eggs and sperm are released into the surrounding water at the same time. (Courtesy of R. F. Head from National Audubon Society)*

**15-20** *The oviparous crocodile produces large numbers of eggs which hatch as miniature immature adults. Fertilization occurs internally but the eggs develop essentially on their own externally, food being supplied within each egg by the yolk. (Jen and Des Bartlett—Photo Researchers)*

**15-21** *Unlike most lower animals Macroperipatus gives birth to live and fully developed young. This "missing link" combines features of annelids—nephridia and ciliated reproduction organs—with those of arthropods—claws and tracheal tubes. (Courtesy of the American Museum of Natural History)*

mechanisms of differentiation make the recognition of a potential mate very easy.

Other adaptations serve to attract mates, so that a mating pair will remain together for at least the time it takes to achieve fertilization, perhaps for an even longer period that includes the care of the young. Behavioral adaptations—stroking; singing; displaying special feathers, color patches on the skin, or other secondary sex characteristics; or the release of odoriferous chemicals—help both to attract mates and to elicit the mating response that allows internal fertilization to take place.

Internal fertilization may be a prelude to egg-laying. This is true of all insects and of all classes of vertebrates except the mammals; these egg-laying animals are described as **oviparous.** Oviparous animals typically produce large numbers of eggs at each mating. Even though the female tries to lay the eggs in some concealed place, they are frequently consumed by predators, and the mortality rate of the unprotected young is also extremely high.

In mammals and some fish and reptiles, internal fertilization has led to another evolutionary advance, the development of the fertilized egg inside the female's body. Mammals are **viviparous** (live-bearing), and therefore they produce only a relatively small number of young at each mating. Many mammalian species extend their protection of the young to the period after birth, when the mother nourishes the young with her milk, and both parents may provide protection against predators.

## Reproductive structures

In animal groups that practice external fertilization (aquatic animals only) reproductive structures are very simple. They consist of the reproductive organ, or **gonad,** which may appear singly or in pairs, and a duct to carry eggs and sperm from the gonad to the environment, where fertilization will take place. In animals adapted to a terrestrial environment, where internal fertilization is the rule, the male develops a copulatory organ that permits transfer of the sperm into the female reproductive tract. The female tract is generally modified in some way for this function. In viviparous mammals, the female reproductive system also includes a specialized organ in which the fertilized zygote will develop before birth.

**Reproductive structures of the male.** The male gonad, or **testis,** is the site of sperm production. Inside the testis are germ cells, cells which are capable of meiotic division, forming haploid gametes. The testis also contains secretory cells, that produce male sex hormones, or androgens. Thus the testis is considered an endocrine gland as well as a reproductive structure.

The inside of the testis is divided into small chambers by folds of epithelial tissue. Within the chambers are small seminiferous tubules, where germ cells are produced. These cells undergo meiosis and become the haploid gametes. These gametes then leave the testis through the sperm duct.

In many lower animals that are adapted for external fertilization, the male reproductive system consists only of testes and a sperm duct. In higher animals, however, there are other organs associated with the testes that serve a reproductive function. There may be a place to store sperm, such as a sperm sac, or (in humans) a set of coiled tubules called the epididymis. Sperm storage permits the discharge of large numbers of sperm at a single time, improving the chances of successful fertilization. It also permits the discharge of sperm to be timed to coincide with successful copulation. Other associated organs, such as the prostate gland and the seminal vesicles, add various fluids to the sperm duct, where they mix with the sperm cells and increase the chances of successful fertilization. The seminal fluid thus produced is a viscous liquid in which the sperm are suspended. It helps the

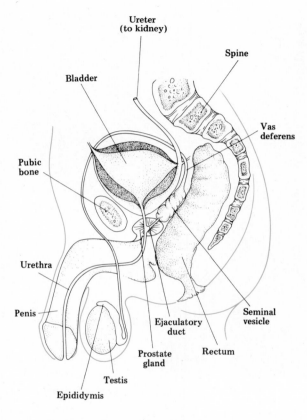

Ureter (to kidney)

Spine

Bladder

Vas deferens

Pubic bone

Urethra

Penis

Ejaculatory duct

Seminal vesicle

Prostate gland

Rectum

Testis

Epididymis

**15-22** *The drawing indicates the major structures of the human male reproductive system. Each testis consists of about 1000 coiled seminiferous tubules, which actually produce the sperm. Sperm is stored in the epididymis. In intercourse, the sperm, suspended in seminal fluid which comes from three glands, enters the urethra from the seminal vesicles, and is forced out into the vagina.*

sperm travel up the female reproductive tract and reach the egg alive. Among the components of the fluid are sugars to provide food for the sperm cells, a chemical buffer that protects the cells against an excessively acidic or basic environment (secretions of the female reproductive tract are often acidic), a mucus to lubricate the female tract, and a compound that stimulates rapid movement of the sperm flagellum, thus helping the sperm to move farther and faster.

In certain species with external fertilization, the sperm duct opens directly to the outside. In salamanders, the male discharges a packet of sperm, in a thick membranous jacket, onto the ground. The female picks up this **spermatophore,** and inserts it into her reproductive tract, where the membrane ruptures and the sperm move up to fertilize the egg. In land snails and many terrestrial arthropods, the sperm duct discharges through the anal opening. The small copulatory organ sometimes present serves only to position the male's cloaca near the female's.

In the higher vertebrates, the male copulatory organ actually discharges the sperm; this type of organ is called a *penis.* The sperm ducts and accessory glands discharge the sperm-containing seminal fluid into the **urethra** or central duct of the penis. In mammals, the urethra is also connected to the urinary bladder, thus allowing this one organ to serve two different functions. There are sphincters at the mouths of the sperm duct and the urinary duct. The closing of one or the other of these sphincters, under the control of the autonomic nervous system, prevents the simultaneous discharge of seminal fluid and urine.

In all animals, gonads, whether male or female, develop within the central body cavity. For homeotherms, with their relatively high body temperatures, this poses a problem, since body heat is great enough to kill the sperm. In many mammals, the solution to this problem has been the suspension of the testes in a sac (the **scrotum**) outside the body. In primates, including man, the testes migrate into the scrotum while the male is still a fetus. In other mammals, the migration takes place only at sexual maturity; and in elephants, it occurs only during the breeding season. Sperm cells are extremely sensitive to high temperatures. In some couples that are having trouble conceiving due to a low sperm count, doctors now advise the husband to switch from briefs to boxer shorts. Tight underwear that holds the scrotum closely up against the body causes many sperm to die or lose their motility. The scrotum is an adaptation that permits the sperm cells to be produced and stored at a temperature several degrees cooler than that inside the body.

**Reproductive structures of the female.** The female gonads, or **ovaries,** develop in the embryo from the same germ cells found in the male gonads. It is the action of certain sex hormones, produced according to the genetic code contained in the genes of the zygote, that determines whether the gonads will become a testis or an ovary.

Anatomical studies reveal the basic similarity of the testis and the ovary. The inner surface of epithelium in the ovary divides and sheds new cells into the ovary. In the male gonad, all these cells can become gametes at the same time; in the female gonad, only one cell at a time will undergo meiosis to produce a haploid gamete. Through the action of hormones, the other cells are prevented from becoming gametes. Instead they form a protective enclosure around the egg, called a **follicle.** The egg matures within the follicle, leaving only when it is ready for fertilization.

The mature egg bursts its follicle and moves from the ovary into the coelom and from there enters a special passageway, the **oviduct;** move-

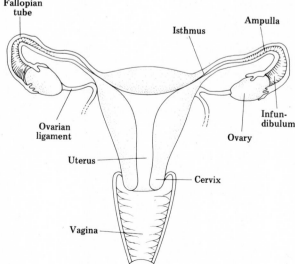

**15-23** *The ovaries in the human female reproductive system are paired almond-shaped organs about three centimeters long. The egg is released from the ovary and travels into one of the two fallopian tubes. The fallopian tubes empty into the uterus. The vagina serves both as a sperm receptacle during coitus and as a birth canal during parturition.*

Fallopian tube

Isthmus

Ampulla

Ovarian ligament

Infundibulum

Ovary

Uterus

Cervix

Vagina

**469**

ment is due to the action of cilia located on the epidermal tissue lining these organs. If fertilization is external, the eggs may simply be covered with a layer of protective (and sperm-attracting) mucus and then expelled through the anus. If fertilization is internal, the female reproductive system develops a number of specialized organs and glands that play a part in the fertilization process; additional reproductive structures have developed in viviparous species.

In the females of oviparous species, there may be a storage receptacle for sperm as well as a gland that secretes the hard protective shell around the egg. In certain snakes and many sharks, there is also a widening of the oviduct in which the fertilized eggs are held while the embryos develop. The eggs hatch inside the uterus, and live young emerge through the external opening of the reproductive tract. This pattern of reproduction, with eggs and live birth, is the **oviparous** pattern.

In viviparous species, fertilization takes place either in the oviduct or in a specialized organ, the uterus. As soon as one sperm enters the egg and fuses with the egg's nucleus, the outer membrane of the cell becomes a barrier that will not permit any more sperm cells to enter, preventing fertilization by two or more sperm that would yield a triploid or polyploid individual. The zygote burrows into the wall of the uterus, where it will develop until it is ready for independent existence.

Inside the uterus, the embryo is nourished by a tissue called the **placenta,** which develops as a specialization of the outer membrane of the zygote. The surface of the placenta contains numerous villi, which protrude into the tissues of the uterine wall. Through these villi, an exchange takes place between the maternal and the fetal blood. As we saw in Chapter 7, fetal blood has an oxygen dissociation curve far to the left of maternal blood, thus allowing the fetus to absorb oxygen. At the same time that oxygen and necessary nutrients diffuse from the maternal blood

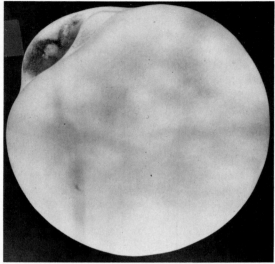

into the fetal blood through the villi, a reverse exchange is taking place, in which the maternal blood absorbs carbon dioxide and metabolic wastes. The placenta does seem to have some ability to "screen" the materials passing from the maternal blood to the fetal blood; experimental studies have shown that many toxic chemicals cannot penetrate the placental barrier. As a result of such studies, the placenta was once thought to be a kind of magic wall, keeping out everything bad while admitting everything good. But there is evidence of placental failure to discriminate between bad and good. A heroin-addicted mother can pass her addiction to her unborn child; pregnant women who took thalidomide passed that drug to their children, with the worst of consequences. Further research is needed before biologists can really understand the mechanism of placental screening.

In addition to serving as a medium of exchange between maternal and fetal blood, the placenta also has an important endocrine function. It secretes hormones that stimulate metabolic functions of the mother during pregnancy

**15-24** *A sperm does not bore its way into the egg. Rather, it is engulfed through an egg cone which comes to surround the sperm. A previously formed fertilization membrane then rises off the surface of the egg to trap the single sperm which has made contact and to prevent other sperm from entering the egg. The sperm leaves its tail at the surface, and the nucleus alone migrates into the cytoplasm toward the egg nucleus. Only after the two haploid nuclei have fused is fertilization complete. (Russ Kinne—Photo Researchers)*

**15-25** *A human pregnancy involves many complex hormonal interactions; the placenta is a major source of endocrine. In this diagram the colored arrows indicate stimulation and the gray arrows indicate inhibition of endocrine secretions.*

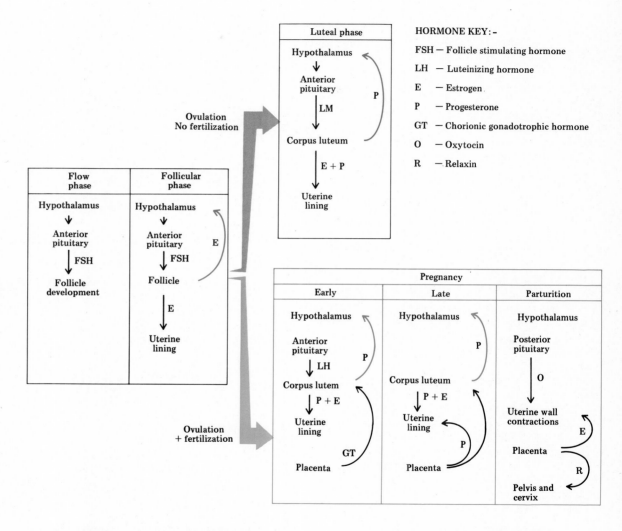

as well as the hormones that prepare both mother and child for the birth process.

The uterus of mammals does not open directly to the outside but rather into a widening of the reproductive tract called the vagina. The vagina is a muscular tube that receives the male sex organ and its sperm ejaculations. The epithelium lining the vagina contains secretory cells that secrete a mucous coating, which helps to lubricate the passage during the insertion of the penis. The mucus is usually rather acidic, with a pH of about 4.5; the adaptive value of this acidic environment probably lies in the fact that it is a protection against bacterial and viral infections. The buffer contained in the seminal fluid helps the sperm cells survive in the vagina.

## Reproductive behavior

In all important life functions, morphological and physiological adaptations are supplemented by behavioral ones. The function of reproduction is no exception in this respect, and in fact, much of observable animal behavior is connected with reproductive processes. Specific behavioral adaptations vary from species to species, but certain broad patterns of adaptation can be picked out. Behavioral adaptations are often categorized on the basis of the time at which they occur— before, during, or after copulation.

**Courtship behavior.** Reproductive behavior that takes place before copulation is referred to as courtship behavior. One function of courtship behavior is the identification of mates of the right sex and species. Where speciation is visually distinct, and sexual dimorphism is present, males and females of the same species can recognize each other by their appearance. But in many cases, behavioral recognition cues must supplement, or substitute for, visual cues. For example, the behavior of sexually mature male crested grebes differs from that of immature males or of females. The mature male displays by erecting his crest and holding his wings outspread to exaggerate the size of his body. Male cichlids (a family of tropical fish) raise their dorsal fins and beat the water with their tails. Such behavioral cues enable potential breeding partners to locate one another, even when they are separated by fairly great distances. Such mechanisms can be very important in the case of species with large hunting territories, for individuals may be widely dispersed.

In predatory species, courtship has another important function; it inhibits the aggressive response customarily evoked by the presence of other individuals within a certain territory. A young male polar bear, for example, will attack all other polar bears, male or female, that try to invade the hunting territory it has staked out

for itself. Such an aggressive response would make mating very difficult, if not downright impossible. During the breeding season, however, male bears alter their behavior toward females, engaging in specific courtship activities that reassure the female that it is safe to approach. In turn, the female's behavior signals the male that her own intentions are also peaceful, so that the male's courtship response will continue to override his aggressive response.

For most vertebrate species, courtship behavior serves as a way to synchronize the reproductive cycles of the male and the female, so that sperm will be produced at exactly the time that the eggs are ready to be fertilized. The sights and sounds of a breeding partner, or the initiation of certain courtship activities, seems to act as a stimulus for the secretion of hormones that in-

**15-26** *Courtship activities among birds are many and varied. Terns present a fish to the female, an act that apparently serves in sex recognition. The female may beg several times before the male actually gives her the fish. If she is not then incubating, the female will return the fish. Courtship rituals and display stimulate hormone production in the female and bring her to a state of reproductive readiness. (U. Böcker—Photo Researchers)*

**15-27** *Natural selection is simply nonrandom reproduction. The battles of bison bulls are a form of selective courtship behavior. The stronger and healthier male will presumably win, and will therefore have access to females and produce offspring. In leaving behind a sample of his superior genetic endowment, he contributes to a change in the gene pool in the direction of greater fitness. (Leonard Lee Rue III—Bruce Coleman)*

winner gains possession of a harem of females. This behavioral adaptation insures that the females will mate with the strongest and healthiest of the nearby males. The amazing migration of salmon traveling upstream to the place of their birth can be considered a specialized kind of courtship behavior that also works as a means of selecting the healthiest individuals for breeding. The weaker individuals never make it back to their spawning grounds.

**Mating behavior.** Behavior that takes place during the time of fertilization or copulation is called mating behavior. In lower vertebrates, certain kinds of behavior help to coordinate the release of the eggs and sperm. For example, the male salamander strokes the female, thus stimulating her to release her eggs when he is present to fertilize them. In fish, the stimulus is often an erratic swimming pattern rather than a stroke.

In vertebrate species with internal fertilization, mating behavior prepares both the male and the female for the act of copulation. The female jaguar, for example, may begin to arch her back and knead the ground with her paws. This behavior stimulates a response in the male, and his penis becomes erect and extruded. He bites the female on the neck, strokes her flanks and then mounts her. The response this behavior elicits in the female causes her to bend her tail sharply, exposing the opening of the genital tract. At the same time, additional mucous secretions help to lubricate the female's reproductive tract so that the male penis can be easily inserted. Thus behavioral signals can elicit both behavioral and physiological responses.

Immediately after fertilization or copulation, behavioral responses of many species serve to diminish the interest of the male in that particular female, although another female may be able to elicit a second mating response. This behavioral inhibition allows a single male to fertilize the eggs of a number of females, a trait which may have survival value if the ratio of

duce the next stage of the courtship. For example, in many female birds the activity of nest building, undertaken during the early stages of courtship stimulates the secretion of the hormones that will cause the female to accept the male during copulation and to lay eggs afterward. Many avian species feature an elaborate courtship ritual, each step of which prepares the way for the next phase of activity. This explains why many birds cannot breed in captivity; the confines of a zoo cage will not permit the performance of all steps in the courtship ritual, and thus the essential hormones are not triggered. Falcons serve as a good example of this problem, for their courtship behavior includes swooping flights and mock chases. Successful breeding of captive falcons is very rare.

In some species, courtship activities also serve a selective function. Among Rocky Mountain bighorn sheep, the males of breeding age engage in fights; the loser is driven away and the

males to females is unbalanced or if male fighting during courtship has driven away some of the males of the group.

**Parental behavior.** Where fertilization is external, there are usually few behavioral adaptations that take place after fertilization; as a rule, both male and female cease to concern themselves with the fate of the coming generation. There are interesting exceptions, among them fish that incubate the eggs in their mouths, and the male stickleback (a fish) that builds a nest for the eggs. But in general, animals that fertilize their eggs externally have no parental instinct.

In fact, many fish will cheerfully eat their own children. For example, unless the male Siamese fighting fish works fast to get the eggs safely in the nest, the female will swim along behind him and gobble them up.

In many oviparous species with internal fertilization, on the other hand, both males and females tend the eggs and feed and protect the young. The most striking examples of devoted parenting can be seen among birds, where the body heat of a parent is needed for the incubation of the eggs.

Species that bear live young tend to have elaborate and long-lasting parental behavior. In

**15-28** *This young snowy owl is gaping for food, a response typical of newly hatched birds. Certain birds may gape initially without apparent external stimulus, but only later do they direct this behavior toward the parent, suggesting that some learning is involved. The behavior serves as a stimulus to elicit feeding behavior from the adult. Caregiving behavior in the parent is reciprocal to care-soliciting behavior in the young. (Eric Hosking, F.R.P.S.—Bruce Coleman)*

some of the higher mammals, parenting may go on for years after the birth of the young. Behavioral biologists believe that in some species (especially primates) courtship activities may serve to establish pair bonds, or ties between the mating male and female, which serve to keep the breeding pair together during the lengthy period of child care. Desmond Morris has suggested that the lack of a breeding season in humans, which permits copulation all year round, may be a mechanism to promote pair bonding, providing two parents to care for the slow-developing young.

## Control and regulation of reproduction

In all vertebrates, the development of reproductive structures and the regulation of reproductive processes is mediated through the carefully timed release of hormones. These control cycles are complex information feedback systems that can adjust to changes in the external environment and the presence or absence of behavioral cues.

### Male development and behavior

In male vertebrates, the testes develop during embryonic growth, but no sperm cells are produced until the male reaches sexual maturity. Maturation begins when the hypothalamus begins to produce a hormone called a releasing factor, whose effect is to stimulate the anterior lobe of the hypophysis. The hypophysis then releases follicle-stimulating hormone (FSH) and luteinizing hormone (LH). These hormones cause developmental changes in the testes, leading to the maturation of these organs. The final step in the chain of control is the production of male sex hormones (testosterones) in the testes.

Testosterone stimulates the development of male secondary sex characteristics—for example, colorful plumage in the bird of paradise, a hairy chest in men—and male sexual behavior. In some cases, the changes, once instituted, are irreversible, but in others, the changes may disappear at the end of the breeding season. For example, once the testes of mammals enlarge and the seminal vesicles are mature, they remain that way for the rest of the animal's life. But in birds, the testes shrink at the end of the breeding season; new growth will be stimulated the following season. The birds' pattern can be regarded as an adaptation to their airborne way of life. Enlarged testes make the bird less streamlined and add an ounce or two to his weight; over the period of a year those disadvantages can be a significant drain on energy resources. Recent studies have shown that gonadal growth in birds is stimulated by light—or more specifically, by changes in the duration and intensity of light. Light is "measured" by the pineal body, and when a significant change is registered, the hypothalamus is stimulated and the entire chain of hormonal control set in action. Because of this control mechanism, breeding often takes place in the spring, so the baby birds will be born when the food supply is largest, the climate is temperate, and the strenuous winter migration still months away.

Testosterone also stimulates male sexual behavior. In laboratory experiments, castrated animals have proved to quickly lose their interest in sex. Not only do they fail to copulate but they also fail to perform any of the male courtship gestures. In humans, the male sex impulse seems to be more complex, and psychological factors are at least as important as physiological factors in stimulating sexual interest.

### Female development and behavior

Female gonads and secondary sex characteristics develop through the same cycle of control mechanisms found in the male. The hypothalamus produces a releasing factor; the hypophysis produces FSH and LH; the gonads then produce the female hormones estrogen and progesterone. Under their influence, the female becomes sexually mature. Since the first several steps of the control mechanism are the same in males and females, developmental biologists were curious about how differentiation was determined. The answer seems to be that it was determined early in the life of the embryo. Significant amounts of either testosterone or estrogen are present in the human embryo by the third month of pregnancy, and this early hormone activity fixes the sex of the individual. By birth, this hormone secretion has stopped, not to resume again until the child reaches the age of puberty.

**Regulation of estrus and menstrual cycles.** Under normal conditions, females are not continuously fertile. Since the development of an egg (and the preparation of the uterus in viviparous species) is metabolically expensive, it is only undertaken at limited intervals. In many wild animals, females have only one period of **estrus,** or "heat," when they are sexually responsive and fertile. In mice, estrus occurs every five days, which explains why a mouse population can grow so explosively in the absence of predators. In humans and many other primates, there is a menstrual cycle rather than a period of estrus. The cycle, which takes place in about one month's time, includes the thickening of the uterine lining and the release of a mature egg. If fertilization does not take place, the lining degenerates and is expelled during a few days of bleeding, the menstrual period.

The menstrual cycle is a good example of the complexity of hormonal control systems. After a menstrual period is ended, the hypothalamus begins to produce the releasing factor. This hormone (actually a peptide) stimulates the hypophysis, which in turn releases FSH. The FSH is carried by the circulatory system, eventually

A

B

15-29 *The appearance of the immature male bald eagle (A) is dramatically transformed by the development of male secondary sex characteristics (B). Testosterone, male sex hormone, stimulates prominent changes in the texture and color of plumage, the eyes, beak, and general bearing. (a: Karl H. Maslowski from National Audubon Society. b: David M. Davenport from National Audubon Society)*

reaching the ovary, where it causes an ovarian follicle to grow rapidly and stimulates it to secrete the female sex hormone estradiol.

Estradiol has several different effects. It causes some growth in the cells lining the uterus; it also acts as a feedback mechanism, stimulating the hypophysis to release more FSH and some LH. The joint effect of these two hormones is to cause ovulation, or the release of the egg. At the same time, the LH, in combination with another hormone produced in the hypophysis, causes the follicle to develop a *corpus luteum*, or secretory structure. The corpus luteum then begins to release progesterone, another female sex hormone.

Progesterone directly stimulates growth of the uterine lining, and causes the cells of the uterus to secrete a nutrient fluid that supports the new cells in the lining. It also serves as a mechanism of negative feedback. The presence of progesterone in the blood inhibits the production of the releasing factor in the hypothalamus. Thus no other follicles will develop, allowing the body to put all its resources in one egg.

Unless the corpus luteum receives an additional supply of LH (this would happen if fertilization took place), it slowly degenerates and ceases to produce progesterone. Menstruation occurs at this point. Since there is no longer any progesterone in the blood, the hypophysis is no longer inhibited, and it begins to produce more of the releasing factor. Thus a new cycle will begin.

An understanding of this complex cycle and the hormonal control mechanisms that regulate it has led to the development of the birth control pill. The pill is a synthetic progesterone; that is, it is basically similar to progesterone but has slight changes in the molecular structure that make it more stable. The effect of the pill is to inhibit the production of FSH and LH and thus prevent ovulation.

## Summary

In asexual and colonial life forms, reproduction involves only mitotic cell division. In sexually reproducing forms it involves some pattern of mitotic and meiotic division and the fusion of haploid gametes into a zygote. As a rule the haploid form is much more prominent in plants than in animals.

The simplest form of asexual reproduction is fission, widely exemplified in the Protista and Monera. Budding, the pinching off of a new, nucleated cell from the parent cell, is found in yeast and, at the organismic level, in hydra and some higher plants. Fragmentation is found in a few, primarily aquatic, multicellular and colonial organisms. Regeneration, the ability to regenerate lost parts or limbs, may extend to the ability, found in starfish, to develop an entire new organism from a severed fragment. In many plants, asexual reproduction is carried out through the production of specialized haploid reproductive cells called spores. Many plants also reproduce asexually by vegetative propagation, root division, and the rooting of leaves.

The possibility of genetic recombination confers a tremendous adaptive advantage upon sexually reproducing species. Forms of sexual reproduction range from the more or less isogamous patterns of Chlamydomonas and Spirogyra, where male and female gametes are similar or identical, through the heterogamous pattern of Oedogonium, to a pattern in which gametes are produced by specialized germ cells, which appears at the level of Volvox. Higher processes are simply elaborations of this last.

In the basic plant reproductive pattern a generally haploid gametophyte generation alternates with a generally diploid sporophyte generation. In the lower plants the gametophyte generation is the more prominent, and fertilization takes place in water or via a film of moisture. As the tracheophytes have become progressively more adapted to a terrestrial environment, the sporophyte generation has tended to become predominant, and nonaquatic mechanisms of fertilization have evolved. Among the Filicinae the gametophyte is an inconspicuous nonvascular plant which can reproduce only in a moist environment. In the seed plants fertilization is typically effected by wind, insects or birds, and the zygote goes through a dormant seed stage.

The gametophyte in seed plants is a small, dependent structure. In gymnosperms the reproductive structures are ovulate (female) and staminate (male) strobili, modified leaves or cones, which undergo cell division to produce, respectively, the archegonia and the pollen. Fertilization is by wind. In angiosperms the reproductive structures are elaborately modified leaves called flowers. Male and female gametophytes develop within the anthers and carpels respectively. The zygote develops within the ovule into the embryo. The fruit, which develops from the ovary, is an exclusive feature of angiosperm plants, and it serves as an extremely efficient mechanism of seed dispersal.

Most animals reproduce sexually, with heterogamous gametes arising in specialized reproductive organs. Hermaphroditic species are characterized by self- or mutual fertilization. Parthenogenesis, the development of a new individual from an unfertilized egg, is largely limited to such arthropods as the social insects, but can be induced artificially in other species.

In marine organisms external fertilization is common, as in the coelenterate Obelia. The efficiency of external fertilization is greatly increased by the evolution of elaborate behavioral mechanisms ensuring the proximity of eggs and sperm.

The move to a terrestrial environment is associated with internal fertilization and the development of sexual dimorphism and elaborate physiological and behavioral mating mechanisms. Species with internal fertilization may be oviparous (egg-bearing) or viviparous (live-bearing).

The basic male reproductive structure is the testis, which contains sperm-producing germ cells. The sperm duct may open directly to the outside, as in species with external fertilization, or may be associated with other structures and glands which aid in reproduction.

The ovaries, or female gonads, develop from the same embryonic germ cells as the testes. At maturity, the egg leaves a protective follicle to move into the oviduct, from which it is expelled from the body if fertilization is external. Internal fertilization takes place in specialized parts of the reproductive tract. In oviparous species the egg may be covered with a protective shell before leaving the body. In oviviparous species the eggs hatch in a specialized region of the oviduct. In viviparous species fertilization usually occurs in the oviduct. The embryo in mammals is linked to the uterus by a physiologically complex tissue, the placenta, which produces an elaborate series of hormones influencing pregnancy, parturition, and lactation.

Higher animals engage in courtship behavior that is important in species recognition, in synchronizing reproductive impulses, and in natural selection. This is followed by the mating behavior that actually leads to conception, and may in turn be followed by parenting behavior after the birth of the young. All reproductive processes are mediated by endocrine control systems, which reach their highest levels of complexity in the estrous or menstrual cycles of the female higher mammals.

# Bibliography

### References

Asdell, S. A. 1964. *Patterns of Mammalian Reproduction*. 2nd ed. Cornell University Press, Ithaca, N. Y.

Benirschke, K., ed. 1969. *Comparative Mammalian Cytogenetics*. Springer-Verlag, New York.

Bullough, W. S. 1961. *Vertebrate Reproductive Cycles*. 2nd ed. Barnes & Noble, New York.

McKerns K. W., ed. 1969. *The Gonads*. Appleton-Century-Crofts, New York.

Nalbandov, A. V. 1964. *Reproductive Physiology*. 2nd ed. Freeman, San Francisco.

Parkes, A. S. 1952, 1956, 1966. *Marshall's Physiology of Reproduction*. Longmans, Green, London.

Rowlands, I. W., ed. 1966. *Comparative Biology of Reproduction in Mammals*. Academic Press, New York.

Van Tienhoven, A. 1968. *Reproductive Physiology of Vertebrates*. Saunders, Philadelphia.

### Suggested for further reading

Barrington, E. J. W. 1967. *Invertebrate Structure and Function*. Houghton Mifflin, Boston.
Part 5 generally and Chapter 18 in particular comprise an excellent discussion of invertebrate reproductive patterns.

Davey, K. G. 1965. *Reproduction in the Insects*. Freeman, San Francisco.
A short but complete introduction, written for the biology student, to the detailed structure and function of insect reproduction.

Frazer, J. F. D. 1959. *The Sexual Cycles of Vertebrates*. Hutchinson University Library, London.
A general treatment of vertebrate reproduction, concentrating on morphological and physiological aspects of the sexual cycles.

Michelmore, S. 1964. *Sexual Reproduction*. Natural History Press, Garden City, N. Y.
An up-to-date and far-ranging introduction to the cellular, behavioral, morphological, and evolutionary aspects of sexual reproduction.

Odell, W. D. and D. L. Moyer. 1971. *Physiology of Reproduction.* Mosby, St. Louis, Mo.
The physiology of the human reproductive organs and their associated hormones, including the processes of reaching puberty, is discussed by two physicians, one a specialist in endocrinology, and the other a gynecologist. The book is well illustrated with electron micrographs, charts, and drawings.

Raven, C. P. 1961. *Oögenesis: The Storage of Developmental Information.*
A very clear, somewhat technical analysis of the formation of the egg in the ovary, exploring the relationship of oögenesis to genetics, embryology, and theoretical biology.

Wilmoth, J. H. 1967. *Biology of Invertebrata.* Prentice-Hall, Englewood Cliffs, N. J.
The discussion of each invertebrate phylum includes a brief but clear treatment of reproduction and embryology.


**Books of related interest**

Jenkins, M. M. 1970. *Animals Without Parents.* Holiday House, New York.
A popularly written survey of asexual or parthenogenetic reproduction in a number of species, ranging from protozoa to mammals.

Milne, L. J. and M. J. Milne. 1954. *The Mating Instinct.* Signet (New American Library), New York.
A perceptive, popularly written treatment of courting, mating, and parental behavior in a wide variety of animals.

# 16
# Developmental biology

The growth of a child into an adult; the continuous growth of a tree; the germination of a seed; the change of a tadpole into a frog; the lobster's regeneration of a lost cheliped: all constitute developmental events.

Developmental biology is the discipline which studies such sequential orderly changes or transformations in the life cycle of organisms. These changes may involve the creation of a new organism from some portion of a parent organism, or they may involve changes in size, replacement of lost parts, and repair of defects. Actually, any progressive and relatively stable changes from one condition in time to another constitutes development.

The study of any developmental process begins with the attempt to describe the events that can actually be seen. These observations establish the basic pattern of the process. The developmental biologist can then challenge the developing system experimentally, working out the patterns of control during the transformation.

Traditionally, there have been two conflicting views of the developmental process. One states that the patterns are preformed, or present from the onset of the transformation. The second view suggests that the patterns may be acquired progressively as a result of the developmental process itself; this is the epigenetic view. Although the concepts of preformation and epigenesis have been conflicting philosophical approaches to development throughout the nineteenth and twentieth centuries, both the patterns appear to be present in all organisms. A more modern view is that development appears to progress as a sum of the actions of the whole organism in relation to the changing external conditions.

Certain developmental events in certain organisms can be shown to proceed along predetermined lines that are apparently an unalterable part of the genome. Yet other experiments have shown that the environment often plays an important part in influencing and shaping developmental processes.

**483**

## Developmental processes

The basic processes of development involve growth, differentiation, cellular interactions, movement, and metabolism.

### Growth

In simplest terms, growth is increase in living mass; this may be either increase in cell size or in number. Single-celled organisms increase in bulk and divide into two to utilize the environment more efficiently. Multicellular organisms increase first in cell number, then in cell size. When the maximum limit of cell number is reached, the organism may still increase in size by adding intercellular material or space. For example, a developing apple fruit increases in volume more than it increases in weight. This difference indicates that air spaces have developed as part of the fruit's growth.

Growth is a measurable phenomenon, and this has probably been its fascination for the biologist. Growth can be measured in units designating increase in size or amount, or as a rate. If the measured increments of a growing organism are plotted on a graph against time, an s-shaped curve is formed. The curve indicates a slow-rising beginning, then a rapid increase in the rate of growth, and a final period of leveling off.

In developing systems, various parts of the organism may grow at differing rates, in a differential growth process. The growth of deer antlers compared to the animal's body weight is an example. In many plants, differential growth is important in **morphogenesis,** or changes in form and structure, and in repair processes. The shape of leaves and flowers is determined by differential growth. The petiole of a leaf develops

16-1 *At the two-month stage, this snapping turtle displays the shell characteristic of the species. Note also the development of the eye, in which the lens and cornea can be clearly distinguished. (Lynwood M. Chase from National Audubon Society)*

at a different rate than the blade. If the two rates are plotted against each other, a straight line results, indicating that both petiole and blade grow proportionally to each other during leaf development.

In regeneration experiments, it was found that growth rates were related to the amount of original tissue lost; the greater the amount of tissue removed, the faster was the growth rate. Spratt (1964) showed that the rate of development varied, according to whether the number of cells present was higher or lower than the normal range. When early chick embryos (in the form of a disc) were cut into smaller pieces, they had an increased cell division rate. If they were fused to form larger than normal discs, they had a decreased division rate. Both rates returned to normal when the discs reached either the correct size for that developmental age or the characteristic developmental age for the cell number. Unicellular organisms confined to a measurable system, such as a flask, also show the same influence of cell number on cell division rates. The environment appears to have an influence similar to the organismic controls of multicellular organisms.

In plants, differential growth plays a role in many transformations. The orientation of mitotic spindles can determine the form a particular fruit can take. If most spindles are oriented perpendicular to the long axis of the floral stem, the fruit will be rounded. When the spindles are parallel to the long axis, the plant will produce an elongated fruit.

Growth appears to be controlled by various secretory products, or hormones. Plant growth is controlled largely by auxins and cytokinins. Auxins stimulate cell elongation; cytokinins stimulate cell division. Their action is based on concentration levels and interactions with other hormones, and they are not as target-specific as animal hormones.

Organ growth stops when a certain size, proportional to the whole organism, is achieved. The factors that result in this balanced development have not yet been identified. It may be due to some balance of the varied metabolic end-products produced by the cells, rather than to any specific growth-controlling substance.

Certain physical factors that help direct developmental growth have emerged. It appears that there is an optimal number of cells needed to achieve a maximum rate of mitosis. When a relatively large number of cells is placed in a culture medium, cell division takes place more rapidly than when only a few cells are grown. But the concentration of cells has a limit that cannot be surpassed even with increased nutrients.

The texture of the substratum on which the cell or organism is growing is another physical factor influencing growth. The inner surface of the vitelline membrane surrounding the developing chick embryo has a texture that stimulates growth, but the outer surface does not allow for growth. If, however, a polypore filter with a pore size of 0.45 microns replaces the inner vitelline membrane, surface growth is also enhanced. Both membrane and filter have a similarity of texture, the growth-controlling factor in the embryo of the chick embryo.

Developing nerve cells and twining vines are controlled by touch contact with their substrate;

**16-2** *Reproductive structures may be conspicuous or barely noticeable. In this pawpaw tree from the Brazilian Amazon, the fruit are a conspicuous feature of the reproducing plant. (Norman Myers—Bruce Coleman)*

this is a form of **thigmotropism.** Physical contact in other instances inhibits growth. Cultured cells cease growing when they group together into a monolayer.

Further modifications of growth can occur by selective use of specific nutrients by different cells at varied times. Thus one cell may have an optimum amount of nutrient necessary for a specific developmental stage to complete its growth. For example, if a fern gametophyte is placed in a medium high in sucrose, it will change from a gametophyte into a sporophyte generation that is haploid. Other cells will vary both as to nutrient and time.

If an organ or tissue is forced to function at a higher than normal level, it will grow to compensate for the increased functional load, increasing cell number and/or size. When a person loses one kidney, the other undergoes this type of compensatory growth by enlarging.

## Differentiation

Discovering what cells normally do during the course of development has proved to be very difficult. If, however, the cells are challenged to see what they do in response to abnormal conditions, it is possible to extrapolate back to the normal conditions. Thus the experimenter discovers a cell's potential capacity for change during sequential transformations, rather than focusing only on the final outcome of the process.

The process of differentiation involves a sequence of transformations resulting in cell specialization. This appears to be a function of selective utilization of only a portion of the genome at any given time.

The original fertilized egg, the zygote, is the whole organism, as well as the first cell of the organism at the beginning of development. At this point, it has the ability to become all the cells of the later organism. Once cell division has occurred, the resulting cells, or **blastomeres,** may or may not have the totipotency (ability to become the whole adult organism) of the zygote. If the blastomeres when separated can adjust to become a complete individual, we say that zygote is capable of regulation. The sea urchin and human are examples of such regulative cells. Other zygotes, such as in the mollusks, cannot maintain totipotency through early cell divisions, and they are nonregulative types. These two zygote types represent different general methods of differentiation, but they are not mutually exclusive. The zygote has inherited a set of preformed guidelines which become manifest by the two-cell stage of development.

In regulative cellular differentiation, the course of development is influenced by the environments in which a cell finds itself. In the early stages of cleavage, simply being on the left or right side is difference enough. If the cells are not disturbed, they will produce all the cells on the left and right side of the organism respectively. Thus regulative zygotes appear to acquire their properties as a result of interaction with their environment.

All developing cells, whether from regulative or nonregulative zygotes, are placed in their own microenvironments throughout development and must interact within these conditions. If the cytoplasm of the regulative egg does not contain preformed guidelines early in development, then it will be the preformed guidelines of the genome that later determines the differentiation process.

Where two cells of different size are produced, the cells often have very different potentials. In aphids, those daughter cells produced during spermatogenesis that lack mitochondria degenerate. In fern gametophytes, the first unequal division of the spore cell produces a rhizoid cell lacking chloroplasts. If some chloroplasts pass into the rhizoid cell they shortly disintegrate. The root hair cell of the grass *Phleum* contains enzymes its sibling does not have and thus develops a hair.

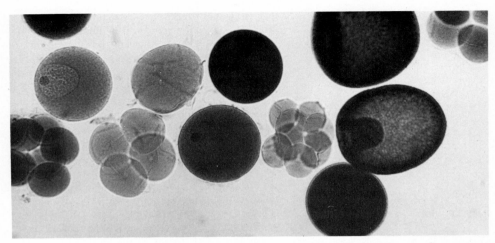

**16-3** *Starfish eggs are shown at several stages of development. An unfertilized egg is shown at the left; at the right, fertilized eggs can be recognized by the dark fertilization membranes. Several embryos have reached the four- and eight-celled stages. (Russ Kinne— Photo Researchers)*

**16-4** *Growth is an important developmental process, for all multicellular organisms are smaller at birth than they are at maturity. This baby octopus (Octopus briareus) is a good illustration; the photo shows it compared with a penny. (William M. Stephens from Omikron)*

The microenvironment of the cell can be shown to influence differentiation. If a piece of tissue from the dorsal side of the amphibian embryo that would have developed into neural plate and then into brain is transplanted during the early stages of development to a prospective ventral skin area, the tissue will develop into skin cells. The reciprocal transplant produces the opposite results.

If cells from a tissue are disaggregated (separated) and then placed into a culture medium, they will reaggregate and reestablish the tissue pattern. If more than one type of cell is placed in the medium, they will not mix together but will reaggregate in separate masses. The cells apparently can identify each other as the same or differing. Mixing the same tissues from different species as mouse and chick produces mosaic tissues. Mixing of cells from different species results in reaggregation by tissue, not by species. Tissue cells, therefore, appear to have specific aggregating properties related to their differentiation, and these properties are similar in different species.

One of the central problems of differentiation is understanding how cells with the same genome make different use of the information. The genome exists within a microenvironment at or near the center of the cell, surrounded by other environments, much like circles within circles. At each step—from the outside of the cell to the genome—a factor or factors appear to be involved in influencing the next microenvironment, until the conditions around the genome result in genetic action. The genetic action then appears to filter back through the cell, resulting in developmental changes. A detailed study of these environmental agents seems to be the way to reveal the control mechanisms.

In some organisms, single factors appear to direct differentiation. The hormone ecdysone affects molting in insects. Vitamin A suppresses keratin synthesis in the human skin. Phenylalanine promotes melanin synthesis. Some single environmental factors have varied effects depending on their concentration. In plants, indoleacetic acid will induce different types of cell specialization depending on its concentration.

## Cellular interactions

The way cells relate to each other during the processes of growth and differentiation is called cellular interaction. An increase in the number of cells, whether in an organism or a culture medium, lead to increasing cell-to-cell communication. This interaction can take place both between neighboring cells and between those opposite ends of the organism or culture medium.

It is necessary for cells in a multicellular condition to maintain some sort of communication in order to remain a synchronized functional unit, and to develop specializations of each tissue. Better understanding of cellular interactions can only come with more study of cell membranes and of metabolites exuded from and captured by cells. Recent work has shown that, through pinocytosis, the cell releases some compounds to the environment and picks others up. Some of the compounds appear to result in cytoplasmic events that seem to be important in cell-to-cell contact.

The number of cells present determines some development actions. If a large piece of dorsal ectoderm is cultured, it will produce nerve cells, but if many small pieces are placed in culture, they will not produce nerve cells. Only when the small pieces are clumped together into a larger mass are they able to differentiate into nerve cells. Some cell interactions can only be completed if sufficient cell numbers are present.

## Movement

The positioning of cells and tissues by differential growth is enhanced by cell and tissue movement.

These migrations in animals are important in the formation of the primary germ layers—ectoderm, endoderm, and mesoderm.

Many examples of movement have been observed. In birds, a rodlike cell aggregation called the primitive streak exhibits striking posterior movement as the embryo elongates. Folding in, in the formation of the neural tube, invaginations in the inner ear vesicle formation, or evaginations in the formation of the eye vesicles, all involve tissue movement. Cells left between the newly formed nerve cord and dorsal ectoderm are called neural crest cells. These cells migrate ventrally and reaggregate to form ganglia and cartilage plates around the nerve cord. Migration of individual amoebae of slime molds aggregate into a multicellular plasmodium.

Within the cell, cytoplasmic movement apparently plays an important role in cellular differentiation. At the time of fertilization, surface cytoplasm or ecotoplasm may migrate or shift position to cover the entire endoplasm. These events redistribute cytoplasmic components that may be important in later development. In the egg of the fern, *Pteridium aquilinum*, DNA moves into the cytoplasm during maturation. Upon fertilization, the egg nucleus forms protrusions that seemingly capture the head of the spermatozoan.

## Development in higher plants

Fertilization of the egg within the female gametophyte begins the development of the flowering plant. The dividing zygote forms an embryo that takes over the space formerly held by the gametophyte. This area, together with the integument is called the ovule; it becomes transformed into the seed.

The first division divides the embryo into two cells. The next division produces the characteristic four-celled embryo. From this stage,

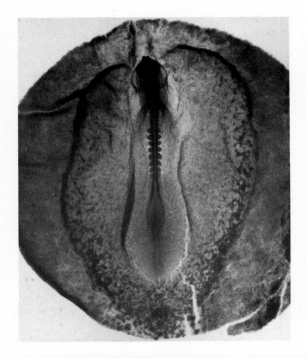

**16-5** *This 26–27 hour chick embryo is shown at a magnification of 30X. Earlier, during gastrulation, involution of cells along the midline of the embryo produces a primitive streak which is clearly visible in the photograph. (Russ Kinne—Photo Researchers)*

**16-6** *The plasmodium of the slime mold* Physarum *produces fruiting bodies when conditions are right. The fruiting bodies then produce spores, which in turn give rise to flagellated gametes. (Hugh Spencer from National Audubon Society)*

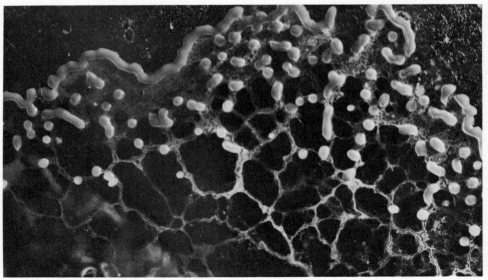

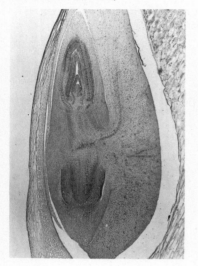

**16-7** *In seed plant development, three major portions of the embryo appear very early in development. The epicotyl, pointing upward, the hypocotyl aimed downward, and the cotyledon to the right. This is an embryo of corn, Zea. (Courtesy of Ripon Microslides)*

the cell or cells at the micropyle become an elongate basal stalk called the suspensor. The cells farthest from the micropyle organize into the embryo proper. The embryo proper continues development until the complete embryo is formed, consisting of the two seed leaves (the cotyledons) and the epicotyl and the hypocotyl.

## Primary growth

The primary meristem is formed very early in the development of the embryo. These tissues develop into the various mature tissues of the plant, but in the apical regions of the organism they remain in the embryonic condition. This continuance of the embryonic condition in certain localized regions throughout its lifetime is one way in which plant development differs strikingly from animal development.

The seed embryo exhibits the basic organization characteristic of the adult plant. The cotyledons are attached and function at the meeting point of the shoot and the root. The shoot axis is mainly confined to the epicotyl, which consists of a tiny bud and two small leaves called the plumule. Below the point of attachment of the cotyledons the axis is the hypocotyl. The lower end of the hypocotyl is called the radicle, and it becomes the primary root of the plant. The apical meristem is the group of cells from which all shoot and root tissue will develop.

While the embryo and endosperm are developing, the ovule matures, becoming a seed by the conversion of the integuments around the ovule into a seed coat. The seed is an important structure, enabling the plant to undergo a dormant stage to pass through adverse conditions. The dormant condition appears to occur during development before seed dispersal. Various triggers lead to the biochemical events of dormancy. These environmental triggers, such as length of photoperiod, low temperature, or chemicals outside the seed, must be counteracted for ger-

mination to occur. Abscissic acid has been shown to prolong seed dormancy. The environmental triggers noted have also been shown to increase abscissic acid in some plants. The gibberellins decrease markedly as the embryo ceases growth, indicating a critical balance between growth promoters and inhibitors. Two factors are necessary to induce germination: a reversal of the environmental agents inducing dormancy; and the addition of a certain amount of moisture. In lettuce, the phytochrome system of conversion induces the production of gibberellin, which in turn causes germination. The induction of a hormone may be the main function of all germination triggers. The spread of water into the seed allows the inductive processes of germination to become chemical events.

As cell division begins again in the process of germination, the dormant embryo begins to push against the seed coat and the radicle emerges. Cell elongation in the hypocotyl region pushes the epicotyl out of the seed coat and above the ground. This elongation process, or etiolation, which takes place in the dark, places the epicotyl in the best position for continued growth. In plants such as the bean, the hypocotyl forms an arch which penetrates the soil surface first. The phototropic response that occurs at this point affects auxin concentrations; the arch straightens up because the cells on the underside elongate at a faster rate than those on the top, nearest the light. The straightening hypocotyl pulls the cotyledons and epicotyl out of the ground to face the light.

The root of the plant is covered with a protective layer of cells called the root cap. The meristem region of the root tip is a zone of cell multiplication; it extends about 1.5 to 3 mm. from the root tip. The new cells begin to differentiate by elongating; their growth extends the root further into the soil. The elongation zone extends about 4 to 10 mm. above the root tip. Once the elongation process is well along, the cells complete their maturation and become specialized functional cells. The process of elongation and differentiation involves the creation of a large vacuole. Rather than increasing the amount of cytoplasm, as happens in animal cells, the main cellular area is taken up by a large water-containing cell vacuole. The differentiation process also includes the cessation of mitotic activity.

The cells do not differentiate haphazardly. The position of the cell at the onset of differentiation determines its fate. If the cell is in the epidermis, it may become a root hair. The cells positioned in the cortex will become parenchyma cells; those within the vascular bundle become vascular elements. This harmony of development by position is similar to the morphogenic fields that define areas of organ development in animals.

The vascular elements in the root begin differentiation with the phloem cells lying just inside the endodermis. Within the phloem bundle, the xylem cells differentiate, and all vascular tissues complete the differentiation process before the other cells surrounding them have completed their elongation. The completely differentiated xylem cell dies, loses its cytoplasm, and functions as a hollow cell wall. The phloem cells lose their nuclei to become functioning sieve cells. Between sieve elements are a series of smaller nucleated cells which remain as companion cells.

Primary root growth results in a central core of xylem surrounded by a series of cylinders of different types of cells. In woody tissues, the layer immediately outside the xylem becomes the secondary meristem, or cambium. This layer increases the girth of the stem by laying down xylem on the inside and phloem on the outside.

The stem differs from the root in lacking the rootcap and endodermis and in the production of leaves. Yet the basic developmental process is the same. Since cell division occurs at the apex of the stem and at the base of each growing leaf, the shoot contains a series of zones of cell division. These meristematic areas are separated

16-8 *In this close-up photograph of a radish seedling, the hypocotyl arch is clearly visible. Notice also the extensive network of fine root hairs just behind the zone of elongation. (Alycia Smith Butler)*

16-9 *The growing stem tip of a corn plant is shown in cross section. The apical meristem can be clearly seen within a sheath of rudimentary leaves. (Courtesy of Ripon Microslides)*

**16-10** *Close-up photograph of magnolia seeds. The large, showy flower has been fertilized, the petals have dropped off, and the seeds formed. They will now remain dormant until temperature and moisture conditions are right, when germination will be initiated. (Jack Dermid from National Audubon Society)*

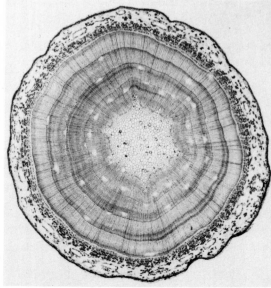

**16-11** *A growing pine stem is seen in cross section, showing the important structural parts. The living cambium is represented by the dark cells just within the perimeter of the stem; the growth of the cambium gives rise to annual growth rings, concentric dark and light rings located between the secondary cambium and the central vascular tissue. (Courtesy of Ripon Microslides)*

from each other by zones of differentiated cells. The resulting arrangement of the stem is unlike that of the root. In dicots and gymnosperms, the vascular bundles are arranged cylindrically around the central pith. In monocots, the bundles are scattered throughout the cortex. The phloem always lies to the outside of the xylem in the vascular bundle. The apical growth zones with their ability to grow indefinitely have no counterpart in animal development. In animals, the regions of cell division are responsible only for a few types of cells; the plant meristem produces the whole complex of cell types making up the organism.

## Secondary growth

Initially, primary growth alone can provide sufficient girth to the plant to maintain its height. But if the plant lives more than one year, increased thickness is necessary to support its additional height; this can take place without cell elongation. The most obvious plants exhibiting secondary growth are the woody plants. Some trees continue to increase their girth for hundreds of years. Herbaceous plants may exhibit some secondary growth after elongation has ceased.

Secondary growth is the function of the cambium layer between the xylem and phloem. In the stem, the cambium is initially confined to the vascular bundles. But in the course of later development, cells in the parenchyma may resume meristem activity and form a continuous sheet of cells with the vascular cambium. The seasonal activity of the cambium results in the so-called annual rings of tree trunks. The cambial cells exist in two forms; some cells are very long, and others are small polyhedral cells in clusters between the elongate cells. The elongate cells produce the xylem and phloem elements. The smaller ones form the rays that transport material laterally in the stem.

As the stem increases in girth, the cambium

**491**

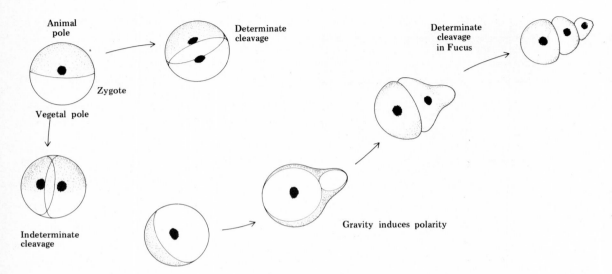

Animal pole

Zygote

Vegetal pole

Determinate cleavage

Indeterminate cleavage

Gravity induces polarity

Determinate cleavage in Fucus

16-12 *Cleavage may be determinate or indeterminate. In this hypothetical example an important substance is concentrated at the animal pole; an initial cleavage which distributes this material equally between the two daughter cells may be indeterminate, while a cleavage across the axis would be determinate. This is because the cell with the important substance would provide its genetic material with an altered cytoplasmic environment, leading to different development. In Fucus, the orientation of the early cleavages is determined by the location of a protuberance which arises on one side of the zygote.*

must also increase to encompass the ever widening xylem. This is accomplished by divisions that occur perpendicular to the stem radius.

Both stems and roots usually have lateral branches, but lateral development in root and stem is quite different. The root branch begins in the cells of the pericycle, as a bud forms which grows out through the root tissue and breaks through the epidermis to continue growth. The primordia cell groups, or **primordia,** that will become the leaves and buds are laid down in the stem apex during primary growth. It appears that the presence of the previous primordia controls the positioning of the next younger primordia.

The lateral buds can remain dormant for extended periods of time, perhaps until hormonal differences trigger germination. Once a bud germinates, its growth is indistinguishable from the main shoot. One of the interesting problems in plant development is how a long-dormant bud initiates development after so many years of developmental inactivity.

The growth of the various parts of the plant obviously depend on the processes of division, elongation, and differentiation of cells. These harmonious events between the different populations of cells indicate that careful regulation in the organism must be an ongoing process throughout the life cycle of the plant. The control seems to be manifested through each cell's relation with its neighbors. The only controls presently understood to some extent are those involving hormones. The explanation of the mechanisms of differential growth remain as an important challenge facing developmental biologists.

## Development in higher animals

Early development in the embryo is directed toward producing enough cells to enable the organism to form the primary germ layers— ectoderm, mesoderm, and endoderm. These layers form the basic ground plan for the future tissues and organs of the organism. The first stage, simple cell increase, is called **blastulation.** The formation of the three germ layers and the embryonic ground plan is called **gastrulation.**

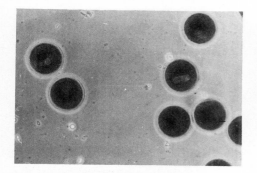

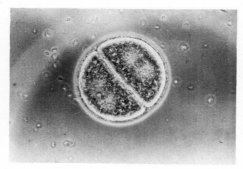

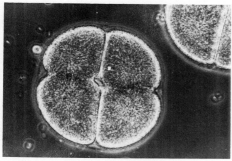

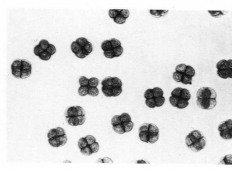

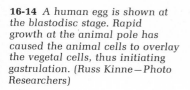

**16-13** *The process of blastulation serve, to increase the number of cells, thereby permitting future differentiation. The pattern of cell division during blastulation varies according to the characteristics of the original egg and the informational guidelines contained in cytoplasm and nucleus. (R. Zulola—Omikron)*

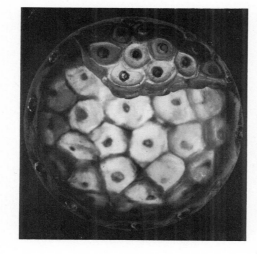

**16-14** *A human egg is shown at the blastodisc stage. Rapid growth at the animal pole has caused the animal cells to overlay the vegetal cells, thus initiating gastrulation. (Russ Kinne—Photo Researchers)*

## Blastulation

The many variations in the type and extent of cell division are related to the amount of storage material or yolk present in the egg. **Holoblastic** zygotes, such as the frog and amphioxus, cleave completely. In amphioxus, the yolk is equally distributed, or homolecithal, and therefore the resulting cells are close to being equal in size. In the frog zygote, there is a great deal of yolk confined toward the vegetal pole. The third cleavage plane will be above the yolk, leading to unequal blastomeres and thus producing a rapidly growing animal pole region and a slower vegetal pole.

Incomplete, or **meroblastic,** cleavage is found in birds and reptiles; their eggs have extremely large quantities of yolk. The yolk material is so massive that the cells cannot divide it. The early cells form from a cytoplasmic disc at the animal pole of the egg, and membranes are formed around the dividing nuclei within the cytoplasmic disc. The resulting divisions form a layer of cells called the blastodisc. The insects exhibit another type of meroblastic cleavage; their eggs have a centrally located yolk. The nuclei divide and migrate around the periphery of the zygote, forming cells around the central yolk.

## Gastrulation

The cleavage pattern thus leads to the blastula type, which in turn affects the process of gastrulation. In homolecithal eggs, the blastula is a ball of cells that invaginates during gastrulation. In the frog type, the cells at the animal pole, which are dividing at a faster rate than the yolk-burdened cells at the vegetal pole, will eventually overgrow the vegetal cells. This process that initiates gastrulation is called epiboly. The discoidal bird and reptile embryos must develop special membranes to cover the yolk. Gastrulation in the cellular disc occurs by the cellular movement called the primitive streak.

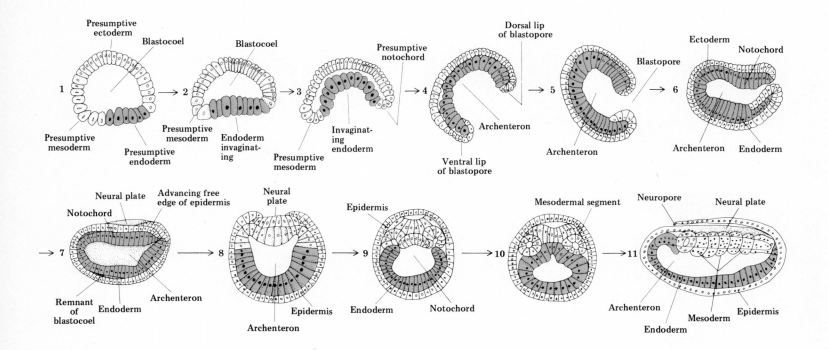

**16-15** *Gastrulation and primary organ formation in amphioxus involves several stages. Drawings 1–6 show different stages of gastrulation; 7 shows neurulation; 8–10 are cross section views of the formation of the primary organ rudiments; 11 is a lateral view of the embryo showing mesodermal segments overlying the notochord and neural plate.*

Once the basic germ layers are formed, they interact with each other to complete the development of the embryo. The ectoderm gives rise to the integument and the nervous system. The chordamesoderm gives rise to the notochord, bone, muscle, vascular system, and connective tissues. The endoderm forms the digestive system and its associated organs, such as the liver and pancreas, and the respiratory system. The posterior opening formed from the process of gastrulation is called the blastopore and becomes the anus in vertebrates. The gastrula cavity, the archenteron, remains as the cavity of the gut.

Dr. H. Spemann, in the early part of this century, attempted to find out whether fragments of the embryo could construct a complete normal adult. He cut salamander embryos in half at various developmental stages. Before gastrulation, both halves developed normal embryos.

After gastrulation, only the half with the dorsal lip of the blastopore developed into a normal embryo. The other half became an amorphous mass of cells. The dorsal lip appeared to have the power to organize the development of the whole embryo, and no other cells seemed to be able to do this.

Spemann confirmed these observations by transplanting pieces of one embryo to other parts of a host embryo. Other parts of the embryo developed into the local tissue of the host site, but the dorsal lip developed into notochord and mesodermal somites regardless of position in the host. The special status of the dorsal lip tissue needed no more confirmation.

Further study revealed that the inductive capacity of chordamesoderm was also positional. The anterior chordamesoderm cells were better inducers of brain formation than more

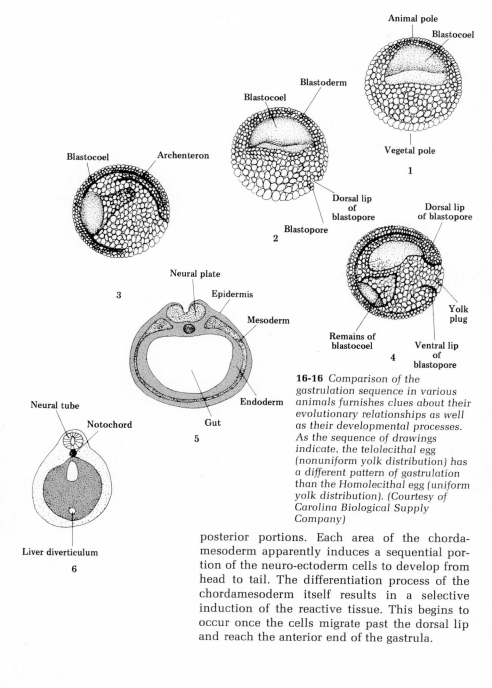

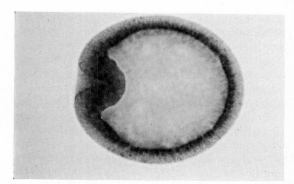

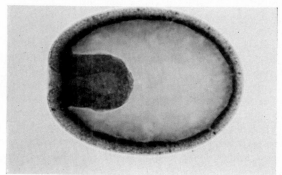

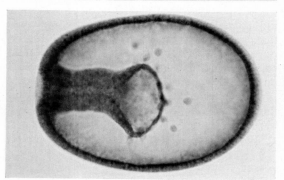

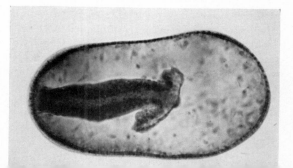

**16-16** *Comparison of the gastrulation sequence in various animals furnishes clues about their evolutionary relationships as well as their developmental processes. As the sequence of drawings indicate, the telolecithal egg (nonuniform yolk distribution) has a different pattern of gastrulation than the Homolecithal egg (uniform yolk distribution). (Courtesy of Carolina Biological Supply Company)*

posterior portions. Each area of the chorda-mesoderm apparently induces a sequential portion of the neuro-ectoderm cells to develop from head to tail. The differentiation process of the chordamesoderm itself results in a selective induction of the reactive tissue. This begins to occur once the cells migrate past the dorsal lip and reach the anterior end of the gastrula.

Interestingly, the search for a specific compound supplied by the chordamesoderm proved fruitless. Apparently, the ectoderm at this stage is so labile that almost any change in the cell environment will induce the next series of transformations. Even though many agents can induce ectoderm development in the embryo, only the chordamesoderm accomplishes these transformations without producing secondary embryos all over the organism. Under the normal pattern, particular transformations occur in regions called morphogenetic fields. These fields of cells are ranges over which a particular path of differentiation will occur, such as a limb bud field. At this level, secondary inducers come into play. These inducers operate in the following way. Tissue I must be present for tissue R to form a specific organ. If tissue I is moved to a new site where the cells are closely related to tissue R, the organ will be constructed at this site.

When two different tissues mutually interact to cause a transformation that would not occur if each were present alone, a synergistic induction takes place. Kidney tubule development is an example of synergistic induction.

Interactions must also stop or inhibit some inductions to maintain regional development in the construction of the embryo. These inhibitions may work at two levels. Competition for a common pool of materials enables development only by those cells that receive the greatest supply. The second level may involve one cell inhibiting another by the material it produces in the nutrient pool.

## Final stages of embryology

At the completion of the early stages of development, the embryos of all vertebrates look very similar. They possess a neural tube, a notochord with somites (mesoderm segments) on either side, four limb buds formed from various somites, and a more or less complete alimentary canal. A

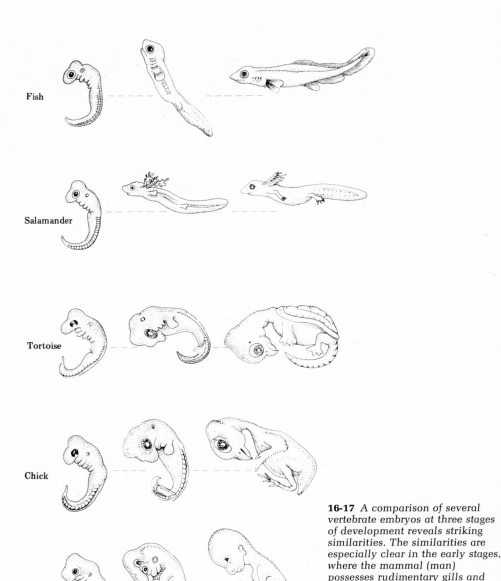

Fish

Salamander

Tortoise

Chick

Man

**16-17** *A comparison of several vertebrate embryos at three stages of development reveals striking similarities. The similarities are especially clear in the early stages, where the mammal (man) possesses rudimentary gills and a tail.*

degree of cephalization ("head-formation") is also present. The anterior region of the neural tube has differentiated into the rudimentary brain, and the eye structures are laid down.

The similarity of vertebrate embryos at the early stage of development and the fact that a mammalian embryo appears to pass through a fishlike, then an amphibianlike, and finally reptilelike stage, led Ernst Haeckel, a nineteenth-century naturalist, to the theory that "ontogeny recapitulates phylogeny." This means that the developmental course of an organism (its ontogeny) repeats its evolutionary history (phylogeny). This is not quite true. Though at certain stages the mammalian embryo has structures in common with the fish embryo, such as pharyngeal pouches, it can in no sense be called a fish; it never actually develops functional gills and is at all times a mammal.

## Maturation

At birth or hatching, the young vertebrate is usually far from mature. It lacks the size, strength, and reproductive capabilities of an adult and may have different nutritional requirements. The final stages of development, however, involve the growth of previously differentiated tissues and organs. The young mammal, for instance, possesses all the elements of a reproductive system, though these are not developed to a functional level. The onset of puberty sees the hormonal-controlled maturation of these already existing structures.

The postembryonic development of invertebrates can present a very different situation. Many mollusks, echinoderms, and arthropods hatch in a larval form quite distinct from the adult. In mollusks and echinoderms, the adults are often sedentary, and the free-swimming larva provides a means of dispersal for the species. The process of metamorphosis which converts the larva to its adult form involves differentiation of new structures, often without a concurrent increase in overall size.

The young of many insects (e.g., wasps, flies, moths, beetles) hatch as larvae or grubs, bearing little or no resemblance to the adults of the species. These larvae may grow many times their original size and then form a pupal stage, characterized by a cessation of movement and eating, and extensive reorganization of its tissues and organs. The adult structures develop from essentially undifferentiated cell discs present in the

**16-18** *Opossum young are relatively undeveloped at birth, as is the case among all marsupial mammals. Each moves about within the mother's pouch until it finds a nipple, from which it takes nourishment until fully developed. At a later stage, the young cling to the mother's fur as she travels about. (Leonard Lee Rue III — Bruce Coleman and National Audubon Society)*

**497**

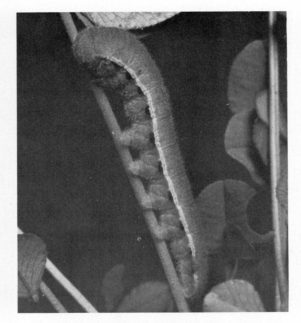

16-19 *Eggs within the egg case (left) have given rise to caterpillars such as the one on the right which has been feeding and growing. Ready to pupate, it enfolds itself in a chrysalis where its tissues undergo a basic reorganization. After a period of weeks it leaves the chrysalis (below), emerging as a mature butterfly. (Photo by Alexander B. Klots)*

larva, using the old larval body as a source of raw materials. Such a pattern of development is called complete metamorphosis.

Other insects, such as locusts and cockroaches, undergo incomplete metamorphosis. The young resemble the adults in many respects, except that structures such as wings and reproductive organs are poorly developed. Development takes place through differential growth of various organs and a series of molts. Each succeeding stage bears more and more resemblance to the adult.

## Cell stability

Once differentiation has been completed, can the cell dedifferentiate or redifferentiate along a new path? This question involves the basic mechanisms that keep a cell in the differentiated state. Certain cell properties, such as shape,

**16-20** *Incomplete and complete metamorphosis are compared: The grasshopper undergoes incomplete metamorphosis, shedding the chitinous exoskeleton with each increase in size. The black swallowtail butterfly,* Papilio polyxenes, *undergoes complete metamorphosis. The caterpillar pupates within a chrysalis and emerges as a mature butterfly, which lives just long enough to breed. The red-spotted newt,* Notophthalmus viridescens *goes through a more extended metamorphosis. The aquatic larva develops into the land stage, the red eft, which after a year or so returns to the water as a mature newt.*

motility, and mitotic activity, appear to be largely controlled by extracellular influences. These cell differences may be modulations related to the substratum or the contiguity of the other cells.

Erythrocytes of mammals, and sperm and egg cells, are intrinsically stabilized cells. Their differentiation is terminal. Some other cells also appear to be irreversibly differentiated. These include the blastomeres of mollusks and annelids, sieve tube elements, companion cells, and root hair cells, among others.

Certain outside influences appear to become heritable parts of the cell, thus maintaining the differentiated state they induce. The bacterium *Agrobacterium tumefaciens* induces normal plant cells to form crown gall tumor cells. These cells can synthesize all growth hormones needed for rapid growth. If all cells are not affected by the bacteria, the gall is called a teratoma. Some of the cells are normal and can form the usual plant structures. If the bacteria are removed from the teratoma and rapid growth is induced, a normal plant condition can redevelop. The normal cells seem to dilute the tumor cell presence. Yet in an infected plant, even when the bacteria

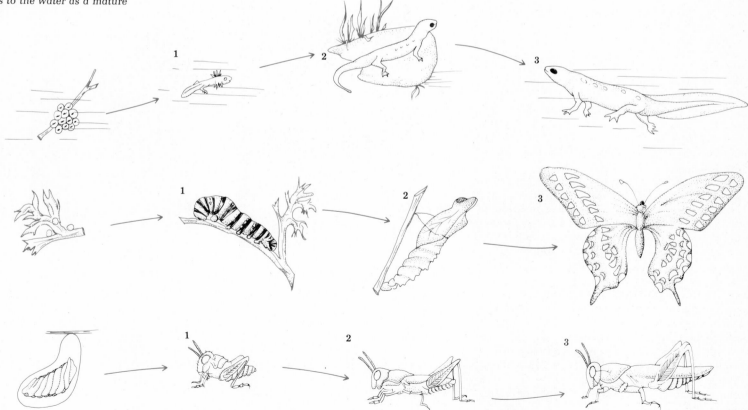

are killed, the tumor continues to grow. Since the tumor cell state continues when the bacteria are removed, it may be that the bacteria bring in bacteriophages which attach to the plant cell genome, forming the tumor cell. The rapid growth in the teratoma might dilute the virus population which allows for the development of normal plants out of the teratoma but not the crown gall.

Some examples of cell stabilization can be found in animals. The pigmented retinal cells of salamander eyes are dependent on the presence of the nonpigmented neural retinal cells. If the neural retinal cells are removed or become degenerate, the pigmented cells lose their pigment and undergo mitotic activity. Once the mitotic activity ceases, they continue to transform into neural retinal cell types.

The differentiated state is maintained even if cell modulations take place. Often when redifferentiation occurs, it is confined within the limits of the potentialities of the source of the cells. Embryonic chick cartilage cells retain their stable phenotype for more than 35 generations in dilute clonal cultures. This shows that the differentiated state is heritable. The mechanisms for maintaining this heritable state elude definition.

Briggs and King in the early 1950s produced a series of experiments that began to shed light on the question of whether or not a differentiated cell retains a complete genome. They transplanted the nucleus from a differentiated frog cell into an enucleated egg cell, using the techniques of microsurgery developed by Chambers and Kopac in *Amoeba*. The transplant experiments resulted in a certain percentage of nuclei with the ability to produce tadpoles. When the nuclei were taken from cells in later stages of development, the percentage of success decreased. Some success was achieved with differentiated endoderm cells in late gastrulas. Nuclear cloning experiments, however, showed that a certain number of nuclei had undergone irreversible differentiation. But the basic problem

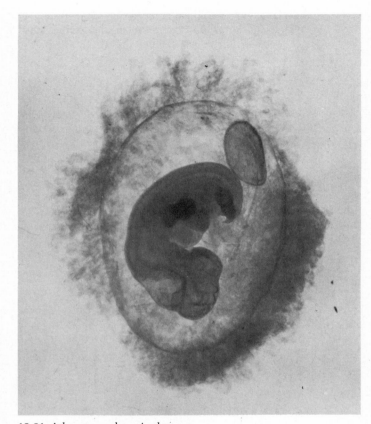

**16-21** *A human embryo is shown at an early stage of development. In in vivo development the embryo would be linked to the mother by the placenta, a tissue which is part of both the maternal and fetal bodies. (Courtesy of Carolina Biological Supply Company)*

appeared to be technique and not the nucleus. Later experiments involved serial transplants of the same nucleus into a series of eggs before finally implanting it into one; this technique seemed to "wash" the nucleus and improved the results. J. B. Gurdon and co-workers, using

*Xenopus* frogs instead of *Rana* frogs, produced adult reproducing frogs from nuclei taken from functionally differentiated intestinal cells in feeding tadpoles. When these nuclei were transplanted to the egg, they had the totipotency to interact with the egg cytoplasm and produce normal development.

These studies and their refinements throughout the 1960s leave us with greater certainty that the genome is essentially intact from zygote to differentiated cell. It appears that the nucleus within the microenvironment of the specialized cell can only receive information that reads the genetic code at a particular sequence step. When returned to the egg, the environment is such that earlier sequences of genetic information can become derepressed and therefore operable.

Gurdon performed another series of experiments to show the influence of cytoplasmic types on nuclear activity. *Xenopus* eggs are capable of active DNA synthesis, but their RNA synthesis is very limited. Blastula cells are capable of active DNA synthesis and cell division. The oöcytes in the ovary of the frog, however, are undergoing very active RNA synthesis during their maturation process. If the nucleus of a blastula cell is transplanted to an egg cell, it continues as before to synthesize DNA, but no appreciable amounts of RNA are synthesized. When the blastula nucleus is transplanted to an oöcyte, it converts over to RNA production, as if it were the original oöcyte nucleus. These experiments add further weight to the concept that the microenvironments in and around the nucleus influence gene activity during development.

It is interesting to note that the degree of cell stability seems to increase with an increase in cell specialization. In one-celled and simple multicellular organisms, changes in the outer environment of the cell or in the microenvironment of the nucleus can bring dramatic changes in the form and function of the cell. But the more specialized a cell is, the more difficult it becomes to induce

any sort of physiological or morphological change without damaging the cell's viability.

## Aging and death

It is often tacitly assumed that development is complete when the mature, adult stage is reached. Yet aging and even death are also a part of the development process.

The process of aging has attracted much attention at the level of both the cell and the organism. If the attainment of maturity completes development, then aging might be considered a result of deterioration of the adult tissues to a point where death occurs. But the rather nebulous concept of "wear-and-tear" is not altogether satisfactory. Aging appears more to be a process of development in itself.

Senescence in leaves of deciduous plants is not a matter of simple deterioration. At the end of the summer, a series of hormonally controlled events results in, among other things, the breakdown of chlorophyll and the red-coloration of the leaf, until the hormone abscissin causes the leaf-fall. Cell death in leaves is not a random process but part of the total development of the organ. In the development of xylem, the final stage in differentiation is cell death!

Cell death is an important sculpturing device for producing the final form of an organ. In the limb of the metamorphosing tadpole, phagocytic cells go to work eating the remains of cells that have committed a form of cell suicide. These dying cells were positioned in regions that would not produce the correct form of the limb. If the phalanges are to be webbed, cell death between the digits is limited. If the limb is to have unwebbed phalanges, cell death between the digits is extensive, allowing for free movement of the fingers or toes. In humans the webbed-fingered condition is probably due to some inactivity of the trigger inducing this selective cell death to occur. In frogs the release of the proteolytic en-

16-22 *The axolotl,* Ambystoma mexicanum, *ordinarily retains the gills in the mature stage. Injection with appropriate hormones can cause a further metamorphosis which transforms the axolotl into a tiger salamander. (A. W. Ambler from National Audubon Society—Permission of Ward's National Science Establishment, Inc., Rochester, New York)*

zymes from the cell lysosomes is triggered by the thyroid hormone which maintains the metamorphosis process. The death factor appears to be some diffusible substance which as yet is unidentified.

These factors may be present for many other regressive processes in development. In sex development, despite our chromosomes, we are all sexually neuter in the early stages, bearing the potential ducts and structures of both sexes. The timing and quantity of the sex hormones produced under the direction of chromosomes induce resorption of the other sex structures. Mishaps in timing and dose can have striking results, inducing the persistence of various types of hermaphroditic conditions. One wonders whether these modulations may have any relation to the development of biological homosexuals.

The human dream of evading the processes of aging and death is an old one, going back even

**16-23** *Several cells from a human tissue culture are shown in the early stages of cell death. The cell cytoplasm is dense and granular, indicating the relative inactivity of the cells. Note the large vacuoles, in which have accumulated various types of metabolic wastes; even the mitochondria have begun to vacuolize. A photo of the same cells at a later stage would show severe disruption of the cytoplasm, fragmentation of the mitochondria, increased nuclear clumping, and a marked increase in the number of visible lysosomes. (Leonard Ross— Omikron)*

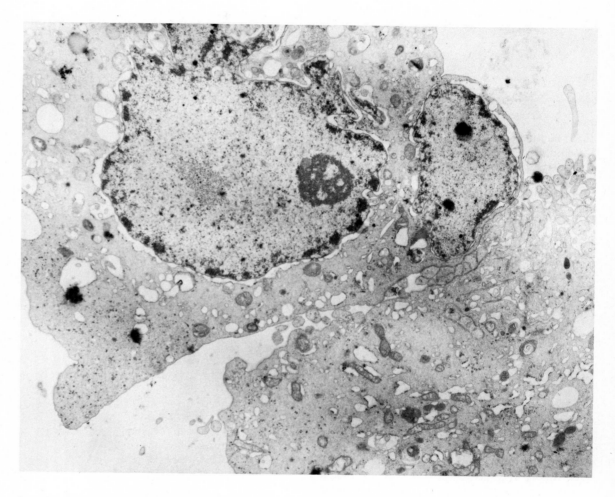

long before the quest for the mythical fountain of youth. Today even responsible biologists are speculating about the possibility of at least retarding many of the aging processes. It is known that in certain kinds of cells, an accumulation of the end products of metabolic reactions serves to trigger the processes that lead to cell death, either by activating a part of the genome or deactivating a repressor. If we could in some way intervene in this chain of action and reaction, for example by changing the microenvironment in which they take place, then we might be able to prevent or retard much of man's aging. Whether or not such an intervention would really be desirable seems a question that must be answered outside the realm of scientific investigation.

# Summary

Development biology is the study of the sequential orderly changes in organisms; these include development of new organisms, seed germination, regeneration, metamorphosis, and aging. The basic developmental processes are growth, differentiation, cellular interactions, and movement.

Growth is increase in living mass; this may be accomplished by increase in cell size or cell number. Differentiation is the process by which cells specialize, losing the ability to undergo certain kinds of future changes but gaining increased efficiency in one particular function. Cells in a multicellular condition remain in contact with one another, in order to synchronize their activities; chemical compounds generally serve as the agents of communication. Migration of cells helps reposition individual cells for further development.

Embryological development in higher plants begins at the first division of the zygote. The three major areas of development are the seed leaves, or cotyledons, the epicotyl, and the hypocotyl. Meristematic tissue of the tip of the shoot and the root divides to produce new cells; behind this zone of division is the zone of elongation in which cell growth takes place. After the cells elongate, they differentiate, forming the various structural and functional tissues of the mature plant. An interesting feature of development in higher plants is the ability of the embryo to remain dormant for long periods before resuming normal development; a similar ability is found in lateral buds.

Embryological development in higher animals begins with the process of simple cell increase, or cleavage. Following blastulation, the process of gastrulation forms the germ layers and lays down the embryonic ground plan. Early cell division, or cleavage, may take place in several different patterns, depending on whether or not the cleavage is complete and where the yolk of the egg is located. The type and plane of cleavage determines later developmental events.

A subject currently receiving much attention and study is the question of cell stability, or the permanence of specialization processes. Experiments with frog embryos suggest that, under the proper cytoplasmic conditions, even rather highly specialized cells can dedifferentiate and then undergo a new course of specialization. It appears that the process of differentiation in no way alters the genome of somatic cells, so that they return the information needed for all other developmental processes.

Both aging and death should also be considered developmental stages, for in multicellular organisms the senescence and perhaps even death of individual cells is part of the overall sequence of maturation.

# Bibliography

### References

Cohen, P. P. 1970. "Biochemical Differentiation During Amphibian Metamorphosis." *Science*. 168: 533–543.

Flickinger, R. A., ed. 1966. *Developmental Biology*. Brown, Dubuque, Iowa.

Gardner, L. I. 1972. "Deprivation Dwarfism." *Scientific American*. 227(1): 76–91.

Maugh, T. H. 1972. "Chalones: Chemical Regulation of Cell Division." *Science*. 176: 1407–1408.

Mithchison, J. M. 1972. *The Biology of the Cell Cycle*. Cambridge University Press, New York.

Saunders, J. W., Jr. 1968. *Animal Morphogenesis*. Macmillan, New York.

Steward, F. C., et al. 1964. "Growth and Development of Cultured Plant Cells." *Science*. 143: 20–27.

Waddington, C. H. 1966. *Principles of Development and Differentiation*. Macmillan, New York.

### Suggestions for further reading

Bonner, J. T. 1963. *Morphogenesis*. Atheneum, New York.
A straightforward introduction to the development of living organisms in general. Using examples from a variety of living forms, the book touches on the physics and chemistry of development, patterns of growth and morphogenic movements, polarity and symmetry, and patterns of differentiation.

Gentile, A. C. 1971. *Plant Growth*. Natural History Press, Garden City, N.Y.
A brief but reliable source of information on the complete growth process in plants. Describes also much recent research on plant growth regulators. Includes appropriate illustrations.

Reiner, J. M. 1968. *The Organism as an Adaptive Control System*. Prentice-Hall, Englewood Cliffs, N. J.
Using recent developments in mathematical biophysics and molecular genetics, the author provides a detailed analysis of development from the origin of polarity in oögenesis through the various embryonic stages. Considers possible molecular mechanisms at each stage and suggests future lines of research.

Sussman, M. 1960. *Animal Growth and Development*. Prentice-Hall, Englewood Cliffs, N. J.
The early embryology of multicellular animals is introduced in lucid terms.

## Books of related interest

Koontz, D. R. 1972. *A Darkness in My Soul*. Daw, New York.
The Artificial Creation Laboratory of the future succeeds in tampering with the genes of the unborn to produce a child who embodies vast potential and creative power. The results prove horrifying and vicious.

Niven, L. 1968. *A Gift from Earth*. Ballantine, New York.
A few men manage to escape the inexorable aging process through constant organ transplantation. However, a problem is created for the donors.

Rosenfeld, A. 1972. *The Second Genesis*. Arena, New York.
A scientific journalist's forecast of man's biological future when the processes of growth and development can be controlled prenatally and man's genetic make-up perfected.

Rugh, R. and L. B. Shettles. 1971. *From Conception to Birth*. Harper & Row, New York.
The text and its striking pictures present an up-to-date account of the development of the new human being day by day and week by week.

Tanner, J. M. and G. R. Taylor and the Editors of *Life*. 1965. *Growth*. Time, New York.
Alternating word and picture essays elucidate the human growth patterns and timetables and explore possible control mechanisms.

# 4

# Organism and its environment

# 17

# The course of evolution

Modern evolutionists have set themselves the task of explaining the origin of life and the developmental relationships of all organisms. Although evolution is certainly a fundamental part of twentieth-century biology, it would be a mistake to believe that interest in the subject began only with the work of Charles Darwin. Some of the earliest recorded writings, such as the clay tablets found at Sumer, deal with the question of where life came from and how the human form developed. Many anthropologists feel that the prehistoric cave paintings of animals, such as those at Altamira or Lorthet, are attempts to establish the nature of the relationship between the man that drew the picture and the animal depicted. Primitive rites often include symbolic reenactments of the origin of life; these are also a feature of modern-day religion.

The earliest explanations of evolution were mythological. An interesting example can be seen in the myth by which the Turtle clan of the Iroquois explained their own presence on earth. According to this story, a group of turtles were forced, because of a prolonged drought, to leave the lake where they lived and look for another. The overland trip under the summer sun made them very hot—so hot that the largest and strongest turtle began thrashing his arms and legs in the air, trying to get out of his stuffy shell. At last he succeeded, upon which he began to stand upright and to assume the characteristics of a man. This was the ancestor of the Turtle clan.

It is interesting to note that certain elements of this myth—the changing climate, the pressing need to exploit a new type of environment, the abandonment of highly specialized adaptations such as a shell—remain a part of the modern explanation of evolution. The desire to find answers to the question of where man came from and how he is related to other animals also remains unchanged. What is new is the amount of evidence that has been collected, and the methods by which the evidence of evolution is interpreted.

## Darwin and evolution

The modern scientific study of evolution really began with the work of Charles Darwin. Darwin (1809–1882) was a superb field biologist and tireless researcher whose work served to crystallize what had previously been only a general tendency to invest the living world with some sort of historical development. Although his original ideas have undergone considerable modification since their inception a century ago, later research has on the whole affirmed Darwin's conclusions, which remain the core of the modern, synthetic theory of evolution.

### Evolution before Darwin

Until the start of the eighteenth century, Western scientists offered very little challenge to the widely accepted theological interpretation of the Biblical account of creation. But by the 1700s, some progressives challenged the notion that every plant and animal was created in finished form during the three Biblical days devoted to living things. One was Charles Darwin's grandfather, Erasmus Darwin. A well-known naturalist, he left several hints throughout his writings that species of animals might have some relationship to each other. They might even compete (to the advantage of some species and disadvantage of others), and new generations might inherit traits from their parents which would help them to survive. But his theories were more or less buried in the great bulk of his work, and his grandson Charles was not thought to be overly impressed with them.

While Erasmus Darwin wrote (largely unnoticed) of gradual changes in animals, George Cuvier was establishing himself as Europe's leading authority on animal development.

Paleontology, the scientific study of fossils, was his field. Cuvier took a dim view of evolutionary theories, asserting instead that a series of earth-shattering catastrophes was responsible for the extinction of species over time. Some of his many disciples believed that these catastrophes were floods, and other maintained that they were volcanic eruptions. But Cuvier and his followers were all convinced that after each catastrophe only a few species remained alive on earth, and the lost fauna were replaced by immigration from other localities. All others were either flood or lava victims, and the paleontologists had the fossil records to prove it.

In 1801, in the face of catastrophism and its formidable evidence, Jean Baptiste de Lamarck advanced a theory of organic evolution. His thesis was that all animals, including man, descended from other species, and that acquired traits were passed on from one generation to the next. A species' physical characteristics were constantly changing, becoming more or less valuable for survival according to how often they were used. The case which he pointed to most frequently (one that still has biology students chuckling) is the giraffe's neck. The early giraffes had relatively short necks proportionate to their bodies. According to Lamarck, it was through stretching progressively higher and higher for leaves growing from tall trees that their necks eventually reached the outlandish length we observe today. Since the long neck was a desirable characteristic for food-getting, the giraffe passed its new trait on to its offspring, and they were also able to reach the higher branches. Unfortunately, Lamarck's giraffe story was intuitive but untrue, and some of his contemporaries (such as Cuvier) were quick to point this out.

But the idea of transmitting advantageous traits was a breakthrough in evolutionary thought —one which Darwin later incorporated into his own writings. Lamarck's theory of evolution also included another useful observation, namely that every living creature is engaged in a kind of

**17-1** *This sketch of Charles Darwin was made in 1849, before he became famous through the publication of* The Origin of Species. *At this time, he was living a quiet life of contemplation in the country, continuing to ponder the significance of the plants and animals he had observed in the Galápagos Islands. (Courtesy of the American Museum of Natural History)*

upward mobility, striving ahead on the evolutionary scale toward greater heights of complexity. As simple organisms at the lowest rung of the ladder move up to slightly higher strata, they are constantly being replaced by new rudimentary forms which appear through spontaneous generation. Like his notion of inheritance of acquired traits, this concept contained the basic developmental theme which would grant him an honorable mention in the development of evolutionary thought.

While some scientists concerned themselves with the progress of life on earth, others were working to trace the evolution of the earth itself. Theological scholars had already established a date for the creation of the earth—4000 B.C.—and this determination was accepted until the late 1700s. Then the geologist James Hutton postulated that the earth had not been formed in one day at all, but had actually been shaped by an ancient progression of natural events. His ideas were virtually ignored, however, during the eighteenth and early nineteenth centuries, as catastrophism was elevated to a doctrine. It was not until Darwin's day that the various biological and geological discoveries were brought together through one monumental insight— Darwin's theory of "descent with modification."

The 22-year-old Charles Darwin seemed an unlikely candidate for scientific achievement. His medical studies had ended abruptly when he fled from an operating room during surgery. Now his family prodded him to pursue the harmless career of a country parson, but Darwin was less than ecstatic about this prospect. His true talents lay in hunting, fishing, collecting insects and other small life forms, and dabbling in biology. Through his wide reading and correspondence he was in touch with the most advanced speculation of his time, which was one of cautious but determined curiosity about all aspects of nature.

Over dinner one evening, Darwin discussed his lack of future plans with sea captain Robert Fitzroy. Fitzroy, weatherbeaten at 26, already had several arduous sea voyages behind him and was planning another. This time his plans included young Darwin. They sailed from Davenport shortly after Christmas of 1831, aboard a ship with the whimsical name H.M.S. *Beagle*. Their destination was the South American coastline. Shortly after they left port, a violent storm lashed the ship. The *Beagle* pitched and rolled, kept afloat only by Fitzroy's skillful (and brutal) handling of his crew. Darwin weathered the storm in his berth, seasick. Eventually the battered ship returned to Plymouth Harbor, but Darwin had already decided to remain aboard for the journey south. One of the few books he took along was Charles Lyell's new *Principles of Geology*. Lyell was a disciple of Hutton; in the book he rejected catastrophism and advanced the theory that the earth had changed gradually over long periods of time.

When the *Beagle* reached the coast of South America, Darwin devoted his time to hunting, fishing, and exploring the interior on horseback. Many things startled him—particularly the great diversity of organisms he found. Noticing that various points along the way were inhabited by totally different species of plants and animals, he collected specimens of virtually every new life form he came across. He also developed an interest in the rich South American fossil beds, keeping in mind Lyell's theories on the age of rocks. When the *Beagle* dropped anchor in the Galápagos Islands off Ecuador, Darwin was ready to make shrewd observations on the unusual organisms he would find.

## The islands of giant tortoises

The Galápagos Islands were a spectacle in themselves. Formed by the aftermath of volcanic eruptions, they were named after their most unusual inhabitants—monstrous tortoises weighing up to 200 pounds each. And the sheer variety of the

**17-2** The H.M.S. Beagle, *aboard which Darwin sailed as naturalist in 1831, was commissioned to spend four years charting the southern oceans. It was during the voyage that Darwin developed his early insights into the process of evolution. (Courtesy of American Museum of Natural History)*

**513**

turtles was astounding. Every island supported a different species of tortoise, all of them lumbering around in search of cacti to feed on. Marine iguanas, resembling prehistoric creatures, scuttled along the shoreline, surviving principally on seaweed.

But what intrigued Darwin the most were the different types of birds he discovered, especially the Galápagos finches. Darwin counted a total of 14 species of finches during his explorations—each species exhibiting characteristics that seemed specialized for unique food-getting needs. Every species of finch fell into one of three categories, according to how it obtained its food. First were the ground finches, living on seeds scattered over the terrain. The six species in this group each possessed a different-sized beak—just large enough to accommodate the size of seed most plentiful in its habitat. Second were the tree-dwelling finches, also six species, which lived in moist forests and fed on insects. The most ingeniously specialized bird in this group was the woodpecker finch—a termite eater. It had evolved the woodpecker's beak but not its long tongue, so this finch carried a twig in its beak to pry termites out of crevices in tree bark. Third was the warbler-like finch—only one species—which ferreted its small insect prey out of bushes.

From his study of Lyell's work and his own observations, Darwin pieced together the lesson of the Galápagos Islands. The islands themselves had obviously emerged from the sea over a long period of time, and whatever life established itself there must also have come from the sea. The plants and animals Darwin found on the islands were considerably different from any organisms he had seen on the mainland. His conclusion was, logically enough, that the range of life forms unique to the Galápagos Islands represented an entirely different "creation" from life on the mainland. And each species of island life was curiously well equipped to take care of its own feeding problems. Would not this suggest that different kinds of organisms had developed in distinct ways, according to the requirements of survival?

Occupied with these thoughts and several trunkloads of fossil samples, Darwin set sail for England.

## The master synthesis

Charles Darwin arrived home determined to explore the implications of the things he had seen. Along with his observations, he had acquired an illness on his voyage that would leave him an invalid. Plagued with chronic headaches and nausea, he settled down with a new wife in the isolated village of Kent. Now he began to formulate his theory, organizing a massive collection of data

**17-3** *These giant tortoises, weighing several hundred pounds, start life at two or three ounces. They grow very rapidly, however, and may increase 25 times in weight and triple in length during their first two years. At one time they were extremely abundant in the Galápagos Islands. But since Pacific sailors regularly stowed them on board their ships as a source of fresh meat, tortoises have become almost extinct. (Miguel Castro—Photo Researchers)*

17-4 *There are a number of land birds in the Galápagos Islands, including the Galapagos hawk. Darwin noted the tameness of the island birds, a characteristic unchanged despite a hundred years of experience with man. This trait is typical of many oceanic island birds where there has been no pressure to develop an instinctive fear of predatory mammals. (Eric Hosking—Bruce Coleman)*

17-5 *Writing about his stay in the Galápagos Islands, Darwin said, "It was most striking to be surrounded by new birds . . ." One of the remarkable birds Darwin saw was the cactus finch, a bird adapted to a specialized niche. It was through his observation of the varying adaptations of the Galápagos finches that Darwin began to formulate his theory of evolution. (Eric Hosking—Bruce Coleman)*

to support his insights. He was thoroughly convinced that since variations existed among individuals of any species, some selective force must operate to make certain characteristics advantageous to survival and other traits detrimental. What he could not fathom was how the selective process actually worked.

Then in 1838 he discovered the gloomy writings of Thomas Malthus—the first author to recognize overpopulation and its consequences. Reverend Malthus' book was called *An Essay on the Principle of Population.* In it he pointed out that human population increases at a phenomenal rate which far exceeds the increase in its potential food supply. The inevitable outcome, according to Malthus, is a frantic struggle for survival, through such mechanisms as a fight

for food, war, and disease with the strong eliminating weaker competitors. This was the insight Darwin had been reaching for.

Applying Malthus' principle of competition to all living things, Darwin concluded that certain small variations in anatomy or physiology may confer on the individual a better chance of survival in the struggle for life. Such individuals could then be expected to produce a larger share of the next generation in the same way that domestic plants and animals were selected by breed. Now Darwin's observations on individual differences fell into a meaningful frame of reference—the principle of natural selection.

But Darwin was in no great hurry to see his theories in print. For 22 years he continued to ponder and refine his theory, gaining a reputation as something of an eccentric. He devoted his time to gathering even more data—staring reflectively at plants for hours on end, receiving small parcels of soil in the mail which he would use to plant obscure varieties of seeds. His closest friend throughout this period was Charles Lyell, the geologist whose work he had absorbed during the voyage of the *Beagle.* Darwin confided in Lyell his plans for a major, encyclopedia-length dissertation on the principle of evolution through natural selection. Lyell suggested to Darwin that he publish right away, since a recent book on Lamarck's theories had enjoyed popular interest, and the time seemed ripe for a definitive work on evolution.

Lyell was right. In the summer of 1858, Darwin received a manuscript in the mail from a younger naturalist named Alfred Russel Wallace with whom he had been in correspondence. The letter inside explained that Wallace had developed a theory on the scheme of evolution while suffering a high fever in Indonesia. Now he had written it into thesis form, and would appreciate it if Darwin would read it for professional criticism. Wallace had arrived independently at Darwin's theory of natural selection. And quite reasonably he had submitted his thesis to the one

**515**

other man who he knew might be working along the same lines.

Wallace and Darwin agreed to read their papers jointly before the scientific Linnaean Society. The Society members listened politely to the presentation, then rushed to discuss the papers with their colleagues. A definite undercurrent of excitement began to develop in scientific circles. Now Darwin was ready to go to press, with a hastily prepared version of his life's work that he titled *An Abstract of an Essay on the Origin of Species.* He considered his manuscript only a preview to the much longer work he was still formulating. His publisher begged Darwin to shorten the title to *The Origin of Species.* Darwin agreed, and his book was released in the fall of 1859. *The Origin of Species* was an instant best-seller—the entire first edition sold out in one day. Darwin congratulated himself on his success and soon forgot about the longer work he had planned, since the scientific community responded to his work with wide acclaim. There was controversy, but not the storm of protest Darwin had envisioned.

Carefully and laboriously, Darwin had outlined his entire theory, notably without direct references to its implications for human ancestry. First, he established that all living things tended to increase in number at a prolific rate—a growth rate that far exceeded the ability of the environment to support them. Then he pointed out that no single group of organisms had overrun the earth. Rather, the size of any population seemed to remain constant over long periods of time. Consequently, some individuals in a population must die off while others survive. This led to the crucial point of his argument—that individuals in a population are not identical but possess different features. And the individuals which exhibit the most favorable variations in relation to their environment were the ones who would survive. They would then pass their advantageous traits along to the next generation, according to the principle of natural selection. What emerges

is a simple definition of evolution—changes in the inherited traits of a population with the passing of each generation.

## The story of the past

The origin and evolutionary history of life is a subject of much speculative interpretation. But like any scientific theory, the theory of evolutionary development must fit all the known facts of the case. In this instance, the facts include the evidence of fossil remains, the established geological time scale, and the evidences of change in the planet itself. It is important to remember that life evolved, not in a vacuum, but in re-

17-6 *Volcanic activity, both submarine and terrestrial, has contributed to the earth's development in a variety of ways. Aside from building up the outer crust with layers of volcanic ash and solidified lava, early eruptions introduced new chemicals into the atmosphere and were an original source of oceanic salt. (H. W. Kitchen from National Audubon Society)*

sponse to a changing environment. It is very likely that a static environment would also be a lifeless one, since the chemical events that led to the origin of life were probably dependent on certain fluctuations in the physical environment.

## The dynamic earth

According to the most widely accepted modern interpretation, the planet earth was formed from a hot mass of gas and dust, or nebula, which swirled around the sun when it was a young star. As the nebula cooled, colliding grains of dust began to stick together, first forming particles and then joining to form planetary bodies. It is estimated that the entire process probably took about 100 million years.

The atmosphere of the planet when it was first formed probably consisted principally of methane ($CH_4$) and ammonia ($NH_4$). At a later date, certain other gases, such as nitrogen and water vapor, escaped from the interior of the earth during volcanic explosions and also entered the atmosphere. The modern-day composition of the atmosphere, however, is due in large part to the activities of living organisms.

The earth's crust consists of a layer of heavy basaltic rock on top of which "floats" a layer of granitic rock; the basalt forms the ocean floor, and the granitic blocks are the continents. Since the cooling of the earth did not proceed in a uniform way, but consisted of periods of cooling alternating with periods of expansion and melting, the earth's crust is folded and wrinkled and unevenly distributed. This unevenness led to the upthrust of ranges and mountains.

It is currently believed that the granitic skin of the earth was once concentrated in a single large landmass, the continent given the name of Pangaea. About 180 million years ago, Pangaea split into two continents, called Laurasia and Gondwana. As shown in Figure 17-7, these continents later split again, forming the present-day continents. With each split, the continents drifted farther apart. Many geologists believe that this continental drift continues. They predict that Northern Africa will eventually bump into Europe's Mediterranean coast, and that all of California west of the San Andreas fault will break off and float away to the northwest, arriving off the Alaskan coast in about 60 million years.

During the same time that the earth's crust was undergoing such dramatic changes, the climate of the earth also changed. Three major ice ages have been recorded: one about 600 million years ago, another about 300 million years ago, and the last a scant 2 million years ago. The movement of glaciers associated with these ice ages changed the landscape, and of course drastically changed climatic conditions. The climatic changes of the second ice age were probably accentuated by the northward continental drift that was taking place at the same time.

## The origin of life

The question of how life began seems to interest everyone. All cultures have myths to explain this perplexing problem; theologians have expounded upon it; and scientists have been investigating it for centuries. It is probably a puzzle that can not be solved empirically; perhaps that is the secret of its intellectual fascination.

The traditional explanation for the origin of life is its creation by a deity. There are two widespread versions of this explanation. One, sometimes called original creation, states that a God or gods initiated life—gave it the divine spark—and then let nature take its course thereafter. This explanation is in no way inconsistent with the known facts about evolution. It would not be disproved even if scientists succeed in creating life themselves; the fact that man can now do it does not mean that God did not do it before. The second traditional explanation is that of special creation. According to this ver-

**517**

**Permian -225 m.y.**

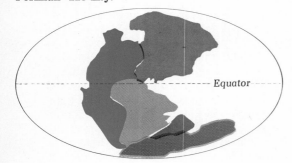

**Late Triassic -180 m.y.**

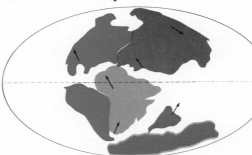

**Late Jurassic -135 m.y.**

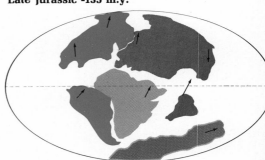

*Equator*

sion, each and every form of life was separately created by a divine being. In other words, God created the English sparrow and then the song sparrow and then the chipping sparrow and then Henslow's sparrow and so on right down the list of species. It is this theory of special creation that has suffered most from the evidence and interpretations of evolutionists, for **speciation** (the creation of a species) can be shown to have occurred many times through the mechanisms of biological evolution. Scientific interpretation of fossil and biochemical evidences does not of course actually disprove the theory of special creation, but it makes that theory more and more unnecessary and thus harder to believe.

The present scientific explanation is that life came from an aggregation of nonliving substances, by means of random events that brought the right combinations of chemicals and energy together. This explanation has the advantage of being philosophically compatible with the scientific outlook, which stresses the randomness of occurrences. It is also a hypothesis that can be tested empirically, which adds to its appeal to scientists.

That life can arise spontaneously from nonlife is not a new idea; it is, in fact, the concept embodied in the doctrine of spontaneous generation. It was the scientists—notably Pasteur—who proved that the apparently spontaneous genera-

tion of bacteria in flasks of nutrient broth was actually due to natural causes and could be prevented by sterilizing the broth. Ironically, scientists now accept a modified version of spontaneous generation as the explanation for the origin of life. There is, however, a crucial difference. Earlier scientists saw the process as ongoing, while the contemporary view holds that it only happened once at the very beginning. The new proposition was put forth by Russian chemist A. I. Oparin in 1924. His work was titled *The Origin of Life*. His theory was that life emerged through a process of chemical evolution that took approximately two billion years to produce the first simple organisms.

**In the beginning.** Life on earth is thought to have arisen three to four billion years ago. The earth's atmosphere was different in those days. Free oxygen and carbon dioxide were almost nonexistent. Since oxygen was only available as a component of water molecules, it could not combine with carbon to form $CO_2$. And because the oxygen was never free, it could not decompose organic compounds as it can now. This was an important consideration in the origin of life; if oxygen had existed in a free state, it would have oxidized existing compounds, breaking them down into simpler components. But hydrogen was available in abundance and in a free form.

**Late Cretaceous -65 m.y.**

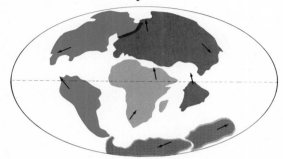

**Present**

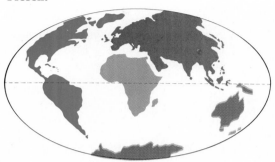

**17-7** *The study of plate tectonics reveals the history of continental drift. Toward the end of the Permian period the earth's continental plates had come together to form a single land mass, Pangaea, surrounded by what is called the Tethys Sea. Different regions of this land mass are called Gondwana and Laurasia. The physiography of the continents since that time has been determined by the drift of these land masses, as shown in the map.*

Thus the earth had a reducing atmosphere, favoring synthesis reactions rather than breakdown reactions. The atmosphere also contained methane and water vapor.

Scientific calculations indicate that the elements all living things require in any large amount were among the most plentiful throughout the solar system. (These findings were arrived at through spectroscopic analysis of light from the sun and planets, coupled with chemical analysis of meteorites which have landed on the earth's surface.) There are six of these major elements in all—hydrogen, oxygen, carbon, nitrogen, sulfur, and phosphorus.

The crucial question is one of explaining how the elements happened to get together in the right combinations. The organic molecules which form an organism are of great variety and complexity. To produce even the simplest kind of life requires a very specific and stable molecular arrangement. How could this incredibly sophisticated fitting of molecules have happened by random coincidence?

Chemist George Wald explains the phenomenon as a matter of simple probability. The probability that a tossed coin will turn up heads is 1/2 on the first try. But the probability that a head will turn up over a series of 10 tosses is much greater—999/1000; the probability of occurrence is increased by extending the number of trials.

**17-8** *The theory of spontaneous generation was finally refuted by Louis Pasteur. Earlier experiments showing that nutritive broth sealed off from the air while boiling does not spoil were discounted by biologists who maintained that boiling made the broth and air above unfit for life. Pasteur convinced them by simply drawing out the neck of his flask rather than sealing it off. Air molecules could pass into the broth but dust and bacterial spores could not. No life was generated. (Rockefeller University Press)*

**519**

So even a fairly improbable event, with a probability of 1/1000, will become almost a sure thing after 1000 trials—again, 999/1000. Eventually the event will be inevitable. Accepting A. I. Oparin's theory that life on earth occurred through a chemical process lasting two billion years, the probability of the right organic elements arranging themselves in all the right combinations to form a rudimentary kind of life is rather high. The time was available, and the event only had to happen once.

**The spark of life.** In Mary Shelley's *Frankenstein*, the deranged scientist assembled a kind of human being with raw materials that his assistant removed from graveyards. His problem was how to turn the huge lump of flesh on his operating table into a living thing; he accomplished this by harnessing electricity and sending a massive jolt through the monster's body. This was an intuitive insight on Shelley's part, because the first life forms must have been produced through the same sort of energy transfer. Even assuming the right combination of elements to make the structure of life, some form of energy was required to turn that structure into an organism which could sustain itself, repair itself, and reproduce.

The fact that an energy input was essential for life was arrived at through experimentation. Early chemists felt that organic compounds could not be synthesized in the laboratory, but in 1828, Frederick Wohler proved them wrong. Wohler managed to produce the organic compound urea through chemical synthesis in the presence of energy, creating a demand for further research. But it was not until 1953 that a conclusive experiment was conducted. Stanley L. Miller, a graduate student, subjected a mixture of water, methane, ammonia, and hydrogen to a series of electrical sparks, duplicating the effect of violent electrical storms during the earth's infancy. After a week, the four inorganic molecules had interacted to form amino acids, the building blocks of

proteins and organisms. In a later experiment, Melvin Calvin exposed a solution of carbon dioxide and water to radiation and produced formic and oxalic acid, both organic chemicals present in man. This study showed that a sufficient amount of radiation could provide the energy to manufacture organic substances. In 1964, Sidney Fox conducted a similar experiment in which he obtained cell-like units called microspheres out of amino acids. This was actual synthesis of polypeptides through radiation energy.

These studies have a sequence of events that might well account for the spontaneous origin of life. The current scientific conclusion is that primitive organisms emerged from an environment referred to as a "dilute organic soup"—bodies of water (probably shallow tidal pools) loaded with the inorganic molecules essential to life. What provided the energy was a combination of ultraviolet rays from the sun, lightning from electrical storms, and dry heat supplied by volcanic eruptions. These forces shaped the simple organic compounds from the "soup" into amino acids, then organisms.

**Chemical events in the creation of life.** Shaking a bottle of salad dressing produces an emulsion; the oil inside will form droplets dispersed through the vinegar. It was this type of action that went on in the "dilute organic soup" three or four billion years ago; the tossing and bubbling of the water caused lightweight substances spontaneously to form droplets called **coacervates,** a colloidal particle enclosed by a membranelike layer, called a hydration membrane.

In the typical primordial pool with a high degree of electrical activity, a number of these droplets were formed. They were relatively stable and, most important, their outer lipid layer made them semipermeable; there were some substances they would admit and some they would not. As the droplets floated passively about on the pool's surface, they would bump into various molecules. Some would be amino

**17-9** *Reconstruction of a Cambrian submarine landscape reveals a scene dominated by algae (left and lower right) and sponges (extreme upper right and top, right of center). Several kinds of worms, brachiopods, and gastropods are present, but arthropods seem to dominate the fauna. Four species of trilobites are shown; they can be recognized by their characteristic three-lobed, segmented exoskeleton. (Courtesy of American Museum of Natural History)*

acids, nucleic acids, fats, and carbohydrates; others would be atoms of sulfur and carbon. Since the coacervates could admit these molecules through their semipermeable outer layer, these chemicals would be mixed together inside, with the possibility of forming more sophisticated proteins, fats, carbohydrates, oils, and nucleic acids. Eventually the new combinations of molecules would become too large for the coacervate to reject through its membrane, so the droplet would grow fatter.

As the plump coacervates drifted about in the water, they often succumbed to the elements and broke up into smaller droplets, many of which tended to disintegrate. But some of them were able to sustain themselves as their "parents" had, by taking in molecules which would chemically combine or react to provide them with additional energy supplies. Different tidal pools were probably producing different kinds of coacervates, with some from each pool floating out to the coastal region of the sea. It was only a matter of time before they were in competition with each other for the available nutrients. The coacervates which were best equipped to screen the environment and absorb energy-producing molecules began establishing their numerical superiority over the others—a primitive form of natural selection.

These original coacervates were heterotrophs that obtained their nourishment through a process similar to fermentation; they did not produce their own food but utilized the energy released in chemical reactions between absorbed molecules. But there was a distinct problem with this method of "food-getting." By removing the high-energy molecules from the immediate environment, the population of coacervates was rapidly doing away with its only resource. Had this process continued, the droplets would eventually have starved themselves to death. But the population was saved by one of their own waste products—carbon dioxide.

As the coacervates became more and more numerous, they released large amounts of carbon dioxide into the water. This provided them

**521**

with an abundance of one of the raw materials they would need to undertake a new form of food production. This new way was photosynthesis. How the coacervates became photosynthetic autotrophs is not known, but somehow as compounds became scarce, some of these colloidal systems "discovered" how to synthesize those in short supply. The fact that they could now manufacture their own food gave them a new lease on life. Now they were able to make organic molecules, such as sugar, from carbon dioxide and water, and a number of other essential molecules with the addition of ammonia and nitrates for oxygen. But they were still dependent upon fermentation as a source of energy. This created grave problems in terms of long-range development of the organism. One problem was efficiency; tremendous amounts of food had to be ingested to release a tiny amount of energy. Another difficulty was the production of waste products, since the wastes of fermentation were always poisonous.

## 17-1
## The origin of life

I   Ammonia, methane, hydrogen, and water vapor subjected to frequent energy bursts from lightning, cosmic rays, etc., resulting in the synthesis of such organic compounds as sugars and amino acids from inorganic substances.

II   Formation of macromolecules as organic compounds collect together in primordial organic "soup."

III   Origin of primary colloidal systems, composed of coacervates, complex large molecules which adsorbed water molecules to form a hydration membrane.

IV   Origin of heterotrophism: some molecules become able to incorporate simpler compounds to enlarge their own structure.

V   Some organisms develop the ability to replicate themselves, at first with many errors, but with growing efficiency.

VI   Origin of autotrophy: as organic compounds become scarce, some colloidal systems develop the ability to synthesize those in short supply. Over time this synthetic ability allowed them to synthesize all the compounds intermediate between carbon dioxide and sugar.

VII   Origin of true cellular structure, more or less contemporaneous with the origin of autotrophy.

VIII   Origin of secondary heterotrophs: this occurred when organisms probably somewhat like dinoflagellates lost the ability to synthesize.

IX   As the autotrophs increase in number, the oxygen they produce begins to accumulate, becoming a major constituent of the atmosphere and forming the ozone layer, shielding the surface of the earth from lethal ultraviolet radiation.

Luckily, the primitive organisms were again saved by a waste product. This time it was oxygen from photosynthesis, which made it possible for them to acquire energy through respiration. By burning sugar, they could dramatically increase the efficiency of energy release. This additional energy permitted them to go beyond mere survival into an increasingly complex organization.

## A history of life on earth

The interpretive history of life on earth is based primarily on the evidence of fossils. A fossil may be defined as the remains of an organism, or its traces, preserved in rock formations by natural processes.

Fossilization may take one of several different forms. It may involve the preservation of the entire organism, as in the case of the woolly mammoth carcasses found perfectly preserved in the frozen gravels of Siberia. Or the hard parts — the bones, teeth, and scales of vertebrates, the shells and chitinous exoskeletons of invertebrates, or the leaves and woody parts of plants — may be preserved, if they are buried quickly enough. Another kind of fossilization occurs when arthropod exoskeletons, or leaves or stems, are buried and subjected to pressures and temperatures great enough to drive off their volatile components, leaving behind only the carbon contained in their organic compounds. Fossils of trilobites, insects, and other shelled animals were commonly fossilized in this manner. Often an organism buried in porous rock will be dissolved away by groundwater, leaving an impression fossil or mold. Insects preserved in amber are an example of another process by which mold fossils can be formed. One of the most interesting categories of fossils is formed when the durable mineral silica has replaced all the organic material of the organism. As in petrified wood, a silicified fossil may preserve the structure of the original organism in exquisite detail. Other types of fossils include coprolites (fossilized excrements), traces of footprints, fin marks, and even raindrops and molds of such artifacts as burrows and tubes.

Any kind of fossilization is the exception rather than the rule. As we can see if we look at the forest floor, or the bottom of the sea, efficient predators crush and scatter their prey, and then scavengers decompose the remains. Only in very rare instances is there left behind any trace that might, if the circumstances are right, undergo some process of fossilization. Students of evolution face an enormous problem as they try to assemble scattered bits and pieces of fossils into some meaningful pattern. The task of drawing a composite of a complete organism is a difficult one, and paleontologist Nicholas Hotton gives us a memorable analogy to dramatize the reconstructive process:

> ... to get a complete picture of evolution is a little like trying to reconstruct *Gone With the Wind* from the scraps on the cutting room floor. Although we may be able to deduce that the story takes place during the Civil War, that it is told from a Southern point of view, and that it is centered in Atlanta, we might not be certain who the major characters are, let alone what the relationship was between Scarlett O'Hara and Rhett Butler. The one advantage the fossil record has over the cutting room floor is that we do know for certain the sequence in time in which things happened.

And yet so great is the time span of the geological past, and so abundant the life on earth, that a fairly continuous, and in some cases a nearly complete record has been left of the passing generations. Detailed interpretation of this record allows us to reconstruct the history of life on earth with considerable accuracy. Allowing for gaps in the record, and misinterpretations of its meaning, we can tell when and where the major groups of organisms arose, flourished, and either passed to extinction or evolved into new forms. In many cases it is possible to determine the environment in which they lived and what changes in that environment caused extinction or further evolution. In fact, so much is known about the course of evolution on our planet that only a summary can be attempted here.

**17-10** *The ant trapped in amber exemplifies the mold fossil. The body parts dissolve away in time, leaving an impression or mold of the original organism. (Courtesy of American Museum of Natural History)*

Chin strap penguins

Bighorn rams

## Territoriality

One of the strongest motivating forces of animal behavior is the drive to establish a protected territory. In animals with social organizations, such as the ant, the territory belongs to an entire population; in many birds and mammals, mating pairs establish the rights to a specific territory in which to make a nest or den; in predatory species, the territory may be a single individual's hunting preserve.

An interesting type of territoriality can be seen in certain types of flocking birds, such as these chin strap penguins, photographed on Deception Island in Antarctica. Note that there is a certain distance between each standing penguin, and that in most cases the distance is exactly the same. If any penguin steps into another's individual distance, he will trigger an immediate aggressive response; the fight will go on until the correct distance is reestablished. The individual distance is actually a small movable territory, the only kind of home the penguin has.

The bighorn rams, found above the tree line on the slopes of both the American and Canadian Rockies, engage in fierce battles every spring. The winner stays in the disputed territory; all the losers must depart. In this instance, the drive to establish a territory is coupled with another powerful motivator, the drive to reproduce, for along with the territory comes a harem of females. Thus the fighting serves the purpose of natural selection.

# Courtship and mating

The activity of courtship seems to bring out the best in the male of the species. It is at that time that his coat is the glossiest, his colors the brightest, and his behavior the most extravagant. The male peacock epitomizes this transformation; the cumbersome tail that hampers his mobility during the rest of the year suddenly become a resplendent attraction as he displays for the hen.

Another kind of display can be seen in the behavior of the male ruffed grouse. Not only does he fan his feathers invitingly, he also drums—beats out a rhythmic call while dancing on a log or cleared spot of ground. The male cockatoo displays by raising his crest to reveal its colored patch; he then courts the female by tenderly touching her beak.

The courtship of the grunion proceeds in accordance with a rigid schedule. Huge schools of these fish swarm in the waters of the Pacific coast during the spring. At the precise moment when the spring tide is highest, they plunge onto the beach, bury themselves in the sand to deposit their eggs and sperm, and then dash to catch the receding tide. In exactly 15 days, when the high tide comes in again, the young will hatch and move out to sea.

The problems of courtship and mating that are comically exemplified by the horseshoe crab have all been avoided by the aphid. The females commonly reproduce by parthenogenesis; the young (born live) are conceived without male help.

Grunion

Horseshoe crabs

Cockatoos

Ruffed grouse

**Aphid giving birth**

**Australian brown snakes**

Peacock

# Nesting

Nests serve to protect the eggs and the young. For this reason, they are usually built in inaccessible or concealed locations. The osprey's nest, built at the top of a tall dead tree that no land animal could climb, is a good example of this habit. The nest of the deer mouse is hidden by an overhanging rock, for the mice also fear flying predators.

In most birds, nest building is highly territorial, and any bird that moves in and tries to build a second nest close to an existing one will be driven away. An exception to this general rule can be seen in the behavior of the cacique, a South American relative of the oriole. These sociable birds build their nests side by side, almost completely covering the tree they have chosen. The males build the nests; when all the nests are finished, the females tour them, inspecting each carefully. The female chooses the nest that seems to her the most appealing, and then she takes as a mate the male that built it. Ornithologists, who have also inspected and measured the nests, are unable to see any difference in them, but the females are very decided in their preferences.

The male stickleback is an unusual nest builder, constructing a little pile of stones and debris to cover the eggs. By extension, the concept of a nest can also include the furry pouch in which the baby kangaroo develops, and the protective shell secreted by the female paper nautilus to hold her eggs.

Grey kangaroos

Osprey and nest

Male stickleback

**Paper nautilus shell**

**Deer mouse with 10-day-old young**

**Nests of cacique**

## Parental care

We tend to associate parental care with birds and mammals, but arthropods may also be devoted parents. An example can be seen in the male water bug that carries the fertilized eggs on his back until they hatch. The scorpion's care of its young extends even farther; after the birth of the live young, the mother carries them around until they have passed through several molts, acquiring a dark exoskeleton that provides protective camouflage.

The cheetah is a solitary predator, and thus the mother must hunt for the family's food all by herself. The one in the photo has just killed a young Grant's gazelle, which she will carry back to the den and share with her cubs.

Baby Canada geese are cared for by both parents. The family will stay together all summer on the lake where the young hatched; in the fall they will migrate as a unit to the winter feeding grounds. Thus each generation teaches the next the patterns of migration.

Elephant mothers and their young live in the larger group of the herd, which may also include the male that is the father of most of the children and several immature males that are probably his sons. Although the care of the young is primarily the responsibility of the female, the larger male is usually nearby to discourage attacks on the baby elephants by lions. Baboons live in similar groups, called troups, with a more advanced kind of social organization, and having a role for each member.

Canada geese

Chacma baboons

**Cheetah and cubs**

**Elephant family**

**Scorpion with young**

**Male water bug**

# Animal technology

It is sometimes said that man is set apart from other animals by his use of tools. There are, however, interesting exceptions to this general rule. Chimpanzees in the wild have been observed to strip the leaves from twigs, making a tool with which to extract termites from their nest. Otters look for sharp stones to drop on bivalves, to break the shell and expose the edible part of the animal; gulls reverse the process and drop the shells on the stone. The woodpecker finch, pictured at right, is another example of a tool user. This bird, which is native to the Galapagos Islands, seizes in its beak stout twigs or the long sharp spines of the cactus; it then uses this tool just as a real woodpecker uses its bill, digging out the insects hidden under the bark of trees.

The beaver is an animal whose technology rivals that of man in at least one respect, that of dam building. Since the last period of glaciation, most of the new lakes in North America have been formed by the action of the beavers. The one in the photograph shown here is towing an aspen branch to add to his dam.

The common garden spider, found throughout much of the United States, is another amazing technologist. The various species of these spiders can be distinguished on the basis of the web they weave; the one shown here is an orb weaver. It has been suggested that the ability to weave such complex webs evolved from the habit of suspending the eggs out of the reach of predators.

Woodpecker finch

Beaver

Orb spider

**17-2**

**Time scale of earth history and evolution**

| Era | Period | Epoch | Characteristic species | Climate and geological events | Years before present (1,000,000) |
|---|---|---|---|---|---|
| Cenozoic | Quaternary | Holocene | modern man | glaciations | .01 |
| | | Pleistocene | early man | glaciations | 2.5–3 |
| | Tertiary | Pliocene | large carnivores | | 7 |
| | | Miocene | abundant grazing mammals | | 26 |
| | | Oligocene | large running mammals; angiosperms dominant | Rocky Mountains raised and eroded | 37 |
| | | Eocene | modern types of mammals | | 54 |
| | | Paleocene | first placental mammals; modern birds | | 65 |
| Mesozoic | Cretaceous | | first flowering plants; climax of reptiles and ammonites, followed by extinction; conifers dominant | | 135 |
| | Jurassic | | first birds; first true mammals; many reptiles and ammonites; cycadeoids (flowering cycads) | | 180 |
| | Triassic | | labyrinthodont amphibians; first dinosaurs; abundant cycads and conifers | | 225 |
| Paleozoic | Permian | | chondrostean fish, cotylosaurs dominant; widespread extinction of marine animals, including trilobites; many insect groups | Allegheny Mountains formed | 280 |
| | Pennsylvanian (Carboniferous) | | great coal swamps; first reptiles | continental glaciation in Southern Hemisphere | 310 |
| | Mississippian | | sharks and amphibians; seed ferns and lycopods | | 345 |

| | | | | |
|---|---|---|---|---|
| Paleozoic | Devonian | first amphibians; fishes very abundant | Appalachian orogeny | 400 |
| | Silurian | eurypterids, ostracoderms; first terrestrial plants | | 435 |
| | Ordovician | brachiopods and first fishes; marine invertebrates dominant; marine algae abundant | | 500 |
| | Cambrian | first abundant fossils of marine life; trilobites and brachiopods dominant | | |
| Precambrian | (period divisions not well established) | Relatively few and primitive fossils | continental glaciations | 600 |

## The Precambrian era

The Precambrian period stretches back in the mists of time. It began unknown billions of years ago, and ended about 600 million years ago. Very little is really known about the life forms that thrived then, or the course of their evolution. Our deductive conclusions about the origin of life constitute the first of four kinds of evidence bearing on the nature of Precambrian evolution. The other three include: (1) traces of organic matter found in Precambrian formations, which usually take the form of layered graphite incorporated into rocks, probably by algalike photosynthetic plants; (2) deductions about the life forms that must have been ancestral to fossils of a later period; (3) recognized fossils of the Precambrian age.

Unfortunately, very few Precambrian fossils have been found. Part of the explanation for their scarcity lies in the nature of the rock records. Few of the Precambrian rocks have survived erosion, and fossils deposited at the edge of the large continents that existed at that time now lie far out on the continental shelf. Another reason for the scarcity of fossils from this period probably lies in the nature of Precambrian life. Biologists believe that the life that existed in this early era floated or swam in the warm, oxygen-rich waters near the surface of the sea; most of these organisms were soft-bodied and left no fossilized traces. In the late Precambrian, this environment may have become crowded, with some forms moving to other, more exciting marine environments. The development of hard external covering of chitin seems to be associated with this move. This hard covering is more likely to fossilize, and thus most Precambrian fossils that have been found date from the latter part of the period.

Among the Precambrian fossils are included the reef-building stromatolites; rod-shaped and coccoid bacteria, some perhaps as much as three billion years old; filaments of blue-green and green algae; algal spores; and the microfossils of dinoflagellates. The few known animals include several genera of cnidarians and two genera of annelids.

**525**

17-11 Diadectes *and related diadectomorph skeletons have been found in Permian deposits in Texas, South Africa, and Northern Russia. They followed an evolutionary trend leading to increased size and certain specializations of the skull and teeth. The quadrate has shifted forward, shortening the jaws. The front teeth are peglike while cheek teeth have broadened transversely. These specializations suggest that* Diadectes *and its relatives were plant-eating reptiles. (Courtesy of the American Museum of Natural History)*

## The Paleozoic era

This era extends from about 500 million to about 225 million years ago; it is subdivided into numerous periods.

**Cambrian period.** In marked contrast to the scarcity of fossil remains from the Precambrian era there is an abundance of fossil evidence of life in the Cambrian period. By the early Cambrian, some very advanced plants and animals appear in the fossil record. Arthropods, the most complex of invertebrate phyla, account for more than 30 percent of the fossils that date from this time. The diversity of species increased significantly through the Cambrian, as organisms spread from isolated colonies to fill the shallow continental seas. Even a hemichordate fossil has been found dating from this period.

Most of the Cambrian plant fossils are algae, including again the stromatolites. Others are species of blue-green, green, and red algae. It seems safe to assume that all plant life was still marine at this time.

Many protist fossils date from this period, including diatoms and some protozoa. The animal phyla represented among Cambrian fossils include cnidarians, brachiopods, bryozoa, phoronidians, gastropods, and annelids. The most numerous and largest Cambrian animals were the

17-12 *Trilobites and brachiopods were among the most active and abundant animals during the Cambrian. Approximately 60 percent of Cambrian fossils are trilobites. They appear here as segmented, three-lobed fossils. Most were bottom dwelling scavengers but some developed into active swimmers. Brachiopods, often called lampshells because they are similar in shape to Roman lamps, were also common in the early seas. Rather than having right and left halves, as in a clam, their shell has a smaller dorsal section fitting into a larger ventral part. (Laurence Pringle from National Audubon Society)*

Trilobita, an extinct class of arthropods characterized by the segmented, three-lobed chitinous carapace for which they are named. The early trilobites reached a length of 20 inches and had as many as 45 segments, suggesting that these arthropods may have evolved from an earlier multisegmented annelid. The lower Cambrian trilobites are found distributed throughout the world, but in the middle and upper Cambrian the trilobites of different regions, for example present-day Africa and South America, began to diverge more widely. Trilobites reached their point of greatest diversity and abundance in the late Cambrian period.

The apparent abundance and diversity of Cambrian life poses an interpretive problem for evolutionists. Why is there such a difference between the late Precambrian and the early Cambrian periods? What permitted such rapid—even explosive—evolution to take place? The answer probably lies in some major geological change that in turn brought about a change in the environment, particularly the climate; the new conditions must have been much more favorable for the evolution of life. Some geologists believe the change was the splitting of a major landmass to form new continents; others believe it was a change in the depth of the ocean floor. This debate will not be settled until new geological evidence is found.

**Ordovician and Silurian periods.** The span of these two periods marks an important transitional phase in the evolution of living things.

Invertebrate animals accounted for the bulk of living organisms at the end of the Cambrian, but by the end of the Silurian land plants and vertebrates began to assume greater importance. This change in life forms was undoubtedly a response to the opening up of new environments, as shorelines altered and the extent of shallow continental seas increased.

Ordovician and Silurian plants are represented mostly by marine algae; this includes not only the blue-green and green algae found in earlier periods, but also the red and brown. A limestone-secreting alga, *Cryptozoon*, was altering the landscape of the ocean floor. Land plants have been reported in the Ordovician formation of Poland, and occasionally elsewhere, but for the most part the landscape was as devoid of plant life as the mountaintops of today.

Among the lower invertebrates were radiolarians, foraminiferans, sponges, and several related groups of corals, which were the major reef builders of the time. Brachiopods reached a climax, or point of greatest number and diversity, in the middle Paleozoic. One type found at that time was an inarticulate, *Lingula*, that is still extant (see Figure A-32). Among the mollusks, cephalopods became prominent by the middle of this period. The modern genus *Nautilus* arose at this time; among its early relatives are included the world's largest invertebrate shells, reaching a length of 17 feet. Among the arthropods, the trilobites, which by now had become highly specialized for a variety of habitats, were declining. Ostracods reached an abundance they have maintained since, and the eurypterids, or "sea scorpions," included members reaching a length of nine feet, making them probably the most effective of Silurian predators and the largest arthropods ever to live. All the major echinoderm divisions appeared in this period.

Ordovician fossils include fragments of an early bony-plated fish, having no jaw, belonging to the class Agnatha; well-preserved, intact fossils date from the Silurian. The most notable of these were the ostracoderms, flat, armored bottom-feeding fish which became abundant in the early Devonian. There is some debate among evolutionists over the significance of the heavy armor of these early fish. Of course the armor would offer protection against predators; but since its development is closely correlated with the move into a fresh-water environment, it is quite possible that the armor's primary survival value was its ability to limit water gain.

527

## ESSAY  Coral reefs

The coral reef community has been called the oldest and most stable of the earth's ecological systems. The reefs of today are directly descended from communities of stromatolites, plants similar to certain modern filamentous blue-green algae. The stromatolites secreted limestone, to lay down the cores of the earliest, simplest reefs about two billion years ago, in the middle Precambrian. At the beginning of the Paleozoic, the stromatolites became associated with spongelike animals, the archaeocyathids, to form the first true reef community.

The composition of the succeeding communities has been altered drastically over geological time. The chart indicates the major changes. The width of the vertical lines reflects the relative abundance of plant and animal groups at different times. The horizontal bands (light color) represent periods of widespread collapse of the reef system, periods which were followed by a resurgence of new, differing types of species, some of which are illustrated. These periods of collapse, lasting between 10 and 60 million years, were associated with widespread changes in the world climate, which in turn were due to continental drift and changes in the physiography of the earth. The modern scleractinian corals, small carnivorous polyps of the phylum Coelenterata, emerged in the Triassic period. In the Cenozoic era, they have diversified into 20 groups, such as the brain coral, which form the backbone of the modern reef.

Perhaps the most interesting lesson the reef can teach us is the value of diversity within an ecological system. As the variations in the widths of lines in the chart show, certain species suddenly become more numerous (and therefore more important to the reef's ecology) and just as suddenly decline. The ecosystem was saved only because of the presence of other organisms that could move into the vacant niches.

The reef community today forms an almost incredibly complex web of ecological niches, including members of almost every animal phylum. The reef has a growing core consisting of limestone secreted by various algae, and of the skeletons of corals, the major reef animals. The corals in turn live in symbiosis with a group of acellular plants, the zooxanthellae. The great bulk of a reef is a fine, sandy detritus held in place by the organisms attached to its surface. Other limestone-secreting animals, from the phyla Porifera, Protozoa, Bryozoa, Echinodermata, Brachiopoda, and Mollusca, contribute to the growth of the reef.

The most conspicuous members of the reef community, however, are not the reef-builders, but such animals as the electric ray, squid, red-bearded sponge, and conch. The parrot fish play an important role in the system. Named for their parrotlike beaks rather than for their gaudy coloring, the parrot fish are reef browsers, nibbling on algae and breaking off bits of the reef limestone. After passing through the gut, this limestone is defecated as fine white sand. It has been estimated that in one acre the parrot fish thus deposit a ton of sand each year.

*(Gifford — Bruce Coleman)*

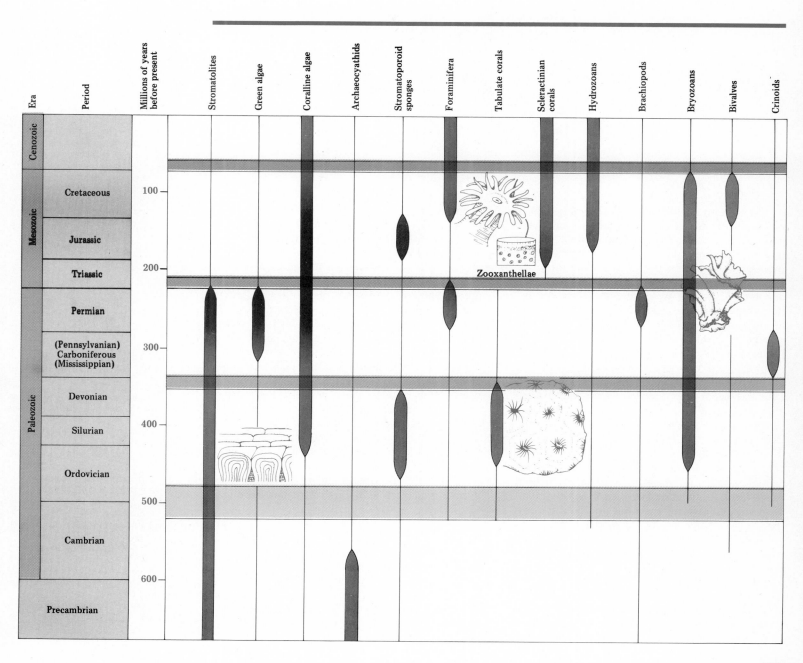

The origin of the vertebrates is a question of considerable interest. The vertebrates probably evolved from a primitive protovertebrate that was streamlined, had segmental muscles, and perhaps looked rather like the modern amphioxus. This animal probably lived in, and was well adapted to, a fresh-water stream. Therefore the origin of vertebrates can be linked to the upthrust of mountains, which led to the creation of such streams. It is believed that this protochordate and the early echinoderms had a common ancestor.

**Late Paleozoic era.** The late Paleozoic era includes a number of different periods: the Devonian (395 to 345 million years ago); the Carboniferous (345 to 280 millon years ago); and the Permian (280 to 225 million years ago). It was a time of rapid evolution, in which the most significant change was the adaptation of both plants and animals to a terrestrial environment. As plants in lowlying estuarial mudflats evolved the characteristics needed for terrestrial life, such as vascular tissue, a cuticle to prevent excessive water loss, and erect growth, they moved slowly onto the land. Their presence permitted animal forms to begin exploiting the terrestrial habitats.

The ancestor of the land plants was probably a green alga. The first plants were homosporous, but some heterosporous forms began to appear in later land plants. This heterospory led to the

evolution of the seed. The dominant terrestrial plants of the Devonian were of the primitive subphylum Psilopsida, spore-bearing plants that grew to a height of about three feet and bore rudimentary leaves. Soon they were replaced by their descendants: lycopsids, or straight-trunked, narrow-leaved "scale trees," reaching a height of 100 feet; sphenopsids, also reaching a height of 80 to 100 feet and represented today only by the horsetail; and pteropsids, the true ferns. These plants grew in abundance on the marshy land surface, which was slow sinking. They were frequently buried in layers of silt that washed over when the water rose. The pressure of the massive upper layers caused heat-producing compression, leading to chemical changes in the buried plants. Their volatile components were given off, leaving behind only a hard black carboniferous "mummy" —the coal beds of today. The conifers, which were also found during this period, remained the dominant terrestrial plants until the rise of the angiosperms in the Cenozoic era.

Late Paleozoic invertebrate life was characterized by an abundance of bryozoa and brachiopods, by the development of early insects, and by a movement of mollusks and arthropods into fresh water. More significant was the spread of the first jawed fish, called the placoderms. The evolution of jaws actually occurred in the late Silurian. This adaptation permitted new types of feeding; since the jawed fish did not have to

**17-13** *The shell and lower jaw of Corythosaurus were very broad and flat in front, earning him the name, "duck-billed dinosaurs." Each jaw was solidly paved with 500 lozenge-shaped teeth which crushed food with a grinding action. The nasal region was also distinguished by unusual adaptations. Premaxillaries and nasals formed a high, helmet-shaped crest on top of the skull, perhaps serving as an accessory air-storage chamber used when the head was submerged under water. (Courtesy of American Museum of Natural History)*

**17-15** *The carnosaurs, typified by Allosaurus of upper Jurassic times, grew to be the largest land-living meat-eating creatures of all time. With reduction of the forelimbs, the activities of predation, killing, and feeding were concentrated in the skull and jaws. The skull became tremendously enlarged and the lengthened jaws accommodated many huge, daggerlike teeth. Greatly enlarged openings in the skull helped to cut down weight while leaving a frame of bony arches for muscle attachment. Allosaurus was admirably adapted to hunting and killing other large reptiles. (Courtesy of the American Museum of Natural History)*

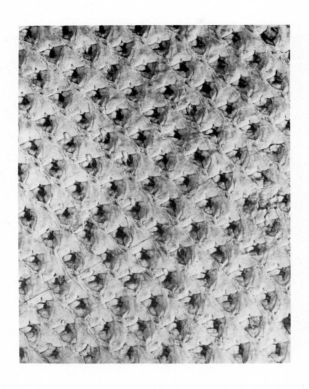

**17-14** *The Devonian lycopsids were represented by* Lepidodendron, *the giant club mosses (no relation to true mosses). All were huge trees up to 120 feet in height. The bark pattern of this fossil* Lepidodendron *suggests why it has been called "the scale tree." Emergences from the outer tissues of the stem probably gave rise to leaves, some of which were 20 inches long. Lepidodendrids also had true roots which may have arisen from branches of ancestral algae that penetrated the soil and branched underground. (John L. Gerard from National Audubon Society)*

depend on sucking alone, they could move from the shallow bottom to other areas of the sea. The placoderms were initially so successful that the Devonian is known as the "Age of Fish." One rather fearsome member of this group was the 30-foot *Dinichthys,* with bony plates and a strong jaw that could crack the armor of other fish. During the late Paleozoic the placoderms were overshadowed by cartilaginous fish, the ocean-dwelling Chondrichthyes, of which the sharks and rays are living representatives. It is interesting to note that the placoderms are the only vertebrate class to have become extinct.

The bony fish also arose during the lower Devonian. They were originally fresh-water fish, but by the end of the Paleozoic they had moved into the sea. Among the advances associated with this group are the protective gill cover, the operculum; a body covering of overlapping scales, very effective in preventing water loss or gain; and an improved jaw suspension that permitted a wider range of feeding habits.

The early Devonian saw the rise of lungfish, with a limited ability to gulp air (represented today by three genera in Africa and South America) and of lobe-finned fishes morphologically similar to the early amphibians. These lobe-finned fishes were thought to have become extinct long ago, but in 1938 one was caught off the eastern coast of Africa. Named *Latimeria,* after the lucky scientist who got the first look at it, it was probably forced up from the ocean depths by a major storm.

As the lobe-finned fishes were forced by seasonal drought out of the shallow waters to which they were adapted, natural selection favored those which were able to leave their old home and move to another body of water. These fishes had stout fins like pillars, able to support their weight on land and serve as a means of locomotion there. In other words, the development of limbs actually helped these animals remain aquatic. It was a group of their descendants, perhaps less successful in finding new

**531**

bodies of water, that gave rise to the amphibians, the first tetrapods. Still linked to their old environment by the necessity of laying their eggs in water, the amphibians evolved many adaptations to terrestrial life, including more highly developed lungs, a stronger skeleton, an improved auditory mechanism, eyelids and tear glands to protect eyes, and loss of scales and gills. These early amphibians ranged in size from two inches to five or six feet long.

A great evolutionary advance took place with the appearance of the first reptiles, the Cotylosauria, or stem reptiles. These stubby-limbed creatures arose from certain amphibians that evolved the ability to enclose their fertilized eggs in a tough semipermeable shell; this egg that could remain viable on land freed the stem reptiles from a dependence on an aquatic environment. From the early cotylosaurs there arose the abundant reptiles of early Permian. Some forms are characterized by a large dorsal fin that may have been used in temperature control. On cold mornings, the membranous fin could be exposed broadside to the sun, absorbing as much heat as possible. As the body reached its optimum temperature, the reptile would turn so that the fin was parallel to the direction of the sun's rays, thus preventing the absorption of heat.

Severe climatic changes, probably caused by the drifting of the continents in a northerly direction, wiped out many widespread terrestrial plants and marine invertebrates, but did not severely affect the terrestrial animals. Insects appear to have steadily increased in number and diversity from Pennsylvanian times to the present. Their evolution and distribution is closely linked to that of the flowering plants. The continents were uplifted at this time, and many geological and climatological changes took place. The drastic changes in environment were fatal to many highly specialized groups, but benign or beneficial to groups either relatively unspecialized or possessed of adaptations that fitted them to the altered environment.

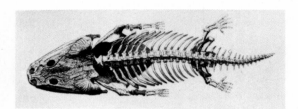

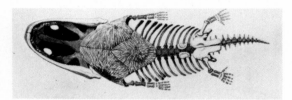

17-16 *The early land-living vertebrates, the ancestral tetrapods, such as Metopias, possessed many characteristics found in crossopterygian fish. The tail still retained the fin rays of these ancestral fish. Teeth had the same complex labyrinthine structure of enamel. The function of the stapes, used in Metopias as a prop to hold jaws and brain case together, was transformed from a bone that formed part of the gill arch in crossopterygians. (Courtesy of American Museum of Natural History)*

17-17 *The development of armor reached an extreme in the large reptiles, Typothorax, of the upper Triassic in North America. Literally encased with heavy plates, these quadrupedal pseudosuchians were almost impregnable to attack. (Courtesy of American Museum of Natural History)*

## The Mesozoic era

The Mesozoic era, or middle period, ran from about 225 million years ago to 65 million years ago. Three major periods are included in this time span: the Triassic (225–190 million years ago), Jurassic (190–135 million years ago), and Cretaceous (135–65 million years ago).

During the Mesozoic era, the ray-finned fish rose rapidly in number and diversity to the dominant position they retain today, by far outnumbering all other vertebrates. Amphibians declined in numbers, to leave only frogs, caecilians, and salamanders surviving. The most dramatic change was the rise of the reptile. If the Paleozoic is known as the age of invertebrates, the Mesozoic deserves to be called the age of reptiles. All the major environments of land, sea, and air were dominated by highly successful reptile groups, but only a few of these survived into the Cenozoic era.

The forerunners of the dominant reptiles were the thecodonts — agile animals two or three feet high that could use their large heavy tails to balance their bodies and run on their hind legs alone. Some thecodonts gave rise to reptiles that look like crocodiles but are only distantly related; others gave rise to the land reptiles, or dinosaurs (from the Greek for "terrible lizard"). These include large four-footed dinosaurs, such as *Brachiosaurus*, an animal taller than a three-story building that attained lengths of 80 feet and weights in excess of 50 tons. The somewhat bizarre adaptations of the dinosaurs catch the imagination. One form had a skull with bones more than 10 inches thick; another had at least 1000 teeth in its jaw. The existence of many herbivorous dinosaurs created a niche for carnivorous predators, such as *Tyrannosaurus rex*, which was 19 feet high and had six-inch teeth. Marine reptiles included the dolphinlike ichthyosaurs, which bore live young, the distantly related plesiosaurs and the serpentlike mosasaurs. The Mesozoic skies were dominated by pterosaurs, flying reptiles such as *Rhamphorhynchus* and *Pterodactylus*. They glided through the air on an enlarged leathery membrane stretched between the elongated fourth digit and the hind legs or body; what is hard to explain is how they ever got up in the air in the first place, since their bodies were too heavy and their wing muscles too weak to permit them actually to take off. It is thought that they may have lived at tops of cliffs, where they simply waited for favorable wind currents.

No satisfactory explanation has been forthcoming for the dramatic extinction of the reptiles at the end of the Cretaceous period. The most likely explanation seems to be that the gradually changing climate created environmental fluctuations to which they were unable to adapt rapidly enough. Climatic changes also severely diminished the gymnosperms on which the herbivorous dinosaurs depended for food. As the herbivores disappeared, so did the carnivores that preyed on them.

The first known bird, *Archaeopteryx*, had evolved by the late Jurassic. This was a reptilian, long-tailed animal the size of a crow, with claws on the forelimbs and a long beak equipped with teeth, both of which features were lost by later birds. The few bird fossils found in Cretaceous rocks reveal types much closer to modern birds. There are many gaps in this group's evolutionary history.

Mammals arose in the upper Triassic. They originated from a group of mammal-like reptiles, the therapsids. A combination of biological characteristics, including a high metabolism, a constant body temperature, fur or hair rather than scales as a body covering, and especially the habit of bearing the young live (although this habit is not limited to mammals alone) and nourishing them with milk, enabled the mammals to succeed the reptiles as the dominant life form. The most important Mesozoic mammals were of seven groups: placentals, marsupials, monotremes, pantotheres, symmetrodonts, multi-

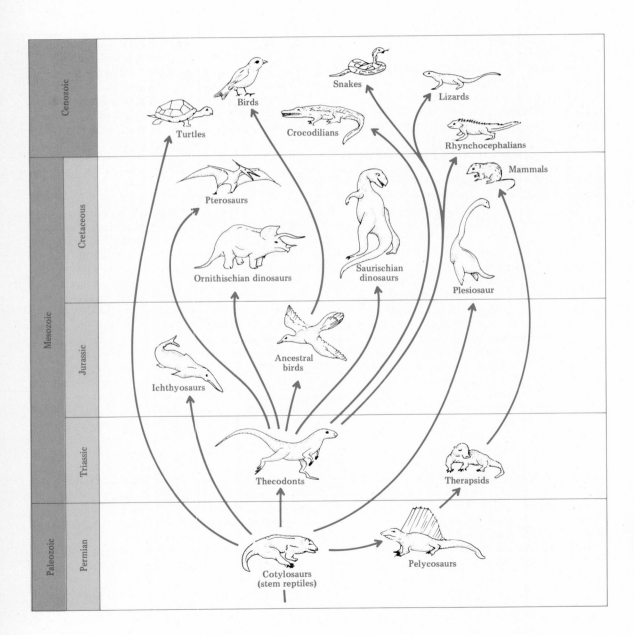

**17-18** From the basal stock, the cotylosaurs, or stem reptiles, a vast array of reptiles evolved during the Mesozoic era. Reptiles of many types dominated the air, land, and sea for well over 100 million years. Some evolved into birds or mammals, but of the rest only a few survived into the Cenozoic.

**17-19** Pterodactylus elegans *is
an early flying form that appeared
during the Jurassic. Several
adaptations to flight are prominent:
a short strong back, strong hind
limbs, a long sternum for
attachment of wing muscles.
Pterodactyls are distinguished by
their enlarged forelimbs with an
extremely elongated digit that
folds back and is attached to a
leathery membrane along the side
of the trunk. By flapping this
structure, the pterodactyls were
able to glide through the air,
although not with the greatest of
ease. (Courtesy of American
Museum of Natural History)*

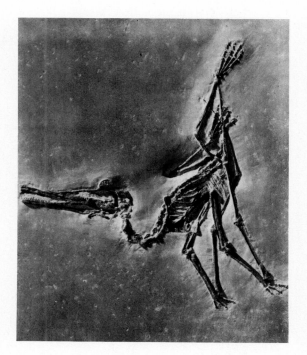

In addition to the horsetails, seed ferns, ferns, and conifers, which had survived from Paleozoic times, the Mesozoic flora included two distinctive groups, the cycads, still represented by the sago palms, and the cycadeoids, now extinct. They included a variety of palmlike forms, reaching a height of 30–40 feet. The cycads reproduced by means of cones, and the cycadeoids by means of the first complex flowers. From a Jurassic ancestor, possibly among the cycadeoids, the first angiosperms arose, to colonize the Mesozoic uplands. The climatic changes that so affected the reptiles during the Cretaceous allowed the angiosperms to expand explosively throughout the rest of the world, giving rise to today's flowering plants and deciduous forests.

## The Cenozoic era

The Cenozoic era covers a time period from about 65 million years ago to the present (although some geologists prefer to divide this span into the Tertiary and Quaternary periods). The Cenozoic era is divided into epochs with rather amusing names, based on the ancient Greek word *kainos*, meaning recent. The oldest epoch is the Eocene, or "dawn of the recent." Then comes the Oligocene ("few recent"), Pliocene ("more recent"), Pleistocene ("most recent"), and finally, in desperation, the Recent.

The Cenozoic is often referred to as the age of mammals. It might just as accurately be called the age of angiosperms, or perhaps even the age of insects, since all three groups spread and flourished in this period. The story of mammalian evolution, however, is one in which we take a personal interest, since it includes the descent of man.

The small mammals of the earliest Cenozoic occupied much the same ecological habitat as the modern insectivores. The extinction of the ruling reptiles provided evolutionary opportunities for mammals to move into vacant habitats.

tuberculates, and tricodonts. The multituberculates, which looked rather like rodents and often reached the size of a woodchuck, were the most successful of Mesozoic mammals, and survived to the Eocene epoch. The monotremes, represented today by *Platypus*, are the least progressive of mammalian orders.

Mesozoic invertebrates underwent a development corresponding rather closely to that of the reptiles. Between the Triassic and Jurassic the ammonites, a large group of mollusks with a shell shaped like a ram's horn, approached extinction, but later evolved into at least 23 specialized and highly successful families. These dominated the invertebrate fauna of the Mesozoic seas, but became extinct for unknown reasons at the close of the Mesozoic. The corals, also common during the Mesozoic, remain important today, as do the other invertebrates that comprise the plankton which form the basis of the marine food chain.

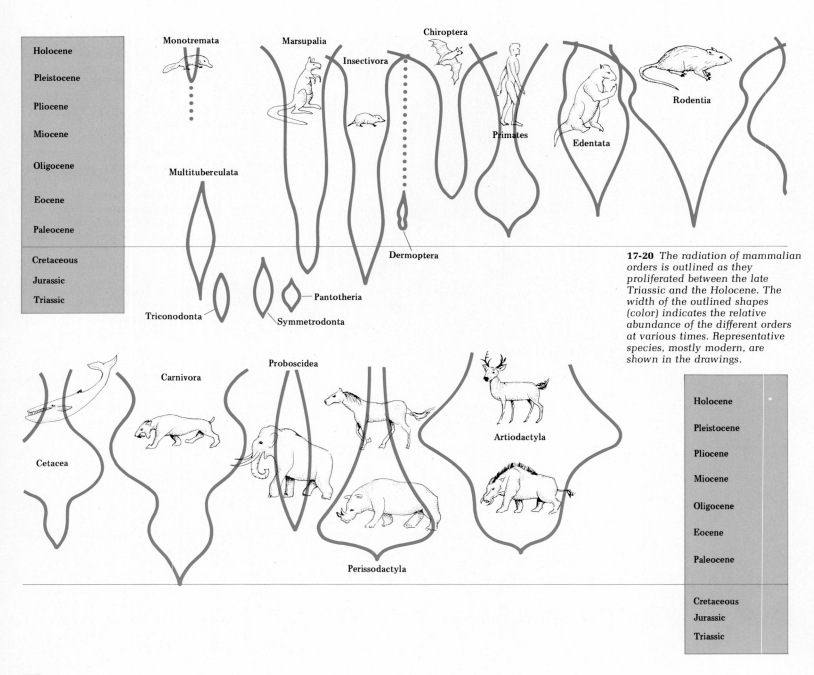

Holocene

Pleistocene

Pliocene

Miocene

Oligocene

Eocene

Paleocene

Cretaceous

Jurassic

Triassic

Monotremata

Marsupalia

Insectivora

Chiroptera

Primates

Edentata

Rodentia

Multituberculata

Triconodonta

Symmetrodonta

Pantotheria

Dermoptera

**17-20** *The radiation of mammalian orders is outlined as they proliferated between the late Triassic and the Holocene. The width of the outlined shapes (color) indicates the relative abundance of the different orders at various times. Representative species, mostly modern, are shown in the drawings.*

Cetacea

Carnivora

Proboscidea

Artiodactyla

Perissodactyla

Holocene

Pleistocene

Pliocene

Miocene

Oligocene

Eocene

Paleocene

Cretaceous

Jurassic

Triassic

In their subsequent evolution, four features are particularly noteworthy:

1. Many mammalian lines evolved from small forebears to larger, even gigantic forms.
2. The early mammals had the five digits characteristic of man and the primates. Others, especially running herbivores, tended to lose the outer digits and modify the claw into a hoof.
3. The teeth of early mammals became modified to reflect diet.
4. The larger mammals have tended to acquire larger brains. In the primates the convoluted cerebrum has come to dominate the rest of the brain.

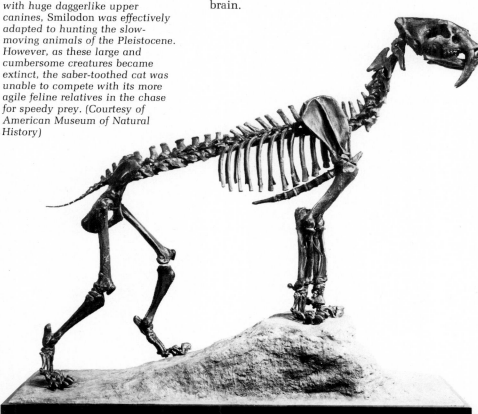

**17-21** *Saber-tooth evolution culminated in the later Pleistocene cat, Smilodon. As large as a modern mountain lion, with huge daggerlike upper canines, Smilodon was effectively adapted to hunting the slow-moving animals of the Pleistocene. However, as these large and cumbersome creatures became extinct, the saber-toothed cat was unable to compete with its more agile feline relatives in the chase for speedy prey. (Courtesy of American Museum of Natural History)*

The marsupials, which arose in the late Cretaceous, are represented today by such animals as the opossum, koala bear, and kangaroo. In South America, where they were abundant until the end of the Pliocene, they became predacious carnivores and remained at the top of the food chain until the end of the Pliocene; competition with placentals from North America may have been the factor that forced them to extinction. The placentals, which arose in the Mesozoic from an insectivorelike stock, were elaborated into an additional 20 orders in the early Cenozoic.

In the later Cenozoic, especially the late Eocene and Oligocene epochs (roughly 40 to 25 million years ago), primitive, unspecialized mammalian orders were gradually replaced by more specialized competitors. Small Paleocene carnivores with specialized teeth and brains evolved into the otterlike forms, the first true carnivores. These diversified into two lines: on the one hand, dogs, raccoons, bears, and weasels; and, on the other, the civets, hyenas, and cats. A distant group of relatives includes marine seals and walruses. A notable development was the saber-tooth cat or "tiger," which reached its maximum size in the late Pleistocene genus *Smilodon*.

The hoofed mammals, or ungulates, are divided into two great orders, the perissodactyls (odd-toed ungulates) and the artiodactyls (even-toed ungulates). An interesting story of perissodactyl evolution begins with the late Paleocene *Hyracotherium*, also called *Eohippus* or "dawn horse." The evolution of the horse has been well studied and is often used as a textbook example of modification through evolution. Unfortunately, it is often assumed that this evolution proceeded directly from *Eohippus* to the modern horse, *Equus*. The evolutionary process was much more complex; the evolution of *Equus* has involved progressive specialization of several features, including size, dentition, hoof form, and range. Evolution after *Hyracotherium* proceeded in numerous and complex lines, including many differences in rate, kind, and

**537**

direction of evolutionary change. Most of the browsers and grazers that arose from *Hyracotherium* are extinct; of these, several branching lines led to the modern horse. Among other perissodactyls, the rhinoceros developed in the Cenozoic from swift, hornless forms including *Baluchitherium*, the largest land mammal ever to live, 25 feet long and 18 feet high at the shoulders. The large artiodactyl order is divided into two groups; Suina, including pigs and hippopotamuses, and Ruminantia, including camels, deer, giraffes, and cattle. Striking examples of these groups were the early "giant pig," Dinohyus, with a skull three feet in length, and the late Pleistocene "Irish elk" (in fact a deer), whose antlers were the largest ever found.

Three other nonprimate groups should be mentioned. The largest living land animals are the African and Indian elephants. These are survivors of a flourishing Eocene lineage of proboscidians that included the browsing mastodons and the grazing woolly mammoths, the largest elephants, which were contemporaries of early man. The edentates, represented today by anteaters, sloths, and armadillos, were abundantly represented in the North American Pleistocene by the giant ground sloth *Megatherium* and the heavily plated armadillo *Glyptodon*. The Cetacea—whales, porpoises, and dolphins—evolved from terrestrial carnivores as early as the Eocene. Today this group includes the blue whale, the largest animal ever to live on earth, and now threatened with extinction.

**Evolution of man.** More than any other aspect of Darwinism, the idea that man is closely related to the anthropoid apes has been responsible for antiscientific and metaphysical attacks on evolutionary theory. Today, overwhelming evidence has been accumulated which seems to establish beyond doubt that the evolution of man and the special attributes that are uniquely human can be explained entirely within the framework of modern evolutionary theory. This evidence is of two kinds: skeletons of early man and his predecessors, and the tools and other artifacts they fashioned and left behind. The latter kind of evidence throws light upon the origins and development of human culture, technology, and society. The skeletal evidence establishes man's biological relation to the other animals; his closest relatives, of course, are the other primates.

Compared to other groups of placental mammals, the order Primates may be considered relatively primitive. This is because features which in other mammalian orders have become highly specialized remain generalized among the primates. Dentition and elements of the skeleton have remained unspecialized, permitting varied and generalized habitat and food habits. The following structural features are common to all the primates, including those ancestral to man:

1. The primates are basically arboreal or tree-dwelling, although some forms, including man, have moved to a ground-dwelling habitat.
2. The sense of smell is less acute, and hearing

17-22 *Progressive specialization of the size, dentition, hoof form, and range of the horse is well documented in the fossil record. Hyracotherium, the original Eocene ancestor, was a browser about the size of a fox terrier with four toes on each front foot and three on the hind feet. Mesohippus, a Miocene grazer with only three toes on the front feet was followed by the slightly larger Merychippus, with a highly developed middle toe and dwindling accessory toes. The modern horse, Equus, is much larger than any of the early horses and has only one toe on each foot. (Courtesy of American Museum of Natural History)*

and vision have become the most important of the senses. The highly developed eyes have moved to the front of the head, permitting binocular vision and accurate depth perception.

3. The areas of the brain associated with vision and hearing have become enlarged; the entire brain is larger and more complex, permitting heightened powers of reasoning and the coordination of rapid and complex reactions. The cerebral hemispheres, the area of sensory integration, have come to dominate the brain, especially in the higher primate forms.

4. The braincase has expanded to accommodate the enlarged cerebral hemispheres.

5. The thumbs have become opposable to the other digits, permitting effective grasping of branches and, ultimately, sensitive and accurate manipulation of objects. The claws have become modified and reduced to nails.

Zoologically, the family Hominidae, to which man belongs, is closest to the family Pongidae, the apes of tropical Africa and Asia. The two can be distinguished by several features. The cranial capacity of modern man is about 1350–1500 cc., compared with a maximum of 750 cc. among the apes. In man the frontal brain is predominant, resulting in a high forehead and a nearly vertical face. In even the early hominid specimens there is evidence that the brain and lower jaw have been modified to permit speech. The curve of the teeth in the apes forms a U-shape, while in man it forms a more open arc; man's molar regions are not parallel, as are the apes', and the canines in man are not prominent. Correlated with these differences are other skeletal differences, associated with man's upright stance, the position of his head, and his characteristic bipedal locomotion. The evidence is that these characters developed in association with one another.

The earliest hominid fossils, known as *Ramapithecus*, appear in upper Miocene and lower Pliocene deposits, especially in East Africa and India. *Ramapithecus* probably moved on all fours and had moved from an arboreal habitat to open country. The dentition suggests that, unlike the modern apes, *Ramapithecus* did not have the ability to strip leaves from branches with his teeth. This in turn suggests a manlike dependence on use of the hands.

The next group of man's ancestors are referred to collectively as *Australopithecines*, or southern apes. These were a group of bipedal near-men combining teeth, pelves, and limbs similar to those of modern man with other features characteristic of the apes. They first appeared in the early Pleistocene, about two million years ago, and some forms persisted for about 1,100,000 years. Some of them used stones to kill animals and made primitive stone implements. Two species are of special interest. The first is *Paranthropus robustus*, a large herbivorous form which failed to develop tool use to

**17-23** Megaceros *(the word means "great horns"), or the Irish elk, flourished throughout parts of Europe during the Pleistocene. Although the horns give this beast an air of ferocity, his flattened teeth, suitable for grinding and macerating, indicate that he was a vegetarian. (Courtesy of American Museum of Natural History.)*

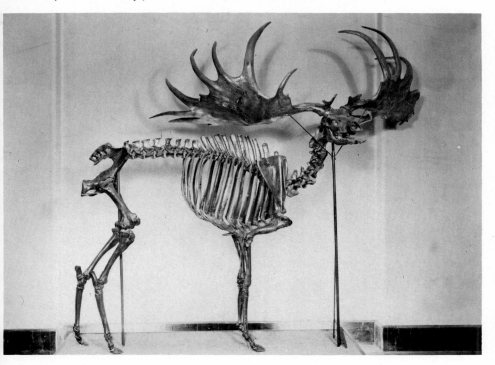

any high degree. The other, *Homo transvaalensis,* was a more adept tool-maker. Both forms were established in the early Pleistocene, 1.7 million years ago, and seem to have coexisted until about 900,000 years ago. The evolution of modern man can be traced back to *H. transvaalensis.*

By all evidence, the cranial capacity of *H. transvaalensis* was only 600 cc., about that of a modern chimpanzee, although the total body size of this hominid was much greater than that of the chimpanzee. The difficult question here is how an animal with such a small brain could become a tool-maker. There seem to be two answers. The first is that the upright posture of the near-men freed their hands, so that the habit of using tools in a major way became an evolutionary possibility. The other is that while the total brain volume was only comparable to that of a chimpanzee, its distribution was more similar to that of modern man, with sensory, motor, and association areas already beginning to dominate the brain structure. The development of even the primitive technology of *H. transvaalensis,* and possibly of the capacity for speech, meant that these forms could cross the threshold into a new adaptive zone, that is, a radically new and hitherto unexplored way of life. The crossing of this threshold set up quite new and intense selection pressures, resulting in the extremely rapid evolution of human characteristics, especially the large, and efficient human brain.

About 800,000 years ago, while *Paranthropus* was still undergoing evolution in Africa and Java, *H. transvaalensis* became extinct and was replaced by a daughter species, *Homo erectus.* This species can be unequivocally accorded the status of man. It had a cranial capacity of about 1300 cc., complex social organization, human morphology, and a relatively advanced technology, including the use of fire. Different races included within this species are: *Pithecanthropus erectus, P. pekingensis* (Peking Man), and *Homo heidelbergensis* (Heidelberg Man). All were characterized by receding forehead and a shallow brain pan, heavy brow ridges, flat nose and a receding or nonexistent chin.

The first specimens of modern man, *Homo sapiens,* appear in the fossil record about 300,000 years ago. In their European form they are called Neanderthal Man, *Homo sapiens neanderthalensis,* after the site of their first discovery at Neanderthal in Germany. Neanderthal Man was rather squat and short-necked, massive and rather brutish in his features, and walked in a stooped position with head thrust forward. (Recent reports suggest the startling possibility that he may have lacked the capacity for speech, but this cannot be determined yet.) Possibly weakened by the extreme cold of the Ice Age, the Neanderthals disappeared after the Fourth Glaciation and were replaced by a tall, erect race of unknown origin, the Cro-Magnons. Cro-Magnon Man, though similar in appearance and intelligence to modern man, was not exactly the same. Rather, modern man, *Homo sapiens sapiens,* seems to have coexisted with the earlier Neanderthals and to have become dominant after their extinction, ultimately absorbing the Cro-Magnons through interbreeding.

Man in his ancestry exhibits a pattern of divergent evolution and extinction exactly similar to that of the other groups of animals discussed earlier. The story of man's past is thus

17-24 *Examining the sequence and height of various strata, the late Dr. Louis Leakey discovered the geologic, animal, and human record of evolution in East Africa. At this site, Olduvai Gorge, australopithecines were first discovered in strata which could be dated absolutely. Volcanic minerals present in the strata made potassium-argon dating possible. The basaltic rock at the bottom of the gorge is approximately two million years old, while the volcanic ash in layers nearer the top average half a million years. (Jen and Des Bartlett — Bruce Coleman)*

**17-25** *Restorations of Pithecanthropus, Neanderthal, and Cro-Magnon based on reconstructed skulls show differences in facial structure and skull size. Pithecanthropus, or Homo erectus, had essentially the same body structure as modern man. He differed mainly in his massive skull, larger teeth and smaller brain (about 75 percent as large as that of modern man). Both Neanderthal and Cro-Magnon man are now classified with our own species, Homo sapiens. Despite his heavy features, Neanderthal man's cranial capacity falls within the same range as our own. Cro-magnon man, one of the last of the Old Stone Age men, appears virtually indistinguishable from a twentieth-century individual. (Courtesy of American Museum of Natural History)*

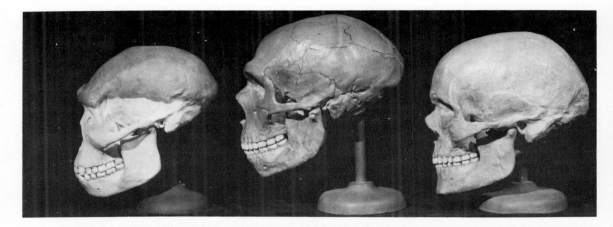

solidly embedded in natural history, and more particularly in evolutionary theory. Yet in one feature—his elaborate culture—*Homo sapiens* resembles no other creature. It has been the rapid evolution of culture and technology as much as his physical evolution that has brought modern man to his dominant position among the mammalian orders. The highest rate of human cultural evolution has been reached only in the last 5000 years, a period that represents only one percent of our history as a species. This nonbiological, cultural evolution has delivered the rest of nature into our safekeeping, as mankind has become ubiquitous and powerful beyond any species previously seen on earth. For good or for ill, the evolution of the future is in our hands.

## Summary

In all times and places people have wondered when and how man and the living world came into being. To answer these questions, modern science has developed the theory of evolution. This theory is associated with the nineteenth-century work of Charles Darwin, but its roots go back to the earlier work of Lamarck and others, and to the work of the great early geologists. The turning point of Darwin's life came on an extended voyage when he visited many regions, especially the Galápagos Islands, whose unique fauna raise fascinating questions about the origin of species. After many years, Darwin published his theory of natural selection.

The story of life begins with the formation of the planet earth from a cloud of gas and dust, the formation of its first atmosphere, the movements of the landmasses, and the advances of oceans and glaciers. The scientific explanation of the origin of life presupposes a primitive atmosphere of methane, ammonia, and water vapor. Between three and four billion years ago organic chemicals, created in the atmosphere by radiation and lightning, accumulated to form a dilute "organic soup." With the passage of time, the molecules formed droplets called coacervates, which already had many of the attributes of living matter. Under the inevitable pressure of natural selection, the coacervates became more developed, changing the atmosphere and acquiring the processes of photosynthesis and respiration. At some point this primitive living matter crossed the boundary between precellular and cellular life forms.

The detailed record of life on earth depends on the systematic study of fossils, remains or traces of organisms preserved in different ways in rock formations. The fossil record allows us to divide the geological past into distinct eras and periods.

The Precambrian era, from the beginning of the earth 4.5 billion years ago to 600 million years ago, is the longest and most poorly documented geological period. The Precambrian witnessed the origin of life and its organization into the major phyla, but little record is left of these events.

Paleozoic era, 600 million to 225 million years ago, is the age of invertebrates and fish. Arthropods especially dominated the early Paleozoic seas, although many other invertebrates were present. By the later Paleozoic, vertebrates, in the form of fish, become important in the fossil record, and primitive plants become widespread on the continents. By the end of the Paleozoic, certain fish had given rise to the amphibians and, through them, to the reptiles.

The Mesozoic era, 225 to 65 million years ago, is called the age of reptiles. Reptiles dominated all environments of land and sea, but few survived to later times. Mesozoic invertebrates were also widespread and numerous, and more advanced plant types superseded the earlier Paleozoic forms. Toward the end of the Mesozoic the first mammals and angiosperms appeared, though they were at first overshadowed by the earlier forms from which they had evolved. Drastic climatic changes contributed to the widespread extinction of the reptiles at the close of the Mesozoic.

The Cenozoic era began 65 million years ago and continues to the present. It is called the age of mammals because it has seen the evolution of mammals to the point where they dominate all terrestrial and many marine habitats. In the last million and a half years this has included the evolution of man from protohominid ancestors to the point where man, in turn, dominates the mammalian orders, and in fact affects the evolution of all life on earth.

# Bibliography

### References

Barghoorn, E. S. 1971. "The Oldest Fossils." *Scientific American.* 224(5): 30–42.

Farrand, W. R. 1962. "Frozen Mammoths." *Science.* 137: 450–452.

Kurten, B. 1969. "Continental Drift and Evolution." *Scientific American.* 226(6): 54–65.

Lack, D. 1953. "Darwin's Finches." *Scientific American.* 188(4): 67–72.

Miller, S. L. 1955. "Production of Some Organic Compounds Under Possible Primitive Earth Conditions." *Journal of the American Chemical Society.* 77: 2351–61.

Newell, N. C. 1972. "The Evolution of Reefs." *Scientific American.* 226(6): 54–65.

Romer, A. S. 1949. *Time Series and Trends in Animal Evolution.* Princeton University Press, Princeton, N. J.

Romer, A. S. 1966. *Vertebrate Paleontology.* University of Chicago Press, Chicago.

West, M. W. and A. Ponnamperuma. 1970. "Chemical Evolution and the Origin of Life." *Space Life Sciences*. 2(27): 225–295.

## Suggestions for further reading

Darwin, C. 1959. *The Origin of Species*. Variorum ed. University of Pennsylvania Press, Philadelphia.
The revolutionary work that established the doctrine of organic evolution. This edition includes all the modifications made by Darwin in the course of the six editions during his lifetime.

Darwin, C. 1962. *The Voyage of the Beagle*. Doubleday (Natural History Library), Garden City.
Based on notes kept during his voyage to South America, this book describes the researches from which Darwin developed his theory of evolution.

Dobzhansky, T. 1951. *Genetics and the Origin of Species*. 3rd ed. Columbia University Press, New York.
An especially helpful presentation on the mechanisms of evolution by a leading authority in the field.

Eiseley, L. 1961. *Darwin's Century*. Doubleday, Garden City.
Masterfully traces the development of evolutionary concepts within the context of Western intellectual history by examining particular achievements and discoveries in geology, biology, and anthropology. Focuses on Darwin's impact on the Victorian world and also treats the challenges that have led to changes in the theory of organic evolution.

Kummel, B. 1961. *History of the Earth*. Freeman, San Francisco.
A descriptive-narrative account of the history of life. Includes names and characterizations of the various groups involved.

Kurten, B. 1968. *The Age of Dinosaurs*. McGraw-Hill, New York.
A noted paleontologist presents a lively treatment of fossil carnivores, late Tertiary and Quaternary stratigraphy, and other aspects of paleobiogeography.

Waddington, C. H. 1961. *The Nature of Life*. Allen & Unwin, London.
A lucid and accurate statement of the problems of heredity written by a leading geneticist.

Wald, G. 1955. "The Origin of Life." *In* Editors of *Scientific American*. *The Physics and Chemistry of Life*. Simon & Schuster, New York.
A biochemical explanation of how a living organism arises from the nonliving. This concise treatment attempts to show that life, as an orderly and natural event, was inevitable on a planet such as ours.

## Books of related interest

Ardrey, R. 1972. *African Genesis.* Dell, New York.
A provocative attempt to account for human aggression by tracing the development of *Homo sapiens* from carnivorous predatory apes.

Clarke, A. C. 1968. *2001: A Space Odyssey.* New American Library, New York.
A fictional and mystical exploration of man's place in the cosmic order.

Greene, J. C. 1961. *The Death of Adam.* New American Library, New York.
An insightful examination of the social and philosophical impact made by evolutionary theory on Western intellectual life and thought.

Montagu, M. F. A. 1961. *Man and Aggression.* Oxford University Press, New York.
An assortment of essays challenging the view of man as an innately aggressive creature descended from killer apes.

Morgan, E. 1972. *The Descent of Woman.* Stein & Day, New York.
Presents a radical alternative to accepted theories of human evolution by proposing that men are descended from a group of female primates (followed by males) who left the trees for the oceans.

Morris, D. 1967. *The Naked Ape.* Dell, New York.
A well-known zoologist makes a startling and entertaining case for man's fundamentally savage nature.

Porter, E. 1971. *Galápagos.* Ballantine, New York. 2 vol.
A master of nature photography has created a beautiful portrait of the islands where Darwin gathered much of the data for his life's work.

# 18

# The principles of evolution

Evolution can be simply defined as a change over time. Yet such a definition, though accurate, leaves much unsaid. An ice cube that is removed from the refrigerator will change in time to a puddle of water; a rock along the seashore will be changed into grains of sand by the ceaseless action of the waves. Is there some essential difference between these changes and the changes that evolutionists are referring to when they say that mammals evolved from therapsid reptiles?

Certainly there are some similarities. One is the need for some interval of time, as evolution takes place only very gradually. Another similarity lies in the gradualness of the change, which takes place in such imperceptible steps that it is impossible to draw a line between any two of the steps, although the change is quite evident when comparing the original form (the stone or the therapsid) with the final product (the sand or the mammal). A third important similarity is that in both cases the process of change is irreversible. Mammals cannot at some later date evolve back into therapsids any more than the grains of sand can be reconstituted into the original rock.

What then are the differences? One important one is that changes in the physical state of matter, such as the changes in the rock, usually proceed in the direction of greater disorganization and higher entropy. Evolutionary change, on the other hand, is not a move toward equilibrium, and in fact very often results in an increased complexity. There is a school of thought that attributes this property seen in biological evolution to some special quality of life—a vital force or drive toward complexity or urge for progress. This philosophical interpretation cannot of course be disproved, but many biologists have pointed out that no such assumption is necessary to explain the course of evolution. In fact, a case can be made for viewing all evolutionary change as a kind of failure rather than any sort of progress. Geneticist Jacques Monod has pointed to *Lingula,* a brachiopod that first appeared during the Ordovician period, as nature's triumph, for it has remained virtually unchanged for all the millions of years since its origin. Monod suggests that evolution, far from being some sort of glorious progress, is the result of error that stopped short just this side of disaster.

The second major difference between evolution and other kinds of change in the natural world is that evolution is adaptive; it changes the population so that it is adapted to its environment. It is important here to recognize the difference between the concept of directional change and that of purpose. Although evolution can be viewed as changes that move toward a certain direction, such as from simple forms to complex ones, there is no purposefulness in these changes. They are the result of selective forces acting on random variations.

The third major difference between our two examples is that in the case of the stone, we are talking about a single individual; in the case of the therapsids, we are talking about a large group of individuals in which there is a gradual percentage change. Biological evolution really means a change in the average of the group rather than a change in any one individual; it is a change in the frequency with which certain genes appear in the population.

18-1 *A picturesque example of the diversity of life forms that is thought to be the result of evolution is this shoebilled stork. A native of the papyrus swamps of northern Africa, this bird feeds largely on shellfish, especially clams. Its bill is well adapted for digging out its prey and cracking open the shells. (William Vandivert)*

## Population genetics

Evolution is really a change in the frequency with which genes appear in a given population. A population is an interbreeding group of individuals; in some cases, it may include all the members of a species, but more often, it includes only those in a certain geographical area. Unlike a species, which is sometimes defined as all those individuals that can interbreed, a population might be defined as those that 'have a good chance of doing so.

All the members of a population share a common gene pool, or total number of different alleles that can be inherited. Each individual member of the population has only a selective portion of the genes within the pool; as Ernst Mayr puts it, "The individual is only a temporary vessel, holding a small portion of the gene pool for a short time. . . . It is the entire effective population that is the temporary incarnation and visible manifestation of the gene pool."

By and large, the gene pool of a population remains relatively stable, and thus the phenotypes of individuals within the population stay within a certain range of variation. However, counteracting the factors that make for stability of the gene pool are certain factors that make for change, through removing certain genes from the pool, or adding new ones, or changing the frequency with which genes appear. Such changes may increase the range of variation among phenotypes, or they may eliminate one extreme or the other of the spectrum of variations.

## Stability factors

Many early geneticists were unaware of the great stability of the gene pool, even though the evidence—the restriction of variation in the phenotypes to a certain range—was right before their eyes. Perhaps they were misled by terms like "dominant" and "recessive," for dominance has a connotation of victoriousness, making it seem as if the dominant gene has somehow defeated all

**18-2** *As this photo of a bog indicates, there can be wide differences in environmental conditions even within a small local area. There are varying degrees of sun and shade, differing amounts of water, probably also differences in soil composition. Such environmental mosaicism helps preserve variation within the gene pool of local populations. (Mary M. Thacher—Photo Researchers)*

the others, who were left with no choice but to disappear. While it is true that the trait controlled by a recessive gene may disappear from the phenotype of a heterozygous zygote, it is definitely not true that the recessive gene disappears from the genotype. In fact, the significance of Mendel's work was that it proved beyond a doubt that recessive traits were neither blended in with the dominant trait, nor submerged by it. Just like the dominant gene, the recessive allele is randomly sorted during meiosis and then passed on to one of the four daughter cells. Its presence may be masked by the dominant allele for a hundred generations, or even a thousand, but it is still there, still sorted during meiosis, still inherited by the next generation.

Because each allele is assorted independently, none is ever "lost" in the process of reproduction. Thus as long as mating is random between members of the population (each of whom possesses a particular fraction of the gene pool) and no mutations occur, and the population is a large one, the gene pool will remain exactly the same from one generation to the next. This conclusion was expressed in 1908 by two research workers at the same time—G. H. Hardy, an English mathematician, and G. Weinberg, a German physician. Therefore it has come to be known as the **Hardy-Weinberg Law.**

The validity of the Hardy-Weinberg Law can be demonstrated by an example. Let us assume that in a hypothetical population the percentages of the genotypes for these alleles are as follows:

CC = 36%     Cc = 48%     cc = 16%

What would be the frequency of gametes produced in such a population? Since CC individuals constitute 36% of the population, they would produce 36% of the total gamete output of the population—all C allele gametes. Likewise, cc individuals, comprising 16% of the population, would produce 16% of the total gamete output—all c allele gametes. Heterozygous individuals, Cc, would produce both C and c allele gametes in

**549**

equal proportions—24% C gametes and 24% c gametes.

So the total output of C allele gametes would be:

    36% (from CC individuals)
    + 24% (from Cc individuals)
    or   60% C gametes

and the total production of c allele gametes would be:

    16% (from cc individuals)
    + 24% (from Cc individuals)
    or   40% c gametes

These gametes could be expected to combine in four possible ways:

    a C sperm with a C egg
    a C sperm with a c egg
    a c sperm with a C egg
    a c sperm with a c egg

The frequency of these zygotic combinations would be the products of the individual frequencies of which each allele occurs in the population:

    for C with C—60% × 60% = 36%
        c with c—40% × 40% = 16%
        C with c—60% × 40% = 24%
        c with C—40% × 60% = 24%

So we see that in this new generation, the frequencies of the three genotypes are exactly the same as they were in the previous generation. There has been no change in gene frequencies, hence no evolution.

The Hardy-Weinberg Law can be expressed in mathematical terms, with the symbol $p$ used for the frequency of one allele, and the symbol $q$ used for the alternate allele. Since the total of their combined frequencies cannot exceed 100 percent, then $p + q = 1$. If we know the frequency of one allele, then we can find the frequency of the other:

$p = 1 - q$. The frequencies of the genotypes within the population relative to the frequencies of each of the alleles can then be expressed by the expansion of the binomial $(p + q)^2$, or $p^2$ (CC) + 2 pq (Cc) + $q^2$ (cc). Since $p + q = 1$, this tells us that the total of the frequencies of each of the genotypes must be 100 percent. Therefore, given either the frequency of p or q or the frequency of the genotype of either homozygote, we can calculate the expected frequency of all other genotypes in the population. Taking the problem used above, assume we knew only the frequency of the cc individuals—16%. That would mean that the frequency of the c allele must be 40% (since 40 × 40 = 16%). Therefore since $q = 0.4$, $p = 1 - q$ or $1 - 0.4 = 0.6$. Now we know the frequencies of the alleles, and if we wish to find the frequency of the heterozygotes in the population we would use the appropriate term, 2 pq or 2 (0.6) (0.4) = 0.48.

**Other stability factors.** Many of the important factors that make the gene pool more or less stable have already been mentioned in earlier chapters on reproduction and genetics. One of these is the accuracy with which DNA replicates, coupled with the fact that errors in the sequence of codons can be rectified simply by cutting out the faulty codon. There are also mechanisms that help prevent errors in chromosomal movement and sorting. Moreover, certain groups of genes tend to remain together even during crossing over of chromosomes; they constitute a sort of supergene, helping to prevent the accidental separation and consequent loss of a single gene in the unit.

Another factor important in maintaining the stability of the gene pool is the fact that individuals heterozygous for two alleles are sometimes larger, stronger, and have more survival advantages than individuals homozygous for either allele. This situation is known to geneticists as **heterosis;** stockbreeders call it hybrid vigor, a desirable condition that they try to

**18-3** *Some natural populations of animals have developed close associations with man. The house mouse,* Mus musculus, *has been man's constant companion since earliest agricultural times, and is found throughout the world wherever there are concentrations of people. (National Audubon Society)*

achieve by crossing different strains. Even in cases where the homozygous genotype produces a phenotype with disadvantageous traits, the superiority of the heterozygote will maintain the allele within the population. This is called **balanced polymorphism;** it seems to be the case with the gene for sickle-cell anemia.

Another factor that prevents the gradual disappearance of some alleles from the gene pool is the great variation in the physical and biological environment. Consider, for example, a field of buttercups. Some may be in full sunlight all day, while others are slightly shaded; most are in a low-lying marshy area, but a few may be on higher ground; some may live in richer or better aerated soil than others. One allele might produce a phenotype well suited to withstanding bright sun; its recessive produces a phenotype that is at a disadvantage under such conditions. But since the environment is varied, with both sunny and shady spots, both alleles will be preserved within the gene pool.

Observations of populations in the wild suggest that preferences in mate selection may also help to preserve certain rare alleles within the gene pool. There have been many reports of females choosing males with rare phenotypes in preference to those with common phenotypes. This was demonstrated in laboratory studies with *Drosophila melanogaster;* it is quite possible that some similar preference factor in humans is the basis of the old adage that opposites attract.

## Sources of genetic change in populations

If the Hardy-Weinberg Law applied to all populations at all times, then there would be no change in the gene pool of any population, and consequently no evolution would ever take place. But in fact, the conditions set forth as part of the Hardy-Weinberg equilibrium principle rarely occur in nature, and evolution does indeed take place. The most important factors working to change the gene pool of a population are mutation, genetic drift, and natural selection.

**Mutation.** The primary source of new genes within a given gene pool is the process of mutation. All species, from bacteria to man show this process. Although genes are relatively stable in structure, any gene can mutate, either spontaneously or under the influence of mutagenic chemicals or radiation. Moreover, as we have seen, in some species there are actually genes that promote mutation; it has been suggested that such genes are present in all species but have simply not yet been identified.

Each gene can be shown to mutate at its own characteristic **mutation rate.** Mutation rates have been calculated for 16 human hereditary diseases; the average is about 30 mutations per million sex cells. The highest of these, 100 per million, is for the mutation causing a certain form of muscular dystrophy. The mutation for aniridia, a condition in which the iris of the eye is absent or rudimentary, arises only five times per million sex cells. While any specific mutation must be considered rare, the great number of genes involved—numbering in the tens of thousands for man—insures that many sex cells carry newly mutant genes.

**Genetic drift.** In a large population, genes that are found in only a small percentage of the total group are nevertheless found in a fairly large number of individuals. But if the population is a very small one, then a rare genotype might be found in literally only a handful of individuals. In such cases, chance may play a great role in determining the fluctuations in frequencies of these genes. If some of those few individuals get sick, or are caught by a predator, or fail to find a mate, or have any one of the accidents that befall a certain proportion of the population every year, then those genes may disappear. On the other hand, if those few individuals by chance escape most of the accidents, the rare genes may become much more prevalent.

This phenomenon of the random effects of chance on the gene pool of a small population is called **genetic drift.** The Hardy-Weinberg Law, like the genetic laws of Mendel, deals with statistical averages, and those averages may not be expressed in a population of a small size, just as the Mendelian laws may not be expressed in an $F_1$ or $F_2$ of only a few individuals. Genetic drift is seen most frequently in geographically isolated populations, such as those on islands; in fact, many of the Galápagos animals that Darwin studied are probably examples of the effects of genetic drift. Evolutionists explain a number of bizarre traits that seem to have no adaptive value in terms of genetic drift.

Because genetic drift is a chance occurrence, it was long considered to be unimportant to the general evolutionary history of life. The recent trend, however, is to accord it more significance. It has been pointed out that genetic drift will fix adaptive traits in a population; if the opportunity presents itself, the population may then radiate out into new habitats. It is believed that this process may have occurred many times, most notably in the radiation of the early hominids throughout the world.

**Natural selection.** Natural selection is the mechanism that Darwin proposed as the basis of evolutionary change. Although the effect of natural selection may be both subtle and complex, the basic concept is really quite a simple one.

**18-4** *An example of a mutation can be seen in albino animals, such as this albino gray squirrel, who owes his color to a mutant gene he inherited. Generally, albinos are less successful than their normally colored relatives, for their color makes them very easy to see and therefore especially vulnerable to predators. (Leonard Lee Rue III from Omikron)*

**18-5** *Movement of the earth's crust has played an important role in evolutionary history, permitting some genera and species to evolve for long periods isolated from competitors. The ringtailed lemur, Lemur catta, is found only on the island of Madagascar. The fauna of Madagascar includes several species unique to that habitat, typifying the effects of genetic drift on small isolated populations. (A. W. Ambler from National Audubon Society)*

**18-6** *In some cases sexual selection may result in exaggerated secondary sexual characteristics, whose only function is to attract a mate. In such cases, such as that of the Chinese silver pheasant shown here, we may wonder if this form of selection is not ultimately inadaptive. Since the pheasant's plumage would seem to make it conspicuous to predators, perhaps predation rates have been low. (T. M. Blackman from National Audubon Society)*

In any population, there is a broad spectrum of inheritable variations in phenotypes. Some of the phenotypic characters may make the organism slightly better adapted to survive and reproduce in its environment, others may make it slightly less adapted. Those that are best adapted will be more successful at reproducing, and many of their offspring will inherit the adaptive trait. Over time, a larger and larger number of individuals within the population will have the adaptive trait. In other words, the population average will move from one point on the spectrum of phenotypes to another.

In popular literature, the phrase "survival of the fittest" is often used as a synonym for natural selection. To some extent, this is a misinterpretation. Obviously the individual must survive if he or she is to reproduce, but it is reproduction that is the crucial point in the context of evolution. A better synonym for natural selection is differential reproduction.

In most cases, the characters that help the individual survive are either the same as, or compatible with, the characters that help him reproduce successfully. There are, however, a few exceptions, and these serve to point up the fact that reproductive success is the basis of natural selection. An example can be seen in the Argus pheasant. The male of this species courts the female by displaying to her the red eyespots on his wing feathers. The larger and more exaggerated the spots, the more likely it is that they will catch the female's eye. Therefore the trend of natural selection has been toward bigger and bigger eyespots on longer and longer wing feathers. The males that are most successful in reproducing have such unwieldy wings that they can barely get off the ground, making them quite vulnerable to predators. In this instance, natural selection works to favor a maladaptive trait solely because it increases reproductive success, even though it may threaten survival.

If you examine closely the members of a natural population, you may conclude that the variations you see are so slight that they seem almost negligible. Does the difference of a fraction of a millimeter in a wing or tail really confer some selective advantage? Several interesting statistical studies were carried out in the 1930s, most notably by Ronald A. Fisher and J. B. S. Haldane, to find an answer to that question. They found that even if a trait proves advantageous in as few as 0.1 percent of cases, it will in fact be selected for survival.

More recent studies indicate that the actual degree of advantage of many traits is much higher than the figure used in those early studies. The results of one of these studies, concerning the peppered moth, are shown in Table 18-1.

The efficiency of natural selection sometimes makes itself felt in spectacular fashion. When DDT was introduced in 1944, it was the most effective pesticide yet developed, but within a few years, DDT-resistant strains of houseflies had evolved in response to this change in the environment. By 1960, over 120 insect species—at least half of them important in public health—had evolved similar resistance to pesticides, and

it has become necessary to develop new formulas continuously to combat this problem. The same difficulty arises in regard to antibiotics; microorganisms rapidly evolve a resistance to such "wonder drugs" as penicillin and streptomycin.

We have spoken of natural selection as a force that changes the composition of the gene pool, but it is important to remember that it is not the genes themselves that are selected. Rather it is the entire set of genes, or **genome,** of the individuals. Successful genomes will become more prevalent and unsuccessful genomes will become less prevalent.

Because it is actually the whole genome, rather than a single gene, against which selection pressure is exerted, an important source of evolutionary change is recombination, or the reordering of genes on chromosomes. Although recombination does not involve any change, or mutation, of individual genes—the gene pool remains the same—it does bring a change in genomes and thus, through natural selection of genomes, also brings about a change in gene frequency.

**18-1**

**Differential mortality of the peppered moth** *(Biston betularia)* **released in different woodlands**

| No. released | | Type of woodland | No. eaten by birds | | Percent recaptured alive | |
|---|---|---|---|---|---|---|
| *Melanic* | *Pale* | | *Melanic* | *Pale* | *Melanic* | *Pale* |
| equal | | grayish | 164 | 26 | — | — |
| equal | | sooty | 15 | 43 | — | — |
| 473 | 496 | grayish | — | — | 6.3 | 12.5 |
| 447 | 137 | sooty | — | — | 27.5 | 13.0 |

From Kettlewell, H. B. D. 1961, "The phenomenon of industrial melanism in Lepidoptera," *Ann. Rev. Entomol.* 6:245–262.

**Migration.** Another source of genetic change in populations is migration, or the movement of specific individuals into or out of the population. This may mean the loss of certain genomes and the addition of new genomes, thus immediately changing the genome of the population. For example, in the pocket gopher of the American southwest, many local races have evolved. One race of the valley pocket gopher, *Thomomys botta* is closely related race to the Baily pocket gopher, which is slightly smaller, inhabits the foothill regions, and has even been occasionally treated as a separate species, *T. baileyi*. Migration of these gophers from one deme to the next causes interbreeding, which leads to slight changes in both genotypes and genomes introducing what are in effect "new" genes into the neighboring population.

## Rate and direction of genetic change

In order to trace the process of evolution, biologists have set up carefully controlled experiments testing the effects of various mechanisms of genetic change in populations. One of the most effective methods was to substitute artificial selection for natural selection.

The principle of artificial selection has long been familiar to man. It is the basis of agriculture and of the domestication of animals, and has been practiced more or less deliberately for many thousands of years. In the opening section of the *Origin of Species*, Darwin discusses artificial selection in the many varieties of domestic pigeon. Darwin's observations convinced him that the 20 or more breeds of pigeon he saw, although so divergent in appearance that they might be taken for distinct species, were all nevertheless descended from the rock dove, *Columba livia*. He was later able to prove in breeding experiments that rock dove characters tended to reappear in hybrids between breeds.

A classic experiment in artificial selection was undertaken by K. Mather and his associates. They used *Drosophila*, and attempted to select for an increased number of bristles on certain abdominal segments. Within 20 generations, bristle number rose from an average of 36 to about 56. At that point, however, the rate of sterility became so high that no further selection breeding was possible. When the laboratory population was then permitted to mate at random, the average number of bristles fell sharply, but it never fell back to the level of 36, and it finally stabilized at about 40. Clearly, the pressures of artificial selection had somewhat altered the frequency of a gene for many bristles within the laboratory population. It is interesting to note a second conclusion that can be drawn from this

**18-7** *Artificial selection can be for any arbitrarily chosen phenotypic character. The Polish hen shown here has been subjected to generations of selection for elegant plumage, and illustrates how dramatically effective selection can be. It no longer bears any obvious resemblance to its wild ancestors and probably could not survive in nature. (Alycia Smith Butler)*

work. Even when selection is unidirectional— exerted for or against a single trait—there is also selection for other traits, since it is the genome rather than the gene that is selected. Thus the selection for increased bristles also served as a selection for sterility. In natural selection, where the selection pressure is exerted in favor of over-all fitness of a complete genome, unidirectional selection is not feasible.

## The effects of evolution

Evolution can be conceptually defined as a change in gene frequency within a given population. But what are the consequences of this change? Three major areas in which the effects of

## ESSAY **Adaptation of the vertebrate foot**

The structural adaptations of the vertebrate foot reveal how animals have responded to the evolutionary challenges of getting food and escaping their enemies in a wide variety of environments. All primitive land animals are cumbersome and waddling walkers, because their limbs are short and stout and are directed outward at right angles to the body. Reptiles adapted for faster movement by taking to a bipedal mode of life. Bipedalism, by freeing the front limbs from contact with the ground, paved the way for the later, adaptive modification of wings.

Certain birds—the penguin, for example—later abandoned flying. Their wings are now used as flippers in swimming, and their webbed feet serve as rudders (Figure A). Although the generalized tetrapod foot has five digits, in the penguin two toes have atrophied. Penguins may walk as far as 60 miles across the ice from the sea to their rookeries (frequently carrying their eggs on their feet). Their plantigrade stance, in which the metatarsals and heel form a large bearing surface flat on the ground is characteristic of slow animals, including primates.

In contrast, the faster spiny mouse evolved a digitigrade stance (Figure B), bearing its weight on the ends of the metatarsals. Digits lying flat on the ground and a permanently elevated ankle provided more spring and enabled this vulnerable species to run quickly and quietly.

As members of the most successful order of living mammals, the spiny mouse and the porcupine (Figure C) demonstrate the great range of adaptation among rodents. They have developed a diversity of limbs, feet, and toes which enable them to move about effectively in a variety of environments. The porcupine foot was a relatively simple though effective adaptation to arboreal life. The nails or claws were developed to assist in clinging to branches or bark.

Various methods of climbing vertical tree trunks are used by different mammals. The slow and deliberate porcupine is aided by granular nonskid pads on the soles of its feet. The gibbon (Figure D), in contrast, is a highly developed arboreal acrobat. Gibbons do not rely heavily on their hindlimbs for locomotion; although they come to the ground occasionally and can run well on their hind legs in plantigrade fashion, the foot is used mainly as an accessory in grasping and for carrying small supplies of food. The toes and sole are extremely elongated, giving a greater surface area for grasping. The big toe, like the thumb, is opposable and represents the final stage in mammalian adaptation for grasping.

evolution make themselves felt are adaptive, diversification, and speciation.

## Adaptation

One major consequence of evolution is **adaptation,** or the change of form, function, or behavior that makes a population better able to survive in its environment. Whenever a factor is present in the environment that favors certain genetic characters, and that factor remains constant over time, there will be a cumulative effect on gene frequency within the population. It seems likely that this process and this alone is responsible for the high degree of adaptation observed throughout the living world. Moreover, as time goes by and new studies are carried out, every physical

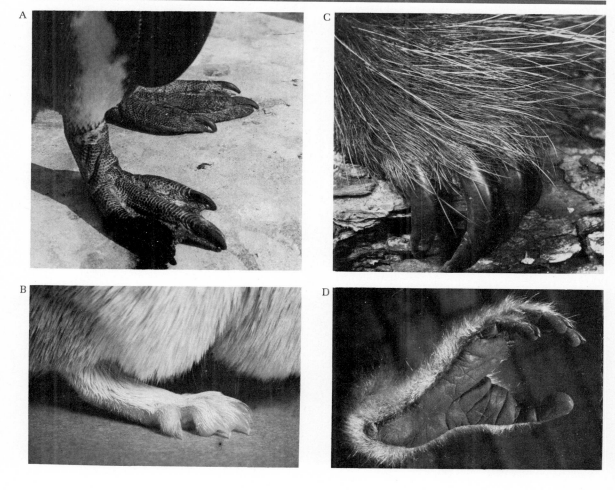

A

B

C

D

and behavioral character studied is shown to have some adaptive value. Natural selection generally operates in favor of maintaining or increasing the integration of an organism with the environment or its adaptation to a way of life. Alternatively, selection may favor changes which will adapt the organism to a different but accessible way of life. The requirements of a local environment are always found, upon close study, to be very exacting, and the gene pool of a deme, as well as gene frequencies within that pool, is always found to be very precisely adjusted to these requirements.

A very well-studied area of adaptation is the survival value of body color and form. One example of such adaptation is **aposemasis,** or warning coloration. Aposemasis is especially common in insects, many of which have evolved coloration which contrasts sharply with their environmental background, rendering them highly visible to potential predators. Aposemasis advertises the unacceptability of the organism; it stings or bites, secretes poisons, or smells or tastes obnoxious to the predator. An example of aposemasis can be seen in the brightly colored larvae of the flannel moth; the spines of this animal are covered with a toxic chemical that causes severe inflammation. Other examples of aposemasis can be found in Color Portfolio 2. After experience with such an obnoxious prey, predators become conditioned to avoid the aposematic species. Some predators, especially among the higher animals, may become so highly conditioned as to avoid all conspicuously colored prey.

Where one species can successfully minimize predation by warning coloration, others can follow. This is the phenomenon of mimicry. In **Batesian mimicry,** the mimic species evolves coloration to resemble a genuinely protected species, which is distasteful to predators. One example of Batesian mimicry can be seen in the flies that mimic wasps, hornets, and bees; among vertebrates, this type of mimicry is found in the

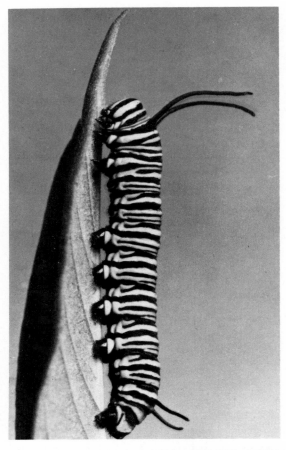

**18-8** *The bright color and vivid pattern of this caterpillar advertise its unacceptability to predators. The larval form of the Monarch butterfly (Danaus plexippus), this insect feeds on leaves of the milkweed plant, and so its body contains the same poisonous chemical. (Photograph by Alexander B. Klots)*

**18-9** *This flower fly is a member of the family Syrphidae, its scientific name is Spilomyia fusca. In body coloration and in behavior, this harmless insect mimics its more dangerous model, the wasp, and therefore many predators give it a wide berth. (Photograph by Alexander B. Klots)*

**18-10** *The walkingstick insect,* Diapheromera femorata, *provides a classic example of protective (cryptic) form. The walkingstick's resemblance to a bare twig is heightened by its tendency to remain nearly motionless for long periods of time during the daylight hours. (Photograph by Alexander B. Klots)*

**18-11** *The green marvel moth,* Agriopodes fallax, *is protected by cryptic coloration. Against the background of the foliose lichen on which it usually rests, the green marvel becomes nearly indistinguishable. (Photograph by Alexander B. Klots)*

Arizona king snake, that mimics the poisonous coral snake. In **Mullerian mimicry,** two or more species, each with its own chemical or behavioral defense mechanisms, come to resemble each other; each species plays the double role of model and mimic. Their similar appearance causes potential predators to treat them as a single group, and therefore fewer of each species are sacrificed in the process of conditioning predators to avoid the group.

A second type of adaptive coloration is **crypsis,** or the "hiding" of an organism due to coloration or form that blends in with its accustomed background. The camouflage can be amazingly successful and widespread. In the Namib Desert of southwest Africa, for example, certain edible rock plants show resemblance to the color of the substrate which is also found in many local animals, including shrews, rodents, larks, lizards, and grasshoppers. Camouflage is often combined with behavioral adaptations which make it more effective. Cryptically colored insects, for example, tend to remain motionless during the day, to select habitats appropriate to their coloring, and to remain in a position in which their coloring is most effective. Caterpillars that resemble twigs typically rest in an extended position which renders their camouflage more convincing. The moth *Automeris io* has evolved a behavior pattern which allows it to combine cryptic with warning coloration. When at rest, this moth assumes a position which exposes the cryptically colored ventral surfaces of the wings. When disturbed, however, it moves the wings to expose two brightly colored "eyespots," at the same time orienting itself so that the eyespots are turned to present the observer with the illusion of large, blinking eyes. This combination of coloration and behavior has been shown to be effective in startling insectivorous birds, permitting the moth to make a hasty departure.

Physiological and morphological adaptations are fully as important as protective coloration.

Flowers, for example, often have characteristic odors, colors, and structures which attract and accommodate the insects which pollinate them; the joint evolution of the Yucca cactus and the moth that pollinates it is a classic case of this type of adaptation-changed physiological characteristics, such as an increased ability to tolerate different pH or osmotic concentration, permitted the adaptation to new types of aquatic environment, such as fresh water.

Many evolutionists have stressed that a major determinant of animal evolution is behavior. Almost all shifts from an accustomed to a new niche are initiated by changes in behavior, ranging from preferred habitat and food selection to mating display. After the behavioral transition has been made, other adaptations are acquired secondarily. The evolution of the brown and long-billed thrashers of the genus *Toxostoma* is a case in point. The evidence points to a gradual transition from arboreal birds resembling the mockingbird. As the thrashers moved more completely into the terrestrial habitat, their toes and bills became increasingly adapted to their new activities, such as scratching in the soil and digging through leaf litter. A selective pressure then arose favoring stronger toes and tarsus and a longer and more sharply curved bill resulting in the thrasher forms observed today.

## Diversification

A second important consequence of evolutionary change has been the diversification of living organisms. The phenomenon of diversification explains the wide variety of forms found within the biosphere at the present time. Populations develop and change, increase in numbers and expand into new habitats or dwindle into extinction, creating new types of environments or filling vacant ecological niches.

One form of diversification is **adaptive radiation,** which occurs when the progeny of a

18-12 *The small moth* Tegeticula alba *pollinates the flower of the Yucca plant. Evolution of insects and flowers has often been linked, with the two organisms becoming more intimately adapted over the course of time. (Photograph by Alexander B. Klots)*

18-13 *Behavioral adaptations are usually as important as structural or physiological ones. The quills of the common hedgehog,* Erinaceus europaeus, *would be useless against predators without the behavioral habit of rolling up into a ball for protection. (Eric Hosking from National Audubon Society)*

**18-14** *One of the mechanisms of diversification is extinction. The California condor is one of the world's largest and rarest birds. The population size is now so reduced that the species is likely to disappear unless special measures are taken to preserve it. This particular individual is photographed in the Los Padres National Forest at the opening to a cave where it probably has its nest. Here the bird is relatively safe from hunters and ecosystem disturbance. (Carl Koford from National Audubon Society)*

**18-15** *The Bering wolf-fish, a native of the cold arctic seas, exemplifies adaptation to food habits. The massive jaw and heavy blunt teeth allow the wolf-fish to crush the large crustaceans which are its prey. (Fred Bruemmer)*

population originally adapted to a restricted habitat radiate out into a diversity of newly available habitats. This may occur when an entire new adaptive zone becomes available for the first time, as when the early Pennsylvanian stem reptiles first moved away from the water. Or the disappearance of competitors may suddenly open a formerly occupied region to adaptive radiation; the explosive radiation of mammals in the Cenozoic was permitted by the extinction of the ruling Mesozoic reptiles. A classic example of small-scale radiation is that of Darwin's finches, in the Galápagos Islands. From a single South American ancestor, several genera of finches have radiated out into diverse ecological niches on the islands. On different islands in the group, different species may fill corresponding roles, and no member of the group is found anywhere but on the Galápagos Islands.

Adaptive radiation seems to be a major feature of evolutionary history. Most radiations were short-lived and of minor importance. But on the time scale of earth history, radiation over the past three billion years, as it was described in Chapter 17, explains the diversity of habitat and adaptation throughout the living world.

Two factors are generally necessary in order for the diversification to take place. One is the existence of some selection pressure that acts to alter gene frequency within a given population. The second factor is the existence of some barrier to **gene flow,** or the free and random movement of genes throughout the entire breeding population. As long as there is a normal gene flow, selection pressures will be expressed throughout the entire population; only if some members of the population are separated will diversification result.

## Speciation

A third major consequence of evolution is **speciation,** or the formation of new species. Each new species may be regarded as a potential evolutionary pioneer, for each species contains a unique gene pool, occupies a unique niche, and therefore has the capacity to divide into local populations which can explore new niches. A population that makes a successful adaptation to a new zone may become the founder of a new taxonomic category. The origin of every class, every order, every phylum can ultimately be traced back to the local population of a species.

How is speciation initiated? In some cases, it may be virtually instantaneous, due to the accident of **polyploidy,** a sudden multiplication of the normal number of chromosomes. Polyploidy is common in the plant kingdom, where it is important in speciation; probably one-third of plant species arose through polyploidy. Among animals, polyploidy seems to be much less common, probably because the presence of the additional chromosomes interferes with the normal process of sex determination, and thus leaves the organism incapable of reproduction. Many polyploid plants are larger and hardier than their diploid relatives; more important is the fact that they tend to be highly variable, and therefore some of them may be preadapted to exploit new habitats.

Spontaneous speciation is quite rare; gradual speciation is much more common. In the simplest basic pattern of evolution, **sequential speciation,** change in gene frequency occurs over time in an entire species group. In other words, one species gradually evolves until, at some point, it is different enough to be called another species.

Populations of any size and type tend to undergo regular sequential evolution. Scarlet tiger moths, for example, were studied in Great Britain over a period of 23 years, with respect to alleles that caused their carriers to have either two large or many small light spots on the anterior wings. Although the frequency of both alleles fluctuated considerably with the years, a trend toward the two-spot allele could be observed; over the 23 years the percentage of the two-spot allele rose from about two to about 10 percent.

**18-16** Speciation can be observed in these two rattlesnakes. (A) is the Eastern diamondback rattlesnake, native to southern Georgia. (B) is the timber rattler, found in temperate deciduous forests. Although the two snakes are rather similarly patterned, close comparison reveals a number of differences. (A by Scott Ransom— Photo Researchers; B by Leonard Lee Rue III from Bruce Coleman)

**18-17** *We tend to think of geographical isolation in terms of barriers like mountains and rivers. Another type of barrier is depth. The fish adapted to live in the lower depths of the ocean are effectively isolated from other fish populations, and they often show the bizarre structure that is the result of genetic drift within a small isolated population. (Peter David — Photo Researchers)*

After millions of years, this accumulation of small changes in gene frequency add up to such change in the group that speciation has occurred.

Perhaps more significant than sequential evolution is **divergent evolution,** in which a species becomes divided into local populations, gene flow is impeded, and the populations become differentiated. Geographic isolation seems to be of major importance in initiating the divergent evolution that results in speciation. It occurs when demes become separated by mountain ranges, bodies of water, canyons, oil pipelines, and other obstacles. For a while, the populations that are separated in space remain potentially capable of interbreeding. Crosses between such **incipient species** can often be produced in the laboratory. With the passage of time, intrinsic **isolating mechanisms** arise in the genotype which prevent further interbreeding, even if the geographic barrier is removed.

## Isolation mechanisms

Geographically isolated demes begin with different initial genes and gene frequencies. Furthermore, as time goes on, different mutations will arise in the isolated populations. Finally, selection pressures are likely to be different in the different habitats, so that different gene combinations will be favored.

The isolating mechanisms that prevent interbreeding of diverging species may be classified either as prezygotic (taking place before possible interspecific fertilization) or postzygotic (taking place after the fusion of gametes that forms a zygote). An important type of prezygotic isolation mechanism is the ecological mechanism that prevents potential mates from meeting. The two diverging populations may occupy different areas or, occupying the same area, may breed at different times, and thus be separated by a temporal isolating mechanism. Animals breeding at the same time may have different mating behavior, in which case the isolating mechanism is described as ethological. Ethological mechanisms comprise the largest and most important group of barriers to gene interchange in the animal kingdom. They include a fascinating array of visual, auditory, chemical, and other factors, which will be discussed in Chapter 19.

Another prezygotic isolation mechanism is mechanical isolation. Here the potential mates may meet and attempt to breed, but cannot successfully complete fertilization. Where the ranges of the Gulf Coast toad and the oak toad overlap in Louisiana, hybridization is prevented by a sharp difference in size between the two species. The "lock and key" hypothesis was developed to describe the isolation of species which arises from structural incompatibility between the sexual or copulatory organs of different species. Some individuals of *Drosophila* may suffer injury or death as a result of interspecific crosses.

When two species are closely enough related to permit gametes to be transferred, gamete

**563**

mortality may prevent hybridization. The acidity of the vagina may be fatal to the entering sperm, or, as in many species of *Drosophila*, an inter-specific "insemination reaction" may cause swelling of the vagina, with consequent death of the sperm. Where these factors are absent, the sperm may still die because it is unable to penetrate the egg membrane.

The second major category of isolation mechanism is the postzygotic type. These include:

1. Zygotic mortality. The fertilized egg may develop irregularly, and may stop developing at any stage between fertilization and birth, due to a number of causes. For example, zygotic mortality may be due to an imbalance between the developmental rates of the parental genotypes, or, among mammals, to a failure to form a placenta.

2. Hybrid inviability. A hybrid may be born, but dies before reaching sexual maturity. Even individuals with somatic hybrid vigor and full sexual capacity may be less successful in courtship than members of the pure species.

3. Hybrid sterility. This is more common in some groups than in others, but it has not, on the whole, been systematically studied except in *Drosophila*. The sterile mule is a well-known example of a vigorously viable, but sterile, hybrid.

4. Selective hybrid elimination. In some cases, the hybrid may be fertile, but it nevertheless suffers selective elimination, because its new combination of characters is in some way less adaptive than those of its parents.

## The modern species concept

There is no definition of a species that is completely satisfactory; an absolute definition is probably unattainable. This is because a central tenet of evolution is that species evolve from races through the gradual accumulation of genetic differences. As long as evolution continues, it

**18-18** *Although some hybrids show increased vigor, many are only marginally viable, and sterility is common. Yet hybridization is an important tool of agriculturalists developing new breeds of stock or strains of plants. This animal is the hybrid offspring of the mating of a wolf and a dog. (Jean and Des Bartlett—Bruce Coleman)*

**18-19** *The elementary evolutionary process explains change at the level of the local population and at all higher levels. The range of genetic variation in the descendant populations exceeds that of the parent population. Selection is the sum total of environmental factors affecting gene frequencies. The overall result of this process is adaptation, an increased efficiency of interaction with the environment.*

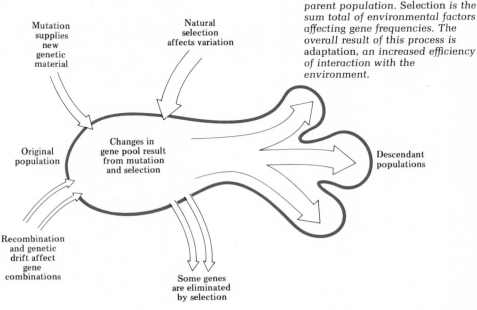

Mutation supplies new genetic material

Natural selection affects variation

Original population

Changes in gene pool result from mutation and selection

Descendant populations

Recombination and genetic drift affect gene combinations

Some genes are eliminated by selection

will always be possible to find forms too distinct to be easily included within one species, but too similar to be regarded as separate species. Such forms must be considered incipient species, and whether they are considered races or species is arbitrary.

"Species" should be regarded as a relative rather than an absolute term. Mayr has succinctly defined species as "groups of interbreeding natural populations that are reproductively isolated from other such groups." Reproductive isolation in the natural state is the central point here. There is little or no gene exchange between contemporaneous species, and this means there is no genetically effective interbreeding. While mules can be produced by crossing horses and donkeys, they are almost wholly sterile, and no gene exchange takes place. Lions and tigers, on the other hand, can produce fertile hybrids in captivity, but their habitats and habits differ so greatly that they do not interbreed in nature.

## Problems in species definition

The biological definition of species was developed to deal with sexually reproducing forms. Asexual forms differ radically from sexual forms, both genetically and evolutionarily, and distinctions regarding interspecific breeding are meaningless when they are applied to asexually reproducing organisms.

In cases where the distinction between two species cannot be made through test breeding, it must be determined experimentally. For example, a culture of several different blue-green algae might be prepared, and then specific nutrients or other environmental factors can be varied, in a series of tests. If the cultures respond differently —one being able to survive without a certain nutrient whereas the other is not—then the two can tentatively be classified as different species, on the basis of this physiological difference.

Fossil species present another problem in species identification. The biological species concept cannot be applied directly to extinct species which may have lived millions of years apart; it cannot even be applied to contemporary species, since we have no way of knowing whether or not gene flow occurred between the two groups. We are fairly certain that at some point the therapsid reptiles gave rise to the early mammals, but it is impossible to say with any certainty exactly when or how this took place. Paleontologists are forced to rely entirely on skeletal criteria, their only data, for their definition of species. Where a gap appears in the fossil record it is assumed that speciation has occurred.

Populations that are at intermediate stages of differentiation pose special problems. The isolation mechanisms which cause species to differ genetically develop very slowly over long periods of time, and only very rarely, if at all, can a definite point of divergence be indicated. Generally, what happens instead is that, with respect to a given character, races vary gradually over a range, in a gradient that is termed a **cline.** Clines arise from the interplay between selection, which tends to make every deme specifically adapted to its immediate environment, and gene flow, which tends to make each deme identical to the others. Because selective factors in the environment, such as temperature, vary along gradients, genetic variation also follows those gradients. The various groups that make up a cline will not, under natural conditions, ever have an opportunity to interbreed, so they are effectively isolated. Yet in laboratory experiments they often prove capable of interbreeding. Clinal variation is extremely common among continental species.

Related to the problem of clinal variation is that of determining speciation in allopatric forms, or populations that occupy different geographical regions. Speciation criteria derived from speciation studies of forms sharing the same area do not apply directly and unequivocally to allopatric speciation.

Why is this difficult question of the species

**18-20** *Are these two animals different species, or do they merely reflect a range of phenotypes found within a single interbreeding population? One way to answer the question is to conduct breeding experiments, to see if the two can produce viable offspring. Yet if they are effectively isolated geographically, by behavioral adaptations, or by breeding time, they are effectively isolated geographically, by behavioral physiologically capable of interbreeding. (Left by Leonard Lee Rue III from National Audubon Society; Right by A. W. Ambler from National Audubon Society)*

concept important? The best answer is to imagine a world without species. Breeding among various living organisms would produce a wide variety of genotypes, some of which would be especially well adapted for particular environments. But since we are imagining a world without specia-

tion mechanisms, the superior gene combination would not survive into the next generation. This, then, is the biological meaning of species: the evolution of a species, with its limited, protected gene pools, imposes limits on genetic diversity and preserves adaptive gene combinations.

## Summary

Although evolution can be simply defined as a change over time, it is more accurate to term it a change in the frequency with which certain genes appear in a given population. This change is irreversible; it refers to a group average rather than to any single individual, and it cannot be attributed to any special drive to progress or complexity.

The concept of the population, or the potentially interbreeding group that is in some way isolated from other similar groups, is central to the modern interpretation of evolution. Each population shares a common gene pool, and the gene pool generally remains relatively stable. Mendel's work with inheritance, which demonstrated that no heredity factor is ever lost or blended or submerged, is one explanation for this stability. As long as mating is random, no mutations occur, and the population is a large one, the gene pool will remain exactly the same from one generation to the next. This is expressed in the Hardy-Weinberg Law; it can be stated mathematically in an equation: $p^2 + 2\,pq + q^2 = 1$. Other stability factors include genetic mechanisms that prevent errors in chromosomal movement and sorting, heterosis, the variation in the environment, and mating preferences for individuals with rare genes.

Yet the gene pool is not always stable; evolution does occur. One reason is mutation. Another is genetic drift, a phenomenon that occurs only in small populations, such as are found on islands. Another is the pressure of natural selection, which permits those individuals with a phenotype slightly better adapted to survive and reproduce, to contribute more offspring to the next generation than other individuals; this gradually brings about a change in gene frequency. Although natural selection is sometimes defined as the survival of the fittest, it is more accurate to call it differential reproduction. Migration also leads to changes in gene frequency.

An important effect of evolution is adaptation, or the change of form, function, or behavior that makes a population better able to survive in its environment. Adaptation may take the form of aposematic or cryptic coloration or form, of physiological characteristics such as the odors of flowers, or of behavioral traits, such as mating displays.

A second effect of evolution is diversification, such as adaptive radiation, which occurs when a group radiates out into new types of habitats. A third effect is the formation of new species, or speciation. For speciation to occur, there must be not only some kind of evolutionary change but also some isolation mechanism at work. Although there are many problems in defining species, the species is an essential evolutionary and ecological unit.

# Bibliography

### References

Cavalli-Sforza, L. L. 1969. "'Genetic Drift' in an Italian Population." *Scientific American.* 221(1): 30–37.

Crow, J. F. and M. Kimura. 1970. *An Introduction to Population Genetics Theory.* Harper & Row, New York.

Eckhardt, R. B. 1972. "Population Genetics and Human Origins." *Scientific American.* 226(1): 94–102.

Jepsen, G. L. and others, eds. 1949. *Genetics, Paleontology and Evolution.* Princeton University Press, Princeton, N. J.

Manwell, C. and C. M. A. Beker. 1970. *Molecular Biology and the Origin of Species.* University of Washington Press, Seattle.

Margulis, L. 1971. "Symbiosis and Evolution." *Scientific American.* 225(2): 48–57.

Mayr, E. 1961. "Cause and Effect in Biology." *Science.* 134: 1501–1506.

Scriven, M. 1959. "Explanation and Prediction in Evolutionary Theory." *Science.* 130: 477–482.

Simpson, G. G. 1965. *Tempo and Mode in Evolution.* Hafner, New York.

### Suggestions for further reading

Dobzhansky, T. 1970. *Mankind Evolving.* Bantam Books, New York.
Discusses in lucid and comprehensive detail the interplay between genetic endowment and environmental factors.

Dubos, R. 1968. *Man Adapting.* Yale University Press, New Haven, Conn.
Drawing on the fields of astronomy, chemistry, geology, and anatomy, this noted microbiologist focuses on the problems of the individual organism in responding adaptively to environmental challenges.

Mayr, E. 1970. *Populations, Species, and Evolution.* Harvard University Press, Cambridge, Mass.
An abridged but thorough inquiry into the nature of species, their population structure, their biological interactions, the multiplication of species, and their role in evolution.

Savage, J. M. 1969. *Evolution.* Holt, Rinehart & Winston, New York.
Objectively surveys various theories proposed to explain the fundamental evolutionary process. Emphasizes the development of isolating mechanisms and their role in evolutionary divergence.

Simpson, G. G. 1971. *The Meaning of Evolution.* Bantam Books, New York.
Examines the concept of evolution and explores from a carefully reasoned scientific view its controversial and cosmic aspects.

## Books of related interest

Aldiis, B. 1961. *The Long Afternoon of Earth.* New American Library, New York.
A nightmarish vision of an earth where plants have destroyed the animal kingdom and of the few humans who struggle for existence.

Clarke, A. C. 1953. *Childhood's End.* Ballantine, New York.
Reflecting on the next step in man's evolution, this noted science fiction writer examines the unexpected tragedy that comes with transformation and perfection.

Heinlein, R. A. 1948. *Beyond This Horizon.* New American Library, New York.
A novel depicting the deliberate attempts of men to destroy a nearly perfected world through modern science.

Huxley, A. 1962. *Brave New World.* Bantam Books, New York.
The classic apocalyptic view of man's evolutionary future.

Simpson, G. G. 1964. *This View of Life.* Harcourt Brace Jovanovich, New York.
A unified collection of essays written by this noted evolutionist on the broader aspects of his explorations, including the problems of objectivity in science, purpose in the universe, and speculation on man's place in the natural order.

# 19
# Behavior

Perhaps man's greatest challenge both now and for the future—as even the most casual observation of human affairs would suggest—lies in obtaining a very real understanding of behavior. Is this a challenge for biology? There is some tendency to conceive of the mind as separate from the body, of behavior as far removed from biology. Yet biology is the study of life; behavior is a fundamental characteristic of animal life, and certainly very much involved in behavior are the patterns of biochemical and molecular events set in motion as an integral part of the response to any stimulus.

Consider, as a simple example, a person who comes from a farm or small community to visit a big city. He moves about in crowded public conveyances, shops in crowded stores, finds himself bustled and bustling, and returns home with great relief and a fervent dislike of the big city. All psychological reaction? But he has reacted biologically: Stressed by the unusual (for him) jostling, sights, sounds, and tempo, his adrenal glands have reacted, pouring out excessive amounts of hormones. There are internal repercussions, not pleasant, from the outpouring. He may intensely dislike the big city not so much because he fears or dislikes crowds, noise and other big city attributes, but more because he dislikes his biological response to them.

## Innate and learned behavior

Every form of behavior, from simple to most complex, is a reaction to a stimulus, which is usually a change in some aspect of the environment. Some behavior is innate, or genetically inherited; some is learned, developing on the basis of experience.

Distinguishing the innate from the learned has been one of the major difficulties faced by behavioral scientists. Early in the twentieth century, a fierce nature-versus-nurture controversy raged. Some psychologists went to an ex-

**571**

treme, dismissing inheritance as having very little, if anything at all, to do with behavior; all was a matter of learning. To biologists, who had studied the nervous and sensory physiology of various organisms and were aware that the potential for their development was inherited, any such dismissal of inheritance had to be looked upon as ridiculous. It seemed clear to them that animals could display only those behavior patterns for which they had suitable nerve pathways and response apparatus; learned behavior patterns were not possible if appropriate pathways were not present. But some biologists also took an extreme position, assigning to inheritance virtually the whole role of instigating in fine detail even the most sophisticated and complex behavior.

Today, it is no longer a matter of "either/or." Credit is accorded both inheritance and learning —and interactions between the two. Inheritance is considered to set a broad pattern within which specific individual behavior may occur. Within those limits, the learning experiences of the individual may determine what the behavior shall be.

In some cases, inheritance may play the prime and virtually only role. The reflex response to some stimulus may be determined solely by the available inherited neural pathways and effectors; you do not need to learn to pull your hand away from a hot stove. But in other cases, where there are many available neural pathways and effectors, the limits set by inheritance are very broad. Learning then determines what the behavior will be; for example, you learn to rest your hand on the counter beside the stove, as a result of your experience of burning your hand.

One well-documented instance of interaction between inheritance and learning is the pecking that takes place almost immediately after baby chicks hatch. The pecking movements— downward head thrust, upward lift, and upward pointing of the beak while swallowing—are inherited. A newly hatched chick pecks at every-

19-1 *Intense crowding is a phenomenon typical of modern urban life. The discomfort many of us feel may be due less to the crowding itself than to our innate biological response to crowding. Other phenomena, such as the speed, noise, and complexity of life in the industrial age, need to be studied in terms of their implications for the biological nature of man. (John Hendry, Jr. from National Audubon Society)*

thing, from small stones to other chicks' feet. Gradually, however, it learns to peck only at grains of corn and other food.

Another interaction example is provided by the European chaffinch which, when raised in isolation, is able to produce a normal chaffinch call but not the typical chaffinch song. Therefore it would appear that the call must be inherited and the song not. Yet when, at about age six months, isolated chaffinches are exposed to a recording of the chaffinch song, they quickly learn to sing it; but they do not learn to sing quite similar songs of other species of birds when these are presented to them on recordings. Chaffinches, then, must learn their song from other chaffinches, but they have an inheritance factor which usually determines that they will learn only the song of their own species. Thus their song can be regarded as both inherited and learned.

Studying inheritance is relatively simple when it comes to such characteristics as flower color or seed shape. Studying inherited behavioral traits is somewhat more difficult. The reason is that a behavioral trait is probably not inherited through a single pair of alleles, but in a pattern of multiple alleles and multiple genes. Another problem lies in separating the effects of inheritance from those of learning. In some instances, this can be done by isolating newborn individuals from others of their species; but of course isolation cannot prevent the individual from learning from his own experience. Yet despite these difficulties, some information has been obtained through carefully controlled behavioral experiments.

When, in captivity, two kinds of lovebirds—peach-faced and Fischer's—are offered sheets of

paper as nesting material, both will proceed to tear the paper into strips. The peach-faced lovebird carries the strips to its nest tucked among its feathers; the Fischer's lovebird carries them in its beak. Peach-faced and Fischer's lovebirds can be mated and their eggs will hatch. And the behavior of the hybrid offspring is notably confused. They will tear paper into strips, a behavior pattern found in both their parents. But they are unable successfully to tuck strips among their feathers; almost invariably the strips drop out. Occasionally, they are able to carry the strips in their beaks, and, after about three years of effort, this becomes their prime mode of transport. Even so, in their confusion, when using their beaks for actual carrying, they still fluff their feathers for the tucking they no longer employ. The clash between two patterns of inherited behavior is never totally ended.

## Innate behavior

Where behavior is largely a matter of some external stimulus triggering a genetically determined response, it is said to be **innate.** Innate behavior is stereotyped; that is, it is almost exactly identical in every member of a population. It includes taxic responses, or direct orientations in response to stimuli, reflexes, and instincts.

### Taxes

A very simple form of adaptive behavior is the taxis, a direct orientation of an organism in respect to stimuli; a taxis is analogous to a plant tropism. For example, *Euglena* will move toward weak light but away from strong sunlight. *Vorticella* will rapidly contract its stalk when touched. *Paramecium*, swimming in a spiral fashion, will halt abruptly and back up when it encounters a droplet of lower pH. Both ants and bees utilize

*19-2 There is a definite difference between the "educated" song of wild chaffinches who have listened to adults of their own species and the "uneducated" song of chaffinches raised in isolation from their own kind. The song of the laboratory chaffinch is a monotonous pattern which retains a recognizable chaffinch quality but is not nearly so elaborate as a normal chaffinch song. It appears, then, that while some singing ability is innate, much of it is learned. (Eric Hosking from National Audubon Society)*

taxic responses in returning to the hive or nest; they continually orient themselves of a constant angle to the source of light; this response is as reliable as a compass.

An early investigator of taxic responses was Jacques Loeb. He found that taxes can be observed even in protozoa. An amoeba will draw back from —i.e., its pseudopodia will flow away from— certain stimuli, such as extreme pH or electrical current. This is an example of a negative taxis.

An illustration of the adaptive value of a taxis is the orientation of the grayling butterfly which always flies toward the sun. Thus its movement is unidirectional, putting as much distance as possible between it and any pursuing predator. If the butterfly is blinded in one eye, it will fly around in circles, making it much more vulnerable. This movement toward the sun is a positive taxis.

## Reflexes

Reflex actions, like taxes, are fixed, stereotyped responses to stimuli. They depend upon the reflex arc; here the connection between behavior and physiology is very clear.

The constriction of the pupils of the eyes in response to intense light and the familiar knee-jerk reflex are examples of reflexes in man. The reflex arc provides a short cut, avoiding the delay that would be involved if impulses first had to travel to the brain and the brain in response then had to send out impulses to achieve action. Many reflexes are protective. And although they may seem like instinctive responses, reflexes operate only so long as stimuli are present, whereas instinctive behavior, once set in motion, continues even if the originating stimuli have disappeared. Another important difference is that reflex behavior affects only a portion of the organism, whereas instinctive behavior involves the entire animal.

Reflexes can be classified into two general

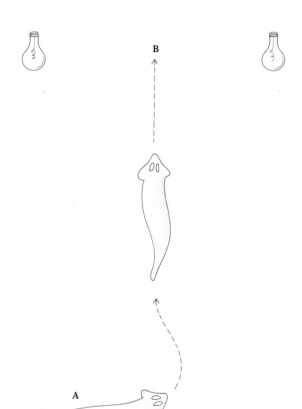

**19-3** *Phototaxis causes a planarian to move toward light. When illuminated by two lights, it then moves toward a point midway between them.*

**19-4** *Oriented movement toward light stimuli is called phototaxis. Grayling butterflies, for example, fly directly toward the sun when pursued by predators, orienting themselves so that each eye receives equal illumination. The bright light partially blinds the pursuer and gives the butterfly a better chance to escape. (Hugh Spencer from National Audubon Society)*

19-5 *Foot-stamping, bowing, and strutting distinguish the courtship behavior of the male sage grouse. This fixed action pattern involves a number of inborn or hereditary movement forms that are specific to the male sage grouse and are as constant as his anatomical characteristics. The unconscious goal of such display behavior is not survival but discharge of the fixed action pattern, perhaps stimulated by the sight of a female grouse. (Joe Van Wormer from National Audubon Society)*

19-6 *The submissive pose is one form of appeasement display common among hummingbirds as well as other species. It is distinguished by the lowered or indrawn head, the crouched position, the fluffing or ruffling of feathers. Such displays are especially characteristic of encounters during the reproductive period. They serve to prevent attack without provoking escape and to reduce the strength of the opponent's attack drive. (Courtesy of the American Museum of Natural History)*

types. Phasic reflexes, such as the knee-jerk, allow quick, brief adjustments. Tonic reflexes permit slow long-lasting adjustments, such as those that maintain posture and equilibrium.

There are complex patterns of reflexes such as those involved in the highly coordinated flexion and extension required for locomotion. For the latter, many segments of the spinal cord and the midbrain as well are involved. Once it was believed that even the most complex behavior, including thought, might conceivably be nothing but sophisticated combinations of innate, conditioned, and acquired reflexes. It is now clear that much more is involved.

## Instinct

The most complex type of innate behavior is instinctive behavior. Most feeding and mating behavior is instinctive, but it should be stressed that these instinctive "urges" are really complex patterns built up of sequences of actions, and we must be careful not to think of such behavior in human terms of purposefulness. For example, we should not conclude that a male bird sings the song of his species because he wants to mate,

even though mating is likely to be the final outcome of the chain of events set in motion by his singing. What really happens is a complicated behavioral process, in which hormonal changes play an important part. His song is likely to be the competitive response to the stimulus of hearing another male sing. The sight of a female during the time he is singing stimulates the next response, which may be some sort of preening or display. There may be a number of courtship activities such as nestbuilding, feeding of the mate, or mock battles with other males or nesting pairs. Each of these actions is undertaken for its own sake; in turn, the actions evoke new behavior in other individuals, and perhaps physiological changes in the courting male as well, that stimulate the next response in the sequence. The very value of such instinctive behavior patterns is that, although the behavior is goal oriented, the individual need not have any sense of purpose.

**Behavior releasers.** No organism can possibly respond to all the stimuli to which it is exposed. It must be selective. It must respond only to a limited number of stimuli, those of significance for it. And such stimuli which are of significance, which are particularly effective in producing behavioral responses, have been termed "releasing stimuli."

As a corollary to the concept of releasing stimuli, ethologists (the biologists and psychologists who study behavior) have proposed an innate releasing mechanism concept. An example can be seen in a favorite subject for study by ethologists: a brightly colored little fresh-water fish, the three-spine stickleback. In its reproductive behavior, a male stickleback goes through a series of instinctive acts—migratory, territorial, fighting, nesting, mating, and parental. Under the influence of increases in gonadal hormones brought on by the increasing daylight of springtime, a male stickleback migrates, moving away from the school in which it ordinarily lives.

575

Guided by temperature, the fish moves to the shallow warm water of a stream or estuary and, upon sighting green vegetation, selects his territory and builds his nest, prepared to defend nest and territory against any other male that might intrude. Now, too, his color changes; his underbelly becomes red, his back a whitish blue. When he sees a cruising school of egg-swollen females, he goes into a zigzag dance; in response, one of the females follows him. Once she is inside the nest, the male prods her tail with his snout; she spawns and then swims away; and the male enters the nest and fertilizes the eggs.

This is complex behavior—and yet a male stickleback, raised in total isolation, is able to perform it precisely. Ethologists suggest that each species has inborn fixed action patterns that are as specific as its physical characteristics. Until a releasing stimulus activates the innate releasing mechanism, the action patterns—the instinctive acts—are held in check by a neural inhibiting mechanism. A releaser may be light, water temperature, vegetation, scent or some aspect or activity of another animal, ranging from the swollen abdomen of the egg-carrying female to the red belly of an intruding male, or even the assumption by another male of a threatening, nose-downward posture.

Studies have demonstrated how specific releasing stimuli can be. Sticklebacks have been aroused to defend their territory against dummy fish—even dummy fish not shaped like sticklebacks—as long as a red belly was painted on. Experiments have shown that a stickleback intruder, dummy or real, provokes a much more vigorous attack if introduced in nose-downward instead of horizontal position.

Similar studies have been carried out with other animals. Baby thrushes, like many other young birds, beg food from their parents by thrusting their heads upward and gaping. Blind when first hatched, the nestlings do their gaping in response to a light touch on the nest or a puff of air. At the age of about one week, a visual

stimulus can release the gaping response. At 10 days, the gaping is no longer straight up but directed toward the head of the parent. And investigators have been able to establish that if the gaping response is to be directed toward the head of a "parent," the head must be about one-third the size of the body.

Herring gull nestlings, instead of gaping, beg food by pecking at a parental bill. Investigators have shown that nestlings will beg from a dummy gull head, but the strength of the response will depend upon the presence of a spot on the dummy's beak and its color. Apparently, the whole configuration of a stimulus or the pattern of stimulation has much to do with the release of the innate response.

**19-7** *Herring gull chicks elicit feeding behavior from their parents by pecking at the red patch on its parent's bill. The parent, in turn, regurgitates food for the hungry chicks. The red patch is one type of "relational" stimulus. The chicks respond not simply to the color of the patch, but to the contrast between the patch and bill color and to the shape of the bill as well. (Gordon Smith from National Audubon Society)*

The concept of the releaser is not without its critics. Some ethologists believe it may be an oversimplification; they also suggest that experimental efforts to determine the nature of the releaser could, in fact, alter the releaser. Some psychologists tend to believe that all behavior involves a mix of inherited capability, experience, and even normal maturation. They suggest that even the almost immediate pecking of a chick upon hatching need not be instinctive behavior. While still in the egg, a chick head rests against the chest and is moved up and down as the heart beats. Not long afterward, the embryo actively begins to nod the head and to open and close the beak at the same time, and by the time a chick is hatched, it pecks—as the result not of instinct but of development. Could all seemingly inherited or instinctive behavior be of much the same origin?

## Learned behavior

Ethologists define the process of learning as a permanent change in behavioral patterns that comes from some individual experience. Although we tend to associate learning with relatively advanced development, learned behavior is by no means limited to vertebrates. Planarian flatworms have been conditioned to associate light with electric shocks; cockroaches and ants can run mazes; and the octopus can learn to uncork bottles to get at its food. Many different types of learning have been demonstrated in animals.

## Imprinting

A highly specialized and limited kind of learning is called imprinting; it was discovered almost 40 years ago by Konrad Lorenz, the Austrian zoologist. Lorenz, at the time, was studying species, such as greylag geese and partridges, whose young locomote on their own very soon after hatching. He noted that a specific attachment of young to parent formed very rapidly after hatching. And his experiments demonstrated that the attachment depends upon the parent's being the first moving object to which the young were exposed. When Lorenz took young animals only a few hours old and had them follow him

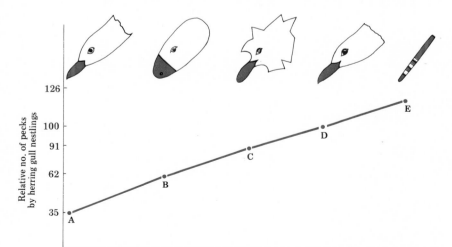

**19-8** *The models of mature herring gull heads were used to elicit pecking (begging) responses from nestlings. The red bill patch (dark color) proved to have the greatest influence in eliciting responses. Some nestlings pecked at the point indicated by the arrow at C. The "superoptimal" bill with greatest contrast, which elicited the strongest response (E), has not evolved because it would be a very inefficient tool for hunting fish.*

before seeing their own mother, they would thereafter look upon him as their parent.

Since then, many investigators have studied imprinting, mainly in birds, particularly ducks and geese. Other birds—including swans, chickens, and turkeys—similarly have been found to follow the first moving object they see after being hatched. Instances of imprinting also have been reported in insects, fish, and some mammals including sheep, deer, and buffalo. The survival value of the following response is that it serves to keep the young close for protection.

Imprinting can occur in only a limited period. If a bird is not imprinted within a day or so after hatching, it may not imprint at all. It is also possible that the lesson of imprinting requires some kind of periodic reinforcement. In recent experiments, it was found that abnormal imprinting—to a parent of another species or an inanimate object—often wears off, perhaps because there is no subsequent reinforcement.

Imprinting may influence later mate selection. Even long before Lorenz coined the term, one investigator had observed that in order to cross two different species of wild pigeons it was first necessary to rear the young of one species under the adults of the other. Upon reaching maturity, the birds thus reared preferred mates of the same species as their foster parents. In more recent studies, ducks and geese have been found to become sexually fixated on objects or on members of other species as the result of imprinting. It was suggested that the reason the female panda in the London zoo failed to mate with the male panda from Moscow was that she had accidentally been imprinted by her human keeper and wanted him as a mate.

## Habituation

Perhaps the simplest type of learning is habituation. With repeated exposure to a stimulus, the animal gradually may decrease its natural

original response; in time the response may disappear altogether. One example is the gradual loss of fear by an animal in a laboratory. In effect, habituation is a dropping of responses that serve no useful purpose—in contrast to much of the rest of learning which involves strengthening of responses that are significant.

## Associative learning

Associative learning is another simple type. It is exemplified by the excited pawing at a door of a dog, ready to go out, when it is shown a leash. The conditioned reflex, made famous by the experiments of Ivan Pavlov, is a type of associative learning. When a dog sees food, it produces saliva. Pavlov found that if he regularly rang a bell just prior to letting a dog see food, after a time the dog would begin to salivate as soon as the bell rang. He also found that while a dog would go on salivating at the bell sound so long as food was duly presented as usual, the salivation would gradually diminish and eventually disappear altogether if the bell was rung regularly

**19-9** *Hawks as well as other birds learn to associate the mother with nourishment. A short time after hatching they respond to the weight of the mother bird alighting on the side of the nest by clamoring for food and turning their heads in the direction of the weight. (Lewis W. Walker from National Audubon Society)*

but the food was withheld. Clearly, reinforcement is important in maintaining a conditioned reflex which can be defined as a response, after training, to a stimulus which previously elicited no response.

Animals also learn by trial and error and by imitation. A newly hatched chick, for example, will peck at virtually anything but by trial and error over a period of time learns to be selective and to do most or all of its pecking at objects it can eat. Imitative learning is exemplified by the chaffinch which, when raised in isolation, can sing only a simple song but when raised with and able to hear other chaffinches sing, imitates them and produces a more elaborate song.

**19-10** *White leghorn chicks have been imprinted by a bantam foster mother. Although this abnormal imprinting may wear off if there is no subsequent reinforcement, the chicks will not be imprinted again later, even to their own mother. This imprinting may influence later mate selection. The leghorns may prefer bantam mates. (Richard Guppy from National Audubon Society)*

## Learning as a physiological process

Learning takes many forms but invariably produces a lasting, sometimes permanent, change in behavior. And changed behavior, from the biological viewpoint, must be related to physiological change, to some alteration in nervous system functioning and perhaps even in nervous system structure. Actually, the idea that experience can produce changes in brain anatomy is not new. Almost 200 years ago, in the 1780s, an Italian anatomist named Malacarne trained one member of each of a pair of dogs from the same litter and one each of a pair of parrots, goldfinches, and blackbirds from the same clutch of eggs, leaving the other member untrained. Upon sacrificing the animals and examining their brains, he found more folds in the cerebellums of the trained than in those of the untrained animals.

But this did not settle the problem of whether or not experience could affect brain anatomy. In the 1950s, armed with new biochemical tools and techniques, some investigators went back to the problem. Typically, they might work with laboratory rats, all of the same genetic strain. At weaning, sets of three males would be taken from each litter and assigned at random. One would go into a standard laboratory colony cage along with a few other rats; the cage was adequate in size, and food and water were always available. A second rat would go into an enriched environment—a large cage occupied by several other rats and furnished with varied objects for play, with a new set of playthings placed in the cage daily. A third rat would go into an impoverished environment, living alone in a cage.

By 1964, investigators had determined through such studies that rats that had spent four to 10 weeks in enriched environments had a greater weight and thickness of cortex than those that had spent the same time in impoverished environments. They also had a greater activity of acetylcholinesterase, an enzyme that rapidly breaks down acetylcholine, the chemical that serves as a transmitter across the synapse between nerve cells. If nerve impulses are to follow each other rapidly, any excess transmitter must be broken down quickly; so the greater quantity of acetylcholinesterase had some significance for rapid nervous connections.

The brain differences produced by an enriched environment are not large, but they appear to be genuine. Yet there was the possibility that the brain changes could be the result not of learning, but of differences in handling, stress, or even change in the rate of maturation. Mark R. Rosenzweig, Edward L. Bennet, and Marian Cleeves, who had carried out many of the original experiments, set to work to look into that possibility.

Could it be that enriched experience simply speeds up maturation? It turned out not to be so. Cortical thickness of control group rats reaches a maximum 25 days after birth, then decreases slightly with age. Yet enriched experience was found to increase cortical thickness even in year-old rats. To many investigators, it now seems that there can be no doubt that experience can produce change in brain anatomy and brain chemistry.

At the University of California, Albert Globus has compared the number of dendritic spines in brain sections of rats exposed to enriched and impoverished environments. Most of the synaptic contacts between nerve cells in the cortex are made through the branchlike dendrites or dendritic spines (small projections from the dendrites). Globus has found more spines in rats exposed to enriched environments. Electron micrographs have also demonstrated that enriched-environment rats have synaptic junctions which are 50 percent larger in cross section.

Investigators have found increased synaptic contact in enriched-experience rats. Some have also found that increased synapse size is associated with a decreased number of synapses, whereas decreased synapse size is associated with an increased number. It seems that learning or memory may be encoded in the brain either by addition of contacts between nerve cells or by removal of contacts, and that both processes may go on simultaneously; an analogy can be drawn to the rewiring of electrical circuits.

If such research ultimately leads to knowledge of how memories are stored in the brain, it could of course help in studies of how to establish conditions that favor learning and memory. But investigators caution that much yet remains to be learned, and that it is difficult to extrapolate from the rat to man.

## Learning and memory

Many attempts have been made to establish the nature of memory. In the first decade of this century, a book was published comparing heredity to memory. The author, Richard Semon, introduced the term "engram" for the physiological correlate of memory. He speculated that both nerve cells and germs cells might have the ability to preserve and transmit engrams, and that the engrams are produced by influences coming from the environment and might constitute a racial memory. Some years ago, during brain surgery, the discovery was made that when some areas of the human cortex were stimulated electrically, past events, even though long forgotten, could be recalled vividly. It appeared that there might, indeed, be engrams—memory items—stored in specific areas of the brain.

Karl S. Lashley at Harvard continued the investigation of memory by training rats and other animals to solve various problems. He then tried removing various portions of the brains, hoping to be able to remove specific engrams. But he could never find any brain speck which he could remove and leave the animal able to remember one thing and unable to remember another. The best he could do was to remove one-fourth or one-third or one-half of a cortex, whereupon the animal no longer remembered one-fourth, or one-third or one-half of what it had learned. Lashley then conceived a theory that memory might be a matter of reverberating impulses going through brain circuits. Yet when he made many cuts through the cortices of rats in an effort to sever connections between various areas and thus break up the circuits, memory still persisted. The theory seemed untenable.

More recent studies by other investigators, however, provide some support for the Lashley hypothesis. If an octopus is placed in a tank with stones at one end, it quickly makes its home among the stones. When a crab is offered at the other end, the octopus moves out from the stones and seizes and devours the crab. If whenever a crab is offered, a white plastic square is moved about near the crab, the octopus will soon come out from the stones at the sight of the square alone. If another shape is introduced into the tank but this one is electrified and produces a mild shock when touched, the ocotpus soon learns to

stay at home when the second shape is introduced but will come out eagerly when the white square is displayed.

Once an octopus has been trained in such fashion, part of its brain can be removed to determine if the removal has any effect on memory. When the vertical lobe at the top of the brain is removed, the octopus no longer remembers what it has learned visually. Moreover, with its vertical lobe gone, an octopus no longer can learn from the usual type of training program in which a training trial is repeated every two hours; yet it can learn when the trials are repeated every five minutes. The vertical lobe appears to be involved in long-term memory; if it is removed, the long-term memory engrams also seem to disappear.

Another approach to the understanding of memory was taken in the 1950s by James V. McConnell. For his subject, McConnell took the planarian. As the worms were gliding along through a trough of water, McConnell turned a light on, then two seconds later administered an electric shock that caused the worms to contract. After 150 such training sessions, the worms contracted to the light alone over 90 percent of the time. After this, the worms were cut in half and allowed four weeks for regeneration, whereupon it turned out that both regenerated head and tail sections had a high degree of retention of what had been learned earlier. Moreover, after the regenerated worms were cut again and a second period regeneration followed, there was still retention of the conditioned response to the light alone. Apparently learning is not confined to one part of the planarian's nervous system but has an effect throughout the system.

Some years after McConnell began his work, Holger Hyden, a Swedish biologist, was studying neurons in rat brains. He made the interesting discovery that they contained large amounts of RNA. Why should brain cells, which do not divide and do not manufacture any sizable quantities of protein, have large amounts of RNA? In further studies with rats, Hyden found that as

**19-11** *Octopuses have been conditioned to respond to white plastic squares associated with crabs and to avoid shapes that deliver a mild shock when touched. They failed, however, to learn the differences between other shapes, such as an inverted V and an M. The handicap was not a failure of memory or of ability to make visual interpretations, but rather a limitation of the eye. The octopus scans an object vertically and horizontally to measure its dimensions, rather than its precise shape. (Ron Church—Photo Researchers)*

they were trained for various tasks, the RNA content of their neurons increased. It appeared that possibly memory might be encoded in RNA.

Was this the reason for memory retention in planarian tails? McConnell recognized that it would be difficult to separate out RNA from one planarian and inject it, totally unchanged, into another. But, since hungry planarians are not averse to being cannibalistic, injections were not necessary. He fed chopped-up sections of trained worms to untrained ones and, for comparison, chopped-up sections of untrained worms to other untrained worms. McConnell was able to report that worms fed on chopped-up educated worms were smarter about showing the conditioned response than those fed chopped-up uneducated worms.

Is memory, then, encoded in RNA? Is it this which is transferred when planarians feed on trained planarians? The question is still an open one. As of now, there are thus two theories of memory: one, the electric circuit; the other, the molecular code.

## Biological rhythms and behavior

Internal rhythms—clocklike regulatory mechanisms—have been found to operate in many animals and in man. Bees can be trained to feed at certain times with great accuracy. Many people wake in the morning at predetermined hours. Even plants have fixed hours at which they open their flowers.

Much periodic behavior appears to be connected with natural rhythmic events, such as the alternation of day and night, the cycle of seasons, and the ebb and flow of tides. Although it is true that such events produce environmental changes—for example, the gray dusk between day and night, the increasing length of day as winter fades and spring comes on—animal behavior often anticipates such changes in the environ-

**19-12** *In plants, growth movements brought about by external stimuli, such as light or gravity, are called tropisms. These flower petals close up as a result of physiochemical changes produced by unequal stimulation of the petal surfaces. Taxes, oriented movements in animals, bear some resemblance to plant tropisms. However, in animals the underlying mechanisms depend upon nervous reflexes rather than hormonally controlled differential growth patterns. (Eschscholtzia Californica)*

ment on the basis of internal changes which may be independent of the environment. Nocturnal animals, for example, including rats and cockroaches, continue to show regular periods of "nightly" activity even when raised under conditions excluding all environmental indications of day and night.

Biological rhythm research is young but rapidly expanding. Investigators have grown rat heart cells in culture and have found regular daily fluctuations in contraction rate. Many biological rhythms in animals are circadian geared to roughly 24-hour periods, but some are longer, some much shorter. The lugworm, for example, has a short-period rhythm. The worm lives in a burrow in muddy sand by the seashore. It sucks in sand at its head end and expels a twist of sand at the other end. In laboratory experiments, it has been shown that every 40 minutes or so, with almost consistent regularity, a worm moves to the tail end of a burrow and pushes out its feces, then returns to the bottom of the burrow to suck in more sand. In the laboratory there are no tides to exert an influence and periods of light and darkness can be controlled. Yet the regular 40-minute cycle goes on, not influenced by any external rhythmical event but rather determined by an endogenous rhythm, one originating within the animal.

Rhythmic fluctuations which affect metabolic activities may, in so doing, determine the degree of sensitivity of animals to various stressful situations. A poison capable of killing most mice at one time of day has been found to kill less than five percent at another time of day. Studies in man have shown that body temperature changes by as much as 2° F. from a low usually at about 4 a.m. to a high at about 4 p.m. It has also been found that there are daily cycles in heartbeat rate, the number of white corpuscles circulating in the blood, brain wave activity, blood pressure, excretion of urinary substances, even in the division of cells. One graphic demonstration of clocks in man is the dulling of mental

acuity in travelers who jet across a succession of time zones. Studies made for the Federal Aviation Agency show that it is the inability of the body to readjust its internal rhythms quickly which generates both mental and physical fatigue.

Internal clocks and their variations between people are now held to explain why some are morning people, "larks," who arise bright and cheerful at early hours, while others are "night hawks" who wake surly but then, later in the day, are at their best while the larks become less vigorous. It has even been suggested, in some seriousness, by Robert Townsend in his book, *Up the Organization*, that "anyone who makes over $150 a week should be allowed to set up his own office hours. If you work better from noon to midnight and your job makes these hours appropriate, you should be able to do it. And if you have a secretary, pick one with the same general metabolism." Some physicians now believe it may be advisable to schedule surgery and even the administration of medication to take advantage of optimum conditions determined by patients' internal rhythms.

Internal rhythms also seem to serve as clocks enabling animals to measure the passage of time. A time sense has been demonstrated in honeybees—important to them since some flowers provide nectar only at certain times of day. Many birds, insects, and crustaceans make long journeys and employ the sun as an aid to navigation; in doing so, they use internal rhythms to allow for the changing position of the sun in the sky.

## Behavior as adaptation to the social environment

Whenever one organism reacts in some way to the presence of another, it may be said to be displaying social behavior. Some social behavior is to be found at almost all levels of the animal kingdom. Virtually all multicellular organisms display social behavior when they mate.

One well-studied form of social behavior in animals is the mating ceremony. Sexual in origin, it also serves to hold flocks together and, by helping to prevent interspecific breeding, maintains species intact. *Drosophila* males, for example, perform elaborate courtship rituals that may last many hours. Although two *Drosophila* species, *pseudoobscura* and *persimilis*, often breed and feed in the same areas, hybrids are rare. Although the two species are physically capable of mating with each other, *Drosophila* females can detect the difference and refuse a male of another species, even though their courtship rituals appear quite similar to human eyes.

**19-13** *The boundaries of individuals are usually set by aggressive encounters between male birds. Once established, the limits are recognized. Fulmars, for example, may bluff or threaten each other at a boundary. They may chase each other for a short distance, but once the chased bird is safely back in his own territory, he in turn becomes the aggressor and drives the other bird back. (Eric Hosking from National Audubon Society)*

Social behavior among animals is displayed in offspring rearing and the protection of the young, and in the establishment of family groups. Some animals—fish, birds, and subprimate mammals—show social behavior, too, in migration, territorial defense, fighting, and in schooling, flocking, and herding.

Insects have remarkably complex social organizations. Examples can be seen in ants with their specialized classes of individuals, their exclusion of outsiders, their milking of aphids, and their cultivation of fungus gardens and in bees with their even more complex classes of individuals and hive organization.

Social living has its advantages for survival. A peregrine will not hesitate to dive on a lone starling but will not attack a flock (possibly because it might injure its wings in the dive among them). Large numbers of animals may be conspicuous to enemies, but there is protection in living in groups. As with humans, animals living in groups generally mature earlier than those living alone.

## Animal communication

In ethology, a branch of study called **zooseinotics,** the science of signs, has been emerging as a dominant theme. Much research is going into animal communication.

It is known that some fishes produce electric fields, and it seems likely that some of the feebler impulses are used for signaling, particularly in species where the frequencies and patterns of discharge are distinct. Not only energy but matter may serve as a message conductor as in honeybees, ants, and termites, where processes of food and water exchange transport information as well as calories. Communication by food exchange is referred to as trophallaxis.

It has been noted that "phatic communion" —a kind of speech in which social bonds are created by mere exchange—is the first verbal

**19-14** *Animals of different species who do not share the same communication signals must resort to blatant shows to make themselves understood. Flying in the face of the established order of dominance, this belligerent mockingbird conveys a warning threat to the great horned owl. The owl may be startled enough by this outrageous and unconventional behavior to withdraw. (G. Ronald Austing from National Audubon Society)*

function acquired by human infants, and it commonly predominates in communicative acts in other species. One example is found in certain shrikes of the genus *Lanius* where each pair tends to develop an individual repertoire of duet patterns to serve for mutual recognition and maintaining contact when visual display is ineffective. Investigators have found that in the mountain gorilla, vocalizations are used in dense vegetation to draw attention to the animal emitting them, to notify others of a special emotional state, and to alert them to watch for gestures which communicate further information. When the distance between individuals decreases, postures and gestures, rather than vocalizations, coordinate behavior within the troop. When dis-

tance is further diminished, the visual language is replaced by a tactile one, such as can be observed between mother gorilla and infant.

It would appear that most, if not all, animals are social beings making alliances which presuppose some measure of communication. Thomas A. Sebeck of the Indiana University Research Center in Anthropology, Folklore and Linguistics has pointed out that protozoa interchange signals. He further suggests that the development of complex multicellular organisms is due to the fact that the component cells can influence one another. Sebeck says:

> Metazoa assemble in various kinds of alliances. It is essential for them to come together to form a temporary tandem for mating, thus enabling their species to continue. When the sexual partners remain together until the offspring appear, we may speak of a family community, that is, a group whose members stay together but become differentiated—a type of process realized in most dramatic fashion by insect colonies. On the other hand, members of a species not necessarily stemming from one mother may come together and become integrated into "interest communites," joined together, for instance, for mutual protection—like a school of porpoises. In all such unions—whether transient or persistent, closed or open, simple or complex—creatures of the same species must locate and identify each other; moreover, they must give information as to what niche they occupy in territory as well as status in the social hierarchy. . . .

## Social dominance

Long before the first human king was crowned, the animal world had its social hierarchies. With remarkable frequency, animal societies show stratified patterns, divisions of members into dominant leaders and subordinates. Such organization, it appears, contributes to the stability of populations and species. Monkeys, for example, organize in bands, with a male "chief" who dominates and has a harem of females. He is in command until challenged and defeated by another male, at which point he may be ostracized and forced to live alone.

Social dominance is the arrangement of a group of animals of the same species into a hierarchy influenced generally by strength, size, weight, health, maturity, previous fighting experience, or hormonal level. The pecking order of birds—in which dominant birds peck more submissive birds—is one form of social dominance which has been studied at length. The pecking order is set up in flocks of hens kept together for a period of time. Usually, one hen in a flock dominates all others and can peck any of them without having them peck back. Another hen dominates and can peck all but the first hen, and so on through the flock. Hens high in the pecking order are readily recognizable; they have a confident air and well-groomed look—understandably enough, since they are first at food, roost, and nest boxes. Hens low in the pecking order show it in their timid moping at the edges of the flock. Often, there are fierce battles while a pecking order is being established but afterwards harmony prevails. Cocks, too, have their own pecking order, but usually one sex does not peck the other.

The general order in animal populations is that all males dominate the females; however, within each sex there is a separate hierarchy. Thus in chickens there is a male-chauvinist hierarchy in which the lowest-ranking male member precedes the highest-ranking female member in the social structure. This arrangement is further expanded with the appearance of offspring. The females are thus accorded dominance over at least one social group. Among jackdaws, which form permanent mating relationships, a female gives up her own status when she mates and acquires the status of the male.

**19-15** *Constantly on guard against predators, the ground-dwelling baboons are organized for defense along lines of dominance and subordination. Among males, dominance is based upon age, strength, and aggressiveness, and the size and condition of canine teeth, a female rises in status merely by giving birth. Despite intense competition for a position of dominance among young males, the baboon troop remains quite stable. (Russ Kinne—Photo Researchers)*

**19-16** *Dik-diks, like many animals, stake out their territory by releasing a strong-smelling secretion. These chemical cues provide information to other members of the species and elicit immediate or delayed responses. They serve variously as attractants, as trail markers, or as warnings to stay away. (Leonard Lee Rue from National Audubon Society)*

19-17 *Two types of home ranges are compared in these maps. On the left, home ranges of eight male box turtles occupying parts of a five-acre plot. Home ranges overlap, because there is no defense of territory. On the right, territories of male song thrushes in two successive years. Notice that three individuals held the same territories in both years. Two others, A and B, were replaced by three new individuals. In general, if an individual succeeds in holding its territory, it will keep the same area throughout its life.*

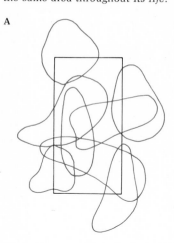

A

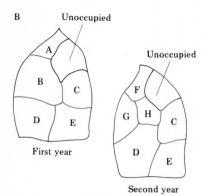

B

Baboon troops are an example of a highly organized society built on social dominance. Dominant baboons are groomed by the others more often, and they take the best feeding and sitting places. Let a dominant baboon come along when a subordinate has occupied his place and the subordinate immediately surrenders the position without argument. When the troop moves, the dominant males stay in the center, closest to the females and infants; the less dominant males serve as the vanguard and also bring up the rear.

Another phenomenon of social structures seemingly related to dominance, but not actually so, is leadership. One need only recall a flock of pigeons circling or geese flying south for the winter to illustrate the presence of a leadership individual. As opposed to social dominance, the leader is one who secures the compliance of the other members of the group without force. His position may be awarded because of age or experience. In some species, the oldest female member is the leader. This is attributed to the habituation of the offspring to receiving sustenance, warmth, and protection from the mother.

## Territoriality

As we have seen earlier in the case of the three-spined stickleback during the mating season, males migrate, select their territory, build their nests, attract females, and defend their territory against intruding males. Such behavior is quite common. Most birds and many fish do the same, often withdrawing in pairs to specific limited areas which the male defends. These private territories provide places where animals can mate, and their owners can feed, in relative safety. While a territory owner will fight off intruders of his own species, he often admits other animals, since their preferences for mates and food are different.

Mammals often mark their territories. Ante-

lope, for example, cover the ends of twigs with a strong-smelling secretion from their facial glands, and the scented twigs serve as ownership notices. Ocelots use piles of excreta; bears and many of the cat family use urine; hippopotamuses spray dung and urine about to mark out their territories.

While challenges to territorial rights do occur, rarely do they result in bloodshed, possibly because an animal away from its own territory seems to be at a disadvantage. In laboratory studies, investigators have alternately moved male sticklebacks into each other's territories and have found that once in rival territory the sticklebacks quickly lose their aggressiveness. A similar phenomenon has been observed in aggressive dogs which may try to attack any visitors to houses where they live but often are docile elsewhere. For that matter, it is a common observation that humans tend to have greater self-confidence in their own homes than elsewhere.

When fighting does break out over territories, losers are allowed either to escape or to withdraw unharmed after they have assumed a ritual posture of submission. When, for example, male grunts fight over territory, each grasps the other by the lips and the first to let go is the loser, who then swims away, defeated but unharmed. Rattlesnakes, capable of killing each other with a bite, instead push their heads against each other when they fight. When two wolves fight, the loser acknowledges his defeat by exposing the most vulnerable part of his body—the jugular vein—to the sharp fangs of the victor. This posture apparently inhibits the other's aggression. One notable exception to the general rule that territorial fights are not fatal is the male dragonfly which will defend a section of river bank against other males and is so aggressive that, after fighting with a trespasser, he may make a meal of him.

Territories vary greatly in size. Meat-eating animals—lions, for example—stake out large territories to permit wide hunting. Grass-eaters usually have much smaller claims. For a European robin, a territory may be about 6 square kilo-

19-18 *The territories of birds have width and depth as well as height. They are frequently circumscribed by obstacles of various sorts—physical, biological, spatial, or psychological. Since few barriers are absolute, their effects on bird distribution are selective. Some species, such as the Old World cattle egret shown here, readily surmount certain barriers to enlarge their range. Perhaps carried by trade winds or a tropical hurricane, this species made its way across the Atlantic from Africa to British Guiana late in the nineteenth century. Perhaps genetic changes in marginal populations of the cattle egret initiated this expansion in range. (Roger Tory Peterson—Photo Researchers)*

meters, but a golden eagle marks out a territory almost sixteen times as great, some 93 square kilometers or 37 square miles.

An interesting type of territoriality can be seen in the relationship between the bitterling, a small fish, and the mussel; for the mussel *is* the bitterling's territory. Their relationship begins when the female bitterling, with her long ovipositor, deposits unfertilized eggs in the mussel's body. The male sheds his sperm outside, and the mussel draws the sperm in through its incurrent siphon, thus fertilizing them. The bitterling embryos then develop within the mussel. The mussel, for its part, sheds its own fertilized eggs into the water after they have developed into

a peculiar parasitic larva called a glochidium. These attach to the gills (or skin) of adult fish (not necessarily only the bitterling) where they develop in parasitic association with the fish. When they metamorphose, they drop off and take up free-living existence on river or lake bottom.

Some animals, including bull seals and birds such as European swallows and American robins, return each year to the same territory, often with the same mate. On the other hand, some—storks, for example—appear to be more attached to nesting territory than to any one mate. If a previous mate fails to return to the territory, a new one is quickly selected.

Population control apparently is a major

advantage of territoriality. Since animals unable to claim a territory are unable to reproduce, and since there are limits to territories that can be claimed, younger animals may have to wait several seasons for their turn at breeding. This could help to prevent overpopulation and outgrowth of food supply. Territoriality may also serve to minimize the passing on of hereditary weaknesses since those without the characteristics required to compete successfully for territory may never have the chance to breed. Territoriality also serves to limit population density in any one area.

## Groups and societies

Groups of animals are not necessarily societies; they can be simply collections of individuals responding in the same way to some environmental feature, such as a source of food. When people stand together under a theater marquee for protection from a rainstorm, they make a group, or aggregation. But as long as all they do is to stand there to keep from getting wet and do not interact with each other, they are not displaying social behavior. Groups of wood lice gathered under the bark of logs share an environment but do not interact socially. However, there is evidence that individuals living together often survive longer than those living independently, and it may be that nonsocial aggregations represent a first step in the evolution of animal societies.

Such societies range from relatively simple to some which, for complexity, rival human society. Birds live in flocks but few do so continuously, most joining for part of the year. Flocks, especially those that will follow migratory paths to winter homes, may be huge. Having joined a flock, a bird, during feeding and roosting, will establish its "individual distance" and will protect the territory about itself.

Birds of a flock may act together in many ways, taking off, wheeling and turning in the air, landing again—each apparently remarkably sensitive to the movement of the others and, indeed, even to what could be called "flight-intention" movements. Let an enemy appear and the flock, acting together, may pursue it and beat it with their wings. It has been observed repeatedly that if food is thrown to pigeons or gulls, flocks quickly gather, consisting not only of nearby members able to hear the cries of those feeding but even of others at a distance. They are apparently able to detect special patterns of movements in those heading for food.

Schools of fish also wheel and turn with remarkable precision, making sharp turns in a

**19-19** *Whirligig beetles are scavengers, feeding on the bodies of insects that fall into the water. Since there is safety in numbers, they do their hunting in packs. They characteristically gyrate in dizzy circles on the surface of ponds or slow-moving streams. If alarmed, however, they dive, carrying a bubble of air at the tip of the abdomen. (Walter Dawn from National Audubon Society)*

**19-20** *Flocking, herding, or schooling is the simplest aspect of social behavior and may occur for a variety of reasons. To be considered social, such an aggregation of animals must at least move toward and remain with each other. The individuals in a flock interact and communicate with each other, timing and coordinating their activities. Perhaps the most obvious advantage of social interaction is increased safety. Birds that feed together also find more food than solitary birds, stimulating each other to feed. As the flock becomes larger and conditions become more crowded, the size of the flock may be limited by the amount of aggressive behavior tolerable among its members. (Alycia Smith Butler)*

fraction of a second and always maintaining perfect formation. The spacing of fish in a school suggests that here again individual distances, or territories, are maintained. Investigators have studied schooling in young whitebait which, immediately after hatching are about one-fourth of an inch long. At that stage, they avoid each other and dart away if approached. Then comes a stage when they still dart away if there are head-on encounters but may swim briefly together if one approaches another from the rear. Gradually, such encounters increase in frequency. By the time the whitebait are half an inch long, they greet each other upon meeting with body waggles and swim away together. By the time they are about three-quarters of an inch long, they have formed miniature schools.

## Insect societies

Among the most extensively studied insect societies are those of the bees. Not all bees live in communal hives. Among some solitary species, the female builds a nest, lays her eggs in it, endows it with a mix of honey and pollen, seals it, and flies off. In what may be considered subsocial species, the female goes a step further, returning to feed her young larvae.

Slightly higher up in terms of social organization are the bumblebees which each spring found colonies. A queen bumblebee builds a two-cell nest from wax exuded from her abdominal surface. In one cell, she lays her eggs; in the other, she stores nectar and pollen. After wax-sealing the egg cell, she settles down on it to hatch her

**19-21** *The thousands of members of a honeybee colony are organized into castes, each with a particular role in the hive. There is a queen, thousands of workers, and at certain times of the year, males or drones. There is even division of labor among the workers. Depending upon their age, some workers feed the brood, forage for food, or stand guard, while others are responsible for ventilating the hive. (C. G. Maxwell—Photo Researchers)*

eggs, then feeds her larvae on the nectar and pollen. About a week after hatching, the larvae spin cocoons; while guarding the cocoons, the mother bee builds more cells, lays more eggs, gathers more nectar. After about two weeks in their cocoons, the larvae have pupated. All female bees, they crawl out and set to work to gather nectar and pollen and care for new offspring, while the mother bee spends almost all of her time in laying eggs. Toward the end of a season, a colony may contain hundreds of bees, and at this point both young males and young queens are produced. The young queens mate; the female workers, males, and old queen die; and the young queens move out, scatter, and hibernate. The next spring they emerge to found new colonies.

The more complex honeybee colony consists of as many as 50,000 individuals living together, under conditions of far greater crowding than possible for humans. An isolated bee is unable to fly when temperature goes below 50°F. and cannot walk if it drops below 45°F. But the honeybee colony survives the winter, its internal environment maintained with remarkable constancy. Temperature and humidity are closely regulated, polluted air eliminated, wastes and foreign objects and dead bodies promptly removed.

A honeybee colony is made up of queen, large complement of worker bees, and for part of the year as many as 3000 drones whose only function is to mate with the queen. Only six or eight actually mate; the rest serve no purpose and are driven out in the fall.

Egg laying begins in January or February. Individual eggs go into separate wax cells. When the larvae hatch, nurse workers feed each larval bee more than 1000 meals daily. Within a week, the larval bees have grown to the point where each fills its cell, which is now sealed with wax by the nurses. After some 12 days of pupation, adults emerge. Within a day or two, the adults go to work. First they serve as nurses, carrying honey and pollen from storage cells to queen, drones, and larvae. After about a week, they graduate to housework. They produce wax, chew it, use it to enlarge the comb. They remove any dead bees from the hive, clean and ready emptied cells for further use, do guard duty at the hive entrance.

Genetically, queen bee and all workers are the same. But while worker and drone larvae are fed honey and pollen, a larva destined to be a queen receives only glandular secretions—royal jelly—produced by young workers. The precise nature of the substance in the jelly that turns a larva into a queen is still unknown.

Pheromones are chemicals secreted by one organism that have effects on other organisms. And the queen bee produces a pheromone, "queen substance," which prevents worker bees from becoming queens or producing eggs that could become queens. The pheromone is passed through the hive by trophallaxis. The substance is made in the glands of the queen's head, and as she licks her body, she coats it with the material which in turn is licked off by attendant workers who are constantly grooming the queen. It is quickly passed on by them to other bees in the hive in the course of food exchanging.

Food exchange may also have the purpose of allowing insects to recognize their fellows. The guards at a beehive examine each bee landing in front of the hive entrance and strangers are quickly turned away or stung if they persist in trying to enter. The odor distinguishing bees of one colony from those of another may be the result of the interchange of food within the colony.

Interesting facts about bee behavior have been noted in observation hives with glass sides. For example, the first bee to find and pick up a foreign object in a hive is usually not the one that flies out of the entrance to dispose of it. As many as five bees participate in removing a single piece of debris, usually grasping it one at a time. Any bee finding an object will carry it at least some distance away from where she is working, and others will carry it on.

Bees do not retire; they keep working until they die. The life span of a bee worker in summer is about six weeks and the death rate in a colony may be more than 1000 a day. Most manage to die away from the hive, apparently a defense mechanism to protect the hive from disease. When a bee does die in the hive, workers carry it to the entrance, and one bee flies away with it, usually making certain to carry it a considerable distance before disposing of it.

Usually, by spring, so many new broods have been raised that the bees separate into two colonies. The queen leaves, taking with her about half the workers, to establish a new hive. In the old hive, with the queen's substance no longer circulating (studies indicate it takes only half an hour before its absence is noticed), some of the

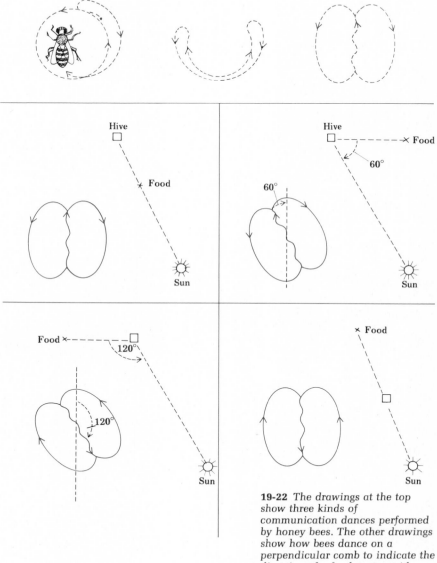

**19-22** The drawings at the top show three kinds of communication dances performed by honey bees. The other drawings show how bees dance on a perpendicular comb to indicate the direction of a food source with respect to the position of the sun. In performing this dance, the bees transform the bearing of the sun into an upward direction.

workers begin to lay eggs. A young queen emerges, goes on a nuptial flight with the drones following her. She mates just once and returns to the hive for a lifetime of egg production, endowed with enough sperm from the single mating to last her for a lifetime of five to seven years.

## Social parasites

Perhaps the most successful social parasite is the European cuckoo. A female European cuckoo busily monitors the nests of other species of birds; she may do this with as many as 50 other species. Once she finds a nest where eggs have been laid and the parents have departed briefly, she promptly throws one egg out of the nest and replaces it with an egg of her own. When the host

parents return, the egg count is as before and, if the cuckoo has chosen the right species as host, her egg closely resembles the others, although it is usually somewhat larger. When a cuckoo chick hatches, it pauses only for a brief rest before proceeding to burrow under each egg or baby chick with which it is sharing a nest and push it out of the nest. The foster parents then have but the one interloping cuckoo chick on which to expend all their feeding efforts.

Studies of American parasitic birds undertaken by Dr. Jurgen Nicolai of the Max Planck Institute for Behavioral Psychology in Bavaria indicate that the birds learn the songs of their hosts as well as those of their own species. A socially parasitic chick, initially exposed to the songs and behavior patterns of its host species, grows up to sing and to behave like its own species, but it also remembers the songs heard in its childhood in a way important for survival. During courtship, a male socially parasitic bird sings snatches of songs of its former host as well as of its own species. In this way he attracts females also reared by the same host species. The result is that males and females adapted to producing eggs of a color suitable for smuggling into a host's nest find each other.

**19-23** *The European cuckoo egg hatches faster than the eggs of most host species, giving the young cuckoo an advantage in growth and in claiming food from the foster parents. When it is about 10 hours old, an instinct appears that causes the blind young chick to "throw overboard" any solid object such as an egg, young bird, or even an acorn, that touches the sensitive shallow depression on its back. The instinct disappears after the nest is cleared of all objects but the young cuckoo. (Eric Hosking—Bruce Coleman)*

## Behavioral change

For animals, as for man, relationships with the environment are complex. The environment exerts many pressures and challenges. To survive and to contribute to future generations, animals must meet the challenges, and behavior may play a part in this adaptation. An example at adaptive behavioral change can be seen in sparrows, which naturally live and nest among trees and hedges in the countryside. To survive as human communities encroach, they have adapted their behavior through natural selection and nest and live in caves and gutters.

**593**

Innate behavior generally is too rigid to allow coping with many new situations. Many otherwise perilous challenges can be met only through learned behavior. When all members of a species have learned a new behavior as an adaptation to an environmental condition, the behavior may become traditional. It may even seem instinctive. For example, one investigator has described a large colony of wild rats living in burrows near ponds and streams heavily stocked with fish. When food was thrown into the water, young fish congregated. The rats learned to congregate, too—at the same time—and they learned not only to recover scraps of food but also to catch fish and to swim well and rapidly for the purpose.

How can new young rats acquire the tradition? Humans can accumulate, maintain, and pass on tradition by the spoken word and by written documents. Humans can also use speech and writing for direct instruction in useful behavior. Animals, however, must depend upon imitation. The young rats had to learn by imitating the learned behavior of older individuals.

## Evolution and behavior

The theory of evolution depends upon the selection of adaptive characteristics, and, clearly, behavior is one of the principal means of adaptation for animals. But much remains to be learned about behavior and natural selection. "There are vast areas of modern biology, for instance . . . the study of behavior," Ernst Mayr has remarked, "in which the application of evolutionary principles is still in the most elementary stage."

Some aspects of the relationship between genetics and behavior seem clear enough. Certainly, there is no doubt about whether a genotype has behavioral effects. It has been demonstrated, for example, that a gene, the Y gene, in *Drosophila* controls body color and also influences the strength and duration of wing vibration, a male sexual display.

19-24 *Pigeons once lived on rocky ledges and fed on wild grasses and grains. But as man encroached, this habitat began to disappear. The pigeons were able to adapt to the new conditions, substituting window ledges for their natural perches, and now they thrive in all big cities. (Alycia Smith Butler)*

For years, of course, animals have been bred for behavioral characteristics—aggressivity in fighting cocks and Siamese fighting fish and docility in laboratory rats, for example—and such characteristics could not have been bred for if they had no genetic basis. It has been established that certain behavioral traits in dogs are determined by specific genes; in man, major effects on behavior, such as feeble-mindedness, have been found to result from metabolic disorders caused by specific genes.

It seems clear enough that behavior is not inherited as such; the fertilized egg shows no activity except that connected with its internal physiological function. Behavior must be developed, and it develops under the influence of both hereditary and environmental influences.

Once it was taken for granted that acquired characters are inherited. Darwin wrote of instinct as inherited memories. He also proposed that instincts were the products of natural selection. In one famous study, rats were trained to escape from a water tank. They had to swim to the escape route. There were two ways they could go. One was through a brightly lit alley; but any attempt to enter this was punished with an electric shock.

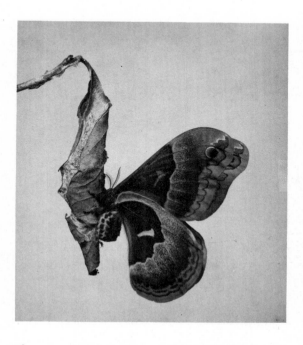

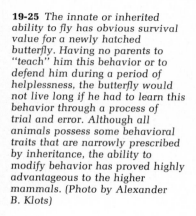

**19-25** *The innate or inherited ability to fly has obvious survival value for a newly hatched butterfly. Having no parents to "teach" him this behavior or to defend him during a period of helplessness, the butterfly would not live long if he had to learn this behavior through a process of trial and error. Although all animals possess some behavioral traits that are narrowly prescribed by inheritance, the ability to modify behavior has proved highly advantageous to the higher mammals. (Photo by Alexander B. Klots)*

The correct way out was dimly lit. Successive generations of rats were trained in this way, and a progressive improvement in ability to perform the task was noted. It seemed that inheritance of acquired characters had been demonstrated experimentally.

But there was an error in the method: failure to maintain a control group, keeping some rats in each generation untrained. When other investigators bred rats for 20 years, training them in a similar manner for 50 generations, the rats showed a progressive improvement with time, but so did those in a control, untrained group. That was true during the first 10 years. Later, both experimental and control rats showed a decline in performance, followed by an improvement once again. These fluctuations have been attributed to uncontrolled environmental variations, especially dietary.

Investigators considering the possibility that genetical assimilation could take a behavioral

response learned by ancestors and turn it into a fixed action pattern for descendants, reason that this might theoretically happen only if the possession of a fixed response provided an advantage over simply possessing the ability to learn. Adaptedness, as some put it, would have to be made to confer more fitness than adaptability.

There are, of course, some instances in which "innateness" has survival value—for example, where a behavior pattern must be nearly perfect the very first time, as in flying by a newly hatched butterfly. But in many other cases, it is advantageous not to be too precisely adapted innately and to leave much to trial-and-error learning. For example, young birds that at first peck at anything and have to learn what is edible have the advantage that they can thus learn individually to specialize in types of food most readily available to them in their particular environment.

Many scientists believe that the trend of evolution, particularly mammalian evolution, has been toward greater individual adaptability and less dependence on fixed patterns, and they see no sound reason for believing in any Lamarckian effect on behavior.

## Summary

The study of behavior is increasingly an area of biological research. Early studies encountered difficulty in distinguishing innate from learned behavior. Innate behavior refers to taxes, reflexes, and instincts. But innate behavior patterns are often modified by learning. Crossbreeding experiments have shed additional light on the interaction between acquired and innate behavior patterns. Instinctive behavior has sometimes been studied in terms of responses to releasers, specific stimuli that activate innate behavioral releasing mechanisms. As an alternative, some psychologists suggest that all behavior involves inherited capability, experience and normal maturation.

Behavior may be studied physiologically. It has been shown that experience affects brain structure, and evidence from brain surgery and elsewhere points to the localization in the brain of some functions of learning and memory. In experiments with planaria, subjects could acquire a certain amount of "training" by ingesting pieces of previously trained worms.

Learning may be of several sorts. One well-studied type of learning is imprinting, in which an attachment, usually to the mother, is formed at a critical period early in life. Other types of learning include habituation, associative learning, and learning by trial and error and by imitation.

A great deal of behavior is adaptive response to physical or chemical stimuli from the environment, such as taxes and reflexes. Other behavior, associated with biological rhythms, seems to be governed solely by some sort of internal "clock." In higher animals and social insects, much of behavior is a response to social stimuli, and results in more or less elaborate communications networks and social systems. Well-studied aspects of this last are social dominance and territoriality. Insect societies are exceptionally complex and stable.

Changes in behavior are usually possible only when the behavior is learned; innate behavior is too stereotyped to be flexible. Evolutionary and genetic studies of behavior are still at a very early stage, but they promise to yield results of profound importance.

# Bibliography

### References

Bullock, T. H. 1970. "Physiological Bases of Behavior." In J. A. Moore, ed. *Ideas in Evolution and Behavior*. Natural History Press, Garden City, N. Y.

Caspari, E. W., J. L. Fuller, W. C. Rothenbuhler, et al. 1964. "Behavior Genetics." *American Zoologist*. 4(2).

Feder, H. A. 1972. "Escape Responses in Marine Invertebrates." *Scientific American*. 227(1): 92–100.

Harlow, H. F. and M. K. Harlow. 1962. "Social Deprivation in Monkeys." *Scientific American*. 207(5): 34–36, 136–146. Offprint no. 473. Freeman, San Francisco.

Hess, E. H. 1972. "'Imprinting' in a Natural Laboratory." *Scientific American*. 227(2): 24–31.

Holldobler, B. 1971. "Communication Between Ants and Their Guests." *Scientific American*. 206(5): 50–59. Offprint no. 1218. Freeman, San Francisco.

Levine, S. 1971. "Stress and Behavior." *Scientific American*. 224(1): 26–31. Offprint no. 532. Freeman, San Francisco.

Morse, R. A. 1972. "Environmental Control in the Beehive." *Scientific American*. 226(4): 92–98.

Watts, R. C. and A. W. Stokes. 1971. "The Social Order of Turkeys." *Scientific American*. 224(6): 44–51. Offprint no. 1224. Freeman, San Francisco.

### Suggestions for further reading

Armstrong, E. A. 1965. *Bird Display and Behaviour*. Dover, New York.
Authoritatively unifies a vast range of material on bird display. Presents a rather detailed and well-illustrated discussion of ritualistic display habits among gannets. Suggests interesting hypotheses on psychological forces fundamental to all levels of animal life.

Dethier, V. G. and E. Stellar. 1961. *Animal Behavior*. Prentice-Hall, Englewood
Cliffs, N.J.
Emphasizing the role of the nervous system as the matrix of behavior, this
exceptionally concise and lucid study examines the mechanisms and evolution of
animal responsiveness.

Frisch, K. von. 1961. *Bees*. Cornell University Press, Ithaca, N.Y.
A straightforward account of the author's noted researches on color sense,
chemical sense, and the language of bees.

Griffin, D. R. 1958. *Listening in the Dark*. Yale University Press, New Haven, Conn.
A study of echolocation clearly elucidates the problem of understanding behavior
in terms of the perceptual environment of animals.

Lorenz, K. 1967. *Evolution and Modification of Behavior*. University of Chicago
Press, Chicago.
The founder of modern ethology examines the concepts of learned and innate
elements in behavior and the nature of phylogenetically adapted behavior patterns
as revealed through the classic deprivation experiment.

Roe, A. and G. G. Simpson. 1958. *Behavior and Evolution*. Yale University Press,
New Haven, Conn.
An encompassing collection of authoritative essays on the evolutionary
significance of behavior and its place in evolutionary theory.

Tinbergen, N. 1970. "Behavior and Natural Selection." In J. A. Moore, ed. *Ideas in
Evolution and Behavior*. Natural History Press, Garden City, N.Y.
In examining the consequences of behavior, this noted zoologist applies to a
number of species the question, "How does the intricate machinery that controls
behavior affect survival and reproduction?"

**Books of related interest**

Bach, R. 1970. *Jonathan Livingston Seagull*. Macmillan, New York. Photographs
by R. Munson.
A charming philosophical fantasy about flight, the sublime, and purpose in the
universe. A beautiful photographic study of flight as well.

Bates, M. 1967. *Gluttons and Libertines*. Random House, New York.
A sprightly and wide-ranging examination of what is "natural" in human behavior.
Aimed at correcting cultural biases on such topics as food and sex, gentility,
clothing, and happiness.

Cousteau, J. and P. Cousteau. 1970. *The Shark: Splendid Savage of the Sea.* Doubleday, Garden City, N.Y.
A personal and wholly engaging tale of a two-year ocean voyage made to study and record the behavior of sharks. Includes full-color photographs.

Goodall, J. van Lawick. 1971. *In the Shadow of Man.* Houghton Mifflin, Boston.
A celebrated field observer's fascinating account of the life of chimpanzees. Records what years of patient watching revealed about birth and death among these closest relatives of man, the ways of chimpanzee mother and child, adolescence, and tool-making. Illustrated with striking photographs.

Lorenz, K. 1972. *King Solomon's Ring.* Crowell, New York.
A noted ethologist shares lively and entertaining observations and insights into the behavior of animals. Delightfully illustrated with the author's original line drawings.

Schaller, G. 1964. *The Year of the Gorilla.* University of Chicago Press, Chicago.
A popular but authoritative account of this primatologist's extensive field research on the behavior of the mountain gorilla.

# 20
# Concepts of ecology

The term "ecology" is today a household word. Headlines about smog, oil spills, poisonous pesticides, industrial water pollution, overpopulation, and other environmental problems have spurred a public outcry in the name of ecology. Because of this response, some effort is being made to curb the most glaring abuses, but often the damage cannot be undone by merely eliminating the cause; the natural balance of the environment must also be restored. Unfortunately man's capacity to alter his environment has far outpaced his understanding of its essential processes.

The science of ecology is attempting to close this gap. Sometimes referred to as "environmental biology," ecology is defined as the branch of science dealing with the interactions among organisms, and the relationship between organisms and their environment. It is a young science, assuming its present direction in only the past decade and a half, when it abandoned the general descriptive approach of natural history to focus on the specific structural and functional relationships between living things in the setting of their environment. Today ecology has become one of the major sciences; some say it is the key to the survival of mankind.

## The nature of the ecosystem

The subject matter of ecology can be broadly divided into two interdependent parts: the **biotic,** or living component, and the **abiotic,** or nonliving component. The biotic component consists of animals, plants, and microbes, interacting in various ways with others of their own kind, with organisms of other species, and with the physical aspects of their environment. The abiotic component consists of all external conditions under which the biotic segment must function, including temperature, rainfall, moisture, wind patterns, currents, sunlight, minerals, and topography. The specific set of conditions surrounding the organism is referred to as its **habitat.**

**601**

Each species in an ecosystem defines its activities in specific ways, including, for example, where it lives, when and how it reproduces, what it feeds on, when it feeds and where, the climatic conditions it favors or can tolerate, and other considerations. The functional role pursued by each individual species in the ecosystem is termed its **ecological niche.** Generally, no two ecological niches within a given ecosystem are identical in all details. For example, two closely related warblers, the cerulean warbler and the black-throated green warbler, live in the beech-maple-hemlock forests of the northern United States. Both are insect eaters, and they may nest very near one another. But the cerulean warbler searches for insects in the upper leaves of the deciduous trees, whereas the black-throated green warbler feeds only in the upper branches of the hemlocks.

In studying the relationships between the abiotic and biotic components of the environment, ecologists concern themselves with specific and self-contained areas in which these components may be identified and observed. Such areas include ponds, meadows, deserts, streams, or lakes; or they may be artificially prepared laboratory situations. They may be large or small, terrestrial or aquatic. Such ecological systems formed by the interaction of a community of organisms and its environment are known as **ecosystems.**

Just what animal and plant populations will comprise an ecosystem is, in part, determined by the physical factors of the environment. Temperature is one important consideration. All living species have a preferred temperature range and a range beyond which they will cease to develop or reproduce. For this reason, desert animals accustomed to temperatures of 45°C. will not be native to ecosystems with daily temperatures of –70°C., such as found in Siberia. In addition to temperature range, temperature variation is important. Short-term variations, for example over a 24-hour period, will be a major factor in determining the

20-1 *Cerulean warblers are usually found in beech-maple climax forests, living in the upper canopy. They feed on insects, both adult forms and larvae. The male of this nesting pair is bringing the female a short snack. (Hugh M. Halliday from National Audubon Society)*

20-2 *The elf owl has evolved various behavioral and physiological adaptations to a desert way of life. Like most desert animals, the owl is small and nocturnal, efficiently conserving his body fluids. He is able to avoid the heat by hunting at night and resting in the holes of saguaro plants during the day. His predatory activities help to keep the desert's prolific rodent population in check. (Roger T. Petersen from National Audubon Society)*

species composition of an ecosystem. This will also be true of sharp seasonal variations. Most animal and plant life is unable to survive extremes in temperature.

Precipitation is another major factor. The amount and distribution of rainfall strongly influences the variety and abundance of vegetation, and consequently of the animal population as well. Seasonal rainfall patterns, alternations between wet and dry spells, flood and drought, are all important in dictating the biotic composition of an ecosystem. So, too, is the amount of snowfall and the length of the period before thaw.

In addition to temperature and precipitation, other climatic influences include wind, which may vary in temperature and velocity, moisture, and atmospheric pressure. But many other environmental conditions do not involve climate. One of these factors is light. The duration of the daylight hours or photoperiod may vary from season to season and from region to region. So may the intensity. These aspects not only influence the variety of species but also affect the behavior of birds, fish, and mammals which respond to light cues.

Biotic composition is also strongly influenced by the nature of the soil in the ecosystem. Its texture, its ability to hold water and air, and its share of various nutrients, such as phosphate, potassium, nitrates, and calcium, are all important factors that influence the composition of the biotic component of that area.

If abiotic factors influence the nature of the biotic composition, the reverse is also true. Soil inhabitants, such as earthworms, break up the soil particles and aerate the soil. The life processes of plants involve the removal of certain substances from the environment and the addition of others. Decomposition of dead animals and plants releases organic and inorganic substances into the soil, enriching its nutrient content. Humidity is influenced by the transpiration from vegetation; wind patterns by tree height and

density; light distribution by foliage; water flow by the dam-building activities of beavers; and rainfall in any specific spot by the trees and foliage sheltering that location.

These are only some of the ways in which the living and nonliving segments of the ecosystem interact to their mutual advantage. Numerous complex relationships exist, involving the availability of food and essential elements, social organization, cooperation and competition, population interdependence, and community development and stability. These relationships are the fundamental concern of the ecologist.

## Flow of energy and materials

At its most fundamental level, the survival of life on earth depends on the utilization of energy. All energy on earth originates from one central and enduring source, the sun. The sun releases enormous quantities of energy in the form of electromagnetic waves. This solar radiation is expended in all directions, so that only about one fifty-millionth of its output reaches the outer atmosphere of the Earth. While this seems a small fraction, it is further decreased by the loss of nearly half through cloud reflection (42 percent) and dust (9 percent). More radiation is also absorbed and scattered by the molecules of gases in the air. And yet even from a tiny fraction of the solar output, the atmosphere, soil, and water are warmed, and the all-important process of energy transfer has its beginnings.

## Food chains

The first and most important step in the conversion of solar energy into the energy consumed by the biosphere begins with the process of photosynthesis. By combining water and carbon dioxide in the presence of sunlight, plants containing chlorophyll are able to carry out the conversion of these low-energy compounds into high-energy food supplies. Other autotrophs, such as chemosynthetic bacteria, can utilize the energy stored in chemical bonds to drive synthetic reactions. Heterotrophs, lacking the ability to make their own food, must consume autotrophs. In this process, the chemical energy stored in the plants is reordered. Some of the energy is utilized in the heterotroph's own metabolic reactions; the rest is released as heat energy.

The transfer of food energy from autotrophs to heterotrophs, with its sequence of eating and being eaten, is known as a **food chain.** All food chains have as their first link some form of autotroph. Because these organisms are the only ones in the chain capable of manufacturing their own food, they are termed **producers;** this is the first trophic level. All other organisms are dependent upon the producers either directly or indirectly. They are therefore appropriately termed **consumers.**

The first consumer link in the food chain following the producer is the plant-eater, or herbivore. A herbivore, an organism that can convert energy stored in plant tissue to animal tissue is at the second trophic level. The major herbivores occurring on land are hoofed animals, some insects, and rodents. In aquatic environments, they consist largely of small crustaceans. Because they comprise the first consumer level, herbivores are often referred to as primary consumers.

Occupying the next link in the food chain, the third trophic level, are those organisms which consume herbivores. These carnivores, or flesh-eaters, since they are a second step away from the producer level, are classified as secondary consumers. Being flesh-eaters, of course, carnivores may also consume carnivores, and these organisms make up other links in the total food chain. The scavengers, that feed on dead plants and animals, are also an important link in the chain.

While the food chain is relatively simple in organization, most natural communities contain far more complex relationships. Food chains generally crisscross and result in the formation of a **food web.** Few consumers restrict their feeding habits to one food source. Plants provide a supply of food to a variety of animal life, and animals are normally consumed by a variety of predators. In addition, some species occupy more than one consumer group, feeding at times on plants and at other times on animals. Man is typical of this group, which is known as omnivores.

One final feeding group that participates in the food chain is the **decomposers.** These are largely bacteria, fungi, and yeast; they decompose the remains and wastes of other organisms. In so doing they convert some organic waste into other useful organic forms and break others down into their inorganic components. The decomposers close the food chain, returning nutrients for the use of the producers.

20-3 *This hyena from the Serengeti plains of East Africa occupies the position of tertiary consumer in its food web. Rarely killing its own prey, the hyena usually scavenges its food from the abandoned prey of larger predators, such as the lion. (Donald Paterson—Photo Researches, Inc.)*

**20-4** The food web shown here is found in midwinter in a Salicornia salt marsh in the San Francisco Bay area. The producers are terrestrial and salt marsh plants. Primary consumers are of several familiar types. Secondary consumers include egrets, shrews, clapper rails, ducks, sandpipers, and the ubiquitous Norway rat. Top carnivores are the marsh hawk and the short-eared owl.

The flow of energy from one trophic level to the next occurs according to the laws of thermodynamics that were discussed in Chapter 1. The law of conservation of energy states that the amount of energy existing in the universe may be transformed or transferred, but it will never increase or diminish; the second law of thermodynamics states that any process involving energy transformation will result in a degradation of energy. In accordance with these two physical laws, the food energy passed on from one trophic level to another occurs in ever-decreasing amounts of transferable resources. Thus the plants in an ecosystem possess more energy than the herbivores which feed on them, and the herbivores possess more than the carnivores, and so on down the chain. It is because of this depreciation of reusable energy that food chains rarely occur in nature with more than three or four, and virtually never more than five, links.

The energy loss experienced at each level is substantial. It has been calculated that energy is reduced in magnitude by 100 from plants to herbivores and by 10 from herbivores to carnivores, and likewise for each succeeding trophic level. Therefore, if we start with 1000 units of light energy fixed by green plants, 10 units would be utilized by herbivores, 1 by carnivores, and only 0.1 for carnivores consuming carnivores. It is therefore apparent that energy flow is a one-way phenomenon, passed on through the community and then dissipated. Without a constant supply of solar energy, life would soon disappear from this planet.

The successive reduction of energy through the system is often represented visually in the form of a pyramid known as a **pyramid of energy.** Because it contains the highest amount of usable energy, the first trophic level is situated at the base of the structure. The apex is formed by the last trophic level, as it has the least energy available. This arrangement will represent energy flow for all communities, although the species which occupy each trophic level will vary from food chain to food chain.

Another characteristic of the food chain which can be demonstrated in pyramid form is the decreasing number of organisms occupying each successive trophic level. The organisms in the producer level will usually be much greater than the number in the next level, with a very small number of carnivores at the apex. This pyramid is known as the **pyramid of numbers.** There is a general tendency toward correlation of size of the organism and its position in the pyramid; the larger organisms are usually found at the top of the pyramid.

In assessing energy flow between ecosystems, comparing numbers of different species is often misleading when one species is as large as an elephant, while the other is as small as a sparrow. More useful is the **pyramid of biomass** which instead considers the total dry weight of protoplasm or bulk of organisms at each level. Again the decrease of energy at each level results in the support of smaller biomass at the next.

**20-5** *The praying mantis, Tenodera sinensis, is a well-known insect predator. Stalking its prey in its still, hunterlike manner, it has been widely introduced to control garden pests. The mantis shown here is consuming a butterfly. (Photo by Alexander B. Klots)*

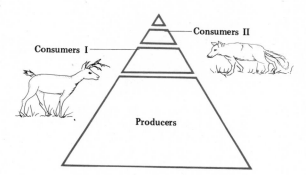

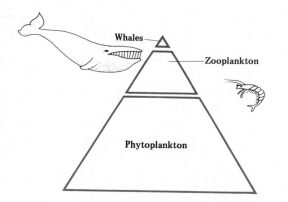

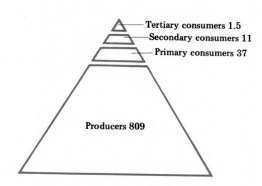

**20-6** *Pyramids of energy (top), numbers (center), and biomass (bottom). As the pyramids are ascended, from producer to primary, secondary, and tertiary consumer, energy, numbers, and biomass undergo parallel diminution. Millions of plankton organisms are required to support one whale (center). The lower pyramid refers to biomass in the aquatic ecosystem of Silver Springs, Florida. The figure refers to grams of dry biomass, the standard form of measurement, per square meter.*

Consumers II

Consumers I

Producers

Whales

Zooplankton

Phytoplankton

Tertiary consumers 1.5
Secondary consumers 11
Primary consumers 37

Producers 809

## Cycles of elements

In addition to food energy, living organisms require the interchange of certain materials essential to their growth and well-being. Included among these materials are some 30 or more naturally occurring chemical elements. Unlike the pattern of food chains, however, elements are exchanged in a cyclical manner—passed from the biotic component to the abiotic component and back again indefinitely. This circular flow of materials is termed the **biogeochemical cycle.**

**Nitrogen.** The atmosphere of the earth is nitrogen-rich, for this element comprises 79 percent of atmospheric composition. In this state, however, the element is virtually useless to living organisms, being consumed and expelled without any significant effect on bodily processes. Before it can be utilized by organisms, it must be converted into another form. This process is carried out in various steps. The first conversion, called **nitrogen fixation,** occurs primarily through the activities of two major groups of microorganisms: (1) symbiotic nitrogen fixers, including bacteria and some fungi, and (2) free-living nitrogen fixers, such as bacteria.

The first group of nitrogen fixers carries out its activities by penetrating the root hairs of leguminous plants (members of the pea family). Once inside, the bacteria greatly increase in numbers causing the root hair to swell into a nodule. Within this structure the nitrogen fixing takes place. Atmospheric nitrogen is reduced to ammonia and then synthesized into organic compounds. Most of these products are incorporated into the plant's cytoplasm as amino acids. Any surplus is excreted into the soil. The second group of nitrogen fixers lives free in soil or water. There they convert nitrogen to ammonia and release it into the medium in which they exist. When these microorganisms die, decomposer organisms also break down the fixed nitrogen in the dead cells into ammonia.

**607**

While some of this ammonia is utilized by various forms of plant life, most higher plants prefer nitrates as their source of nitrogen. The production of this material is also dependent on bacteria. Two different groups of bacteria act in successive steps to reduce ammonia to nitrites and to nitrates through oxidation. This process is known as **nitrification.**

The nitrates which result from the nitrification process are released into the soil, and are subsequently assimilated by plants into organic nitrogen compounds. When the plant is eventually consumed by carnivores, the cycle may begin its return flow. When the plant or carnivore dies or when the carnivore excretes waste products, decomposers again act on the organisms or its wastes to convert their nitrogen content to ammonia. They then utilize some of it in their metabolism or release it for use by nitrogen-fixing plants. This process is called **ammonification.**

To carry out the return of nitrogen to the atmosphere, other bacteria serve to counter the action of nitrifying bacteria. Through a process of denitrification, these organisms convert nitrogen compounds back to gaseous nitrogen and release it into the air.

**Carbon.** In contrast to nitrogen, carbon possesses a relatively simple cycle. Its reservoir is also the atmosphere, but much smaller quantities exist than the abundant nitrogen. Only 0.03 to 0.04 percent of the content of the air is carbon dioxide. However, the ocean is believed to contain more than 50 times this quantity dissolved in its waters.

The first step in the carbon cycle begins with photosynthesis and the conversion of carbon dioxide to carbohydrates. Once stored in plant tissues, the carbon compounds are consumed by herbivores and converted into other chemical forms. Each successive consumer in the food chain causes them to undergo further breakdown and resynthesis.

One of the ways in which carbon is returned

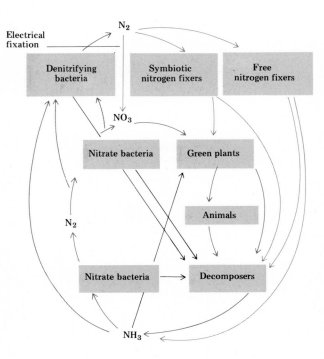

*20-7 The nitrogen cycle is stable and self-regulating. Amounts of available nitrogen remain fairly constant with respect to large ecosystems or to the biosphere as a whole.*

*20-8 A root nodule containing symbiotic nitrogen fixers is seen in cross section (75×). Certain nitrogen-fixing bacteria occur in the nodules of legumes where they live in symbiosis with the roots. These bacteria release as much as 90 percent of the nitrogen they synthesize into the host plant's cytoplasm. The surplus is excreted into the soil. Thus, not only can legumes grow well in nitrogen-poor soils by manufacturing their own supply, but such plants actually enrich the soil in which they grow. (Hugh Spencer from National Audubon Society)*

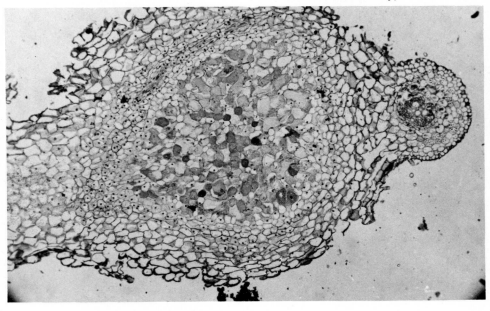

**20-9** *The carbon biogeochemical cycle makes this element available for life processes through photosynthesis, respiration, and decomposition. More recently man's use of large-scale combustion processes has started to influence the natural carbon cycle.*

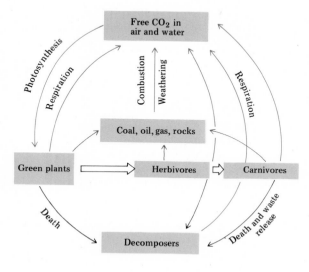

to the atmosphere occurs when bacteria and fungi decompose animal wastes or the protoplasm of dead plants and animals. In breaking down these materials into simpler substances, some carbon dioxide is released. Most of the carbon substances are added to the earth's raw materials, ending up as part of limestone or clay or as coal or petroleum. The combustion of these products or volcanic activity again release the carbon dioxide into the atmosphere.

**Phosphorus.** Phosphorus is an element of primary importance to biological processes. Unfortunately, the natural element is often in scarce supply. The main sources of phosphorus are rocks and natural phosphate deposits. To free the element into the cycle, these sources must undergo such environmental phenomena as leaching and weathering, as well as such man-made processes as mining. Some of the phosphorus is utilized through the food chain and ultimately is released for recycling through death and decomposition. Much of the element, however, is carried by the waterways to the sea, where, settling deep into the sediment, it becomes temporarily or permanently inaccessible.

Through ground disturbances, some of the material may rise to a water level in which tiny one-celled plants, or **phytoplankton,** are able to absorb it. They, in turn, are eaten by microscopic aquatic animals, or **zooplankton,** which may utilize and excrete the phosphorus. Some may also resettle into the sediment. Only a small portion is returned to land by the deposits of fish-eating birds or through fish harvesting.

**20-10** *Compared to nitrogen, phosphorus is a rare element; in natural waters nitrogen is about 23 times as abundant as phosphorus. Phosphorus is, however, a vital element. What the diagram alone does not make clear is that the return of phosphorus to the land, and thus to the cycle, does not keep up with the loss of phosphorus to ocean sediments.*

## The species as an ecological unit

As we have seen from the study of food-energy chains and nutrient cycles, species must depend on other species and on the environment for support of many life processes. But species must also depend on their individual members. In their social and functional relationships, whether cooperative and beneficial or antagonistic, the interactions of members of a species are important ecological influences.

## Intraspecific relations

Intraspecific relationships (relationships between members of the same species) may be either cooperative or competitive. Often, a group of plants or animals withstand environmental stress better than single animals and plants. We see this kind of cooperation occurring in stands of pines on the Maine coast. The group of pines can better withstand the action of the wind and the effect of dehydration than a single pine growing alone in the same environment. Groups of trout can withstand a given dose of poison introduced into the water better than isolated individuals. Experiments found that individual fish could withstand a larger dose of poison if they were in water previously occupied by other fish. They concluded that mucus and other secretions given off by the fish aided in counteracting the poison. A hive or cluster of bees can withstand temperatures that would kill all the bees if they were isolated and alone.

Grouping does not only occur in response to extreme environmental stresses. Some species must exist in groups no matter what the condition of their environment. If they are not in groups, the species dies. Colonial birds fail to reproduce in isolation.

There is also intraspecific competition for food and space and mating partners. Normally intraspecific competition for the limited resources of the ecological niche does not lead to extinction, but instead to a stable population. The members of the species may adjust to intraspecific competition in a variety of ways. One type of interest to sports fishermen was discovered when fishery officials introduced new

bluegills into a lake already containing bluegills. To their surprise, fishermen stopped catching any bluegills at all. The ecologist brought in to analyze this singular situation discovered that the number of bluegills in the lake had increased but the size of the individual bluegills had decreased. Thus, the total biomass of bluegills had remained constant. But because the fish became so small, sport fishermen were unable to catch them using conventional hooks.

In extraordinary cases, intraspecific competition may lead to the near extinction of a species. When population has increased to a critical level, the species members may exhaust and destroy the food supply for future generations. A familiar example occurred when the Eastern whitetailed deer reproduced geometrically when its predators were destroyed and hunting was controlled. During the winter season, the majority of the population starved to death. This dramatic example contradicts the usual case, in which intraspecific competition results in homeostasis of the population.

## Interspecific relations

In addition to relationships within a specific population, ecologists are also concerned with relationships among the various populations in the biotic community. These relationships may be competitive or cooperative; they may be indirect in nature, such as relations within the food chain, or immediate and direct, with necessary and permanent contact between two species. They may range from the most beneficial to the most harmful.

**Symbiosis.** A term which refers to a situation in which two species interact is called **symbiosis;** the term more literally means "living together." The advantages obtained through these relationships may involve food, protection, shelter, transportation, or support, and may range from

**20-11** *The clownfish and the sea anemone enjoy the type of symbiosis known as mutualism. While the anemone's tentacles are lethal to most animals, the clownfish enjoys immunity to their poison and is therefore able to live among them in safety. The sea anemone benefits from the relationship because the brightly colored clownfish attracts would-be predators, which become the anemone's prey. (A. Giddings— Bruce Coleman)*

**20-12** *The pearlfish, Carapus, enjoys a commensal relation with the sea cucumber, Actinopyga. Living within the cloaca of the sea cucumber, the pearlfish derives its shelter and nourishment from its host, but does it no harm. The echinoderm is apparently unaffected by this form of symbiosis. (Courtesy of the American Museum of Natural History)*

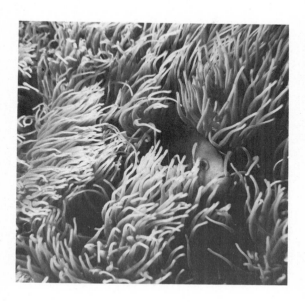

mutual gain to potential death as a result of the relationship.

The most beneficial species interaction because it involves an exchange necessary to the survival of both species is called **mutualism.** One example of this relationship was illustrated previously in the interaction between nitrogen-fixing bacteria and leguminous plants. Another example of mutualism can be seen in the clown-fish *(Amphiprion)* that often rests among the tentacles of a sea anemone. This fish, which is immune to the sting of the anemone, obtains a protected spot to rest. The anemone benefits because the calm presence of the clownfish lures other fish within striking range of its stinging tentacles. At times dependency between two organisms is so complex that they appear as one organism. This is true in the case of plants called lichens, in which the algae member and the fungi member have become so interdependent as to be referred to by botanists as a single species. The turbellarian worm, *Convoluta roscoffensis,* also exemplifies intimate mutualism. This tidal resident has numerous cells of microscopic algae actually incorporated into its tissues. The algae remove metabolic wastes, that might otherwise accumulate and poison the worm; these wastes supply needed nutrients for the algae.

In some cases, the mutualism is obligatory; one or both of the symbionts is unable to survive independently. This is true, for example, of the individuals that make up a lichen. In other instances, the two species benefit one another, but are not essential for the other's survival. This casual mutualism is found in the relationship between certain sea anemones and hermit crabs. The crabs attach the anemones to their shells. Thus the anemones provide protection and camouflage for the crab, while they receive transportation and food from the leavings of their host.

Another type of symbiotic interaction is called **commensalism.** This occurs when one species receives benefit from an association, while the other is neither benefited nor harmed. An example can be seen in another kind of turbellarian, belonging to the genus *Bdelloura.* This worm lives in the gill hooks of the horseshoe crab; it does not feed on the tissues of its host but finds there a shelter from which to search for the rotifers that constitute its chief food.

A third type of symbiotic relationship is one in which one species benefits by the association, while the other is harmed. This is known as **parasitism.** The parasite usually derives its nourishment from the body of the host, either externally—from skin, hair, feathers, or scales— or internally from cells, tissues, tubes, or ducts of the host. Some parasites live on the skin, but suck blood from their victims.

Parasites may be fungi, worms, protozoans, bacteria, lice, or fleas. They are usually small, and rely on the host both for their food supply and for a habitat. Some common forms of parasites especially troublesome to man are tapeworms, pinworms, hookworms, flukes, lice, and mites. Most of these parasites have successfully adapted themselves to their hosts so that the host is able to support the intruder in a state of toleration. Obviously it would not be to the parasite's advantage to kill the host, but most parasitized organisms are not healthy. The alien organisms often indirectly cause the death of the host, by weakening it so that it becomes vulnerable to a predator or disease.

A final symbiotic category is **amensalism,** a condition in which one species inhibits the other but receives no benefit itself. For example, certain species of fungi produce by-products which are toxic to bacterial species. The process in which one organism is capable of destroying another or inhibiting its growth is exploited in medicine in the form of antibiotics.

**Predation.** Another interspecific relationship, similar in some ways to parasitism, is predation. This is the relationship between organisms in which one captures and feeds upon another.

**20-13** *The term symbiosis refers to any system wherein two species live together and interact, although its use is sometimes restricted to those instances where both species benefit from the interaction. The leech and turtle depicted here live in a form of symbiosis called parasitism: the parasite benefits to the detriment of the other, host, species. (Lynwood M. Chace from National Audubon Society)*

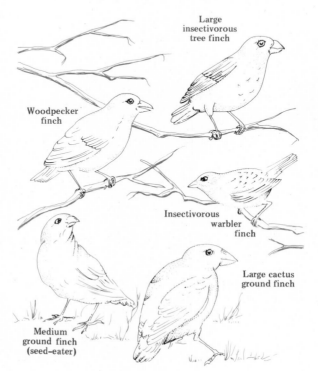

**20-14** *Gause's Law states that no two species can remain in competition in precisely the same ecological niche. Darwin's finches exemplify this law: As interspecific differences have evolved, so have corresponding differences in habitat. The beaks especially serve to distinguish the seed-eaters, for example, from the insectivorous warbler or woodpecker finches.*

Large
insectivorous
tree finch

Woodpecker
finch

Insectivorous
warbler
finch

Large cactus
ground finch

Medium
ground finch
(seed-eater)

Both parasitism and predation involve the utilization of one organism by another for sustenance. There are, however, several generalizations about the differences that may be noted:

1. A parasite is usually smaller in size than a predator.
2. A parasite lives on or in its host, so that the host is both a source of food and a habitat, while a predator is free-living and relies on its prey only for food.
3. Parasites have generally higher reproductive rates than predators.
4. Parasites often have greater host "specificity" than most predators, becoming genetically more specialized in structure and physiology to fit into their host environment.
5. The predator and prey have independent lives until the act of predation takes place; parasites often cannot survive apart from their host.

These differences between predators and parasites are only generalized rules of thumb; many animals do not fit neatly into these separate categories. The leech is a good example. Except when it is feeding, it lives an independent life; moreover, many leeches can feed on other little animals if they cannot find a suitable host for a meal of blood.

Most predators kill their prey, but some, including various insect species, only consume a portion of their prey without killing it. Predator species also vary in the diversity of the organisms they will utilize as prey, in their adaptability to alternative food sources during periods of low supply, in their preferences for hunting grounds, and in their hunting ability, among other factors.

## Interspecific competition

No two species are ever in direct competition in the same ecological niche. This principle is expressed in Gause's Law, which states that competitors of different species with exactly the same requirements cannot coexist. In general, this law has held up well, although recently some studies have indicated that there may be coexistence over time; for example, weather conditions in the summer may favor one species, whereas the winter favors another for the same niche. It is difficult to say whether this really constitutes an exception to Gause's Law or whether it merely indicates that the concept of ecological niche should be redefined to include seasonal climate as a factor.

Although there is not direct competition between species for the same ecological niche, there may well be competition for some environmental resource which is in short supply. For example, during a prolonged drought in the African grasslands, the small inland lakes or water holes may dry up, causing great competition for the remaining water. Thus the giraffe and the wildebeest, for example, may be in direct competition for water, even though they occupy completely different ecological niches; in this one requirement the two species are overlapping.

The shorter the supply of the limited resource, and the more similar the need of the two species, the greater the competition between the two will be. In its severest form, this competition may result in decreases in the numbers of one or both species groups, decreases in the rates of growth and reproduction by individual competitors, and perhaps even ultimate extinction.

Fortunately, extinction is not the only possible outcome of direct interspecific competition. To protect itself from extinction, the less successful competitor may choose to abandon its home range and emigrate to another locality; for example, the giraffe herd may move to another area where there is more water, even though it means accepting the risk of increased vulnerability to predators. If it remains within its own home range, the competitor may be forced to settle on another less desirable but acceptable substitute from among its marginal alternatives; for example, it may switch to a food that is

**613**

harder to find or more difficult to digest but at least enables the animal to survive. Over a period of time, the species may through natural selection undergo an evolutionary adaptation to limit or eliminate the overlap and to allow it to adopt a new ecological niche; thus the competitive characteristics are displaced.

Many ecosystems have evolved to a relative state of complementary niches in which competition is at a minimum. Even among herbivores that favor the same plants, one may prefer the leaves of the plant, or the flower, roots, or stems exclusively. Or the competitors may favor plants at different stages of maturity or moisture content. The adaptations of aquatic species may enable them to live within the same narrow area but at different heights or depths. Time of day, season, or year at which species utilize the same resources may vary.

Competition may exist among plants as well as among animals. In fact, the phenomenon is far easier to observe in balanced relationships among plant life. Competition for light, space, water, and nutrients results in plants of different heights, root depths, varying rainfall requirements, or different cycles of growth all occupying the same area in relative harmony.

An interesting example of competition among plant species is that of the antibiotic inhibitors produced by shrubs in the California chaparral. Chemicals produced by two species of shrubs, *Salvia leucophyla* and *Artemisia Californica*, inhibit the growth of herbaceous plants. These toxins are produced in the leaves, and may accumulate in the soil during the dry season to such an extent that when the rainy season comes, germination or growth of seedlings is inhibited in a wide belt around each shrub group. There are very few herbs in the mature chaparral. Antibiosis is not, of course, limited to higher plants. Numerous examples among microorganisms are known. Penicillin, the bacterial inhibitor produced by a species of bread mold, is one example.

20-15 *Wildebeest, Connochaetes taurinus, congregate to drink at Ngorongoro Crater in Tanzania. Among the most active of antelopes, the wildebeest often herds with the zebra of the East African plains. (Mark Boulton from National Audubon Society)*

20-16 *Uniform spacing allows individual saguaros to maximize the limited water and nutrient supply available to them. Able to store much water in their tissues, saguaros can successfully resist long periods of drought. The absence of leaves, of course, minimizes water loss through transpiration. (Russ Kinne—Photo Researchers)*

**20-1**

**Ecological equivalents in four coastal zones**

| Niche | Upper east coast | Upper west coast | Gulf coast | Tropical |
|---|---|---|---|---|
| Grazer on intertidal rocks | Periwinkle (*Littorina, littorea*) | *L. danaxis, L. scutelata* | *L. irrorata* | *L. ziczac* |
| Bottom-feeding carnivore | Lobster (*Homarus*) | King crab (*Limulus*) | Stone crab (*Menippe*) | Spiny lobster (*Palinurus*) |
| Plankton-feeding fish | Atlantic herring, alewife | Pacific herring, sardine | Menhaden, threadfin | Anchovy |

Ecological niches are similar in type from ecosystem to ecosystem. In each, however, the niche may be taken over by different species. Thus the buffalo graze in Africa, the kangaroo in Australia, and cattle in North America. Species occupying the same niche in different ecological systems are termed **ecological equivalents.**

## The ecology of populations

A population is a group of organisms of the same species occupying a given geographical area. To comprehend those factors that determine the number of organisms in a population and their distribution, ecologists must probe the regulating mechanisms of birth, death, and survival, as well as the various operational population checks.

The total population numbers in a given area for a particular time period are known as its **density.** This can be expressed as the number of organisms per volume of unit area or as biomass per volume or per unit area. Density is an important consideration in studying the extent of interaction between individuals of the same and different species.

The size of a population is determined in large measure by its **natality,** or rate of production of new individuals by birth, and its **mortality,** or rate of individual loss through death per unit of time, and the difference between the two rates. If the number of births increases, and the number of deaths decreases or stays the same, population increases; if the reverse occurs, population decreases.

A second important factor influencing population density is the migration of individuals. **Emigration,** or the moving away from an area with no return, occurs generally as a result of overcrowding, food shortages, competition for resources, and various other limiting environmental situations. **Immigration,** or the inward movement to an area, can greatly increase population density. Having abandoned the situation in which its numbers could not increase successfully, the species often undergoes a boost in reproductive power. In some cases there may be two-way migration, with a movement away and a later return to the same area. Most common in

**615**

birds and fish, migration is usually seasonal, triggered by changes in weather or temperature. Migrating populations, therefore, undergo a seasonal or periodic flux in density.

When a species first colonizes a particular area, it begins to multiply in accordance with environmental factors until it reaches a point of balance with available resources and then continues to fluctuate around this level. When the population has reached the maximum size that can be supported on a sustained basis over a long period of time by the resources of a particular habitat, the environment is said to have reached its **carrying capacity.** It is a basic rule of ecology that no species may exceed the carrying capacity of its environment.

A number of mechanisms are at work to limit population size within carrying capacity. In 1840 Justus Liebig, a botanist, stated that the factor, or influence, of the environment that tends to stop the growth or spread of an organism is a **limiting factor.** This was originally thought of as a lower limit, but with Shelford's Law of Tolerance an upper threshold or ceiling was incorporated in the concept. The factor of which the limit (upper or lower) is first reached is often referred to as the primary limiting factor or the **critical factor.** A limiting factor for the eastern spadefoot toad is sandy soil, but the critical factor may well be soil moisture. A critical factor may vary by day or season; it may be predation or parasitism.

Population density is not only determined by the size of the population; it is also determined by its distribution. One factor that must be considered in measuring population density is **spacing,** or the distance between individuals. Dispersal of organisms may be random, uniform, or clumped, of which the last is most common. The phenomenon of territoriality also contributes to uniform distribution among certain species. Random spacing, on the other hand, is sometimes found among those plant groups which are characterized by random seed dis-

persing mechanisms. The most common form of spacing, however, is illustrated by the numerous instances of flocks, schools, colonies, swarms, and herds following the familiar clumped patterns. Many plant species also assume a clumped distribution. All such aggregates have points of optimum density after which overcrowding occurs.

**Cycles,** the regular fluctuations of some populations, also influence population density. Although the cause of rhythmic rises and falls in the population of some species is not fully understood, it is known that they typically occur in three- to four-year cycles or in nine- to 10-year cycles. The most well-known of these concerns the lemming, which undergoes such massive

**20-17** *Vast herds of barren ground caribou migrate during July across interior Alaska to their summer feeding grounds. Caribou form large aggregations for the purposes of traveling and feeding together where tundra vegetation is abundant. During the winter when food is scarce, they disperse to forage in isolated groups. (Charlie Ott from National Audubon Society)*

swings in population as to experience first over-population, then severe underpopulation, only to recover and undergo the same cycle. Two theories on cause have been posed. The first relates cycles to change in the physical environment, in the ecosystem, or in the population itself; while the second terms them purely random population fluxes brought about by random changes in the environment.

## Population checks

Once a given population has achieved balance with the carrying capacity of its environment, a great many situations may occur to throw off

**20-18** *The population of koala bears is self-regulated through competition for food. Since koalas eat one kind of leaf exclusively, the density of the koala population is dependent on the amount of eucalyptus leaves available to them. Less successful competitors may die or migrate to new habitats containing the requisite eucalyptus trees. (R. Van Nostrand from National Audubon Society)*

**20-19** *This flood in the basin of the Yellowstone River is an example of a density-independent limiting factor. Periodic floods, sometimes caused by overgrazing or logging, cause a cutback in populations of both plant and animal species, and the phenomenon is not affected by the density of the affected populations. (Courtesy of Carolina Biological Supply)*

this balance. In order to understand the dynamics of population control, ecologists often consider the theoretical situation in which unlimited resources are available to a particular species. Obviously if no natural checks exist on the species, the population can increase indefinitely. Under ideal conditions, the maximum population growth rate is termed the **biotic potential.**

One example of such a situation takes a single bacterium as a starting point. Within each 20-minute interval, the bacterium divides, as do its offspring. Thus if all progeny survive, within a day and a half mankind would be knee-high in these organisms. In only one additional hour, mankind would be drowning in them. The same point can be illustrated by a pair of houseflies. In a period of four months this set of organisms could produce nearly 20 billion billion descendants. Fortunately, unlimited population growth is impossible, owing to an elaborate system of natural checks and balances, known as **environmental resistance.** Through these devices, nature keeps its species populations at comparatively stable density.

Natural population checks are of two types: **density independent** (those forces outside the population which remain constant despite population size) and **density dependent** (those forces within the population that vary in intensity with population density). The combination of these forces serves to influence virtually every population fluctuation.

The fact that population density is influenced by external or density-independent factors is evident each season. As changes in precipitation or temperature occur, various species undergo rapid increase in numbers, as, for example, the familiar housefly or mosquito. As the season moves on, these favorable factors diminish and are replaced by unfavorable factors which serve to reduce the population. Unusual weather conditions, such as severe winters, extraordinarily heavy or light rainfall, and unseasonably hot or cold spells, are some of the causes of population

**617**

reduction or growth. Animals and plants starve, freeze, dehydrate, drown, have their habitats destroyed, or become exposed to predation through such natural climatic occurrences. Obviously increases or decreases in population size caused by such conditions occur without regard to present population numbers.

Man is also a density-independent factor. At times he has destroyed habitats and exterminated species through careless timber harvesting, hunting for hides, feathers, and trophies. Most recently he has added the careless or ill-advised use of pesticides and herbicides to his deadly arsenal.

In addition to external population checks, ecologists have discovered that each species living in a healthy environment is the self-regulator of its own density. Under normal conditions, a homeostatic balance is reached between the population and its environmental resources. If the population exceeds these resources or falls well below them, various density-dependent effects are triggered which operate to restore the balance.

One such factor is competition. As population numbers in a given area increase, available resources must be shared in ever-diminishing portions. When these portions become insufficient or virtually unavailable, competition accelerates both among members of the same species and members of different species sharing the same or a similar niche. When the scarce commodity is food, competition takes two forms: (1) **scramble competition,** in which each individual receives an insufficient share of the resource and wastage and high mortality result, and (2) **contest competition,** in which each successful competitor receives sufficient food, and only the smaller numbers of unsuccessful contestants die or are forced to seek alternative food supplies or new habitats. When space is the scarce commodity, fighting among members of the species occurs, and territorial animals increase their watchful defense at their own area. Unsuccessful

**20-20** *The lion is perhaps the best-known predator of the East African savanna. This one is feeding on a cape buffalo calf which it has killed. After such a meal, a lion may not hunt for several days. (Leonard Lee III —Omikron)*

competitors fail to find nesting areas or mating situations. Both forms of competition often result in large-scale emigration.

Severe overcrowding may also result in physiological and psychological stress among the population. Such conditions are believed to affect the endocrine systems of the organisms and to act as behavioral checks. In laboratory tests with mice, severe crowding was found to delay or halt sexual maturation, to increase fetus mortality, to reduce fertilization, and to cause inadequate lactation. Numerous examples have also occurred in natural settings where overcrowding causes stress that may lead to derangement, physiological shock, or failure of adrenal and pituitary function; the result is severe population cutbacks. A recent paper suggests that this may be part of the reason for the rapid extinction of the dinosaurs. Measurement of the shells of fossil dinosaur eggs revealed that shells were twice as thick at the start of the Cretaceous period as they were near its end; overcrowding may have led to physical stress that prevented the production of viable eggs.

Dense populations may also be drastically reduced through the agent of epidemics. Not only do the large numbers facilitate spread of disease, but experiments show that normal physiological responses are inhibited by crowding. Studies show that decreased antibody formation and weakening of other disease defenses increase the population's susceptibility. As a result, the rampant spread of disease may cause large population cutbacks. Lower susceptibility also affects parasitism, and a higher number of organisms become victimized as population numbers increase.

Predation also serves as a density-dependent factor in population checks. Increases in prey populations attract predators. Availability of a single food source often results in a tendency on the part of predators to concentrate on this source to the exclusion of the other food varieties in their diet. This ultimately results in an increase in predators and severe population cuts in the prey species. With the reduction in its food source, predator populations also undergo population reduction.

An example of the way these various checks and balances operate can be seen in the deer population at the Kaibab Plateau in Arizona. A census of the Kaibab deer taken in 1907 revealed a population of 4000 deer, although the carrying capacity of the area was estimated at 30,000. In the next 15 years, a concerted, if somewhat misguided, effort was made to increase the deer population by removing their predators—wolves, coyotes, and pumas. By 1925, the deer population had increased to 100,000, a far larger number than the environment could support. The available food supply was depleted and over the next two winters, 60,000 deer died of starvation. Disease and parasites also took their toll, so the herd was soon reduced below 10,000.

## Dynamics of ecological communities

We have seen that organisms interact with organisms and species with species, but populations also interact to create a stable ecosystem. The biotic community undergoes numerous stages of development, ranging from the simple to the complex, with the population of each stage leading to the opening of new niches, permitting additional species to enter the system. The stability of an ecosystem is based on an orderly and predictable sequence of changes. The orderly progression from one community stage to another is termed **ecological succession.**

## The nature of succession

As an open field gradually evolves into a forest, a series of intermediate steps are taken—from

*20-21 Normal predation may be essential in permitting a population to approach the carrying capacity of its environment. In the first quarter of the century, predators were removed from the Kaibab Plateau on the north rim of the Grand Canyon in Arizona in order to allow the deer population to expand. This has become a classic example of mismanagement: in the following years the deer population increased to about 100,000, but was then decimated by starvation.*

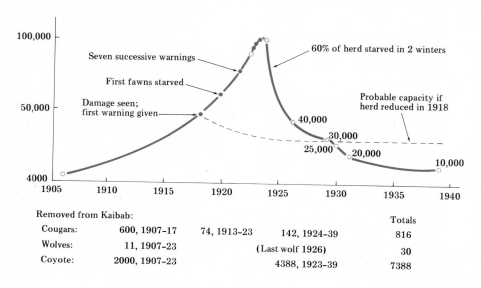

| Removed from Kaibab: | | | | Totals |
|---|---|---|---|---|
| Cougars: | 600, 1907–17 | 74, 1913–23 | 142, 1924–39 | 816 |
| Wolves: | 11, 1907–23 | | (Last wolf 1926) | 30 |
| Coyote: | 2000, 1907–23 | | 4388, 1923–39 | 7388 |

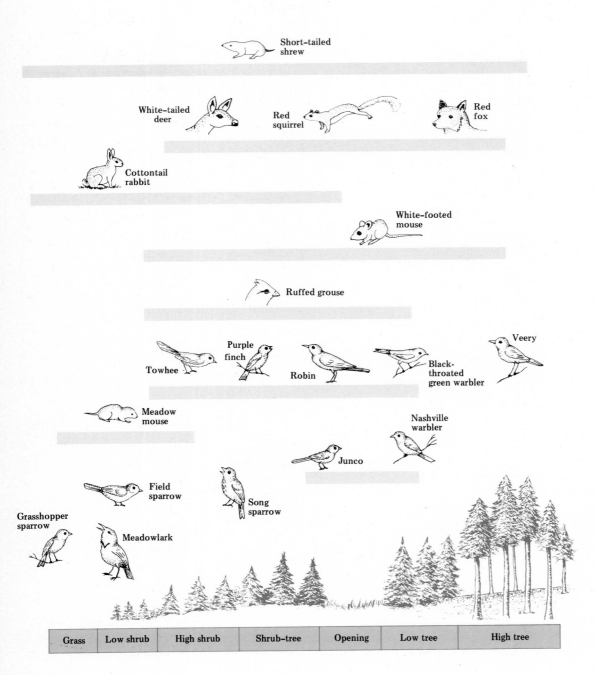

**20-22** *Wildlife succession is shown in a conifer plantation in central New York State. Some species are common to all stages, while others appear and disappear with changes in the density and height of the vegetation.*

grass, to shrubs, and finally to trees. Each of these developmental stages in the succession process is known as a **seral stage.** The entire progression of seral stages is termed a **sere.**

In each seral stage, dominance of certain species first rises and then declines and is superseded by a seral stage more compatible with the new physical properties which its successors have introduced. Each stage thus serves to alter the environment which it originally found favorable in ways which favor other populations. Ultimately these seral stages may progress to a final permanent stage known as a **climax community.** However, in some cases, a community remains in a fairly stable state at a seral stage just preceding the climax; this is called a subclimax community.

The net result of ecological succession is an increase in symbiosis among members of the community, nutrient conservation, and stability. The overall function is to achieve as large and diverse an organic structure at climax as is possible within the limits set by the available energy input (sunlight) and the prevailing physical condition (water, climate, soil). The increasing size and complexity of the biological community enables it to approach at climax a state of homeostasis or self-regulation. Homeostasis reduces the effect of environmental change upon the community.

Development that begins in an area not previously occupied by a community (a lava flow for example) is known as **primary succession.** The succession of plants and invertebrates on the sand dunes of Lake Michigan illustrates this type of succession. The lake was once larger than it is now. As it retreated, it left successively younger dunes. The communities first to establish themselves on the dunes were grasses, willows, cottonwoods, and cherry trees; and animals like white tiger beetles, burrowing spiders, and grasshoppers. These communities were displaced by the jack pine forest community. Locusts, digger wasps, and ants appeared, and burrowing

**20-23** *Tom's Swamp was changed by generations of plants that had succeeded each other in cycles beginning long before the Revolutionary War, when it was named for the man who disappeared there with his oxen. Water plants filled up the pond until they formed a mat of roots and peat thick enough for water-loving tree species, such as spruce, to grow upon. As the soil gradually accumulated, drying out the area, hardwood trees began replacing the evergreens. (Alycia Smith Butler)*

spiders and white tiger beetles died out. The jack pine forest gives way to the black oak dry forest. Ant lions, flat bugs, wire worms, snails and six new species of grasshoppers appeared. Black oak gives way to oak and oak-hickory moist forest. Millipedes, camel crickets, centipedes, betsy beetles, snow bugs, and earthworms become dominant. The climax community is a beech-maple forest. The oak-hickory forest invertebrate species are retained and a new species of locust and seven new species of snails now thrive. Thus, the first community is followed by a series of different forest communities, each with a changing animal population. Although it began upon dry and infertile sand, the development of the community results in moist and rich beech-maple forest. The soil is deep humus in contrast to the sand. The original inhospitable sand dunes are completely transformed by a succession of communities. It has been estimated that it took 1000 years to reach a forest climax on the dunes of Lake Michigan.

Development that begins in an area from which a community has been removed is called **secondary succession.** In 1917 Shantz described the succession which occurred on the abandoned wagon roads used by the pioneers crossing the grasslands of the central United States. There are four stages to the ecological succession on the wagon tracks. First, there is an annual weed stage. It is followed by a short-lived grass stage. Short-lived grass gives way to early perennial grass. A climax grass is reached in 20 to 40 years. While the species that comprise the seral and climax communities of this grassland differ from those that comprise the communities of the Michigan dunes, the overall pattern is the same. A succession of communities transforms initial environment until a stable climax community is established.

When a climax community is reached, it will be characterized by a wide variety of species, complex food webs, stable social organization, and a self-perpetuating structure. As such, it may endure for hundreds of years, unless it is entirely destroyed through such major environmental disturbances as fire, disease, severe drought or storms, overgrazing, or industrialization.

Fully developed climax communities are generally categorized into a limited number of broad biotic units called biomes. A **biome** is a group of communities characterized by a distinctive type of vegetation and climate. The major continental biomes include the desert, tundra, deciduous forest, tropical rain forest, taiga, and grasslands.

## Biogeography

The geographic distribution of living organisms is a subject of great interest to ecologists. It would seem that the appearance of various species of plants and animals in different parts of the world would be a readily explainable phenomenon. Species occur in unlikely distributions across the face of the earth and scientists seek the reasons why.

## Dispersal and colonization

The first step in studying the theories of biogeography is to comprehend the methods of dispersal by which species move from area to area, and what accounts for their successful colonization of their new habitat. There are various means by which organisms transport themselves to new locales. Among the chief means for small organisms are the wind and tides. Airborne insects, spores, and seeds, as well as small invertebrates, may be carried great distances by wind, especially during storms. Strong tides may also carry both aquatic and terrestrial plants for miles; while floating tree limbs may house unwittingly seafaring organisms.

For larger animals, the main means of dis-

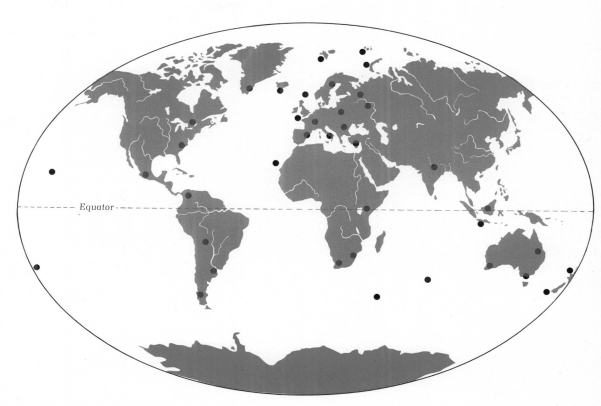

**20-24** *The dispersal of plant and animal life may be by means of land, water and air. The ecologist creates dispersal maps (above) indicating distribution of species. These maps aid him in the study of emigration, immigration and migration of various species over select periods of time.*

persal is through self-locomotion. Although great distances may not be traversed by this means, over many generations widespread distribution may occur. Along with these animals will also travel seeds and larvae caught in their fur or feathers. Man, too, by introducing new species into various parts of the world, is an important agent of dispersal.

Despite the sometimes random nature of dispersal, certain species are never found to be inhabitants of given areas. Successful colonization is dependent on various factors which govern the presence or absence of all animal and plant species. The first consideration is the problem of accessibility. Large desert masses, major bodies of water, or mountain ranges are natural barricades for many species of land animals. So, too, may be an open expanse of land for species fearful of exposure. Small organisms traveling by

airborne or waterborne routes are far more likely to experience broad dispersal than are such organisms as large terrestrial mammals.

But even if one were to remove the problem of access by transporting the species by mechanical means, successful colonization still depends on other factors. To survive in its new habitat, a species must be physiologically suited to it. The potential colony must have an adequate food supply and an acceptable climate, as well as meet other environmental requirements. This will generally occur if the new habitat is similar to the one formerly occupied. If survival is accomplished and reproduction takes place, the colony is likely to be successful.

One additional factor, however, will still be necessary. Within the potential community, the new species must be able to contend with competition for resources. This means that the species

**623**

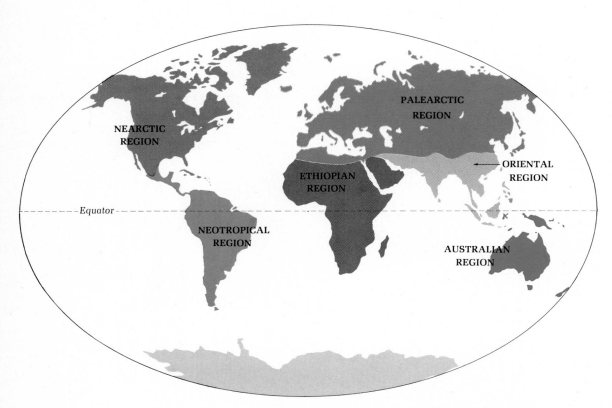

NEARCTIC
REGION

PALEARCTIC
REGION

ORIENTAL
REGION

ETHIOPIAN
REGION

Equator

NEOTROPICAL
REGION

AUSTRALIAN
REGION

will either have to occupy an available niche or will need to compete with established members of the community. Functional relationships must be achieved in relation to the environment and to other populations in the same area.

## Biogeographical regions

According to the principles of successful colonization just described, a species must have physical accessibility to a geographical region in order to establish a population in that area. Curiously, however, closely related species are found distributed upon continents and across natural barriers that would seem insurmountable. One logical hypothesis for this odd distribution states that shapes of continents, land links, and climates were previously far different from those

existing today. Continents that were connected now stand severed; frigid zones were temperate. Dispersal was therefore possible through normal means.

Data on worldwide animal and plant distribution as well as fossil records and other biological and geological evidence indicate that the earth may be divided into six major biogeographical regions. Each covers a broad continental landmass and is separated by major natural barriers, such as a desert, ocean, or mountain range. These are the Australian, the Neotropical, the Oriental, the Ethiopian, the Palearctic, and the Nearctic. Each is distinguished by its major animal species.

**Australia.** Australia has the greatest number of endemic species because of its relative isolation from the rest of the world. The montremes, the spiny anteater and the platypus live here. The

**20-25** *The world's faunal regions, each of which embraces several or many biomes, are indicated on the map. A faunal region may be distinguished by the prevalence of characteristic animals, for example, the African elephant in the Ethiopian and the Indian elephant in the Oriental faunal regions.*

duck-billed platypus is an egg-laying mammal. While it suckles its young, it does not have true breasts, only tiny sievelike pores. To nurse, the female lies on her back and contracts her stomach muscles to form a groove or tunnel out of which the young scoop up the milk. The platypus feeds in the water. It uses its bill to dredge for worms, crayfish, and grubs. The platypus thus combines two characteristics—egg-laying and nursing—that were thought to be incompatible in one species before it was discovered.

Australia has more and more different species of marsupials than any other biogeographical area. The koala bear lives in the Australian forest. It feeds upon the leaves of the eucalyptus tree, another endemic species. It is a marsupial because after its young are born they must mi-

**20-27** *Although the wallaby looks rather like its relative, the kangaroo, it is a smaller animal and fits into a somewhat different ecological niche. Both are marsupials, and both graze on the grasslands of the open plains. They are great hoppers and leapers, like many grasslands herbivores. (Gordon Smith from National Audubon Society)*

**20-26** *The duck-billed platypus, a reptilelike mammal that lays eggs yet secretes milk is exclusive to Australia and New Zealand. A curious blend of reptilian, mammalian, and unique traits, the platypus belongs to the monotremes, a small group of mammals that diverged from the main lineage too early to be considered ancestral to modern mammals. (Tom McHugh—Photo Researchers)*

grate to a pouch upon the stomach where they remain for about six months. Wallabies are the junior members of the kangaroo family. They are herbivores and feed in the morning and evening. They are the ecological equivalent to deer in North America. Another Australian marsupial is the brush-tail possum. It is a night feeder that eats fruits and nuts. Its diet and habitat are similar to those of several rodent species in the United States. It is itself the principal food source for the tree goanna lizard, an endemic and carnivorous reptile. Other marsupials in Australia are the kangaroo, the wombat, the Tasmanian devil, and the sugar glider.

**625**

**South America.** South America, because its only terrestrial attachment to another continent is through the thin neck of Central America, has developed many species of its own. The vicuña and guanaco are two grazing animals of the Argentine pampas. Their ancestor is the camel, which emerged in North America in the Eocene and evolved entirely in the western hemisphere before migrating to Asia. The rhea, or American ostrich, is the largest bird in the western world (four feet high) and flightless. The capybara, a four-foot rodent, lives in swampy sections of the pampas and is an agile swimmer.

The gallery forests of the Brazilian jungle are a microcosm of endemic life. Lianas and epiphytes (air plants) grow upon the tall trees that crown the forest. The three-toed sloth clings to branches of the lower trees. Night monkeys hunt insects among the trees. The tayara, a weasel-like carnivorous mammal, climbs the trees to hunt wild

honey, or a bird's nest and its young. On the floor of the forest the tapir, related to the rhinoceros, browses the rivers and lakes, drawing aquatic plants into its mouth with its mobile proboscis. The jaguar feeds on the tapir.

**20-29** *The capybara, the world's largest living rodent, is indigenous only to South America. It is a primary consumer (herbivore), and belongs to the class of rodents known as caviomorphs, which also includes the guinea pig and chinchilla. Before the late Pliocene such lines were able to evolve free of competition from other rodents on the island continent of South America. (Karl Weidmann from National Audubon Society)*

**20-28** *The rhea is one of the dominant birds of the plains. Flightless because its wings are vestigial, the rhea is a strong runner, and avoids predators in this way. Although superficially similar to the ostriches, of the order Struthioniformes, the rheas are grouped into their own order, the Rheiformes. (Photo Researchers, Inc.)*

**20-30** *The tapir is native to South America, and is found from Colombia to northern Argentina. Although it looks rather fierce, and has the reputation of being a bad-tempered sort, the tapir is strictly a vegetarian. It is distantly related to the African rhinoceros. (A. W. Ambler from National Audubon Society)*

## Summary

Ecology, or "environmental biology," is the branch of science dealing with the interactions among organisms and the relationship between organisms and their environment. For ecological purposes the habitat is divided into the living biotic component and the nonliving abiotic component. The term "ecological niche" indicates the functional role of a species in its ecosystem; the ecosystem is a specific and self-contained area in which the ecological features and interactions can be described. The precise composition of a particular ecosystem will depend on climatic, physiographic, and geological factors, but these in turn may be modified by biotic factors.

Ecology studies the flow of energy and materials within ecosystems, through analysis of food chains or webs. Food webs involve interaction between producers, primary and secondary consumers, and decomposers. The second law of thermodynamics requires that food chains involve energy loss which can be shown graphically by pyramids of energy, numbers, and biomass. Food webs also involve biogeochemical nutrient cycles, such as the nitrogen, carbon, and phosphorus cycles.

In environmental studies, the species is considered as an ecological unit. Intraspecific cooperation or competition has important ecological consequences. Interspecific relations include symbiosis—mutualism, commensalism, parasitism, and amensalism—and predation. Interspecific competition is limited by Gause's Law, but may have important ecological or even evolutionary consequences.

Population ecology is the study of species density, environmental carrying capacity, limiting factors, and population cycles. In practice this comes down to the study of biotic potential on the one hand and environmental resistance on the other. Populations interact to create a stable ecosystem; in the course of their interaction the biotic community undergoes an orderly ecological succession through seral stages to result in a climax or subclimax community. Climax communities are stable, complex, and self-perpetuating, and are generally classed within broad biotic units called biomes. The study of species distribution within the major biomes is the complex subject of biogeography, and is a subject of great interest to the ecologist.

# Bibliography

## References

Cole, L. C. 1958. "The Ecosphere." *Scientific American*. 198(4): 83–92. Offprint no. 144. Freeman, San Francisco.

Eisenberg, J. F., N. A. Muckerhirn, and R. Rudran. 1972. "The Relation Between Ecology and Social Structure in Primates." *Science*. 176: 863–875.

Hutchinson, G. E. 1970. "The Biosphere." *Scientific American*. 223(3): 44–53. Offprint no. 1188. Freeman, San Francisco.

McVay, S. 1966. "The Last of the Great Whales." *Scientific American*. 215(2): 13–21. Offprint no. 1046. Freeman, San Francisco.

Talbot, L. M. 1970. "Endangered Species." *BioScience*. 20: 331.

Turner, F. B., ed. 1968. "Energy Flow and Ecological Systems." *American Zoologist*. 8: 10–69.

Westhoff, V. 1970. "New Criteria for Nature Reserves." *New Scientist*. 46: 108–113.

Woodwell, G. M. and H. H. Smith, eds. 1969. *Diversity and Stability in Ecological Systems*. Brookhaven Symposia in Biology No. 22, New York.

## Suggestions for further reading

Bates, M. 1960. *The Forest and the Sea*. Vintage Books, New York.
An early and classic introduction to the principles of ecology. Carefully and sensitively examines a diversity of ecological communities.

Boughey, A. S. 1968. *Ecology of Populations*. Macmillan, New York.
Provides a statement of the principles underlying relations between living organisms and their environment. The mechanisms for changing relationships and the problems involved in studying them are examined.

Kormody, E. J. 1965. *Readings in Ecology*. Prentice-Hall, Englewood Cliffs, N. J.
A representative selection of papers covering a wide range of ecological literature, from Linnaeus and early natural history to modern concepts of populations, communities, and ecosystems.

Milne, L. and M. Milne. 1971. *The Arena of Life*. Doubleday, Garden City, N. Y.
An outstandingly illustrated and extremely up-to-date text on the dynamics of
ecology. Examines the solar energy cycle, the many webs of life, physical
challenges, and the pyramids of life in major biomes.

Odum, E. P. 1959. *Fundamentals of Ecology*, 2nd ed. Saunders, Philadelphia.
One of the earliest but still appropriate texts on the principles of ecology. Also
emphasizes description and application. Remains an excellent reference.

Whittaker, R. H. 1970. *Communities and Ecosystems*. Macmillan, New York.
An introductory text touching on a variety of topics including the nature of
communities, their relationship with the physical environment, species diversity,
gradient analysis, major communities in relation to climate, chemical ecology, and
biogeochemical cycling.

**Books of related interest**

Farb, P. and the Editors of *Life*. 1963. *Ecology*. Time, New York.
A sensitively photographed and simply expressed study of the relationships
among living things.

Krutch, J. W. 1949. *The Twelve Seasons*. Sloane, New York.
A distinguished nature writer reflects on the unity and diversity of life as it is
revealed in each month of the year.

Laycock, G. 1969. *The Alien Animals*. Ballantine, New York.
The principles of ecology are revealed by examining case histories of animals that
have been transplanted into alien environments with tragic results.

Leopold, A. 1949. *A Sand County Almanac*. Ballentine, New York.
An established classic and now even more timely plea for an ecological conscience.
Describes the seasonal changes of a region and shows how the delicate natural
balance can be ruined easily by man's destructive violence.

Porter, E. 1962. *In Wildness Is the Preservation of the World*. Sierra Club, San
Francisco.
Exquisite photographs by Porter complemented by selections from Thoreau's
*Journal*. Joseph Wood Krutch has provided an introduction.

Storer, J. H. 1953. *The Web of Life*. New American Library, New York.
A popular and concisely written introduction to the principles of ecology.

# 21
# The biome

The ecological process of succession results in a fairly stable community, the climax, which then remains relatively unchanged over long periods of time. Together with the climax community will be others—either intermediate in stages of succession or more or less stabilized at an earlier seral stage because of special local conditions. Such a group of communities dominated by a typical terrestrial climax community and shaped primarily by the climate of the region is known as a **biome.** Biomes differ, because the local climate determines a characteristic community of vegetation, which in turn supports a particular assemblage of animals. Among the major types of biomes are the tundra, taiga, grassland, desert, and several varieties of forest.

The general composition of a biome type is remarkably uniform; the plants and animals of a tundra biome, and their interrelationship, are much the same whether the tundra is in Ireland, northern Russia, or the Canadian northwest. Still there may be significant local differences. For example, one vast area of coniferous forest may be dominated by Engelmann's spruce, while in another area there is mostly black and red spruce. Oak-hickory may predominate in one region of deciduous forest and beech-maple may be the important association in another. Recent disturbances in one ecosystem, such as floods, population fluctuations, or the introduction of a new species, may also create changes that lead to local variations in biome types.

The delicate equilibria characterizing the biomes have come about through complex interactions between the living inhabitants and their environment. Bison and prairie dogs were probably as important in maintaining the grasslands of yesteryear as were periodic fires. Beavers have created many of the lakes and ponds on the North American continent. Trees too are important to the ecosystem, for within a deciduous forest they moderate temperature and humidity while enriching the soil. Moisture is maintained in Pacific coast soil by redwoods that are able to condense water droplets from the constant fog that enshrouds that region.

In the discussion that follows we shall look at some of the biomes found in the western hemisphere. The character of the vegetation in corresponding biome types around the world will be much the same as in these samples chosen for study, but ecologically equivalent animals consist of different species.

## Tundra

Circling the globe in a broad belt at the edge of the Arctic Ocean, or south of the polar ice cap, is the tundra, the most northerly of the biomes. Its southern boundary has been variously defined as the Arctic Circle, the tree line, or the line connecting all the points on the globe that have an average temperature of about 10°C. in the month of July. The tundra extends throughout much of northern North America, Eurasia, and the coastal regions of Greenland.

The tundra is remarkably uniform regardless of geographic location. This uniformity is due in large part to the common and fairly recent evolutionary origins of the plants and animals in this biome. There simply has not been enough time for the evolution of large numbers of varying organisms.

The most obvious distinguishing characteristic of the tundra is the lack of erect trees. Because it is treeless, it seems at first glance to be a depressingly bleak landscape. But the gently rolling terrain of the arctic tundra possesses a powerfully awesome beauty. The biotic characteristics of this biome are the result of its harsh environmental conditions, most importantly the intense cold of the long winters, the nearly incessant wind, and the permanently frozen ground. Other factors influencing the tundra biome include the short growing season, the poor soil, and the marked climatic difference between the summer and the winter seasons.

During the winter there is little daylight and temperatures may hover as low as −70°F. The mean January temperature is −30°F. Although the summers are short, the days are long; in areas such as Tromso, Norway, the sun never falls below the horizon between May 21 and July 23. The warmth of the summer sun melts the upper few inches of permafrost, creating a soggy landscape riddled with thousands of shallow puddles or bogs. With the low annual precipitation (generally less than 10 inches), and the permafrost making water unavailable for much of the year, it is no wonder that the tundra is often referred to as a cold semidesert. Despite the scant precipitation, however, the air is too cold to absorb much moisture, and since the ground is always frozen, water cannot drain down through the soil. Thus water is conserved, trapped between the permafrost and the cool upper air.

Since the tundra is of relatively recent origin, having developed as the glaciers of the last ice age retreated, the soils have scant organic matter and are characterized as either brown loams or rocky and gravelly sands. The constant freezing and thawing makes the soil extremely unstable and hazardous to plant growth.

21-1 *The arctic or shrub willow is a particularly hardy species, able to withstand constant disturbance of the soil, buffeting by the wind, and abrasion from wind-carried particles of soil and ice. Willows, along with birches, occupy well-drained sites where the soil is of relatively fine material. They form part of the arctic "snow patch" community, small pockets which occur where wind-driven snow collects in shallow depressions and protects the plants beneath. (Photo by Alexander B. Klots)*

The vegetation that dominates the tundra are the lichens, such as the ubiquitous reindeer "moss," true mosses, grasses, and sedges. Shrubs such as bilberry, bearberry, and leatherleaf are common. Here and there may be found patches of herbaceous plants, such as arctic poppy and saxifrage. Most tundra plants are perennials and may take years before flowering, if they flower at all. The few trees that manage to survive are either dwarfed or grow horizontally along the ground. Examples are the arctic willow, birch, and alder.

Animals have only two choices if they are to survive the cold of the tundra. They may migrate south, as do most of the birds and the larger mammals. Or, as in the case of lemmings, they may burrow beneath the snow and enter an extended period of torpor.

Adaptations to the cold climate can also be seen in the size and shape of tundra animals. The species of bears and foxes found in the tundra are somewhat larger than their relatives to the south. This observation is expressed in Bergmann's Law, which states that homeotherms tend to become larger in cold climates. The survival value of this adaptation is that a larger animal has more internal bulk and less surface area to radiate vital heat. Allen's Rule, which states that extremities become shorter in colder climates, also seems to be generally true. For example, the ears of one arctic fox are much shorter than those of the red fox found in the more southerly forest biomes. This adaptation also helps to conserve heat.

The largest herbivores of the tundra are the barren ground caribou and the musk ox. The tendency of large herbivores of open areas to be running animals is well illustrated by the caribou, which depends on its speed to escape its chief predator, the wolf. Caribou are on the move the year round. During the spring they leave the forests to the south and migrate northward, feeding mainly on lichens. They are forced to move constantly, since they are bothered by clouds of mosquitoes and blackflies. Later, as summer is ending, they begin their southern trek back to the forests to mate and breed and to spend the winter.

At one time, large herds of musk oxen roamed the tundra, but by 1917 they had fallen in numbers to an estimated 500 due to excessive hunting. Now, protected by law, the musk ox seems to be staging a slow comeback. About the size of a cow, the musk ox is well adapted to withstand cold with its double fur coat. The long shaggy outer guard hairs cover an undercoat of short, thick downy wool. Moreover, musk oxen, being gregarious animals, tend to group together in herds of as many as 50 individuals, so that the collective breath exhalations produce a miniature "greenhouse effect," warming the air about them. When threatened, adults form a defensive circle with calves inside.

The giant of the western North American tundra is the barren ground grizzly. The grizzly is an opportunist in search of almost any animal or even plant food. Like the musk ox, its numbers have declined drastically because of overhunting.

Few animals are as closely identified with the tundra as the lemming. This small furry rodent is a primary consumer, living on tundra grasses and their roots. Like most other tundra residents, the lemming's population is cyclic, rising rapidly and then crashing suddenly approximately every four years. It is a staple food item for a number of predators such as the snowy owl and the arctic fox. Another important herbivore is the arctic hare, whose white fur lends a measure of protective camouflage during the winter from its predators, the ermine and the snowy owl, both of which also have white coats so they can get close to their prey without being seen.

The willow ptarmigan, so called because of its preference for willow, is superbly endowed against predation and cold. During the summer this pigeon-sized bird is spotted brown and matches the lichen-covered rocks; in winter it changes its regalia to snow-white. The ptarmigan is well camouflaged on the ground but quite visible in the air, where it is preyed on by jaegers and gyrfalcons. Therefore it rarely flies, but remains motionless almost until it is stepped on. In the winter, a frightened ptarmigan may dart right into a snowdrift to escape a predator. The ptarmigan's feathers completely cover its feet, thus preventing heat loss from its extremities.

Unlike the ptarmigan, many of the tundra's summer inhabitants are visitors, the migratory birds which may fly thousands of miles from their winter quarters. Many take up temporary residence in coastal areas of the tundra to avail themselves of the abundant plant and insect life, and also to mate and breed.

Succession in the tundra follows a cyclic course unlike that in other biomes. As one plant community succeeds another, thickening layers of vegetation create more shade, which gradually reduces the amount of heat that actually penetrates to the soil. This causes the permafrost to creep gradually toward the surface until it pushes against the roots of the established plant life. As the permafrost nears the surface, there is less summer meltwater and the ice kills the roots. When the plants die, the exposed soil is eroded by water and wind. With the surface raw, the way is once more open to an invasion of the pioneer community of lichens and succession begins anew. The total cycle takes place in about 50 years.

The tundra is both the simplest and the least productive of all the biomes. Thus, the populations of many herbivores, such as the lemming and the arctic hare, move through cycles of explosion and crash; this, in turn, affects the population of predators that depend upon them. The fact that the ecosystem of this biome depends on only a few producer species and primary consumers make it an extremely vulnerable and unstable ecosystem. In recent years fires have inflicted serious damage to the lichen ground cover. Coupled with this has been the great harm done by excessive hunting of the musk ox, grizzly

21-2 *The "grizzly" bear is so called presumably because he is grizzled-coated and grizzly-looking. Despite their lumbering appearance, grizzlies are swift and powerful, can climb trees and swim well. They eat enormous quantities of almost any kind of animal or vegetable matter and appear immune to almost everything except natural disasters and man. (W. J. Schoonmaker from National Audubon Society)*

21-3 *Recurrent population crises among lemmings have produced some rather far-fetched stories about mass suicide attempts in which hundreds of pathetic little rodents "periodically patter to their doom" from the mountains by casting themselves en masse into the ocean. The lemming population does, however, outreach its food supply every three or four years. At population peak, they emigrate outward in all directions from their habitat, showing some erratic behavior that usually proves suicidal. (Photo by Alexander B. Klots)*

*21-4 The willow or rock ptarmigan is the resident bird of the tundra. Unlike most birds, it stays through most of the winter. If conditions prove too difficult, however, this species may also retreat to the margin of the taiga. The ptarmigan has three distinctive changes of plumage in adaptation to the change of season. In summer, they are brown; in autumn, gray; and in winter, white. (Pat Witherspoon from National Audubon Society)*

and polar bears. The recent discovery of enormous deposits of oil near Prudhoe Bay on the North Slope of Alaska poses a new threat to the tundra.

## Taiga

At the southern edge of the tundra, a sprinkling of scraggly wind-swept trees marks the transition zone with the taiga, or boreal forest. This northernmost forest, dominated by thick stands of conifers, circles the polar area, ranging over much of Siberia, Scandinavia, Canada, and Alaska; it forms an irregular band roughly 800 to 900 miles wide. This biome type is also found hundreds of miles to the south along the spines of such mountain systems as the Rockies.

As a result of the planing action of moving glaciers, the topography of the taiga is fairly flat. Characteristically, there are many small lakes. The climate is only slightly less severe than that of the tundra. Winters are bitterly cold, with temperatures averaging always below zero, but summers are warmer and of longer duration than in the tundra. The growing season is approximately three months, and summer highs range between 50° to 70°F. Precipitation in the taiga is evenly distributed throughout the year, and varies from 15 to 40 inches annually, depending on the region. Fog is frequent and the humidity is usually high. In the more northerly regions there is some permafrost. A continuous winter cover of snow and the windbreaking action of the trees tend to moderate the freezing temperatures. Because of the cold, decomposition and decay are slow, so there is a thick accumulation of leaf litter on the forest floor.

The dominant trees of the taiga are evergreens; in the North American coniferous forests, spruce, tamarack, fir, jack pine, and hemlock are the important needle-leaved trees. Because their growth is so dense, and because of the thick carpet of leaves, there is practically no shrub growth under the trees. Deciduous trees such as aspen, birch, and willow are typically found growing around the edges of acid bogs in the taiga. Other bog inhabitants are insectivorous plants (the sundew, for example) and quaking mats of sphagnum moss.

If spruce is the dominant plant of the taiga, then certainly the moose is the dominant large herbivore. For this reason, the biome is sometimes referred to as the "Spruce-Moose" biome. The moose, usually depicted as ugly and ungainly, is well adapted to running swiftly on deep snow with its broad hooves. In late winter, when food becomes scarce, it stamps on the snow to pack it down, creating a hard runway that

21-5 *Lakes, ponds, and bogs are numerous in many regions of the taiga and provide a variety of habitats. There is a gradual decrease in the variety of animal and plant life northward from the deciduous forest toward the tundra. As conditions of life become more severe, trees become smaller and more stunted, giving way to shrubs and herbs. In the central taiga, dominated by evergreens, there is little undergrowth. (Courtesy of the American Museum of Natural History)*

enables it to reach choice morsels of willow or birch. In summer, the moose is usually found browsing in glens of aspen, birch, and willow, from which it can wade into a pond. There, up to its shoulders in water, it may feed on the succulent plant life, such as pond lilies and sedges, that it dredges from the bottom.

Other large herbivores commonly found in the taiga are the elk or wapiti, the caribou, and the whitetail and mule deer. Numerous small herbivorous mammals include the snowshoe hare, the northern flying squirrel, the red squirrel, and the red-backed vole. As its name implies, the snowshoe hare has broad, furry feet, enabling it to bound easily over deep soft snow. Unlike the

arctic hare which remains white the year around, the snowshoe exchanges its white coat for brown during the summer months.

The most important large predators of the taiga are the timber wolf, Canada lynx, and red fox. Lesser predators include the fisher, marten, and wolverine, and such prized fur-bearers as the mink, ermine, otter, and short-tailed weasel.

The timber wolf, which once ranged over all the North American taiga, is now restricted to Canada and Alaska; it has largely disappeared from the continental United States because of the campaign waged against it by farmers and ranchers, who consider the wolf a threat to their livestock. Until very recently, most western

**21-6** *The Alaskan moose is a powerful animal and a notorious fighter. Once the young bulls have developed their horns in the fall, they immediately begin looking for adversaries. Moose browse for leaves and bark and submerge themselves completely in rivers to obtain their favorite food, water plants. They are exceptionally strong swimmers and have been observed crossing wide arms of the sea. (Charlie Ott from National Audubon Society)*

**21-7** *Wolves are highly adaptive creatures that can accommodate themselves to a variety of living conditions. They sometimes live alone, sometimes in family groups, and at other times in packs. They are inveterate travelers. Individuals or whole populations may migrate if their food supply itself fluctuates. Except during breeding season when they usually raise their young in permanent lairs, they are basically nomadic. (Tom McHugh —Photo Researchers)*

states paid bounties for dead wolves. Wolf packs hunt moose, caribou, and deer, and play a valuable role in maintaining healthy and stabilized herds of these herbivores. The animals that are successfully downed by wolves are the old, sick or weak, or very young. Where wolves are absent, the moose populations build up to such numbers that the vegetation is devastated. Despite this knowledge, wolves are now being decimated in Canada by professional hunters.

Chief predator of the snowshoe hare is the Canada lynx, a medium-sized cat with a bob-tail. Its huge paws make it as adept as its prey in bounding over soft snow. Population size of the lynx, not unexpectedly, follows that of the hare, but with a lag phase.

Birds such as the Canada jay, the red crossbill, and the goshawk may be regarded as indicators of the taiga, as they are endemic to this biome. The crossbill may frequently be seen hanging from spruce or pine cones, using its peculiarly shaped bill to extract seeds. With its short wings, the goshawk is able to maneuver adroitly among tree branches as it pursues small birds and rodents. Other birds that are typical of this biome are the great horned owl, the Hudsonian chickadee, and a variety of siskins.

Some mammals, such as the hoofed herbivores, remain active throughout the winter. Other mammals, such as bats, ground squirrels, and woodchucks, survive the winter by hibernation. Their body temperatures drop drastically, and their metabolic processes slow down. The black bears retreat to a den where they enter a period of torpor, or deep sleep. Since there is no drop in body temperature, bears do not hibernate, contrary to popular belief.

While homeotherms are well able to survive the cold, poikilotherms are poorly adapted to life in the taiga, since they cannot keep their body temperatures high enough to survive the cold winters. Consequently, reptiles are rare, but a few amphibians such as the wood frog and the Hudson Bay toad are widely distributed here.

**637**

As a result of disturbances (most of them either directly or indirectly inflicted by man), the taiga has undergone some changes in recent decades. Overhunting of predators prized for their pelts, or killed simply for sport, have brought about shifts in population density of small rodents as well as other mammals. Another disturbance to the taiga's ecosystem is the increasingly heavy logging, particularly where whole areas are completely cleared of trees. Not only is this devastating to wildlife, but it also results in the erosion of valuable topsoil. Fires, caused either by lightning or by man, have destroyed entire ranges of slow-growing lichens in the Alaskan taiga. Such areas are quickly invaded by rapidly growing birches and willows. This change in vegetation has brought about a corresponding change in the animal population. There are fewer caribou (they eat lichens) and more moose (they eat willows).

Fires have been a mixed blessing to the balance of the ecosystem. Over the short-term period, fire obviously results in widespread destruction of wildlife and loss of habitat. However, it may be that large stretches of taiga are still dominated by coniferous trees rather than the broad-leaved trees because conifers are better able to recover from the disastrous effects of fire.

## Deciduous forest

The temperate deciduous forest is situated to the south of the taiga; there is often a transition zone in which the two biomes intermix. The deciduous forest is not as continuous or extensive as the taiga, nor are its boundaries so clearly defined. In the western hemisphere it sprawls over much of the eastern United States, extending some distance beyond the Mississippi River. In the eastern hemisphere it occurs in parts of western Europe and eastern Asia. Although the biome types of the two hemispheres share few species in common, the same genera are often represented in both, and many ecological equivalents can be found.

The climate of the deciduous forest is much milder than that of the taiga. Precipitation is generally high, averaging 40 inches annually, but ranging from 25 inches in the western United States to over 60 inches in the east. The summers are hot, and the growing season varies from four to six months. Winters are cold, with temperatures ranging between −20° and 30°F. Snow cover is rarely continuous for more than two or three weeks. The acid topsoil is very well developed, with a very high fraction of organic content or humus. The combination of long mild summers, relatively high precipitation, and rich soils are the important factors that shape the deciduous forest.

Characteristically, the broad-leaved trees of a temperate zone forest show a blaze of color in the fall and lose their leaves. It is the decomposition of the thick carpet of dead leaves and logs that eventually produces the rich spongy humus. This process is carried on by a large and varied array of reducer organisms. A great diversity of species, both plant and animal, make up the deciduous forest. Forest growth is layered vertically, and each layer is composed of a characteristic group of species. Thus, there is a canopy of tall trees, an understory of shorter trees, a shrub layer, an herb layer, and the layer of the upper soil. Much of the animal life also follows this stratification.

Among the common species of trees are the oaks, maples, hickories, birch, beech, sweetgum, and tulip trees. Conifers, such as pines, cedars, and hemlocks, are also indigenous to the deciduous forest biome. Flowers, fruits, seeds, and leaves offer a rich source of food for the primary consumers of the forest. Hollow trunks and tree crowns provide shelter and nesting areas for birds and mammals. Animal life is far more varied than in the taiga, and the complex food webs make for much greater stability.

**21-8** *The nomadic habits of the red crossbill, in contrast to the geographic fixity of most species, enables it to survive even widespread failures of its food crop. It has been found that where spruce cone seeds, the staple of crossbills, were abundant, the birds were also plentiful. Where the cone crop failed, the birds were rare. (Hugh M. Halliday from National Audubon Society)*

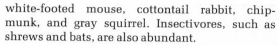
The dominant large herbivore of the eastern deciduous forest is the whitetail deer. Because of elimination of predators, the whitetail has often undergone upsurges in population. These explosions often culminate in serious destruction of vegetation and mass starvation of the deer. The wolf, black bear, and porcupine were once abundant throughout the deciduous forest; the wolf and porcupine are now rare in the United States, but in some areas of the northeast the black bear (a legally protected species) appears to be returning.

Among the many smaller mammals that are common in the deciduous forest are the opossum, the raccoon, the striped skunk, the red fox, and the longtail weasel. Numerous in individuals and in species are such small herbivores as the white-footed mouse, cottontail rabbit, chipmunk, and gray squirrel. Insectivores, such as shrews and bats, are also abundant.

The beaver, the largest rodent in North America, plays an important part in the ecosystem of the deciduous forest. The beaver exerts its influence through its dam-building activities which account for the creation of many new ponds and lakes. When the pond or lake becomes silted in and overgrown with vegetation, the beavers move on to another area. Eventually, the pond or lake is replaced by a highly fertile meadow. Because of its activities, the beaver has produced millions of acres of fertile meadow interspersed among the deciduous forest.

The canopy and understory of the deciduous forest support a large population of birds—mostly warblers, tanagers, thrushes, and woodpeckers. Predators, such as hawks and owls, are common; they prey mostly on rodents and small birds. Gallinaceous or fowl-like birds, such as the ruffed grouse, are fairly abundant residents of the forest floor. At one time the wild turkey was extremely abundant in upland woods. But since the days of the early European settlers, it has gradually been decimated to such an extent that by 1920 it was virtually extinct. Ironically, it was saved in part by the industrial revolution. As more people began to leave rural areas to seek work in city factories, hunting pressure began to ease. Its numbers were also increased by the Pennsylvania Game Commission, which began to raise large numbers of truly wild turkeys. As a result, they are now beginning to reappear in fair numbers in several states.

Because of the mild and moist climate of the deciduous forest, there is an abundance of amphibians and reptiles. The spring peeper and the gray treefrog are frequently heard calling from trees in spring and summer. The black-masked wood frog, which is restricted entirely to the forest floor, is a good indicator species of the deciduous forest biome. Under rotting logs and rocks and in streams can be found many kinds of

**21-9** *Working in cooperating family communities in highly disciplined fashion under the guidance of experienced individuals, beavers perform extensive engineering feats. However, the whole process is directed at a single objective—construction of a safe and well-stocked winter home in which to raise their young. (Karl Maslowski—Photo Researchers)*

salamanders. Reptiles, too, are represented by an impressive number of species of snakes, turtles, and lizards. Among the snakes common throughout most of the eastern forest of the United States are the ribbon snakes, ring-neck, king and water snakes, copperhead, and timber rattler.

Along the Atlantic coastal plain is an ecosystem sometimes regarded as a distinct biome, the Southeastern evergreen biome. Because it is dominated by pines, it is often referred to as piney woods or pine barrens. This represents a unique ecosystem characterized by sandy soil, fairly dry conditions in the slightly elevated portions, and acid bogs in the lower portions. The dominant evergreens may be pitch pine, loblolly, longleaf, or slash pine. These are usually associated with several species of scrub oak. Because of the layer of pine needles, the forest floor is generally clear. Some of the animals that may be considered indicator species of the pine barrens are the pine snake, the fence lizard, the pine barrens treefrog, and the pine woods treefrog.

In the United States, the deciduous forest has suffered more ecological disturbance than the biomes to the north. Virtually nothing remains of the grand primeval forests that were here when the Dutch settlers arrived. Although the Indians had brought about some change with their practice of burning large areas to flush game, large-scale changes began only as the white colonists expanded westward. All virgin timber was cleared for cultivation purposes, or to provide building materials or fuel. The forest of today is almost all secondary growth—only a pale shadow of the magnificent woodlands of the seventeenth century.

As a result of man's influence, the composition of the vegetation has been changing, and wildlife populations have been drastically disrupted. Until the turn of the century, the American chestnut was one of the largest as well as the most common tree. Roughly a third of the timber was of this species. Then, about 1900, a

**21-10** *Turtles have adapted successfully to almost every type of biome. Although frequently cited as examples of slowness and stupidity, their adaptations for heavy protection at the expense of mobility have served them well. The most successful group, the cryptodires, are today found throughout the world—in swamps and rivers, forests, plains, deserts, and the sea. (Karl Weidmann from National Audubon Society)*

**21-11** *The pine snake of the Southeastern evergreen biome is colored in a way that blends perfectly with the splotchy light of his environment. The pines provide a fine perch for sunning. (Leonard Lee Rue—Bruce Coleman)*

**21-12** *Wild oats grass, a form of sod grass, still covers the lower open rolling foothills of the California Sierra Nevada. The grasses, with 10 thousand species, make up one of the largest families of plants. They are the most widely distributed seed plants, ranging from arctic regions to tropics, from marshes to deserts, and from mountains to lowlands. (Verna R. Johnston from National Audubon Society)*

fungus was accidentally imported from Asia on nursery plants. Known as chestnut blight, it has resulted in the destruction of nearly all American chestnuts. Young saplings may still be found throughout the woods, but they are girdled by a fungus and die after reaching a height of six to 10 feet. Similarly, the American elm is now vanishing from the American scene. This magnificent tree, once so common not only in forests but also as a shade tree of town and city streets, has fallen victim to Dutch elm disease, a fungus transmitted by a tiny bark beetle. Unquestionably, the disappearance of the chestnut and elm have transformed the deciduous forest. It has profoundly affected the insects, birds, and small mammals that depended on these trees. It has also reduced the complexity of this ecosystem, and therefore its stability.

## Grassland

The grassland biome is found within the temperate zone, where moisture is insufficient to support trees, yet adequate to support a continuous carpet of vegetation. This biome is represented in parts of Africa, Australia, the U.S.S.R., and North America. It is estimated that more than 40 percent of the earth's surface was once covered with grass.

The climate of the prairie features cold winters and hot dry summers. Rainfall extremes are between 10 and 40 inches per year, but averages are commonly within the 16–30 inch range. Because there is no natural windbreak, the winds blow almost continuously. The evaporation rate is high in the hot dry air, and winds exert a drying effect. The alkaline soils are characterized by black or brown colors, high fertility in the topsoil layer, and poor drainage due to an underlying hardpan with a high calcium content.

When the grassland prairie was still intact, two factors were important in maintaining grasses as the dominant type of vegetation. Probably the most important of these was the occurrence of periodic fires. These tended to favor the faster growing grasses over the slower growing shrubs and trees. With the removal of shrubs, competition is reduced and grasses tend to grow back with increased vigor from underground rhizomes and rootstocks. Fire also clears the soil surface of matted dead grasses which bind up valuable

**641**

nutrients and suffocate new growth; after a fire, the nutrients are recycled, returned to the soil. When grasslands are completely protected from fire, they begin to deteriorate.

The second important factor in maintaining grasses as the dominant type of vegetation was the role of the buffalo or bison. Their grazing action cropped the vegetation, but since they were constantly on the move, damage due to over-grazing was prevented. The constant trampling of the ground by their hooves probably served to compact the soil so that water could not penetrate deeply enough to support the roots of trees or shrubs.

Before frontier days, the prairie of the midwest was the most extensive biome in North America and an ecosystem in delicate balance. In the eastern portion where rainfall is higher, tall grasses were dominant; the more westerly and drier regions supported short grass prairies. To learn which species of grass populated the undisturbed grasslands of yesterday, a handy museum to visit is an untended prairie cemetery. Common species of the tall grass prairie were tall bluestem, slough grass, and Indian grass. Short grass prairies included blue grama grass, bluegrass, and mesquite grass.

Three species of large herbivores formerly dominated the prairie: the bison, the pronghorn antelope, and wild horse. Again, as with the caribou of the tundra, we note the evolutionary adaptation toward grazing and a running habit of large-hoofed mammals in an open environment. Grassland dwellers, both predators and prey, tend to be fast runners, and the fastest animal in the world—the cheetah, which can reach a speed of 70 m.p.h.—lives in the African grasslands. Speed obviously helps adapt an animal to life in the wide-open spaces.

Associated with the large dominant herbivores were numerous species of small herbivores, such as the prairie dog, the ground squirrel, pocket gophers, and the jackrabbit. Many of these illustrate the evolutionary tendency of small

**21-13** *This pronghorned antelope is just one of many varieties. Most members of this numerous group are exclusively African. All are adapted to life in open country—grassland, savannah, or scrub. They combine fleetness with sociability, grazing in herds where there is safety in numbers and many eyes are better than one. (Allan D. Cruickshank from National Audubon Society)*

**21-14** *Certain grassland birds, because of the dense grass and lack of trees for singing perches, have developed conspicuous flight songs that advertise territory and attract mates. Others, such as this prairie chicken, have extensive ritual courtship dances. (Mary M. Tremaine from National Audubon Society)*

herbivores in open areas to burrow. The prairie dog, a large social rodent, built spreading subterranean "towns" and was ubiquitous throughout the grassland biome. In Texas alone, a rough census of the early 1900s placed this species at approximately 800 million. The prairie dog feeds on shoots and seeds. Its constant burrowing action served to aerate the soil, and by mixing it with droppings, it also enriched the soil. The same burrowing habit can be seen in small mammals that live on the grasslands of other continents, such as the tuco tuco in South America and the hamster in Asia. There is even a species of owl on the western prairies that lives in burrows.

Another evolutionary tendency of grasslands residents is the habit of leaping. This can be seen in the American jackrabbit, the Australian kangaroo, and the Asiatic jerboa. This adaptation permits herbivores to see above the tall grass and catch sight of approaching predators.

## Changes in the grassland biome

Within the span of little over a century, dramatic changes have been wrought on the grassland by the white man. Where once a variety of natural grasses supported a complex web of wild herbivores and their predators, there now exists a monoculture of grains and domesticated livestock. Native animals have either declined drastically or have been totally eliminated from many areas. In the early nineteenth century, the total population of buffalo or bison herds was estimated to be between 45 and 60 million head; they now number somewhere under 15,000 individuals. Similarly, the pronghorn antelope, once numbering more than 50 million, has been reduced to less than 400,000. It, however, continues to occupy portions of its former range in the west and the southwest. Herds of wild horses, once quite common, are now extremely rare.

Man has helped to accelerate a noticeable evolutionary trend toward the extinction of large herbivores; in the past 12,000 years, about three-fourths of all the large herbivores in North America have disappeared.

The great herds of native herbivores have been replaced by domesticated cattle, goats, and sheep. Often too many are crowded onto a rangeland. Since they are confined, they exceed the carrying capacity of the land. In some areas damage by overgrazing has been severe. Usually, the soil is exposed; erosion follows and grasses are replaced by prickly pear and mesquite trees.

Settlers on the prairie regarded prairie dogs as an unnecessary nuisance; not only did they create holes which broke many a horse's leg, but they were consuming forage grass which, the ranchers believed, deprived their cattle of food and therefore took dollars out of their pockets. Large predators, such as the wolf and the coyote, were (and still are) also considered undesirable, since they turned to preying on livestock as their natural prey became more scarce. With the help of government agencies, cattlemen and farmers began a campaign with poisons and traps against prairie dog and predator. Now the prairie dog has been largely eliminated from its former range. To some extent it has been replaced by the ground squirrel. The wolf, too, has been eradicated from American prairies. The black-footed ferret has become extremely rare. Although enormous numbers of coyotes have been killed, that animal has miraculously managed not only to maintain its hold, but even to extend its range.

The American prairie today is an ecosystem out of balance. Present wild inhabitants of this highly disturbed community are such small mammals as the ground squirrel, jackrabbit, and various species of mice and rats. Important predators are the coyote and badger, which sometimes cooperate to catch jackrabbits. Reptilian predators, such as the bullsnake and the prairie rattler, are also important. A large number of lizards, frogs, and toads are supported by the

**643**

numerous species of grasshoppers, crickets, and spiders.

It might be added that the grassland biome types of Australia and Africa are also undergoing changes. Kangaroos, the ecological equivalents of the bison, are being slaughtered by Australians, many of whom often brag about the number of 'roos they have shot. The African veldt has been steadily shrinking, and many of its native species are endangered. A hopeful sign for the future is the fact that governments of African nations are beginning to set aside large tracts of grasslands as natural parks, where hunting is not permitted and man made environmental changes are regulated.

## Desert

The desert biome covers vast tracts of Australia, Africa, Asia, and North and South America. Most important of the climatic factors shaping deserts is the minimal amount of precipitation over the region. Deserts are usually defined as areas in which the annual rainfall averages 10 inches or less. In more temperate zones, the scant rainfall is often the result of a "rain shadow" effect cast by a mountain chain on its leeward (away from the direction of the wind) side. As prevailing winds drive moisture-laden air up one side of the mountain, it becomes cooler and loses its moisture-holding capacity. By the time it clears the crest, all moisture has been wrung from it. The desert of the American southwest lies in the "rain shadow" of the western mountain chains.

In the United States, the desert biome is found in the lower elevations of the western and southwestern states. Two types are recognized: the cold desert with its cooler winter temperatures (it may even have snow on the ground) and the hot desert with high temperatures throughout the year. Examples of cold desert are found in Nevada, Oregon, and Utah, while

**21-15** *Coyotes are essentially wolves that have become closely adapted to the open grasslands. However, the ranges of the coyote and wolf overlap, and anatomically they are quite similar. Coyotes are predators and scavengers. They feed extensively on rodents and rabbits, helping to keep these prolific species in check. They are quite shy and retiring and almost never dangerous to man. (Pat Kirkpatrick from National Audubon Society)*

**21-16** *The black-footed ferret, or polecat, ranges from Montana to Texas, feeding mainly on prairie dogs. Anatomically they might be described as larger, slower moving weasles. Unlike weasles, however, they do not hunt in packs, and they are less wanton in their killing. Ferrets have been trained to work with falcons to hunt rabbits and rodents. The males, especially, make delightful pets. (Edward Bonn from National Audubon Society)*

**21-17** *The curious and picturesque spine-studded Joshua tree grows exclusively in the more favored sandy soiled areas of the Mojave Desert. A member of the Yucca family, the Joshua tree has developed narrow bladelike leaves to cut down transpiration as a means of conserving water. The tree flowers briefly with brilliant purple or white blossoms and helps to break up the almost unrelieved monotony of the Mojave, the least diverse desert area of North America. (Al Green & Assoc.—Bruce Coleman)*

typical hot deserts occur in Arizona, California, New Mexico, Texas, and northern Mexico. Daytime summer temperatures of deserts typically remain over 100°F.

The popular picture of the desert, seen so often in grade B movies, is that of miles of sand dunes without so much as a blade of vegetation anywhere. While this may be true of the Sahara, the largest of deserts, it is by no means typical. Far from being barren and lifeless, most deserts support a rich flora, and, although it may not be too obvious during the day, a tremendous number and variety of animals as well.

Precipitation in the desert is not evenly distributed, but falls mostly during a rainy season. Along with the scant rainfall, deserts are characterized by very low humidity and drying winds. Soils are typically alkaline. Often just below the surface in hot deserts there is a cement-like layer of hard soil with a high clay content, or caliche, that impedes drainage. When storms occur, the water cannot soak into the ground because of the caliche, and flash floods result. Because of the desiccating effect of the climate, dead plant and animal material dries out before any decay or decomposition can occur; consequently, the desert soils contain very little organic material.

The major problems of both plants and animals living in hot deserts is escaping the effects of excessive heat and preventing dehydration. Water problems are the most critical for plants. Not only must plants obtain and conserve enough water to remain alive, but they must be physiologically ready to flower and produce seeds whenever the rains arrive. Indeed, herbaceous plants must germinate, grow, flower, and produce seeds, all within a very brief span of time if the next generation is to survive. Immediately following rains, the desert floor is often carpeted with a variety of flowers.

Shrubs and trees exhibit a number of adaptations to scarcity of water in the desert. One obvious adaptation that strikes the visitor to any desert is the natural spacing between plants, almost as if each plant were deliberately planted at a uniform distance from its neighbor. Many desert plants exude substances from their roots that actually inhibit the growth of other plants. Each plant thereby insures for itself sufficient soil space to absorb enough moisture. Furthermore, root systems of many cacti are shallow and sprawling to take advantage of even light rains. Many desert plants have fewer stomata than their relatives that live in moister biomes, and often the stomata are sunken below the surface of the epidermis to reduce moisture loss through transpiration. The ocotillo and palo verde, among other desert plants, shed their leaves, thus avoiding transpiration loss altogether. Desert plants often have green stems so that photosynthesis can occur without leaves; this is true of the palo verde and most cacti. Some cacti, such as the saguaro of the Sonoran desert, have pleated stems enabling them to store water by expanding when it rains.

**21-18** *The bobcat is native to the desert biome, although unfavorable conditions there sometimes force them into other biomes. Bobcats sometimes mate with domestic cats, but the kittens rarely survive. (Alfred M. Bailey from National Audubon Society)*

Since plants are unable to move to escape the lethal effects of extreme heat, they must utilize other devices to cut the effect of the sun's rays. A typical adaptation for this purpose is a covering of spines. These serve to protect them against grazing animals, but even more importantly, they also deflect a considerable amount of heat and radiation. An extreme instance of this may be seen in the old man cactus, in which the spines have been modified into a thick covering of long white hairs. Other plants, such as the totem pole cactus, have bizarre contours that probably serve to scatter radiation. Coloration also is important. Many cacti are light in color,

**21-19** *The collared peccary is the more common and less predacious of the two known varieties of peccary. They have scimitar-formed tusks in both jaws and sharply pointed cheek teeth. When angry or alarmed they habitually clap their jaws together in warning. (Leonard Lee Rue from National Audubon Society)*

**21-20** *Like most desert animals, the western ground gecko avoids the heat of day and does its serious hunting (for insects) at night. Although cold-blooded, the gecko is actually able to regulate its body temperature by behavior. Geckos expose themselves to the sun to get warm and retreat to shade when they get too hot. They are thus able to maintain a surprisingly constant body temperature throughout their period of activity. (Robert H. Wright from National Audubon Society)*

often grayish or gray-green; these light colors tend to absorb less heat than darker colors.

Animals of the desert must deal with the same problems of too much heat and too little water; but they cope in different ways than do plants. Although we tend to regard many desert animals as highly adapted to desert conditions, the most successful desert animals actually spend much of their time evading extreme desert conditions.

Since animals are capable of movement, their best solution for the problem of heat is behavioral; they simply change their location when soaring temperatures approach lethal limits. Except for a few birds and insects, all activity is suspended during the "midday siesta." Larger animals, such as coyotes and bobcats, seek what little shade may be available. Many homeotherms practice a type of evaporative cooling simply by opening their mouths and panting.

The collared peccary, or javelina, utilizes the same device as some of the spiny cacti to shield it from excess heat. It is no accident that peccary fur is so shaggy and unkempt. Its rough hairs form a thick boundary layer that serves to slow the rate of heat transfer between the environment and its skin.

Reptiles and small mammals either bury themselves in sand or seek underground burrows during the day. Desert floor temperatures may

reach a fiery 180°F in midafternoon, but with each inch of depth, there is a drastic drop. Only a few inches beneath the surface, temperatures may be a relatively comfortable 90°F.

The best way to avoid desert heat is to be nocturnal. Most desert animals are "night folk," and it is during the night that the desert really throbs with life. The kangaroo rat forages for seeds. The rattler and kit fox hunt for kangaroo rats. An elf owl emerges from its hole in a giant saguaro to swoop down on a large spider. Bats and gecko lizards search for insects; solpugids (they are a kind of arachnid, related to spiders) and scorpions are to be found everywhere. Many animals, such as the mule deer and the spotted skunk, take their nightly walk to the water hole, where each waits its turn.

In addition to avoiding the killing heat, desert animals must also be able to obtain and conserve water. Water conservation is accomplished in different ways by different animals, for not all animals meet their water needs by drinking. As we have seen in Chapter 8, the kangaroo rat obtains its water metabolically from its seed diet, and remains in a humid subterranean chamber during the day to avoid water loss by evaporation. Reptiles conserve water by their covering of scales that prevents evaporation. Insects are protected against evaporation by their chitinous exoskeleton, and they conserve water by excreting a dry paste of uric acid rather than a liquid urine.

Today the desert biome also shows signs of man-made change. Although to many of us, the changes barely seem perceptible, they are every bit as profound as those that swept the American prairie. Motorcycles and dune buggies now shatter the silence in formerly isolated areas; the litter that is the hallmark of our industrialized civilization is beginning to mount. Biotic changes, however, are the most profound. In Arizona, for example, the magnificent giant saguaro, which still form bizarre forests, appears headed toward extinction. This slow-growing cactus requires

**647**

approximately 75 years before an arm is produced, and it may live as long as 250 years. During its early tender years of growth, the saguaro requires the shade dispensed by larger "nurse plants." Two important factors are stacking the deck against survival of the saguaro. One of these is the overgrazing by cattle which remove the essential nurse plants. The second factor is the decline in the coyote population. The coyote, whose preferred prey is the packrat, has been so decimated in the desert area that packrat populations have soared. Packrats like to nibble on the succulent young saguaro seedlings, and under this withering attack, the saguaro is simply failing to produce enough surviving seedlings. This example provides a vivid instance of the intricately interwoven threads of the ecological web.

## Tropical rain forest

Of all the biomes none is more productive, more varied, more highly organized and complex than the tropical rain forest. Because it is so rich in numbers of species and so complex, it is also the most stable of ecosystems. This biome is largely confined within the limits of the Tropics of Cancer and Capricorn. Sprawling tracts of tropical rain forest are to be found in central Africa, in India, and in southeast Asia. In the New World it occurs only in Central America and in the Amazon and Orinocan Basins of South America.

The tropical rain forest climate is characterized by uniformity. There are no seasons and little fluctuation of conditions. Unlike the tundra, there is little change in day length throughout the year. Temperatures are generally mild, and the high rainfall, at least 80 inches, is evenly distributed. Depending on the region, the rainfall may be as great as 500 inches in a year, and for all regions frequent torrential thunderstorms are characteristic. In regions where rainfall is be-

**21-21** *With so much vegetation crowded together in the tropical rain forest, competition for light has led to the development of trees with very tall trunks that can use light in depth. Typically these trees are supported by buttresses. The dense canopy of trees allows little light to penetrate, and vegetation at ground level in a rain forest tends to be open and limited to grasses. (Norman Myers—Bruce Coleman)*

**21-22** *A bewildering variety of orchids adorn the trees of tropical rain forests. They have adapted very successfully to lack of light on the forest floor by evolving aerial roots. These are not in contact with the ground at all and permit the orchids to grow where light is available. (Elsie M. Rodgers from National Audubon Society)*

**21-23** *Toucans are among the most curious of rain forest canopy birds. In flight they appear to trail after their own gaudy bills. Like canopy animals in general, they are brightly colored and extremely sociable. They sometimes travel in bands composed of many different species. (A. W. Ambler from National Audubon Society)*

tween 50 and 80 inches there is a definite dry season; in these areas may be found tropical deciduous forests. The air is saturated with humidity and moisture drips constantly from vegetation. Soils in tropical rain forests are typically yellow or red clays, rich in some nutrients, but deficient in certain minerals. This is due to the heavy rainfall which causes leaching of these minerals from surface layers. Many tropical regions have soils that are rich in iron and aluminum.

The tropical rain forest of the Amazon Basin of South America is a fairly typical example of this biome type. Green is the all-pervasive color wherever one turns. But contrary to the image fostered by pulp magazines and Tarzan movies, the thick, impenetrable tangle of jungle growth is not to be found within a mature rain forest. That type of growth may be found at the edge of man-made clearings or along river banks. The floor of the interior, however, is wet, mild but not hot, and remarkably uncluttered. Because of the excessive humidity there is a luxuriant growth of mold.

The tropical rain forest, like the temperate deciduous forest, is stratified vertically. Making up the canopy are tall trees which average between 100 and 180 feet, although an occasional emergent may be taller than 200 feet. Most of the sunlight and heat that falls on a tropical rain forest is absorbed by the canopy, and relatively little reaches the forest floor. Since there are no dry seasons, the trees are not deciduous, and different species are in bloom throughout the year. The variety of vegetation is staggering; there may be 3000 species of trees alone. These include the strangler fig, various palms, ironwood, and cecropia trees. A characteristic of rain forests are the thick woody vines, called lianas, which are draped from branches. Also growing on branches are numerous "air plants" (plants with no underground roots) such as orchids, of which there are thousands of species, and bromeliads, which are related to the pineapple.

Much of the animal life in a tropical rain forest is adapted primarily to existence in the canopy. New World monkeys with their long spidery limbs and prehensile tails are superbly adapted to grasping branches and moving through the trees. The prehensile tail is also found in other tropical animals, such as opossums, lizards, and snakes.

Flitting among the branches are colorful birds such as parrots, toucans, honey creepers, and tanagers. Birds and insects in particular illustrate another axiom of life in the tropics: species tend to be more brilliantly colored.

In keeping with Bergmann's Law, many mammals tend to be smaller than those in more temperate zones. Interesting exceptions to the law can be seen in the coypu and capybara, giant rodents. Poikilotherms, such as reptiles and arthropods, are not subject to Bergmann's Law, and they tend to be larger than temperate zone relatives. Because of the mild temperatures and high humidity, the climate of the rain forest is ideal for frogs and toads. The arboreal frogs have adhesive toe discs that enable them to cling to vegetation. These frogs often lay their eggs in the water that collects in bromeliads or on vegetation overhanging ponds.

There is a teeming scavenger industry on the forest floor. Various insects, such as beetles, termites, and ants, are constantly on the prowl for dead vegetation. Periodically, a brown column of army ants embarks on a foraging raid, sweeping the forest clean as they go. This constant activity of scavengers accounts for the relative absence of litter on the forest floor. This, in turn, means less plant decay and decomposition, and consequently scant humus in the soil.

The extraordinary productivity of the tropical rain forest has prompted many to suggest that here is the answer to the world's food problem. All we have to do, they say, is just clear the forest and plant various food crops. From time to time such experiments have been tried, but most of them have ended in failure. When the trees are cleared, the iron-rich soils are exposed

**21-24** *Allen's Rule is well illustrated by comparing the size of ears and snout in an arctic fox (left) and a gray fox (right). Along with all other appendages, the ears and snout of animals in cold climates tend to be reduced. The more compact the body, the less likelihood of freezing. (Leonard Lee Rue—Bruce Coleman)*

to the direct sunlight. This brings about certain chemical changes in the aluminum and iron oxides, usually transforming the soil into a rock-like substance. The problems of rain forest agriculture were highlighted in the 1960s, when the Brazilian government attempted to establish a farming colony at Iata, in the Amazon Basin. The land was cleared and crops were planted. During the first year, production was good. Then heavy rains washed away valuable topsoil, leaving the lower layers fully exposed to the sun. Production plummeted, and within five years the cleared land was covered with slabs of hard earth.

## Big city biome

In addition to the natural biomes, some biologists, such as A. B. Klots, have recognized an artificial, man-made biome, the "big city biome." Found in dense urban centers, particularly port cities, or those near the seacoast, this biome is the result of man's increasing mobility. Exotic plants and animals (and diseases) have been introduced inadvertently with shipments from foreign countries; in some instances, they have been deliberately introduced. Many introduced species find themselves in an environment so hostile they cannot survive, but others are able to flourish and spread.

Some of the transplanted, now cosmopolitan, species have flourished because of their broad adaptability to various environmental conditions. This would include the ability to subsist on a wide variety of foods, or on foods similar to those in their original environments. In some instances they are able to proliferate because of an absence of natural predators. Many species in the big city biome have evolved a commensal relationship with man, and some live intimately with him in his own home.

The climate found in the big city biome is similar to that of surrounding areas of the particular city. The air, however, usually ranges from "unacceptable" to "unhealthy." Soot, grime, haze, and smog are either constant or frequent. Garbage, trash, and filth are often to be found in back alleys or side streets; it might be considered analogous to the leaf litter on the forest floor.

Asphalt streets, brick and concrete buildings, empty lots, and city parks are the new habitats of the big city biome plants and animals. One of the plants that is extremely successful is the Ailanthus or "tree of heaven." This tree, referred to in the novel, *A Tree Grows in Brooklyn*, can thrive in a tiny patch of hard-packed soil or a chink in the sidewalk. Any empty lot littered with cans and bottles will be found to support a thriving crop of ragweed, common plantain, lady's thumb, knotweed, dandelion, and day-flower.

The dominant biped of this biome, of course, is *Homo sapiens*. A species which equals—perhaps exceeds—man in individual numbers is the Norway rat. This highly adaptable, prolific animal is a familiar, if unwelcome, housemate of central city dwellers. It thrives in sewers, apartment buildings, waterfront areas, and empty lots, feeding on garbage or any filth. Another rodent, more widespread in human dwellings, is the common house mouse. These rodents fill an important ecological niche, that of the scavenger; this is the same role they played in the natural environments where they lived before they came to the city. Their presence creates another niche, that of the predator one step up the food chain. This niche is filled by wild cats and dogs.

Two birds that characterize the big city biome are the domestic pigeon and the house sparrow. Pigeons are common in park areas, on sidewalks, and on building ledges, as any person with a new hat can ruefully testify. House sparrows, which were intentionally introduced from England by the Brooklyn Institute in 1853, underwent a population explosion thereafter, and spread quickly across the country and into Canada and Mexico.

Many insects are also typical of the big city biome. The large American roach, *Periplaneta*, and the smaller German and Oriental roaches are abundant in many homes and apartments. Other common insects include the housefly and the silverfish. Insects that are most often seen in empty lots and park areas are the cabbage butterfly, green and blue bottle flies, horseflies, and Japanese beetles.

## Fresh-water environments

Fresh water in a standing body is best represented by the lake, which can be divided into three zones or regions. The first is the shallow water **littoral zone,** where sunlight penetrates to the bottom. The second region, the **limnetic zone,** is the upper layer of deep water; some light penetrates there. The third region is the **profundal zone** of bottom and deep water where no light penetrates. Each zone has a characteristic plant and animal population.

Rooted seed plants thrive in the littoral zone. In the shallowest water, cattails and bullrushes proliferate. The leaves of these plants project above the surface of the water. They obtain carbon dioxide for photosynthesis from the air, but other raw materials are obtained from beneath

**21-25** *Cockroaches are among the oldest of insects and have flourished as scavengers since the Carboniferous period. There are about five thousand species living, mostly in tropical forests. Undoubtedly some species have always lived with man. Roaches thrive on almost anything (including soap and the starch used in binding this book). In fairness they must be classed as a nuisance rather than a danger, since they do not carry any human disease and do not bite or sting. (Photo by Alexander B. Klots)*

the water. Water lilies grow in deeper water, with leaves floating on the surface. In the deepest water of the littoral zone, the seed plants are completely submerged. Their leaves tend to be thin and finely divided, in order to provide maximum surface area through which to absorb carbon dioxide and other nutrients from the water.

Algae also thrive in the littoral zone. The blue-green algae are essential members of the lake environment. They fix molecular nitrogen into nitrates, fertilizing the water in the same way that nitrogenous bacteria fertilize the soil. They may, however, become superabundant in a polluted lake like Lake Erie. The nitrates which they produce are then present in unnaturally high concentrations, giving the lake water an unpleasant taste and a noxious smell. The overabundance of blue-green algae is a prime indicator of polluted water.

All of the animal phyla are also represented in the littoral zone. Pond snails, rotifers, flatworms, dragonfly nymphs, and midge larvae rest on, or are attached to, the stems of large plants. Snails and midge larvae are primary consumers, feeding directly upon plant material. Dragonfly nymphs are carnivorous secondary consumers. Two insects of the littoral zone should be familiar to anyone who has observed a lake. The first is the whirligig beetle. Its eye, well adapted to an aquatic environment, is divided into two parts; the top half is for seeing above the water, and the bottom half is for seeing in the water. The second familiar insect is the water strider; it walks on the water, held up by the surface tension.

Frogs, salamanders, turtles, and water snakes are vertebrate forms which live almost exclusively in the littoral zone. Adult frogs are secondary consumers, but tadpoles are primary consumers that feed upon algae and other plant material. Fish move freely among all of the zones of a lake, but most species spend a great deal of their time in the littoral zone. Some species—the bream and sunfish, for example—establish terri-

**21-26** *The dragonfly nymph is a voracious eater. It can consume its own weight in insects in just two hours. To capture its food, the nymph shoots out its hooked lower lip and draws the prey back into its mouth. Dragonflies imprisoned without food may resort to self-cannibalism and begin eating their abdomens. (Russ Kinne— Photo Researchers)*

**21-27** *Fresh-water insects have been extremely successful. Most are found in ponds and streams where they lead a double life, spending only part of their lives in the water. These insects have developed a variety of means to get oxygen from the water. This water strider, though truly aquatic, lives on rather than in the water, relying on surface tension for support. Other species have variously developed gills, air tubes, or means of carrying air bubbles with them under water. (N. E. Beck, Jr. from National Audubon Society)*

tories there; others, like the large mouth bass, move throughout the lake but use the littoral zone as a feeding ground.

The vegetation of the limnetic zone is mostly microscopic and therefore does not attract the attention of the casual observer. Yet these microscopic forms—algae and diatoms—usually exceed rooted plants in food production per unit area. Since these plants are at the bottom of the food chain, their abundance determines the abundance of all other life in the zone.

Much of the animal life of the limnetic zone is also microscopic. Rotifers and copepods are two common examples. They are filter feeders that strain bacteria and algae out of the water; these organisms therefore graze the plant life of the limnetic zone in much the same way that cattle graze grass. Other animals are predators, feeding on the herbivores. Fish may also be found in this zone, but they generally go to shallower waters to feed.

The deepest region of the lake is the profundal zone. Since there is no light, plants cannot grow; the inhabitants of this zone depend upon the limnetic and littoral zones for basic food materials. The variety of life in the profundal zone is not great, but it is nevertheless important. Its major function is to recycle primary nutrients upon which plants depend for their growth. The bacteria and fungi of the profundal zone close the circle of food production and use in the lake. They break down decaying plant and animal materials into their basic chemical constituents. These constituents are then used by the photosynthetic plants of the littoral and limnetic zones.

Another type of fresh-water environment is the stream. It is characterized by two sections: the running water and the standing pool. Free-floating plankton are present only in slow-moving water and standing pools. Since the majority of life forms must withstand moving currents, they are specially adapted for attaching themselves to their habitat. Green algae, such as *Cladophora*, have long filaments which enable them to trail over moving waters. Aquatic mosses cling to stones.

The invertebrate organisms which inhabit streams are commonly bottom dwellers, although burrowers such as clams are numerous in standing pools. Many stream animals are equipped with hooks and suckers, to fix them in the current. The caddis and waterpenny (larvae of the riffle beetle) are some examples. The caddis actually spins a net which it uses both as a shelter and as a trap for food. Snails and flatworms are among the animals that have sticky undersurfaces to help them cling to rocks.

Stream animals are shaped to offer little resistance to water flow; they are characteristically streamlined and flat. Flat stonefly and mayfly nymphs can slide beneath rocks for security. These forms also exhibit a helpful taxic response in which they instinctively point upstream and swim against the current. The larger grazers and predators, the fish, have the same kind of streamlined silhouette.

## Marine environments

The **estuary** or tidal zone occurs where fresh water meets the sea. It is a partially enclosed coastal body of water. Organisms dwelling in an estuary must have wide degrees of tolerance for varying conditions such as temperature, salinity, and tidal movement. The stress of living in this habitat is balanced by its unusually abundant food supply; large numbers of clams, mussels, and oysters populate the estuarine community.

Plant forms in an estuary include seaweeds and grasses, microphytes and phytoplankton. Blooms of red-pigmented dinoflagellates, or red tides, are sometimes seen in coastal waters. It is suspected that organic pollution can bring about this condition. Red tides can be toxic, resulting in large fish kills.

The **neritic** biome or continental shelf is

**21-28** *The number and variety of marine and shore birds is conspicuously large. They inhabit mud flats, sandy beaches, and rocky shores in abundance. Gulls are among the most successful of these birds, adapting even to the conditions of man's harbors. One recent study has shown that the size of the gull population is directly related to the amount of human garbage available to them. (Carl Frank—Photo Researchers)*

the second marine zone, located beyond the intertidal area. It is dominated by diatoms and dinoflagellates, which make up the primary link in the food chain. The diatoms are more common in colder waters, whereas dinoflagellates are characteristic of warmer ones. Close to the shore the algae known as seaweeds grow. There are not only green seaweed but brown and red as well; these other colors are useful in obtaining light for photosynthesis at greater depths.

There is a large population of sessile animals on the shelf. These include some kinds of clams, mussels, and barnacles, and filter feeders. Bur-

rowing sea crabs and worms feed by devouring decaying matter which filters into the holes where they live. Marine snails (periwinkles) attach themselves to rocks.

Fish are both grazers and predators on the continental shelf. In the Atlantic the killifish and flounder commonly prey upon bottom-dwelling and burrowing forms. Flounders and rays are particularly well adapted to this way of life, with a flat shape and an ability to change color that camouflages them on the ocean floor. The herring family, including the popular sardine and anchovy, strain their food through gill rakers.

The large predacious fish are exemplified by the shark. The behavior of ocean fish is quite different from that of fresh-water fish. Rather than burying their eggs on the bottom, ocean fish lay their eggs in the water and leave them unattended. Often these eggs have special balancing or floating devices attached to them, such as droplets of oil. Ocean fish move about in groups or schools, as a means of protection in the shelterless ocean. The movement of so many individuals confuses the predator, so that he is unable to track any single target.

Marine birds play a distinctive role in the cycling of nitrogen and phosphorus. Since their prey come from the sea, these birds remove elements from the ocean and deposit upon the land where they breed. The rich guano deposits containing nitrates and phosphates are a good example. Shorebirds which are seen in the tidal zones are cormorants, sea ducks, and pelicans. Farther out in the neritic zone there are petrels and shearwaters.

The final phase of the nutrient cycle in the neritic zone takes place in the marine sediment. Here anaerobic bacteria, such as sulfate reducers and methane bacteria, create the upward-moving gases employed by plants. These fertilizing bacteria supply the basic nutrients to the community.

The third region of the marine environment is the open, or deep, sea. Three factors which determine the presence or absence of deep sea life are salinity, temperature, and the low concentration of nutrients. In the open sea salinity varies only narrowly, and so the organisms found there tend to be poor osmoregulators. As a general rule, there is a low concentration of nutrient factors in this region, making it relatively scarcely populated. But the relative infertility of the open sea is correlated with an incredible variety. Although population density is low, the number of different populations is high.

Algae are the most important plant group of the open sea. Seed plants are unimportant. Insects are absent. Most of the life is free swim-

ming or floats in the upper layer. Life is concentrated in the **euphotic zone,** the zone of effective light penetration, which extends from one to 200 meters down. Beneath the euphotic zone is the **aphotic zone,** where not enough light penetrates for photosynthesis.

As in the limnetic zone of the lake, the plant producers of the open sea come in small packages. Until the early twentieth century, biologists were unable to capture and study the most important of these photosynthetic algae, for the mesh of the nets that they were using to capture them was not fine enough. Recent studies indicate that tiny nannoplankton, green flagellates from 2 to 25 microns in size, account for 90 percent of the photosynthesis which occurs in the sea.

The zooplankton which eat the phytoplankton are of a corresponding size. Medium-sized

**21-29** *The fate of marine gammarids, relatives of shrimps, lobsters, and crabs, is to be eaten. Tiny drifters rather than swimmers, they are a staple in the diet of large fish, cephalopods, and cetaceans. Along with other species of tiny crustaceans, they make up 70 percent of the animal plankton. (Courtesy of the American Museum of Natural History)*

animals are either plankton feeders or feed upon decaying plant materials. Large shrimplike "krill" are important food chain links at this level of animal size, since they are a dietary staple for many larger fish. Most large animals are carnivores. There are, however, a few large and strictly herbivorous animals like the whalebone whales which feed only on plankton.

Most zooplankton and fish in this region are either transparent or blue, making them invisible in the open sea. Other adaptations to life in the open sea are protective spines and fat droplets or air bladders that help maintain buoyancy. Both spines and external air bladders also serve to increase the surface area of an animal and thus provide more surface for the inward diffusion of important but scarce dietary supplements like vitamin $B_{12}$.

Oceanic birds, independent of the land, are found in great diversity on the open sea. Petrels, albatrosses, frigate birds, and some species of terns and boobies seek land only during the breeding season. Despite appearances to the contrary, these birds are not free to fly anywhere they want. They must stay near concentrations of the krill and the fish that they eat; thus, the abundance and location of seabirds indirectly depends upon the abundance of the plankton that are the bottom of the food chain.

The density of life in the open sea decreases with depth. While the number of living forms in the depths is small, there are many different types. No photosynthesis occurs here; instead food must be transmitted from the euphotic zone. Three mechanisms account for this transport. First, there is a constant rain of organic debris from above. Second, organic matter is transported from coastal zones as nutrients flow out from river mouths and down into the depths. Third, dissolved organic matter spontaneously forms into particles which drop to the ocean floor.

The animals of the depths show many specialized adaptations to this environment.

Since large areas of the ocean floor are covered with ooze, many bottom animals have spines, stalks, or other means of support. Although not enough light for photosynthesis penetrates to this depth, some light does, and many deep sea fish have enlarged eyes for seeing in the dimness. Some produce their own light with luminescent organs. The lantern fish and the hatchet fish are examples of this type. Another adaptive characteristic of deep sea fish is an enormous mouth which enables them to swallow prey many times their own size. Meals are few and far between in the depths, so it is important to be able to eat enough food for weeks on end.

**21-30** *Perhaps three-quarters of the deep sea fish have light-organ systems. Many of these species are distinguished by a pattern of round, pearly light-organs studding their trunk. Light producing cells are activated by a nerve. The light is reflected forward by a silvery backing and focused through a lenslike thickening of the scale overlying the light organ. (Gene Wolfsheimer from National Audubon Society)*

## Summary

A biome is a group of communities dominated by a particular climax community shaped primarily by the regional climate, latitude, altitude, and topography. Biomes, generally quite uniform in composition, may vary locally, especially because of recent disturbances, such as fire, or detailed interactions between the biota and the environment. Similar but geographically separated communities are known as biome types, and equivalent niches in biome types are generally occupied by animals called ecological equivalents. Biomes of the western hemisphere, going from the poles toward the equator, include tundra, taiga, deciduous forest, grassland, desert, and tropical rain forest.

The tundra, the most northerly, holarctic biome, is characterized by intense cold, high winds, permanently frozen ground, long summer days, and low precipitation. The biota of the tundra is relatively simple. The dominant vegetation is lichen, mosses, grasses, and shrubs. Such herbivores as lemmings, hares, caribou, and musk oxen feed upon these and are in turn the prey of the arctic fox, wolf, and grizzly bear. The ptarmigan is a typical tundra bird. The simplicity of the tundra ecosystem makes it extremely unproductive and unstable.

The taiga, or boreal forest, forms a belt 800 to 900 miles wide, dominated by thick stands of conifers, such as spruce. Typical taiga is flat and contains many lakes. Large herbivores of the taiga include the dominant moose, elk, caribou, and deer; small herbivores include snowshoe hares, squirrels, and voles. Predators include the timber wolf, lynx, red fox, wolverine, and weasel. The Canada jay, red crossbill, and goshawk are typical taiga birds.

The deciduous forest is a more southerly biome with less well-defined borders. It occurs in warmer regions of higher precipitation, and is characterized by a great diversity of broad-leaved plants and temperate zone animals. Common species of trees include oaks, maples, and birches, and the forest herbivores, such as the dominant whitetail deer, feed on abundant flowers, fruits, seeds, and leaves. Before man's intervention, wolf, black bear, and porcupine were abundant. Other forest mammals include opossum, raccoon, mice, rabbits, squirrels, and the beaver, whose dams are an important force in shaping the forest habitat. Reptiles and amphibians are represented in the forest by frogs, salamanders, snakes, turtles, and lizards. All levels of the forest canopy support a rich variety of bird species.

Grassland appears at approximately the latitude of the deciduous forest, but is largely limited to areas with high winds and an average annual rainfall of 16 to 30 inches. It once formed an extensive and delicately balanced ecosystem throughout the American midwest, dominated by bison, pronghorn antelope, and wild horse. Associated with these were prairie dogs, ground squirrels, gophers, and jackrabbits.

Today the original herbivores are largely replaced by domesticated cattle, goats, and sheep.

Two kinds of deserts, hot and cold, occur in low-lying regions of the American southwest where rainfall is below 10 inches annually. Dominant plants are either such forms as saguaro, bursage, and creosote bush or, in the cold Great Basin Desert, sagebrush. Desert mammals include coyotes, bobcats, kit fox, jackrabbit, and collared peccary. Other animals include elf owls, geckos, and scorpions.

The most varied, productive, and complex ecosystem is the tropical rain forest. Ecologically stratified, the rain forest embraces a tremendous number of species, including strangler fig, palms, and lianas among the plants, and monkeys, toucans, lizards, frogs, and tapirs among the animals.

In addition to these natural biomes, some authorities recognize an additional "big city biome," characterized by man-made structures, among which a limited number of plant and animal species, many of them regarded as pests, are able to gain a foothold and thrive.

The two types of aquatic environments are fresh water and marine. The three zones of the fresh-water lake are the littoral zone, the limnetic zone, and the profundal zone. Each zone is characterized by flora and fauna adapted to the environmental conditions there. A second type of fresh-water habitat is the stream. Many stream-dwellers have special adaptations that help them resist the swift current.

The estuary is often a habitat that serves as a transition zone from fresh-water to marine environments. The three zones of the true marine environment are the coastal zone, the oceanic zone, and the deep sea zone.

# Bibliography

## References

Bormann, F. H. and G. E. Likens. 1970. "The Nutrient Cycles of an Ecosystem." *Scientific American.* 223(4): 92–101. Offprint no. 1202. Freeman, San Francisco.

Cooper, C. F. 1961. "The Ecology of Fire." *Scientific American.* 204(4): 150–159. Offprint no. 1099. Freeman, San Francisco.

Dunbar, M. J. 1968. *Ecological Development in Polar Regions.* Prentice-Hall, Englewood Cliffs, N. J.

Irving, L. 1966. "Adaptations to Cold." *Scientific American.* 214(1): 94–101. Offprint no. 1032. Freeman, San Francisco.

Love, R. M. 1970. "The Rangelands of the Western U.S." *Scientific American.* 222(2): 89–96. Offprint no. 1169. Freeman, San Francisco.

Marks, P. L. and F. H. Borman. 1972. "Revegetation Following Forest Cutting." *Science.* 177: 914–916.

Moir, W. H. 1972. "Natural Areas." *Science.* 177: 369–400.

Stearns, F. 1970. "Urban Ecology Today." *Science.* 170: 1006–1007.

Went, F. W. 1955. "The Ecology of Desert Plants." *Scientific American.* 192(4): 68–75. Offprint no. 114. Freeman, San Francisco.

## Suggestions for further reading

Bates, M. 1952. *Where Winter Never Comes,* Scribner's, New York.
A noted biologist describes in general but sound terms the characteristics of men and nature in the tropical environment. Focuses alternately on man's physical and cultural adaptation to the tropics and on the distinctive characteristics of tropical forests and seas.

Carson, R. 1955. *The Edge of the Sea*. New American Library, New York.
Vividly and carefully examines how the rhythms of tides and surf affect the
typical community of plants and animals that inhabit each of three basic types of
seashore: the rocky shore, sand beach, and coral reef.

King, H. G. R. 1970. *The Antarctic*. Arco, New York.
A nontechnical guide to the Antarctic as a whole—the ice cap, the land, islands,
climate, and the remarkable life on ice, rock, and sea.

Larson, P. 1970. *Deserts of America*. Prentice-Hall, Englewood Cliffs, N. J.
Explores the symbiotic relations and dynamic adaptations which preserve and
perpetuate life in the desert.

Nierung, W. A. 1966. *The Life of the Marsh*. McGraw-Hill, New York.
An ecological study of North American wetlands and a prognosis of their future.
Includes many photographs and line drawings of fresh- and salt-water marshes,
bogs, and swamps.

Stonehouse, B. 1971. *Animals of the Arctic: The Ecology of the Far North*. Holt,
Rinehart & Winston, New York.
A thoughtful and readable text liberally illustrated with paintings of animals and
birds, maps and striking photographs. Presents the life of the Arctic and Subarctic
regions, focusing on animals.

**Books of related interest**

Barlow, E. 1970. *The Forests and Wetlands of New York City*. Little, Brown, Boston.
An elegant description through photographs, old maps, and essays of six "green
islands in an urban sea."

Caulfield, P. 1970. *Everglades*. Sierra Club/Ballantine, New York.
Evocative pictorial essay capturing the subtle diversity of this vanishing wilderness
area. Accompanied by passages from the writings of P. Mathiesson.

Costello, D. F. 1969. *The Prairie World*. Crowell, New York.
A generalized book about the history, ecology, landscape, climate, plants, and
animals of the prairie.

Gibbons, E. 1970. *Stalking the Blue-Eyed Scallop*. McKay, New York.
A field guide and manual for preparing seacoast foods written by a delightfully
original authority on nature and food.

Hay, J. 1963. *The Great Beach*. Ballantine, New York.
A noted naturalist describes the diversity of life on Cape Cod's vast outer beach.

Leopold, A. S. and the Editors of *Life*. 1962. *The Desert*. Time, New York.
Presents a large body of information in especially clear and lucid terms. Touches on the creation of deserts, life patterns in arid lands, the adaptations of plant and animal, the eternal problem of water, and the taming of the desert.

Ley, W. and the Editors of *Life*. 1962. *The Poles*. Time, New York.
Discusses the present and past ecology of the Polar regions dominated by repeated shifts in the balance of ice and water. Also examines the role and problems of the Arctic as a natural laboratory and hypothesizes about the coming boom in the Arctic.

# 22

# Man and the environment

Photographs of earth taken from the moon have reminded us with dramatic impact that we live on a tiny, vulnerable, and finite speck in space. The resources of our "spaceship earth" are limited, as is its capacity for absorbing wastes. It is the only planet we have and should some nuclear or ecological catastrophe overtake it, we would be unable to move to another. Even if we could be transported by some modern Noah's ark, there is no other planet within our solar system even remotely as hospitable to life as we know it on earth. As earthlings, then, we are completely dependent on our planet's life support systems.

We have seen that any living creature has two important characteristics: it requires energy and it maintains adaptive homeostasis. Populations or communities exhibit essentially the same features as individual organisms. An individual organism obtains its energy by ingesting food, and a natural community requires constant inputs of solar energy. The rate at which energy is utilized by individual organism or community is indicative of its metabolic rate. We have also seen that natural communities, like organisms, tend to achieve a balance with their environments, a sort of dynamic homeostasis. More diversified communities are more stable since they suffer fewer, less drastic fluctuations. Since they are in harmony with their abiotic environment, such communities are more balanced homeostatically than simpler communities.

In addition to solar energy, an industrialized society requires enormous supplements of energy inputs derived from fossil or nuclear fuels. Such a society, compared with a more primitive one, is operating at a very high rate of metabolism. It consumes more resources, and, inevitably, creates more wastes. This is consistent with the laws of thermodynamics, and the human population, living within the closed system of "spaceship earth," is no less bound by these laws than a single organism.

A cardinal principle of ecology is that no

species may exceed the carrying capacity of its environment. But *Homo sapiens* is currently behaving as though he were a species apart, for that is precisely what he is in danger of doing. An important aspect of carrying capacity is the ability of an environment to recover from damage by, and absorb the wastes of, a community. Today the capacity of natural systems around the world to absorb the wastes of humanity and industry and cleanse themselves is being strained to the breaking point. No other species but man has had such a profound effect on the environment of this planet. As we shall see, many of these changes are altering the state of nature in which we evolved. We thus are in danger of exceeding our homeostatic boundaries.

Another ecological law is that no species can continue for very long to increase its numbers at the expense of other species. As the number of humans has soared, the numbers and kinds of other species have declined sharply. It remains to be seen how much longer we can test the parameters of this law before we have to face up to its consequences.

22-1 *The depressing sight of a fish, dead of oxygen starvation, floating on the hazy waters of an algae-filled lake, is becoming more and more common as the condition of our water deteriorates. At this stage, the ecosystem is in severe danger, but the damage is not yet irreversible. (Alfred Eisenstaedt)*

22-2 *The human population remained at moderate numbers throughout most of prehistory and history. Fluctuations are indicated for such events as the Black Death in Europe in the thirteenth century, which temporarily diminished the population but did not really retard its growth. The dramatic increase, approaching four billion in this century, is a phenomenon of the industrial age.*

## Population growth

At present, world population stands at approximately 3.8 billion people. According to some estimates, it will reach some 7 billion by the end of this century if present rates of increase continue. This, of course, is a projection, not a prediction. Many things could change in 27 years.

For hundreds of thousands of years, the human population was small and increased very slowly, almost imperceptibly, because of diseases and famines. Like other species, it was in homeostatic balance with its environment. Then, as a result of an interplay of changes brought about by the Industrial Revolution and advances in technology and medicine, an upsurge began in the eighteenth century.

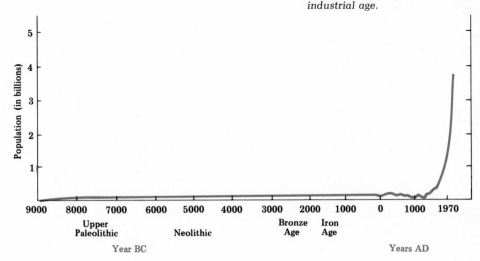

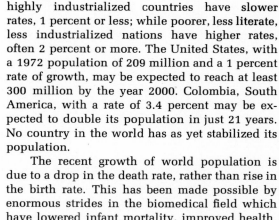

A concept particularly useful in demonstrating this acceleration is that of "doubling time," the time required for a population to double itself. Estimates of numbers are necessarily rough and approximate, but it is believed that 2000 years ago, world population was between 200 and 300 million people. By 1650, it had grown to approximately 500 million. In 1850 it had reached 1 billion, and by 1930 it stood at around 2 billion. With present population at 3.8 billion, it appears virtually certain that in 1975 it will reach 4 billion. From Christ's time, then, the first doubling to a half billion people required 1650 years. The next doubling time was 200 years; the next was 80, and the last will have occurred in only 45 years. This represents a drastic telescoping in doubling time.

World population is now increasing at a rate of 2 percent per year. That means that the current doubling time is approximately 35 years. Rates of increase vary, of course, for different countries.

In general, the wealthier, more literate, more highly industrialized countries have slower rates, 1 percent or less; while poorer, less literate, less industrialized nations have higher rates, often 2 percent or more. The United States, with a 1972 population of 209 million and a 1 percent rate of growth, may be expected to reach at least 300 million by the year 2000. Colombia, South America, with a rate of 3.4 percent may be expected to double its population in just 21 years. No country in the world has as yet stabilized its population.

The recent growth of world population is due to a drop in the death rate, rather than rise in the birth rate. This has been made possible by enormous strides in the biomedical field which have lowered infant mortality, improved health, and extended the human life span. At the moment there is no indication that birth rates will soon come into balance with death rates and thereby stabilize population. If they should, stabilization would not occur right away. In the United States, for example, even if the birth rate is reduced to replacement level (two children per couple), population would continue to grow for another 70 years because of the youth of that population and because of its long life expectancy.

## Human ecology

Malthus was the first to clearly articulate the relationship of population growth to food production. He noted that if food supply is adequate, human population would be limited only by such checks as war, disease, and "moral restraint." Many times in the past, critics have argued that Malthus must have been wrong since population continues to rise. Malthusian law has thus been assumed by many to have been "proved" wrong. Recently, however, it has been gaining renewed supports from many ecologists.

Those that reject Malthus today apparently downgrade the significance of a number of facts.

22-3 *As the size of the human population increases, and the trend toward urbanization creates more areas of high population density, the phenomenon of crowding becomes more widespread. Behavioral studies have indicated that overcrowding, such as may be experienced in major cities, can lead to psychological and even physiological symptoms of stress. (John Hendry, Jr. from National Audubon Society)*

Tidal pool

## Balanced ecosystem

The tidal pools found along the ocean's edge are microcosms—each a small balanced ecosystem. All the links in the tidal pool food chain are shown in the photo at left.

The primary producers in this system are the algae. The large branching plant growing on the blue mussel shell is *Chondrus crispus,* or Irish moss. Another kind of red algae, of a much smaller size, is the coralline algae that shows up as pink patches on the shell. No doubt many microscopic green algae are also floating in the waters of the pool.

The next link in the chain, the primary consumers, are the filter feeders that live on the algae. The mussel is one such organism; it pumps water in through one siphon, strains out all the organic material, and pumps it back out again. The acorn barnacle, a member of the genus *Balanus,* is also a filter feeder. *Balanus* is well adapted to life in the tidal zone, for it can survive exposure to the air for hours at a time when the tide recedes. The barnacles and mussels feed on all kinds of organic debris and therefore play an important role in the recycling of nutrients.

The secondary consumer in the tidal pool is the starfish, a carnivore that feeds on the mollusks. The starfish is probably not a permanent resident of this pool; he has merely dropped in for dinner. Having eaten his fill, he likely to move back out with the flow of the next high tide.

# Biomes

The biomes on these pages represent two climatic extremes—the cold tundra and the hot desert. These two biomes, both of which can be found in North America, are not as dissimilar as one might think, for they share the problems of low rainfall, poor soil, and continually fluctuating environmental conditions.

A typical view of the Mojave desert shows how far from barren that biome really is. The large saguaro cactus dominates the landscape; notes of color are added by the goldenhead and the vivid red flowers of the Indian paintbrush. A fearsome predator of the desert biome is the Gila monster, a carnivore with teeth modified for the injection of poison venom. The niche of scavenger in this biome is filled by the vulture; its sharp curved beak is an adaptation to this way of life.

In the tundra, the strong winds and the weight of winter snows prevent the growth of upright trees; all vascular plants in this biome grow horizontally, parallel to the ground. At first glance, the tundra landscape may seem monotonous, but a closer look reveals a variety of life forms. The rocks are covered by lichens, some of them quite brightly colored, and many plants flourish along the melting edge of a snowdrift. Animals such as the snowshoe rabbit and the arctic ground squirrel are so well camouflaged that they blend into the scenery; but the musk oxen catch even the casual eye, impressive in their threatening solidarity.

Mojave desert

Egyptian vulture

Gila monster

**Musk oxen**

**Arctic ground squirrel**

**Tundra**

**Snowshoe rabbit**

# Land use and misuse

When man sets out to "conquer" nature, he sometimes damages his environment so severely that he destroys the ecological balance of an entire area. One example of this problem can be seen in the bauxite mining process that peels the soil from acres of ground. It will take centuries before such a pit can support life again. The problems of succession after strip mining has changed the landscape are evident in the photograph of an old slag heap in Kentucky. Although some ground cover has been established, all the new trees have been killed by the sulphur released by the coal slag. The aerial photo taken over Vietnam shows another kind of destruction of the environment, in the cratering of a lowland forest. Damage to an aquatic biome is evidenced by the pulp mill wastes floating on the surface of the Columbia River.

Yet man's intervention need not be harmful. The famous Ifugao rice terraces, located in the Philippines, exemplify land use that strengthens the ecosystem. The hillside terraces are said to be more than 2000 years old and have been built and rebuilt over the centuries. The slope of the terraces prevents erosion; it also creates new habitats, adding variety and therefore more stability to the ecosystem. Another pattern of use can be observed in the Ohio farm scene; a notable feature is the diversity which has been retained here. Instead of clearing every tree in sight, the farmer has left some of the woodland undisturbed.

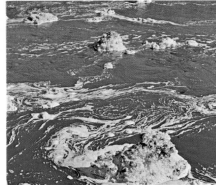

**Pollution on the Columbia River**

**Cratering in Vietnam**

**Strip mining**

**Bauxite surface mining**

**Rice terraces**

**Cattle grazing**

**Vineyard**

**Farm land**

# Food chain in the woods

An example of a complex food chain can be seen in almost any woodland area of the United States. Such areas are likely to contain both a deciduous forest and a fresh-water pond, and the two biomes are linked together to form one integrated ecosystem.

The grass, weeds, and water plants are the primary producers in this system, and the herbivores are the primary consumers. This first step in the progression is represented by the Viceroy butterfly that is feeding on the nectar of the *Potentilla* flower.

The butterfly may in turn fall victim to another insect, the predatory damselfly, which lies in wait so well concealed on the green leaves. The damselfly itself may then become the prey of the green frog that is hiding by the edge of the water.

The next step in the chain is vividly portrayed by the garter snake that is trying to swallow the green frog. Although it appears that the frog is almost as large as the snake, the snake can easily stretch its jaws and accommodate its prey. But it may have made a mistake in starting with the legs rather than the head. The snake will have to loosen his grip to make the next swallow, and if the frog times it just right, he can take advantage of that instant and make a leap for safety.

At the top of the food chain in this area is the hawk. The broad-winged hawk shown here feeds on snakes and rodents.

Damselfly
Woodland pond

**Viceroy butterfly**

**Green frog**
**Garter snake and green frog**

**Broad-winged hawk**

# The planet earth

Even when seen from a distance of 22,000 miles in space, our earth remains a remarkably beautiful planet. In this photo taken from a NASA satellite, the entire continent of South America is clearly visible, as are portions of North America and Europe, and a speck of Africa. To the north, the Greenland icecap is just barely distinguishable, but the continent of Antarctica is entirely blanketed under a heavy cloud cover.

The picture also gives some indication of the earth's weather. The thick clouds that stretch over central North America from the Great Lakes to Mexico is a cold front moving eastward, bringing typical November weather. A tropical storm is swirling over much of South America.

Seeing the earth from space in this way has helped make us all more conscious of the fact that our planet is really one large ecosystem. "Ownership" of parts of the land or water by private individuals or national governments is purely a social convention; the biological reality is that the earth is a whole and that it belongs only to itself.

When the earth is viewed from this distant perspective, man himself is given a new context. No longer master of all he surveys, he becomes another of the earth's dependents. His fate—even his continuing existence—is seen to be inextricably linked to the survival of the ecosystem earth.

The earth

Most important of these is that more than half of the world population now alive is either hungry or suffering from malnutrition. While many of the approximately 12,000 people who die each day because of starvation have succumbed more because of political and economic dysfunctions than through the inability of the earth to produce enough food, there are, for whatever reason, critical food problems with our present population of 3.8 billion. These problems will certainly be dwarfed by those facing the world in the year 2000 if its population reaches 6 or 7 billion.

For the poorer nations, providing enough food for their people hinges, for the most part, upon a successful grain crop each year. But vagaries in rainfall all too frequently result in poor crops or no crops. Famines afflicting hundreds of thousands continue to occur. Virtually all arable land in the world is now under cultivation. Soils, already poor because of centuries of cultivation that removed nutrients without adequate refertilization, are being farmed more intensively.

There are two prominent, divergent views of the future. Many agronomists say that massive famines resulting in the death of millions are now inevitable. Some food experts believe that these famines may be only two or three years away, and that they will last for many years. Others believe we are on the brink of a "Green Revolution" which will make starvation and hunger a thing of the past. They point to the development of new varieties of "miracle grains" which grow faster and result in a manyfold increase in production per acre. But there may be many serious obstacles to realizing a Green Revolution. New strains of hybrid grains are often susceptible to insect and fungus attack, so they require the application of tremendous amounts of fungicides and pesticides. These new strains also require inputs of chemical fertilizers, water, and machinery for planting and harvesting. These biological considerations raise the problem of how sufficient amounts of money can be made available to develop these new grains and to acquire the machinery to plant, nurture, and harvest them.

## Resource depletion

Many of the resources we require for our way of life are considered "nonrenewable," in the sense that they may have taken millions of years to form, but can be rapidly consumed. For all practical purposes, these resources are finite. While some of them may be recovered by recycling, mostly at great expense in money and energy, the second law of thermodynamics decrees that there will nevertheless be a loss. Coal and oil, for example, are highly organized states of matter of low entropy. As these are oxidized they become more disorganized; entropy has increased sharply. They are converted from an energy-rich state to an energy-poor state. So it is in the use of most materials. Entropy rises so sharply and the molecules become so scattered or chemically changed that reconstitution is virtually impossible.

Probably the most important single factor underlying the temporary expansion of the earth's carrying capacity and supporting the population explosion has been the availability of large amounts of energy. This has made possible the Industrial Revolution. Until now, most of this energy has come from the combustion of the fossil fuels — petroleum, coal, and natural gas. This combustion has created undesirable and unforeseen side effects, primarily air pollution.

When the Industrial Revolution began, population was small, and it was thought that the supply of fossil fuels was literally unlimited. Now, however, we are beginning to see a threat of depletion. The search for crude oil by all industrialized countries has become more intense, and deposits are being sought miles offshore and in the Arctic tundra. According to some estimates, 80 percent of the world's petroleum reserves will be gone within 50 years. We have estimated coal reserves of perhaps 125 years, although if current usage rates remained static, these reserves could well last some 2300 years.

Recycling is often offered as an answer to this growing dilemma. While it is a valuable practice, and one that is sure to grow in usage, it does have some limitations. Some metals, such as silver and mercury, are irretrievably lost in the refuse of photographic film and discarded or broken thermometers. Others, such as aluminum and steel, can be salvaged, but it is often more costly, particularly in energy requirements, than using raw ores. Another trend, in addition to recycling, is toward digging deeper and using diminishing grades of ore. These require correspondingly greater amounts of capital and energy to mine and refine.

Perhaps substitutes will be found for many if not most of these minerals. But it should be kept in mind that many are elements with unique properties that generally cannot be duplicated. Only cobalt, for example, can be used to make

permanent magnets. Mercury is the only metal that is liquid at room temperature.

Unlimited cheap new sources of energy would do much to alleviate some, but not all, of the critical problems now looming. The still-to-be-developed fast breeder reactor should extend our plutonium reserves. But atomic plants will continue to pose a small but definite risk of contaminating the environment with radiation. Furthermore, there are the growing problems of thermal pollution and the storage of long-lived radioactive wastes. Fusion reactors, which would operate on the virtually unlimited hydrogen atom, would certainly produce vast amounts of inexpensive energy, but aspects of its technology are still unknown, and a prototype power plant is probably at least 30 years in the future.

The total global picture, then, is one of rapidly dwindling nonrenewable resources, including the fossil fuels. Whether sufficient sources of other means of energy can be found quickly enough is uncertain. As semi-industrialized countries become more industrialized, the drain on, and competition for, remaining stocks of resources and fuel will worsen.

**22-4** *The general effect of intense and persistent smog is summarized in an extract from the Clean Air Act (88th Congress, 1963):*

*. . . growth in the amount and complexity of air pollution brought about by urbanization, industrial development, and the increasing use of motor vehicles, has resulted in mounting dangers to the public health and welfare, including injury to agricultural crops and livestock, damage to and the deterioration of property, and hazards to air and ground transportation. (Dave Repp from National Audubon Society)*

## Environmental deterioration

Given our patterns of consumption, today's material resources become tomorrow's garbage. The point of recycling, of course, is to take such "garbage" and use it as tomorrow's resources. If our present garbage were merely litter, junked cars or old washing machines, there would be little to worry about. But environmentalists now monitoring changes in our life support systems are not very encouraged by their findings. Read any newspaper or magazine on any day; watch the newscast on television and count how many items relate to pollution. These are not isolated nor merely local incidents. They are part of a deepening and accelerating global pattern. Much of it involves not litter or junk but organic and chemical wastes that are returned to us in the air we breathe, the water we drink, and the food we eat. Little is known concerning the effects of many of these wastes on living things, of how some of these chemicals interact with each other to produce effects many times more deadly than would be expected by the addition of their separate effects. All we can see are the results. We have no real knowledge of how long our life support systems can continue to absorb these effluents before some begin to break down.

## Land and soil

In times past, when the human population was much smaller, settlements in new territories were often selected because of the presence of rich fertile soil. With the acceleration of the Industrial Revolution, more people migrated to these settlements, and they gradually became densely populated cities. Gradually, prime farming land was covered with buildings and pavement. This trend is continuing today. Many can still recall when much of northern New Jersey supported vegetable farms. Today, farms have been replaced by private homes, lawns, and shopping centers. Los Angeles in the 1920s was as important for its citrus groves and farmland as it was for its silent movies. Today, groves and farms have disappeared, and Los Angeles is a mass of housing developments, industrial plants, freeways, and urban sprawl. In that city 70 square miles are being paved over each year. Iowa loses one percent of its farmland each year. Due to creeping urbanization, at least 1600 square miles of open land are being paved in the United States each year. This means that the amount of arable land per person is falling. In 1800 there were 104 acres for each individual in the United States; by 1900 there were only 25 acres per person, and now this has shriveled to 2.6 acres.

What of the soil on remaining farmland?

What is not being lost by erosion is being contaminated.

Whenever the natural vegetation cover is removed from an area, imbalances occur and deterioration of the soil begins. Much of our valuable topsoil is lost as a result of the practice of clear-cut logging. With the cover of trees stripped away, rains wash much of the humus away into the rivers.

Obviously, we must raise crops and livestock if we are to produce food. But few activities of modern man are as destructive to the environment in general and to the soil in particular as agriculture. Clearing of land removes the diverse community of native plants and therefore destroys or drives away the natural community of animals. Cultivation, even with the most modern management methods, inevitably involves some loss of topsoil through wind and water erosion. Hundreds of thousands of tons of farm soil are washed into the oceans each year. In addition, intensive cultivation removes important minerals, thereby decreasing soil quality. The pages of history are filled with examples of cultures and civilizations that faded following intensive farming and soil depletion. The Mayan civilization which flourished in Yucatan centuries ago is a prime example. In many regions around the world, gardens have been transformed into deserts. Mali and Ghana, Egypt and much of the Sahara, were once fertile and green. These are now parched, spent areas, even though scientists do not believe that the climate of these regions has changed significantly.

Modern agriculture requires the use of millions of tons of artificial fertilizers and pesticides. Pesticides have been used because replacement of a natural, diverse and therefore stable community with a simple community of one or two crop species presents a vast dinner table to pest species. The natural checks and balances inherent in a varied system are lacking, and there is nothing to prevent an explosion of pest numbers. At worst, crops may be wiped out com-

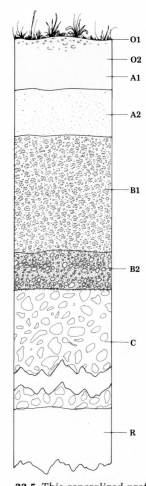

**22-5** *This generalized profile of the soil includes a greater number of horizons than is usually found in any one soil. The O horizons are composed of loose, matted, or decomposed organic matter. The A horizon is the one of greatest biological activity. B is the horizon of maximum accumulation of clay materials or of iron and organic matter. C is weathered material, which may be either like or unlike the material from which the soil was originally formed. R is consolidated bedrock.*

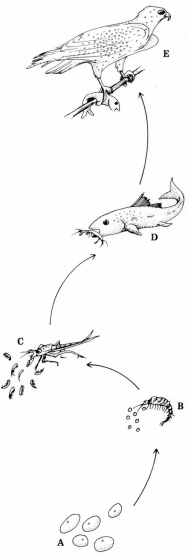

**22-7** *A food chain, such as this one involving a single-celled plant, Gammarus (a small amphipod), a dragonfly nymph, a fish and an osprey, may result in the progressive concentration of any nonexcretable substance (color) such as DDT.*

pletely. At best, fruits and vegetables are riddled with holes in which the insect larvae may still be wriggling. So vast amounts of chlorinated hydrocarbons, such as DDT, have been sprayed each year.

**22-6** *One of the ways in which man has chosen to interfere with the workings of the natural ecosystem is by removing certain links in the food chain, especially the top predators. Although farmers intend only to protect their own livestock, the killing of coyotes upsets the balance of predator and prey within the grassland biomes, with far-reaching effects. (C. T. Hotchkiss from National Audubon Society)*

But there are unexpected catches to their use. For one thing, some of the most effective of the chlorinated hydrocarbons are persistent; that is, they resist being broken down to simpler, less harmful molecules by natural agents of decay. The half-life of DDT residues is 10 years. Another drawback is that chemicals do not remain where they are applied. They are carried by wind or water runoff to other ecosystems. There they become incorporated in tiny amounts within producer organisms. But as the energy is transferred along the food chain, the pesticide concentration

increases manyfold with each transfer. Animals at the top of the food pyramids have the highest concentrations. Only recently have we begun to learn how these pesticide concentrations are harming top carnivores. In addition to frequent mass fish kills in many American and European rivers, we are witnessing the decline of such birds as the brown pelican, the bald eagle, and the peregrine falcon. It has now been established that buildups of DDT cause hormonal imbalances in birds that result in the production of eggs with progressively thinner shells. Under the weight of the mother's body, many of these eggs are crushed during incubation.

Ironically, the pest species that the chemicals are intended for keep coming back. Many insect pests have had millions of years of natural selection to cope with various plant poisons. When man sprays, he is speeding up that selective process. The result is more resistant strains of pests. In addition, the parasite or predator that naturally keeps the pest in check often is wiped out because it has evolved no defense against these poisons. Thus we have a vicious cycle: the more pesticides used, the more resistant the pests and the more pesticides needed. Not one species of pest has been eliminated by use of pesticides. On the contrary, we are creating mutant species. In southern California, for example, two species of mosquitoes have become virtually immune to DDT. One of these, *Aedes nigromaculus*, is a vicious stinger that attacks in swarms. The other, *Culex tarsalis*, is much more worrisome since it transmits human encephalitis and has foiled all insecticides. How many more such mutants are in the making is anyone's guess.

But the pesticide problem is only one of several now worsening as a result of the intensification of modern agriculture. Another problem involves the use of chemical fertilizers.

Farmers have found that in nitrogen-depleted soils, crop yield can be improved dramatically with applications of synthetic nitrate fertilizers. Within the past 25 years, there has been a 12-fold

**669**

increase in their use. But with repeated applications of artificial nitrate fertilizers, soils lose their natural capacity to regenerate nitrogenous compounds. The geochemical cycle is disrupted. It is as if the soils have become "hooked." As in the case of pesticides, the more synthetic fertilizers used, the more required.

Not all these chemicals are picked up by the plants. Excess nitrates seep into ground water, or are carried by runoff into ponds or lakes, where they cause other problems. In some regions of the world high nitrate content in well water has resulted in methemoglobinemia, a disease that often attacks infants. Intestinal bacteria convert nitrates to nitrites, which are absorbed into the bloodstream. Hemoglobin has a stronger affinity for nitrites than oxygen, and consequently the infant begins to suffer from an acute lack of oxygen. Serious outbreaks of this disease have occurred in the Central Valley region of California as well as in Illinois, Missouri, and Wisconsin. The problem of excess nitrogenous compounds in the ecosystem is further intensified by the excretions of a growing livestock population. In the United States there are 50 million pigs, 38 million cattle, 350 million chickens in addition to a variety of other fowl, sheep, and goats. Many individuals are crowded into feedlots, and their excretory products become highly concentrated in the soil.

Many other widespread human activities are having an adverse impact on the land. Farming in arid regions requires intense irrigation which, because of the high rate of evaporation, causes salts to be pulled to the surface where they concentrate. The soil then becomes too saline to support normal crops. This process of salinization undoubtedly converted rich farming cultures into the Syrian desert, ended in dismal failure a project to make the Afghanistan desert bloom, and is becoming a serious problem in many regions of the southwestern United States.

Strip mining has ravaged hundreds of thousands of acres in Kentucky and elsewhere in Appalachia. Stripped of its topsoil and "overburden," the scarred land remains barren of any vegetation for many years. Acids percolating from the rubble find their way into ground water and streams where they poison all aquatic life.

## Water

Water is a prime necessity of life. For many animals and plants, it is an important limiting factor. Although, like soil, it has been described as a renewable resource, at present the hydrological cycle is not replacing pure water as fast as it is being consumed or contaminated. In some regions, water shortages have begun to appear and in others they are looming. In arid or semiarid regions where irrigation is practiced, water sources are being depleted at a rate many times faster than they can be recharged. Because millions of years were required to build up underground water supplies, the pure water they contain is known as "fossil water." In some coastal regions, removal of this water is resulting in its replacement by salt water. In west Texas, New Mexico, and Arizona, the drop in the water table is becoming critical. In some of these areas water is being pumped out 140 times as fast as it is trickling in. In other regions, clearing of forests removes the valuable watersheds that drain into reservoirs and rivers. Damage here is long-lasting because the water-storing humus layer that is stripped away requires many hundreds of years to replace.

Water that is not being used or squandered is being rapidly polluted. Pollutants come from industrial and technological sources, human and livestock wastes, and runoff of chemicals applied to the soil. Many creeks, rivers, lakes, and bays in the United States have now been seriously contaminated. Some rivers in the United States and Europe have become open sewers. Among the worst are the Rhine, the Potomac, the Hudson, and the Cuyahoga. The Cuyahoga in Ohio has

**22-8** *Following the Industrial Revolution and the invention of the indoor toilet the amount of supplemental organic material added to waterways and lakes increased dramatically. It is estimated that 20 percent of U.S. sewage is still flushed untreated into the nearest watercourse. Raw sewage disturbs the biogeochemical circulation of an aquatic ecosystem by vastly increasing the accumulation of organic matter and the demand for oxygen. (Gordon S. Smith from National Audubon Society)*

**22-9** *The recurring red tides of Tampa Bay have brought sudden death to thousands of fish, including these black drum fish washed onto the beach at Tampa. The red tides seem to occur in summer as the result of an aberration in local algae. The noxious bright red algae may cover the surface of the water for several hundred miles, killing most of the fish in the vicinity. (R. F. Head from National Audubon Society)*

ents which bring about "blooms" of blue-green and green algae. When the huge masses of algae die, bacterial decay robs the water of its natural oxygen content, killing fish and other aquatic life. This not only results in an overwhelming stench but also causes eutrophication, a process that accelerates natural succession. Thus a lake that might normally last over a thousand years may reach senility and death in 30 or 40 years. The recent use of phosphate-laden household detergents has added to this problem.

At one time, it was believed that it would be impossible to pollute such large bodies of water as the Great Lakes. Today, Lake Erie, which receives the discharges of the Cuyahoga and other equally filthy rivers, has been pronounced "biologically dead." This is a sad end to a lake that not too many years ago produced millions of pounds of whitefish yearly and provided swimming recreation for thousands of people. Lake Michigan appears to be headed in the same direction.

One of the side effects of our growing demand for electrical energy has been contamination of our waterways with excessive heat—or thermal pollution. Most steam-electric plants, both fossil- and nuclear-fueled, employ river water as a coolant and discharge hot water further downstream. It has been shown that some thermal loading is beneficial to fish life, but excessive thermal loading is extremely destructive. In some instances, heated water acts as a barrier that prevents such cold-water fish as salmon from reaching their spawning streams.

All over the world, rich industrialized countries are concerned about chemical contamination of water, and poor countries are worrying about contamination of water by human wastes. Can pollution cleanup reverse the trend? Although billions are now being spent in the United States, and many billions more are needed, the fact is that water quality worsens each year. Simply "cleaning up" the water will not restore the conditions and complex communities that

been so burdened with flammable wastes that it has been declared a fire hazard and has actually caught fire several times along one stretch.

The nitrogenous compounds and phosphates that originate in farms and feedlots is also killing ponds and lakes. These chemicals serve as nutri-

**671**

existed previously; in many cases the damage done to aquatic life forms is irreparable.

## Air

The pollutants we discharge into our soils and water are matched by the effluents released into the air. Some 200 million tons of contaminants are emitted by autos, jets, and industry each year. These contaminants consist of two components: particulates, some of which precipitate as soot; and various noxious gases that include oxides of carbon, sulfur, and nitrogen, and a variety of hydrocarbons. More than 60 percent of this nation's air pollution is produced by its 83 million cars, and the major pollutant is carbon monoxide.

Smog, burning eyes, and coughing have become a familiar way of life for inhabitants of big cities. Several ingredients and conditions go into the making of smog. Nitrogen oxides emitted mainly from autos give the air that brown tinge so familiar to jet passengers approaching Los Angeles. Under sunlight, a photochemical reaction results in which nitrogen oxides react with waste hydrocarbons from gasoline to form PAN (peroxyacetyl nitrate), the cause of smarting eyes and probable lung damage. PAN and ozone are the two most toxic components of smog. Another set of smog ingredients are the oxides of sulfur, of which some 24 million tons are released by American industries each year. Sulfur dioxide is an unstable compound which tends to oxidize into a sulfuric acid mist. In recent years, acid rains of extremely low pH have been recorded in Sweden and Chesapeake Bay. Samples of these rains contained large amounts of sulfuric acid. Mix together vast quantities of carbon monoxide, nitrogen and sulfur oxides, PAN and sulfuric acid, and you have a fairly lethal "soup" indeed. In cities situated in pockets or basins, this smog may be trapped for many days by stagnant air held by temperature inversions that are formed when a layer of cool air is covered by a layer of warm air.

In the Los Angeles basin, the smog level is so acute that it is no longer profitable to grow lettuce and spinach. Smog is also killing more than a million trees in southern California. Pines exposed to smog suffer a decline which is indicated by a yellowing of needles; the ponderosa pine is particularly susceptible.

If air pollution is harmful to plants, it is to be expected that it would affect the health of animals and human beings. No actual proof is yet at hand, but certain correlations are evident. The incidence of bronchitis, emphysema, and lung cancer are rising rapidly everywhere; a relative rarity 30 years ago, emphysema now kills 25,000 people a year. Deaths from all respiratory diseases are up 50 percent. It has been estimated that in walking the streets for a full day, a New Yorker inhales the equivalent of close to two packs of cigarettes.

There are currently two prominent theories that regard air pollution as having an effect on global climate. However, they are mutually contradictory in terms of the direction of the change they predict. The first theory is an attempt to explain the fact that during the first half of this century there was a global warming trend. Thus, any burning or combustion must produce at least carbon dioxide and water vapor as waste products. Together, when released into the atmosphere, they create a "greenhouse effect" by trapping much of the heat energy produced when the light from solar radiation is reflected from the ground. Since the carbon dioxide content of the world's atmosphere has increased by some 10–12 percent since the Industrial Revolution began, it is theorized that this increase is responsible for the warming trend.

The second theory attempts to account for the fact that we have been experiencing a cooling trend since the end of the Second World War. It is believed that this has occurred because of a sharp rise in the atmosphere's particulate load

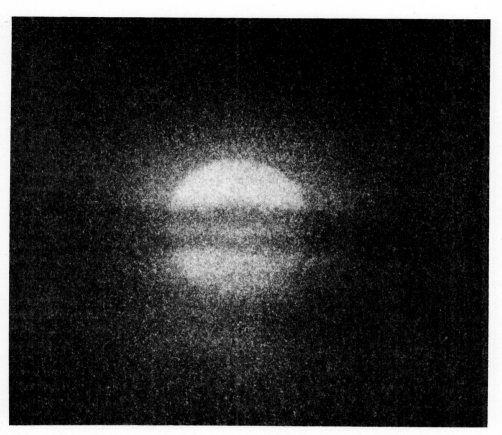

**22-10** *The presence of particulate matter in the air may have a dramatic impact on the weather, reducing the amount of solar radiation reaching the earth's surface and initiating cloud condensation. A five percent increase in the world's cloud cover would so drop temperatures that the earth would be plunged into another ice age. During the 1960s there was, in fact, a slight decrease in the earth's temperature and ice coverage in the North Atlantic reached a peak in 1968. (John Hendry, Jr. from National Audubon Society)*

since particulate matter or atmospheric dust blocks out solar radiation.

Scientists doubt that the two trends will cancel each other out, but they are unable to agree on which trend will dominate. One could precipitate another ice age, while the other could result in the inundation of the world's coastal regions when melting ice caps bring about a rise in sea level. As a matter of fact, the sea level has risen three inches in the past eight years. This represents an enormous rise in so short a time. Whether it is due to a subsidence of land because of removal of subsurface water and petroleum, or to a melting of glaciers, is unknown.

## Oceans

The oceans have been described as the "ultimate garbage pail" since they eventually receive so much of the excrement, garbage, and industrial wastes of the world and have such a huge capacity. The land-locked shallow Baltic Sea has become so polluted and oxygen-deficient that it is approaching Lake Erie's fate. The once sparkling Adriatic is in similar condition, as is the Mediterranean. Alarm is spreading among local populations and tourists because of outrageously polluted beaches in France and Italy. Less than 15 percent of Italy's beaches are now free of pollution. Thor Heyerdahl reported after his most recent crossing that a continuous stretch of at least 1400 miles of the Atlantic Ocean is polluted by asphaltlike lumps of oil.

No wonder that marine biologists are shocked by the rapid biotic and chemical changes they are now witnessing in the open oceans. What has them concerned is not so much the detritus of discarded dolls, driftwood, and debris, but rather the lethal and largely invisible mixture of industrial wastes compounded of oil, pesticides, detergents, herbicides, heavy metals, and the chemical by-products now gushing out of the rivers of many nations. In addition, a considerable amount of wastes is flushed from the bilges of ships, leaked from broken pipelines, or sinking tankers.

The oceans have always provided vast quantities of protein for the peoples of the world. Today, sophisticated fishing fleets equipped with the finest sonar locating devices are in fierce competition with each other. Whether because of growing pollution, or overfishing, or both, certain fisheries are disappearing rapidly. The sardine, once abundant off California and the west coast of France, is now rare. Near Marseilles, out of 13 species of food fish plentiful before World War II, 9 have disappeared, and the remaining 4 are scarce. Off the New Jersey coast, the annual catch of such food fish as fluke, weakfish, and

**673**

bluefish has dropped from more than 2 million pounds to 907 thousand pounds from 1965 to 1970. Menhaden, formerly an abundant source of fishmeal for animal food, has shown a sharp decline. Within the last decade, the catch of menhaden off the New Jersey coast has dropped from 330 million pounds to 31 million pounds per year. Whaling, once the mainstay of New England, has now been abandoned by all nations except Russia and Japan. Jacques Cousteau estimates that life in the oceans has diminished by 40 percent in 20 years.

Some of the poisons expelled by industrial man have now begun to return to him on his dinner table. Particularly worrisome are such dangerous heavy metals as cadmium, lead, and mercury. Like DDT, mercury becomes greatly concentrated as it moves its way up the food pyramid. Unsafe levels of mercury have been discovered not only in tuna and swordfish, but in many large fresh-water fish as well. A heavy diet of mercury-laden fish could induce severe poisoning, causing brain damage.

Of all the chemicals man's activities are now adding to the oceans, pesticides, such as DDT, DDD, and DDE, may prove to be the most ubiquitous. Although originally applied to the land, they ultimately accumulate in the oceans. According to a report by the National Academy of Sciences, as much as 25 percent of the DDT compounds produced as of 1971 may have been transferred to the sea. The amount of DDT compounds in marine organisms is estimated to be less than 0.1 percent of total production, yet this small amount has produced a demonstrable impact upon the marine environment.

Because of these chemicals, food chains are shortening and top carnivores are being eliminated. There is a shift from diversity toward larger populations of a few species of small animals and plants with fast rates of reproduction. Recently, evidence has been found that photosynthesis as well as reproduction in algae may be inhibited by pesticides. There is now

22-11 *The accumulation of debris on beaches is one of the most highly visible forms of solid waste pollution. Almost all wastes contain recoverable substances. Even garbage can be sorted and recycled. It only remains for an economically feasible recovery method to be applied. In this case, the Miami resort owners may find it economically profitable to endorse and help subsidize such efforts. (Maurice E. Lanore from National Audubon Society)*

22-12 *This oil-soaked gannet is the victim of "accidental" pollution resulting from oil spillage. Unable to fly or to get food efficiently, many such birds have been found washed up on California beaches either dead or seriously ill. (George Komorowski from National Audubon Society)*

growing alarm that the diatoms, which generate more than half of the earth's atmospheric oxygen and which are the main food base for all ocean life, may be in jeopardy. Scientists are unable to agree on what elimination of diatoms may do to our oxygen supply. Some believe it would remain virtually unchanged for 500 years even if all diatoms should vanish tomorrow; others claim there would be a sharp dip in atmospheric oxygen within a short period of time.

The world is looking to the oceans to supply increasing amounts of protein for its growing millions. But as overexploitation continues and pollution worsens, the capacity of the oceans to produce must surely suffer. At one time it was thought impossible that the vast oceans could be seriously damaged by pollution. Now some marine biologists are warning that if nothing is done to curb the present trend, the oceans will be dead in 50 years or less.

## Alteration of ecosystems

We have seen that stability of an ecosystem hinges upon its diversity or complexity. Simple ecosystems are marked by drastic changes or fluctuations in populations. Yet, as in the case of

the oceans, the basic pattern around the world is one of simplification. As species decline or become extinct, communities become simpler. Thus, by replacing a tract of natural grassland with a wheat field, man has created a simple ecosystem in which homeostatic balance is more fragile.

One vital ecosystem now being rapidly destroyed is the salt marsh. Salt marshes have been aptly described as the nurseries of ocean life. The coastal mud flats where rivers meet the ocean support a unique community of plant life that nourishes resident invertebrates and the larvae and fry of much of the ocean's fisheries. A key producer here is the cord grass or salt hay of the genus *Spartina*. When this plant dies, it is broken down by bacterial decay and becomes detritus. *Spartina* detritus provides the nutrients for microscopic plants and a direct source of food for many worms, crustaceans, and bivalve mollusks. The microscopic plants provide food for microscopic animals, and fish depend on these, as well as on worms and crustaceans. Smaller fish serve as food for larger fish, which, in turn are preyed upon by birds and mammals. So much of the vast food web of the oceans depends on the integrity of the salt marshes.

Today, less than 30 percent remains of the salt marshes that lined much of the east coast of the United States a century ago. These wetlands have been drained, bulldozed, and filled, to be replaced by housing developments and industrial plants. Much of the remaining marshes are polluted by the outflow of rivers and pesticides sprayed to control mosquitoes.

As population increases, pressure mounts to develop or exploit not only salt marshes but also all remaining natural areas. Forests are cleared, grasslands are cultivated, resorts intrude in formerly wild regions. This push of people, farms, and dwellings into natural areas, either drives out native animals because of habitat destruction, or results in outright extinction. The triple threat of overexploitation, pollution, and

**22-13** *Between 1800 and the early 1900s, millions of alligators were killed each year for their fashionable hides. The alligator has managed to survive despite a dwindling range and more than two centuries of unrestricted killing. Formerly 'gators were found from North Carolina to the tip of Florida and westward into Texas as far as the Rio Grande. Today they are limited largely to established refuges in Georgia and Louisiana. There they are provided with the proper environment—a coastal swamp crisscrossed by waterways, an abundance of aquatic and other food, and sunny spots for basking. (Allan D. Cruickshank from National Audubon Society)*

ecosystem alteration is taking its toll of animal species.

Since the extermination of the dodo in the 1600s, some 225 species of animals have become extinct. More than 75 percent of these are directly attributable to mankind. Gone are the lesser moa, the passenger pigeon, the heath hen, and the great auk. The rate of extermination of mammals has increased 55-fold. Nearly 1000 species are now regarded as endangered. The largest animals in existence, the great blue whales, head this list; also in great peril are the California condor, the Australian crocodile, the cheetah, the Indian lion and snow leopard, and the symbol of the United States, the bald eagle. Saving these and other animals presents a formidable challenge to biologists.

Each species is inextricably bound to countless others by direct and indirect ecological threads. The removal of any species link by extinction disrupts the integrity of the entire community fabric. In Chapter 21, we saw how the decline of the saguaro cactus is related to the removal of predators. We are beginning to unravel similar complex interrelationships, which in some instances touch man directly. Who, for example, would have suspected that the successful campaign to eradicate the South African hippopotamus would result in an increased incidence in the debilitating disease, schistosomiasis? It was discovered that hippos not only keep river silt churned up but also walk single file from the river, thereby creating natural irrigation channels. With the hippos removed, the rivers silted up and the overflow, no longer restricted to channels, created widespread flooding. The resulting changed conditions favored the proliferation of water snails that serve as the host for the schistosoma parasite.

The extinction of a species, then, should not be regarded merely as an aesthetic or moral issue, but more importantly, as the removal of a vital link that erodes the health and stability of the community, and ultimately affects man. There

can be little doubt that as the population of the human race continues to increase, more and more animals will be forced to the brink of extinction.

## The ecological meaning of growth

This negative impact of man on his physical and biotic environment increases as his numbers increase. The carrying capacity falls and the quality of life decreases. In backward countries with low levels of industrialization, the permanent damage done to the environment is correspondingly low. But the greatest strain on life support systems is inflicted by technological man.

22-14 *Cheetahs have been semidomesticated for centuries in India where they are admired for their beauty, swiftness and unexpected tamability. The animals are caught full-grown and wild. They take about six months to train and become completely docile although regularly allowed to hunt. Once they have knocked their prey down, they wait for their owners to actually kill the beast. Cheetahs have proved successful in killing coyotes in the American Southwest. However, since they are worth at least $5000 a pair, they are not in widespread use. (Mark Boulton from National Audubon Society)*

materials and unavoidable contamination of the environment.

## Views of the future

For many years, conservationists and scientists have been warning that resource depletion and global pollution must quickly be slowed and halted. They have pointed to human population increase as the major factor underlying these problems. Recently, three independent world organizations have issued statements urging stabilization of population and industrialization.

In England, a document entitled, "Blueprint for Survival," supported and endorsed by 34 eminent scientists, was published in *The Ecologist* in 1972. Its recommendations, oriented to Britain but encompassing the entire planet, recognize that one of the keys to a stabilization of economic and ecologic conditions is population stabilization.

In the United States, a Commission on Population Growth and the American Future empanelled by President Nixon to study the impact on environment by population growth, has published similar findings.

The most persuasive argument, however, was embodied in a book, *The Limits to Growth*, prepared by a team of 17 scientists. Sponsored by the Club of Rome, an organization of scientists, educators, and industrialists, this system-dynamics study has identified man's worsening predicament as the most serious problems of all times. Using data of recent conditions and trends, the scientists have produced computer models which project a drastic collapse in both population and industrial capacity within 130 years unless there is an immediate slowdown and halt in growth of each. This collapse, according to the models, would result from environmental deterioration, famines, or resource depletion, or combinations of these. At very least, according to the study, stabilization of population and

**22-15** *The whooping crane has all but disappeared from the North American taiga and tundra. Among large birds it is estimated that 69 percent of the cause for present rarity is due to human factors. Of this, 32 percent results from hunting and 26 percent from habitat disruption. (O. S. Pettingill, Jr. from National Audubon Society)*

Highly industrialized nations enjoy a high standard of living, but their fast rates of metabolism require ever larger infusions of energy and raw materials. As a result, they generate prodigious quantities of industrial pollution.

It has been estimated that, merely to keep up with its present growth rate of one percent per year, the United States would have to build a city the size of Tulsa every 30 days. This is one reason that the demand for electrical energy in the United States doubles every decade. Not only must new electrical plants be built, but also schools, hospitals, sewage facilities, homes, roads, and shopping centers. The high metabolic rate of industrialized countries, given current industrial practices, means swift depletion of

economic growth would buy necessary time, a commodity which is now running out. Stressing that there are simply no panaceas, this study clearly identified exponential growth as the prime factor responsible for our predicaments. The study emphasizes that zero population growth must be accompanied by enormous outlays of capital for pollution control and cleanup, and for means to increase food production.

These studies, particularly *The Limits to Growth*, are not without their critics. Both the assumptions upon which these models are based and the data they use for their projections have come under critical attack, often justifiably so. What the critics have failed to do, however, is provide any studies, using what they consider to be appropriate methodologies and data, that lead to alternative, more optimistic, conclusions. Rather, this tendency has been to fall back upon the assumption that our social and technological systems will prove to be self-correcting before the advent of ecological disaster.

**22-16** *Man's frightening power to alter his environment makes him the custodian of the future. Only through an informed and intelligent approach to the rest of the natural world can man hope to preserve the fragile ecosystem of the planet. (Grant Haist from National Audubon Society)*

## Summary

The resources of our planet are finite and its capacity for absorbing wastes is limited. Although it is a basic principle of ecology that no species may long exceed the carrying capacity of its environment, man persists in challenging the parameters of this law.

One alarming trend is the rate of growth of the human population. Presently standing at 3.8 billion people, the world population may reach 7 billion by the start of the next century. The current growth rate of two percent per year means a cut in doubling time; it now takes only 35 years for the population to double. The population increase is largely due to a drop in death rates rather than to a rise in birth rates.

The rate of population growth raises the question of whether or not the earth can provide food for so many people. Although current hunger problems are those of distribution rather than production, it is undeniable that the amount of arable land is limited.

Another important area of concern is the rate of resource depletion. So large a population entails rapid resource use. We are using fossil fuels and mineral resources that are, for all practical purposes, irreplaceable. Although recycling will provide some relief for this problem, the process requires an additional energy input.

The problem of environmental deterioration is also widespread. Soil is eroded, misused, and polluted with pesticides, herbicides, and fertilizers. Lakes and rivers are polluted and often undergo eutrophication, leading to the biological death of the aquatic ecosystem. The air is filled with a haze of chemicals and particulate matter. Even the oceans are beginning to show signs of ecological stress.

Man's ability to alter ecosystems—particularly his ability to simplify them— has weakened the homeostatic balance of many natural systems. This not only leads to the extinction of other species, it threatens the whole fabric of life on the planet. Only immediate and informed action can avert the worsening of this situation.

# Bibliography

## References

Berelson, B. 1969. "Beyond Family Planning." *Science*. 163: 533–543.

Darling, F. F. 1964. "Conservation and Ecological Theory." *Journal of Ecology*. 52(suppl.): 39–54.

Disch, R. 1970. *The Ecological Conscience*. Prentice-Hall, Englewood Cliffs, N. J.

Holcomb, R. W. 1969. "Oil in the Ecosystem." *Science*. 166: 204–206.

Langer, W. L. 1972. "Checks on Population Growth: 1750–1850." *Scientific American*. 226(2): 92–99.

Leighton, P. A. 1961. *Photochemistry of Air Pollution*. Academic Press, New York.

McElroy, W. D. 1969. "Biomedic Aspects of Population Control." *BioScience*. 19: 19–23.

Ridker, R. G. 1972. "Population and Pollution in the U.S." *Science*. 176: 1085–1090.

White, L. 1967. "The Historic Roots of Our Ecologic Crisis." *Science*. 155: 1203–1207.

Wilson, B. P., ed. 1968. *Environmental Problems: Pesticides, Thermal Pollution & Environmental Synergisms*. Lippincott, Philadelphia.

## Suggestions for further reading

Borgstrom, G. 1969. *Too Many: The Biological Limitations of Our Earth*. Macmillan, New York.
A bold discussion of the limits of food production that has provoked controversy among scientists.

Boughey, A. 1971. *Man and the Environment*. Macmillan, New York.
An amply illustrated introductory text in human ecology organized as an ecological account of the origins and emergence of urban technocracy.

Carson, R. 1962. *Silent Spring*. Fawcett, Greenwich, Conn.
An exacting scientist makes a forceful appeal against the destructive use of chemical pesticides which are disrupting the delicately balanced scale of the natural world.

De Bell, G., ed. 1970. *The Environmental Handbook*. Ballantine, New York.
  Assorted essays from a variety of contributors who examine the meaning of ecology
  and some major environmental problems. Suggestions for concrete lines of action
  and a helpful bibliography are included.

Dubos, R. 1968. *So Human an Animal*. Scribner, New York.
  Focusing on the crucial prenatal and early postnatal influences, a respected
  microbiologist considers the effects that environment and events exert on human
  development. A plea for a humanistic science of man dedicated to studying the
  effects of urbanization and technology on the body.

Whiteside, T. 1970. *Defoliation*. Ballantine, New York.
  An up-to-date account of the use of herbicides and an assessment of their effects.
  Includes supporting technical reports.

## Books of related interest

Adams, A. and N. Newhall. 1960. *This Is the American Earth*. Sierra Club,
  San Francisco.
  A stirring photographic portrait that dramatically contrasts images of wilderness
  and wanton technology in America.

Asimov, I. 1972. *The Gods Themselves*. Doubleday, Garden City, N. Y.
  Extraterritorial parties unite to oppose man's relentless material expansion and
  incursions into the natural order.

Bromfield, L. 1970. *Malabar Farm*. Ballantine, New York.
  A journalistic account of how one man and his family managed the return to a
  simple rural life by rebuilding a house, cultivating the soil and creating a working
  organic farm.

Dobzhansky, T. 1967. *The Biology of Ultimate Concern*. Meridian Books, New York.
  Consists of essays written by this noted evolutionary biologist to thoughtfully
  examine some philosophical implications of various biological and anthropological
  theories on man's place in the natural order.

Ehrlich, P. 1968. *The Population Bomb*. Ballantine, New York.
  Vivifies the dimensions of overpopulation and its effects.

Fuller, B. 1970. *Operation Manual for Spaceship Earth*. Pocket Books, New York.
  Challenging the need for specialization and traditional solution to overpopulation,
  this noted environmental designer calls for "comprehensively commanded
  automation" to solve the environmental crises.

Herbert, F. 1965. *Dune*. Ace, New York.
  Dramatically depicts how men wrest a living and evolve a social system adapted
  to an overwhelmingly hostile planet of harsh deserts.

# A taxonomic survey

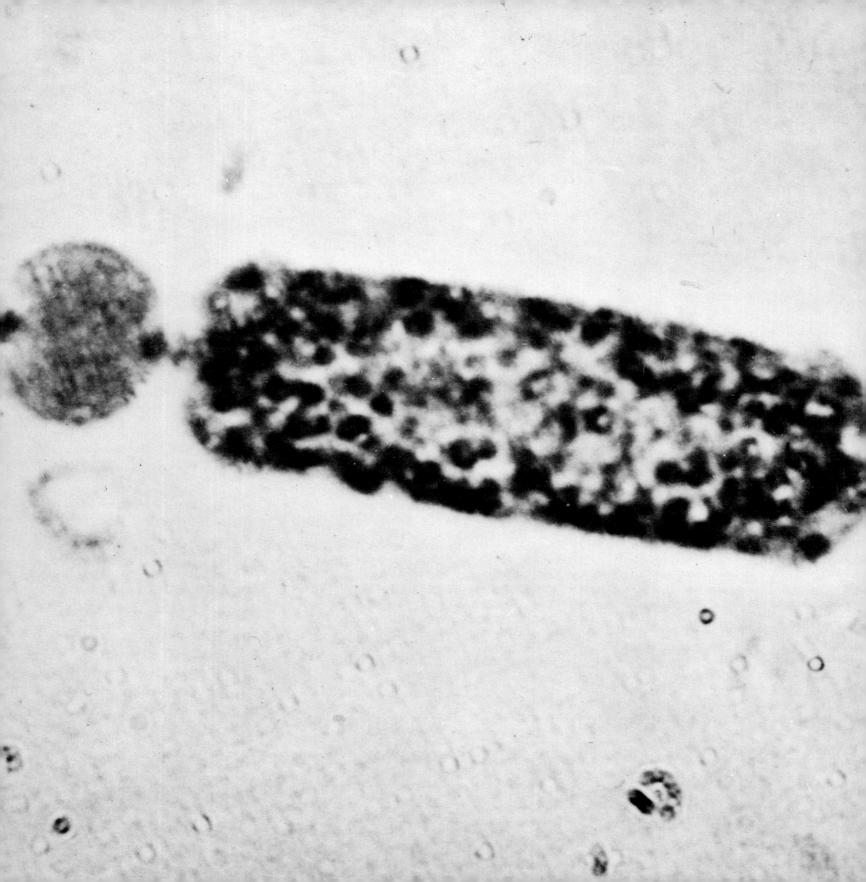

# Kingdom Monera

The Kingdom Monera embraces a staggering variety of bacteria and blue-green algae. Monerans have the most primitive level of specialization imaginable; they are surpassed in complexity by all other life forms. Members of the Monera lack the cellular features common to most other organisms: they are single or colonial procaryotic cells.

The monerans lack a nuclear membrane; the genetic material is distributed throughout a central area of the cytoplasm called the nuclear area. Although all the blue-green algae and a few of the bacteria are photosynthetic, none of them has chloroplasts; their photosynthetic material consists of protein units containing various pigments. These units are called chromatophores, and they occur at random in the cell.

Many monerans are heterotrophic, obtaining their food through absorption. The autotrophic species, such as the blue-green algae, produce a unique cyanophytan starch. The chemosynthetic species of bacteria perform oxidation-reduction reactions to obtain energy, and some of them can carry on these reactions without the presence of free $O_2$. Reproduction generally takes place by simple fission or budding, although protosexual activities, such as conjugation, transformation, and transduction, have been observed in some bacteria. Many monerans do not move at all. The ones which do utilize simple flagellar movement or employ a gliding movement.

Most monerans live as individuals, but in some species, cells clump together to form colonies. The majority of bacteria subsist on organic material and cause decay.

# Phylum Schizomycetes

This group, which includes all bacteria species, is diverse in nutrition, habitat, and ecological niche. Most reproduce by fission.

### Class Myxobacteria

These rod-shaped, thin-walled organisms are known as the slime bacteria because they secrete a slimy substance that surrounds the cells. The secreted material creates a slippery layer on solid surfaces, allowing the bacteria to creep or glide along. The same substance also permits the formation of colonial groups and fruiting bodies. Typical genera are *Myxococcus* and *Chondromyces*.

A number of environments support the versatile Myxobacteria cells, which can degrade dead bacteria, fungi, or cellulose materials. The gliding bacteria often inhabit rotting vegetation, feces, or water that provides a home for other kinds of bacteria.

### Class Rickettsiae

There has been some question as to whether the Rickettsiae can be considered organisms at all, let alone forms of bacteria. Some biologists have preferred to classify them as viruses, since they are much smaller than any other bacteria and require living hosts for survival. But their forms—spherical or rod-like—suggest a close relationship to other bacteria. This group is now thought to be intermediate between viruses and bacteria.

Rickettsiae are intracellular parasites on such arthropods as fleas, ticks, and lice. Many are pathogenic for man and other mammals, and are transmitted by bites of the arthropod vectors. One species of tick transmits the carrier of Rocky Mountain spotted fever. Another transmits epidemic typhus fever.

# Phylum Schizophyta

This phylum includes monerans that are propelled by simple flagella, and also many that are nonmotile. Bacteria are the smallest free-living organisms and survive almost everywhere. There are probably more of them than of all other life forms put together.

## Class Eubacteria

Members of this class are commonly known as "true bacteria," accounting for the three types most important to other organisms—parasitic (sometimes pathogenic) species, decay-causing types, and the nitrogen-using bacteria. The majority of eubacteria are unicellular, although some species create colonies. The cocci especially may gather in pairs (*Diplococcus*), four-celled clumps (*Tetracoccus*), chains (*Streptococcus*), clusters (*Staphylococcus*), or cubes and packets (*Sarcina*). The plane of cell division determines how a colony will take shape. Members of this class tend to be heterotrophic, and many of these are saprophytic, although a few are parasitic. Some are autotrophs, primarily chemosynthetic. The photosynthetic species have chromatophores containing various kinds of the chlorophylls. Many kinds of metabolic pathways appear in this group. Some eubacteria cannot exist in the presence of oxygen, others absolutely require it, and a few forms can carry on with or without it.

The activities of symbiotic bacteria may or may not benefit their hosts. The ones that have earned the most notoriety are the pathogenic species. *Salmonella* causes typhoid fever in people who drink contaminated water. Another genus—*Escherichia*—is used to test water for *Salmonella* contamination. Other bacteria are important in both food processing and food spoilage. Some species help to make pickles and sauerkraut through their fermenting action. *Streptococcus* sours milk, whereas its close relative *Staphylococcus* can cause food poisoning.

The nitrogen-using bacteria are especially important in ecological cycles. Classified as nitrifiers, dentrifiers, and nitrogen-fixing, they supply plants and animals with nitrogen that would otherwise be unavailable. *Nitrobacter* oxidizes nitrites to nitrates; *Azotobacter* fixes atmospheric nitrogen, forming nitrogenous compounds; *Rhizobium* can also fix free nitrogen.

## Class Actinomycota

Some species in this class are distinguished by filaments which cause them to bear a superficial resemblance to certain kinds of fungi. Their filaments tend to branch out, so they are commonly known as "branching bacteria."

Members of this class exhibit a wide range of asexual reproduction, including fragmentation, budding, and fission. Some form conidia and motile sporangiospores. They usually are nonmotile.

The branching bacteria are important to man as the source of several antibiotics, such as streptomycin, aureomycin, and actinomycin. Among the genera in this class are *Streptomyces* and *Actinomyces*.

| Class Spirochetae | Spirochetes are flexible, spiral-shaped, largely parasitic bacteria adapted for a high level of mobility. *Treponema pallidum* is the cause of syphilis in man, and the long-term deterioration of syphilis patients is due to movement of *Treponema* throughout the body. *Treponema pertenue* is the spirochete responsible for yaws. A third member of this group, *Leptospira*, is a pathogenic genus which lodges itself in healthy teeth and gums. Free-living spirochetes may inhabit stagnant water and sewage. |
|---|---|

# Phylum Cyanophyta

Cyanophytes are blue-green algae and are considered closer to the bacteria than to the other algae. Representatives of this phylum occur in a spectrum of colors that ranges through yellow, red, blue, and green. This color variation is due to the fact that each contains a different combination of pigments. Chlorophyll produces a green hue; phycocyanin causes a blue tint; phycoerythrin often results in red, and carotene produces a yellow color.

Blue-green algae have a distinct cell wall of cellulose; in addition, some may also have chitin in the cell wall. Inside, the central area is colorless and the outer region contains the pigments, although the two are not sharply separated. Many blue-green algae form colonies. The difference in colony type provides a basis for distinguishing them from one another. Often they are held together in groups by secretion of a slimy envelope around each cluster, but some blue-green algae form more tenacious colonies than others. *Gleocapsa*, as one example, retains its envelope intact even after dividing, so colonial arrangements are not disturbed.

Cyanophytes are chiefly residents of aquatic environments, and populate most bodies of water, damp soil, and marshlands. (Waterbed owners have often reported algal growth when they neglected to add chlorine to the water.) Climate is seldom a problem for blue-green algae, since they seem to flourish in any extreme. Cyanophytes are not particularly motile. There are about 1500 species of blue-green algae, but as far as is presently known, none of these is capable of sexual reproduction.

# Kingdom Protista

Until recently, members of this kingdom were the subject of some controversy among scientists. Since many of these organisms have cellulose cell walls and carry on photosynthesis, most biologists included them in the Plant Kingdom. But there remained the problem of certain animal-like features, such as motility. Finally it was decided that because of their unique features, the protista should be placed in a kingdom of their own.

The protists are generally unicellular or colonial-unicellular organisms, although the category includes some multinucleate forms. They are eucaryotic cells, with nuclear membranes, mitochondria, and often plastids. Nutritional methods are various, including photosynthesis, absorption, phagocytosis, and sometimes combinations of these. Their reproductive cycles are diverse, and sexual reproduction is universal. Some are entirely nonmotile, but many others use flagella, cilia, or amoeboid movement to achieve motility. Many of the flagellated forms and all ciliated forms possess a cellular covering called a pellicle. When the pellicle is thin, the protist can change its body shape; if the pellicle is thick, its shape is fairly rigid.

# Phylum Euglenophyta

This phylum represents a small group of aquatic organisms inhabiting marine environments and fresh water. They may have a cell wall and one or two hairlike flagella. They are especially numerous in waters with a high nitrogen content, such as stagnant ponds, polluted streams, and sometimes wet barnyards. Some are able to change shape; others have a rigid pellicle that keeps the shape constant. An important plantlike feature of euglenoids is their ability to synthesize their own food. Not all species are photosynthetic, and all are incapable of manufacturing the vitamin $B_{12}$ that they require for growth. Some photosynthetic species are known also to ingest solid food when it is available. A starchlike substance, paramylon, is stored as a food reserve. Reproduction is usually by binary fission, although a few species are known to reproduce sexually. *Euglena* is a popular subject for experimental purposes since it exhibits both plant and animal characteristics, grows rapidly, and is simple to cultivate. Locomotion is by means of one whiplike flagellum. It also has a structure called an eyespot—a primitive photosensitive organelle that directs its movements toward light. *Euglena* is photosynthetic, but it can also form cells which lack chlorophyll and thus cannot photosynthesize. This can happen both under natural conditions and in laboratory cultures grown in total darkness.

*Euglena* has a colorless relative, *Astasia,* which is so close in form that the two are called a "species pair." Aside from similarity of appearance, *Astasia* has an elastic pellicle, flagellum, and proplastids rather than chloroplasts. *Phacus* is a third member of the euglenoid group that shares many of the same features, but it is unable to change its shape because it is enclosed by a rigid pellicle. *Colacium* is the only euglenoid known to form colonies.

# Phylum Pyrrophyta

This phylum includes some 1900 species of generally marine organisms known as dinoflagellates and cryptomonads. Among the dinoflagellates, most are acellular, free-swimming flagellates, but certain species are colonial. Some pyrrophytes have wall coverings called periplasts similar to the euglenoid pellicle, while others have the more common cellulose wall. Those that have only periplasts, such as *Gymnodinium,* are said to be "naked" or unarmored. The rest are armored forms, such as *Glenodinium,* which has rather flimsy sculptured plates, or *Gonyaulax,* which carries around a heavily armored plate with spiny projections, large horns, and papillae. The pyrrophytes move by means of flagella; most of them are biflagellate species with two dissimilar flagella occurring side by side. The most interesting locomotion occurs in the armored forms,

692

which have their flagella projecting through large pores in the cellular plates. One is wrapped around the cell, completely encircling it, while the other fits in a longitudinal groove called the sulcus. The one that encircles the cell rotates to propel the organism forward, while the other acts as a kind of rudder.

The cryptomonads are photosynthetic, usually having brown or yellow chromoplasts. These free-living forms are found in both fresh-water and marine habitats, many of them symbiotic with radiolarians and corals. *Cryptomonas* is the most common genus, with two dissimilar flagella arranged as in the dinoflagellates, a flattened shape, and a thick wall containing cellulose. It reproduces by longitudinal cell division.

## Phylum Chrysophyta

This phylum is divided into three classes—the yellow-green algae, golden-brown algae, and a third group of algal forms known as diatoms.

### Class Xanthophycae

This class consists of about 400 species of yellow-green marine and fresh-water algae. The yellow-green algae have cell walls of cellulose and sometimes silicon. Inside the cell are a number of green or yellow chloroplasts which give the organisms their distinctive coloration. All have central nuclei, and some have central vacuoles. Although most of them are photosynthetic autotrophs, there are a few heterotrophs characterized by the lack of any pigmentation at all. And some of the more pigmented species have been observed taking in food particles, like their heterotrophic relatives. Most xanthophyceae are unicellular, such as the common *Tribonema*, but a few, like *Botrydium*, are multicellular. These algae reproduce asexually by fission in most cases, but some (*Botrydiopsis*, for example) carry on sexual reproduction. They can be found in the scum floating on the surface of stagnant ponds, in spring water, or even on trees.

### Class Chrysophyceae

Members of this class are the golden-brown algae, including both fresh-water and marine species. Acellular or multicellular, with cell walls made of cellulose, their golden hue comes from the presence of carotene pigments in addition to chlorophyll in the chloroplasts. These organisms lack central vacuoles, but they do store food in the form of globules of fat and leucosin. Except for a very few colorless species, all of them are photosynthetic. This numerous and diverse class includes unicellular amoebic forms like

*Chrysamoeba,* unicellular flagellates such as *Mallomonus,* and colonial species such as *Dinobryon.*

| | |
|---|---|
| **Class Bacillariophyceae** | This class includes the diatoms. These are acellular species with elaborately shaped cell walls made of silicon, which may exhibit pores or ridges. Most are autotrophic, but some heterotrophs among them store food in fatty deposits within the cell. Diatoms are generally sedentary, since only certain species (called the pennate diatoms) are capable of moving at all. Most diatoms are free-living organisms; common genera are *Pinnularia* and *Navicula.* Others form loose aggregations joined together by the sheath material secreted from their pores; *Asterionella* forms a colony in the shape of a star. Fission is the usual mode of reproduction, and this sometimes occurs at a fantastic pace, resulting in a full-fledged bloom of new diatoms. They can be found in stagnant ponds, flowing streams, and marine waters. They are among the most numerous of marine algae. |

# Phylum Fungi

This phylum consists of organisms that lack chlorophyll and store their food in some form other than starch. They are commonly found in soil and decaying organic matter. Of the two subphyla, Myxomycophyta includes the slime molds and Eumycophyta, the true fungi. A characteristic of the true fungi is a mycelium, a mass of branched filaments. In some species the branches, or hyphae, grow as long as 50 feet. The mycelium itself cannot move, but the cytoplasm inside can be seen to undergo streaming.

# Subphylum Myxomycophyta

These are the slime molds. All are characterized by heterotrophic nutrition and unusual life cycles.

| | |
|---|---|
| **Class Myxomycetes** | This class includes the slime molds found in moist wood, soil, feces, and decaying vegetation. Their life cycles are fairly complex: an amoebalike stage, the plasmodium, is most conspicuous. The plasmodium is multinucleate, lacks a rigid cell wall and feeds by phagocytosis. Under certain conditions, the plasmodium produces fruiting bodies. Sporangia arising from |

the fruiting body produce spores, which upon germination give rise to flagellated gametes. The resulting zygote develops into a new plasmodium. *Echinosteliales* is a genus whose tiny plasmodium produces only one sporangium. *Trichiales* is a colorful slime mold whose spores are yellow or red.

| | |
|---|---|
| **Class Acrasiae** | Members of this class are known as the cellular slime molds, which pass through a lengthy amoeboid assimilative stage. In this state, they cannot be distinguished from a true amoeba. The organism settles on a substrate and begins reproducing by fission. It may remain in this stage for its entire lifetime if conditions are suitable. If not, it assumes a second stage, that of aggregation. Here several amoebae converge to form a "slug" which migrates for a period whose duration is determined by heat and light conditions. Finally the "slug" transforms into a stalk with a knob at the tip. Spores formed in the knob are released and give rise to new amoeboid individuals. *Dictyostelium* is a well-known representative of this group, undergoing all the changes discussed. |
| **Class Labyrinthuleae** | This group contains fresh-water and marine slime molds with an assimilative stage. In *Labyrinthula* (the most common member), this stage involves a naked cell with only one nucleus. |

# Subphylum Eumycophyta

These are the true fungi, characterized by mycelial growth and heterotrophic nutrition. Most are saprophytic but a few are parasitic. The method of spore formation distinguishes one class from another.

| | |
|---|---|
| **Class Phycomycetes** | The algal fungi are considered to be primitive forms. Members of this group are chiefly aquatic. They reproduce asexually by spores, or sexually by fusion of motile flagellated gametes. In most forms, the life cycle involves a long haploid stage. *Saprolegnia* is a fresh-water and soil-dwelling genus that is generally saprophytic, but some species parasitize plants or marine animals. *Rhizopus*, the black bread mold, differs from aquatic phycomycetes in that it does not produce flagellated gametes. Its reproductive spores are dispersed in the air rather than through water. *Philobus* is another common genus found exclusively on the feces of herbivores. |

## Class Ascomycetes

The sac fungi represent a diverse group of terrestrial and marine forms. They go through reproductive stages that include formation of a spore sac, the ascus, typically containing eight spores. Several members of this class are important to us for economic and medicinal reasons. The most primitive Ascomycetes are yeasts, like *Saccharomyces cerevisiae*, familiar to us as baker's or brewer's yeast. The *Penicillium* fungus is the source of penicillin. *Neurospora* has been especially useful to experimenters in genetic research. The common *Ergot* is a different matter entirely, and it is usually viewed with ambivalence. On one hand, it is used to make drugs for the treatment of hemorrhaging. But on the other, it was once responsible for a fatal disease known as St. Anthony's Dance, and has been used in recent years to make a hallucinogenic drug.

## Class Basidiomycetes

The club fungi include organisms that produce basidia, club-shaped reproductive structures. Mushrooms and toadstools are well-known members of this class. They can reproduce asexually by means of conidia, but this is less common than sexual reproduction involving plasmogamy and karyogamy. Some Basidiomycetes are called rusts—such as *Cronartium ribicola*, or white pine blister rust. *Puccinia* is also a rust that grows on gooseberries and currants. Then there are the smuts, such as *Ustilago*, which parasitize vascular plants. Other fungi in this group are wood-decaying forms, such as *Poria* and *Fomes*. *Lycoperdon* is a saprophytic puffball that can usually be found in moist, shady woods.

   The vegetative mycelia of the Basidiomycetes contain haploid nuclei and grow just beneath the surface of the soil. When conditions are right—just after a rain, for instance—some rapidly form the fruiting bodies known to us as mushrooms or toadstools. These fruiting bodies bear the basidia, inside of which two haploid nuclei fuse. This is followed by meiotic divisions which give rise once again to haploid nuclei.

# Phylum Protozoa

The Protozoa comprise a large group of organisms, including 30,000 different species and an incalculable number of individuals. Their dispersal is enhanced by their ability to encyst; individuals roll themselves into balls and secrete shell-like cysts when climatic conditions become oppressive.

The phylum is organized into two subphyla. The Plasmodroma have only one type of cell nucleus. They reproduce sexually by forming gametes. Locomotion is by pseudopodia or flagella. The other subphylum is the Ciliophora, which employ cilia for locomotion. Ciliophora have two types of nuclei and a fairly complex internal structure. All fresh-water protozoans have contractile vacuoles.

## Class Flagellata (Mastigophora)

Members of this class may or may not have pseudopodia, but all are equipped with one or more flagella for locomotion. Their flagella may also function as food-getting and sensory structures. Some flagellates are colonial and others free-living, inhabiting both water and soil. A few are parasites. They live in several species of animals, and can cause serious diseases in man. Flagellates reproduce by binary fission, feed on small organisms which they digest in food vacuoles, and move by beating their flagella in a rapid whip-like motion.

Some important flagellates are the various members of the genus *Trypanosoma*—slender parasites that enter vertebrate hosts through bloodsucking or biting. Among other diseases, species of *Trypanosoma* cause sleeping sickness, a condition common in Africa. *Chilomonas* is a saprozoic flagellate inhabiting foul waters.

## Class Sarcodina (Rhizopoda)

These protozoa are ordinarily free-living, with pseudopodia for locomotion and food-getting. Perhaps the most famous member of this class is the amoeba. The amoeba surrounds food with its pseudopodia and digests it in a food vacuole. Reproduction is by binary fission. Another common member of the Sarcodina is *Globigerina,* which secretes an exquisite calcareous shell to house the living occupant. The shell falls to the bottom of the ocean when the animal dies; collectively these shells form a sediment that covers about 35 percent of the ocean floor. Another member of this class is *Entamoeba histolytica,* which causea amoebic dysentery in humans.

## Class Sporozoa

All representatives of this class are internal parasites. Adults have no external organelles of locomotion. They reproduce in great quantity through fission, and they have fairly complex life cycles. Rapid reproduction accounts for the impressive number of sporozoa; they represent the largest group of animal parasites. They are responsible for various diseases of vertebrates, including malaria in man. The organism producing malaria is known as *Plasmodium,* which attacks red blood cells.

A subgroup of the Sporozoa is the Cnidosporidia. Most parasites in this group live in fish or arthropods, but a few inhabit the muscular tissue of birds and mammals. They are all characterized by possession of a complex spore, and are equipped with one or more filaments which enable them to cling tenaciously to their hosts.

# Subphylum Cilophora

This group includes all the ciliates and suctorians, which are grouped into the single class Ciliata. These organisms are relatively specialized, with cilia for movement and a system of organelles performing various individual functions. Their locomotion often involves a singular spiraling motion created by fast-beating cilia. They typically reproduce by binary fission, but most ciliates also engage in conjugation. *Paramecium* engages in conjugation and reproduces by transverse fission. It ingests bacteria and smaller protozoa. *Stentor* is an intriguing ciliate with a trumpet shape, and may actually be a host itself to a suctorian called *Sphaerophrya*. *Didinium* is a barrel-shaped ciliate that voraciously feeds on paramecia. *Vorticella* is a sessile ciliate; its stalk can contract into a tight spiral.

Suctorians have a varied life cycle. In their early stages, they are ciliated. New individuals are formed by budding and simply swim away from the parent. At maturity, the suctorian attaches itself to the bottom and its cilia disappear. The body of the adult has numerous cytoplasmic projections which act like sucking tentacles to catch and feed on floating prey. A typical suctorian is *Acineta tuberosa*.

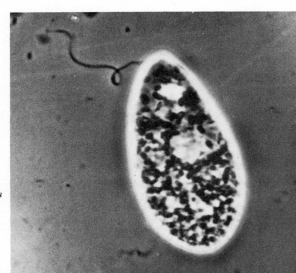

Flagellated *Euglena gracilis*

myxomycetes *Cribraria*, a slime mold

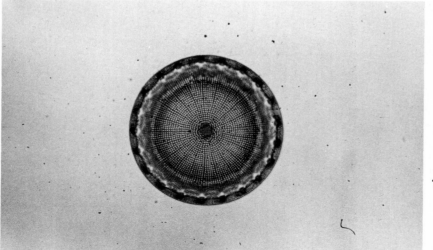

The diatom *Arachnoidiscus orientalis*

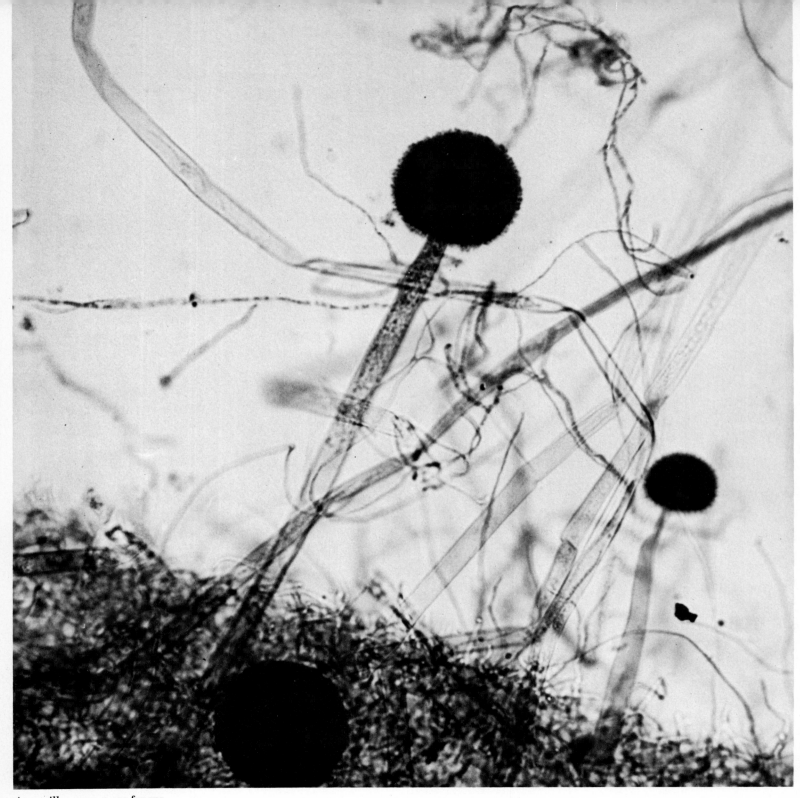

*Aspergillus zoox,* a sac fungus

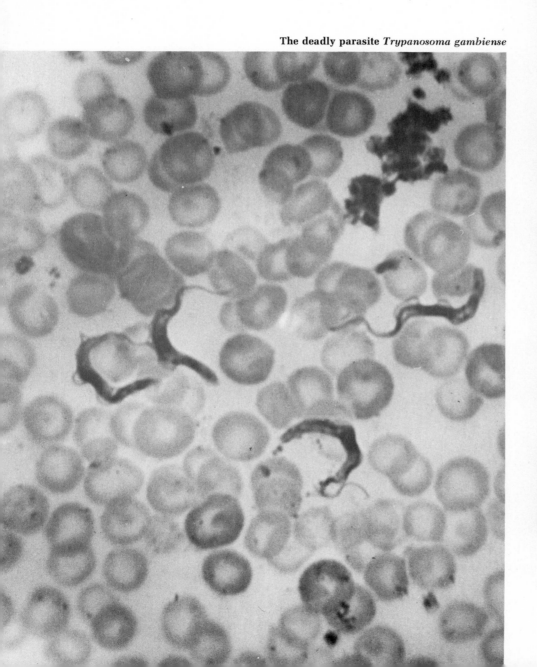

The deadly parasite *Trypanosoma gambiense*

Shelf fungi on a tree

701

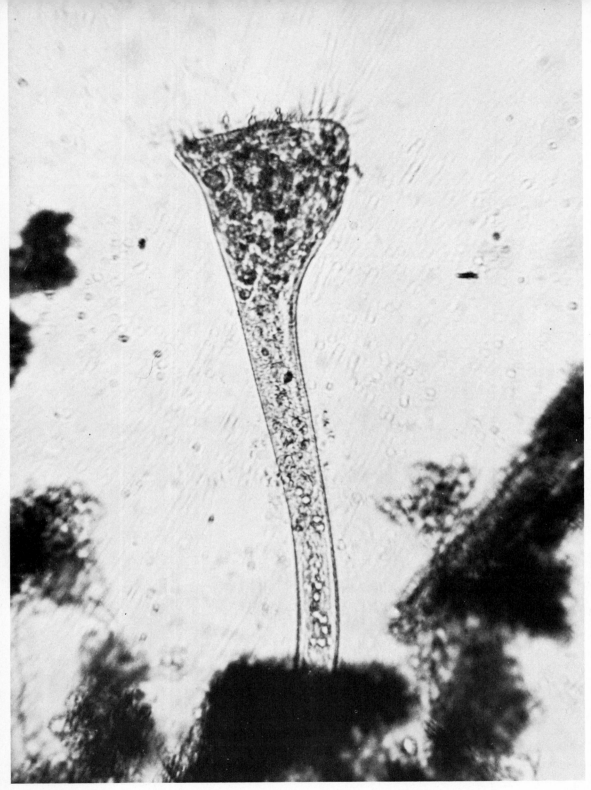

*Stentor*, a ciliate

# Kingdom Plantae

All animals on this planet, including man, rely on plants for their survival, for photosynthesis is the basis of all food energy. The Plant Kingdom contains over 250,000 different species alive today, and countless others known as fossils or still undiscovered. Some existing plants are acellular or syncytial, but most are multicellular organisms made up of eucaryotic cells. Generally plants are holophytic, containing photosynthetic pigments in plastids. But a few forms have turned to heterotrophic nutrition rather than producing their own food. Most members of this kingdom are sedentary or nonmotile types, anchored to a substrate. They display a wide range of structural differentiation from simple organs of photosynthesis, anchorage, storage, and support in lower plants to increasingly specialized photosynthetic, vascular, and covering tissues. Their reproductive methods include both sexual and asexual processes with cycles of alternating haploid and diploid generations. The haploid stage tends to be progressively reduced as the plant's organization becomes more complex.

# Phylum Rhodophyta

This category covers the red algae—distinguished from all other forms of algae by their characteristic color. Most of them are marine, usually found along rocky seacoasts, where they typically abound either in deep water or near the low tide mark. Other marine forms inhabit coral reefs, and calcification of the cell wall of certain of these forms contributes to the formation of the reef. Still other red algae are fresh-water species living in cool, swift-flowing streams. The rhodophytes are fairly small plants, rarely exceeding three feet. The cell wall which surrounds the cell membrane is differentiated into an inner cellulose portion and an outer pectin layer. The wall may be calcified, in some cases containing substances such as agar. Inside the cell, organization is relatively simple in most forms, but in more advanced groups, cells may have more than one nucleus. Their distinctive red hue comes from a mixture of pigments in their plastids—phycobilin pigments, a number of carotene and accessory pigments, and chlorophyll.

The red algae store their manufactured food in a polysaccharide form called Floridian starch; it is practically identical in some species to the starch found in blue-green algae, and in other species very much like potato starch.

Aside from a lack of flagella, an important feature of red algae is their method of reproduction. Fertilization occurs when a nonmotile female gamete fuses with a nonflagellated male gamete that has drifted with the current. The product of the fusion, which takes place inside the female plant, is a diploid zygote which then undergoes mitosis, giving rise to several carpospores. When these are released, they grow into diploid plants, when then produce haploid spores by meiosis. The spores in turn develop into the male and female gametophytes. *Polysiphonia* is a representative member commonly found on the Atlantic coast; it is branched and filamentous.

# Phylum Phaeophyta

The brown algae are a diverse but well-defined group. They are almost all marine forms, growing best along the shorelines of cold seas; most attach themselves to the rocks there. They owe their brown coloration to an accessory pigment called fucoxanthin, which totally masks their green pigments in most species. Hues run from golden-brown to almost black. Most of the brown algae cells have only one nucleus; nucleoli may be present. They store the polysaccharide laminarin.

In appearance, these algae vary from simple branched filaments to massive seaweeds. Many resemble vascular plants because they seem to have a root, stem, and leaf. However, that is not what they are; these parts are known to botanists as holdfast, stipe, and blade. *Sargassum* has miniature air cushions that keep it floating near the surface. It is found in huge quantity in warmer parts of the Atlantic.

There are essentially three types of brown algae, and each can be exemplified by one specific member.

*Ectocarpus* is an example of the simplest type—a filamentous plant that is branched and attaches to other larger brown algae, resembling tiny hairs. This variety is isogamous; the male and female gametes are identical in appearance. The zygotes produced by fusion of gametes grow into diploid plants, which in turn give rise to biflagellate haploid zoospores (through meiosis) that grow into haploid gametophytes.

*Laminaria* is an organism representative of the kelp group of brown algae. It also undergoes alternation of generations, but its generations are dissimilar or heteromorphic. The blade of the diploid plant forms meiosporangia. Once meiospores are released, they germinate and produce very tiny male and female haploid plants. One cell of the female plant develops into an oögonium which forms eggs. Biflagellate sperm are produced in the male plants and make their way to the female gametophyte. Fusion takes place, and the zygote develops into a diploid sporophyte. *Laminaria* is cultivated extensively in the Orient, where it has been used for some time as a food plant.

*Fucus*, is a third type of brown algae, popularly known as rockweed. It is commonly found along rocky coasts where it inhabits intertidal areas. Unlike the other brown algae, *Fucus* does not undergo an alteration of multicellular generations. The haploid phase is restricted to the gametes and does not give rise to separate plants.

# Phylum Chlorophyta

This group consists of the green algae. Some are motile, with two or four flagella; others are not. Starch is the common food reserve. The green algae are one of the largest groups of algae; they are subdivided into two classes.

## Class Chlorophyceae

This class contains the organisms referred to as true green algae. It is by far the larger class, and so diverse in structure and function that taxonomists further divide it into smaller groups that are separated from one another on the basis of evolutionary trends, with several distinct lines in evidence. One of the most primitive is the volvocine line, which includes the simplest forms of green algae. This branch reaches its highest development in *Volvox*.

*Chlamydomonas* is the simplest of the volvocine plants; it is a unicell with two flagella. It sometimes forms nonflagellated spores that, when environmental conditions become harsh, mire themselves in a gelatinous matrix and divide repeatedly to produce a mass of new cells. When conditions become favorable again, the cells grow new flagella and disband, leaving the common matrix. Colonial forms probably evolved from individuals such as *Chlamydomonas*. The simplest type of colony can be seen in *Gonium*, which consists of four, eight, or sixteen cells. Each cell has two flagella, an eyespot, chloroplast, contractile vacuole, and cell wall. There is an evolutionary trend for colonies to arrange themselves in a spherical or ellipsoid group; there is also a trend toward increasing the number of individuals in the colony with each new generation. The most complex form in this particular line is *Volvox*, which forms a hollow spherical colony that increases its numbers rapidly. Certain cells within a *Volvox* colony undergo mitotic divisions to form daughter colonies. The colony may also reproduce sexually, with certain cells specializing to form eggs and sperm.

The tetrasporine line of development probably began as a variation of *Chlamydomonas*, with a trend toward loss of motility. Its members are distinguished by possessing only one nucleus per cell. A typical representative is *Ulothrix*, a fresh-water filamentous form that produces vegetative cells, spores, and gametes. Gametes are produced by mitotic division. They fuse to form a diploid zygote, which rests until conditions are ripe for meiosis to occur. When it divides, four haploid meiospores result, each of which then divides by mitosis to produce a new filament.

A third major line is the chlorococcine line. Individual cells are often multinucleate. A primitive genus in this line of development is *Chlorella*, a unicellular form. *Codium*, which is common in intertidal regions on the west coast, attaches itself by a holdfast and forms velvety ropelike branches. It only reproduces sexually.

*Spirogyra* is a member of a group called the conjugales, which is distinct from all the other green algae. They probably branched off very early in algal evolution because they have little in common with other green algae. None of this group produces flagellated cells at any time, but many feature amoeboid movement. *Spirogyra* is particularly unusual because each individual cell in its unbranched filament contains a helical chloroplast. It reproduces asexually by fragmentation. Sexual reproduction occurs by conjugation.

**Class Charophyceae**

Included in this class is a small group of fresh-water, filamentous, branched plants. They are commonly known as stoneworts, usually found attached by their rhizoids to rocks on the bottoms of ponds and streams. Oögamous sexual reproduction is typical. In *Chara*, the oögonium contains a single egg covered by special twisted cells. These coronal cells serve to protect the egg.

# Phylum Bryophyta

Members of this phylum bear little resemblance to other plant forms. The "parasitic" sporophyte generation is dependent on the gametophyte. Since water is required for sexual reproduction, many bryophytes live in wet or moist situations. About 9000 different species of these primarily terrestrial plants are divided into three classes.

## Class Hepaticeae

Liverworts are small organisms with thin flat thalli, notched at the tip and lobed like the liver of animals. The liverwort sporophyte is differentiated into three segments—the capsule, seta, and foot. There is no localized meristem. As the sporophyte develops, mitotic divisions occur in all parts of the plant; the sporophyte matures when the divisions cease. The gametophyte clings to its substrate by rhizoids, always growing horizontally (flat against the ground). The thalloid variety, for example *Conocephalum*, is usually found in the soil, branching off into a dichotomous pattern. *Lophocolea* is a "leafy" liverwort which looks like a moss, except for its horizontal pattern of growth. *Marchantia* is a common liverwort that has achieved a fairly high level of organization. It has scales, rhizoids, and air chambers. It has developed a special asexual reproductive modification called gemmae cups.

## Class Anthocerotae

The hornworts are a widespread class that resemble thalloid liverworts in appearance. They are found in damp earth virtually everywhere. The sporophyte has a distinct capsule and basal foot separated by a meristem. The meristem is capable of rapid and constant growth activities, but is usually inhibited by environmental conditions. The hornwort gametophyte is a flattened plant whose thallus appears to have fluted borders. It has several pores leading into hollow, mucus-filled chambers, and often supports large groups of blue-green algae inside. *Anthoceros* is perhaps the most common hornwort.

## Class Musci

Mosses are the bryophytes commonly found on rocks splashed with water, moist fields, and woods—even in our lawns and gardens. Mosses also have an uncanny ability to stay dormant but viable for long periods of time. The length of the dormant period depends heavily on factors such as light, soil, and temperature. The gametophytes grow from spores in two specific stages. The first stage is the protonema. Then buds form on the protonema, each giving rise to a new leafy gametophyte. The zygote in the archegonium grows into a sporophyte plant. The sporophyte plant remains attached to the gametophyte by its foot and sends up a long slender stalk (seta) bearing a spore-producing capsule that is often covered by the remnant of the archegonial wall. This cover drops off, exposing the operculum which covers the peristome, an opening ringed by teeth. Movements of these teeth are brought about by changes in humidity and result in dispersal of spores within the capsule. *Sphagnum*, or peat moss, has long been ecologically useful in loosening clay soil to promote growth of other plants. *Polytrichum* is a moss whose sporophyte capsules have a hairy covering.

# Phylum Tracheophyta

This group of plants is distinguished from all others by the presence of xylem and phloem tissues. All tracheophytes probably arose from a common ancestor, although the seedless and seeded forms might represent two different evolutionary branches. They have made a remarkably successful transition to land, and are found in virtually all terrestrial habitats. In all tracheophytes, the sporophyte is the conspicuous long-lived generation. The tracheophytes are divided into five subphyla.

# Subphylum Psilopsida

This group, known from fossil evidence to have arisen long ago, consists today of only two genera — *Psilotum* and *Tmesipteris*. *Psilotum* is found in both hemispheres, in both tropical and subtropical regions. *Tmesipteris* is limited to the damp, shaded island habitats of the Eastern Pacific and Australia. Although the two differ greatly, both are characterized by rhizoids on their underground stems, which grow erect branches that become aerial stems. True roots and leaves are absent. *Psilotum*, known as the whisk fern, has xylem in the form of separate bundles around a pith, and *Tmesipteris* has the same arrangement with the addition of a layer of metaxylem encircling the bundles. The stem has stomata on its surface, for gas exchange, and contains chlorophyll. Both psilopsids have essentially similar reproductive methods. Gametophytes grow from spores formed on the large sporophyte plants. Each gametophyte produces both eggs and flagellated sperm. Once zygotes have formed, they germinate while still inside the archegonium, send up aerial shoots, and begin to grow into new sporophyte plants.

## Subphylum Lycopsida

These are the club mosses, which are often used as decorations at Christmas. They look like mosses but have vascularized organs. They are characterized by true roots, stems, and leaves. Their leaves are usually tiny, and are arranged spirally. Each leaf has a central vein. Sporangia are produced on certain leaves (which look no different from the other leaves) called sporophylls. The spore-bearing structure resembles a tiny club, the reason for the common name given this group. Some plants are homosporous; other club mosses are heterosporous. Two representative members of this subphylum are *Lycopodium* and *Selaginella*. *Lycopodium* is commonly known as ground pine, since it grows on the floor of coniferous forests and somewhat resembles pine branches. *Selaginella* is heterosporous, and its sporophylls are clustered together in aggregations called strobili.

## Subphylum Sphenopsida

These are the horsetails, which bear the name *Equisetum*. They are the last remaining genus of their group, although two fossil genera were extremely abundant in the Devonian period. They are characterized by slender stem with distinct nodes, giving them a jointed appearance, and leaves that are arranged in whorls on each node. The epidermal cells have silicified walls, and stomata occur in the epidermis. Roots are fibrous and arise from subterranean rhizomes. In some horsetails, spores are formed in cones at the end of specialized shoots called sporangiophores. In other sphenopsids, vegetative shoots give rise to strobili at their tips. These release spores that then grow into tiny green gametophytes with rhizoids that penetrate the moist substrate.

## Subphylum Pteropsida

This category includes the ferns—a highly successful group of vascular plants. Ferns, like the higher seed plants, have leaves with branching veins.

## Class Filicinae

Ferns have undergone considerable development of their leaves, the most conspicuous part of the plant. All have roots, stems, and leaves, and most members have rhizomes. The leaves of ferns are often large and flattened with a number of lobes. Subdivisions within the whole leaf (frond) are pinnae, which may be further divided into pinnules. Not all fern leaves are compound; some ferns, such as *Camptosorus* the walking fern, have simple non-divided leaves.

Ferns reproduce asexually by sporangia that are either independent of one another or fused (in which case they are called synangia), and they may be grouped into clusters known as sori in some plants. All living land-dwelling ferns are homosporous; only the water ferns such as the Marsileales follow a heterosporous pattern. The sporogenous tissue divides meiotically and produces a fairly large number of spores. Each spore germinates into a prothallus, the gametophyte plant. It can sometimes be seen on the ground beneath the fern, looking like a heart-shaped plant. Most fern gametophytes are homothallic, producing both types of gametes but not at the same time. Water is required for fertilization, since the sperm must swim to the egg. The zygote formed develops into an embryo, consisting of a foot, primary root, leaf, and stem, while still in the prothallus. As it develops into a young sporophyte plant, the gametophyte withers and dies.

# Subphylum Spermopsida

These plants are characterized by the seeds they bear, and it is this feature that has made them so successful in terrestrial environments. The seed-bearing habit is believed to have evolved from heterosporous ancestors. The spermopsids are divided into two major classes, the Gymnospermae and the Angiospermae.

## Class Gymnospermae

This class was once considered a fairly homogeneous group. But recent evidence suggests that the grouping may have been too hasty, and now they are differentiated from one another as four different subclasses.

## Subclass Ginkgoae

has only one surviving species, *Ginkgo biloba*, popularly known as the maidenhair tree; at least 15 genera flourished during the Mesozoic period. This unusual tree has been in demand with gardeners for some time. In fact, it was introduced to the Western world as a curious inhabitant of Chinese temple gardens. It is extremely rare in the wild, and there was a flurry of excitement when lonely trees were suddenly discovered growing wild in the hinterlands of China. There is some controversy as to whether this was a genuine wild species or simply the product of escaped seeds from someone's garden.

The *Ginkgo* retains the primitive features of motile sperm and terminally borne seeds. It produces male and female gametes on separate trees. Young trees grow in a pyramidal shape until the side branches begin to dominate over the apical growth, producing a rounded profile.

While the *Ginkgo* tree bears certain similarities to cycads (particularly in gametophyte development), biologists have been unable to trace the evolutionary line that gave rise to the *Ginkgo*.

---

## Subclass Cycadae

is another group that has declined markedly, leaving few survivors. Giant cycads formed great forests during the Carboniferous period, but today there are only nine genera left, all found in tropical and semitropical climates. The cycads that are still living bear a resemblance to palm trees and tree ferns. The fronds are pinnate forms with leaflets arranged in two lateral rows. The trunk is covered by leaf scars which can be used to determine the age of the tree. The age of some individual members of the genus *Macrozamia*, an Australian genus, is estimated to be over 1000 years. The leaves of the cycad are very large, often over three feet long. All the cycads are dioecious, with seeds developing in strobili (in *Macrozamia*, the cones may weigh as much as 70 pounds). Growth of new plants may be very slow; the common North American cycad, *Zamia*, which is found in Florida, may take several years to develop a characteristic cycad appearance.

---

## Subclass Gneteae

is a rather strange collection of gymnosperms that seem to have little in common with other gymnosperms. Scientists have suggested that they represent a transitional phase toward the angiosperm group, although there is no evidence confirming this hypothesis. They make up three diverse genera: *Ephedra*, *Welwitschia*, and *Gnetum*.

*Ephedra* is found in warm climates, flourishing on sandy deserts or rocky mountains. It is a profusely branched shrub, with leaves greatly reduced in size. They are so short that most photosynthesis occurs in the branches or stem.

*Welwitschia* was named after the botanist who discovered and collected it. It is found only in one small desert region near the southwestern coast of Africa in a habitat no larger than 100 miles square. It needs little water, has two strap-shaped leaves that may grow to a length of 10 feet, and a giant taproot. The leaves live as long as the plant, an average life span of 100 years. Its evolutionary background is a matter of wild speculation, and there is a law against its collection by nonscientific aficionados.

*Gnetum* includes about 30 species growing in the tropics. Most are vines but a few are shrubs. Most species occur in Asia and Indochina, but some are found in Central and South America. Like the *Ginkgo*, it has both long and short shoots, with leaves occurring only on the short branches.

## Subclass Coniferae

refers to the conventional group of gymnosperms that we are most familiar with, such as the pine, sequoia, cypress, and redwood. They are chiefly trees, (a few are shrubs) and are the dominant plants in most forests. Largest is the *Sequoia*, which can reach a height of 300 feet. Most exhibit spiral or whorled branching. Their leaf patterns are quite diverse. Some, such as the pines, lose their leaves as the branches age, but this is not the case in most. The epidermis of conifer leaves is heavy cutinized. *Picea, Cedrus,* and *Podocarpus* all have the needle type of leaf associated with Christmas trees. Linear leaves are the rule with *Taxodium* and *Abies,* and these are generally flat. *Podocarpus*' leaves are reduced to the point that they seem reluctant to leave the stem. The least common leaves are found in *Araucaria*—broad, flat structures with a network of veins.

Although most conifers are evergreen, retaining their leaves at least two years, many, such as *Larix,* are deciduous. In the central area of the leaves are one or two vascular bundles. There is usually a thick epidermis with sunken stomata. In most conifers, the reproductive mode involves pollen grains and ovules developed in strobili or cones. *Pinus* and *Tsuga* produce winged pollen granules, but *Sequoia*'s pollen is wingless. In the cypresses, the strobilus is a fleshy structure resembling a berry.

## Class Angiospermae

These, of course, are the flowering plants, which are divided into the Subclasses Monocotyledoneae and Dicotyledoneae. The division is based on number of cotyledons present in the embryo, presence or absence of cambium, arrangement of vascular bundles, venation of leaves, and number of flower parts.

## Subclass Dicotyledoneae

is a diverse group consisting of many orders. In all dicots, the embryo has two seed leaves. Floral parts come in 4s or 5s. Leaves have netlike venation, and stems have a prominent cambium.

714

## Order
## Magnoliales

is the most primitive group, composed of woody members. The magnolias are usually found in north temperate regions. The tulip tree and sweet bay tree are also classed in this group.

## Order
## Rosales

members are shrubs, herbs, and trees; it is the second largest order of flowering plants. The apple (*Malus*), strawberry (*Fragaria*), and rose (*Rosa*) are among its representatives. In addition to other fruits (blackberries and raspberries), cockspur, and hawthorn are also placed in this group.

## Order
## Leguminales

includes trees, herbs, and shrubs. Herbs of this order are of great ecological importance, because their roots may contain nodules in which nitrogen-fixing bacteria live. This order contains many of our common vegetables such as peas (*Pisum*), beans (*Phaseolus*), and even peanuts (*Arachis*). The honeylocust, red bud, and sensitive plants are other group members.

## Order
## Salicales

includes only the willows and poplars. Both petals and sepals are usually lacking; only sepals appear in some species, and these are reduced.

## Order
## Fagales

consists of shrubs and trees. Some families include trees with deciduous leaves, such as the beech (*Fagus*) and the chestnut (*Castanea*). The birch, alder, oak, and hazelnut trees all belong to this group.

## Order
## Cactales

contains the woody, spiny, or succulent plants that we recognize as cactus plants. Although we think of such plants as the saguaro and cholla as being desert dwellers, this order is not restricted solely to hot dry environments.

## Order
## Ericales

includes the heath family. Many of these plants are found in the arctic and subarctic zones, as well as the more southerly temperate zones. Rhododendron, wintergreen (*Gaultheria*), and mountain laurel (*Kalmia*) are familiar examples; blueberries, cranberries, and huckleberries are often harvested commercially.

## Order
## Berberidales

includes the poppy barberry and crowfoot family. The crowfoot family is the largest of the order; familiar plants classified in this group include the anemone, columbine, larkspur, peony, and buttercup.

## Order
## Caryophyalles

contains about 1400 species, mostly herbs. Representatives include carnations (*Dianthus*) and baby's breath (*Gypsophila*), as well as many common weeds such as chickweed.

## Order
## Saxifragales

is a herbaceous group that includes several garden species. They are usually found in cold or temperate regions. Common representatives are currant and gooseberry, as well as the ornamental hydrangea. A more exotic member of this order is the carnivorous Venus flytrap.

## Order
## Umbellales

These are herbs for the most part, but include some woody members too. They are a diverse group, since they include some of our favorite vegetables and some of our most dangerous poisons. The carrot (*Daucus*) and celery (*Apium*) are representatives, but so are poison hemlock (*Conium*) and *Aethusa*, known to some poisoned fools as "fool's parsley." Anise, once believed to be a potent aphrodisiac, is also included in this order. The group gets its scientific name from the fact that the flowers are often grouped in clusters that hang from the stem like little umbrellas.

## Order
## Asterales

contains a varied assortment of plants. An interesting feature of all is that the same head may contain two sorts of flowers. The common lettuce (*Lactuca*) and the more expensive artichoke (*Cynara*) belong to this order, as do the chrysanthemum and sunflower, and those annoying weeds, ragweed and dandelions.

## Order
## Solanales

contains both herbs and climbing plants, including a number that wind up as products we use daily. *Nicotiana*, better known as tobacco, belongs to this family, as do belladonna and nightshade. The tomato (*Lycopersicum*) and Irish potato are less dangerous members.

## Order
## Bignoniales

contains both ornamental plants and weeds. *Antirrhinum*, the snapdragon, and *Veronica* are two familiar representatives; the common broadleafed weed, mullein, is a less popular member of this order.

## Order
## Lamiales

is a large order that contains seven families; many of our familiar wild flowers belong to this group, such as verbena, forget-me-not, and hound's-tongue. It also includes parsley, sage (*Salvia*), rosemary (*Rosemarinus*), and thyme (*Thymus*).

## Subclass
## Monocotyledoneae

includes plants with flower parts in 3s. Leaves have parallel venation, and the scattered vascular bundles of the stem lack cambium.

## Order
## Liliales

includes a group of herbaceous plants with underground storage stems—corms, rhizomes, or buds. Leaves occur in different forms, and may be reduced to scales in some species. Fruits may be capsules or berries. The tulip (*Tulipa*), lily (*Lilium*), and lily of the valley (*Convallaria*) all belong to this order; vegetables placed here include leeks and asparagus.

## Order
## Orchidales

includes some of the most beautiful plants in the world. Their leaves are simple and fleshy, and the flowers are often irregular, with petals of different sizes and shapes. In orchids, the middle petal is modified to serve as a landing field for insects. The lady's slipper that grows wild in wooded areas of the United States is a very attractive member of this group.

## Order
## Graminales

These plants represent man's most valuable and essential source of food. One family (Graminae) includes rice (*Oryza*), wheat (*Triticum*), Barley (*Hordeum*), corn (*Zea*), and rye (*Secale*). Forage grasses, such as timothy, Bermuda grass, fescue, and brome grass, also belong to this order, and the seeds and stems are a staple food for many other mammals and some small birds.

*Fucus*, or rockweed

*Marchantia,* a thallose liverwort

The green alga *Draparnaldia*

Sphagnum moss with sporangia

Meadow spikemoss, *Selaginella apoda*

*Polypodium virginanum,* **the rock cap fern**

*Ginkgo* leaves

*Sequoia gigantea* from California

**Passion flower blossom**

**Wild calla,** *Calla palustris*

# Kingdom Animalia

The characteristic most frequently associated with the term "animal" is the ability to move by choice from place to place; thus animals have developed senses, such as vision, smell, and taste, to help them determine where to move next. Some of the organisms classified in this kingdom lack one or more of the features we usually associate with animal life, such as locomotion, and the total range of traits that bring these diverse forms under one categorical heading is not always observed in every organism. Animals are multicellular living things whose eucaryotic cells lack rigid walls, plastids, and photosynthetic pigments. Animals are, therefore, always heterotrophic, and they usually feed themselves by ingestion. That process most often involves an internal cavity that carries on digestion. But some forms absorb their food rather than swallowing it, and quite a few members do not have an internal cavity for storage of food while it is being digested. Higher forms have developed rather complex sensory, motor, and neural systems. Animals practice both sexual and asexual reproduction. Haploid forms are rare.

Representatives of this group are so diverse that they present a bewildering array. Modern classification schemes divide them into three Subkingdoms—Agnotozoa, Parazoa, and Eumetazoa.

# Subkingdom Agnotozoa

This group contains only a single phylum, the Mesozoa. Mesozoa lack both tissue differentiation and an internal digestive cavity. Instead, they have surface cells to carry on ingestive and absorptive activity. These tiny organisms move by beating cilia.

# Phylum Mesozoa

Mesozoa are minute parasitic animals. For one order, Orthonectida, the hosts are marine invertebrates. Their structure is simple but their life cycle is complex. The wormlike body of the adult consists of an outer layer of ciliated cells surrounding an internal mass of either sperm or eggs. Sperm from the male are released and penetrate the body of the female. The ciliated larvae so produced infect the tissue spaces of a new host.

    The phylogenetic position of the Mesozoa is enigmatic. Many zoologists believe that they are degenerate flatworms. Others believe that they are an early offshoot of the metazoans and have a long history of parasitism.

# Subkingdom Parazoa

# Phylum Porifera

The Porifera, or sponges, are sessile or bottom-dwelling forms. They have no anterior end. Their symmetry is primitively radial, but most are irregular. They lack a mouth or digestive cavity. The body is organized around a system of water canals or chambers. They are generally marine, though a few inhabit fresh water.

| Class Calcarea | Members of this class of sponges have spicules (spicules serve the same function of supporting the soft tissues as do the skeletons of vertebrates) composed of calcium carbonate or silicon. Most calcareous sponges are restricted to shallow water. |
|---|---|

| Class Hexactinellida | These sponges are commonly known as glass sponges. Their name is derived from the fact that their six-pointed spicules are often fused to form a glasslike skeleton. The glass sponges are the most symmetrical of sponges. Their shape is like that of a cup or vase. In contrast to the Calcarea, the Hexactinellida are chiefly deep-water sponges. Most live between 900 and 2700 feet deep. Species of *Euplectella*, Venus's flower basket, have an interesting commensal relation with certain species of shrimps. A young male and female enter the sponge and after growth cannot escape. Their entire life is spent in this sponge prison. They feed on plankton brought in by the sponge's water currents. |
|---|---|
| Class Demospongiae | This class includes the most common and familiar North American sponges. They are all marine and range in habitat from shallow to deep water. The skeleton of this class is variable. It may consist of silicious spicules or spongin fibres or a combination of both. A few of the genera, like *Oscarella*, lack any skeleton at all. They may be the most primitive of the Demospongiae. Two genera of this class, *Spongia* and *Hippospongia*, are of commercial value. The skeleton is composed only of spongin fibres. They are gathered and the living tissues are allowed to decompose in water. The remaining undecomposed skeleton of spongin fibers is then washed and bleached. |

# Subkingdom Eumetazoa

Members of this subkingdom are more complex than members of the Subkingdom Agnotozoa. They have organs, a mouth and a digestive cavity.

# Branch Radiata

This branch contains the phyla Cnidaria and Ctenophora. They are grouped together because both display a fundamental radial symmetry and a similar body plan. They both possess a digestive cavity with a mouth as the only major opening to the exterior; the cells lying between the digestive cavity and the outer body surface are not highly organized. Many are attached to the substrate. Those that are free swimming are at the mercy of water currents.

# Phylum Cnidaria

This phylum includes hydras, jelly fish, sea anemones, and corals. Members of the phylum possess two basic structural features. There is an internal space for digestion called the gastrovascular cavity. This cavity lies along the polar axis of the animal and opens to the outside at one end to form a mouth. A mouth and digestive cavity allow these animals to engulf larger particles of food than the sponges. A circle of tentacles surrounding the mouth aids in the capture and ingestion of food.

The body cavity consists of three basic layers: an outer layer of epidermis, an inner layer lining the gastrovascular cavity, and between these two layers a layer called the mesoglea. The mesoglea ranges from a noncellular membrane to a thick jellylike mucoid material. The cnidaria are histologically primitive. A wide range of cell types compose the epidermis and gastrodermis, but they are not grouped to form tissues. Of the five different types of cells that compose the epidermis, the cnidoblast type deserves special mention. The cnidoblast is a round or ovoid cell. One end of the cell possesses a short bristlelike process exposed to the surface. The interior of the cell is filled with a special stinging structure called the nematocyst, which is used for anchorage, defense, and the capture of prey. When the bristlelike process of the cnidoblast is stimulated, the permeability of the cell wall is decreased. Water flows into the cell; pressure builds in its interior until the cell bursts and the nematocyst is shot out. Different species of cnidaria possess from one to four types of nematocysts. Hydra, for example, possesses four types arranged in rows along its tentacles.

There are no special organs for respiration and excretion. Gaseous exchange occurs through the general body surface. The nervous system is primitive.

## Class Hydrozoa

Most species of this class pass through both a polypoid and medusoid stage in their life cycles. Some forms, like the hydra, are solitary polyps, but the majority are colonial. In hydra, reproduction occurs by budding. The colonial forms reproduce by budding as well, but the buds remain attached to the parent body. The resulting collection of polyps is called a polypoid colony. Most hydroid colonies are only a few inches or more in length.

Hydroid medusae are usually small. As in all medusoid forms, the mesoglea is extremely thick. The medusae are carnivorous. All medusae reproduce sexually. The larvae which result from this sexual reproduction may have a free-swimming existence of several hours and then attach to an object and develop into a hydroid colony. *Obelia, Pennaria,* and *Syncorne* display this alternation between medusoid and hydroid stages in their life cycles.

The majority of hydroids do not produce a free-swimming medusoid progeny. Instead, the medusae remain attached to the parent hydroid.

## Class Scyphozoa

Scyphozoans are the cnidaria most often referred to as jellyfish. Similar to the hydrozoan *Obelia* in appearance, the jellyfish's mouth is lined with ciliated grooves leading into the gullet. Gastric filaments secrete enzymes which digest prey that enter the stomach. There is a complicated water vascular system made up of branched canals; water enters the mouth and travels first through the gastric pouches inside and then through the canals. The water carries digested prey and oxygen throughout the body. Thus the system fills the double role of a digestive and circulatory system. A common jellyfish is *Aurelia*. It is shaped like a shallow saucer. A fringe of tentacles surrounds its mouth. Four large gonads shaped like horseshoes are located at the midpoint of the body. The life stages of *Aurelia* are typical of most jellyfish. The medusa reproduces sexually and the resultant larvae develop into a polyp. The polyp grows and develops grooves which ultimately become layers of new jellyfish. The young leave the old polyp and grow into adults. The largest jellyfish, *Cyanea*, is more than 4 meters in diameter, with tentacles as much as 30 meters long.

## Class Anthozoa

This class includes the sea anemones and corals, all of which are noted for the lack of a medusa stage in their life cycle. They are solitary or colonial coelenterates which only produce polyps. The sea coral (*Alcyonium*) is a colonial form, sometimes known as "Dead man's fingers," which attaches itself to stones below the surface of the water. *Alcyonium* polyps are delicate organisms which shrink from the slightest stimulus. They are equipped with tentacles which are located in an area about the mouth called the oral disk. The anemones are solitary individuals. They have no skeleton. They receive their name from their resemblance to a flower. They have tentacles armed with stinging nematocysts which are used to stun prey. Although the sea anemone is not a colonial organism, individuals tend to congregate in tidepool crevices, and some of them seem to remain in the same place indefinitely. Others are quite mobile, and are capable of moving about 18 inches in 24 hours by means of a slime-covered disk. Some anemones are especially sensitive to light. They contract when the light becomes too intense. These anemones can grow only in the dark. Others, like the large green anemone of the eastern Pacific, require light since their tissues contain a mutualistic green alga.

# Phylum Ctenophora

This is a small phylum of marine animals commonly known as sea walnuts or comb jellies. Ctenophores are radially symmetrical and the general body plan is somewhat similar to the cnidarian medusa. The more primitive ctenophores, like *Mertensia*, are spherical or ovoid in shape. The body is divided into equal sections by eight ciliated bands. These bands, called comb rows, are characteristic of the ctenophores. The animal usually maintains a vertical position but can point downward and right itself when tilted by water currents. A long tentacle extends downward on each side of the ctenophore. The tentacle secretes a sticky material used in catching prey. The ctenophores are carnivorous, feeding on small planktonic animals. The food is caught on the tentacles and wiped into the mouth. Ctenophores are noted for their luminescence.

## Class Tentaculata

This class includes the *Mnemiopsis*, or sea walnut, which is quite common along our Atlantic coast. It swims by means of eight rows of cilia, which resemble combs, and cilia around the mouth also serve to send small food particles inside. Sticky tentacles are encased in pouches at opposite ends of the body; they may be stuck out or withdrawn at will. Each tentacle contains cells analogous to the nematocysts found in coelenterates. Each cell, or colloblast, consists of a sticky head which attaches itself to potential prey. The tentaculata swim feebly through the action of their comb plates. Each comb plate beats in succession, and ciliar movement propels the organism ahead. Under most circumstances, the eight rows of comb plates will beat in unison, but disturbing stimuli will throw them off the beat. The sea walnuts (named because of their nutlike size and shape) are hermaphroditic forms, with both male and female gonads located inside the body cavity, and much self-fertilization probably occurs. When *Mnemiopsis* is exposed to illumination, it will light itself up in protest. *Ctenoplana* is a second member of this class. It has lost some of its swimming comb plates, and thus maintains a flattened appearance. Unlike the sea walnut, it does not have the property of illumination when annoyed by an obnoxious light stimulus.

# Branch Bilateria

The animals of this branch are bilaterally symmetrical. Bilateral symmetry is correlated with their locomotor movement. One end is directed forward and a constant surface is directed toward the substratum.

## Grade Acoelomata

The name for this grade was chosen because its members lack any sort of body cavity. The simple mouth is located on the midventral side of the body. Acoelomates, even when they feed on small animals like protozoa, do not pass the prey into a digestive cavity, but rather into the internal mass of nutritive cells.

## Phylum Platyhelminthes

This is the most primitive phylum of bilaterally symmetrical animals. The members of the phylum are dorsoventrally flattened and have a solid, acoelomate structure. The mouth is the only opening to the digestive apparatus. Protonephrida are the excretory or osmoregulatory organs. The reproductive system is hermaphroditic.

| | |
|---|---|
| **Class Turbellaria** | This class consists of free-living worms. They are unsegmented and possess cilia and a fairly well-developed gut. In most species, a concentrated mass of nerve tissue near the eyes takes the form of a pair of cerebral ganglia; this is rudimentary cephalization. Turbellarians, usually only about two inches long, are found in both terrestrial and aquatic habitats. They are hermaphroditic, mating through mutual exchange. They move by a combination of undulation and ciliary action. They receive their name from the beating action of their cilia, which creates turbulence in the water around them. The most familiar turbellarian is *Dugesia*, which is a recent favorite of psychologists because it is the simplest, most dull-witted animal capable of learning. |
| **Class Trematoda** | Included in this class are the parasitic flatworms popularly known as flukes. All are parasitic, with suckers appearing around the mouth and sometimes on the ventral surface. Like turbellarians, they exhibit a well-developed gut and mouth. They lack cilia, and their epidermis is lost during development. They have a cuticle covering the outside of the body. Some trematodes go |

731

through life with only one host, while others visit two or even three hosts in the course of their life cycle. The larvae often live in snails, while the adults parasitize vertebrates. *Diplozooan* is a parasite living in the gills of minnows, attached by suckers during the larval stage. In order to reproduce, it must somehow meet another larva; if they do, they fuse immediately and develop into adults locked in a state of permanent copulation. Like other flatworms, trematodes are hermaphroditic. *Polystomum* attaches to tadpoles in the larval form and then matures when the tadpole undergoes metamorphosis; the adult lives in the frog's urinary bladder. *Gyrodactylus* fastens itself to fresh-water fish with huge suckers modified with hooks. *Schistosoma* is a fluke with a forked tail that lives in human blood. Unlike most flatworms, the sexes in this genus are separate, and the male carries the smaller female around in a groove in his body.

**Class Cestoda**

This group includes the tapeworms—flat creatures that inhabit the intestines of many vertebrates, including man. The nervous system is greatly reduced. The tapeworm's nutrients are simply absorbed from the host's intestine; there is no mouth or digestive system. Adults lack cilia and epidermis, but the body is protected by a cuticle. The body consists of a variable number of sections called proglottids; each contains both male and female reproductive organs. *Taenia* is a common tapeworm that parasitizes man. It adheres to the intestine with four massive suckers and a system of hooks located on the head. Once attached, the maturing posterior proglottids containing the eggs are shed into the host's intestine.

# Phylum Aschelminthes

This is a diverse group of pseudocoelomate worms. All have a complete digestive tract and a tube-within-a-tube body plan. There are six different classes in this category.

**Class Rotifera**

Included in this class are the rotifers, tiny microscopic creatures which are all characterized by a crown of cilia at the anterior end. This crown (which resembles a rotating wheel) serves to keep the rotifer moving and to drive food toward its mouth. Rotifers are unsegmented wormlike creatures that usually live in fresh water; some are sessile, with adhesive disks at the posterior end to attach to the substrate. Oddly enough, in some species males are rarely found. Females usually reproduce parthenogenically; males are produced only at certain times of the year when sexual reproduction occurs. *Conochilus* is a colonial rotifer that congregates with others of its kind and swirls through the water like a revolving sphere.

## Class Gastrotricha

This group encompasses microscopic shore-dwellers, both fresh-water and marine. Typically, there is an epidermis which secretes a cuticle. Cilia on the head sweep food into the mouth. The digestive tract is complete, with a mouth at one end and anus at the other. These animals are often difficult to distinguish from rotifers found in the same habitat. Gastrotrichs lack the rotifer's crown of cilia, but do locomote on the bottom by cilia. In the fresh-water gastrotrichs, only females are known. *Chaetonotus*, which has a spiny cuticle, and *Macrodasys*, a marine form, are common representatives.

## Class Kinorhyncha

These organisms are tiny, ringed creatures which live in mud and shallow water, where they feed mostly on detritus. They possess a retractile head, a nervous system with dorsal ganglia, a midventral nerve cord, and a cuticle. Cilia are lacking. Spines surrounding the mouth give this group its name. Members of this class include *Echinoderes* and *Semnoderes*.

## Class Priapulida

This class includes a group of marine worms that live in the mud. Sexes are separate, and the larva greatly resemble rotifers. *Priapulus* and *Halycriptus* are the best-known members.

## Class Nematoda

This is a group of slender unsegmented worms. Nematodes are found in huge numbers everywhere, and may be either aquatic or terrestrial. They are characterized by the total absence of cilia, an elastic cuticle, a nervous system that includes a bilobed brain and central nerve trunks, and separate sexes. They are known collectively as the round worms. *Wuchereria* is an example of a parasitic form transmitted from insects to humans, causing the disease called elephantiasis, characterized by enormous swelling of the arms and legs.

## Class Nematomorpha

This is a group of parasites known as the horsehair worms. They have a thin cuticle and absorb food from their arthropod hosts through the body wall. Sexes are separate. *Gordius* is a frequently observed nematomorph that infects the body cavity of crickets, forcing its way in with a boring organ characteristic of its larval form. *Paragordius* is a similar form that infects beetles.

# Phylum Annelida

This group includes all the segmented worms. The name of the phylum means "ringed," in reference to the segmented appearance of its members. Each segment is a functional unit. In this phylum, there is a true coelom, a circulatory system with blood, and nephridia as excretory organs. There are four classes of annelids.

## Class Archiannelida

These are tiny marine organisms with a relatively rudimentary level of organization. They have retained the characteristics of larval forms, such as cilia and a nervous system within the epidermis. *Polygordius* is an archiannelid with an elongated body, cylindrical in shape, exhibiting tentacles and cilia. Its circulatory system has also undergone reduction through time. Only one genus, *Protodrilus*, has setae, or shortened bristles on each segment.

## Class Polychaeta

Members of this class are segmented forms which show less evolutionary reduction than other annelids. They have numerous setae, paired appendages, and parapodia. Gonads, rather than being restricted to one body segment, are arranged at random over the body to permit external fertilization. Each segment has two sets of antagonistic muscles. Members of this group include the feather-duster worm (*Sabella*), the clam worm (*Nereis*) and the parchment worm (*Chaetopterus*).

An interesting pattern of reproductive behavior is found in the polychaete palolo worm, which lives around Pacific coral reefs. Every individual reproduces within a two-hour period on the same night of the year. The posterior half of each worm breaks off and swims by itself to the surface. There it bursts, releasing either eggs or sperm, and cross-fertilization takes place.

## Class Oligochaetae

These annelids are the common earthworms and their close relatives; all exhibit reduction in setae. They are always hermaphroditic, with both male and female gonads located in anterior segments. Eggs are laid in a cocoon, and cross-fertilization is the rule. There are two major forms of oligochaetes— the earthworms, larger and adapted to burrowing, and the aquatic forms, which are both smaller and more rudimentary in structure. The earthworms' importance to the ecological balance of our planet is immense; as they burrow through the soil they aerate it and create drainage passages, thus making plant growth possible. Charles Darwin once estimated that in 10 years, earthworms turn over enough soil to form a layer five centimeters thick over the entire earth.

| | |
|---|---|
| **Class Hirudinea** | This class contains the most highly specialized annelids of all, the leeches. The leech has 32 body segments headed by a prostomium, equipped with a sucker surrounding the mouth. *Acanthobdella* is a leech that feeds exclusively on salmon. It has no anterior sucker, but does have a posterior sucker derived from the last body segment. *Glossiphonia* is a fresh-water leech that feeds on mollusks, and *Hirudo medicinalis* is a sharp-toothed form that was once employed by doctors to bleed their patients (possibly to make the ailment seem desirable by comparison). |

# Phylum Sipunculoidea

Phylum Sipunculoidea is the name of an intriguing group known as the "peanut worms," because of their shape and size. There are about 250 species in this phylum. A typical peanut worm features a long curved structure at the front end of the body. This flexible tube, or introvert, looks rather like an elephant's trunk. The mouth is at the tip of the introvert, and the entire structure is retractable. Peanut worms have a twisted digestive tract and a dorsal anus; their circulatory system is rudimentary.

# Phylum Arthropoda

Characterized by a jointed chitinous exoskeleton and jointed appendages, members of this phylum literally swarm the earth with their staggering numbers; over 80 percent of all living things are arthropods. Distinguishing characteristics of the phylum include the presence of a hemocoel, a metameric structure, a ventral nerve cord, and a periodic molting, to shed the old exoskeleton.

# Subphylum Chelicerata

The arthropods in this subphylum have no antennae and the first pair of appendages, the chelicerae, is equipped with pincerlike organs. There are four pairs of walking legs, and the body is divided into two parts.

## Class Xiphosura

This group contains the horseshoe crab *Limulus*, which is often observed creeping or swimming along the shorelines of our Atlantic coast. *Limulus* is a well-armored organism; it is not a true crab, but shares enough features with scorpions and spiders to be classified in this subphylum. Some of the appendages are modified for capturing prey; others are used primarily for locomotion but also are used for handling food. The habitat of the horseshoe crab is the soft mud or sand of the ocean floor.

## Class Arachnida

Members of this class usually conjure up feelings of distaste or apprehension, for these are the world's spiders, ticks, mites, and scorpions. They superficially resemble insects but distinguish themselves by having only two body regions, and four pairs of legs rather than three. Instead of chewing mouthparts, they have a system of two chelicerae in front of the mouth, with clawlike poisonous fangs. In place of antennae, arachnids use sensory bristles which cover the body and perform the same function. A highly successful group, the arachnids have become adapted to a wide variety of environments. Their ancestors are believed to have been aquatic, though the majority of modern forms are terrestrial.

Mites are tiny, fairly simple arachnids that often parasitize plants and other animals, including man and his house pets (even the parakeet). They usually live on the surface of the host, although some may be free-living. Ticks are considerably larger and adapted to sucking blood. They parasitize a number of domestic animals, to whom they may transmit disease; cattle are frequent victims. Spiders make up the largest group of arachnids. *Argiope* is known for its beautiful orb webs. After mating, the female *Argiope* will often wrap her mate in web-silk and feed on him. Scorpions are arachnids that live on the ground and hunt at night, feeding largely on insects or small vertebrates. A solid carapace covers the cephalothorax, and the terminal segment ends in a sting. *Centruroides* is an example of this group. *Mastigoprocteus* is a nocturnal arachnid often called a whip scorpion in reference to its whiplike abdominal appendage. They are also known as "vinegaroons" because of their habit of spraying a fluid rather like acetic acid when disturbed.

---

| | |
|---|---|
| **Class Pycnogonida** | Known as the sea spiders, these chelicerates appear to be all legs. Most of the body is made up of the cephalothorax; the abdomen is greatly reduced in size compared to the cephalothorax. Most are less than 10 millimeters in body length, but one giant species, *Colossendeis colossea*, has a body two inches long and leg span of two feet. They crawl about the ocean floor, feeding on hydroids, bryozoans, sponges, and other marine animals, grasping the prey with their chelicerae and ingesting bits of soft food. The males, at mating time, gather up the fertilized eggs and carry them glued to their anterior legs, until they hatch. Because of several peculiar characteristics—for example, the segmented cephalothorax—the evolutionary relation of this group to the other chelicerates is uncertain. |

# Subphylum Mandibulata

These arthropods all have a pair of preoral antennae. The second pair of somites bears a pair of jaws or mandibles, and the next two somites bear the maxillae, or mouthparts.

---

| | |
|---|---|
| **Class Crustacea** | is an extremely diverse group with a very wide range of creatures, features, and habitats. Crustaceans are primarily aquatic animals, distinguished by having two pairs of antennae. Many crustaceans pass through a free-swimming larval stage and then a brief post-larval "adolescence" before they become sexually mature. |

---

| | |
|---|---|
| **Subclass Cephalocarida** | is a recently discovered group consisting of only four known species. The first genus, *Hutchinsoniella*, was found in 1955 in Long Island Sound. This organism is now thought to be the most primitive type in its class. It is very small; eyeless; retains a carapace that reaches back to cover the first thoracic segment; and has 10 pairs of limbs. It feeds on detritus, using its appendages to suspend bottom deposits in the water so particles can be swept into the mouth. |

---

| Subclass Branchiopoda | includes fresh-water forms that feed on plankton or detritus. The free-living branchiopods exhibit a carapace over the body, and flexible trunk appendages. Filter-feeding is the rule here, and branchiopods have developed special mechanisms for it, such as fine setae to filter food particles from the water. Reproduction may be either sexual or parthenogenic; in some species males are unknown. *Daphnia* is a representative of the group known as water fleas. *Cyzicus* is a representative of the clam shrimps. |
|---|---|
| Subclass Ostracoda | includes a group of free-swimmers that differ very little from each other except in the size of their appendages. They have a bivalve carapace and only two pairs of trunk limbs. They are all small — about one millimeter long — and are best represented by the *Cypris* which is both very mobile and omnivorous. Ostracods are filter-feeders that secrete mucus to help trap food particles. They are primitive crustaceans; fossil records indicate many extinct forms. |
| Subclass Copepoda | includes a group of organisms that are quite disparate in appearance, ranging from skillfully organized free-swimmers to sometimes limbless parasites. They feed chiefly on diatoms and are themselves an important part of the plankton. *Cyclops* is a common form that is pear-shaped, with long antennules but shorter antennae, bristled mandibles, and thoracic limbs that tend to move together while swimming. It locomotes by using its long antennae as oars and utilizes the swimming limbs of the thorax to accelerate its pace. *Calanus* is a more primitive species found in the plankton. Both of these organisms obtain food by filter-feeding, but *Cyclops* can also seize food particles. |
| Subclass Cirripedia | consists of the barnacles, filter-feeding organisms that are the strangest crustaceans of all, different in many respects from all other members of this class. They have a calcareous covering of overlapping plates which led early taxonomists to classify them as mollusks. They are sessile and lack eyes. The members of one group of barnacles, including *Lepas*, hang from a projection of the head known as the peduncle and, as anyone who has owned a boat can tell you, are stubbornly resistant to leaving their chosen moorings. *Balanus* is an example of a sessile form with no peduncle; it feeds by means of protruding appendages that function as a kind of net to trap small food particles. |

## Subclass Malacostraca

contains familiar crustaceans such as the shrimp, crayfish, lobster, and crab. They exhibit a tremendous diversity of form. The pistol shrimp (*Crangon*) has chelae modified to snap together forcefully enough to stun prey by sound waves. Its relative *Stenopus* is a tropical marine species that sports a white body striped with iridescent bands of red, orange, blue, and purple. The lobster that we are fond of (*Homarus*) will live and die as a dark green organism unless it is caught and boiled; then its carapace develops the bright red coloration we associate with it. *Crayfish* are similar to lobsters and are capable of the same defense mechanism—suddenly reversing course when approached by a predator and propelling themselves in the opposite direction. This has proved a limited safeguard in the face of modern fishing methods. The great family of crabs includes a number of intriguing species. The fiddler crab (*Uca*) is named for the giant pincer which it keeps poised as if it were about to grab the nearest fiddle and play. It is commonly found near estuaries, digging burrows in the intertidal zone and roaming over the mud at low tide. *Macrocheira*, the giant spider-crab found in eastern Asian waters, may measure three meters in the span from claw to claw.

## Class Chilopoda

These are the centipedes. They prefer damp places, such as the ground underneath rocks or logs. Although they are carnivorous, and sometimes bite people, they present little threat to man, and some are actually responsible for killing off some of the insects that we find most obnoxious—cockroaches, silverfish, and the like. *Lithobius* is a representative exhibiting the common features. It has a long flattened body of several segments (most of which carry a pair of walking legs and one of which is armed with a pair of poison fangs or claws for killing prey), a long pair of segmented antennae, maxillae and mandibles, and noncompound eyes. Centipedes are notoriously shy of light, and are prone to head for the nearest rock when it becomes too bright for them.

## Class Diplopoda

These are the millipedes, which really do not have as many legs as you may have been led to believe. Like the centipedes, they have a head that consists of six segments. But they have two pairs of legs per body segment, which means that the average millipede can claim about 70 or so. The millipedes live in the same type of environment as the centipedes—under rocks or decaying logs—but they are scavengers rather than predators. *Polydesmus* is distinguished by its dorsal exoskeleton, which flares out to the sides, and by its ability to secrete a cyanidelike chemical for defense. *Spirobolus* slightly resembles a small snake.

| Class Pauropoda | This class consists of only one small group of soft-bodied organisms found in tropical to temperate regions; they are superficially similar to diplopods. *Pauropus* (a representative member) has a head segment and 11 trunk segments, and no eyes, though sense organs called pseudoculi are present on the head. Like the millipedes, they live under leaves and logs and feed on vegetation. |
|---|---|

| Class Symphyla | Members of this class are found in most parts of the world (except for very cold climates), living in soil and ground litter. Resembling tiny centipedes, they range in size from 2–10 mm. The trunk has 12 segments, each bearing a pair of legs, and it is covered by tergal plates. The last trunk segment bears a pair of spinnerets. The young hatch with six to seven pairs of legs. *Scutigerella immaculata* is a representative member. |
|---|---|

| Class Insecta | For the past 200 million years, this single class has established itself as the most successful group of organisms on earth. Insects have managed to inhabit virtually every ecosystem that can sustain life at all. They are characterized by having a segmented body—head, three thoracic segments, and 9 to 11 abdominal segments. Three pairs of legs and two pairs of wings are usually attached to the thorax. Mouthparts consist of a pair of mandibles, a pair of maxillae, and the labium. There is one pair of antennae. They undergo metamorphosis as part of their development process. Respiration is tracheal, and excretion takes place with the aid of Malpighian tubules. |
|---|---|

### Order Thysanura

includes small, wingless, primitive creatures with long jointed antenna, mouths adapted for chewing, and abdomens that consist of 10 segments. The household genus is the *Lepisma*, or the silverfish, which apartment dwellers often discover scuttling out of faucets or drains, or worse yet, eating books and clothing. *Machilis* is a relative found at the seashore. Some thysanurans live in decaying logs or piles of dead leaves.

### Order Diplura

includes small elongated organisms hiding under rocks or in the soil. These insects have abdomens of 11 segments. They lack eyes and wings and have only primitive chewing mouthparts. Their antennae are prominent and articulated. *Campodea* is a representative.

## Order
## Protura

consists of tiny insects that thrive in decaying organic matter. Lacking the usual features of wings, antennae, or eyes, their forelegs are used as tactile organs. They possess primitive chewing mouthparts and have pseudoculi on the head. *Acerentulus* is a North American genus.

## Order
## Collembola

are commonly known as springtails—forms often lacking compound eyes and wings. *Isotoma* and *Achorutes* are two genera that live near the seashore. Their chewing mouthparts are equipped with sharp mandibles.

## Order
## Ephemerida

contains the mayflies, which have large membranous wings. The nymphs are aquatic and have chewing mouthparts. Adults, which have vestigal mouthparts, die soon after mating. Thus the mayflies spend most of their lives as nymphs. One example, *Chloen,* is streamlined to permit rapid escapes from fish that prey on the nymph, and *Ephemera* is adapted for burrowing.

## Order
## Odonata

includes the dragonflies and damselflies—large predators with strong chewing mouthparts. The narrow net-veined wings of dragonflies are held straight out from the body when the insect is at rest. As adults they are fast-moving hunters, but the nymphs are aquatic forms that are relatively slow moving. The dragonflies have thick bodies, while damselflies are delicate and slender.

## Order
## Orthoptera

includes the grasshoppers, crickets, walking sticks, mantids, and cockroaches. These insects are characterized by two pairs of wings, the front set slightly leathery. The front wings are not actively used in flight but serve as a cover for the hindwings. Many orthopterans are herbivores that sometimes endanger crops.

## Order
## Dermaptera

covers the earwigs, a small group of tropical omnivorous insects that come out only at night. They have mouths adapted for chewing. Wings are tiny and functionless. They are elongated insects that resemble certain beetles.

---

## Order
## Embiaria

contains a group of tiny insects known as webspinners. These small slender insects are equipped with silk glands and spinnerets located on the forelegs. They live in communal webs and are herbivorous. Females are always wingless, but the males of most species have wings.

---

## Order
## Plecoptera

contains the stone flies. Some have prominent mandibles and antennae, with well-developed wings which are folded over the back when they are at rest. They have large heads with moderately sized compound eyes; ocelli are also usually present. Young stages, or nymphs, possess gills, an adaptation to aquatic life. Representative genera are *Isoperla* and *Capnia*.

---

## Order
## Zoraptera

is made up of small insects that display nine-jointed slender antennae and live in groups under rotting logs. *Zorotypus* is a typical example. Its wings are somewhat reduced, and they are lost after mating. Its mouthparts are modified for grasping and chewing.

---

## Order
## Corrodentia

consists of insects that feed on lichens and other plants. *Peripsocus* is a tree-dwelling genus with chewing mouthparts. *Atropus* is the scourge of all paper products, leaving torn wallpaper and books in its wake. Its common name is book louse.

---

## Order
## Mallophaga

contains the biting lice. These small, wingless and flattened insects are ectoparasitic, living on birds or mammals, and they often cause great damage to domestic fowls. Eyes are reduced, and appendages are modified to form grasping claws. They feed on hair, skin, and feathers.

---

## Order
## Thysanoptera

includes the thrips. A few are carnivores, but most are herbivores that attack fruit with mouthparts modified for piercing and sucking. The pea is a frequent victim of *Kakothrips*, which lays its eggs in the stamen so its nymphs can begin feeding on the plant immediately. *Taeniothrips* is a close relative that feeds on pears. Thrips have a proboscis and two pairs of nearly veinless wings (fringed with hair).

## Order
## Anoplura

is made up of the sucking lice, ectoparasitic on mammals with claws for clinging to their hosts and mouthparts that pierce the skin and suck their host's blood. They puncture the skin of their host with stylets, then inject a salivary fluid, finally ingesting the blood through a specialized pump mechanism. *Pediculus* is the common body louse, which lays its eggs on human hair or clothing in great quantity and makes its host thoroughly miserable. The bites cause itching and swelling. *Phthirius* is the crab louse that lives in pubic hair; the head louse prefers the hair on the head.

## Order
## Hemiptera

is such a tremendous order it is necessary to divide the members between two subgroups. Suborder Homoptera, recently placed under Hemiptera, includes the forms that have a beak arising from the back of the head and two pairs of membranous wings. It includes the aphids (*Aphis*), leaf hopper cicadas (*Cicada*), and plant lice. Generally these insects are herbivores attacking a vast range of plants. The Hemiptera are distinguished by having a beak arising from the front of the head, and a leathery first pair of wings modified for covering the membranous hindwing. These are the true bugs, which may be land-dwellers or aquatic. *Plesiocoris* is an old enemy of the willow tree that has lately taken to apple trees and fruit bushes. *Notonecta* is an example of an insect with specialized features for an aquatic existence; its strong legs are outfitted with hairs which, moving in unison, can "row" the animal.

## Order
## Neuroptera

contains the alder fly (*Sialis*), dobsonfly, snakefly (*Raphidia*), and lacewing (*Chrysopa*). Adults have chewing mouthparts and two similar pairs of membranous wings. The alder fly has two large pairs of wings, which are held over the back when resting as a kind of roof. The larvae are predaceous aquatic forms equipped with gills. The dobsonflies and snakeflies are small insects with bodies elongated (to the point in the snakefly that it seems serpentine). The head juts forward and tapers toward the thorax.

## Order
## Coleoptera

is the largest order of insects. It includes all the beetles, a large group displaying chewing mouthparts. The front wings are hardened and fold over the rear wings for protection; the rear wings, if present, are membranous. The tough exoskeleton is often brightly colored. Beetles occur virtually everywhere—on plants, in the water, and in the soil. They are particularly abundant in environments of decaying plants, mounds of dung, or fungi. Among the many diverse forms, *Coccinella* is one of the few well liked because it eats insect pests. Many of the beetles eat plants, such as the legendary boll weevil (*Anthonomus*) which manages to devastate cotton crops shortly before they are harvested. Others, such as *Calandra*, invade silos and deplete grain stores. Some are comical at a distance, such as the *Aphodius* or dung beetle and the fat, gruff-looking *Scarabaeus*, or scarab beetle.

## Order
## Hymenoptera

includes certain groups that have evolved a high degree of social organization. They all have mouthparts adapted for chewing or sucking, usually two pairs of membranous wings and a large thorax and abdomen connected by a narrow waist called a pedicel. Some members of this group are parasites; ichneumon wasps and chalcids lay eggs in other insects. Communication is visual; bees perform a dance to indicate the location of food. An interesting feature is the social organization exhibited in the ants (*Formicoidea*), honey bees (*Apis*), and wasps (*Vespa*). The bee colony, for instance, is so specialized that even workers' tasks are divided. They are placed, according to age, as nurses to care for the larvae, ventilators that vibrate their wings to keep the hive cool and livable, scavengers and cleaners to handle housekeeping, and foragers to go out looking for nectar and pollen.

## Order
## Mecoptera

consists of the scorpionflies, a small group characterized by two pairs of similar wings and chewing mouthparts. *Panorpa* is a common scorpionfly, laying its eggs in the soil to hatch grublike larvae. This group gets its name from the fact that the tip of the abdomen in males greatly resembles a scorpion's sting.

## Order
## Siphonaptera

contains fleas, wingless insects adapted to a parasitic existence. Unlike many parasites, they have retained their compound eyes. They have strong claws for attaching themselves to their hosts, and sucking mouthparts. Their legs are adapted for jumping, so they can leap from one host to another. They seem to have a facility for dropping eggs where their hosts cannot find them until it is too late. By then, the adult will have developed (it takes only a month in the common *Pulex*) and will be ready to lay more eggs. With their piercing and sucking mouthparts, they feed on blood. The most dangerous flea is *Xenopsylla* of Asia, which transmits the plague bacillus to man from its primary host the rat.

## Order
## Diptera

includes the true flies, gnats, and mosquitoes, which we have come to dislike for coveting both our food and our bodies. They are distinguished by a head with proboscis of mouthparts adapted to lapping or sucking and biting, one pair of wings (the second pair reduced to short balancing organs, or halteres) and feeding patterns that range from a preference for nectar to blood-sucking. *Glossina*, the tsetse fly, transmits sleeping sickness to man, and several other dipterans are also disease carriers.

## Order
## Trichoptera

refers to a group of insects identifiable by their hairy wings that fold over the back when at rest; they are commonly called caddisflies. Most species have aquatic larvae that build cases of materials like sand or bits of shells, but others may spin webs to catch their prey instead. This more aggressive caddisfly is exemplified by the *Hydropsyche*.

## Order
## Lepidoptera

contains the soft-bodied moths and butterflies. Mouthparts in this order are modified to form a coiled tube for sucking nectar. Wings are covered with tiny brightly colored scales. The forewings are larger than the hindwings. The larvae are typically herbivores.

# Phylum Pentastomida

This is a group of elongated parasites with two pairs of hooked appendages usually found behind the mouth. They have neither respiratory nor circulatory systems. *Linguatala* is the most common member of this phylum. Their larvae, already equipped with the characteristic claws, always begin their parasitic existence on herbivores. When their hosts are finally eaten by carnivores, the young *Linguatala* grow to adulthood on their new hosts.

# Phylum Tardigrada

These tiny organisms, commonly called water bears, have four pairs of stubby legs equipped with claws, but lack both circulatory and respiratory systems. While the tardigrades may appear somewhat comical, with short, flattened bodies and rather clumsy movement, they have an uncanny ability to survive unfavorable climatic conditions. Like rotifers, they can actually shrivel up and dry out from lack of water, then absorb enough water during the next rain to revive and go on about their business. Typical members of this phylum are *Macrobiotus* and *Echiniscus*. They usually live in mossy areas or the sediment that collects in rain gutters. The organism can feed itself by piercing plant cells with its stylets and sucking out the contents through contractions of the muscular pharynx.

# Phylum Bryozoa

These are the *Ectoprocta*, or moss animals, tiny colonial and sessile organisms that lack excretory and circulatory systems. They are found encrusted on rocks in fresh or salt water. This and the next two phyla are collectively described as lophophore-bearing animals. The individual zooids enclose themselves in a gelatinous or calcareous case open only at one end. The lophophore is an extended horseshoe-shaped or coiled fold with projecting tentacles, surrounding the mouth. The tentacles are covered with cilia whose beating action drives tiny plankton toward the mouth.

# Phylum Phoronidea

Members of this group also possess lophophores. Included are about 15 species of tiny, wormlike organisms which secrete chitinous tubes in which they live. Most of them belong to the genus *Phoronis*. All are marine, and adults are found either buried in sand or attached to the substrate. Only in the larval form are they free-swimming.

# Phylum Brachiopoda

This phylum includes animals that look remarkably like ancient oil lamps. They resemble the shelled marine bivalves, but their symmetry is dorsal-ventral rather than lateral. During the Paleozoic era, brachiopods were abundant; now there are only some 250 species living on the shallow ocean floor, permanently attached to rocks by a stalk.

**Class Inarticulata**

Included in this group are the brachiopods which have shells without hinges. They also lack internal skeletal supports for the lophophore. *Lingula* is one genus often found in shallow water, and bears a superficial resemblance to a clam.

**Class Articulata**

This class accounts for other brachiopods, which have hinges and live in deep water, such as *Magellana* and *Terebratula*. Both of these species have ventral shells larger than their dorsal shells, with a posterior beak known as the umbo.

# Phylum Mollusca

This is the second-largest phylum in the animal kingdom. Mollusks are soft-bodied animals with a muscular foot and glandular mantle. A radula, or rasping organ, is also found in most mollusks. The success of this group is reflected in the variety of habitats they have invaded; mollusks are found in fresh water, marine, and terrestrial habitats. There are presently about 80,000 species, divided into six classes.

**Class Amphineura**

Members of this group include the chitons, sluggish marine forms which exhibit bilateral symmetry and have an anterior mouth and a posterior anus. Eight overlapping calcareous plates cover the body dorsally. These mollusks have a reduced head (without eyes, tentacles, or other features usually associated with a head) that is tucked under the mantle. The mantle itself may contain a number of spicules, of different types, which may also come together to form an actual shell. The chiton has a flattened foot and moves slowly. It inhabits the bottom of shallow waters, where it creeps over rocks and feeds on algae. *Chaetoderma* is one member of this group with gills, but most amphineurans, such as *Neomena*, do not have them. *Craspedochilus* looks very much like a long limpet. It lives under rocks in shallow water, and attaches itself to them so tenaciously that it is difficult to pry from the rock surface. It leaves its usual spot every now and then to forage, always returning to the same place after feeding on algae it has scraped from rocks.

## Class Monoplacophora

This class contains a small group of primitively segmented animals that closely resemble members of the annelida. Originally a large class accounting for a great number of fossils, these organisms were thought for a time to be extinct. Now about 10 species are known to exist. *Neopilina* was found in the early 1950s. It has a bilaterally symmetrical form with a well-developed external shell. Feeding on radiolarians, *Neopilina* has a radula and a single flattened foot located beneath its five pairs of gills.

## Class Gastropoda

This is a large group that includes the snails, sea slugs, and their relatives. They live in a wide range of habitats. Although most gastropods are shelled, some, such as sea slugs, or the nudibranchs, are shell-less. Most of the 35,000 species are marine forms, but several are adapted to fresh water and land. In some snails, the gill chamber has been modified to form a sort of lung. The descendants of some terrestrial species have actually returned to aquatic life; they must surface periodically for air.

Gastropods have definite heads with tentacles; nervous systems complete with cerebral, pleural, and pedal ganglia; a heart with one ventricle and two atria; two kidneys; and a radula. A single flat foot moves them along at the proverbial snail's pace. *Helix* is a very common gastropod; *H. aspera* is the garden snail, and *H. potamia* is edible and often regarded as a delicacy. *Buccinum* is a marine gastropod that is more active and carnivorous, feeding on living and dead animals. *Murex* is also carnivorous and is called the oyster drill. *Littorina*, the periwinkle, is a noncarnivorous snail that lives at the high water mark and spends most of its time out of water. The periwinkle shows a high degree of coiling in its shell, whereas its relative, the abalone, shows very little coiling.

## Class Scaphopoda

Members of this class are the tusk shells. A strictly marine group, each has a long tubular calcareous shell open at both ends, with one end usually smaller than the other, and bilaterally symmetrical. The cone-shaped foot is used for burrowing, and several structures called captacula surround the mouth.

There is also a radula, a nervous system with cerebral and other ganglia, and paired kidneys. The captacula or tentacles are used in seizing food. This can be seen in the genus *Dentalium*. The single foot juts out for use as a burrowing tool, and the shell into which it retracts has a hole at each end to allow water inside when the animal is burrowing into sand.

## Class Pelecypoda

This group consists of the bivalve mollusks—animals with shells consisting of two symmetrical valves. The two are almost identical in size and shape and are hinged together on one side. They can be shut tightly through the action of large, specialized adductor muscles. There is no head, radula, eyes, or tentacles. A mouth is located between the labial palps. The wedge-shaped foot is adapted to burrowing in mud and sand. The gills in the mantle cavity are covered with cilia and are the main food-getting organs. The lower posterior edge of the mantle cavity opens into a ventral siphon and above it is a dorsal siphon, allowing a constant stream of water to flow through. Pelecypods are filter-feeders, using mucus and cilia to separate food from waste substances in the water. It has been estimated that an oyster filters about three liters of water in an hour. Most pelecypods are sedentary. One exception is the scallop, which can move quite rapidly through the water by opening and closing its shell in a variation of jet propulsion. Some well-known pelecypods are the clams, oysters, and mussels.

## Class Cephalopoda

This class contains an interesting group of animals bearing little resemblance to other mollusks. Rather than lying passively along shorelines, the cephalopods are vigorous hunters adapted to fast movement and a predatory life style. Members of one branch of this group, including *Sepia* and *Nautilius*, retain an external shell. Other cephalopods, such as the squid and octopus, do not. In this last form, the shell has become reduced and internal. The foot is divided into eight arms in the octopus and 10 in the squid. The octopus exhibits a number of fascinating adaptations, among them jet-propulsion swimming and a parrotlike beak. Squids, which include the largest living invertebrates, have an unusually well-developed nervous system with a complex brain.

# Subgrade Enterocoela

This includes all the animal phyla in which the mouth is formed, not from the blastopore (the primitive opening in the archenteron), but from a second opening which arises later in development. This subgrade includes only five phyla, of which two, Echinodermata and Chordata, are major groups.

## Phylum Chaetognatha

The arrow worms are marine animals, each having an elongated body divided into head, trunk, and tail. They have well-developed lateral and tail fins and are excellent swimmers. They lack circulatory, respiratory, and excretory systems. Eggs are laid in the sea, and their offspring develop directly into adults. *Sagitta* is a well-traveled planktonic animal that lives in the upper hundred feet of the sea. It occurs in great numbers off the coast of California.

## Phylum Echinodermata

This is a strictly marine phylum embracing animals with limey endoskeletons and spines; it includes starfish, sea urchins, sea cucumber, and sea lilies. An unusual characteristic of these creatures is their water vascular system that includes special organs, the tube feet, which function analogously to the tentacles of other marine invertebrates. Echinoderms exhibit bilateral symmetry as larvae and radial symmetry as adults—a characteristic unique to this group. There are five classes in the phylum.

### Class Crinoidea

This class includes the crinoids and sea lilies. Most members of this group are known only as fossils, and most of the remaining 80 species live deep in the sea. But some types, such as *Antedon*, inhabit shallow water. *Antedon* is known as the feather star. This crinoid begins life attached to a stalk, then breaks away from its mooring when mature and swims away by flapping five plumelike arms. It has a stomach, esophagus, and intestine for digestion, and a water vascular ring around its mouth. *Ptilocrinus* is a crinoid that inhabits deeper water, also leaving its stalk behind as an adult. *Metacrinus* is one of the few representatives of this class that remains permanently attached by its stalk.

### Class Asteroida

Members of this group are star-shaped or pentagonal organisms. They are equipped with a well-defined mouth leading to an esophagus, which opens into a saclike stomach. Starfish are quite aggressive in their feeding habits, attacking mostly bivalve mollusks. The usually spiny skin covers a skeleton of calcareous plates. One of the most interesting things about starfish is that they evert their stomachs into their bivalve prey, enveloping the soft parts and digesting them outside of the body. From a central disk radiate a number of arms, anywhere from five to 20. The skeleton is composed of interlocking plates, or ossicles. Protruding from these ossicles are protective spines. *Astropecten* is an Atlantic representative of the class found on the sandy sea bottom. *Brisinga* is a deep-sea sea star with long slender arms.

## Class Ophiuroidea

Members of this class are known as "stars"—brittle, serpent, and basket. They exhibit an exceptional kind of movement utilizing their long flexible arms, which are mounted on the body's central disk in such a way that they may move freely. Ophiuroids have only very reduced and nonfunctional tube feet, but the mouth has five movable plates that act like teeth to chew food. The arms also function to push small animals into the mouth. They are protected from harm by armored skeletal plates. Brittle stars have an incomplete digestive tract, for there is no intestine or anus. Serpent stars are named for the snakelike movement of their arms, which tend to occur in some multiple of five. They are also known as brittle stars, since their arms break off when seized by man or some other enemy. They can regenerate new arms very quickly. *Ophiurus* and *Ophiothrix* are common members of this class.

## Class Echinoidea

Representatives of this group resemble starfish whose arms have been severed. Their spines, unlike those of the starfish and brittle stars, are independently movable. The endoskeleton is generally fused and nonflexible, but may have porelike openings. Digestive mechanisms are highly developed, since the echinoids feed mostly on vegetation and require a longer period of digestion to deal with the cellulose. The mouth contains three jaws and leads into the esophagus.

Sea urchins and sand dollars are the most familiar members of this class, looking very much like large living burrs flaunting sharp spines. (These spines operate as aids to locomotion, as well as protection.) The early development of the sea urchin is used as a representative example in embryology. Similar early stages occur in many other animals.

Sand dollars are flattened members of this class with short spines. They feed by swallowing sand to filter out the organic materials.

| Class Holothuroidea | These are the armless echinoderms called sea cucumbers. They exhibit a secondary bilateral symmetry and contain minute ossicles embedded within the body wall. A well known form is the sea cucumber, a thick-bodied variety with a leathery skin. They are plankton feeders, with a fascinating repertoire of defense mechanisms. Some can throw out their entire digestive tract when attacked or irritated; they can soon regenerate new organs. Other types can toss off a system of slimy threads which entangle the predator while the cucumber makes its escape. Sea cucumbers are gathered and dried in great quantity to make "trepang"—a substance used for soup in Asian countries. |
|---|---|

# Phylum Pogonophora

Known as the beard worms, members of this phylum are marine animals with anterior tentacles. The dorsal nervous system is embedded in the epidermis. Lacking mouth, digestive tract, or anus, these animals are filter feeders, trapping food with long hollow tentacles on which it is presumably digested externally. There is a closed circulatory system. They live in chitinous tubes in the mud of the ocean floor. *Polybrachia* and *Oligobrachia* are common representatives of this phylum.

# Phylum Hemichordata

This phylum includes worms with dorsal and ventral nerve cords. They are soft-bodied and segmented. All hemichordates are marine animals. They are divided into two classes, both characterized by a proboscis and paired gill slits.

| Class Enteropneusta | This class consists of the acorn worms, which are free-living animals with several gill slits and a straight gut. They are burrowers living in tubular structures which are secreted by the proboscis, a lobe which vaguely resembles an acorn. *Balanoglossus* is the best known acorn worm, and *Saccoglossus* is a related genus. |
|---|---|
| Class Pterobranchia | This group includes sessile hemichordates with two distinguishing features: gills are much reduced or nonexistent, and the gut is a U-shaped organ. *Cephalodiscus* and *Rhabpopuleura* are common representatives of this class. |

*Cephalodiscus* has a single pair of gill clefts on the front of its trunk  Both produce reproductive buds from an opening on the belly; buds of *Cephalodiscus* tend to break off and swim freely away, while those of *Rhabdopleurus* remain with the parent to form a colony.

# Phylum Chordata

The chordates are a varied group of animals, able to inhabit virtually any ecosystem and support themselves by a variety of self-maintenance techniques. Underlying their adaptive diversity is a basic uniformity of design. Diagnostic features are the tubular dorsal nerve cord supported by the notochord, and pharyngeal gill slits, which may be lost during development in higher forms.

# Subphylum Urochordata

These are the tunicates — saclike creatures named for their outer covering, a kind of envelope made of cellulose and tunicin. The tunicate obtains its food by ciliary action, sweeping it into a gigantic pharynx which fills most of its body. A heart, a specialization of the pericardium, is connected to a system of blood spaces. There is a reduced central nervous system of round ganglia, with receptor cells and nerve fibers. Its basic purpose seems to be to function as a reflex apparatus, producing movements that protect the organism from disturbing stimuli (intense light, for example).

Most tunicates are hermaphroditic; fertilization occurs externally in solitary tunicates and internally in the colonial forms, producing a fishlike tadpole. Tunicates also possess remarkable regenerative abilities, and can often reproduce simply by budding.

There are roughly 2000 sessile tunicate species, and about 100 of them are adapted to a oceanic life style. The sessile forms inhabit the bottom of the sea, either as solitary individuals, such as the *Ciona*, or as colonial aggregations. Such genera as *Clavelina* and *Amaroucium* require high salinity in the water they inhabit; one genus, *Mogula*, is able to live in brackish water.

Of the oceanic species, certain individuals are distinguished by their great mobility. Some that inhabit warm water possess circular bands of muscle which contract to force water through their body; this type of jet propulsion can whisk them along at high speed. *Doliolum* and *Salpa* are such genera. *Doliolum* and *Salpa* are individual forms, but some tunicates are colonial. *Pyrosoma* is one genus that forms an elongated cylinder-shaped colony. *Pyrosoma* also has the fascinating ability to light itself up; when large masses of the colonies become luminescent, it is possible to read by their light. This bioluminescence may have a protective function. Other tunicates live only on plankton. *Appendicularia* and *Oikopleura* build houselike structures around themselves with a skin secretion known as oikoplastic epithelium.

## Subphylum Cephalochordata

These are the amphioxus, or lancelets. Amphioxus nourishes itself by ciliary feeding. Although it can swim freely, it lives at the sandy bottom of shallow water and spends most of its time with its body buried in the sand. Further adaptations to this way of life include an impressively well-developed pharynx and an asymmetrical form which is most notable in the six species of *Aymmetron*.

Most species of amphioxus fall into the genus *Branchiostoma*. *Branchiostoma lancelatum* is a somewhat flattened, elongated organism that tapers to a point at both ends. About two inches long and translucent, it looks at first glance rather like a fish. It is not adapted to fast movement, and seldom swims freely. It obtains its food from a current of water created by the beating cilia, collecting the particles in the large pharynx. The circulatory system can be considered as a prototype of all chordate circulation. Unlike the tunicates, cephalochordates exhibit metamerism, or repetition of parts. In *Branchiostoma* this is particularly noticeable in the arrangement of the muscles.

## Subphylum Vertebrata

Members of this subphylum are characterized by a cranium and a backbone composed of vertebrae. The brain is connected to sensory receptors, multichambered hearts, and complex motor organization.

**Class Agnatha**

This group includes the earliest known vertebrates. Surviving forms differ now from other vertebrates in their lack of jaws. The few living species include the hagfish and the lamprey.

Lampreys are eel-like fish that are scaleless and lack paired appendages. They inhabit temperate waters virtually anywhere. The lamprey begins life as a fresh-water larva buried in the mud; at this stage it is strikingly similar to the amphioxus. The larva metamorphoses into an adult with an elongated body and a sucking mouth. The adult attaches itself to a fish, which it then tears into with its rasplike tongue. The digestive system includes an esophagus, intestine (no stomach), liver, gall bladder, and bile duct (but no pancreas). The lamprey moves like an eel and reproduces sexually. Representative lampreys include *Petromyzon* (all the larger species of sea-dwellers) and *Entosphenus* (those which inhabit western North America).

Like the lampreys, the hagfish (*Myxine*) have fully adapted sucking mouths. They all live in the sea, but unlike other vertebrates, their blood is isotonic to sea water. Hagfish are as physically unattractive as their name would suggest, and are considered an unusual breed of scavenger. They only attack prey which are helpless (fish which are dying or already dead). Hagfishes often bore right through the skin and remain inside the body of the host while devouring it. What remains of the fish is nothing more than an empty sack of skin and bones.

---

## Class Chondrichthyes

These are the cartilaginous fish—a small but successful group of marine predators. This group is characterized by the absence of bone tissue. Their skeleton is cartilaginous, in some cases hardened by a deposit of calcium salts. This lack of bone is not a sign of primitiveness, for the ancestors of this group had bony skeletons. The group includes sharks, rays, skates, and sawfish.

Of the different varieties of shark, only one group poses a threat to man; individuals in this group, such as *Carcharodon*, may grow to be about 20 feet long. The rest are inclined to feed upon cephalopods, crustaceans, bivalves mollusks, or bony fish. In one species, the basking shark, the predaceous life style has been replaced by filter feeding on plankton through combs, known as "rakers," on the gills. Some sharks are distinguished from others by a spine located in front of each dorsal fin. One of these is the dogfish, *Squalus*. Another is the *Alopias*, or thresher shark. Individual thresher sharks cooperate for feeding purposes. As a group, they thrash at their prey with whiplike tails and drive them into narrow shoals where they can be seized easily. A unique characteristic of all sharks is their skin, which is studded with tiny scales or dentricles, sharp pointed structures analogous to teeth.

Skates and rays have become singularly well-adapted for life at the bottom of the oceans. They move by means of undulating waves of their fins, and have blunt teeth for feeding on invertebrates. *Raja* is an Indian and Atlantic genus that would tickle the fancy of racing car stylists. Its flat body is flared at each side where the pectoral fins are attached, and the tail is an elongated rod equipped with a tailfin configuration.

One last group of Chondrichthyes are the rat-fishes, or *Chimaeras*. Their mouths, unlike those of other chondrichthyes, are small and surrounded by lips, and they have large plates of teeth attached to jaws. Their digestive systems are simple, and they are small-particle feeders.

## Class Osteichthyes

The bony fishes are a tremendously successful class, existing in both salt and fresh water and accounting for an enormous number of individuals. They are characterized by an operculum over the gills, bony dermal scales, and often, a swim bladder. Their intricate sensory receptors keep them attuned to events in the environment. These fish represent a sizable portion of man's food supply, with a good 63 million metric tons of them being taken from the waters of the world by fishermen every year.

## Subclass Sarcopterygii

consists of bony fish with paired fleshy-lobed fins, and nostrils directly connected to the mouth cavity. Only four surviving genera are known in this subclass.

### Order Crossopterygii

contains only the coelacanth. It is a deep-bodied fish with a three-lobed tail. It is large, dark blue in color, and is covered with prominent scales. Once thought to be extinct, an occasional coelacanth (of the genus *Latimeria*) has been brought up from great depth in the ocean. These fish are sometimes called "living fossils" since they represent a group that first appeared in the Devonian period.

### Order Dipnoi

consists of the lungfishes. The three surviving genera in this order are all fresh-water dwellers: *Protopterus* (living in large lakes of Africa), *Neoceratodus* (inhabiting the rivers of Australia), and *Lepidosiren* (from the rivers of South America). They are closer to amphibians than other fish in two respects; they have fins adapted to walking on the bottom, and they all have a specialization of the swim bladder that serves as a lung during periods when the water becomes stagnant.

## Subclass Actinopterygii

includes the "higher" bony fishes. These ray-finned fish are divided into three infraclasses on the basis of scale type and composition of skeleton. The actinopterygii include the dominant modern fish, represented by the species we usually see and eat.

| | |
|---|---|
| **Infraclass Chondrostei** | contains the bichir and sturgeon. *Polypterus* (bichir) is basically a Paleozoic form that still survives. It retains an air-bladder very much like a lung (a trait it shares with the Order Dipnoi), a body covered by rhomboidal scales, and a spiral valve inside its intestine which tends to suggest that the bichir resembles all early actinopterygii. The *Acipencer* (sturgeon) is also an ancient genus, but with more cartilage and fewer bones than its ancestors. Sturgeon can grow to be quite large, and are in great demand for its eggs, which we consume as caviar. |
| **Infraclass Holostei** | includes the bowfin and garpike—the only holosteans which survived beyond the Cretaceous period. *Lepisosteus* (the garpike) is a primitive appearing, large-scaled fish. Its air bladder is used as a respiratory organ when the fish surfaces. *Amia* (the bowfin) has thin scales, a symmetrical tailfin, and other features resembling those of the teleosts. |
| **Infraclass Teleostei** | includes the largest and most successful group of bony fish. Characteristics include a shortened tail, cycloid scales and a skeleton formed entirely of bone. Two major structural patterns are found among teleosts. One pattern includes rather soft fins and smooth scales; examples are the salmon (*Salmo*), trout, and herring. The perciform pattern features very spiny fins and rough-textured scales. Examples are the perch, bass, and tuna. |
| **Class Amphibia** | At some time during the Devonian period a group of primitive coelocanths made the transition to land and gave rise to the amphibians. The move was made possible by a number of adaptations, particularly the evolution of pentadactyl limbs from fins for movement on land. Their skin became moist and slimy, and modified for respiration; scales eventually disappeared. As the first tetrapods, the amphibian vertebrates became adapted to four-legged locomotion.<br><br>Although this class is successful in moist or wet areas, its members have never been able to inhabit the vast range of ecosystems occupied by some other vertebrates. The problem is that amphibians cannot breed and repro- |

duce without being close to fresh water. Many species which have ventured into drier areas depend on ponds or temporary rain pools for breeding.

Contemporary amphibians are divided into three subclasses. Urodela, which includes the newts and salamanders, resemble fish in some ways. Anura, the frogs and toads, have developed strong leg muscles and move about by hopping or jumping. Apoda are the caecilians; they are tropical forms which are blind, legless, and move only by burrowing into the ground.

## Subclass Urodela

includes species in which the adults have retained their tails. They are fairly close to primitive amphibians in other respects. Adult and larval forms are usually indistinguishable from one another, and adults typically maintain features associated with aquatic life. The salamanders run a gamut of life styles from the almost totally terrestrial to the fully aquatic. *Necturus* is an aquatic form that walks along the bottom of fresh-water streams and rivers. Closely related to the salamanders are the newts. *Triturus*, a newt common in Great Britain, typifies the life cycle of this group. The gilled larvae are aquatic; they then lose their gills and develop into a second immature stage (commonly known as a eft) which is terrestrial; several years later, the adult newt returns to water permanently. The most common North American newt belong to the genus *Notophthalmus*.

## Subclass Anura

consists of frogs and toads. They are in many ways the most versatile amphibians. They live in a variety of aquatic and terrestrial habitats. Among their adaptations to the terrestrial environment are the ability to carry on cutaneous respiration and the evolution of a protective eyelid. Frogs and toads are preyed upon by a number of enemies, particularly birds and snakes, but they can often elude these predators thanks to their powerful jumping legs. Young anurans begin life as aquatic tadpoles before making the transition to land and adulthood. *Bufo* is a successful land-dwelling toad found all over the world, but even this species returns to water every year during the breeding season. *Rana* is a very successful genus of frog, completely cosmopolitan in distribution. *Rana* includes the so-called "true" frogs (the ones which we recognize as frogs in North America). Some amphibians have developed adhesive pads on their toes, enabling them to climb trees. The tree-dwellers include representatives of the genus *Hyla* (see Color Portfolio 1).

## Subclass Apoda

refers to the blind, limbless creatures resembling earthworms which burrow into tropical soil. Most are terrestrial. *Ichthyophis* survives on land due to its development of a shorter tail and external sex organs. *Typhlonectes* is a water dweller with sensory tentacles in place of eyes.

## Class Reptilia

Reptiles are characterized by dry scaly skin, often well-developed teeth, a four-chambered heart, and in many species, claws. Reptiles are generally adapted to warm climates, since they have no internal mechanisms to regulate body temperature. Instead they must adjust to changes in their environment. When the weather is too cold, they remain inactive or find warm spots in the sun. When it becomes too hot, they move to cooler places under rocks or underground. Each species has an individual temperature range that cannot be exceeded, especially at the upper limit, if they are to survive, and many reptiles must remain inactive during hot spells.

Unlike the skin of amphibians, reptilian skin is dry and contains no glands. The thick scales help prevent cutaneous water loss. Some reptiles utilize skin coloration for camouflage. Since the females lay thick-shelled eggs that can remain viable in a terrestrial environment, there is no need to return to the water each year during the breeding season.

Many reptilian forms, such as the dinosaurs, are extinct. Surviving groups fall into four orders: Rhynchocephalia, Crocodilia, Chelonia, and Squamata.

### Order Rhynchocephalia

decimated by the gradual extinction of most species, has only one surviving member —the lizardlike tuatara, *Sphenodon*. And even this, a native of New Zealand, is an endangered species. *Sphenodon* is about two feet long, has a pineal eye, and eats both insects and other vertebrates. Fertilization of eggs always takes place internally. Reptiles were the first group to evolve a penis. The tuatara is the sole survivor of a group that can be traced back more than 180 million years into the past. Scientists are now attempting to save the few that are still in existence.

### Order Crocodilia

is made up of crocodiles and alligators. They are descendants of a reptile group that were ancestral to such diverse forms as the dinosaurs and the birds. Modern crocodiles (*Crocodylus*) closely resemble their ancestors of the Triassic period, exhibiting the same adaptive features. These include the ability to keep their mouths open under water, due to a valve which closes off the respiratory passage; webbed hind feet; a skin covering that can close the outer ear; and a powerful tail. They walk on all fours, but their front legs are somewhat shorter than the hind legs. They lay hard-shelled eggs in protected sandy areas or nests. The alligator is a smaller crocodilian that inhabits this continent exclusively.

## Order Chelonia

exhibits one of the most intriguing protective mechanisms among living reptiles, that is, the hard shell worn by its members. These are the turtles and tortoises, which flaunt coverings made up of several bony layers topped by outer plates of horny material. Once withdrawn into its shell, a chelonian is virtually indestructible. There are some 200 species of chelonia that include both terrestrial and marine forms. Members of this order possess fairly well-developed brains. Through evolution they have lost their teeth. Both carnivorous and herbivorous species rely on ridged jaws and horny beaks. Modern chelonians include *Chelydra*, or the snapping turtle, and the brightly-colored *Terrapene*, the box turtle.

## Order Squamata

includes the highly successful lizards and snakes. Both have a double copulatory organ, the hemipenes, which is eversible. Lizards generally have four limbs, and there are usually five digits bearing claws per limb. Since some lizards may have only two forelimbs, such as *Bipes*, and others (e.g., the glass "snake") may have none; they may be mistaken for snakes. Eyelids of lizards are usually movable, allowing them to close their eyes. Lizards feed primarily on insects; some feed on other invertebrates, and a few are herbivorous. Snakes lack limbs altogether, but some, such as the boas and pythons still retain vestiges of pelvic girdles and hindlimbs. Snakes do not have movable eyelids, and so their eyes are fixed in an unbroken stare. However, the eye is protected by a specialized transparent scale. Unlike the lizards, whose lower jaws are fused anteriorly, snakes have their lower jaws joined by a ligament. The hind part of the lower jaw is hinged in such a way that it can be swivelled to permit enormous distension of the mouth. This is a necessity since all snakes swallow their food whole. Snakes simply do not chew, but depend on their powerful digestive secretions to break down their prey. As there is very little room to spare in the slender body of a snake, it is not surprising that the right lung is either reduced in size or missing altogether. Some snakes have fangs associated with poison glands and are capable of delivering a fast-acting neuro- or hemotoxin. Rattler poison is slower since it attacks the blood;

but cobra poison is very fast-acting since it attacks the nervous system. Snakes are extremely valuable predators of temperate, semitropical, and tropical ecosystems.

## Class Aves

Stemming from the same ancestry that produced the dinosaurs, the birds have developed into an extremely successful highly active group. Their ability to fly, due to their covering of feathers, allows them to avoid enemies and change their habitats when unfavorable conditions arise. Their large brains are associated with high intelligence. They are homeothermic and maintain a high body temperature. This permits them to live successfully in cold environments. Color patterns are often protective in the species that are likely to be preyed upon. Flight speed varies up to a record of 100 miles per hour set by the swifts. The birds are divided into three superorders.

### Superorder Palaeognathae

includes the large birds which do not fly. Once regarded as more primitive than birds capable of flight, they are now believed to have the same ancestor as their flying relatives. In fact, some flying birds go through a palaeognathaeanlike stage of development. The ostrich *Struthio* is the largest bird in existence, now found only in Africa and Southern Asia. The New Zealand kiwi, or *Apteryx*, comes out only at night to feed on insects and worms native to New Zealand. Other members of this group are the cassowary, emu, and rhea.

### Superorder Impennae

includes the birds which were once capable of flight, but lost the ability when they became adapted to an aquatic existence. The most common member is the penguin (*Spheniscus*), in which the wings have been modified as flippers and webbed feet. Penguins live in the Southern Hemisphere only, and must come ashore to breed. The emperor penguin is a large form which breeds during winter on ice floes of the antarctic shore.

### Superorder Neognathae

embraces all the modern birds familiar to us, and many exotic ones as well. Collectively they represent an extremely diverse array of over 25,000 species and subspecies. Of the nine different orders, the Passeriformes account for half of all known bird life.

## Order
## Procellariiformes

contains the birds which have become modified for oceanic life and soaring flight. They lay only one large egg. Members of this order include the petrels (*Fulmarus*), shearwater (*Puffinus*), and the albatross (*Diomedea*) so beloved of sailors. Birds of this order are characterized by the amazing distance of their migrations.

## Order
## Anseriformes

is represented by ducks (*Anas*), geese, and swans (*Cygnus*). These birds are also specialized for life in the water. They survive on a variety of foods including plants, mollusks, and fish, which they take in with the help of flattened bills. Ducks and swans build nests on land and lay many eggs.

## Order
## Falconiformes

includes birds that are efficient predators, feeding on smaller birds, fish, and living, dying, or dead mammals. They hunt by day, and are in the habit of laying their eggs in inaccessible locations — cliffs, treetops, and protected plots of ground. The American buzzard (*Buteo*), hawk (*Neophron*), and eagle (*Haliaeetus*) are all members of this group.

## Order
## Galliformes

includes many birds that have the dubious distinction of being sought out by man for food and sport. These are the game birds — land-dwelling, grain-eating, and often, slow to take flight (which helps to account for their popularity with hunters). Sadly, many species have been greatly decimated for this reason. Their distribution is world-wide, however, and they remain a moderately successful group. The turkey (*Meleagris*), partridge (*Perdix*), and grouse are North American representatives of this order.

## Order
## Charadriiformes

encompasses the wading birds, which live on the ground in marshes or other watery places. Normally in great abundance around any seashore, they are predators and scavengers on beaches and the shoreline. The sandpiper (*Erolia*) and auks and puffins (family Alcidae) belong to this order, but the most familiar member is the seagull (*Larus*).

## Order
## Columbiformes

includes members that are familiar to any city-dweller. Here we find the pigeon (*Columba*) which has been successful in overrunning many metropolitan areas. But pigeons were originally cliff-dwellers with impressive flight capabilities; they feed on fruit or grain. The dodo (*Raphus*) was a large whimsical-looking pigeon that became flightless and was exterminated by man during the 1600s.

## Order
## Strigiformes

contains only the owls—predators that hunt at night and swallow their small prey whole. Their eyes are large and they enhance their visual capacity by rotating their heads. They generally detect their prey through highly sensitive ears. Their flight is stealthy, with loosely arranged feathers making no noise to give them away. The barn owl (*Tyto*) and eared owl (*Asio*) are representatives.

## Order
## Apodiformes
## (Micropodiformes)

includes the fastest bird in the world—the swift, which is capable of flight at speeds up to 100 miles per hour. The swifts' wings are long and they have large mouths enabling them to snap up and catch their insect prey. They lay their few eggs in nests. The active, colorful hummingbirds are also members of this group; they are adapted to hovering flight.

## Order
## Passeriformes

is the largest and most recognizable avian order, known collectively as the perching birds. They are usually small, and have four toes on each foot, three in front and one in back, adapted to a perch. Many emit species-specific songs associated with many aspects of behavior. The eggs of passeriform birds exhibit a spectrum of colors and markings. The colonial rooks and jackdaws (family Corvidae) are the largest and probably the most intelligent birds. Finches (family Fringillidae) helped Darwin develop his theory of evolution. Sparrows (*Passer*), warblers (*Sylvia*), and blackbirds (*Turdus*) are familiar on this continent. Swallows (*Hirundo*) are fast-flying insectivores which catch their prey in mid-air. And tree-creepers (*Certhia*) have developed long bills useful in digging up insects buried in the bark.

## Class Mammalia

Mammals evolved from a group of mammal-like reptiles, the Synapsida, which dominated the land during the Permian period (before the "great age of reptiles" in the Jurassic and Cretaceous periods). The early mammals survived until the end of the Cretaceous as small, relatively inconspicuous animals. Since that time, they have undergone explosive adaptive radiation, entering very diverse types of habitat, from the tundra to the tropics, the desert to the ocean.

All mammals possess milk glands, and only mammals possess them. The jaw is comprised of a single fused dentary bone and the brain, particularly the forebrain, is relatively large; there are three small auditory bones in the middle ear. A muscular diaphragm separates the thorax from the abdominal cavity. Like birds, mammals are homeotherms, and possess a four-chambered heart which completely separates the oxygenated and deoxygenated blood. To assist thermoregulation, the mammalian body is covered in varying degrees by hair. Except in the monotremes, mammals are viviparous, and most develop a placenta of embryonic and maternal tissues in the uterus.

The many physiological and behavioral advantages conferred by thermoregulation and a well-developed cerebrum (as well as the adaptations which allow protection and nourishment of both the embryo and the young) have enabled mammals to colonize successfully most of the earth, including some of its most inhospitable regions.

The Class Mammalia is divided into two subclasses—Prototheria and Theria.

## Subclass Prototheria

is the province of only one remaining order, the Monotremata. The individual survivors are the duck-billed platypus (*Ornithorhynchus*) and the spiny anteater (*Tachyglossus*), which are the only mammals that still lay eggs. Both are Australian forms, and are probably an offshoot of the mammals of the Mesozoic era. They have a single dentary bone in the lower jaw, but the young platypus retains flattened teeth unlike any other animal in this order. The monotremes' vertebrae resemble those of reptiles more than those of other mammals; their limbs and girdles are also reptilian in character. Their skeletons strongly suggest features exhibited by early reptiles, and this might indicate that the monotremes' mammalian characteristics were produced by parallel evolution. The prototherians' most singular trait is egg-laying; their eggs are large, flexible, and yolky with off-white shells.

Their primitive traits notwithstanding, the monotremes have developed a number of highly adaptive features. Hair is usually present as a single coat, but the anteater retains an interesting combination of spines on its back and hair on its belly. The platypus is well-suited to an aquatic existence, with its dorsal nostrils and webbed feet. The spiny anteater has a long snout modified for seizing insects, and sharply clawed feet for digging up ants' nests.

## Subclass Theria

includes the mammals more familiar to us. These are the placental and marsupial animals, with the eutherians (placentals) well established as the dominant group.

### Order Marsupialia

includes the marsupials, the ranks of which have dwindled considerably in recent years, partly because of competition with placentals. They are similar to other mammals except for their adaptations for offspring care. Some features suggest an early divergence from the eutherian stock and a convergent evolution that resulted in many forms similar to placental mammals. The embryo is born extremely young, in an undeveloped state, and completes its development in a ventral fur-lined pocket, the marsupial pouch. The embryo itself is equipped with well-developed forelimbs and nervous system enabling it to climb into the pouch. The opossum (*Didelphis*) differs only slightly from the early marsupials of the Cretaceous period. Modern opossums live in trees aided by a prehensile tail and eat insects at night. Other marsupials are carnivorous, including the Tasmanian devil (*Sarcophilus*) which may be extinct. Still others are vegetarians; the koala bear, or *Phascolarctos*, lives exclusively on eucalyptus leaves. Wallabies and kangaroos (family Macropodidae) are large marsupials common to Australia.

### Order Insectivora

consists of moles and shrews which are nocturnal animals with features reminiscent of the earliest mammals. They are small, quite primitive, and are believed to be the most interesting and most ancient of living placentals. Many of them hibernate in winter, and some may build nests for their young. Shrews (family Soricidae) may be insectivorous or omnivorous, terrestrial or aquatic, with highly specialized incisors. Moles (family Talpidae) are modified for burrowing, and have relatively weak eyes. The hedgehog (*Erinaceous*) and tree shrew (*Tupaia*) are also members of this order.

### Order Dermoptera

is represented by a single genus. The so-called flying lemur (*Galeopithecus*) which does not fly and is not a lemur, probably arose from insectivorous ancestors. It is modified for a kind of skydiving, with a parachuting membrane stretching from neck, limbs, and tail. This enables it to glide from branch to branch. This genus appears to have several features in common with bats, and it has sometimes been classified in Order Chiroptera.

## Order Chiroptera

contains the bats, the only mammals capable of true flight. Their wings are developed as folds of the skin manipulated by elongated digits on both hands. Bats can also climb and crawl, although they are incapable of walking. They hang upside down when not flying. Since bats navigate by the echoes of their own high-pitched cries, their ears are large and extremely sensitive.

Some common representatives of this order are the big-eared bat (*Plecotus*) and the whiskered bat (*Myotis*). Vampire bats are alive and well in Central and South America.

## Order Primates

is the group to which we belong, so we are naturally most interested in this order. Basically arboreal, the primate group has developed many traits associated with organisms which were once ground-dwellers but became adapted to life in the trees. Since tree-dwellers cannot hunt by smell, eyes and ears become important mechanisms for food-getting and navigation.

The primates' eyes are forward-facing, which enabled them to evolve stereoscopic and binocular vision to help navigate among trees. Along with changes in the sensory organs, the brain became larger and more complex. Skeletal and muscular systems became specialized for in jumping and swinging maneuvers. The opposable thumb developed as a useful grasping digit. Omnivorous eating habits were accompanied by modifications of the teeth, and although fewer offspring are produced, they receive longer and better parental care. In their natural environment, many primates exhibit intricate social patterns which permit stable, well-protected family groups.

The primate order is divided into two suborders: Prosimii, which includes lemurs and tarsiers, and Anthropoidea, which includes monkeys, apes, and man. Lemurs are considered to be the most "primitive" members of this group, with characters reminiscent of the early primates of 50 million years ago. They are nocturnal tree-dwellers adapted for omnivorous, insectivorous, or herbivorous feeding. Equipped with large bushy tails, they seem almost squirrel-like in appearance. The tarsiers, represented by the single genus *Tarsius*, occupy a position between the lemurs and the anthropoids. *Tarsius* is a tree-dweller that comes out at night to feed on insects. Small, with enormous round eyes, the tarsier is well adapted to a nocturnal existence.

In the anthropoids, the cerebral hemispheres reach the greatest development, overhanging the cerebellum and medulla. The superfamily Platyrrhina includes all

New World monkeys, which have probably been isolated from other anthropoids since the Eocene epoch. Typically they have widely separated nostrils and a prehensile tail. Examples are the capuchins (*Cebus*), spider monkey (*Ateles*), and howlers (*Alovatta*). The superfamily Catarrhina includes the Old World monkeys. They are narrow-nosed primates, often with very short tails (for example, rhesus monkeys and mandrills). As with hominoids, menstruation seems to be universal in this group. Most catarrhines are arboreal but some, such as the baboons (*Papio*), are ground dwellers and walk on all fours. The superfamily Hominoidea includes the great apes and man. Examples are the chimpanzee (*Pau*), orangutan (*Pongo*), gibbons (*Hylobates*), and of course, *Homo* himself.

## Order
## Edentata

is represented by members that are insectivorous, as were the original primates. Edentata are found only in the New World. Included here are the anteaters (*Myrmecophaga*), the three-toed sloth (*Bradypus*), and the armadillo (*Dasypus*). They all have a long snout and tongue for removing insects from holes and trees, but they have no teeth. The three-toed sloth, forsaking its insectivorous heritage, has turned to a herbivorous diet.

## Order
## Pholidota

includes the scaly anteaters (*Manis*) of the Old World. They share the traits exhibited by the edentata—long tongue and snout, no teeth, and a diet that consists solely of insects. They may live on the ground or in trees, and are nocturnal animals.

## Order
## Lagomorpha

consists of small, highly mobile creatures with teeth specialized for gnawing on vegetation. They are the rabbits (*Oryctolagus*), hares (*Lepus*), and pikas (*Ochotona*). Rabbits and hares are famous for another specialization—a development of the hindlegs enabling them to hop away from predators. Rabbits are also proficient at burrowing, and their reproductive prowess is legendary. Lagomorphs are closely related to rodents.

## Order
## Rodentia

includes the rodents of the world, a tremendously successful group. Most rodents are vegetarians, except for such species as the rat (*Rattus*) which will gnaw on flesh if it becomes available. A distinctive feature of the rodentia is the dentition, which is a highly developed set including chisel-like incisors with a well-honed cutting edge. Two sets of these incisors, one on each jaw, work against each other. As well as using them for chewing food, some species are able to gnaw through obstructions (even wood and metal) in their paths. Squirrels (*Sciurus*) are rodents, as are beavers (*Castor*), gophers (*Geomis*), and kangaroo rats (*Dipodomys*),

## Order
## Cetacea

includes the mammals with a fluke or flipperlike tail as well as finlike appendages, which have adapted to an aquatic life. The blue whale (*Balanoptera*) is the largest animal in existence. This great size is made possible by the buoyancy of the sea water, which helps support the enormous body weight. The naked smooth skin and highly streamlined shape are adaptations to swimming. Propulsion is achieved by an up-down motion of the tail. Physiological adaptations permit the whale to remain under water for over 30 minutes at a time, and the residue of vapor that is tossed off on emerging from the water is the familiar "blow" spout. The larger whales live on plankton and small crustaceans (krill), but the killer whale (*Orcinus*) will eat virtually anything that crosses its path, including birds, seals, and other whales. Dolphins (*Delphinus*) are also cetaceans, and so are porpoises (*Phocaena*). Their highly evolved behavior is of great interest to scientists, and includes communication by sound, social organization, and an impressive capacity to learn.

## Order
## Carnivora

consists of mammals that are chiefly meat-eaters. The order includes many highly efficient hunters. Brains are well developed; dentition is characteristic, with large canines and shearing molars. Well-known carnivorans include the tiger, wolf, hyena, and fox. The skunk is also in this group, and has evolved a unique and well-known mechanism of defense, its obnoxious spray and warning coloration. Bears (*Ursus*) are carnivorans that have an omnivorous diet. The badger, raccoon, and the sealion are all carnivorans.

## Order
## Tubulidentata

is the order of the aardvark (*Orycteropus*), a virtually unclassifiable mammal of dubious origins. Roughly the size of a young pig, it has a long snout and feeds exclusively on ants and termites. Its most unusual feature is its teeth—rootless and tube-shaped. The aardvark inhabits the area from South Africa to Sudan.

## Order
## Proboscidea

includes the elephants, the largest living land mammals. There are two genera; the African (*Loxodonta*) and the Asian (*Elephas*), which are distinguishable from one another on the basis of tusk and ear size. Aside from their great size (*Loxodonta* may be almost 12 feet high), their food gathering mechanisms are noteworthy. Food is collected by the trunk and shoved into the huge mouth which is equipped with specialized grinding teeth. Though distantly related to the Orders Sirenia and Hyracoidea, the proboscids occupy a relative isolated taxonomic position.

## Order
## Hyracoidea

contains small herbivores known as coneys (*Procavia*). Similar to the rabbit in life style, coneys live in Africa and the Middle East. They cannot burrow, but are very agile and active. Their flattened, hooflike feet are much like those of large hoofed mammals.

## Order
## Sirenia

includes the sea cow (*Halicore*) and the manatee (*Trichechus*); they are herbivorous mammals adapted to a marine life. Manatees live in shallow waters along seacoasts of America and Asia. The sea cow is found off the coasts of the Indian Ocean and South Pacific.

## Order
## Perissodactyla

is known collectively as the odd-toed ungulates—a mixed assembly of horses, zebras, tapirs, and rhinoceri. The horse provides paleontologists with an almost flawless evolutionary pedigree—from the tiny *Eohippus* of Eocene times to the contemporary *Equus*. Tapirs (*Tapirus*) are small, nocturnal herbivores which are relatively unchanged from their ancestors of some 20 million years ago. *Rhinoceros* is a bizarre giant member of this group that is fairly specialized with its thick coat of armorlike skin. Its formidable appearance matches its unpredictable and dangerous temperament.

## Order
## Artiodactyla

encompasses the even-toed ungulates, an even more diverse group, ranging in form from the pig to the giraffe. They are distinguished by a digestive system in which the stomach is divided into a number of chambers. Some species are equipped with special systems of scent glands, employed in setting territorial limits and establishing sexual and social contacts. The domestic pig (*Sus*) has a fascinating group of relatives that includes the wild boar and a modern but primitive-looking derivative known as the warthog (*Phacochoerus*). The hippopotamus is specialized for a semiaquatic life style, and can remain submerged for five minutes. Another group of artiodactyls are the ruminants, such as the camel (*Camelus*), peccary (*Pecari*), and deer (*Cervidae*). They have all developed the habit of chewing a cud. The giraffe is particularly specialized for grazing in high branches, and the Barbary sheep has become adapted to grazing on rocky, mountainous terrains. But the ruminants familiar to most people are the bovids—cattle and sheep, which are an important souce of food for mankind.

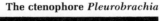

The ctenophore *Pleurobrachia*

*Nereis,* the segmented sand worm

*Phagocata,* a freshwater planarian

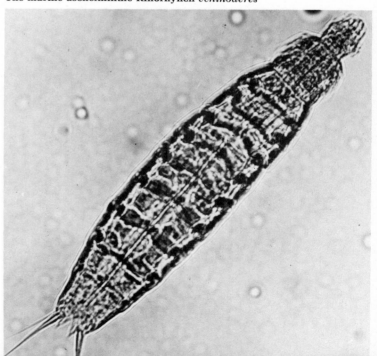

The marine aschelminthe Kinorhynch *echinoderes*

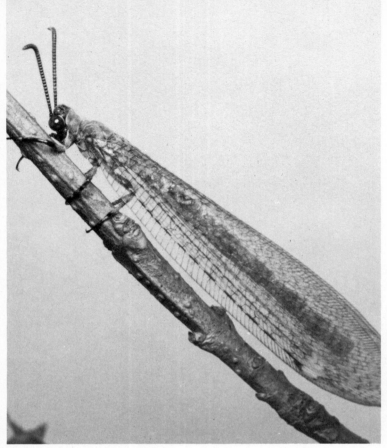

The adult ant lion *Myrmeleon*

The wood tick, an arachnid

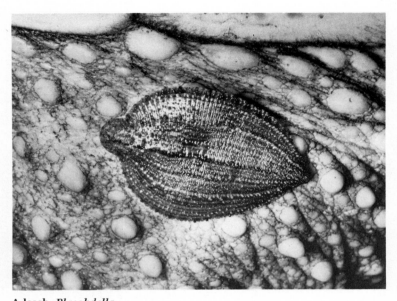

A leech, *Placobdella*

*Curculio,* the acorn weevil

Cloud forest millepede

The centipede *Lithobius*

*Alaus oculatus* or eyed elater, a click beetle

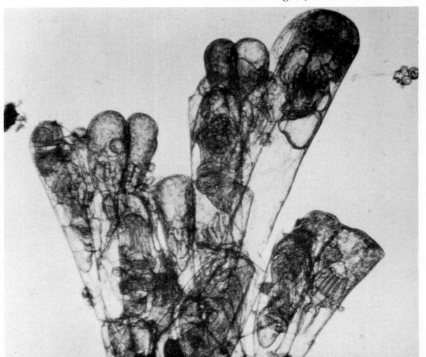

The wood-boring sawfly *Tremex columba*

The brachiopod *Terebratalia transversa*

Colonial *Bugula*, marine moss animals

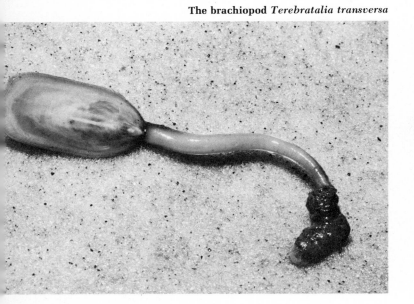

Luna moth

*Lingula antina*, a lamp shell

Chiton, a simple marine mollusk

Tusk-shelled mollusk, *Scaphella junonia*

Squid

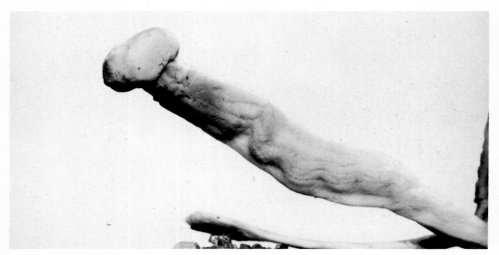

The marine acorn or tongue worm

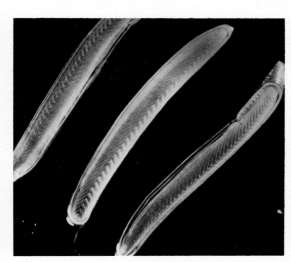

Amphioxus *Branchiostoma*, a primitive chordate

The sand dollar, *Encope michelini*

A tunicate, or sea squirt

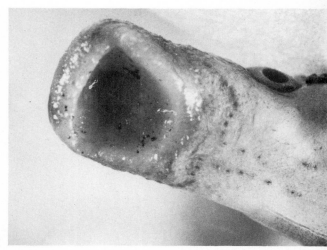

Mouth of the river lamprey *Petromyzon fluviatilis*

*Astrophyton*, the basket starfish

Yellow headed jawfish at nest

Shark

The double-wattled cassowary

Man-o-war birds

The tuatara of New Zealand

**Red-shouldered hawk**

Male yellow-shafted flicker

The Tasmanian devil, a marsupial

A male woodcock

Elephant shrews

The lesser mouse-lemur

**Blacktail jackrabbit**

**The white-handed gibbon**

Three-toed sloth

Polar bear mother with twins

786

**Bottle-nosed dolphin, a marine mammal**

A sea cow or manatee

Young aardvark, East Africa

**Giraffes**

# Appendix credits

Flagellated *Euglena gracilis* (Walter Dawn, from National Audubon Society); The diatom *Arachnoidiscus orientalis* (Carlton Ray, Photo Researchers); *Myxomycetes cribraria*, a slime mold (Walter Dawn, from National Audubon Society); *Aspergillus zoox*, a sac fungus (Russ Kinne, Photo Researchers); Shelf fungi on a tree (Alycia Smith Butler); The deadly parasite *Trypanosoma gambiense* (Walter Dawn, from National Audubon Society); *Stentor*, a ciliate (R. F. Head, from National Audubon Society); *Fucus*, or rockweed (Hugh Spencer, from National Audubon Society); The green alga *Draparnaldia* (Hugh Spencer, from National Audubon Society); *Marchantia*, a thallose liverwort (Hugh Spencer, from National Audubon Society); Sphagnum moss with sporangia (Russ Kinne, Photo Researchers); Meadow spikemoss, *Selaginella apoda* (Hugh Spencer, from National Audubon Society); *Polypodium virginianum*, the rock cap fern (John H. Gerard, from National Audubon Society); *Ginkgo* leaves (Hugh Spencer, from National Audubon Society); *Sequoia gigantea* from California (Philip Carpenter, from National Audubon Society); Passion flower blossom (Russ Kinne, Photo Researchers); Wild calla, *Calla palustris* (Laurence Pringle, from National Audubon Society); The ctenophore *Pleurobrachia* (Jack Dermid, from National Audubon Society); *Phagocata*, a freshwater planarian (Walter Dawn, from National Audubon Society); The marine aschelminthe kinorhynch *Echinoderes* (Walter Dawn, from National Audubon Society); *Nereis*, the segmented sand worm (Walter Dawn, from National Audubon Society); A leech, *Placobdella* (John H. Gerard, from National Audubon Society); The wood tick, an arachnid (G. Ronald Austing, from National Audubon Society); The centipede *Lithobius* (N. E. Beck, Jr., from National Audubon Society); Cloud forest millepede (Walter Dawn, from National Audubon Society); *Curculio*, the acorn weevil (Photograph by Alexander B. Klots); The adult ant lion *Myrmeleon* (Photograph by Alexander B. Klots); *Alaus oculatus* or eyed elater, a click beetle (Photograph by Alexander B. Klots); The wood-boring sawfly *Tremex columba* (Photograph by Alexander B. Klots); Luna moth (Photograph by Alexander B. Klots); Colonial *Bugula*, marine moss animals (Walter Dawn, from National Audubon Society); *Lingula antina*, a lamp shell (Russ Kinne, Photo Researchers); The brachiopod *Terebratalia transversa* (Russ Kinne, Photo Researchers); Chiton, a simple marine mollusk (Russ Kinne, Photo Researchers); Tusk-shelled mollusk, *Scaphella junonia* (William M. Stephens, Photo Researchers); Squid (Russ Kinne, Photo Researchers); *Astrophyton*, the basket starfish (Carleton Ray, Photo Researchers); The sand dollar, *Encope michelini* (William M. Stephens, Photo Researchers); The marine acorn or tongue worm (Russ Kinne, Photo Researchers); A tunicate, or sea squirt (Russ Kinne, Photo Researchers); amphioxus *Branchiostoma*, a primitive chordate (Walter Dawn, from National Audubon Society); Mouth of the river lamprey *Petromyzon Fluviatilis* (Sdeuard C. Bisserot, F.R.P.S.); Shark (Elgin Ciampi, from National Audubon Society); Yellow headed jawfish at nest (Russ Kinne,

Photo Researchers); The tuatara of New Zealand (Tom McHugh, Photo Researchers); The double-wattled cassowary (Russ Kinne, Photo Researchers); Man-o-war birds (Allan D. Cruickshank, from National Audubon Society); Red-shouldered hawk (G. Ronald Austing, from National Audubon Society); A male woodcock (Leonard Lee Rue III, Bruce Coleman Inc.); Male yellow-shafted flicker (Leonard Lee Rue III, Bruce Coleman Inc.); The Tasmanian devil, a marsupial (R. van Nostrand, from National Audubon Society); Elephant shrews (Russ Kinne, Photo Researchers); The lesser mouse-lemur (Russ Kinne, Photo Researchers); The white-handed gibbon (Arthur Ambler, from National Audubon Society); Three-toed sloth (Paris Match, Photo Researchers); Blacktail jackrabbit (Leonard Lee Rue III, Bruce Coleman Inc.); Bottle-nosed dolphin, a marine mammal (Russ Kinne, Photo Researchers); Polar bear mother with twins (Sven Gillsater/TIO, Bruce Coleman Inc.); Young aardvark, East Africa (Jen and Des Bartlett, Photo Researchers); A sew cow or manatee (Russ Kinne, Photo Researchers); Giraffes (Jeanne White, from National Audubon Society).

Kingdom openers for appendix: pg. 684 Omikron; pg. 691 Omikron; pg. 706 Omikron.

# Portfolio credits

## Portfolio 1

Oedogonium—*Roman Vishniac*; Desmid—*Roman Vishniac*; Colonial green algae—*Roman Vishniac*; British soldiers—*Ken Brate-Photo Researchers*; Sea anemone and barnacles—*Roman Vishniac*; Portuguese man-of-war—*Roman Vishniac*; Green sea anemone—*Tom Myers-Photo Researchers*; Slime mold—*Roman Vishniac*; Beavertail cactus—*C. G. Maxwell from National Audubon Society*; Indian wheat—*Verna R. Johnston from National Audubon Society*; Cycad—*Russ Kinne-Photo Researchers*; Cypress trees—*Aaron O. Wasserman*; Horsetails—*Charles E. Mohr from National Audubon Society*; Mascarene bottle palm—*Robert C. Hermes from National Audubon Society*; Porcupine—*Woodrow Goodpaster from National Audubon Society*; African horned chameleon—*George Porter from National Audubon Society*; Spice bush butterfly larva—*Alexander B. Klots*; Sea slug—*Aaron O. Wasserman*; Yellow spotted turtle—*Alycia Smith Butler*; Female copepod—*Roman Vishniac*; Hyla andersonii—*Aaron O. Wasserman*; Ceratophrys calcarata—*George Porter from National Audubon Society*; Hyla ebreccata—*Aaron O. Wasserman*; Agalychnis callidryas—*Aaron O. Wasserman*

## Portfolio 2

Sphinx caterpillar—*Ray Glover from National Audubon Society*; Rat-tailed maggot—*Alexander B. Klots*; Long-tailed salamander—*Alexander B. Klots*; Mangrove trees—*George Holton-Photo Researchers*; Bee on goldenrod—*Alexander B. Klots*; Lodgepole pine pollen cones—*Winton Patnode-Photo Researchers*; White birch catkin—*Jerome Wexler-Photo Researchers*; Green Orchid—*Lt. Col. Jack C. Novak-Photo Researchers*; Raccoon and wild grapes—*Karl H. Maslowski-Photo Researchers*; Bursting milkweed pod—*C.G. Maxwell from National Audubon Society*; Dandelion—*Lynwood Chace from National Audubon Society*; Burdock seeds in bobcat fur—*Lynwood Chace from National Audubon Society*; Red-bellied squirrel—*Arthur W. Ambler from National Audubon Society*; White-tailed deer—*Leonard Lee Rue III from National Audubon Society*; Agalychnis—*Aaron O. Wasserman*; Geometrid moth—*Alexander B. Klots*; Cilix glaucata—*Alexander B. Klots*; Lacewing larva eating aphids—*Alexander B. Klots*; Geometrid larva—*Alexander B. Klots*; Thorn insect—*Robert C. Hermes from National Audubon Society*; Ptarmigans—*C. Vibe*; Saw-toothed elm caterpillar—*Alexander B. Klots*; Milkweed leaf beetle (adult)—*Alexander B. Klots*; Milkweed leaf beetle (larva)—*Alexander B. Klots*; Monarch butterfly—*Alexander B. Klots*; Viceroy butterfly—*Alexander B. Klots*; Currant moth—*Alexander B. Klots*; Puss moth larva—*Alexander B. Klots*; Catocala concumbens—*Alexander B. Klots*; Leaf hopper—*Alexander B. Klots*;

Saw-whet owl — *G. Blake Johnson from National Audubon Society;* Caiman — *Russ Kinne-Photo Researchers;* Male horsefly — *Alexander B. Klots;* Female horsefly — *Alexander B. Klots*

## Portfolio 3

Chin strap penguins — *Bjorn Bolstad-Photo Researchers;* Bighorn rams — *Phil Farnes from National Audubon Society;* Grunion — *Tom McHugh-Photo Researchers;* Cockatoos — *Bucky Reeves from National Audubon Society;* Horseshoe crabs — *Alexander B. Klots;* Ruffed grouse — *Elsa & Henry Potter from National Audubon Society;* Aphid giving birth — *Alexander B. Klots;* Australian brown snakes — *Tom McHugh-Photo Researchers;* Peacock — *Farrell Grehan-Photo Researchers;* Grey kangaroos — *Harold J. Pollack from National Audubon Society;* Osprey and nest — *Alycia Smith Butler;* Male stickleback — *Dwight Kuhn from National Audubon Society;* Paper nautilus shell — *Lee E. Battaglia-Photo Researchers;* Deer mouse with 10-day-old young — *Cordell Anderson from National Audubon Society;* Nests of cacique — *Karl Weidmann from National Audubon Society;* Canada geese — *W. Munro-Photo Researchers;* Chacma baboons — *Cyril Toker from National Audubon Society;* Cheetah and cubs — *Mark N. Boulton from National Audubon Society;* Elephant family — *Alouise Boker from National Audubon Society;* Scorpion with young — *Anthony Mercieca from National Audubon Society;* Male water bug — *Alexander B. Klots;* Woodpecker finch — *Miguel Castro-Photo Researchers;* Beaver — *Leonard Lee Rue III from National Audubon Society;* Orb spider — *James Carmichael Jr.-Photo Researchers*

## Portfolio 4

Tidal pool — *Jeanne White from National Audubon Society;* Mojave desert — *Richard Weymouth Brooks-Photo Researchers;* Egyptian vulture — *Mark N. Boulton from National Audubon Society;* Gila monster — *Bucky Reeves from National Audubon Society;* Musk oxen — *Fred Bruemmer;* Tundra — *Fred Bruemmer;* Arctic ground squirrel — *Fred Bruemmer;* Snowshoe rabbit — *William Jahoda from National Audubon Society;* Pollution on the Columbia river — *John V.A.F. Neal-Photo Researchers;* Strip mining — *Arthur Tress-Photo Researchers;* Cratering in Vietnam — *Arthur H. Westing;* Bauxite surface mining — *Bjorn Bolstad-Photo Researchers;* Rice terraces — *Jules Bucher-Photo Researchers;* Cattle grazing — *Maurice and Sally Landre from National Audubon Society;* Vineyard — *Joe Munroe-Photo Researchers;* Damselfly — *Alycia Smith Butler;* Viceroy butterfly — *Alycia Smith Butler;* Woodland pond — *Alycia Smith Butler;* Green frog — *Alycia Smith Butler;* Garter snake and green frog — *D. Mohrhardt from National Audubon Society;* Broad-winged hawk — *John B. Holt Jr. from National Audubon Society;* The earth — *NASA*

# Glossary

**abscissic acid:** A plant hormone, once known as the "dormancy hormone," which acts as an antagonist to auxins, gibberellins, and cytokinin.

**acid:** A substance which, upon being dissolved, introduces an abundance of $H^+$ into water; dissociates to yield hydrogen ions; and has a pH of less than 7; reacts with base to form a salt.

**ACTH:** A pituitary secretion, the adrenocorticotropic hormone, which stimulates secretion of cortisol by the adrenal cortex.

**actin:** One of two proteins found in the muscles of higher animals.

**action potential:** The entire depolarization sequence characteristic of a neuron during a nervous impulse.

**active transport:** The diffusion of molecules through the cell membrane against the concentration gradient with attachment to "carrier" molecules and ATP energy.

**adrenal:** One of a pair of double glands lying close to the kidneys, which secrete adrenalin, noradrenalin, and (in the adrenal cortex) a large number of steroids.

**adsorption:** The clinging of tiny particles of a gas or liquid to the surface of a solid body or large colloidal particle.

**aldosterone:** Hormone that increases the rate at which sodium ions are reabsorbed in the distal tubule of the nephron of the mammalian kidney; regulates the final adjustment in the urine's composition and concentration.

**all-or-none response:** The principle that under given conditions the response of a nerve or muscle fiber to a stimulus at any strength above the threshold is the same.

**alleles:** Alternate forms of genes that may be found at the same locus on homologous chromosomes and are responsible for hereditary variation.

**alpha helix:** The shape often produced by a spiral twisting of straight-chained polypeptides with hydrogen or disulphide bonds between adjacent levels of the spiral.

**alveoli:** Saclike compartments within the lung which facilitate gas exchange; consist of epithelial cells and a capillary into which the oxygen moves.

**ambient temperature:** Temperature of air or water in which an animal lives.

**amino acid:** An organic acid containing an amino group and an acid carboxyl group that serves as building blocks of protein.

**anaphase:** The stage in mitosis, following metaphase, in which the daughter chromosomes move apart to opposite ends of the cell.

**angstrom unit:** One ten millionth of a millimeter; abbreviated by the symbol Å.

**anion:** A negatively charged ion.

**antagonist:** A muscle which acts in opposition to another.

**antidiuretic hormone (ADH):** Hormone affecting the rate of water reabsorption in the collecting ducts of the nephrons in the mammalian kidney; its presence in the blood makes the membranes of the cells lining the collecting duct more permeable to water.

**aorta:** The main body artery which receives freshly oxygenated blood and pumps it into the arterial system.

**aortic arches:** Loops in the simple circulatory system of the earthworm which contract and act as the pumping mechanism which propels the fluid; rudimentary heart.

**archegonium:** The female reproductive organ of ferns, mosses, and other plants in which eggs are produced and fertilized.

**arteries:** All the large blood vessels leading away from the heart.

**arterioles:** The thinner branches of arteries, usually not larger than 0.1 mm. in diameter.

**aster:** The starlike structure of fibers radiating out from the centriole of an animal cell during mitosis and meiosis.

**atom:** The smallest possible particle of matter into which a chemical element can be broken down and still retain its characteristic identity.

**atomic number:** A number uniquely associated with each element and equivalent to the number of protons in its nucleus.

**atria:** The two upper chambers of the heart.

**autodigestion:** The breaking down of an organism's body cells by its own digestive enzymes.

**autonomic nervous system:** Collective name for the neurons controlling all muscles and organs not subject to voluntary control.

**autosomes:** All chromosomes other than the sex chromosomes.

**autotrophs:** Organisms which manufacture energy-containing compounds from inorganic raw materials; includes green plants, most algae, and certain pigmented bacteria.

**auxins:** A number of related plant hormones which function in the elongation and differentiation of newly formed cells and which regulate the movement of water in plant tissues, causing turgor changes.

**axon:** A very long, slender cytoplasmic projection of a neuron, which conducts impulses to the dendrites of adjacent neurons: may reach a length of several feet.

**basal body:** A structure formed from a centriole which moves toward the end of a cell to form cilia or flagella.

**basal metabolism:** The rate at which an animal uses energy under normal, nonstressful conditions; determined by measuring the amount of oxygen used by an organism.

**base:** A substance which, upon being dissolved, introduces an abundance of $OH^-$ in water; dissociates to yield hydroxyl ions; has a pH of more than 7; can react with an acid to form a base.

**basement membrane:** Fibrous substance that separates epithelial cells from underlying tissue.

**basilar membrane:** A membrane stretched across the cochlea and adapted to sound reception by the fact that it supports a great number of hair cells.

**bicarbonate ion ($HCO_3^-$):** An important component of all digestive secretions in intestine; ions react with the acid chyme and gradually make it alkaline, a necessity for the subsequent steps in digestion and absorption.

**bile:** A liver secretion of salts which functions in digestion by emulsifying fats, rendering them susceptible to digestion by an aqueous solution of enzymes.

**biogenesis:** The concept that life arises only from prior life.

**biogeochemical cycle:** The circular flow of materials from the biotic component to the abiotic component and back again.

**biomass:** The part of a given habitat consisting of living matter, expressed either as the volume of organisms per unit volume or the weight of organisms per unit area of habitat.

**biome:** A group of communities characterized by a distinctive type of vegetation and climate.

**bivalent:** Pertaining to a genetic unit formed by two similar or identical chromosomes.

**bond resonance:** A process whereby electrons switch back and forth between ionic and covalent bending.

**Bowman's capsule:** One of the two principal parts of the nephron; a cup-shaped receptacle in which filtration takes place.

**Brownian movement:** The constant, random movement of molecules according to their thermal energy.

**budding:** A form of asexual reproduction characteristic of certain acellular organisms, in which a new individual develops as a bud from the older and larger parent.

**buffer system:** A system which resists or cushions change in pH when a strong acid or base is added; consists of a weak acid and a salt of that acid produced in an easily reversible combination reaction.

calorie: The amount of heat required to raise one kilogram of water 1° C.

cambium: A layer of meristematic tissue between the phloem and the wood (xylem) which produces new phloem on the outside and new xylem on the inside in stems and roots, it is responsible for growth rings in wood.

capillaries: The extensive network of very small vessels between the arterial and venous systems; usually about 8–10 microns in diameter; provide the surface for diffusion, thus are the vessels that supply body tissues with nutrients and remove waste products.

carbohydrase: An enzyme which hydrolyzes the bonds of polysaccharide molecules.

carbohydrate: An organic compound in which hydrogen and carbon typically occur in a ratio of 2:1.

cardiac muscle cells: The cells of the heart muscle, which are characterized by an unusually large number of mitochondria to aid in continual contraction.

carnivore: An organism which feeds only on animals.

carrying capacity: The maximum size of a population that can be supported on a sustained basis by the resources of a particular habitat.

cartilage: An elastic connective tissue composed of widely spaced, spherical cells embedded in a flexible matrix; found in the nose, ears, and discs between bones.

catalyst: A chemical substance which speeds up the rate of chemical reaction without becoming a part of the end-product and without itself being changed.

catalyzed transport: The diffusion of molecules through the cell membrane, and their transport to an area of lower concentration just within the cell cytoplasm by highly concentrated "carrier" molecules and osmosis.

cation: Anion with a positive charge.

cell body: An enlarged part of the neuron containing the cell nucleus.

cell membrane: The living outer membrane of a cell that serves as a selective barrier to molecules entering and leaving the cell.

cell theory: The assumption basic to modern biology, that cells are the basic structural unit of all living matter.

cellulose: An insoluble polysaccharide that composes the cell wall of plant cells.

centriole: A cylindrical body located adjacent to the nucleus which furnishes spindles for cell reproduction; usually essential to mitosis in animal cells.

centromere: A specialized structure in the chromosome to which a spindle fiber is attached.

cerebellum: A large outgrowth of the medulla in some higher vertebrates, with important functions in coordination.

cerebrum: The convoluted and largest part of the vertebrate, and especially the mammalian, brain; the center of sensory and motor function.

chemoreceptors: Specialized epithelial cells which function as the receptors for smell and taste.

chemosynthesis: A form of autotrophic nutrition in certain bacteria which obtain energy for manufacturing carbohydrates from oxidizing inorganic material, and from converting atmospheric nitrogen into organic compounds.

chitin: The hard substance forming the heavy external skeleton of the arthropod; necessitates the development of special respiratory organs such as gills and tracheae.

chlorophyll: A conjugated, green, light-trapping pigment which is essential as an electron donor in photosynthesis; the chlorophyll electron is returned to the chrolophyll molecule to reenter the cycle.

chloroplasts: Chlorophyll-containing plastids which are used to trap light and donate electrons in photosynthesis.

chromatid: One of two identical strands into

which a chromosome splits longitudinally prior to cell division.

**chromatin:** Readily stainable substance within the cell's nucleus containing the hereditary material, DNA.

**chromoplasts:** Pigment-containing plastids in plant cells.

**chromosomes:** Threadlike structures found in the cell nucleus that carry the genes in fixed patterns; the hereditary material within the cell nucleus; the number is constant for each species.

**cilium:** Short, hairlike projection that aids one-celled organisms in movement.

**circadian rhythm** (from the Latin *circa die*): One of a number of metabolic or behavioral cycles with a period of about one day.

**cistron:** A genetic unit of physiological function, consisting of a large, complex molecule; usually the nucleotide chain that determines a polypeptide chain.

**coacervates:** A reversible aggregation of emulsoid particles consisting of liquid masses or droplets enclosed by a membrane-like layer of lipid molecules.

**cochlea:** A receptor organ, located deep in the skull, which is the actual organ of hearing in vertebrates.

**codon:** A unit in the cistron controlling the synthesis of a single polypeptide unit of the genetic code.

**coenzyme A:** A coenzyme used in the aerobic oxidation of pyruvic acid.

**collagen:** Fibrous protein found in connective tissues and secreted by fibroblasts; one-third of body protein is in the form of collagen.

**collenchyma:** A slightly specialized varient of parenchyma with thick, elastic primary cells often present as support in maturing plant tissue.

**columnar cells:** Rectangular epithelial cells that function in much the same way as cuboidal cells.

**combination reaction:** The chemical bonding of two substances to form a third.

**commensalism:** A relationship between two biological species in which one species benefits

from the association, while the other is neither benefited nor harmed.

**companion cells:** Phloem cells that contain nuclear material and develop an unusually large number of organelles concerned with metabolic activities; they function to aid sieve tubes.

**compound:** A union of two or more different kinds of ions or atoms that form a different substance when held together by chemical bonds.

**concentration:** A measure of the amount of dissolved substance or number of molecules per unit volume.

**concentration gradient:** The strength of density difference between regions of molecular diffusion.

**cone cell:** A relatively insensitive visual cell, responsible for visual acuity and color vision.

**consumers:** Organisms, largely animals, that utilize other organisms as their food source.

**corpus luteum:** A secretory structure formed in the ovary following ovulation whose function is to secrete progesterone if fertilization occurs.

**countercurrent principle:** Concept whereby water always flows through the gill in a direction opposite to the flow of blood within the gill, assuring a high rate of oxygen absorption; also found in capillary exchange.

**covalent bond:** A chemical bond which is formed when two electron clouds overlap and each atom shares one or more electron with the other.

**cristae:** Folds formed by the inner membrane of the mitochondria.

**crossing over:** The interchange of corresponding parts of homologous chromosomes through accidental entanglement during the early stages of cell division.

**cuboidal cells:** Square-shaped epithelial cells located beneath squamous cells, or the exterior of internal organs; may have secretory function, or may synthesize enzymes and hormones.

**cuticle:** A waxy coating on the outside of plant cells.

**cyclic AMP (cAMP):** Cyclic-3′5′-adenosine monophosphate, a compound related to ATP, and which seems to be involved as a "second messenger" in hormonal activity.

**cyclic phosphorylation:** A cycle in which an excited chlorophyll electron is picked up and passed through a series of energy-reducing cytochrome molecules for the purpose of bonding ADP and a phosphate group, thus syntheisizing ATP.

**cyclosis:** The movement of cytoplasm around and around the cell, between the large central vacuole and the cell membrane, thus distributing organic molecules throughout the entire cell.

**cytochrome:** One of a group of conjugated, iron-containing proteins which serve as hydrogen carriers in aerobic respiration.

**cytokinesis:** The changes in the cell cytoplasm which occur during such stages of- its development as mitosis, meiosis, and fertilization.

**cytokinin:** A plant hormone, structurally a modified nucleotide, which modifies plant development in a number of ways by stimulating or altering cellular RNA.

**cytoplasm:** A general name for complex chemical matter of a cell encompassing the area between the cell membrane and the nucleus.

**cytoplasmic streaming:** The flowing of cytoplasm from one portion of the cell to another, distributing organic molecules throughout the entire cell.

**ecdysone:** The molting hormone, secreted by the small prothoracic glands in arthropods.

**ecosystem:** An ecological system formed by the interaction of a community of organisms and its environment.

**ecological equivalents:** Species occupying the same functional niche in different ecological systems.

**elastin:** Proteins secreted by fibroblasts that constitute the basic substance of elastic tissue; fibers contained in the connective tissue of blood vessels, especially in arteries, which allow the vessels to stretch and contract in response to pressure.

**electron:** A negatively charged particle that orbits around the atomic nucleus.

**element:** One of 92 naturally occurring fundamental kinds of matter composed of one kind of atom.

**endergonic reaction:** A chemical reaction absorbing more energy than it releases.

**endocrine gland:** One of a large number of ductless glands that secrete hormones.

**endoplasmic reticulum (ER):** A membrane network within the cytoplasm continuous with either the nuclear or cell membrane, or both, which may serve as a transport and/or screening membrane between the nucleus and the cell.

**endoskeleton:** The bone and cartilage which form the supportive structure of vertebrates and some echinoderms.

**engram:** A structural change in the nervous system caused by a specific experience which is believed to be the physiological basis of memory.

**entropy:** A measure of the randomness or disorder within a system.

**enzyme:** A protein molecule produced in plant and animal cells that causes changes in other substances by catalytic action.

**epiderm:** The outer cells of animals or the roots, stems, and leaves of green plants which act as a protective cover for the inner cells.

**epistasis:** A type of nonallelic gene interaction in which one pair of genes has a dominant effect over other pairs of genes.

**epithelium cells:** The protective cells of animals; found on outer skin areas, digestive tract linings, and blood vessels.

**equilibrium:** The stage in a chemical change at which the rate of combination (forward rate) will be exactly equal to the rate of decomposition (reverse rate).

**erythrocytes:** The red blood cells which carry oxygen throughout the body.

**eucaryotic cells:** Cells containing a nucleus, nuclear membrane, and specialized organelles.

**euploidy:** Variations in the number of sets of chromosomes in an organism, usually as a result of the fertilization of one egg by two sperms or by a failure of meiosis.

**exergonic reaction:** A chemical reaction that releases more energy than it absorbs.

**exohormone:** A chemical control agent whose effect and mode of action are very close to that of a hormone, except that it functions socially, and travels from one individual to another.

**exoskeleton:** The hard outer covering, shell, or cuticle found in mollusks and arthropods which acts with the muscles to produce movement.

**expressivity:** The extent to which a particular gene produces its effect on a group of organisms.

**eyespot:** A light-sensitive area found in many primitive animals, **e.g., *Chlamydomonas.***

**fermentation:** An anaerobic process in which carbohydrates or derivatives are decomposed by living microorganisms.

**fibroblasts:** The cells that secrete proteins such as collagen and elastin to form connective fibers like those found in tendons and ligaments.

**filtrate:** Fluid filtered from the blood by Bowman's capsule.

**fission:** The simplest form of asexual reproduction, in which the body of the parent organism splits into two new organisms.

**flagellum:** A long, whiplike projection that aids in movement in one-celled organisms.

**flame bulbs:** The primitive excretory mechanisms in freeliving flatworm (*Turbellaria*) which consist of hollow tubules lined with cilia.

**fluorescence:** An emission of light (radiation) from a substance that has absorbed light from another source.

**gametangium:** The specialized structure of plants in which the gametes are produced.

**ganglion:** A cluster of neuron cell bodies.

**gastrovascular cavity:** The central body cavity of lower invertebrates which serves the double function of an incomplete digestive tract and a means of internal transport.

**gel:** Semisolid state of a colloidal system.

**genetic load:** The proportion of harmful genes within a population, which if too high may endanger the existence of a species.

**gene pool:** The group of genes existing in all the individuals within any particular interbreeding population.

**gibberellins:** A number of closely related plant hormones capable of inducing germination, growth, flowering, and pollen development.

**gill:** The respiratory system of aquatic vertebrates; a projecting organ supported by a series of bony arches and consisting of absorptive surfaces called secondary lamella.

**glomerulus:** The bulblike network of blood capillaries, held by Bowman's capsule, which branches off from the renal artery.

**glucose:** The most common simple (monosaccharide) sugar and the chief fuel substance for most organisms.

**glyceraldehyde-3-phosphate (PGAL):** A compound formed when PGA is phosphorylated by ATP, and then reduced by $NADPH_2$; occurs in the dark reaction of photosynthesis and glycolysis.

**glycerol:** A 3-carbon compound which may combine with fatty acids to form a fat.

**glycolysis:** A process of anaerobic respiration in which glucose, starch, or glycogen is broken down to pyruvic acid.

**Golgi apparatus:** A cytoplasmic organelle which may play a role in the packaging of cell secretions.

**gonads:** The reproductive organs, testes in the male and ovaries in the female, which produce gametes (egg and sperm) and the sex hormones, testosterone in the male and estrogen and progesterone in the female.

**grana:** Tiny structures appearing to consist of stacks of disks within chloroplasts.

**guard cells:** Long, narrow cells bordering the opening of the stoma and facilitating its opening to allow water evaporation and oxygen exchange with the outside atmosphere; the stoma and guard cells together are called the stomatal apparatus.

**guanosine diphosphate (GDP):** A coenzyme

essential to energy production within the Krebs citric acid cycle.

**guanosine triphosphate (GTP):** A reduced form of GDP which reacts with ADP to form ATP in the Krebs citric acid cycle.

**guttation:** The exudation, or forcing out, of drops of water from the leaves of green plants, due to root pressure.

**habitat:** The native environment in which an organism lives or grows.

**hair cell:** A sensory cell, evolutionarily derived from specialized ciliated epithelium, which is the basic functional unit in all sound reception.

**haploid:** Having half the normal diploid number of chromosomes; most animal gametes contain a haploid number of chromosomes.

**heartwood:** Area of old xylem, no longer active in transport, in the center of the trunk of an old tree.

**hematocrit:** The percentage of total blood volume that is represented by red blood cells; determines the blood's viscosity.

**hemoglobin:** A conjugated protein, containing four atoms of iron, found in red blood cells, which forms bonds easily and permits oxygen and carbon dioxide transport.

**hepatic portal vein:** Vein connecting digestive tract with the liver; water soluble nutrients enter this vein and are transported to the liver where distribution of nutrients is regulated. The hepatic portal vein is unique in having capillaries at both ends.

**herbivore:** An organism which feeds only on plants.

**hermaphroditism:** A condition in which the organism possesses both the male and the female reproductive organs.

**heterotrophs:** Organisms that must obtain their energy-containing compounds in the cells of other organisms; any organism without chlorophyll.

**heterozygous:** A condition in which the organism contains dissimilar pairs of genes for any given

hereditary characteristic and will therefore not breed true to type.

**histamine:** A hormonelike chemical agent released by damaged tissue as a defense against infection.

**homeostasis:** The tendency of a system to maintain internal stability in the face of external change owing to the coordinated response of its parts to any situation or stimulus tending to disturb its normal function; e.g., regulation of body temperature, adjustment of rate of respiration and circulation.

**homeotherms:** Group of animals which maintain a constant body temperature despite changes in the ambient temperature.

**homozygous:** A condition in which the organism contains identical pairs of genes for any given hereditary characteristic and will therefore breed true to type.

**hormone:** One of a number of chemical agents produced in specific body tissues (usually in the endocrine glands) and carried to its site of activity by bulk circulation.

**hydrogen bond:** A weak bond produced by the polar attraction between positively charged hydrogen atoms on one part of a molecule and negatively charged atoms of oxygen or nitrogen on another part of the molecule.

**hydrolysis:** The chemical decomposition of one molecule into two by the addition or insertion of water. May be used as a means of digestion.

**hydrostatic skeleton:** The supportive structure, necessary for movement, which is made up of body fluids.

**hypertonic:** Characteristic of a solution which contains a greater number of osmotically active solutes than the cell it surrounds; cell will lose water and may die of dehydration.

**hypophysis (pituitary):** The three-lobed "master gland" lying just beneath the vertebrate brain which dominates the activity of the endocrine system.

**hypothalamus:** The part of the vertebrate forebrain linking the nervous and endocrine control systems by regulating the hypophysis; also the site of homeostatic and basic emotional functions.

**hypotonic:** Characteristic of a solution that contains fewer osmotically active solutes than the cell it surrounds; cell will swell and may burst.

**imbibition:** The process of water absorption by colloids; a mechanism which helps bring water into the xylem of roots.

**imprinting:** A highly specialized form of learning which occurs rapidly and very early in life.

**induction:** A process that initiates the activity of the operon by the binding of repressor substance by a substrate chemical.

**instinct:** A form of innate behavior or tendency to action characteristic of a given biological species.

**intercalary disks:** Cellular junctions within cardiac muscle; the adjacent surfaces are so tightly wedged together as to give the junctures the appearance of an independent structure.

**interkinesis:** In meiosis, a brief period of reproductive inactivity following telophase of the first division.

**interphase:** A stage which occurs before cell division has taken place or after it has been completed.

**invagination:** A local infolding of the cell membrane to enclose groups of molecules for transport to the interior of the cell; more generally, any folding or protruding inward.

**ion:** An atom or atoms with an electric charge.

**ionic bond:** A process by which electron clouds overlap and one atom releases one or more electrons to the other atom.

**isomers:** Compounds identical in molecular formula but different in structural arrangement. Isomers exhibit different chemical and physical properties.

**isoosmotic concentration:** The maintenance of a balance between the osmotically active solutes of body cells and the fluids surrounding them.

**isotonic:** A characteristic of two solutions having equal osmotic pressures; no flow of water will take place when a membrane separates them.

**isotope:** One of a number of forms of a chemical element with a differing number of neutrons and atomic weight, but having the same chemical properties.

**karyokinesis:** The series of changes that occur in the nucleus of a cell in the process of division.

**kinetic energy:** A system's energy in motion.

**lactic acid:** The end-product of fermentation in aerobic cells forced to function under anaerobic conditions.

**lateral line system:** A series of specialized hair cells located along the sides of most fish, and serving as extremely sensitive receptors for pressure and low-frequency sound waves.

**leucocytes:** White blood cells whose function is to destroy alien cells such as bacteria or virus.

**leucoplasts:** Unpigmented plastids which are the site of synthesis of starch from simpler sugars.

**lipase:** An enzyme which breaks down the bonds of lipid molecules.

**lipid:** A common organic fat or fatlike compound that functions in living organisms structural components, fuel, and storage materials.

**loop of Henle:** Long, thin U-shaped portion of the nephron tubule through which the filtrate fluid passes and is concentrated before the final adjustments in the solute and water content of urine are made.

**lysosomes:** Small, membranous sacs that enclose enzymes used in decomposing cellular materials.

**macronutrient:** An element required in large amounts by a living thing.

**Malpighian tubules:** Excretory organs of insects consisting of tubules branching off the digestive tract between the stomach and the intestine; so structured to complement the insects' open circulatory system.

**mass:** A measure of the quantity of matter present in an object.

**mass number:** The total number of protons and neutrons in a given atom's nucleus.

**matter:** Any substance which has mass and occupies space.

**medulla:** The main part of the vertebrate hindbrain, actually the enlarged end of the spinal cord.

**megaspore mother cell:** A large, asexually produced spore from which a female gametophyte develops.

**meiosis:** The process of nuclear division in germ cells, in which the number of chromosomes is reduced from the diploid number to the haploid number.

**membrane:** A thin, pliable partition that both separates and organizes chemical reactions in a cell.

**membrane potential:** The difference in voltage between the inside and the outside of an axon; alteration of the membrane potential is responsible for the propagation and transmission of nervous impulses.

**meristem:** Unspecialized plant tissue that is capable of rapid division.

**metaphase:** The stage in mitosis, following the prophase, in which the duplicated chromosomes are arranged on the equatorial plane of the spindle fibers.

**microfilaments:** Structures 40–50 $\mu$ in diameter which perform the contractile function in cellular movement; contain a contractile protein similar to actin.

**micron:** A unit of microscopic measurement equal to one-thousandth part of a millimeter; abbreviated by the symbol $\mu$.

**micronutrient:** An element required only in tiny amounts by a living things; also called a trace element.

**microspore mother cell:** A small, asexually produced spore from which a male gametophyte develops.

**microtubules:** Organelles with an elongated, tubular nature, that occur in cilia and parts of the cytoplasm; involved in spindle formation in cell division.

**mitochondria:** Organelles within the cytoplasm that serve as sites for cellular respiration.

**mitosis:** The exact reproduction of the original cell through cell division, involving the formation of chromatin into long threads, their separation into segments, their lengthwise division, and the halves coming together in two sets to form the nucleus for a new cell.

**mixture:** Two or more different substances that are in physical proximity but chemically uncombined.

**mole:** The number of grams equal to the molecular weight of a compound.

**molecule:** A particle that contains two or more identical or different atoms bonded together chemically.

**motor neuron:** A neuron that leads away from the CNS to cause response in an effector.

**mutation:** A change in the form, qualities, or nature of the offspring from their parent type brought about by a change in the hereditary material from the parent.

**mutualism:** A symbiotic relationship between two biological species in which both benefit from the association.

**myelination:** The insulation surrounding an axon.

**myofibrils:** Structures, about one micron in diameter, making up the muscle fiber; in turn composed of smaller protein filaments, myosin and actin.

**nastic movements:** Movements due to changes in turgor; these are fast and reversible.

**nerve:** A number of neurons organized into a bundle.

**nerve net:** An interconnected circuit of nerve cells radiating throughout the body of lower invertebrates such as *Hydra*.

**nephridium:** Rudimentary excretory organ in the leech and the earthworm consisting of a network of tiny tubules contained within a ciliated sac, actually a specialized part of the central body cavity.

**nephron:** One of about a million units fundamental to the mammalian kidney; produces urine.

**neurons:** Nerve cells engaged in conducting electrochemical impulses; composed of a cell body, dendrites, and axon.

**neutron:** A fundamental particle within the nucleus of the atom which has no charge.

**niche:** The functional role pursued by each individual species in an ecosystem.

**nicotinamide-adenine-dinucleotide (NAD):** A coenzyme necessary for the alcoholic fermentation of glucose and the oxidative dehydrogenation of other substances.

**nicotinamide-adenine-dinucleotide phosphate (NADP):** A coenzyme which oxidizes water molecules and is reduced to $NADPH_2$ in the light reaction of photosynthesis.

**noncyclic phosphorylation:** The oxidation of water and the reduction of NADP (to $NADPH_2$) in the light reaction of photosynthesis.

**nucleic acid:** An organic acid composed of joined nucleotide complexes; the two main types are deoxyribose nucleic acid (DNA) and ribose nucleic acid (RNA).

**nucleolus:** A small oval structure within the cell's nucleus where RNA is synthesized.

**nucleus:** A body present in all eucaryotic cells which seems to be the cell's control center; also the central body of an atom.

**ommatidium:** One of a large number of independent visual units comprising the arthropod compound eye.

**omnivore:** An organism which eats all types of food indiscriminately.

**operon:** A unit consisting of the operator gene together with the structural genes it controls.

**organ:** A structure composed of several tissues associated to perform one or several functions.

**organelles:** Specialized bodies within cytoplasm of a cell; involved in various aspects of metabolic activity.

**osmoconformity:** The condition in which the internal osmotic pressure of a cell conforms to

that outside the cell; marine algae and lower marine invertebrates are osmoconformers.

**osmoregulation:** A system evolved by higher animals to regulate their internal osmotic pressures allowing them to travel through environments having different pressures.

**osmosis:** The process in which water diffuses through a semipermeable membrane from a side containing a lesser concentration of particles to a side containing a greater concentration; diffusion continues until the concentration of particles is equal on both sides.

**osmotic pressure:** The force exerted by the cell content against the outer cell membrane.

**osteoblasts:** Supportive bone-forming cells that secrete collagen.

**oviparous:** Designating animals that produce eggs which develop and hatch after leaving the body of the female.

**ovoviviparous:** Designating animals that first produce eggs which develop and hatch in the uterus and then bring forth live young; sharks are an example of ovoviviparous animals.

**oxygen dissociation curve:** A mathematical curve showing the propensity of hemoglobin to load and unload oxygen molecules, when temperature and pH are kept constant.

**Pacinian corpuscle:** A specialized receptor bulb lying relatively deep within the body and functioning in the reception of "deep touch" or pressure.

**parasitism:** A relation between organisms in which one symbiont is helped and the other is harmed.

**parathyroid:** One of a number of small endocrine glands found attached to or embedded in the thyroid, and whose secretion, parathormone, is essential to the regulation of the body's calcium-phosphate balance.

**parenchyma:** The cells of green plants that may contain chlorophyll or function in the storage of starch.

**parthenogenesis:** A form of reproduction in

which eggs develop into new organisms without fertilization.

**passive transport:** An osmotic process in which water and dissolved mineral molecules or ions may pass through a cell membrane in response to the concentration gradient; no expenditure of energy is required.

**penetrance:** The frequency with which a particular gene produces its effect in a group of organisms.

**peptide:** A combination of from two to about 100 amino acids within a molecule.

**peristalsis:** The progressive wave of contraction and relaxation of the alimentary canal, or any tubular muscular system, by which the contents are forced through the system.

**phagocytosis:** An activity in which large groups or particles are engulfed by the cell membrane and digested.

**phenotype:** The physical appearance of an organism; all organisms distinguished by the same physical characteristics.

**phloem:** Cylindrical cells which specialize in conducting dissolved organic materials to various parts of the plant.

**phosphoglyceric acid (PGA):** A 3-carbon compound that appears as the first intermediate in photosynthesis; metabolic intermediate in glycolysis.

**phosphorylation:** The addition of a phosphate group to a compound.

**photoperiodism:** The regulation of some functions in plants and animals according to variations in period and amount of ambient light.

**photosynthesis:** The process of manufacturing carbohydrates out of carbon dioxide and water by using light energy and chlorophyll.

**pineal gland:** A small endocrine gland associated with the thalamus, which is suspected to play a major role in regulating sexual development and long-range cycles in vertebrates; it is sensitive to light.

**pinocytosis:** A process by which small groups of molecules are transported across the cell membrane by enclosing the molecules in a membranous envelope.

**pit organ:** A heat-detecting receptor organ found on the heads of snakes that feed on small mammals.

**placenta:** The organ within the uterus of most mammals through which the fetus receives nourishment and eliminates wastes.

**plastids:** Specialized cytoplasmic organelles which are the site of synthesis.

**platelet:** A component of the blood which initiates clotting by releasing thromboplastin when a vessel has been damaged.

**poikilotherms:** Group of animals that tend to take on a body temperature similar to the ambient temperature; for example, fish, reptiles, and amphibians.

**polarity of molecules:** A condition arising from an uneven distribution of charge; molecules have a positive and negative end.

**polypeptide:** Many amino acids linked together by peptide bonds.

**potential energy:** The stored energy of a system.

**precursor:** A molecule similar to a hormone in configuration and composition which may be stored in the gland until needed, when it is rapidly altered to produce the related hormone.

**primary wall:** Thin, cellulose coating surrounding a plant cell.

**procaryotic cells:** Simple cells that contain no distinct nuclear membrane or organelles; believed to be the first form of life.

**producers:** Organisms, largely plants, that are capable of manufacturing their own food from inorganic substances.

**proprioreceptors:** A number of receptor organs which monitor the internal environment and are essential to all homeostatic regulation.

**prophase:** The first stage in mitosis in which the chromosomes contract and the nuclear membrane disappears.

**prostaglandin:** One of a large number of chemical control agents resembling hormones, but synthesized by a wide variety of body cells.

**protein:** Any of a class of complex nitrogenous substances composed of numerous amino acid molecules.

**protease:** An enzyme which catalyzes the digestion of protein.

**proton:** One of the fundamental particles of the nuclei of all atoms; carries a positive electric charge.

**pseudopod:** The "false foot" or temporary projection of the protoplasm of an amoeba (or of a cell) serving as an organ of locomotion.

**pyruvic acid:** A fundamental intermediate in protein and carbohydrate metabolism in the cell.

**quantasomes:** Beadlike granules found on the thylakoid membranes in chloroplasts; probably contain chlorophyll and enzymes necessary for photosynthesis.

**quantum:** An elemental unit of energy; a "packet" of energy.

**receptor:** A spot or organ in an organism sensitive to light, heat, and chemical or physical changes in the environment.

**reduced nicotinamide - adenine - dinucleotide (NADH):** A reduced form of NAD that is essential for the oxidative processes of glycolysis.

**reduced nicotinamide - adenine - dinucleotide phosphate (NADPH$_2$):** A reduced form of NADP which is used in the dark reaction of photosynthesis to synthesize carbohydrates from carbon dioxide.

**reflex arc:** A basic stimulus-response pathway, linking a receptor and an effector.

**refractory period:** A period after response, usually about 5 to 10 milliseconds, during which a neuron displays diminished irritability or none at all.

**releaser:** Any stimulus which is particularly effective in producing behavioral responses in a given organism.

**residual volume:** The amount of air which cannot be squeezed out of the lungs; the difference between vital capacity and total capacity.

**retina:** The tissue in the rear of the eye comprised of the light-sensitive rods and cones.

**rhodopsin:** The principal visual pigment in land vertebrates.

**ribosomes:** Granules within the cytoplasm that contain RNA and are the site of protein synthesis; may be attached to the surface of endoplasmic reticulum.

**rod:** A highly sensitive visual cell, registering light but not color.

**root pressure:** The phenomenon where roots act like a pump forcing water through the xylem in the absence of upward pull.

**sapwood:** Rings of xylem, lighter than the heartwood, which conduct water for the tree.

**sarcolemma:** The membrane of the muscle fiber which exhibits strong electrical polarization; it is depolarization of this membrane which initiates contraction.

**Schwann cell:** A nerve cell that wraps itself around the axon; serves as a form of insulation and aids in conduction.

**sclerenchyma:** A supportive plant tissue composed of cells with greatly thickened walls; varients include fibers and stonecells.

**secondary wall:** A thick, rigid, often waterproof wall of the plant cell laid down after the primary wall.

**semicircular canals:** A proprioceptive organ; a specialized part of the vertebrate inner ear, which monitors the equilibrium and movement of head and body.

**sere:** A stage in an ecological succession.

**serum:** Plasma which cannot clot because of the absence of fibrinogen.

**sexual dimorphism:** The existence of physical differences between the males and females of the same species.

**sieve plate:** The perforated end-wall present in sieve tube cells.

**sieve tubes:** The functional units of phloem that, upon maturity, lose their specialized organelles, and usually are without nuclei.

**smooth muscle cells:** Contracting, involuntary muscle cells which line the walls of the intestinal tract, the stomach, and blood vessels.

**social dominance:** The arrangement of a group of animals of the same species into a hierarchy, influenced generally by strength, size, weight, health, maturity, previous fighting experience, or hormonal level.

**sodium-potassium pump:** An active transport mechanism which maintains a relatively high concentration of sodium ions outside the neural membrane, thus maintaining the membrane potential.

**solution:** A mixture of one or more substances evenly dispersed.

**solvent:** The dissolving medium of a solution.

**sphincters:** Small valves which permit and prohibit the flow of blood through a particular vessel; controlled by the nervous system.

**spindles:** Threadlike structures that extend from centrioles during mitosis and meiosis, and guide the movements of chromosomes; present in both plant and animal cells.

**spontaneous generation:** A now-abandoned theory which assumed that living organisms could come into being without preexisting cells; definitely disproven in 1858.

**sporophyte:** The spore-bearing stage of those plants which alternate between asexual and sexual generations; in higher plants, the sporophyte generation is dominant.

**squamous cells:** Flat epithelial cells which are usually the outermost protection of animals.

**statocyst:** An invertebrate equilibrium proprioceptor, consisting of a freely moving mineral particle within a cilia-lined sensory sac.

**stele:** The central cylinder of vascular tissue in the stem, root, and leaves of vascular plants.

**stimulus:** An alteration in the physical or chemical environment capable of detection by a receptor.

**stomata (s., stoma):** Specialized tiny openings, surrounded by two guard cells, scattered throughout the epidermal surface of the leaves of all vascular plants which facilitate the process of gas exchange.

**striated muscle cells:** The long, thin fibers comprising voluntary muscle and showing vertical bands, or striations.

**strobili:** Conelike reproductive structures found in most gymnosperms which contain clusters of specialized, spore-bearing leaves.

**succession:** The orderly and predictable replacement of one community of organisms by another until a climax stage is reached.

**summation:** The cumulative effect of muscle fibers contracting individually to accomplish the smooth contraction of the whole muscle.

**suspension:** A mixture in which the dispersed molecules are so large, they tend to settle to the bottom without dispersing evenly.

**symbiosis:** The living together of two dissimilar organisms.

**sympathetic nervous system:** The system of control that dominates autonomic nervous function in times of stress.

**synapse:** A junction between the axon of one neuron and the dendrite of the next.

**synapsis:** The conjugation of homologous maternal and paternal chromosomes during the early stages of meiosis.

**synaptic bulb:** A bulge at the end of an axon which permits transmission of a nervous impulse to an adjacent dendrite through release of a chemical transmitter substance such as acetylcholine.

**systole:** The powerful contraction of the ventricles which pumps blood out of the heart and into the arteries.

**taxis:** Movement of an organism in a particular direction in response to an external stimulus.

**telophase:** The final stage in mitosis in which new nuclei are formed.

**tendon:** The structure that attaches muscle to bone, allowing flexible movement where bones are hinged or jointed.

**tetanus:** The smooth, sustained contraction of

the muscle by the cumulative summation of individual twitches.

**tetrad:** A pair of homologous chromosomes which behave as a single unit throughout the first phases of meiosis.

**thalamus:** In higher vertebrates, that part of the forebrain that integrates all sensory input and relays information to other parts of the forebrain and back down the spinal cord.

**thorax:** The central body segment of the insect to which wings are attached.

**threshold:** The point at which a stimulus is of sufficient intensity to begin to produce an effect.

**thymus:** An endocrine gland, especially prominent in childhood, whose main function is to establish the body's immune system.

**thyroxine:** A modified amino acid, the principal secretion of the thyroid glands, whose presence affects the basal metabolism and body temperature.

**tidal volume:** The normal volume of air that moves in and out of the lungs with each breath.

**tissue:** A group of specialized cells that are similar in structure and function.

**tonoplast:** Membrane that encloses the vacuole and keeps its contents from diffusing into the cell.

**tracheae:** A number of thin tubes in the body of small land-dwelling arthropods which carry air from the atmosphere to the internal tissues, thus facilitating respiration.

**tracheid:** A cylindrical conducting type of xylem cell that is tapered at both ends, and that has thin spots through which solutions may move from one tracheid to another.

**translocation:** The movement of water through vascular tissue of plants.

**transpiration:** The loss of water vapor during gas exchange in the green plant, leading to constant need for water replacement.

**trituration:** The mechanical breakdown of food as by the churning of the stomach.

**tropism:** An orientation of an organism in response to an external stimulus; the growth movement induced is irreversible, and the direction of the movement is directly related to the direction of the stimulus.

**twitch:** One brief contraction of a muscle fiber.

**tympanum:** A fine membrane, stretched across the path of sound waves, serving as an organ of sound reception in insects.

**unit membrane:** The structural unit of two lipid and two protein layers that is thought to form many membranous cell structures.

**vacuole:** A membrane-lined pocket of fluid within a cell.

**vagus nerve:** Regulates rate of respiration through a reflex reaction, usually coordinated through the medulla.

**valence:** A measure of the capacity of an element to combine with another.

**vascular tissue:** In plants, the combination of xylem and phloem cells, specialized to transport water and inorganic ions and dissolved molecules of sugar and amino acids.

**veins:** Vessels that carry blood back toward the heart.

**vena cavae:** The two large veins that return blood to the heart from the body; the superior vena cava drains the upper body and the inferior, the lower body.

**ventilation:** The pumping of the respiratory medium over the epithelial tissues which absorb oxygen.

**ventral:** Running along an organism's lower side.

**ventricles:** The two lower chambers of the heart; the pumping muscles comprise the walls.

**venules:** Vessels that pick up blood from the capillaries.

**vesicles:** Tiny pockets of secretions, pigments, or cell fluid.

**vessel element:** A large type of xylem cell with end-walls that are partially or entirely dissolved.

**villi:** Minute, fingerlike, vascular processes on mucous membranes of the small intestine, where they aid in absorbing nutriment.

**vital capacity:** The amount of air that can be

exhaled after the lungs have been filled to their maximum.

**vitamin:** An organic substance essential in small quantities to normal metabolism; it must be present in the diet, since the organism cannot synthesize it for itself.

**viviparous:** Designating animals that bring forth living young rather than eggs.

**xylem:** Cells that conduct water and minerals from the roots to the leaves. Upon maturity, the cells die and, in bulk, form the wood of trees.

**zygote:** The cell produced by the union of two gametes.

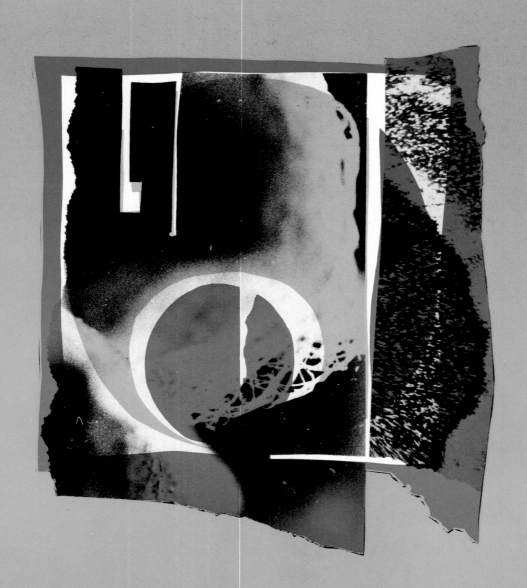

# Index